Introduction to Corrosion Science

E. McCafferty

Introduction to Corrosion Science

E. McCafferty
Alexandria VA 22309
USA

ISBN 978-1-4419-0454-6 e-ISBN 978-1-4419-0455-3
DOI 10.1007/978-1-4419-0455-3
Springer New York Dordrecht Heidelberg London

Library of Congress Control Number: 2009940577

Printed on acid-free paper

Springer is part of Springer Science+Business Media (www.springer.com)

Preface

This textbook is intended for a one-semester course in corrosion science at the graduate or advanced undergraduate level. The approach is that of a physical chemist or materials scientist, and the text is geared toward students of chemistry, materials science, and engineering. This textbook should also be useful to practicing corrosion engineers or materials engineers who wish to enhance their understanding of the fundamental principles of corrosion science.

It is assumed that the student or reader does not have a background in electrochemistry. However, the student or reader should have taken at least an undergraduate course in materials science or physical chemistry. More material is presented in the textbook than can be covered in a one-semester course, so the book is intended for both the classroom and as a source book for further use.

This book grew out of classroom lectures which the author presented between 1982 and the present while a professorial lecturer at George Washington University, Washington, DC, where he organized and taught a graduate course on "Environmental Effects on Materials." Additional material has been provided by over 30 years of experience in corrosion research, largely at the Naval Research Laboratory, Washington, DC and also at the Bethlehem Steel Company, Bethlehem, PA and as a Robert A. Welch Postdoctoral Fellow at the University of Texas.

The text emphasizes basic principles of corrosion science which underpin extensions to practice. The emphasis here is on corrosion in aqueous environments, although a chapter on high-temperature oxidation has also been included. The overall effort has been to provide a brief but rigorous introduction to corrosion science without getting mired in extensive individual case histories, specific engineering applications, or compilations of practical corrosion data. Some other possible topics of interest in the field of corrosion science have not been included in accordance with the goal to keep the material introductory in nature and to keep the size of the book manageable. In addition, references are meant to be illustrative rather than exhaustive.

Most chapters also contain a set of problems. Numerical answers to problems are found at the end of the book.

Finally, the author wishes to recognize the various mentors who have graciously shaped his professional life. These are: Dr. J. B. Horton and A. R. Borzillo of the Bethlehem Steel Corporation, who introduced the author to the field of corrosion; the late Prof. A. C. Zettlemoyer of Lehigh University, who taught the author the beauty of surface chemistry while his Ph. D. advisor; the late Dr. Norman Hackerman, postdoctoral mentor at the University of Texas; the late Dr. B. F. Brown and M. H. Peterson of the Naval Research Laboratory; and Prof. James P. Wightman of the Virginia Polytechnic Institute and State University, a "surface agent extra-ordinaire" with whom the author has spent an enjoyable and exciting sabbatical year.

The author is also grateful to Harry N. Jones, III, James R. Martin, Farrel J. Martin, Paul M. Natishan, Virginia DeGeorgi, Luke Davis, Robert A. Bayles, and Roy Rayne, all of the Naval Research Laboratory, who helped in various ways. The author also appreciates the kind assistance of

A. Pourbaix of CEBELCOR (Centre Belge d'Etude de la Corrosion), C. Anderson Engh, Jr., M.D. of the Anderson Orthopaedic Clinic, Alexandria, VA; Phoebe Dent Weil, Northern Light Studio, Florence, MA; Erik Axdahl, and Harry's U-Pull-It, West Hazleton, PA.

Finally, the author wishes to thank Dr. Kenneth Howell, Senior Chemistry Editor at Springer, for his encouragement and support.

Washington, DC
2009

E. McCafferty

Contents

Chapter 1
Societal Aspects of Corrosion

We Live in a Metals-Based Society

Residents of industrialized nations live in metal-based societies. Various types of steel are used in residential and commercial structures, in bridges and trusses, in automobiles, passenger trains, railroad cars, ships, piers, docks, bulkheads, in pipelines and storage tanks, and in the construction of motors. Aluminum alloys find a variety of uses ranging from aircraft frames to canned food containers to electronic applications. Copper is used in water pipes, in electrical connectors, and in decorative roofs. Chromium and nickel, to name just two more metals, are used in the production of stainless steels and other corrosion-resistant alloys.

In addition, metals are also used in various electronic applications, such as computer discs, printed circuits, connectors, and switches. Metals are even used in the human body as hip or knee replacements, as arterial stents, and as surgical plates, screws, and wires. See Fig. 1.1.

Metals also find use as coins of daily commerce, in jewelry, in historical landmarks (such as statues), and in objects of art.

There are 85 metals in the Periodic Table. Whatever be their end use, all common metals tend to react with their environments to different extents and at different rates. Thus, corrosion is a natural phenomenon and is the destructive attack of a metal by its environment so as to cause a deterioration of the properties of the metal. Figure 1.2 shows an example of a harsh corrosive environment in which the deck of an aircraft carrier is splashed and sprayed by seawater so that both the structural metals and electronic components in the aircraft may suffer corrosion.

Why Study Corrosion?

There are four main reasons to study corrosion. Three of these reasons are based on societal issues regarding (i) human life and safety, (ii) the cost of corrosion, and (iii) conservation of materials. The fourth reason is that corrosion is inherently a difficult phenomenon to understand, and its study is in itself a challenging and interesting pursuit.

These reasons are discussed in more detail below.

Corrosion and Human Life and Safety

On December 15, 1967, the "Silver Bridge" over the Ohio River linking Point Pleasant, West Virginia, and Kanauga, Ohio, collapsed carrying 46 people to their deaths [3]. See Fig. 1.3. The cause of the failure was due to the combined effects of stress and corrosion.

E. McCafferty, *Introduction to Corrosion Science*, DOI 10.1007/978-1-4419-0455-3_1,

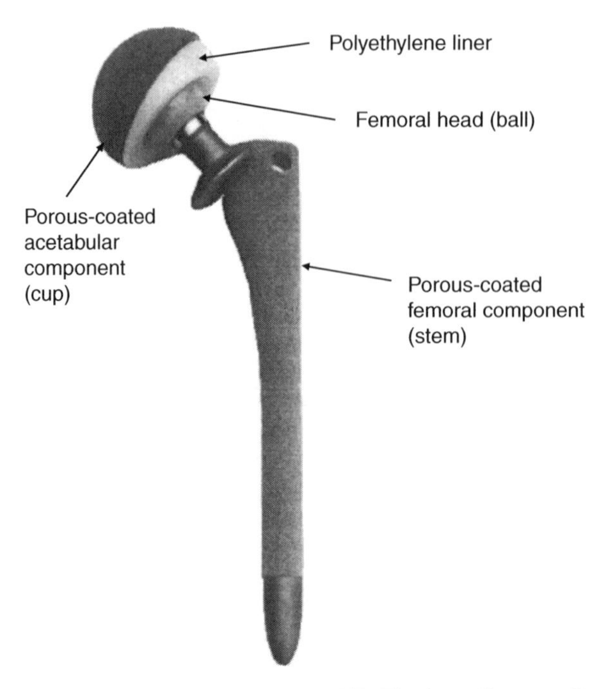

Fig. 1.1 An artificial hip used in hip-replacement surgery [1]. Titanium alloys or alloys of cobalt–chromium–molybdenum are currently used in artificial hips because these alloys resist corrosion in the human body. Figure courtesy of Dr. C. Anderson Engh, Anderson Orthopedic Clinic, Alexandria, VA

Fig. 1.2 An example of a harsh corrosive environment [2], in which both the structural and electronic metals in the aircraft will be subject to corrosion

On June 28, 1983, a 100-ft long section of a bridge span on a major US Interstate highway collapsed [4]. Three people died and three more were seriously injured. The cause of the collapse was due to the corrosion of a support pin.

On April 28, 1988, the cabin of a commercial airliner en route from Hilo to Honolulu, Hawaii, suddenly disintegrated, and a flight attendant was tragically lost and 65 passengers were injured [5]. The cause of the problem was the combined effects of metal fatigue and corrosion.

Fig. 1.3 On December 15, 1967, the Silver Bridge connecting Ohio and West Virginia over the Ohio River collapsed, and 46 people lost their lives. The cause of the collapse was stress-corrosion cracking [3]

In February 2001, a nuclear power plant in Ohio was closed for routine maintenance; and a "pineapple-sized hole" caused by corrosion was found in the reactor's lid [6]. A report issued by the Nuclear Regulatory Commission stated that the plant had been operating on the brink of a potentially devastating nuclear accident.

In July 2005, a 50-ft pedestrian bridge at a shopping mall collapsed on top of a box truck, which was demolished. (The collapse occurred after normal business hours, so no one was injured.) The structural failure was attributed to the corrosion of large metal bolts which connected the bridge to a parking garage and to the stores in the mall [7].

In December 2008, a massive water main break in suburban Washington, DC, unleashed a torrent of water that forced dramatic rescues of trapped motorists [8]. This incident raised anew concerns about the safety of the aging infrastructure in the United States.

These are but a few examples as to how corrosion can impact human life and safety. Other examples where safety is a primary concern involve the structural integrity of pipelines, storage tanks, boilers, pressure vessels, and aircraft engines.

Of recent concern in the United States is the state of the nation's aging infrastructure. There are approximately 583,000 bridges in the United States, and 15% of these are structurally deficient due to corroded steel or steel reinforcement [9].

In Washington, DC, in November 2007, a freight train crashed through a barrier and crossed a railroad bridge which was supposed to be closed. Ten railroad cars were derailed, including six which fell into the Anacostia River. A spokesman for the railway corporation said that "We have not experienced this level of corrosion on a bridge this age" [10].

Another safety issue concerns the disposal of nuclear wastes. The United States plans to establish a nuclear waste repository beneath a mountain at Yucca Mountain, Nevada. The nuclear wastes are to be contained in casks of a nickel-based alloy, and the repository is to be free from radiation loss for a period of 10,000 years. As of 2009, however, corrosion scientists and engineers were grappling with the problem of predicting the extremely long-term corrosion resistance of the nickel alloy casks at high temperatures and in the presence of chloride salts which could leach into rainwater passing from the surface of the mountain to the underground repository [11].

A different safety issue is the buildup of certain toxic ions in solution due to corrosion. The corrosion of chromium to produce Cr^{6+} ions and the use of chromates as surface treatments to prevent corrosion are issues of recent concern. Contamination of water piping systems by the corrosion of cadmium components in the water delivery system has also been under scrutiny. Galvanized steel pipes (zinc coated) usually contain some cadmium as does the solder used to join them together [12]. The toxic effect of dissolved lead ions in water pipes constructed of lead metal has long been recognized and will be a potential problem as long as lead piping already in place continues in service. In the winter of 2004, toxic levels of dissolved lead ions were found in the drinking water in Washington, DC, homes which were serviced by lead pipes [13]. The problem arose from a change in the water treatment procedure which inadvertently disturbed the protective oxide film on lead, thus allowing the underlying metal to corrode and discharge lead ions into the drinking water. Orthophosphates are being added to the water supply in order to re-form the protective oxide film on the interior of the lead pipes.

Corrosion of water pipes constructed of copper can sometimes produce the phenomenon of discolored "blue water" due to dissolved Cu^{2+} ions [14, 15]. Concentrations of Cu^{2+} in excess of 2 mg/L can produce a bitter metallic taste.

Economics of Corrosion

In 1978 a comprehensive landmark study was carried out on the economic effects of metallic corrosion in the United States [16]. The results of this study were that the total cost of corrosion in the United States for the year 1975 was the staggering total of $70 billion or approximately 5% of the Gross National Product for that year.

The figure of $70 billion now seems small compared to the results of a more recent study conducted between 1999 and 2001 [9], which places the annual direct cost of corrosion in the United States at $276 billion. This is approximately 3% of the Gross Domestic Product for the period of study.

There have been various previous studies on the economic loss due to corrosion carried out at various times in various industrialized nations [17]. The results have always been consistent. Corrosion consumes 3–5% of the Gross National Product of that particular nation.

The 1978 report divided the $70 billion cost of corrosion into avoidable costs and unavoidable costs. Avoidable costs are those which could have been reduced by the application of available corrosion control practices. Unavoidable costs are those which require advances in new materials and in corrosion technology and control. The avoidable costs in the 1978 study were about 15% of the total cost of $70 billion.

In addition, in the 1978 study, corrosion costs could be divided into direct costs and indirect costs. The following are some examples of direct costs listed by Uhlig [18]:

1. Capital costs – cost of replacement parts, e.g., automobile mufflers, water lines, hot water heaters, sheet metal roofs.
2. Control costs – maintenance, repair, painting.
3. Design costs – extra cost of using corrosion-resistant alloys, protective coatings, corrosion inhibitors.

Examples of indirect losses are as follows [18]:

1. Shutdown – of power plants and manufacturing plants. (See the example of the nuclear power plant mentioned earlier.)

   A second example is provided by a widespread corrosion problem which occurred in August 2006 and forced a shutdown of an oil pipeline in Alaska [19]. Sixteen miles of pipeline had to be

replaced. The oil leakage caused environmental damage and the pipeline closure sent the price of oil higher.
2. Loss of product due to leakage – leakage of pipelines due to corrosion.
3. Contamination of product – In 1991, 100,000 residents of Northern Virginia were surprised to find that sediment and rust from a temporary pumping station caused discoloration of their drinking water [20]. Public officials claimed no health risks were involved, but the prospect of drinking rusty water was unpleasant and disconcerting. Another example of contamination is provided by food spoilage due to the corrosion of containers.

Corrosion and the Conservation of Materials

Corrosion destroys metals by converting them into oxides or other corrosion products. Thus, corrosion affects the global supply of metals by removing components or structures from service so that their replacement consumes a portion of the total supply of the earth's material resources. Environmentalists are interested in conserving our supply of metals not only to conserve minerals but also to reduce the amount of solid materials at landfills or recycling centers. See Fig. 1.4. In addition, extension of the service life of a metal product or component forestalls additional manufacturing or processing, thus decreasing emissions of greenhouse gases.

Fig. 1.4 (*Top*) Corrosion of vintage US automobiles. (*Bottom*) A view on the storage, recycling, and reclamation of used automobiles. Photographs courtesy of Harry's U-Pull-It, West Hazleton, PA

There are two important issues concerning the world's supply of metals. The first of these is the question as to what total supplies of various metals (or their ores) actually exist in nature. Over the past 30 years, there have been various assessments as to the earth's reserves of important metal ores.

Table 1.1 1975 and 1995 estimates of the global reserves of various metals

	1975 Estimate of years of supply [19]	1995 Estimate of years of supply [20]
Aluminum	185	162
Iron	110	77
Nickel	100	43
Molybdenum	90	–
Chromium	64	–
Copper	45	22
Zinc	23	16

Selected results from two studies [21, 22] are given in Table 1.1, which shows that there is a finite limited supply of the metals listed in the table.

Of course, these studies are estimates, and these estimates will change as new supplies of ores are located, as the demand for a given metal changes, and as recycling efforts intensify. In addition, prolonging the service lifetime through the development and use of more effective corrosion-resistant alloys or improved corrosion control measures will stretch the supply of the earth's natural resources.

A second issue regarding the world's supply of metals is the geographical location of certain ores and minerals. Industrialized nations (primarily the United States, western Europe, and Japan) import 90–100% of their total requirements for chromium, cobalt, manganese, and platinum group metals from additional sources [23]. These metals have been referred to as "critical materials." The largest use of chromium is in the manufacture of stainless steels; the largest use of cobalt is in high-temperature alloys.

In order to develop a national self-reliance in regard to these critical materials, there is a continuing interest in the development of new corrosion-resistant and oxidation-resistant alloys containing substitutes for chromium and cobalt. This goal is a formidable challenge for corrosion scientists and engineers, and research in this direction still continues.

In 2006, China announced that it plans to build strategic reserves of various minerals including uranium, copper, aluminum, manganese, and others that the country urgently needs [24].

Finally, corrosion takes its toll on national landmarks, works of art, and historical artifacts [25–27]. See Fig. 1.5. Preservation and restoration of these objects are an important part of material conservation. For example, the Statue of Liberty was found to have suffered extensive corrosion in the iron structure, which supports its copper skin as well as perforation of copper in the torch area [27]. A massive restoration project was completed in 1986.

Despite all the problems with corrosion mentioned so far, corrosion is sometimes (but not usually) good. Rust possesses an attractive reddish brown hue so that protective layers of rust can be attractive in outdoor settings. Figure 1.6 is a photograph of an attractive rust-colored giant watering can, which can be found in a garden center south of Alexandria, VA. More about the unusual beneficial aspects of corrosion is given in Chapter 17.

The Study of Corrosion

In addition to the importance of corrosion discussed above, the study of corrosion is in itself a challenging and interesting pursuit. Corrosion science is an interdisciplinary area embracing chemistry, materials science, and mechanics, as shown in Fig. 1.7. The study of aqueous corrosion processes

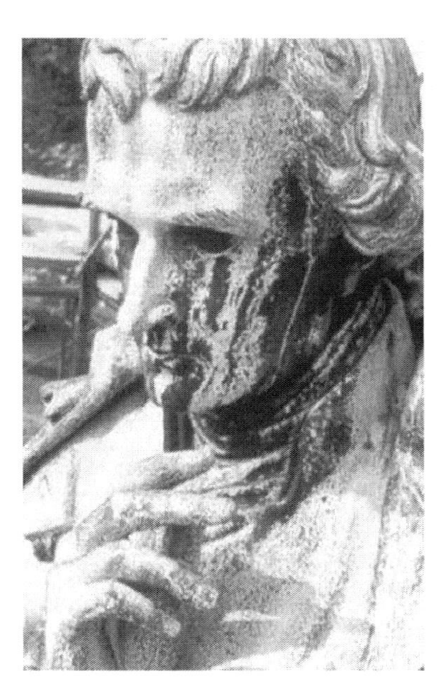

Fig. 1.5 Details of the statue of Thomas Jefferson from the *Washington Monument* (1858), by Thomas Crawford and Randolph Rogers, Virginia State Capitol, Richmond, VA, showing corrosion damage caused by industrial pollution. Photograph courtesy of Phoebe Dent Weil, Northern Light Studio, Florence, MA

Fig. 1.6 Photograph of a rust-covered outdoor work of art near Alexandria, VA

involves the intersection of chemistry and materials science. But the science of mechanics must be added to understand mechanically assisted corrosion processes, such as stress-corrosion cracking and corrosion fatigue.

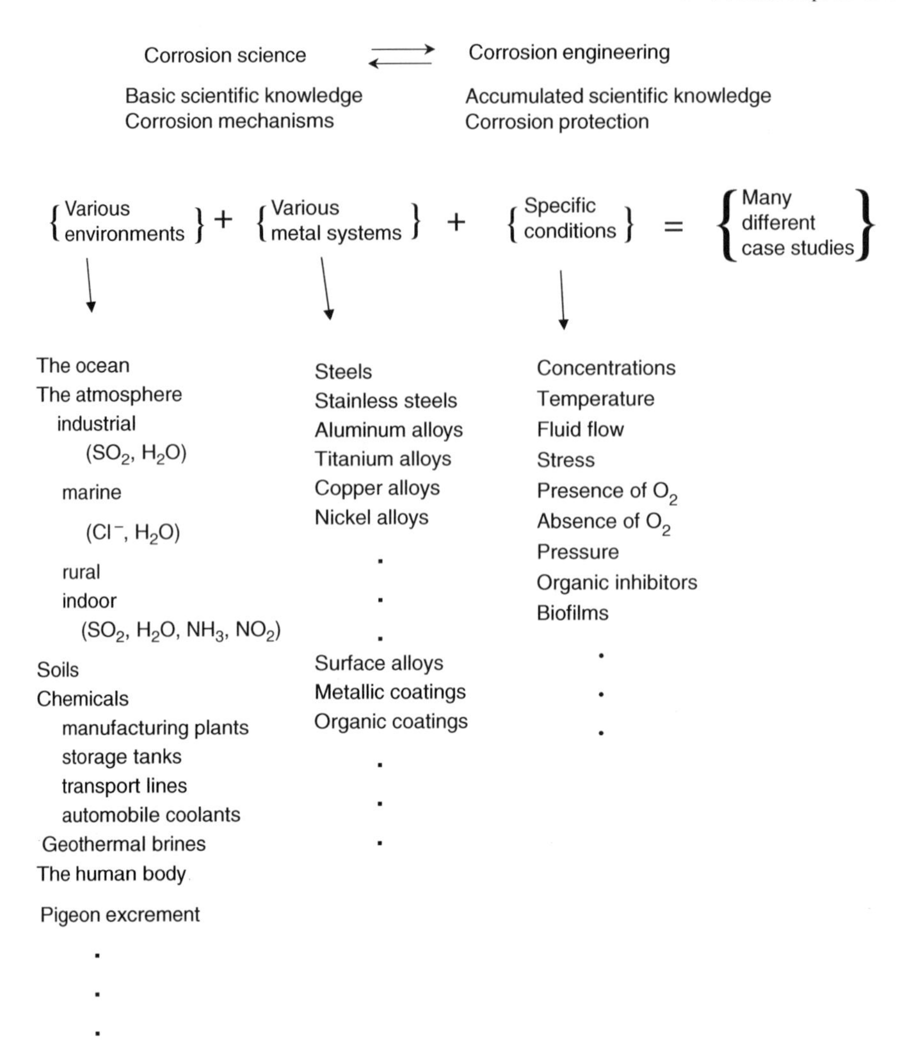

Fig. 1.7 Various disciplines involved in corrosion science

Other relationships exist between the various disciplines, as shown in Fig. 1.7. A blend of mechanics and materials science can address dry fracture processes, and mechanics and chemistry can study chemico-mechanical processes such as adhesion and wear, among others. But again, the science of chemistry must be included to understand environmentally assisted fracture, i.e., stress-corrosion cracking and corrosion fatigue.

Corrosion Science vs. Corrosion Engineering

Corrosion science is directed toward gaining basic scientific knowledge so as to understand corrosion mechanisms. Corrosion engineering involves accumulated scientific knowledge and its application to corrosion protection. Ideally, corrosion science and corrosion engineering complement and reinforce each other, but it has been the author's observation that most workers in the field of corrosion settle into one camp or the other. The most effective corrosionists are those who understand both the science and the engineering of corrosion.

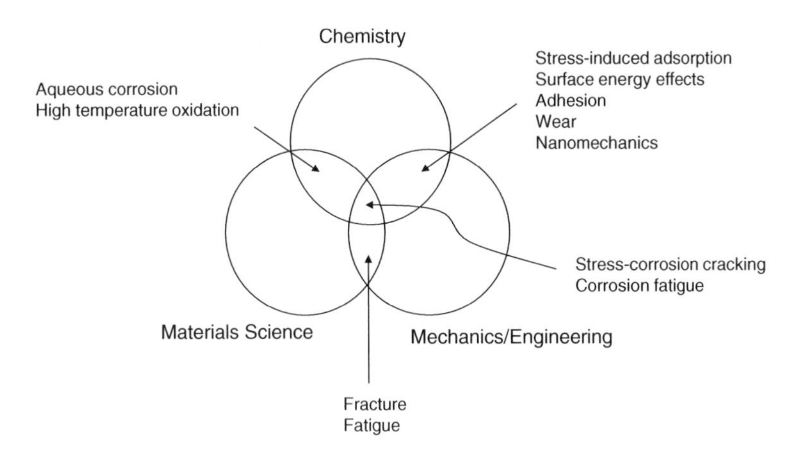

Fig. 1.8 Schematic relationship between corrosion science and corrosion engineering and the large number of variables which can be operative when corrosion occurs

Figure 1.8 shows schematically the relationship between corrosion science and engineering in addition to pointing out the many various factors which can be present in any given situation. The large number of possible various environments in which metals are used plus the large number of possible metals which can be used (with or without protective coatings) plus the large number of possible specific conditions of use generate a very large number of individual case studies. Corrosion science aids corrosion engineering by providing connections between various case studies. In addition, the understanding of corrosion mechanisms can lead to possible new corrosion-resistant alloys, better surface treatments, and improved corrosion control measures.

This text emphasizes the corrosion science approach but will also include extensions to practice.

Challenges for Today's Corrosion Scientist

Based partly on the introductory material in this chapter and partly on more detail to follow in subsequent chapters, several important timely challenges to the corrosion scientist can be listed. These are the following:

1. The development of protective surface treatments and corrosion inhibitors to replace inorganic chromates, which are environmentally objectionable.
2. An improved conservation of materials through the development of corrosion-resistant surface alloys which confine alloying elements to the surface rather than employing conventional bulk alloying. (Initial research strides have been made in this direction, as will be seen in Chapter 16).
3. The formulation of a new generation of stainless steels containing replacements for chromium and other critical metals.
4. An improved understanding of passivity so as to use our fundamental knowledge to guide the development of alloys having improved corrosion resistance.
5. Understanding the mechanism of the breakdown of passive oxide films by chloride ions and subsequent pitting of the underlying metal.

6. The development of "smart" organic coatings which can detect a break in the coating and automatically dispatch an organic molecule to the required site to both heal the coating and inhibit corrosion.
7. The ability to predict the lifetime of metals and components from short-term experimental corrosion data.

It is suggested that the reader refer back to these challenges as he or she proceeds through this text. Perhaps the reader can provide additions to this list.

Problems

1. Examples of corrosion can be found in everyday life. Describe one example which you have seen. Is this particular instance of corrosion primarily an example of wastage of materials, an economic loss, or a safety issue? Or is this example a combination of some of these factors? Note: many photographs of corrosion can be found on the Internet. If you are not familiar with an example of corrosion, select one photograph from the Internet and then complete this problem.
2. Based on your everyday experience, name one method of corrosion protection which you have observed in use.
3. Ordinary garbage cans are often constructed from galvanized steel (a coating of zinc on steel). What direct costs and indirect costs of corrosion are involved if you need to replace such a garbage can with a similar one because your old one is no longer useable due to severe corrosion?
4. Various studies on the annual cost of corrosion always conclude that corrosion amounts to 3–5% of a nation's Gross National Product, no matter in what year the study was undertaken. Does this mean that corrosion science and engineering are not making any headway?

   Note: Before answering this problem, ask yourself the following additional questions: (a) Can you think of any industries which would not exist without the development of corrosion-resistant alloys or corrosion control measures? (b) Are metals being required to perform in increasingly severe environments? (Recall Fig. 1.2). (c) Would you like your automobile to have a longer lifetime before its paint system fails and is overtaken by rusting?
5. Refer to Fig. 1.8 which illustrates the interdisciplinary nature of corrosion. What additional formal disciplines of study would be useful in expanding our knowledge of corrosion?
6. Increasing the corrosion resistance of a metal part or structure so as to make the metal piece last longer is one way to conserve the earth's supply of metals. What other practices can be undertaken to help stretch our natural supply of metals?

References

1. "Hip Replacement Patient Information Booklet", Anderson Orthopedic Institute, Alexandria, VA, p. 1.
2. Front cover of "AMPTIAC Quarterly", (Advanced Materials and Process Technology Information Analysis Center), *AMPTIAC Quart.*, *7* (4), (2003).
3. B. F. Brown, "Stress Corrosion Cracking Control Measures", National Bureau of Standards Monograph 156, Washington, DC (1977).
4. "National Transportation Safety Board Report HAR-84/03", June 28, 1983, Web site http://www.ntsb.gov, July (2004).
5. "Aging Airplanes", S. Derra, *R & D Magazine*, p. 29, January (1970).
6. "FirstEnergy Faced String of Difficulties", Washington Post, p. E1, August 19 (2003).
7. "Walkway Collapse Shuts Laurel Mall", Washington Post, p. B4, July 2 (2005).

8. “Tense Rescues Follow Massive Water Main Break”, washingtonpost.com, December 23 (2008).
9. G. H. Koch, M. P. H. Brongers, N. G. Thompson, Y. P. Virmani, and J. H. Payer, “Corrosion Costs and Preventative Strategies in the United States”, *Mater. Perform.*, *42* (Supplement), 3, July (2002).
10. “At Accident Site, a Bridge Too Far Corroded”, Washington Post, p. B2, November 15 (2007).
11. “DOE Defends ‘Hot’ Repository Design”, *C & E News*, p. 19, May 31 (2004).
12. Excerpted from E. M. Haas, “Staying Healthy with Nutrition”, Web site http://www.healthy.net, July (2004).
13. “Agencies Brushed Off Lead Warnings”, Washington Post, p. A1, February 29 (2004).
14. Australian Water Association, “Fact Sheet: The problem of blue-water and copper pipes”, July (2004).
15. D. J. Fitzgerald, *Am. J. Clin. Nutr.*, *67* (5), 1098S, (Supplement S) (1998).
16. L. H. Bennet, J. Kruger, R. I. Parker, E. Passiglia, C. Reimann, A. W. Ruff, H. Yakowitz, and E. B. Berman, “Economic Effects of Metallic Corrosion in the United States”, National Bureau of Standards Special Publication 511, Washington, DC (1978).
17. J. Kruger in “Uhlig’s Corrosion Handbook”, R. W. Revie, Ed., p. 3, John Wiley, New York (2000).
18. H. H. Uhlig and W. R. Revie, “Corrosion and Corrosion Control”, Chapter 1, John Wiley, New York (1985).
19. “Pipeline Closure Sends Oil Higher”, Washington Post, p. A1, August 8 (2006).
20. “Rusty water rattles N. Va. residents”, *Fairfax J.*, p. A5, April 29 (1991).
21. J. J. Harwood, *ASTM Stand. News*, *3* (1), 12 (1975).
22. T. E. Norgate and W. J. Rankin, “The role of metals in sustainable development”, in “Proceedings, Green Processing 2002, International Conference on the Sustainable Processing of Minerals”, p. 49, Australian Institute of Mining and Metallurgy, May (2002).
23. W. L. Swager in “Battelle Today” Vol. 18, p. 3, June (1980).
24. “China to Build Up Mineral Reserves, Ministry Says”, Washington Post, p. D10, May 17 (2006).
25. B. F. Brown et al., “Corrosion and Metal Artifacts”, National Bureau of Standards Special Publication 479, Washington, DC (1977).
26. R. Baboian, *Mater. Perform.*, *33* (8), 12 (1994).
27. N. Nielsen, *Mater. Perform.*, *23* (4), 78 (1984).

Chapter 2
Getting Started on the Basics

Introduction

What is Corrosion?

Corrosion is the destructive attack of a metal by its reaction with the environment. A more scientific definition of corrosion will be given later in this chapter, but the description just provided is a good working one. As seen in Chapter 1, there are very many different specific environments which are possible, depending upon how the particular metal is used. The most general case is that in which the environment is a bulk aqueous solution. For atmospheric corrosion, the aqueous solution is a condensed thin-layer rather than a bulk solution, but the overall principles are, for the most part, the same.

Note that the word "corrosion" refers to the degradation of a *metal* by its environment. Other materials such as plastics, concrete, wood, ceramics, and composite materials all undergo deterioration when placed in some environment; but this text will deal with only the corrosion of metals.

The word "rusting" applies to the corrosion of iron and plain carbon steel. Rust is a hydrated ferric oxide which appears in the familiar color of red or dark brown. See Fig. 2.1. Thus, steel rusts (and also corrodes), but the non-ferrous metals such as aluminum, copper, and zinc corrode (but do not rust). The term "white rust" is often used to describe the powdery white corrosion product formed on zinc. The "white rusting" of sheets of galvanized steel (zinc-coated steel) is a frequent problem if the sheets are stacked and stored under conditions of high relative humidity. Condensation of moisture between stacked sheets will often lead to "white rusting".

Physical Processes of Degradation

Metals may undergo degradation by physical processes which occur in the absence of a chemical environment. Physical degradation processes include the following:

Fracture – failure of a metal under an applied stress.
Fatigue – failure of a metal under an applied repeated cyclic stress.
Wear – rubbing or sliding of materials on each other.
Erosion or cavitation erosion – mechanical damage caused by the movement of a liquid or the collapse of vapor bubbles against a metal surface.
Radiation damage – interaction of elementary particles (e.g., neutrons or metal ions) with a solid metal so as to distort the metal lattice.

E. McCafferty, *Introduction to Corrosion Science*, DOI 10.1007/978-1-4419-0455-3_2,

Fig. 2.1 An example of red rust showing the corrosion of a ship near the waterline. Photograph courtesy of James R. Martin, Naval Research Laboratory, Washington, DC

Environmentally Assisted Degradation Processes

Each of the physical degradation processes above can be assisted or aggravated in the presence of an aqueous environment. Thus, corresponding to each of the degradation processes listed immediately above are environmentally assisted counterparts, as given in Table 2.1. In each case, metal degradation is intensified by the conjoint action of the physical process and the chemical environment.

Examples of each of these environmentally assisted processes are also given in Table 2.1, and more detail is provided in Chapter 11.

Table 2.1 Physical degradation processes and their environmentally assisted counterparts

	Environmentally assisted process	Example of environmentally- assisted process
Fracture	Stress-corrosion cracking	Stress-corrosion cracking of bridge cables, of landing gear on aircraft
Fatigue	Corrosion fatigue	Vibrating structures, such as aircraft wings, bridges, offshore platforms
Wear	Fretting corrosion	Ball bearings in chloride-contaminated oil
Cavitation erosion	Cavitation corrosion	Ship propellers, pumps, turbine blades, fast fluid flow in pipes
Radiation damage	Radiation corrosion	Increased susceptibility of stainless steels to dissolution or to stress-corrosion cracking [1, 2]

Electrochemical Reactions

Corrosion is an electrochemical process. That is, corrosion usually occurs not by direct chemical reaction of a metal with its environment but rather through the operation of coupled electrochemical half-cell reactions.

Half-Cell Reactions

A half-cell reaction is one in which electrons appear on one side or another of the reaction as written.

If electrons are products (right-hand side of the reaction), then the half-cell reaction is an oxidation reaction.

If electrons are reactants (left-hand side of the reaction), then the half-cell reaction is a reduction reaction.

Anodic Reactions

The loss of metal occurs as an anodic reaction. Examples are

$$Fe(s) \rightarrow Fe^{2+}(aq) + 2e^- \quad (1)$$

$$Al(s) \rightarrow Al^{3+}(aq) + 3e^- \quad (2)$$

$$2Cu(s) + H_2O(l) \rightarrow Cu_2O(s) + 2H^+(aq) + 2e^- \quad (3)$$

where the notations (s), (aq), and (l) refer to the solid, aqueous, and liquid phases, respectively. Each of the above reactions in Eqs. (1), (2), and (3) is an anodic reaction because of the following:

(1) A given species undergoes oxidation, i.e., there is an increase in its oxidation number.
(2) There is a loss of electrons at the anodic site (electrons are produced by the reaction).

These ideas are illustrated schematically in Fig. 2.2.

The following reaction is also an anodic reaction:

$$Fe(CN)_6^{4-}(aq) \rightarrow Fe(CN)_6^{3-}(aq) + e^- \quad (4)$$

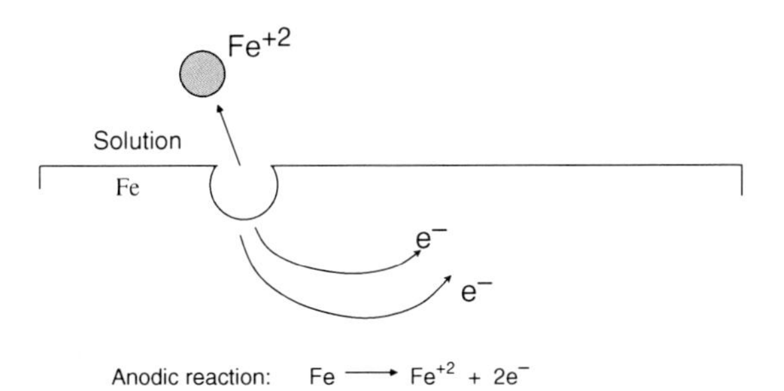

Fig. 2.2 Example of an anodic reaction – the dissolution of iron

The oxidation number of the Fe species on the left, i.e., in the ferrocyanate ion, is +2, and the oxidation number of Fe in the ferricyanate ion on the right is +3. Thus, there is an increase in oxidation number. In addition, electrons are produced in the electrochemical half-cell reaction, so Eq. (4) is an anodic reaction. By the same reasoning the following is also an anodic reaction:

$$Cr^{3+}(aq) + 4H_2O \rightarrow CrO_4^{2-}(aq) + 8H^+(aq) + 3e^- \tag{5}$$

Although Eqs. (4) and (5) are anodic reactions, they are not corrosion reactions. There is a charge transfer in each of the last two equations, but not a loss of metal. Thus, not all anodic reactions are corrosion reactions. This observation allows the following scientific definition of corrosion:

Corrosion is the simultaneous transfer of *mass and charge* across a metal/solution interface.

Cathodic Reactions

In a cathodic reaction

(1) A given species undergoes reduction, i.e., there is a decrease in its oxidation number.
(2) There is a gain of electrons at the cathodic site (electrons are consumed by the reaction).

An example of a cathodic reaction is the reduction of two hydrogen ions at a surface to form one molecule of hydrogen gas:

$$2H^+(aq) + 2e^- \rightarrow H_2(g) \tag{6}$$

This is the predominant cathodic reaction in acidic solutions. See Fig. 2.3 for a schematic representation of this reduction reaction.

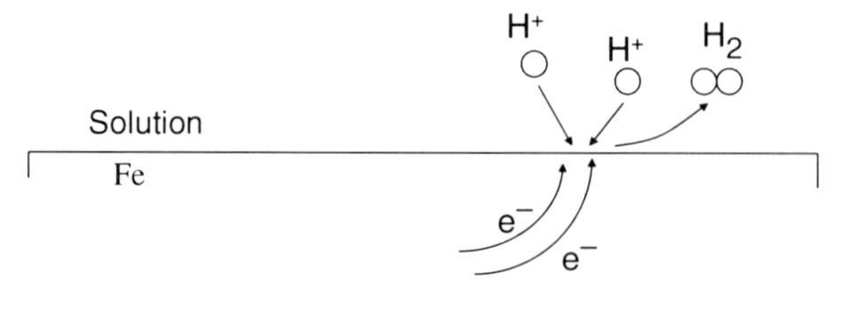

Fig. 2.3 Example of a cathodic reaction – hydrogen evolution on iron immersed in an acid solution

Another common cathodic reaction is the reduction of dissolved oxygen to hydroxyl ions, a reduction reaction which occurs in neutral or basic solutions.

$$O_2(g) + 2H_2O + 4e^- \rightarrow 4OH^-(aq) \tag{7}$$

Coupled Electrochemical Reactions

On a corroding metal surface, anodic and cathodic reactions occur in a coupled manner at different places on the metal surface. See Fig. 2.4, which illustrates this behavior for an iron surface immersed in an acidic aqueous environment. At certain sites on the iron surface, iron atoms pass into solution as Fe^{2+} ions by Eq. (1). The two electrons produced by this anodic half-cell reaction are consumed elsewhere on the surface to reduce two hydrogen ions to one H_2 molecule.

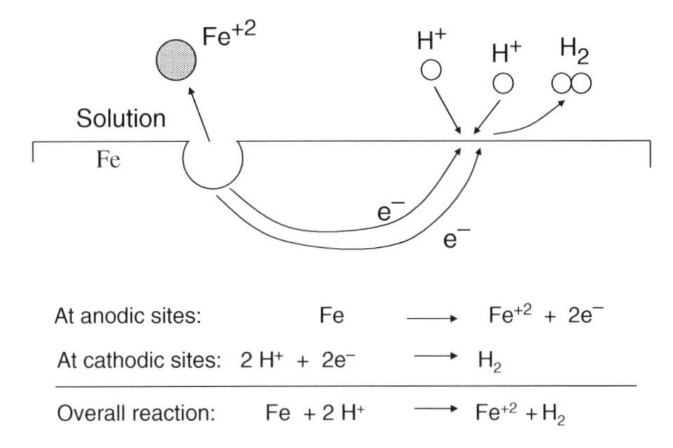

Fig. 2.4 Coupled electrochemical reactions occurring at different sites on the same metal surface for iron in an acid solution. The electrons lost by the oxidation of Fe atoms are consumed in the reduction of two H^+ ions to form hydrogen gas (H_2)

The reason that two different electrochemical half-cell reactions can occur on the same metal surface lies in the heterogeneous nature of a metal surface. Polycrystalline metal surfaces contain an array of site energies due to the existence of various crystal faces (i.e., grains) and grain boundaries. In addition, there can be other defects such as edges, steps, kink sites, screw dislocations, and point defects. Moreover, there can be surface contaminants due to the presence of impurity metal atoms or to the adsorption of ions from solution so as to change the surface energy of the underlying metal atoms around the adsorbate. Some of these effects are illustrated in Fig. 2.5.

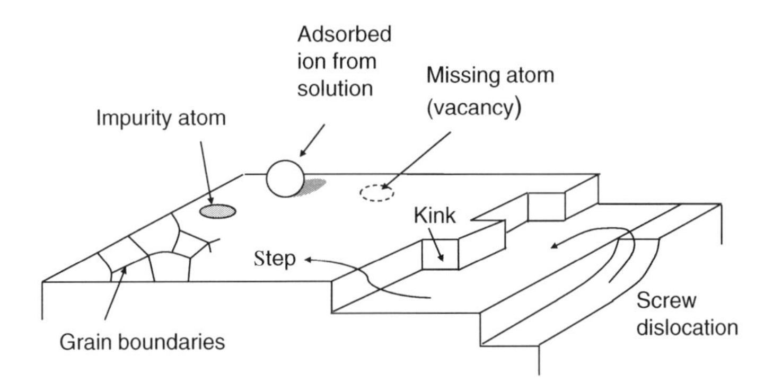

Fig. 2.5 The heterogeneous nature of a metal surface showing various types of imperfections

Metal atoms at the highest energy sites are most likely to pass into solution. These high-energy sites include atoms located at the edges and corners of crystal planes, for example. Stressed surfaces also contain atoms that are reactive because they have a less stable crystalline environment. When a metal is cold worked or shaped, the metal lattice becomes strained, and atoms located in the strained regions tend to go into solution more readily than do atoms in unstrained regions. Once the process of metal dissolution process begins, a new energy distribution of sites is established. Then, the positions of anodic and cathodic surface sites change randomly with time so that the overall effect is uniform corrosion of the metal.

The overall chemical reaction is thus the sum of the two half-cell reactions. For the process shown in Fig. 2.4:

At the local anodes

$$Fe(s) \rightarrow Fe^{2+}(aq) + 2e^{-} \tag{1}$$

At the local cathodes

$$2H^{+}(aq) + 2e^{-} \rightarrow H_2(g) \tag{6}$$

The overall reaction is the sum of these two half-cell reactions:

$$Fe(s) + 2H^{+}(aq) \rightarrow Fe^{2+}(aq) + H_2(g) \tag{8}$$

Figure 2.4 illustrates the four conditions which are necessary for corrosion to occur. These are the following:

(1) An anodic reaction
(2) A cathodic reaction
(3) A metallic path of contact between anodic and cathodic sites
(4) The presence of an electrolyte

An electrolyte is a solution which contains dissolved ions capable of conducting a current. The most common electrolyte is an aqueous solution, i.e., water containing dissolved ions; but other liquids, such as liquid ammonia, can function as electrolytes.

Figure 2.6 illustrates the coupled electrochemical reactions for an iron surface immersed in a neutral or a basic aqueous solution. Figure 2.7 schematically shows the continuation of these reactions en route to the formation of hydrated ferric oxide (rust).

A Note About Atmospheric Corrosion

The need for the presence of an electrolyte as a condition for corrosion to occur is illustrated by the phenomenon of atmospheric corrosion, i.e., the corrosion of metals in the natural outdoor atmosphere. Vernon [3] observed that a *critical relative humidity* exists below which atmospheric corrosion is negligible and above which corrosion occurs. Figure 2.8 shows Vernon's results for iron exposed to water vapor containing 0.01% SO_2 [3]. It can be seen that the corrosion of iron occurs above 60% relative humidity. (The critical relative humidity is typically 50–70% for most metals.)

The critical relative humidity is the condition where multimolecular layers of water vapor physically adsorb from the atmosphere onto the oxide-covered metal surface. (Physical adsorption involves the accumulation of a gas or a liquid on a solid surface due to the secondary short-range forces.) Oxide films on iron can contain various iron oxides, including Fe_2O_3. Figure 2.9 shows adsorption isotherms for water on α-Fe_2O_3 [4], in which it is seen that two or more layers of adsorbed

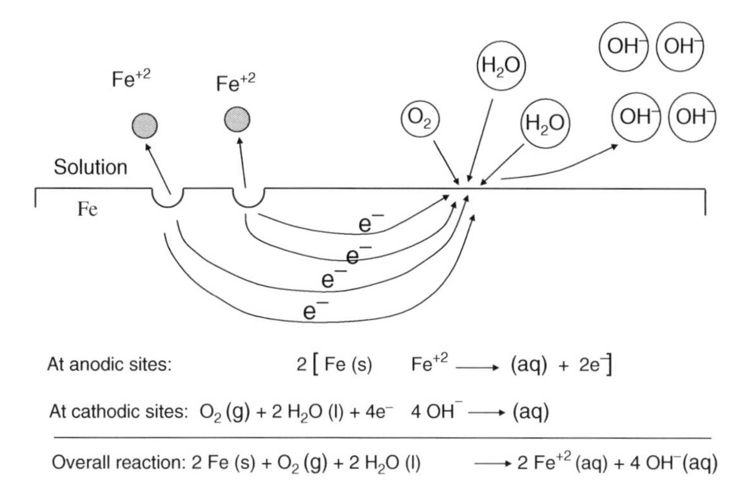

Fig. 2.6 Coupled electrochemical reactions occurring at different sites on the same metal surface for iron in a neutral or a basic solution

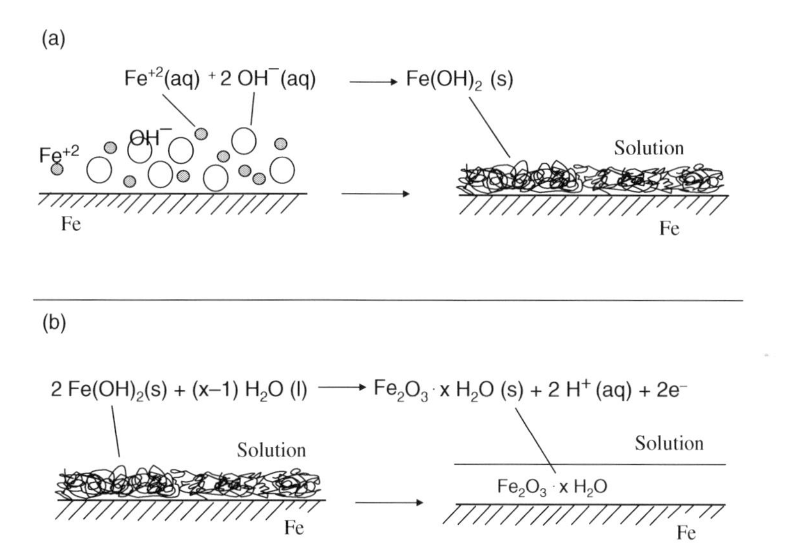

Fig. 2.7 The continuation of reactions initiated in Fig. 2.6. (**a**) The precipitation of ferrous hydroxide on the iron surface. (**b**) The conversion of ferrous hydroxide to a hydrated ferric oxide

water are formed at relative humidities greater than 60%. Thus, the thin layer of electrolyte required for atmospheric corrosion to occur is provided at or above this critical relative humidity.

Secondary Effects of Cathodic Reactions

Cathodic reactions do not involve the loss of the metal substrate and thus are not per se corrosion reactions. However, cathodic reactions are very important for two reasons. First, as seen above, the cathodic reaction is coupled to the anodic reaction so that impeding the cathodic reaction will also impede the anodic reaction. Similarly, accelerating the cathodic reaction will also accelerate the anodic reaction.

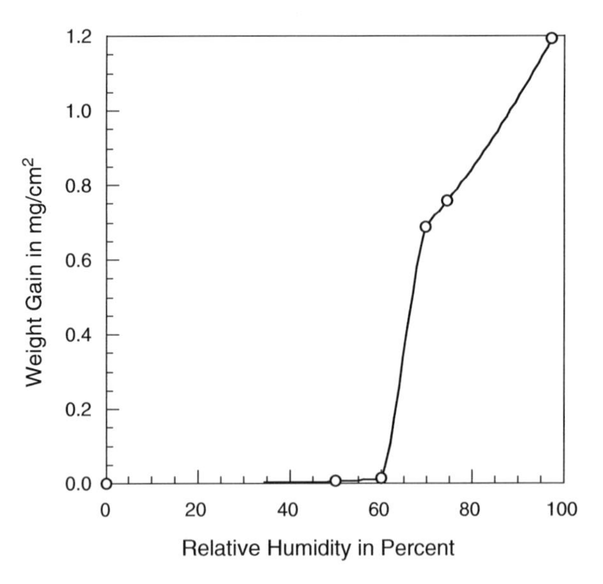

Fig. 2.8 Corrosion of iron in air containing 0.01% SO_2 after 55 days of exposure showing the effect of a critical relative humidity (approximately 60%). Redrawn from Vernon [3] by permission of the Royal Society of Chemistry

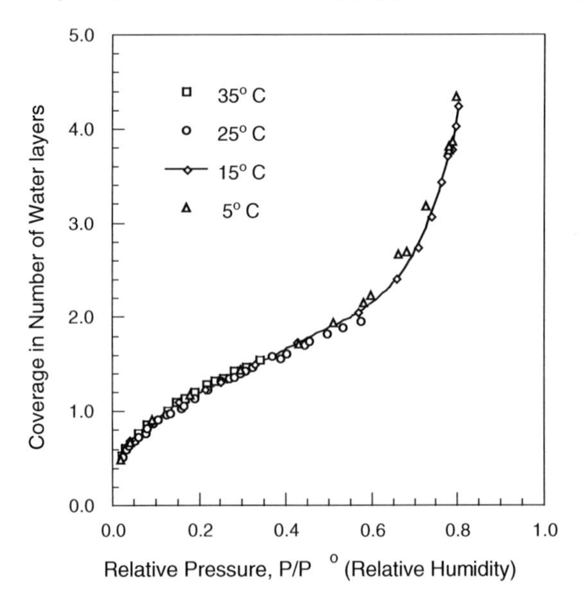

Fig. 2.9 Adsorption isotherms for water vapor on α-Fe_2O_3 [4] showing that multimolecular layers of adsorbed water are formed at a relative humidity of 60% and higher

In addition, cathodic reactions may induce corrosion through secondary effects caused by the products of the cathodic reaction. In stress-corrosion cracking (considered in Chapter 11), the narrow confines of the stress-corrosion crack limit the exchange of dissolved metal ions with the bulk electrolyte, as shown in Fig. 2.10. Thus, metal cations, in this case Fe^{2+} ions, accumulate within the stress-corrosion crack and are then hydrolyzed to form hydrogen ions:

$$Fe^{2+}(aq) + H_2O(1) \rightarrow FeOH^+(aq) + H^+(aq) \tag{9}$$

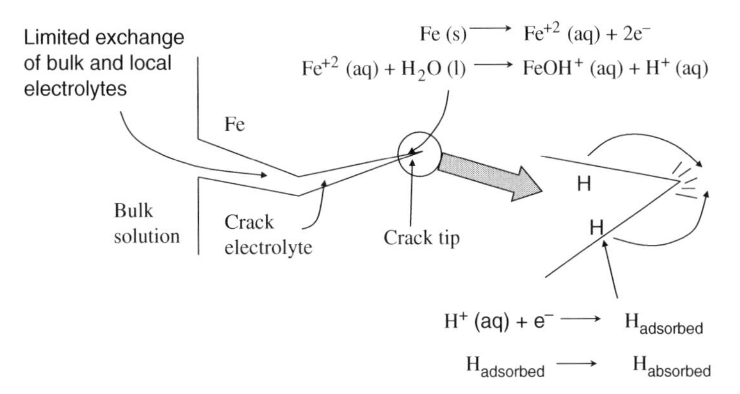

Fig. 2.10 Crack tip reactions can produce hydrogen atoms available for migration into the metal at stressed regions ahead of the crack tip

The local environment within the crack tip becomes acidified due to the production of hydrogen ions. As discussed earlier, in acidic solutions, the major cathodic reaction is the reduction of hydrogen ions. Thus, hydrogen ions produced within the crack can be reduced to form hydrogen atoms which adsorb on the metal surface:

$$H^{+}(aq) + e^{-} \rightarrow H(adsorbed) \tag{10}$$

Some of these hydrogen atoms then migrate into the stressed region ahead of the crack tip (rather than combining to form hydrogen gas). The presence of hydrogen atoms in stressed areas promotes the growth of the stress-corrosion crack by the process of hydrogen embrittlement, as discussed in Chapter 11.

In neutral or basic solutions, the major cathodic reaction is the reduction of dissolved oxygen, as given by Eq. (7). Note that hydroxyl ions are produced by this cathodic reaction. In a thin-layer electrolyte, such as that which exists in atmospheric corrosion, the continued production of OH^- ions and their accumulation will cause an increase in the pH of the thin layer of solution. (See Problem 2.12 at the end of this chapter.) If the pH increases, i.e., the solution becomes more alkaline, then the corrosion behavior of the underlying substrate can be altered. Aluminum, for example, has a low corrosion rate at pH 7, but the corrosion rate increases dramatically with increasing pH, as shown in Fig. 2.11 [5].

Three Simple Properties of Solutions

The pH is a measure of the acidity (or alkalinity) of a solution and is defined as

$$\mathrm{pH} = -\log[\mathrm{H}^{+}] = \log \frac{1}{[\mathrm{H}^{+}]} \tag{11}$$

where $[H^+]$ is the concentration of hydrogen ions in solution. Neutral solutions have a pH value of 7.0, while acid solutions have pH values less than 7.0 and alkaline (basic) solutions have pH values greater than 7.0. Most solutions have pH values between 0 and 14, but lower and higher values are possible. (For example, the pH of 12 M HCl is –1.1.)

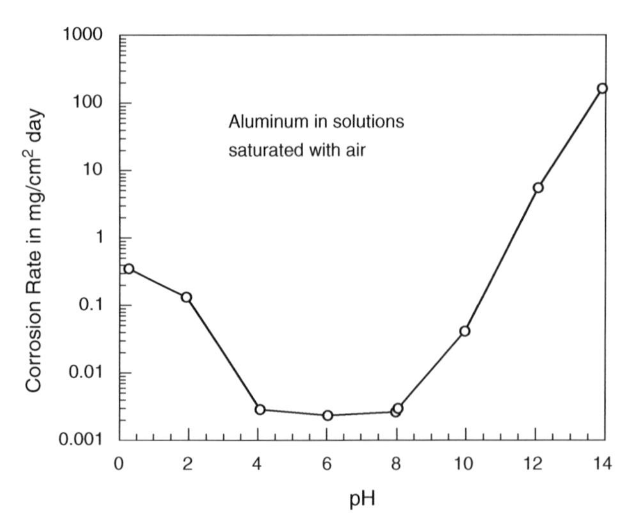

Fig. 2.11 Aluminum has high corrosion rates for both acidic (low pH) and basic (high pH) solutions [5]. Reproduced by permission of ECS – The Electrochemical Society

The concentration of hydrogen ions and of hydroxyl ions (OH^-) in aqueous solutions are related by the following equation:

$$K_w = [H^+][OH^-] = 1.0 \times 10^{-14} \tag{12}$$

where K_w is called the ionization constant for water.

When an ionic solid (acid, base, or salt) dissociates into ions in solution:

$$A_xB_y(s) \rightleftharpoons xA^{y+}(aq) + yB^{x-}(aq) \tag{13}$$

the solubility product K_{sp} is defined by

$$K_{sp} = [A^{y+}] \times [B^{x-}]^y \tag{14}$$

where $[A^{y+}]$ and $[B^{x-}]$ are the concentrations of the dissolved ions. If the ion product $[A^{y+}]^x\ [B^{x-}]^y$ exceeds the tabulated value of K_{sp}, then precipitation of solid A_xB_y occurs. Otherwise, precipitation does not occur.

Example 2.1: The corrosion of aluminum produces Al^{3+} ions in solution. Write an expression for the solubility product of aluminum hydroxide. If the concentration of Al^{3+} ions in solution is 1.0×10^{-6} M, will aluminum hydroxide precipitate at pH 9.0? The tabulated value for K_{sp} for aluminum hydroxide is 1.3×10^{-33}.

Solution: The equilibrium between $Al(OH)_3$ and its ions is

$$Al(OH)_3(s) \rightleftharpoons Al^{3+}(aq) + 3OH^-(aq)$$

and K_{sp} is given by

$$K_{sp} = [Al^{3+}][OH^-]^3$$

where $[Al^{3+}]$ and $[OH^-]$ are the concentrations in moles per liter of Al^{3+} and OH^-, respectively. At pH 9.0

$$9.0 = -\log[H^+]$$

or $[H^+] = 1.0 \times 10^{-9}$ M. The concentration of OH^- is then obtained from the following equation:

$$(1.0 \times 10^{-9})[OH^-] = 1.0 \times 10^{-14}$$

or $[OH^-] = 1.0 \times 10^{-5}$. Then

$$[Al^{3+}][OH^-]^3 = (1.0 \times 10^{-6})(1.0 \times 10^{-5})^3 = 1.0 \times 10^{-21}$$

This product is much greater than K_{sp} for aluminum hydroxide so that $Al(OH)_3$ is precipitated under the conditions given.

The Faraday and Faraday's Law

As stated earlier, the process of corrosion involves simultaneous charge transfer and mass transfer across the metal/solution interface.

The unit of charge is the coulomb, which is the product of the current and its time of passage. The unit of mass is, of course, the gram, which can be obtained from the number of equivalents of metal lost and the equivalent weight of the metal.

The link between charge transfer and mass transfer is the Faraday (F):

$$F \equiv \frac{96,500\,\text{C}}{\text{equivalent}}$$

Faraday's law states that the mass (w) of metal corroded is given by

$$W = \frac{ItA}{nF} \tag{15}$$

where I is the current in amperes, t is the time in seconds, A is the atomic weight of the metal, and n is the number of equivalents transferred per mole of metal. For example, $n = 2$ for the anodic reaction

$$Fe(s) \rightarrow Fe^{2+}(aq) + 2e^-$$

whereas $n = 3$ for

$$Al(s) \rightarrow Al^{3+}(aq) + 3e^-$$

The various simple concepts associated with Faraday's law are assembled for convenience in Table 2.2. It should be noted that all Faraday's law problems can be worked through by using simple units analysis.

Table 2.2 Useful information regarding Faraday's law

Current in amperes $= \dfrac{\text{charge in coulombs}}{\text{time in seconds}}$

There are $\dfrac{96,500\,\text{C}}{\text{equivalent}}$.

For the anodic half-cell rection $M \rightarrow M^{n+}$, there are $\dfrac{n\ \text{equivalents}}{\text{mole M}}$.

Example 2.2: Plain carbon steel immersed in seawater has a uniform corrosion rate expressed as a penetration rate of 5.0 mpy (mils per year, where 1 mil = 0.001 in.). The density of iron is 7.87 g/cm^3. The atomic weight of Fe is 55.8 g/mol.

(a) Calculate the weight loss after 1 year.
(b) Calculate the corresponding corrosion current density in microamperes per square centimeter assuming that the corrosion current is given by

$$Fe \longrightarrow Fe^{2+} + 2e^-$$

Solution: (a) The weight loss is given by

$$\left(\frac{5.0\,\text{mils}}{\text{year}}\right)\left(\frac{0.001\,\text{in}}{\text{mil}}\right)\left(\frac{2.54\,\text{cm}}{\text{in.}}\right)\left(\frac{7.87\text{g}}{\text{cm}^3}\right) = \frac{0.10\,\text{g}}{\text{cm}^2\ \text{year}}$$

(b) The corrosion current density is:

$$\left(\frac{0.10\,\text{g}}{\text{cm}^2\ \text{year}}\right)\left(\frac{1\ \text{year}}{356\ \text{days}}\right)\left(\frac{1\ \text{day}}{24\,\text{h}}\right)\left(\frac{1\text{h}}{3600\,\text{s}}\right)\left(\frac{1\ \text{mol}}{55.8\,\text{g}}\right)\left(\frac{2\ \text{equiv}}{\text{mol}}\right)\left(\frac{96{,}500\,\text{C}}{\text{equiv}}\right) = \left(\frac{1.10\times 10^{-5}}{\text{s}\,\text{cm}^2}\right)$$

or

$$\left(\frac{1.10\times 10^{-5}\,\text{A}}{\text{cm}^2}\right)\left(\frac{1\times 10^{6}\,\mu\text{A}}{\text{A}}\right) = \frac{11\,\mu\text{A}}{\text{cm}^2}$$

Units for Corrosion Rates

Common units for the corrosion current density include microamperes per square centimeter, milliamperes per square centimeter, and amperes per square meter. Various units have been used when the mass loss is the experimentally observed variable. These include grams per square centimeter per day and mdd (milligrams per square decimeter per day). Sometimes the corrosion rate is given as a uniform penetration rate. Units include ipy (inches per year), inches per month, and mpy (mils per year, where 1 mil = 0.001 in.). The units millimeters per year and micrometers per year have also

Table 2.3 Relative severity of corrosion rates

Relative corrosion resistance	Corrosion rate (mils per year, mpy)	Corresponding corrosion current density from Faraday's law (in $\mu A/cm^2$)		
		Aluminum[a]	Iron[b]	Lead[c]
Outstanding	< 1	2.3	2.2	0.85
Excellent	1–5	2.3–12	2.2–11	0.8–4.3
Good	5–20	12–47	11–43	4.2–17
Fair	20–50	47–180	43–109	17–42
Poor	50–200	180–470	109–430	42–170
Unacceptable	> 200	> 470	> 430	> 170

[a] $Al \rightarrow Al^{3+} + 3e^-$
[b] $Fe \rightarrow Fe^{2+} + 2e^-$
[c] $Pb \rightarrow Pb^{2+} + 2e^-$

been used. Collections of corrosion rate data are available for various metals and alloys in different environments [6–10].

A compilation of corrosion rate data by NACE International [6] provides a qualitative ranking of the severity of corrosion rates in terms of the units mils per year. See Table 2.3, which also lists corresponding electrochemical corrosion rates in microamperes per square centimeter calculated from Faraday's law for aluminum (a light metal), iron (a transition metal), and lead (a heavy metal). The electrochemical corrosion rate depends, of course, on the atomic weight of the metal and the oxidation state of the metal ion, but in general, corrosion rates greater than about 100 $\mu A/cm^2$ are considered unacceptable.

Rigorous metric units, such as nanometers per second, have not held wide appeal to corrosion scientists or engineers.

Uniform vs. Localized Corrosion

There are two major types of corrosion: uniform corrosion and localized corrosion. Uniform corrosion and three forms of localized corrosion are illustrated schematically in Fig. 2.12.

In uniform corrosion, the metal is attacked more or less evenly over its entire surface. No portions of the metal surface are attacked more preferentially than others, and the metal piece is thinned away by the process of corrosion until the piece eventually fails. Examples include the corrosion of zinc in hydrochloric acid and the atmospheric corrosion of iron or steel in aggressive outdoor environments. In these cases, localized anodes and cathodes exist and operate as discussed earlier. However, the positions of these localized anodes and cathodes change with time and "dance" all over the metal surface so that the overall effect is that the metal is attacked uniformly. Figure 2.13 shows an example of uniform corrosion.

In localized corrosion, local anodes and cathodes also exist, but their positions become fixed so that corrosion proceeds on established portions of the metal surface. An example of localized corrosion is seen in Fig. 2.14, which shows that corrosion is limited to certain fixed locations on the metal surface. The three most prevalent forms of localized corrosion are (i) pitting, (ii) crevice corrosion, and (iii) stress-corrosion cracking. In pitting, the metal is attacked at certain fixed sites on the metal surface where the otherwise protective oxide film breaks down locally, usually due

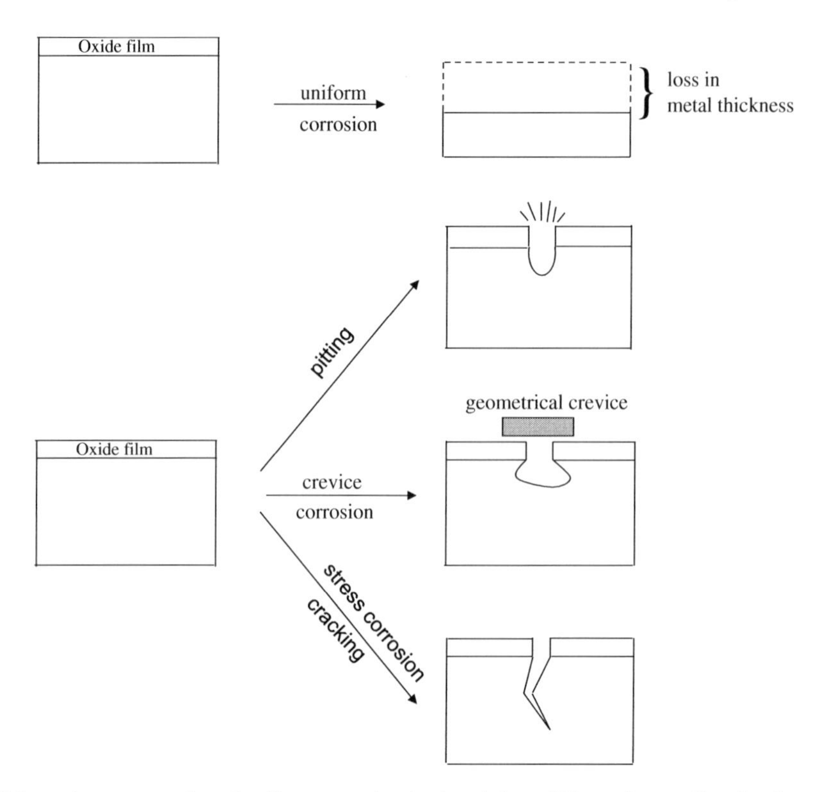

Fig. 2.12 Schematic representation of uniform corrosion (*top*) and three different forms of localized corrosion

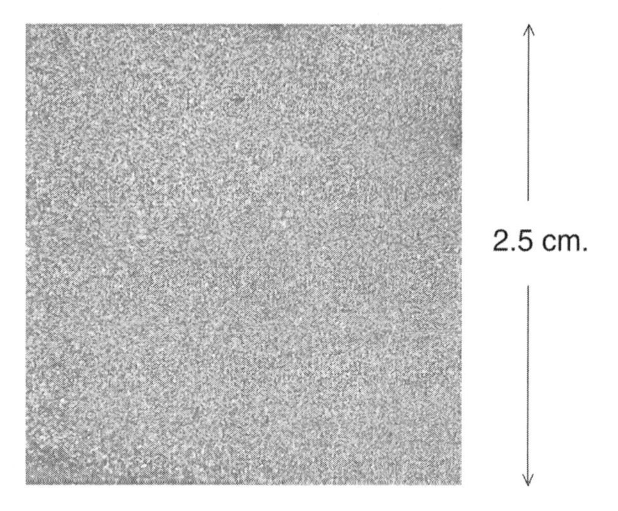

Fig. 2.13 The uniform corrosion of zinc after immersion in hydrochloric acid

to the action of chloride ions. In crevice corrosion, the metal within narrow clearances undergoes localized attack. In stress-corrosion cracking, the combined action of an applied stress and a chemical environment causes the initiation and propagation of cracks in the metal.

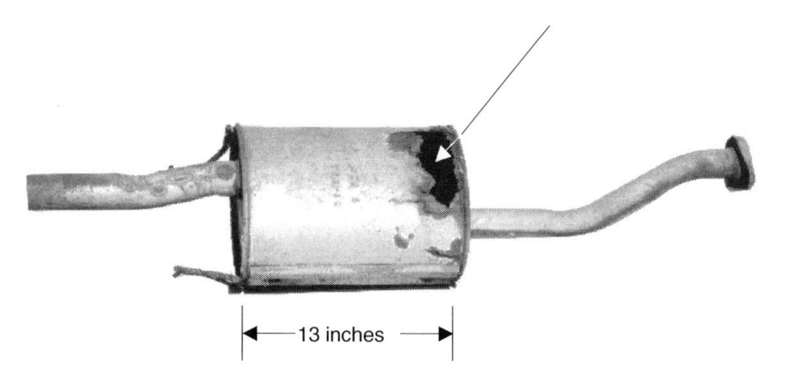

Fig. 2.14 An example of localized corrosion, showing a hole in an automobile muffler caused by the combined action of hot exhaust gases, condensed moisture, and road spray consisting of water and de-icing salts. For an example of localized corrosion on a much finer scale, see Fig. 10.43

Each of these three forms of localized corrosion shares the common feature that there is a geometrical constraint on the system. In the case of pitting corrosion, the geometrical constraint is the cap of corrosion products above the propagating pit. For crevice corrosion and stress-corrosion cracking, the geometrical constraints are the narrow dimensions of the crevice or the crack itself. More thorough discussions of these three types of localized corrosion are given in Chapters 10 and 11.

The Eight Forms of Corrosion

Fontana and Greene [11] have conveniently classified the various types of corrosion into eight forms. The eight forms of corrosion are as follows:

(1) Uniform attack (or general corrosion)
(2) Crevice corrosion
(3) Pitting
(4) Stress-corrosion cracking
(5) Galvanic corrosion (two metal corrosion)
(6) Intergranular corrosion
(7) Selective leaching (dealloying)
(8) Erosion corrosion

We have already mentioned four of these forms in the previous section of this chapter. More details on uniform corrosion are given in Chapters 6, 7, and 8, crevice corrosion and pitting are discussed further in Chapter 10, and stress-corrosion cracking is considered in Chapter 11.

Galvanic corrosion occurs when two metals are in mechanical or electrical contact. In a corrosive environment, one of the metals acts as an anode and undergoes corrosion, while the second metal acts as a cathode and remains unattacked. Galvanic corrosion is discussed in more detail in Chapter 5.

Intergranular corrosion is the pronounced localized attack that occurs in narrow regions at or immediately adjacent to grain boundaries of an alloy. Type 304 stainless steel (which contains 18% Cr and 8% Ni as well as small amounts of carbon) is subject to intergranular corrosion if the stainless steel is heated to the temperature range of 425–790°C (and then cooled). The stainless steel is said to be sensitized and is susceptible to intergranular corrosion. During sensitization, carbon diffuses to

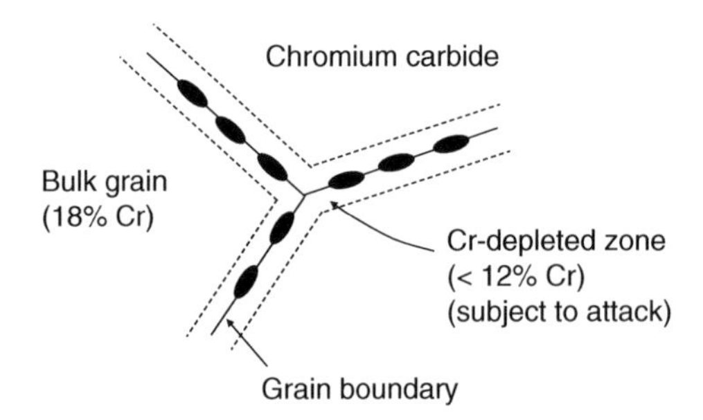

Fig. 2.15 Schematic representation of sensitized stainless steel. After Myers [12]. Chromium-depleted zones adjacent to grain boundaries are susceptible to intergranular corrosion

the grain boundaries where it combines with chromium to form chromium carbide precipitates (such as $Cr_{23}C_6$). This process depletes chromium from the areas in and adjacent to the grain boundaries so that these regions locally contain less than the 12% Cr required for a stainless steel. Thus, localized corrosion occurs in certain aqueous environments in the form of intergranular corrosion, as depicted in Fig. 2.15.

Selective leaching or dealloying is the preferential removal of one element from a solid alloy by corrosion. Examples include the preferential removal of zinc from copper zinc alloys (dezincification) and the preferential removal of iron from gray cast iron (graphitic corrosion) [11]. (Gray cast iron contains carbon in the form of graphite.)

Erosion–corrosion is caused by the mechanical action provided by the movement of a corrosive liquid against the metal surface. This form of corrosion is discussed further in Chapter 11.

Problems

1. Are the following degradation processes strictly physical processes or do they also involve an environmentally assisted component?

   (a) The cleaning of a metal piece by sandblasting.
   (b) Damage to the interior of a pipeline used to transport an abrasive slurry of coal.
   (c) The deterioration of steel reinforcing bars in concrete bridges.
   (d) The fracture of an artificial hip constructed of an alloy of cobalt and chromium when in service in the human body.

2. When pure zinc undergoes corrosion in aerated hydrochloric acid, what is the anodic half-cell reaction? Which cathodic half-cell reactions are possible? What is the overall reaction in each case?
3. Type 430 stainless steel is an alloy of iron and chromium. Suppose that this alloy is used in flowing seawater (pH 8.0) which contains dissolved oxygen. Write the possible anodic half-cell reactions. What is the cathodic half-cell reaction?
4. The corrosion of copper in water occurs by the reaction

$$2Cu(s) + O_2(g) + 2H_2O(l) \rightarrow 2Cu^{2+}(aq) + 4OH^-(aq)$$

   (a) Separate this overall reaction into its two half-cell reactions.

(b) If the concentration of dissolved Cu^{2+} ions is 0.001 M and the pH is 9.0, would the following reaction occur?

$$Cu^{2+}(aq) + 2OH^-(aq) \rightarrow Cu(OH)_2(s)$$

The solubility product for $Cu(OH)_2$ is $K_{sp} = 2.2 \times 10^{-20}$.

(c) When the reaction in (b) is at equilibrium, what is the effect of decreasing the pH on the precipitation reaction? (HINT: recall Le Châtelier's principle.)

5. (a) Derive an expression to convert the corrosion rate in grams per square decimeter per day to the corrosion rate in mils per year (1 mil = 0.001 in). The correct expression will contain the density of the metal. (b) Then derive a second expression to convert the corrosion rate in mils per year to the corrosion rate in millimeters per year.
6. The weight loss of an aluminum alloy corroding in a solution of hydrochloric acid was observed to be 0.250 g/cm^2 after an 8 h immersion period [13]. What is the corresponding anodic current density in milliamperes per square centimeter, assuming that all the corrosion is due to the following anodic half-cell reaction:

$$Al \rightarrow Al^{3+} + 3e^-$$

The atomic weight of Al is 26.98 g/mol.

7. In a short-term laboratory test, the corrosion rate of nickel in boiling 10% phosphoric acid was observed to be 154 mils/year [14] (1 mil = 0.001 in.). The density of nickel is 8.90 g/cm^3 and its atomic weight is 58.7 g/mol.

(a) What is the mass loss per square centimeter after a 1 h immersion period?
(b) What is the corrosion current density in milliamperes per square centimeter?

8. Electronic materials corrode in the presence of condensed water vapor and atmospheric pollutants. The corrosion rate of gold was measured to be 435 mg/m^2 day in an environment consisting of 95% relative humidity and SO_2 and NO_2 pollutants [15]. The atomic weight of gold is 197.0 g/mol and its density is 19.3 g/cm^3.

(a) What is the corresponding corrosion current density in amperes per square meter for the following anodic reaction:

$$Au \rightarrow Au^{3+} + 3e^-$$

(b) What is the penetration rate in millimeter per year?
(c) Based on the penetration rate calculated above, how long will it take to corrode through a gold film which is 0.5 mm in thickness?

9. A warehouse is to be constructed in an industrial city near the seacoast. The roof of the warehouse is to be built of sheets of galvanized steel (an outer coating of zinc on an underlying steel substrate). The outdoor atmosphere near the warehouse contains SO_2 gas and airborne salt particles. When water vapor condenses on the roof, a thin layer of an electrolyte which contains mixed SO_4^{2-} and Cl^- will form. The roof will corrode by the reaction

$$Zn \rightarrow Zn^{2+} + 2e^-$$

Suppose that you have made a chemical analysis of condensed electrolytes in this geographical region and can simulate the condensate with a bulk electrolyte having the appropriate concentrations of SO_4^{2-} and Cl^-. In your laboratory you measure the steady-state corrosion rate of Zn in this electrolyte to be 0.50 μA/cm^2. Galvanized roofs are expected to last 20 years before they are replaced. What is the minimum thickness in mils of zinc coating that is required for this roof to last 20 years? The atomic weight of Zn is 65.4 g/mol, and its density is 7.14 g/cm^3.

10. If metallic iron in an aqueous solution corrodes at the rate of one atomic layer per second, what is the corrosion current density in milliamperes per square centimeter corresponding to:

$$Fe \rightarrow Fe^{2+} + 2e^-$$

The radius of an iron atom is 0.124 nm.

11. If zirconium and hafnium each separately have the same corrosion current density in a given corrosive environment, which of the two metals will suffer the greater weight loss? Explain or show why. Assume that

$$Zr \rightarrow Ze^{4+} + 4e^-$$

$$Hf \rightarrow Hf^{4+} + 4e^-$$

12. In the atmospheric corrosion of aluminum, suppose that the cathodic reduction of oxygen

$$O_2 + 2H_2O + 4e^- \rightarrow 4OH^-$$

occurs for a period of 48 h at a current density of 25 μA/cm^2 in a thin electrolyte film which is 165 μm in thickness [16]. What is the resulting pH in this thin layer of electrolyte if the total electrode area is 2.0 cm^2? If the initial pH was 7.0, would you expect the final pH to cause a change in the corrosion rate? Answer this question by referring to Fig. 2.11.

13. Walk around your campus or drive around your town or locality and identify three instances of corrosion. Describe what you see. Is each individual instance an example of uniform corrosion or of localized corrosion?

14. The occurrence of anodic and cathodic sites on the same surface can be demonstrated by the following experiment [17]. In a small petri dish containing a 3% NaCl solution, immerse a plain iron nail which has been thoroughly cleaned with abrasive paper just prior to its immersion. Add 5 ml of 5% potassium ferricyanide solution and 1 ml of a 1% phenolphthalein solution in alcohol. When anodic areas develop on the nail, they will appear as a blue color. The blue color reveals the presence of ferrous ions. Cathodic areas which produce hydroxyl ions, as per Eq. (7), will give a red color in the presence of the acid–base indicator phenolphthalein.

    (a) On what part of the nail do the anodic areas appear? Explain why they appear at that location.
    (b) Suppose that the nail is bent in half before immersion. Would the bent part of the nail produce anodic or cathodic regions? Explain your answer.

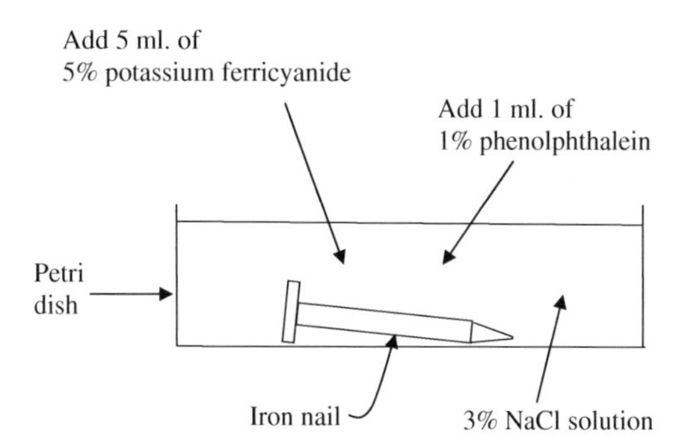

References

1. S. M. Bruemmer, E. P. Simonen, P. M. Scott, P. L. Andresen, G. S. Was, and J. L. Nelson, *J. Nucl. Mater.*, *274*, 299 (1999).
2. S. M. Bruemmer, B. W. Arey, J. I. Cole, and C. F. Windisch, Jr., *Corrosion*, *51*, 11 (1995).
3. W. H. J. Vernon, *Trans. Faraday Soc.*, *31*, 1668 (1935).
4. E. McCafferty and A. C. Zettlemoyer, *Discuss. Faraday Soc.*, *52*, 239 (1971).
5. M. J. Pryor and D. S. Keir, *J. Electrochem. Soc.*, *105*, 629 (1958).
6. D. L. Graver, Ed., "Corrosion Data Survey, Metals Section", NACE, Houston, TX (1985). (Available as an electronic book as "Corrosion Survey Database" by Knovel Corporation, Norwich, NY, 2002).
7. P. A. Schweitzer, "Corrosion Resistance Tables", Marcel Dekker, New York (1995).
8. D. J. De Renzo, Ed., "Corrosion Resistant Materials Handbook", Noyes Data Corporation, Park Ridge, NJ (1985).
9. E. Rabold, "Corrosion Guide", Elsevier Publishing Co., Amsterdam (1968).
10. S. C. Dexter, "Handbook of Oceanographic Engineering Materials", John Wiley, New York (1979).
11. M. G. Fontana and N. D. Greene, "Corrosion Engineering", Chapter 3, McGraw-Hill, New York (1978).
12. J. R. Myers, "Fundamentals and Forms of Corrosion", Air Force Institute of Technology, Wright-Patterson Air Force Base, Ohio, October (1974).
13. E. McCafferty, *Corrosion*, *54*, 862 (1998).
14. W. Z. Friend, "Corrosion of Nickel and Nickel-Base Alloys", p. 51, John Wiley, New York (1980).
15. J. F. Henriksen and A. Rode in "Corrosion and Reliability of Electronic Materials and Devices", R. B. Comizzoli, R. P. Frankenthal, and J. D. Sinclair, Eds., p. 79, The Electrochemical Society, Pennington, NJ (1994).
16. I. L. Rozenfeld, "Atmospheric Corrosion of Metals", p. 65, NACE, Houston, TX (1972).
17. International Nickel Company, "Corrosion in Action", p. 31, International Nickel Company, New York (1977).

Chapter 3
Charged Interfaces

Introduction

Interfaces form at the physical boundary between two phases, such as between a solid and a liquid (S/L), a liquid and its vapor (L/V), or a solid and a vapor (S/V). There can also be interfaces between two different solids (S_1/S_2) or between two immiscible liquids (L_1/L_2). This chapter will consider two special interfaces, the solution/air interface and the metal/solution interface. A brief discussion of the solution/air interface is useful in pointing out some general properties of interfaces. The metal/solution interface is, of course, of paramount interest to the study of corrosion.

This chapter shows how a potential difference originates across a metal/solution interface and discusses the concept of the electrode potential for use in subsequent chapters.

Electrolytes

The Interior of an Electrolyte

An electrolyte is a solution which contains dissolved ions capable of conducting a current. The interior of an electrolyte may consist of a variety of charged and uncharged species.

Consider an aqueous solution which contains the following species:

(1) H_2O molecules
(2) Na^+ ions
(3) Cl^- ions
(4) Organic molecules (which may be present as impurities, biological entities, or may be intentionally added as a corrosion inhibitor).

In a single water molecule, the angle between the two O–H bonds is 105°, and the oxygen atom is more electronegative than the hydrogen atom, so the oxygen end of the molecule contains a partial negative charge. Thus, the water molecule is a dipole, as represented by the vector shown in Fig. 3.1(a). In the interior of liquid water, water molecules are oriented randomly in all directions at any given time, as shown diagrammatically in Fig. 3.1(b). Thus, there is no net electrical field in the interior of liquid water.

In a sodium chloride solution, the electrolyte contains an equal concentration of Na^+ ions and Cl^- ions. In any volume element of solution, there is an equal number of positive and negative ions, and these ions are randomly distributed, as shown in Fig. 3.2. Moreover, these ions are in constant motion migrating through the solution in a random walk. Thus, there is no net charge within any volume element of solution due to the existence of dissolved ions.

E. McCafferty, *Introduction to Corrosion Science*, DOI 10.1007/978-1-4419-0455-3_3,

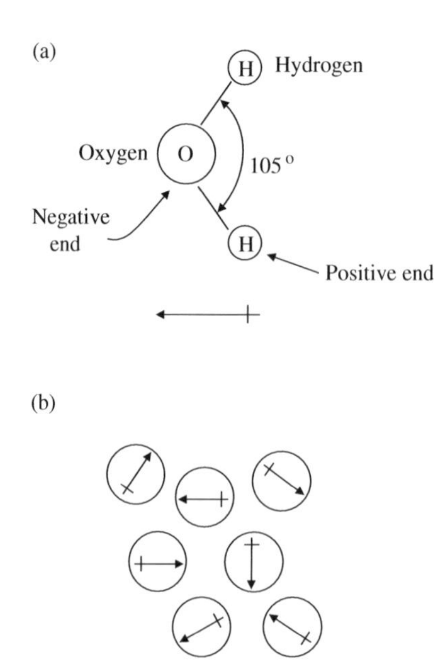

Fig. 3.1 (**a**) The water dipole. (**b**) In the bulk, liquid water consists of an array of randomly oriented dipoles, so the net charge is zero

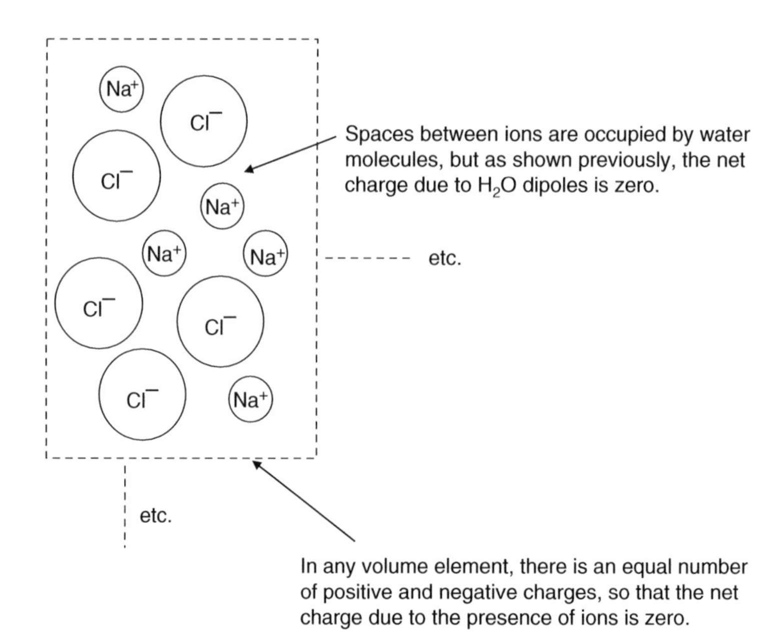

Fig. 3.2 A volume element of sodium chloride solution showing the distribution of ions

This statement needs to be examined more carefully by considering the character of water molecules in ionic solutions. Water molecules in the immediate vicinity of positive or negative ions are attracted toward the charge on the ion. The charge on the ion orients these nearest water molecules with the appropriate end of the dipole pointing toward the ion, as shown in Fig. 3.3. Due

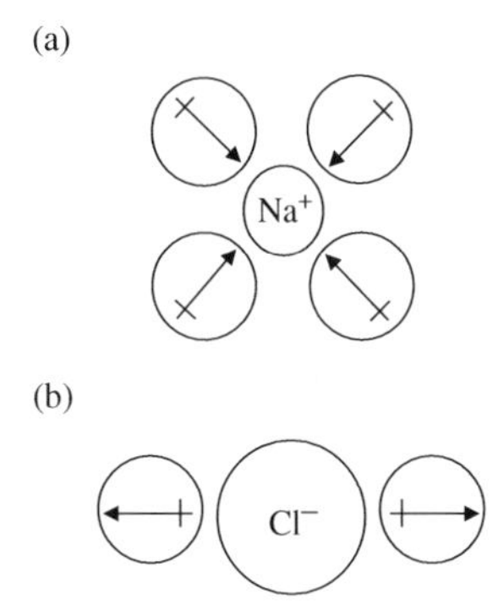

Fig. 3.3 Primary waters of hydration for (**a**) Na^+ ion, (**b**) Cl^- anion. Primary hydration numbers are from Bockris and Reddy [1]

to these ion–dipole forces, a certain number of water molecules become attracted to the central ion. Such water molecules are called primary waters of hydration, and their number usually varies from 1 to 5, depending on the specific ion. Located just outside the primary sheath of oriented water molecules is a secondary region of partially ordered water molecules, called secondary waters of hydration, which balance the localized oriented charge which has developed in the primary water sheath. Thus, the overall effect of ionic hydration is that there is no net charge due to ionic hydration. It can easily be shown that the number of water molecules located in primary water sheaths is a very small percentage of the total number of water molecules in solution. See Problem 3.1 at the end of this chapter.

Outside the region of hydration, the "spaces" between dissolved ions are occupied by bulk water molecules, but we have already established above that there is no net charge for a random distribution of water dipoles. The same situation exists in the bulk of an aqueous solution, so there is no net charge in the interior of an electrolyte due to the existence of water dipoles.

Organic molecules dissolved in solution usually contain functional groups, such as –COOH, which assist in their solubility. Proteins, for instance, contain both –COOH and –NH groups. Suppose that the molecule $CH_3(CH_2)_{10}COOH$ (dodecanoic acid) is contained in solution. Decanoic acid is a weak acid, so in aqueous solution, not all molecules completely ionize, but a certain fraction of the dissolved molecules dissociate into hexanoate ions and protons:

$$CH_3\,(CH_2)_{10}\,COOH(aq) \rightleftharpoons CH_3(CH_2)_{10}COO^-(aq) + H^+$$

The ions produced by this dissociation are free to migrate throughout the solution, and they distribute themselves in a random fashion so that the dissolved organic molecule does not impart any localized charge to the interior of the electrolyte.

Interfaces

Encountering an Interface

Eventually an aqueous solution must terminate in a boundary with another phase. This common boundary is called an interface. In an aqueous solution contained in a laboratory beaker or a test cell, one interface is with the vapor phase above the solution, i.e., the solution/air interface. Even enormous volumes of solution terminate in interfaces. For example, the ocean, which is 35,840 ft (10,924 m) at its deepest level, forms an interface with air at its uppermost level.

The natural world is composed of a variety of interfaces. These include liquid/vapor, liquid/liquid, solid/liquid, solid/vapor, and solid/solid interfaces. The interface of major interest to us is the metal/solution interface. This interface will be treated in more detail after solution/air interfaces are first considered.

The Solution/Air Interface

The properties of a surface region are different than the properties of the bulk. In the bulk of a solution, each ion or molecule is surrounded in all directions by other ions or molecules so that their time-averaged arrangement is the same throughout the interior of the solution. At the surface, however, ions or molecules do not have neighbors distributed in all directions. Figure 3.4 illustrates the situation for a water/air interface. Water molecules exist in the vapor part of the interface due to the vapor pressure of liquid water, but the concentration of these gaseous water molecules is much less than the concentration of water molecules in the liquid. Thus, there is an imbalance of forces for molecules located in the surface region. This unbalance results in a net force inward into the liquid, and this net inward force is the origin of the surface tension of the liquid.

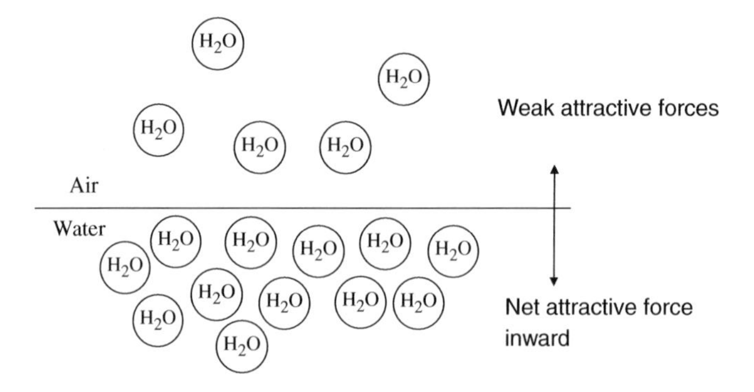

Fig. 3.4 Water molecules at the water/air interface and the origin of surface tension

A similar situation exists for aqueous solutions. With a sodium chloride solution, for example, there is a complete absence of Na^+ or Cl^-ions in the vapor-phase component of the interface. Thus, water molecules near the surface interact not only with interior water molecules in the bulk solution but also with interior Na^+ and Cl^- ions. The net result is that the surface tension of sodium chloride solutions and other strong electrolytes increases slightly with the concentration of the dissolved salt [2].

Because surface ions and molecules experience different chemical neighborhoods than those located in the bulk, there can be a tendency for certain species in solution to preferentially accumulate (i.e., adsorb) near the interface. Such "surface-active" species serve to reduce the surface tension of the solution, as shown in Fig. 3.5. (For a liquid, the surface tension is identical to the surface free energy. The concept of free energy is discussed further in Chapter 4.) There is no experimental evidence to suggest that either Na^+ or Cl^- ions are surface active. However, certain organic molecules such as carboxylic acids, amines, and alcohols are indeed surface active.

Figure 3.6 illustrates the random orientation for $CH_3(CH_2)_{10}COOH$ molecules dissolved in the bulk of the solution. The zigzag portion of the molecules represents the non-polar hydrocarbon tail.

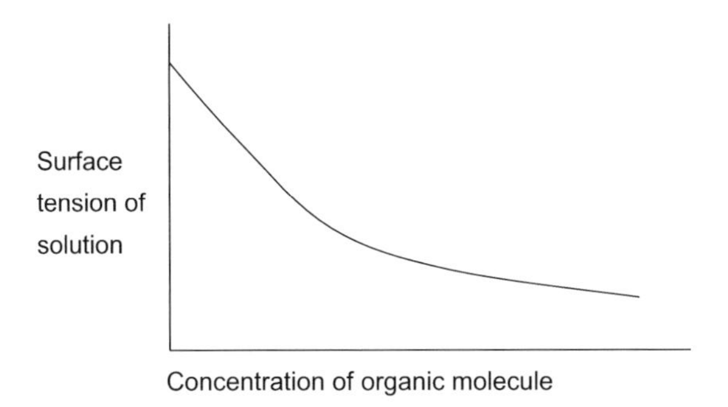

Fig. 3.5 Reduction in the surface tension of a solution by a dissolved surface-active agent (schematic)

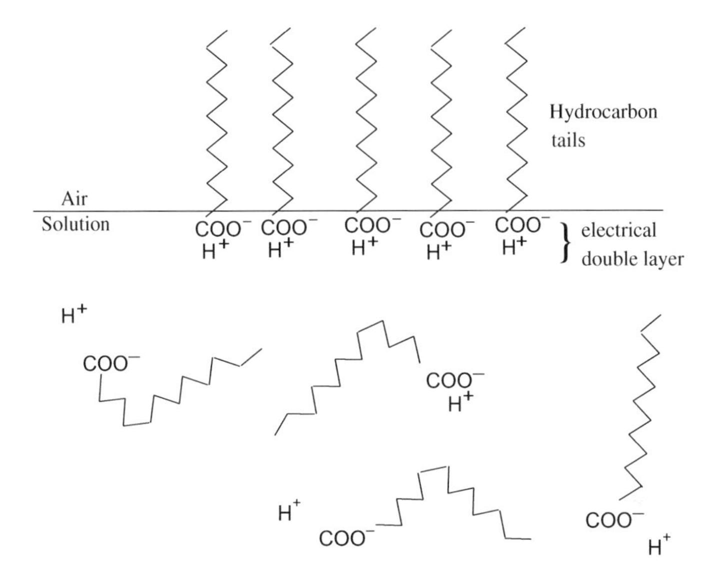

Fig. 3.6 Orientation of $CH_3(CH_2)_{10}COOH$ molecules at the solution/air interface and the formation of an electrical double layer

At the surface, however, the $-COO^-$ polar head group is hydrophilic and is located in the aqueous side of the interface. The non-polar hydrocarbon tail extends outward into the gaseous phase. These surface-active molecules form an oriented layer one molecule in thickness (a *monolayer*).

It can be seen that an oriented monolayer of a carboxylic acid at an interface, as shown schematically in Fig. 3.6, produces an accumulation of negative charges at the aqueous side of the solution/air surface. This array of negative charges is balanced in solution by nearby positive H^+ counterions, and the effect is to establish an *electrical double layer* at the solution/air interface.

The Metal/Solution Interface

Immersion of a metal into a solution creates another type of interface, the metal/solution interface. This interface is much more complicated than the solution/air interface for several reasons.

First, the metal is a conductor of electricity. By connecting external leads to the metal (without immersing these leads into the aqueous solution) and then connecting the metal under study to an external device, we can supply electrons to the metal side of the metal/solution interface, or we can extract electrons from the metal side of the interface. Thus, the metal side of the interface can be charged negatively or positively, respectively.

Second, chloride ions (and other inorganic anions) which are not surface active at solution/air interfaces are adsorbed at metal/solution interfaces.

Third, the water molecule itself is adsorbed at metal/solution interfaces. Moreover, the water molecule being a dipole is oriented at the interface, as shown in Fig. 3.7.

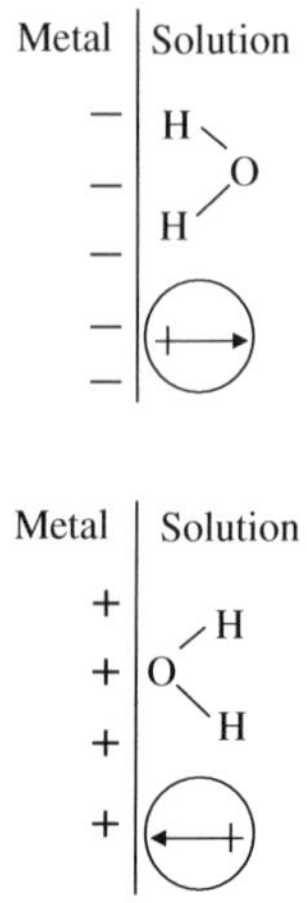

Metal | Solution

Metal | Solution

Fig. 3.7 The orientation of water molecules at a metal/solution interface. *Top*: the "flop-down" orientation of the water dipole. *Bottom*: the "flip-up" orientation [3]

Fourth, the metal/solution interface is not always a stable one. If the metal corrodes, then the interface is neither chemically nor geometrically stable. Under freely corroding conditions, the metal surface supports both local anodic and local cathodic processes, as discussed in Chapter 2. Under these conditions, the metal/solution interface is a hubbub of activity.

Metal Ions in Two Different Chemical Environments

Metals contain closely packed atoms which have strong overlap of electrons between one another. A solid metal therefore does not possess individual well-defined electron energy levels as are found in a single atom of the same material. Instead a vast number of molecular orbitals exist which extend throughout the entire metal, and there is a continuum of energy levels. The electrons can move freely within these molecular orbitals, and so each electron becomes delocalized from its parent atom. The situation is sometimes described as an array of positive ions in a Fermi sea of electrons, as depicted in Fig. 3.8.

The process of corrosion may be thought of as the transfer of a positive ion from the metal lattice into solution. In the metal lattice, the positive ion is stabilized by the Fermi sea of electrons. In solution, the positive ion is stabilized by its water of hydration. This idea is illustrated schematically in Fig. 3.9. To effect this transfer, however, the positive ion must pass outward through the electrical

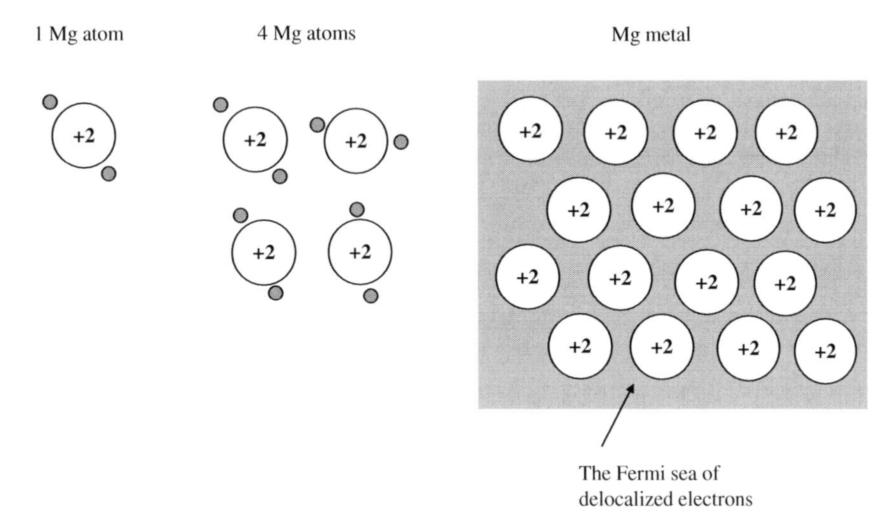

Fig. 3.8 Schematic illustration of valence electrons for 1 Mg atom, 4 Mg atoms, and the Fermi sea of delocalized electrons for solid magnesium metal

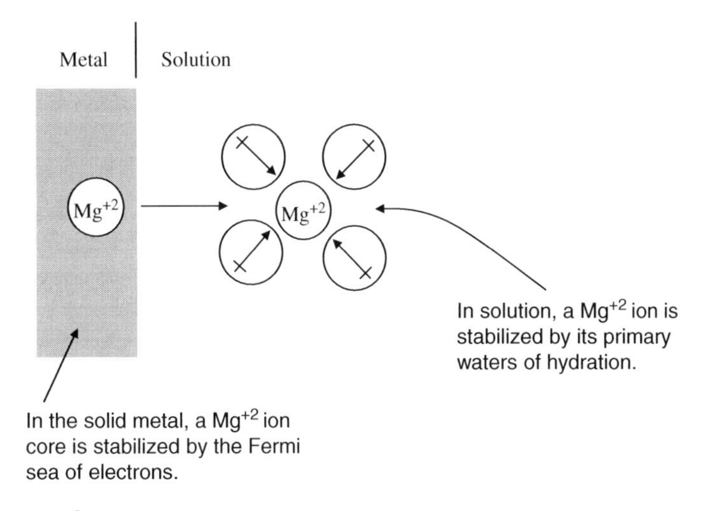

Fig. 3.9 Stability of a Mg^{2+} ion in two different environments

double layer which exists at the metal/solution interface. Anions, such as Cl^-, which can assist this process, must travel from the solution and enter the electrical double layer to interact with the metal surface.

The Electrical Double Layer

The electrical double layer is an array of charged species which exist at the metal/solution interface. The metal side of the interface can be charged positively or negatively by withdrawing or providing electrons, as seen in Chapter 2. The charge on the metal side of the interface is balanced by a distribution of ions at the solution side of the interface. Our view of the electrical double layer has evolved through the development of various models, as discussed below.

The Gouy–Chapman Model of the Electrical Double Layer

Figure 3.10 shows a metal having a positive charge, which is partially balanced in solution by a diffuse layer of negative ions. In this diffuse layer (called the diffuse part of the double layer), ions are in thermal motion, but there is an overall increase in the concentration of negative ions within this layer so as to partly balance the positive charge on the metal side of the interface. This view of the electrical double layer is called the *Gouy–Chapman model.*

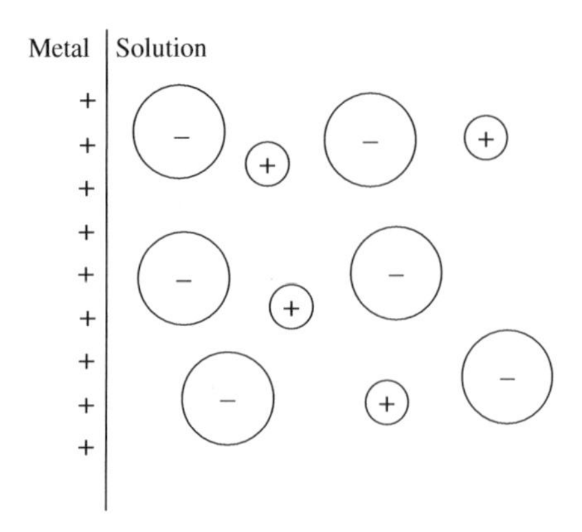

Fig. 3.10 The Gouy–Chapman model of the electrical double layer at a metal/solution interface

The Electrostatic Potential and Potential Difference

As seen in Fig. 3.10, there is a net charge within the diffuse part of the electrical double layer. This is in marked contrast to the interior of the solution, which, as shown earlier, does not exhibit a net charge. This difference in local chemical environments leads to the concept of the electrostatic potential.

The *electrostatic potential* (at some point) is the work required to move a small positive unit charge from infinity to the point in question. This is a thought experiment for which

(1) the positive test charge is small enough not to perturb the existing electrical field;
(2) the work involved is independent of the path taken.

The *potential difference* (between two points) is the work required to move a small unit positive charge between the two points, as shown in Fig. 3.11. The potential difference, PD, between A and B is given by

$$\mathrm{PD_{BA}} = \phi_{\mathrm{B}} - \phi_{\mathrm{A}} \tag{1}$$

where ϕ_{B} and ϕ_{A} are the electrostatic potentials at points B and A, respectively. The potential difference, PD, has the units of joules per coulomb, or volts.

In Fig. 3.10, suppose that a unit positive test charge travels from some point A in the interior of the electrolyte, which is neutral in charge, to some point B within the diffuse double layer, where there is a net charge. There is a potential difference between the two points, and it is the electrical double layer which gives rise to this potential difference.

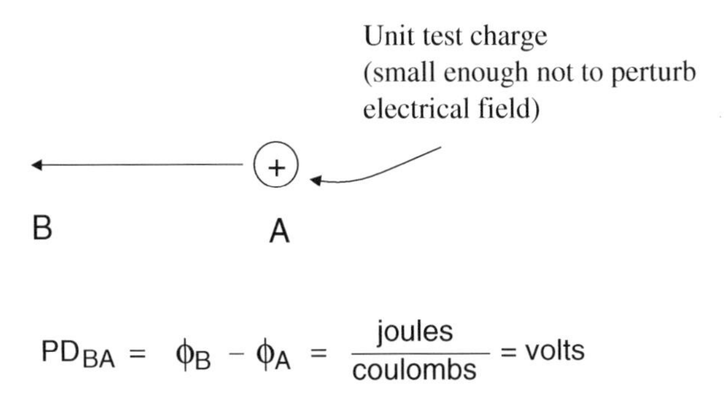

Fig. 3.11 A positive unit test charge and the definition of a potential difference

The Stern Model of the Electrical Double Layer

The Stern model takes into account adsorption of anions or cations at the metal surface. Figure 3.12 shows the case for the adsorption of anions. The distance of closest approach of the ion is its radius, and the plane through the center of these adsorbed ions is called the *Helmholtz plane*. The excess charge at the metal surface is balanced in part by ions located in a Gouy–Chapman diffuse double layer, which exists outside the Helmholtz plane.

A typical potential difference across the Helmholtz plane is of the order of 1 V. The thickness of the Helmholtz layer is about 10 Å (1 Å = 10^{-8} cm). This amounts to a field strength of 1 × 10^7 V/cm. This is a very high field strength and is, of course, the consequence of having a localized charge confined within the narrow region of the interface.

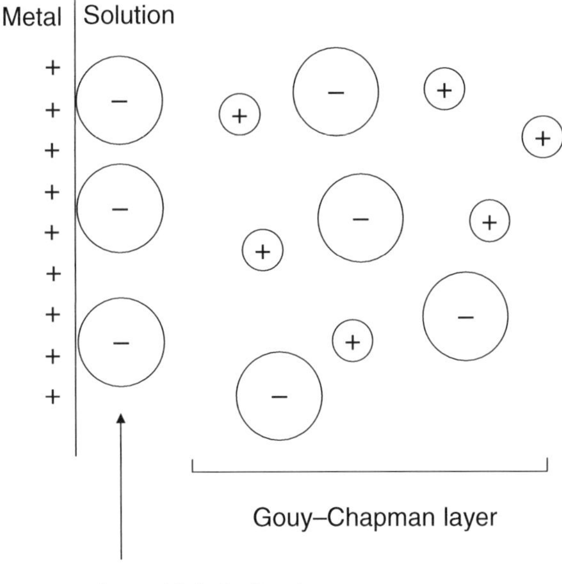

Fig. 3.12 The Stern model of the electrical double layer at a metal/solution interface

The Bockris–Devanathan–Müller Model of the Electrical Double Layer

The Bockris–Devanathan–Müller model is the most recent and most rigorous model of the electrical double layer [4]. This model of the electrical double layer retains all the features of the Stern model and in addition embraces two important considerations.

First, this model takes into account the adsorption of water molecules at the metal/solution interface. For a positive charge on the metal side of the interface, water molecules are oriented with the negative ends of their dipoles toward the metal surface, as shown in Fig. 3.13. This model recognizes that water molecules and ions in solution compete for sites on the metal surface. The adsorption of the chloride ion, for example, may be considered a replacement reaction in which Cl^- ions replace water molecules adsorbed at the metal/solution interface. The plane through the center of these adsorbed ions is called the *inner Helmholtz plane* in this model.

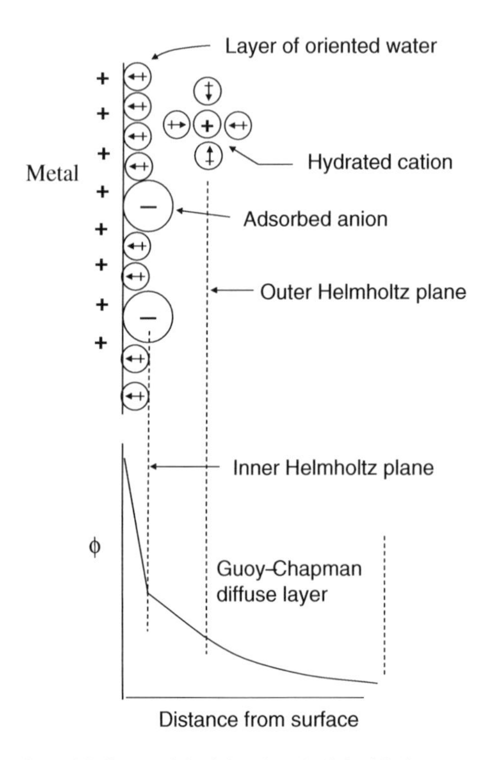

Fig. 3.13 The Bockris–Devanathan–Müller model of the electrical double layer at a metal/solution interface [4]

The second added feature of this model is that the charge introduced by the adsorption of anions at the metal surface is balanced in part by counterions of the opposite charge. These counterions are not adsorbed on the metal surface, but exist in solution, and have associated with them their waters of hydration. The plane through the center of these counter ions is called the *outer Helmholtz plane*.

According to the Bockris–Devanathan–Müller model, water molecules adsorbed at the metal/solution interface have a dielectric constant of 6, and water molecules in the outer Helmholtz plane have a dielectric constant of 30–40. The dielectric constant of water molecules outside the diffuse double layer and in the bulk is the usual value of 78.

Significance of the Electrical Double Layer to Corrosion

The significance of the electrical double layer (edl) to corrosion is that the edl is the origin of the potential difference across an interface and accordingly of the electrode potential (to be discussed in the next section). Changes in the electrode potential can produce changes in the rate of anodic (or cathodic) processes, as will be seen in Chapter 7. Emerging (corroding) metal cations must pass across the edl outward into solution, and solution species (e.g., anions) which participate in the corrosion process must enter the edl from solution in order to attack the metal.

Thus, the properties of the edl control the corrosion process. The edl on a corroding metal can be modeled by a capacitance in parallel with a resistance, as shown in Fig. 3.14. In the simple equivalent circuit shown in Fig. 3.14, the double layer capacitance C_{dl} ensues because the edl at a metal surface is similar to a parallel plate capacitor. The Faradaic resistance R_P in parallel with this capacitance represents the resistance to charge transfer across the edl. The quantity R_P is inversely proportional to the specific rate constant for the half-cell reaction. The term R_S is the ohmic resistance of the solution.

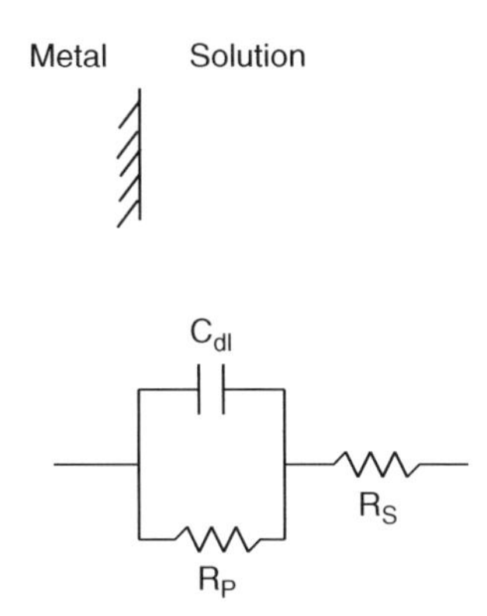

Fig. 3.14 Simple equivalent circuit model of the electrical double layer. C_{dl} is the double layer capacitance, R_P is the resistance to charge transfer across the edl, and R_S is the ohmic resistance of the solution

---

Example 3.1: The double layer capacitance of freely corroding iron in 6 M HCl has been measured to be 34 μF/cm^2 [5]. If the thickness of the entire double layer is 100 Å, what is the average value of the dielectric constant within the electrical double layer?

Solution: The formula for the capacitance C of a parallel plate condenser is

$$C = \frac{\varepsilon A}{d}$$

where ε is the dielectric constant, A is the area of the plates, and d is the distance between plates. Thus

$$C_{dl} = \frac{\varepsilon}{d}$$

where C_{dl} is the double layer capacitance per unit area. The farad is the SI (System Internationale) unit of capacitance, but in the cgs (centimeter−gram−second) system of units, the capacitance must be expressed in statfarads, there being 1.113×10^{-12} F/statF. Thus

$$\left(\frac{34 \times 10^{-6}\text{F}}{\text{cm}^2}\right)\left(\frac{1\,\text{stat F}}{1.113 \times 10^{-12}\,\text{farads}}\right) = \frac{\varepsilon}{(100\text{Å})\left(\frac{1.00\times 10^{-8}\,\text{cm}}{\text{Å}}\right)}$$

The result is $\varepsilon = 31$. Note that this value for the dielectric constant lies between the value 6 for oriented water dipoles adsorbed on the metal surface and lying in the inner Helmholtz plane and the value of 78 for bulk water.

Electrode Potentials

The Potential Difference Across a Metal/Solution Interface

An earlier section discussed the concept of the electrostatic potential (at a fixed point) in terms of a thought experiment. Recall that the electrostatic potential is the work required to move a small positive unit charge from infinity to the point in question. The potential difference (between two points) was also discussed. What happens if we attempt to measure the potential difference across a metal/solution interface in the laboratory? The idea of conceptually moving a test positive unit charge across the metal/solution interface is not of much practical help to us in the laboratory.

We quickly realize that in order to measure the potential difference across the metal/solution interface of interest, we must create additional interfaces [3]. These new interfaces are necessary in order to connect the metal/solution interface of interest to the potential-measuring device so as to complete the electrical circuit. See Fig. 3.15. The metal of interest is designated as M. A second metal which forms a second metal/solution interface is a reference metal, designated by "ref." The required properties of this new interface will be given later. The metal M_1 connects the metals M and "ref" to the potential-measuring device. S and S′ are two points in solution, just outside the electrical double layers on M and "ref", respectively.

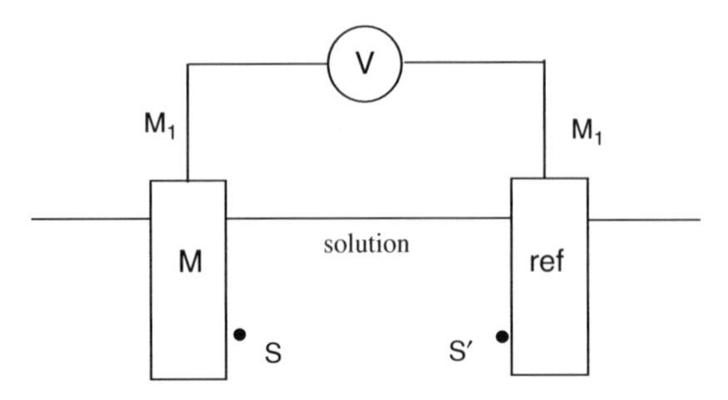

Fig. 3.15 In order to measure the potential difference across the metal/solution interface of interest (M/S), an additional interface must be created using a reference metal "ref" [3]

Consider a point located in solution just outside the electrical double layer on the metal of interest (M). The electrostatic potential has some value ϕ_S. For a second point just inside the interior of the metal M, the electrostatic potential has a value ϕ_M. Accordingly, as one moves inward across the M/S interface, the electrostatic potential will change from value ϕ_s to ϕ_M. Let us continue in a closed path across all interfaces writing the changes in electrostatic potential as we proceed. The sum total of changes in electrostatic potential must be zero by Kirchhoff's law. That is [3]

$$(\phi_S - \phi_M) + (\phi_M - \phi_{M1}) + V + (\phi_{M1} - \phi_{ref}) + \left(\phi_{ref} - \phi_s'\right) + \left(\phi_s' - \phi_s\right) = 0 \tag{2}$$

As shown earlier, there is no net charge in the interior of solution, so that $\phi_S = \phi_{S'}$. Also, by Eq. (1), $(\phi_M - \phi_s) = PD_{MS} = PD_{M/S}$, where the notation M/S refers to the interface formed between metal M and solution S. Then, Eq. (2) can be rewritten as

$$PD_{S/M} + PD_{M/M1} + V + PD_{M1/ref} + PD_{ref/S} = 0 \quad (3)$$

Thus

$$V = -PD_{S/M} - PD_{M/M1} - PD_{M1/ref} - PD_{ref/S} \quad (4)$$

The terms $PD_{M/M1}$ and $PD_{ref/M1}$ are small and can be neglected. These two terms $PD_{M/M1}$ and $PD_{ref/M1}$ can also be removed by extending the metals M and "ref" up to the potential-measuring device. In addition, $PD_{S/M} = -PD_{M/S}$. Thus, Eq. (4) becomes

$$V = PD_{M/S} - PD_{ref/S} \quad (5)$$

Relative Electrode Potentials

The potential difference across a metal/solution interface is commonly referred to as an *electrode potential*. Equation (5) clearly shows that it is impossible to measure the *absolute* electrode potential, but instead we measure the *relative* electrode potential in terms of a second interface. That is, we measure the electrode potential vs. a standard reference electrode.

The hydrogen electrode is universally accepted as the primary standard against which all electrode potentials are compared. For the reversible half-cell reaction

$$2H^+ \text{(aq)} + 2e^- \rightleftharpoons H_2\text{(g)} \quad (6)$$

In the special case

$$2H^+(\text{aq}, a = 1) + 2e^- \rightleftharpoons H_2(\text{g}, P = 1 \text{ atm}) \quad (7)$$

the half-cell potential is arbitrarily defined as $E^\circ = 0.000$ V. The superscript means that all species are in their standard states, which is unit activity for ions and 1 atm pressure for gases. For dilute solutions or solutions of moderate concentration (approximately 1 M or less), the activity can be approximated by the concentration of the solution.

What Eq. (5) means in terms of the discussion given in the previous section is that $PD_{ref/S}$ is defined to be zero for a *standard hydrogen electrode* which satisfies the conditions in Eq. (7). A standard hydrogen electrode (SHE) is shown schematically in Fig. 3.16.

Thus, by measuring electrode potentials relative to the standard hydrogen electrode, a series of standard electrode potentials can be developed for metals immersed in their own ions at unit activity. See Fig. 3.17. Note that strictly speaking, the term "relative standard electrode potentials" should be used; but the word "relative" is understood and is usually dropped.

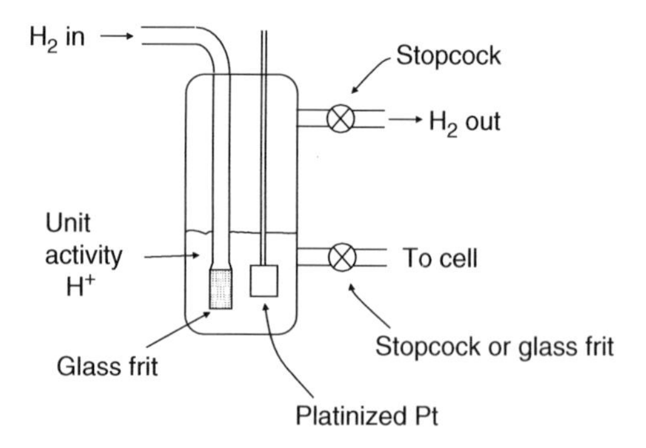

Fig. 3.16 A standard hydrogen reference electrode (SHE)

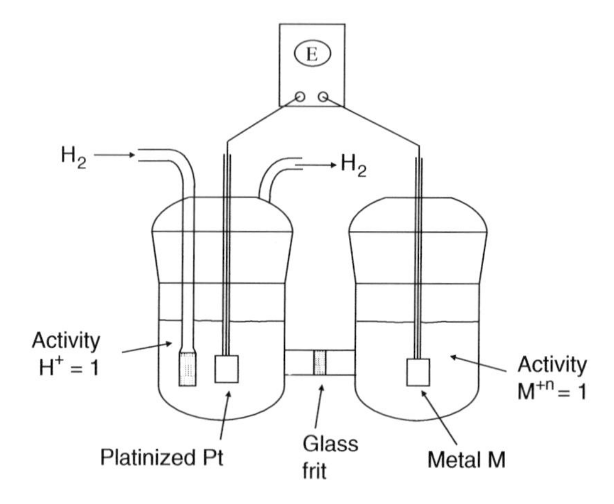

Fig. 3.17 Experimental determination of a standard electrode potential for some metal M using a standard hydrogen reference electrode

The Electromotive Force Series

An ordered listing of the standard half-cell potentials is called the electromotive force (emf) series. See Table 3.1 [6]. Note that all of the half-cell reactions are written from left to right as reduction reactions. This practice follows the Stockholm Convention of 1953 but is a source of some annoyance to corrosion scientists, who are primarily interested in the reverse (i.e., oxidation) reaction. (Different compilations [7–9] of standard electrode potentials may list slightly different values for some half-cell reactions, but these variations are usually of the order of millivolts and should not be considered to be significant.)

Notice that the sign of the standard electrode potential in Table 3.1 ranges from positive to negative values, passing through the value 0.000 which is defined as such for the hydrogen reduction reaction. The most positive potential listed in Table 3.1 is for

$$Au^{3+} + 3e^- \rightleftharpoons Au \qquad E^\circ = +1.498\,V$$

Table 3.1 Standard electrode potentials at 25°C [6]

Reaction	E^0(V vs. SHE)	
$Au^{3+} + 3e^- \rightarrow Au$	+1.498	Noble
$Pt^{2+} + 2e^- \rightarrow Pt$	+1.18	↑
$Pd^{2+} + 2e^- \rightarrow Pd$	+0.951	
$Hg^{2+} + 2e^- \rightarrow Hg$	+0.851	
$Ag^+ + e^- \rightarrow Au$	+0.800	
$Cu^+ + e \rightarrow Cu$	+0.521	
$Cu^{2+} + 2e^- \rightarrow Cu$	+0.342	
$2H^+ + 2e^- \rightarrow H_2$	0.000	
$Pb^{2+} + 2e^- \rightarrow Pb$	−0.126	
$Sn^{2+} + 2e^- \rightarrow Sn$	−0.138	
$Mo^{3+} + 3e^- \rightarrow Mo$	−0.200	
$Ni^{2+} + 2e^- \rightarrow Ni$	−0.257	
$Co^{2+} + 2e^- \rightarrow Co$	−0.28	
$Cd^{2+} + 2e^- \rightarrow Cd$	−0.403	
$Fe^{2+} + 2e^- \rightarrow Fe$	−0.447	
$Ga^{3+} + 3e^- \rightarrow Ga$	−0.549	
$Ta^{3+} + 3e^- \rightarrow Ta$	−0.6	
$Cr^{3+} + 3e^- \rightarrow Cr$	−0.744	
$Zn^{2+} + 2e^- \rightarrow Zn$	−0.762	
$Nb^{3+} + 3e^- \rightarrow Nb$	−1.100	
$Mn^{2+} + 2e^- \rightarrow Mn$	−1.185	
$Zr^{4+} + 4e^- \rightarrow Ze$	−1.45	
$Hf^{4+} + 4e^- \rightarrow Hf$	−1.55	
$Ti^{2+} + 2e^- \rightarrow Ti$	−1.630	
$Al^{3+} + 3e^- \rightarrow Al$	−1.662	
$U^{3+} + 3e^- \rightarrow U$	−1.798	
$Be^{2+} + 2e^- \rightarrow Be$	−1.847	
$Mg^{2+} + 2e^- \rightarrow Mg$	−2.372	
$Na^+ + e^- \rightarrow Na$	−2.71	
$Ca^{2+} + 2e^- \rightarrow Ca$	−2.868	
$K^+ + e^- \rightarrow K$	−2.931	↓
$Li^+ + e^- \rightarrow Li$	−3.040	Active

One of the most negative standard electrode potential in Table 3.1 is for

$$K^+ + e^- \rightleftharpoons K \qquad E^o = -2.931\ V$$

From our common sense knowledge of the chemical world, we know that metallic gold oxidizes with difficulty. This is the basis for the use of gold as a monetary standard and for jewelry and artwork. Thus, the forward (reduction) reaction is favored, and the backward (oxidation) reaction is not favored. Similarly, we know from experience that potassium is a very reactive metal and in fact must be stored under kerosene to keep it from reacting with water vapor in the atmosphere. In this case, the forward (reduction) reaction is not favored, and the backward (oxidation) reaction is favored.

The spontaneity of electrochemical reactions will be taken up more quantitatively in Chapter 4 and 5. However, for the present it can be appreciated that metals located near the top (positive end) of the emf series are more chemically stable than metals located near the bottom (negative end) of the series. Said simply, metals near the top of the emf series are less prone to corrosion. *But the following limitations must be recognized:*

(1) The emf series applies to pure metals in their own ions at unit activity.
(2) The relative ranking of metals in the emf series is not necessarily the same (*and is usually not the same*) in other media (such as seawater, groundwater, sulfuric acid, artificial perspiration).
(3) The emf series applies to pure metals only and not to metallic alloys.
(4) The relative ranking of metals in the emf series gives corrosion tendencies (subject to the restrictions immediately above) but provides no information on corrosion rates.

Metals located near the positive end of the emf series are referred to as "noble" metals, while metals near the negative end of the emf scale are called "active" metals.

The electrode potential for a half-cell reaction for a metal immersed in a solution of its ions at some concentration other than unit activity is related to its standard electrode potential (at unit activity) by the Nernst equation. See Chapter 4.

It should be noted that in some older texts and data compilations, standard electrode potentials are written for half-cell oxidation reactions, and the signs for $E°$ listed in Table 3.1 are then reversed.

Reference Electrodes for the Laboratory and the Field

Although the standard hydrogen electrode (depicted in Fig. 3.16) is the reference electrode against which electrode potentials are defined; this reference electrode is not commonly used in the laboratory. The hydrogen electrode is somewhat inconvenient to use as it requires a constant external source of hydrogen gas. (Hoare and Schuldiner [10] have developed a more compact hydrogen reference electrode which employs hydrogen gas discharged in a palladium wire, but this reference electrode has not found much use in corrosion studies.)

Instead of using the standard hydrogen electrode, other reference electrodes are commonly used in the laboratory. The saturated calomel electrode (SCE) has long been used, especially in chloride

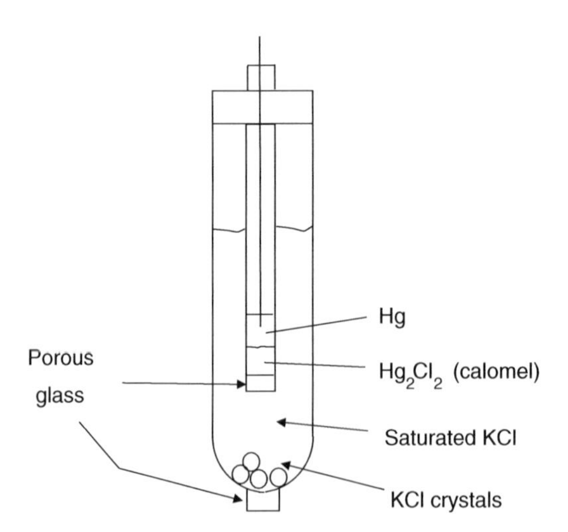

Fig. 3.18 A saturated calomel reference electrode

solutions. Its construction is shown in Fig. 3.18, and its half-cell reaction is

$$Hg_2Cl_2(s) + 2e^- \rightleftharpoons 2Hg(s) + 2Cl^-(aq, sat'd)$$

The electrode potential under these conditions at 25°C is $E = +\,0.242$ V.

Another reference electrode in wide use in the laboratory (usually in chloride solutions) is the silver–silver chloride reference electrode, and its half-cell reaction is

$$AgCl(s) + e^- \rightleftharpoons Ag(s) + Cl^-(aq)$$

In its most common form, the silver–silver chloride electrode consists of a solid AgCl coating on a silver wire immersed in a solution of 4 M KCl plus saturated AgCl, as shown schematically in Fig. 3.19. The electrode potential under these conditions at 25°C is $E = +\,0.222$ V.

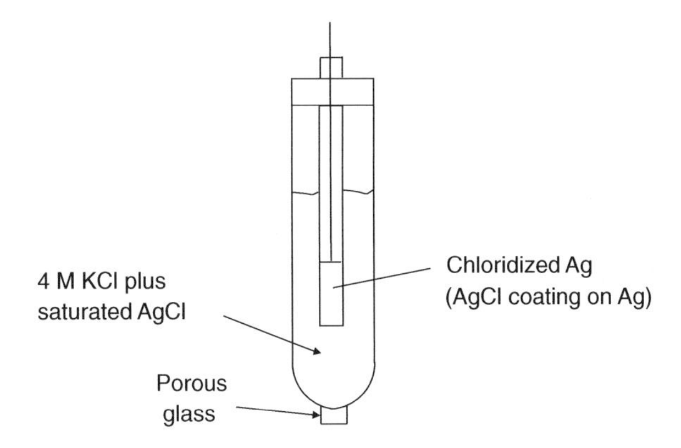

Fig. 3.19 A silver/silver chloride reference electrode

Peterson and Groover [11] have developed a silver–silver chloride reference electrode for use in natural seawater. The electrode consists of a solid mass of AgCl in contact with an Ag wire and

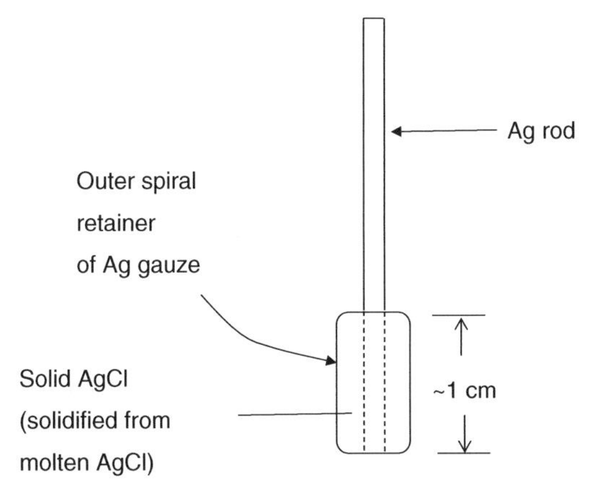

Fig. 3.20 The Peterson–Groover silver/silver chloride reference electrode for use in natural seawater [11]

held in place with an outer retainer of silver gauze. The electrode itself contains no aqueous filling solution but is immersed directly into the ocean, which then provides the source of the Cl^- ion (approximately 0.6 M Cl^-). See Fig. 3.20. This electrode maintains a steady half-cell potential which can be then calibrated vs. the standard silver–silver chloride electrode. The Peterson–Groover reference electrode has been used with much success in measuring the electrode potentials of test specimens in seawater and in various marine applications, such as measuring the electrode potentials at various locations around the hull of a ship. In the latter application, the Peterson–Groover reference electrode may be delivered to the desired location at the end of a fishing line.

There are various other reference electrodes which are used in various aqueous solutions. For example, the mercury–mercurous sulfate electrode is used in sulfate solutions to avoid contamination by Cl^-, and the mercury–mercuric oxide electrode is used in alkaline solutions. For more details, specialized texts on experimental electrochemistry should be consulted [12]. Values for the half-cell potentials of several reference electrodes are given in Table 3.2 [13, 14].

Table 3.2 Electrode potentials of various reference electrodes [13, 14]

Reference electrode	Potential (V vs. SHE)	Conversion to SHE
Standard hydrogen electrode	0.000	–
Saturated calomel	+ 0.242	E vs. SCE = E vs. SHE – 0.242
Silver/silver chloride (saturated)	+ 0.222	E vs. Ag/AgCl = E vs. SHE – 0.222
Peterson–Groover silver/silver chloride (in seawater)	+ 0.248	E vs. Ag/AgCl (in seawater) = E vs. SHE – 0.248
Copper/copper sulfate (saturated)	+ 0.316	E vs. $Cu/CuSO_4$ = E vs. SHE – 0.316
Mercury/mercurous sulfate	+ 0.615	E vs. Hg/Hg_2SO_4 = E vs. SHE – 0.615

The copper–copper sulfate reference electrode is used commonly in the field to measure the potential of buried structures such as pipelines or tanks. The construction of this reference electrode is shown in Fig. 3.21. A porous wooden plug provides the electrolyte path between the reference

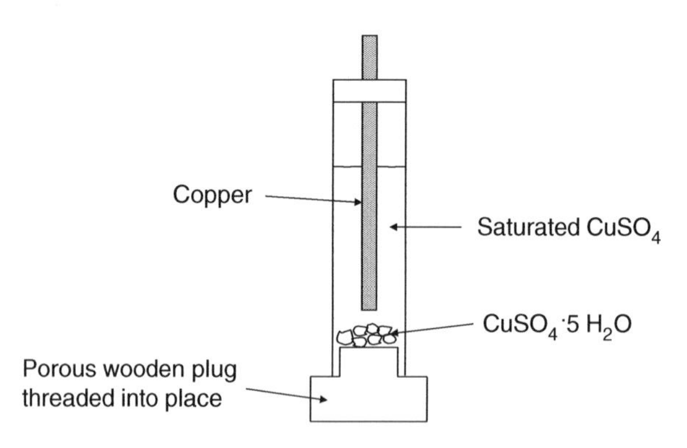

Fig. 3.21 The copper/copper sulfate reference electrode for use in soils

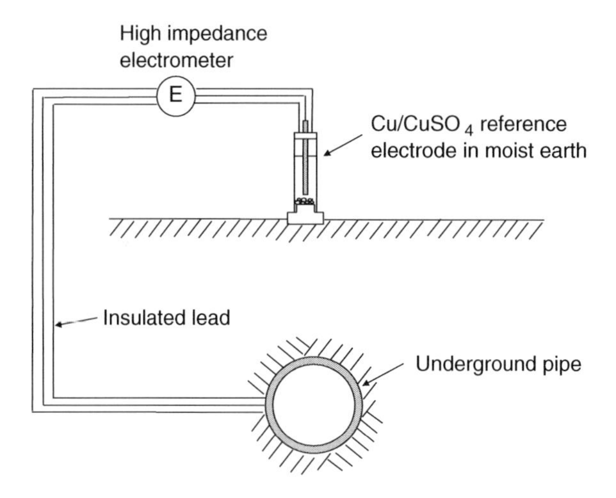

Fig. 3.22 Measurement of the electrode potential of a buried pipe using a copper/copper sulfate reference [15]

electrode and the moist soil. This reference electrode is used more for its rugged and simple nature than for its high precision. Figure 3.22 gives a schematic diagram showing the measurement of the potential of an underground pipe [15].

Example 3.2: An electrode potential was measured to be –0.500 V vs. $Cu/CuSO_4$. What is this electrode potential on the SCE scale?

Solution: From Table 3.2

$$E\,\text{vs. SCE} = E\,\text{vs. SHE} - 0.242$$

Also

$$E\,\text{vs. Cu/CuSO}_4 = E\,\text{vs. SHE} - 0.316$$

Subtracting the second of these equations from the first gives

$$(E\,\text{vs. SCE}) - \left(E\,\text{vs. Cu/CuSO}_4\right) = -0.242 + 0.316 = +0.074$$

Thus

$$(E\,\text{vs. SCE}) - (-0.500) = 0.074$$

or

$$(E\,\text{vs. SCE}) = -0.426\text{V}$$

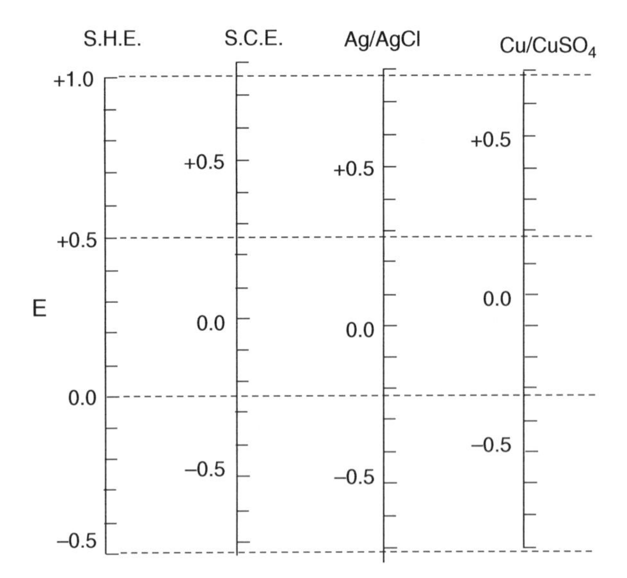

Fig. 3.23 A graphical aid for conversion between reference electrode scales

Using the information in Table 3.2, graphs may be constructed for converting the electrode potential on one reference scale into an electrode potential on another scale, as in Fig. 3.23.

All reference electrodes whether they are used in the laboratory or in the field have the same common features. These are the following: (i) the half-cell potential must be constant, (ii) the half-cell potential should not change with the passage of a small current through the reference electrode, and (iii) the half-cell potential must not drift with time. These three conditions are met if there is an excess of both reactants and products in the half-cell reaction.

It should be noted that the use of any reference electrode introduces a liquid junction potential at the liquid/liquid interface between the test solution and the filling solution of the reference electrode. Such liquid junction potentials are caused by differences in ion types or concentrations across a liquid/liquid interface. The liquid junction potential can be minimized by the proper choice of reference electrode, i.e., use of a saturated calomel electrode in chloride solutions. Liquid junction potentials are usually small (of the order of 30 mV) [12] and are included in the measurement of electrode potentials.

Measurement of Electrode Potentials

Electrode potentials are usually measured with an instrument called an electrometer, which has a high input impedance. The input impedance can be as high as 10^{14} Ω so that for a potential drop of 1 V across the electrical double layer, an extremely small current of 10^{-14} amp will flow in the measuring circuit. This current is too small to interfere with the electrode reactions occurring at the metal/solution interface or to change the electrode potential of the reference electrode. Thus, the electrode potential of the metal under study will not be altered during the measurement.

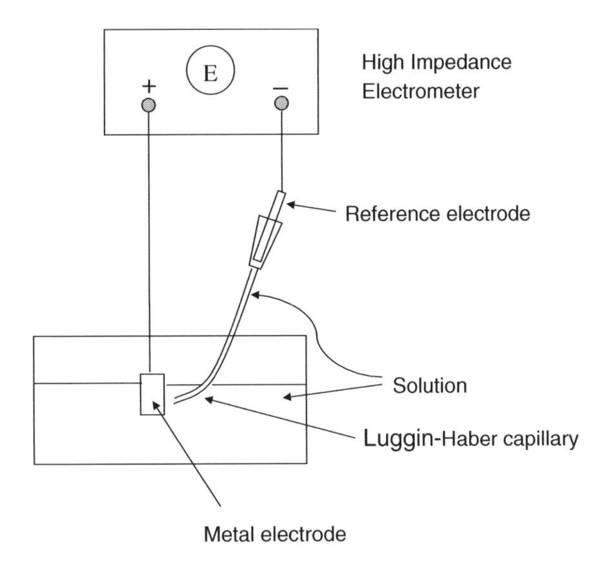

Fig. 3.24 A simple cell for measuring electrode potentials in the laboratory

Figure 3.24 shows an experimental arrangement for the measurement of an electrode potential using a saturated calomel reference electrode. The reference electrode is in essence brought up near the surface of the metal of interest by means of a Luggin–Haber capillary, which contains the same solution as in the test cell. This arrangement allows measurement of the electrode potential near the metal surface rather than at some point in the bulk of the solution. Barnartt used such a capillary which was 0.2 mm in outside diameter and about 0.1 mm in inside diameter, with the capillary placed several millimeters from the electrode surface [16]. If the Luggin–Haber capillary is located too close to the sample surface, there is a danger that the electrode surface will be shielded.

The IR drop in the aqueous solution contained between metal surface and the reference electrode is usually small for most electrolytes of practical interest.

Before the advent of the electrometer, electrode potentials were measured by null methods, such as the use of a potentiometer, but such techniques are only of passing interest now.

Problems

1. Calculate the percentage of water molecules in a 0.10 M NaCl solution which are tied up as primary waters of hydration, given the following information. The primary hydration number for the Na^+ ion is 4 and for the Cl^- ion is 1 [1]. The density of 0.10 M NaCl is 1.0431 g/ml, and the molecular weight of NaCl is 58.45 g/mol.
2. If adsorbed chloride ions occupy the inner Helmholtz layer of the electrical double layer, what is the field strength in V/cm across the inner Helmholtz plane (IHP) when the potential drop across the entire double layer is 0.50 V? Assume that 90% of the potential drop is across the IHP. Also assume that the IHP is located at the center of the adsorbed Cl^- ions. The radius of the Cl^- ion is 1.81 Å.
3. (a) Calculate the double layer capacitance for the Helmholtz layer on a metal if the dielectric constant within the Helmholtz layer is 10 and the thickness of the Helmholtz layer is 15 Å.

(b) Calculate the double layer capacitance for a Gouy–Chapman layer of dielectric constant 50 and thickness 100 Å. (c) Use the Stern model of the double layer to calculate the total double layer capacitance, assuming that the inner Helmholtz layer and the Gouy–Chapman layer have capacitances in series.

4. In the arrangement shown below, two different metals M_1 and M_2 are short-circuited by an external connection. Show that $PD_{M1/S} = PD_{M2/S}$. Also write an expression to give V in terms of the quantity $PD_{ref/S}$. *Hint*: Apply Kirchhoff's law to each of the three cycles a, b, and c shown in the figure. For simplicity, assume that the potential difference across M_1 and its external connecting wire can be neglected. Make the same assumption for M_2 and for "ref" and their respective connecting wires. Also assume that the potential difference between two different points in solution is negligible.
5. A thin platinum wire immersed directly into solution is sometimes used as a "quasi-reference electrode" [11]. Advantages of using this electrode are its simplicity and elimination of the need of a Luggin–Haber capillary. (a) What are some disadvantages? (b) What do you think about using an immersed zinc wire as a reference electrode?

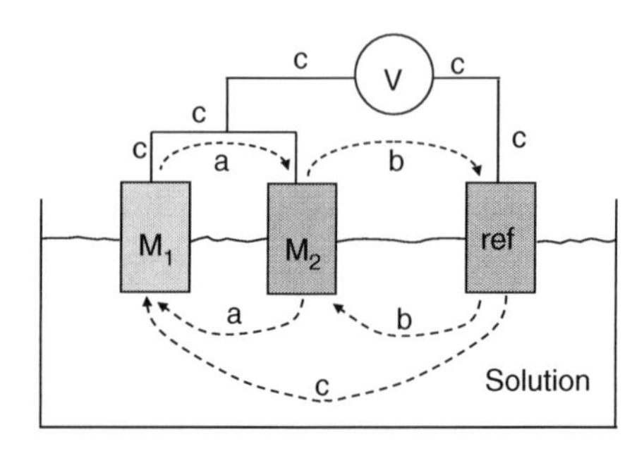

6. The standard electrode potential for tin

$$Sn^{2+}\,(aq) + 2e^- \rightleftharpoons Sn(s)$$

is −0.138 V vs. the standard hydrogen electrode (SHE). What is the value of this standard electrode potential vs. the following:

(a) the saturated calomel reference electrode?
(b) the Ag/AgCl (saturated) reference electrode?
(c) the $Cu/CuSO_4$ (saturated) reference electrode?

7. The operation of a reliable conventional standard reference electrode in the laboratory usually does not present problems. However, it is often desirable to measure the electrode potential of a metal in an outdoor natural environment over an extended period of time (months to years). What types of natural environments might cause problems in the operation of a reference electrode over a long period of time (so as to require modifications to the reference electrode design). Explain how these problems arise.
8. One of the following statements is not true. Which *one* statement is it?

(a) Zinc and chromium have similar standard potentials in the emf series. Thus, these two metals are certain to have similar electrode potentials in seawater.
(b) Standard electrode potentials are reduction potentials.
(c) The liquid junction potential for a solution of 0.1 M HCl separated by porous diaphragm from a solution of 0.1 M NaCl is less than the liquid junction potential between 0.1 M HCl and 0.1 M Na_2SO_4.

9. Only one of the following statements is true. Which *one* statement is it?

(a) The standard electrode potential for copper immersed in unit activity Ni^{2+} ions is the average of the standard electrode potential for nickel immersed in unit activity Ni^{2+} and copper immersed in unit activity Cu^{2+}.
(b) Use of a Luggin–Haber capillary can reduce the IR drop between a metal electrode under study and the reference electrode.
(c) Any two metals in the same Group (i.e., column) of the Period Table will have standard electrode potentials having the same sign.

10. The following laboratory data were taken for iron in 6 M HCl with and without a straight-chain organic amine added as a corrosion inhibitor:

Concentration of inhibitor (M)	C_{dl} (μF/cm^2)	Corrosion rate (μA/cm^2)
0	33	2,000
8×10^{-4}	22	1,400
2×10^{-3}	17	1,000
2×10^{-2}	8	200

Comment on the possibility of measuring the double layer capacitance of a test coupon as a corrosion rate monitor in a steel tank storing waste HCl with an added amount of the inhibitor to reduce corrosion of the storage tank walls.

11. Organic amines as in Problem 3.10 reduce the corrosion rate by adsorption at the metal/solution interface (as is discussed in Chapter 12). Why is there a corresponding decrease in the double layer capacitance? *Hint*: Consider the effect of the organic molecule on the dielectric constant within the edl and on the thickness of the edl.
12. Sketch a figure of the Bockris–Devanathan–Müller model of the edl when the charge on the metal side of the interface is negative instead of positive, as was the case in Fig. 3.13.

References

1. J. O'M. Bockris and A. K. N. Reddy, "Modern Electrochemistry", Second Edition, Vol. 1, Chapter 2, Plenum Press, New York (1998). Also, First Edition, Vol. 1, Chapter 2 (1977).
2. A. W. Adamson, "Physical Chemistry of Surfaces", p. 74, John Wiley, New York (1990).
3. J. O'M. Bockris and A. K. N. Reddy, "Modern Electrochemistry", Vol. 2, Chapter 7, Plenum Press, New York (1977).
4. J. O'M. Bockris, M. A. V. Devanathan, and K. Müller, *Proc. R. Soc.*, *274*, 55 (1963).
5. E. McCafferty and N. Hackerman, *J. Electrochem. Soc.*, *119*, 146 (1972).
6. P. Vanysek in "CRC Handbook of Chemistry and Physics", D. R. Lide, Ed., pp. 8–21, CRC Press, Boca Raton, FL (2001).

7. W. M. Latimer, “The Oxidation States of the Elements and Their Properties in Aqueous Solutions”, Prentice-Hall, New York (1952).
8. M. Pourbaix, “Atlas of Electrochemical Equilibria in Aqueous Solutions”, National Association of Corrosion Engineers, Houston, TX (1974).
9. A. J. Bard, R. Parsons, and J. Jordan, Eds., “Standard Potentials in Aqueous Solutions”, Marcel Dekker, New York (1985).
10. J. P. Hoare and S. Schuldiner. *J. Phys. Chem.*, *61*, 399 (1957).
11. M. H. Peterson and R. E. Groover, *Mater. Prot. Perform.*, *11* (5), 19 (1972).
12. D. T. Sawyer and J. L. Roberts, Jr., “Experimental Electrochemistry for Chemists”, Chapter 2, Wiley Interscience, New York (1974).
13. H. H. Uhlig and R. W. Revie, “Corrosion and Corrosion Control”, Chapter 3, John Wiley, New York (1985).
14. D. G. Ives and G. J. Janz, Eds., “Reference Electrodes: Theory and Practice”, Chapters 2 and 3, Academic Press, New York (1961).
15. J. R. Myers, “Fundamentals and Forms of Corrosion”, Air Force Institute of Technology, Wright-Patterson Air Force Base, Ohio, October (1974).
16. S. Barnartt, *J. Electrochem. Soc.*, *108*, 102 (1961).

Chapter 4
A Brief Review of Thermodynamics

Introduction

This chapter will review briefly the principles of thermodynamics needed to consider corrosion from a thermodynamic point of view. In particular we are working toward the treatment of galvanic corrosion and the construction and analysis of Pourbaix diagrams. As will be seen in Chapter 6, Pourbaix diagrams are useful diagrams to summarize concisely the corrosion behavior of a given metal.

Thermodynamic State Functions

A state function is one whose properties depend only on the present state of the system and not on the path taken to get there. (By the "state" in thermodynamics we mean having a certain set of descriptors, rather than the specific physical state of the system). There are five common state functions: (i) the internal energy E, (ii) the entropy S, (iii) the enthalpy H, (iv) the Helmholtz free energy A, and (v) the Gibbs free energy G.

Internal Energy

The *internal energy* E of a system is the total energy contained within the system. The internal energy includes contributions from both kinetic energy and potential energy sources. These include translational, rotational, and vibrational motion of atoms. Forces within an atom, such as the attraction of an electron to the positively charged nucleus, as well as intermolecular forces between molecules (such as van der Waals forces) are included. The internal energy also includes gravitational forces, magnetic forces, and electrical forces.

The fact that E is a state function means that when the system changes from some condition (condition 1) to another condition (condition 2), the change in internal energy is given by

$$\Delta E = E_2 - E_1 \tag{1}$$

Because both E_2 and E_1 are state functions, their change $(E_2 - E_1)$ does not depend on the path taken between 1 and 2. An analogy is provided in Fig. 4.1. The potential energy difference between the elevation of the second floor of a building and the first floor of that building depends only on the heights of the two floors and not on whether a person climbed the stairs to the second floor from the first or rode the elevator.

E. McCafferty, *Introduction to Corrosion Science*, DOI 10.1007/978-1-4419-0455-3_4,

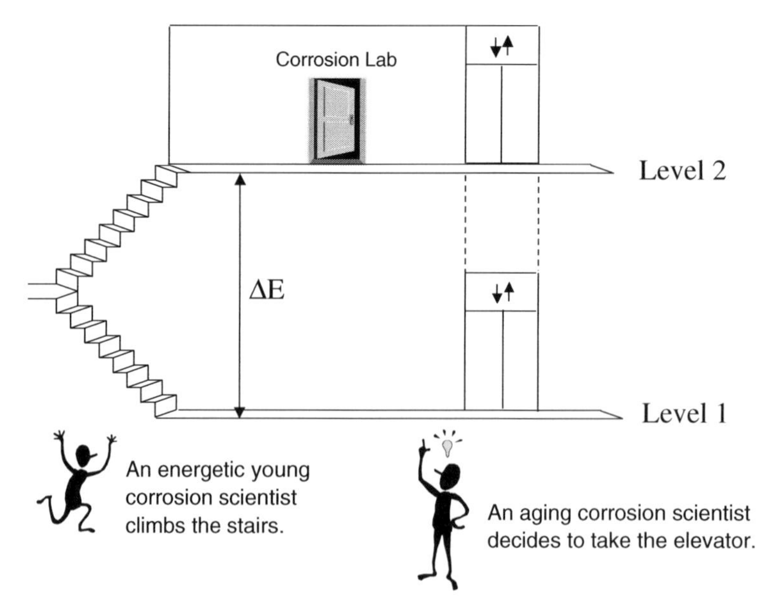

Fig. 4.1 The difference in potential energy between the second and first floors of a building depends only on the heights of the two floors and not on the path taken to reach the second floor

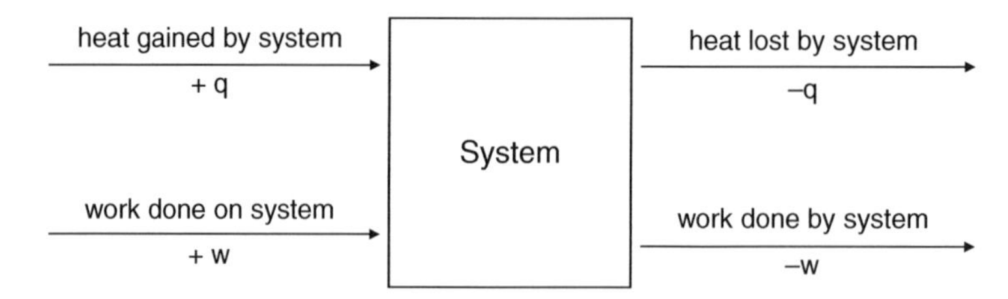

Fig. 4.2 The heat q and work w associated with a thermodynamic system

The *first law of thermodynamics* states that for small changes in internal energy

$$dE = q + w \tag{2}$$

where q is the heat absorbed by the system and w is the work done on the system. The signs of both q and w are important and are illustrated in Fig. 4.2. It is important to note that while the internal energy E is a state function, the quantities q and w are not. Heat and work are not contained in a system but are observed when a system changes from one state to another.

Entropy

The *entropy* S is defined by

$$dS = \frac{q}{T} \tag{3}$$

where T is the Kelvin temperature (degrees centigrade plus 273.1°C). The entropy S is also a state function and has the properties that

(1) $\Delta S = S_2 - S_1$
(2) $dS = 0$ for a reversible system
(3) $dS > 0$ for an irreversible system

The entropy is a measure of the disorder of a system, with the entropy increasing for an irreversible system. This observation is the basis for the familiar statement that the entropy of the universe is increasing.

Enthalpy

The *enthalpy H* is defined by

$$H = E + PV \tag{4}$$

where P and V are the pressure and the volume of gases involved. The change in enthalpy ΔH between two states measures the heat of a reaction at constant pressure. That is, at constant pressure, the heat q_p (the subscript indicates constant pressure) is given by

$$q_p = \Delta H \tag{5}$$

Corrosion reactions usually occur under the conditions of a constant pressure (i.e., atmospheric pressure).

Helmholtz and Gibbs Free Energies

The *Helmholtz free energy A* is defined by

$$A = E - TS \tag{6}$$

and the *Gibbs free energy G* by

$$G = H - TS \tag{7}$$

Both A and G are state functions. Their significance is as follows:
At constant temperature T and volume V, the system is at equilibrium when $dA = 0$.
At constant temperature T and pressure P, the system is at equilibrium when $dG = 0$.
Most corrosion reactions occur at constant temperature and pressure so that the Gibbs free energy G is the appropriate measure of equilibrium for corrosion scientists and engineers. At constant temperature and pressure, Eq. (7) gives

$$\Delta G = \Delta H - T\Delta S \tag{8}$$

where ΔG, ΔH, and ΔS are the changes in Gibbs free energy, enthalpy, and entropy, respectively, when the system undergoes a change from one state to another.

Free Energy and Spontaneity

The notions of reversibility, equilibrium, and spontaneity are related. A reaction is reversible if it can proceed in either direction. A reaction is at equilibrium if the rate of the forward reaction is the same as the rate of the reverse reaction. A reaction is spontaneous if it proceeds in a given direction without any external influence.

For the dissolution of iron in acids at room temperature, as considered in Chapter 2, the overall reaction

$$Fe(s) + 2H^+(aq) \rightarrow Fe^{2+}(aq) + H_2(g)$$

is spontaneous at 25°C because it proceeds as written above if a piece of iron is immersed in a hydrochloric acid solution. In addition, the reaction reaches an equilibrium in which the anodic half-cell reaction

$$Fe(s) \rightarrow Fe^{2+}(aq) + 2e^-$$

proceeds at the same rate as the cathodic half-cell reaction

$$2H^+(aq) + 2e^- \rightarrow H_2(g)$$

However, the overall reaction is not reversible because we cannot treat a ferrous salt with hydrogen gas at 25°C and produce solid iron.

Solid iron can be produced by the reduction of Fe_2O_3 under special conditions and in a process which requires an intensive energy input. The reduction of iron ore (Fe_2O_3) at elevated temperatures over 1000°C and in the presence of carbon (as coke) and $CaCO_3$ (limestone) is the basis for the production of iron from iron ore. Another way of viewing corrosion, then, is that corrosion is the thermodynamic process by which metals revert to their natural form as ores. See Fig. 4.3.

The change in Gibbs free energy ΔG is a powerful indicator of spontaneity. Figure 4.4 illustrates the behavior of the change in Gibbs free energy ΔG for the case of equilibrium, spontaneous, and non-spontaneous reactions. A chemical reaction (or a coupled electrochemical reaction):

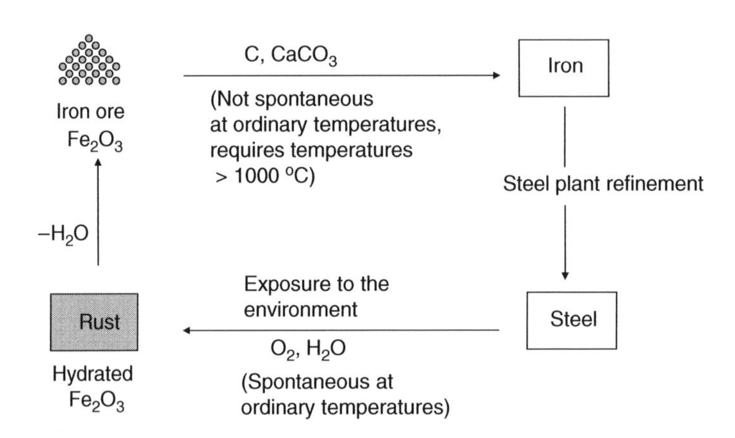

Fig. 4.3 A thermodynamic cycle for Fe_2O_3

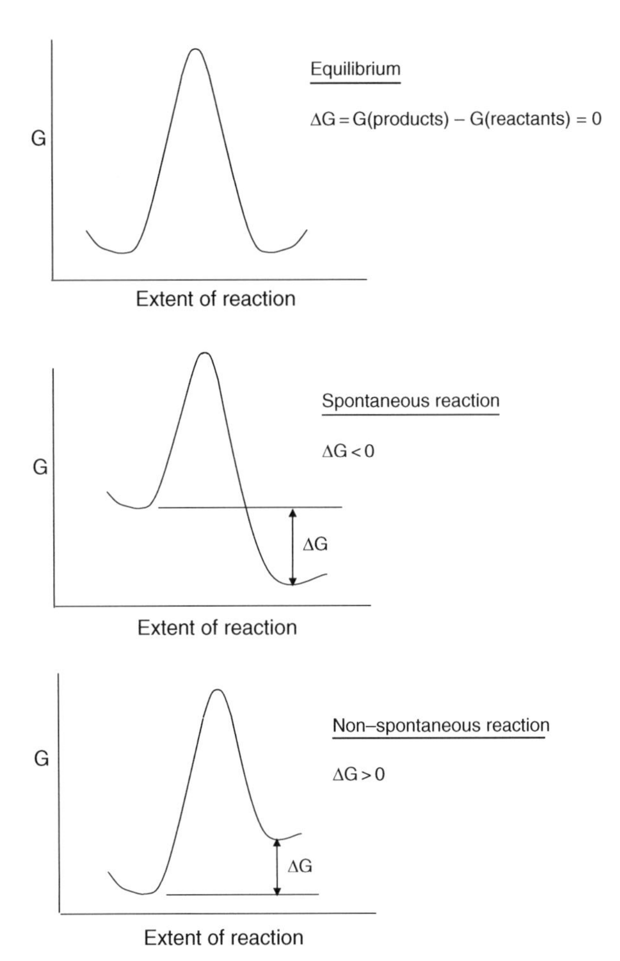

Fig. 4.4 The change in Gibbs free energy ΔG for the case of equilibrium, spontaneous, and non-spontaneous reactions

(1) is at equilibrium if the free energy change ΔG is zero,
(2) proceeds spontaneously if ΔG is negative,
(3) is not spontaneous if ΔG is positive.

Relationships Between Thermodynamic Functions

From Eqs. (4) and (7)

$$G = H - TS = (E + PV) - TS \tag{9}$$

Taking differentials

$$\mathrm{d}G = \mathrm{d}E + P\,\mathrm{d}V + V\,\mathrm{d}P - T\,\mathrm{d}S - S\,\mathrm{d}T \tag{10}$$

Use of Eq. (2) gives

$$dG = q + w + P\,dV + V\,dP - T\,dS - S\,dT \tag{11}$$

From Eq. (3), $q = T\,dS$, and also $w = -P\,dV$ (the minus sign is needed for work done on the system). Then Eq. (11) becomes

$$dG = T\,dS - P\,dV + P\,dV + V\,dP - T\,dS - S\,dT \tag{12}$$

or

$$dG = V\,dP - S\,dT \tag{13}$$

Equation (13) applies to a closed system, i.e., one in which there is no mass transfer in or out of the system. Suppose that an ideal gas is present. Then $PV = RT$, where R is the ideal gas constant. (See Table 4.1.) Then, substitution for $V = RT/P$ in Eq. (13) gives

$$dG = \frac{RT}{P}dP - S dT \tag{14}$$

At constant temperature T

$$dG = \frac{RT}{P}dP \tag{15}$$

The free energy change in going from a standard state (denoted by the superscript zero) to any new state is given by

$$\int_{G^o}^{G} dG = RT \int_{P^o}^{P} d\ln P \tag{16}$$

Integrating gives

$$G - G^o = RT\ln(P/P^o) \tag{17}$$

But the pressure in the standard state is 1 atm (see Chapter 3). Then Eq. (17) becomes

$$G = G^o + RT\ln P \tag{18}$$

Table 4.1 Units for the ideal gas constant R and some additional conversion factors

R	
0.082058 L atm mole^{-1}K^{-1}	P is in atm
62.364 L torr mole^{-1}K^{-1}	P is in torr
8.314 J mole^{-1}K^{-1}	P is in Pa, V is in m^3
1.987 cal mole^{-1}K^{-1}	P is in Pa, V is in m^3

1 cal = 4.184 joules (J).
1 J = 1 V C = 1 N m.

The Chemical Potential and Standard States

The total energy of a system can be changed by

(1) thermal energy $+q$ or $-q$
(2) work $+w$ or $-w$
(3) by adding or subtracting various chemical substances to the system.

If the above third possibility occurs and the system contains n_i moles of various species i, then Eq. (13) must be extended to

$$dG = VdP - SdT + \sum_i \mu_i dn_i \tag{19}$$

where μ_i is the *chemical potential* of species *i* defined as

$$\mu_i = \left(\frac{\partial G}{\partial n_i}\right)_{T,P,n_j} \equiv \overline{G}_i \tag{20}$$

That is, the chemical potential is the change in Gibbs free energy with respect to a change in the chemical composition of a given chemical species, at constant temperature *T*, constant pressure *P*, and constant composition of the other chemical species which are present. The chemical potential is also called the partial molar Gibbs free energy and is also given the symbol G_i.

More About the Chemical Potential

For a system containing *i* components, Eq. (18) becomes

$$\mu_i = \mu_i^o + RT \ln P_i \tag{21}$$

For solids and liquids, Eq. (21) is written as

$$\mu_i = \mu_i^o + RT \ln a_i \tag{22}$$

where a_i is the activity of the *i*th species and

(1) for solids, $a = 1$.
(2) for liquid H_2O, $a = 1$.
(3) for other liquids and for ions, the activity *a* is usually replaced by concentration *C*. This is a valid approximation for dilute solutions.
(4) for gases, the activity *a* is replaced by pressure *P*.

Consider a chemical reaction

$$\nu_1 A_1 + \nu_2 A_2 + \cdots . \rightarrow \nu_1{}' B_1 + \nu_2{}' B_2 + \cdots \tag{23}$$

where ν_1 refers to the number of moles of reactant A_1, ν_2 refers to the number of moles of reactant A_2, etc., and the primes and B′s have similar meanings for the products. When all the reactants and products are in their standard states, then the standard free energy change ΔG^o is given by

$$\Delta G^o = v'_1\mu^o(B_1) + v'_2\mu^o(B_2) + \cdots - [v_1\mu^o(A_1) + [v_2\mu^o(A_2) + \cdots] \quad (24)$$

or in short-hand form

$$\Delta G^o = \sum_i v_i\mu_i^o(\text{products}) - \sum_i v_i\mu_i^o(\text{rectants}) \quad (25)$$

Tables exist for values of μ^o for various chemical species [1–4]. The standard chemical potential is also called the standard free energy of formation $\Delta G_f{}^o$ and is listed as such in some tables.

The standard chemical potential μ^o (i.e., the standard free energy of formation, $\Delta G_f{}^o$) is zero for pure elements.

Also, μ^o (H^+ (aq)) = 0. This is a consequence of defining $E^o = 0.000$ V for the standard hydrogen electrode. See Problem 4.1 at the end of this chapter.

A Note About Units for ΔG^o *or* ΔG

Materials scientists and chemists have traditionally expressed values of ΔG^o in units of calories per mole or kilocalories per mole (kcal/mol). (One calorie is the energy required to raise the temperature of one gram of water by 1°C). However, in SI (System Internationale) units, the proper measure of energy is the joule. (One joule is defined as one volt coulomb). There are 4.184 J/cal.

Either system of units may be used. Pourbaix's classic atlas of corrosion thermodynamic data [1] gives values for the chemical potential μ^o in terms of calories per mole. More recent compilations of thermodynamic data give values of μ^o (or $\Delta G_f{}^o$) in terms of kilojoules per mole (kJ/mol).

Example 4.1: Is the following reaction spontaneous at 25°C when each of the reactants and products are in their standard states:

$$Zn(s) + 2H_2O(I) \rightarrow Zn(OH)_2(s) + H_2(g)$$

given the following thermodynamic data for 25°C [1]: $\mu^o(Zn(OH)_2)(s)) = -75{,}164$ cal/mol; $\mu^o(H_2O\ (l)) = -56{,}690$ cal/mol.

Solution: The standard free energy change for the reaction above is given by

$$\Delta G^o = [\mu^o(Zn(OH)_2(s)) + \mu^o(H_2(g))] - [\mu^o(Zn(s)) + 2\mu^o(H_2O(l))]$$

The chemical potential of a pure element is zero so that $\mu^o(Zn\ (s)) = 0$ and also $\mu^o(H_2\ (g)) = 0$. Thus

$$\Delta G^o = [-75{,}164 + (0)]\text{cal} - [(0) + 2(-56{,}690)]\text{cal}$$

$$\Delta G^o = -75{,}164\,\text{cal} + 113{,}380\,\text{cal}$$
$$\Delta G^o = +38{,}126\,\text{cal}$$

The sign of ΔG^o is positive so that the reaction is not spontaneous and the immersion of zinc into water under standard conditions does not produce hydrogen gas.

The Free Energy and Electrode Potentials

It was shown in the previous chapter that an electrode potential exits across the metal/solution interface, as seen schematically in Fig. 4.5.

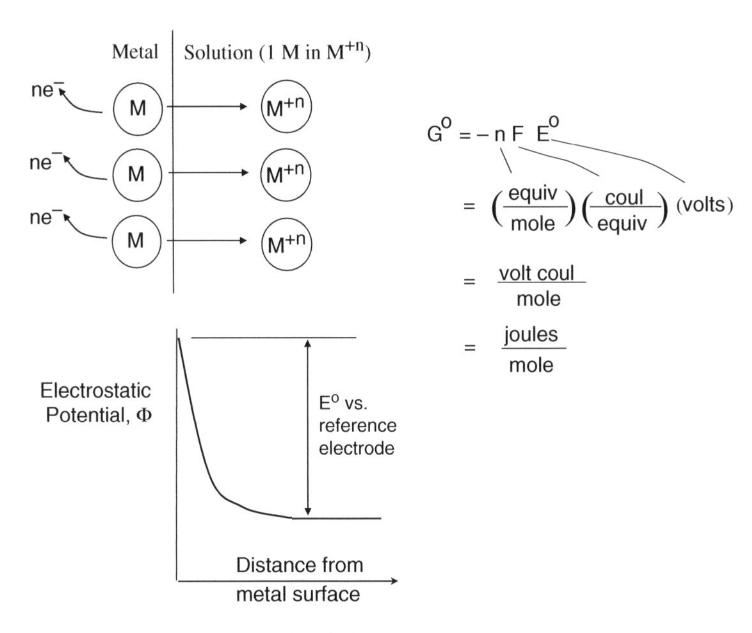

Fig. 4.5 The free energy change for an electrochemical process

The free energy change for an electrochemical process when all the reactants and products are in their standard states is given by

$$\Delta G^{\mathrm{o}} = -nFE^{\mathrm{o}} \tag{26}$$

where n is the number of electrons transferred, F is the Faraday, and E^{o} is the electrode potential. Figure 4.5 also shows that the units resulting from the term nFE are in joules (or calories), as required. The negative sign in Eq. (26) is required to make spontaneous electrochemical reactions have a negative value of ΔG, as is required. In the convention employed here, the number of electrons transferred (n) is always positive. The Faraday F is, of course, positive.

In the general case where reactants and products are not all in standard states

$$\Delta G = -nFE \tag{27}$$

Example 4.2: Chromate inhibitors act on steel surfaces as follows:

$$2\mathrm{CrO_4^{2-}(aq)} + 2\mathrm{Fe(s)} + 4\mathrm{H^+(aq)} \rightarrow \mathrm{Cr_2O_3(s)} + \mathrm{Fe_2O_3(s)} + 2\mathrm{H_2O(l)}$$

Is the above electrochemical reaction spontaneous when each of the reactants and products is in their standard states, given that the standard electrode potential for the overall reaction is +1.437 V vs. SHE?

Solution:

$$\Delta G^{o} = -nFE^{o}$$
$$\Delta G^{o} = -(+n)(+F)(+1.437) < 0$$

Thus, the reaction is spontaneous as written. Note, for our purposes here, we do not need to know the value of n but only that it is positive. We can determine that $n = 6$ by separating the overall electrochemical reaction into its two half-cell reactions:

$$2Fe(s) + 3H_2O(I) \rightarrow Fe_2O_3(s) + 6H^{+} + 6e^{-}$$

$$2CrO_4^{2-}(aq) + 10H^{+}(aq) + 6e^{-} \rightarrow Cr_2O_3(s) + 5H_2O(l)$$

The sum of these two half-cell reactions is the overall reaction given above. More detail on electrochemical cells is given in Chapter 5.

The Nernst Equation

As noted in Chapter 3, standard electrode potentials E^{o} apply only to the situation where a metal is immersed in a solution of its own ions at unit activity. This condition is rarely encountered in corrosion reactions. The Nernst equation allows calculation of the half-cell potential for some other concentration in terms of the standard electrode potential. The derivation is given here.

Consider a general electrochemical reaction

$$a\mathrm{A} + b\mathrm{B} + ne \rightleftharpoons c\mathrm{C} + d\mathrm{D} \tag{28}$$

where a is the number of moles of reactant A, b is the number of moles of reactant B, etc. From Eq. (25)

$$\Delta G = \sum_{\mathrm{i}} \nu_{\mathrm{i}}\mu_{\mathrm{i}}\,(\text{products}) - \sum_{\mathrm{i}} \nu_{\mathrm{i}}\mu_{\mathrm{i}}\,(\text{rectants}) \tag{29}$$

or

$$\Delta G = c\mu_{\mathrm{C}} + d\mu_{\mathrm{D}} - a\mu_{\mathrm{A}} - b\mu_{\mathrm{B}} \tag{30}$$

But from Eq. (22)

$$\begin{aligned} \mu_{\mathrm{C}} &= \mu_{\mathrm{C}}^{o} + RT\ \ln\ \mathrm{a_C} \\ \mu_{\mathrm{D}} &= \mu_{\mathrm{D}}^{o} + RT\ \ln\ \mathrm{a_D} \\ \mu_{\mathrm{A}} &= \mu_{\mathrm{A}}^{o} + RT\ \ln\ \mathrm{a_A} \\ \mu_{\mathrm{B}} &= \mu_{\mathrm{B}}^{o} + RT\ \ln\ \mathrm{a_B} \end{aligned} \tag{31}$$

Substitution of Eq. (31) in Eq. (30) gives

$$\Delta G = c[\mu_{\mathrm{C}}^{o} + RT\ \ln\ a_{\mathrm{C}}] + d[\mu_{\mathrm{D}}^{o} + RT\ \ln\ a_{\mathrm{D}}] - a[\mu_{\mathrm{A}}^{o} + RT\ \ln\ a_{\mathrm{A}}] - b[\mu_{\mathrm{B}}^{o} + RT\ \ln\ a_{\mathrm{B}}] \tag{32}$$

Grouping terms gives

$$\Delta G = [c\mu_C^o + d\mu_D^o - a\mu_A^o - b\mu_B^o] + \left[RT\ \ln \frac{a_C^c a_D^d}{a_A^a a_B^b}\right] \tag{33}$$

But from Eq. (25)

$$\Delta G^o = c\mu_C^o + d\mu_D^o - a\mu_A^o - b\mu_B^o \tag{34}$$

so that Eq. (33) becomes

$$\Delta G = \Delta G^0 + RT \ln \frac{a_c^c a_D^d}{a_A^a a_B^b} \tag{35}$$

With $\Delta G = -nFE$ and $\Delta G^o = -nFE^o$, Eq. (35) becomes

$$E = E^o - \frac{2.303RT}{nF} \log \frac{a_C^c a_D^d}{a_A^a a_B^b} \tag{36}$$

which is the Nernst equation. This equation is very useful in the analysis of electrochemical cells and in the construction of Pourbaix diagrams. These subjects are treated in the next two chapters. At 25°C, Eq. (36) can be written as

$$E = E^o - \frac{0.0591}{n} \log \frac{a_C^c a_D^d}{a_A^a a_B^b} \tag{37}$$

Standard Free Energy Change and the Equilibrium Constant

The equilibrium constant for Eq. (28) is

$$K = \frac{a_C^c a_D^d}{a_A^a a_B^b} \tag{38}$$

so that Eq. (35) can be rewritten as

$$\Delta G = \Delta G^o + RT \ln K \tag{39}$$

At equilibrium, $\Delta G = 0$ so that Eq. (39) becomes

$$\Delta G^o = -RT \ln K \tag{40}$$

or

$$\Delta G^o = -2.303\,RT \log K \tag{41}$$

This equation is useful because it is the link between the standard free energy change and the equilibrium constant for a reaction.

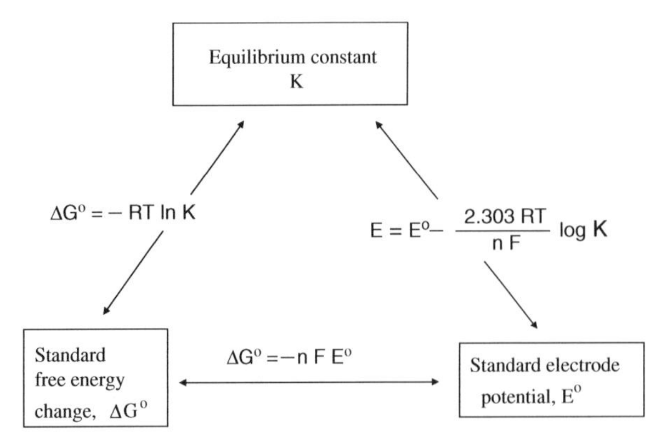

Fig. 4.6 Summary of relationships between the standard free energy change, the standard electrode potential, and the equilibrium constant for electrochemical reactions

Figure 4.6 summarizes relationships between the standard free energy change, the standard electrode potential, and the equilibrium constant for electrochemical reactions.

A Quandary – The Sign of Electrode Potentials

Throughout this chapter we have been adhering to a system of thermodynamics commonly called the American sign convention [4, 5]. We write

$$Zn^{2+}(a = 1) + 2e^- \rightarrow Zn \quad E = -0.762\,V$$

but we write

$$Zn \rightarrow Zn^{2+}(a = 1) + 2e^- \quad E = +0.762\,V$$

These are thermodynamic half-cell potentials, but what is the sign of the electrode potential for the *physical* electrode of an actual sample of zinc metal immersed in an actual solution of unit activity of Zn^{2+} ions?

At the physical electrode, there is an equilibrium between zinc metal and zinc ions so that zinc ions pass into solution at the same rate at which they are reduced back onto the metal surface, as depicted in Fig. 4.7. So what is the electrode potential for the physical electrode? Is it the average of

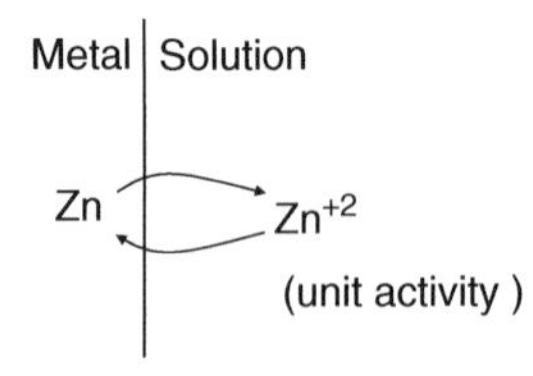

Fig. 4.7 The Zn/Zn^{2+} electrode at equilibrium. The rate at which Zn goes into solution is equal to the rate at which Zn^{2+} ions are reduced to solid Zn

the reduction and oxidation potentials? This cannot be, for the average potential would be 0.000 V, and this is the standard electrode potential for the reduction of hydrogen ions to hydrogen gas. Moreover, the physical electrode potentials for *all* metals listed in the emf series would also be 0.000 V, so actual electrode potentials would be a meaningless concept.

Note that in Fig. 3.24, the metal electrode under study is connected to the positive terminal of the electrometer and the reference electrode is connected to the negative terminal. This is a convention adapted by the American Society of Testing and Materials [6], and this arrangement gives the proper sign of the measured electrode potential. The proper sign for the actual physical electrode potential of zinc in contact with a solution of unit activity Zn^{2+} ions is negative. (A simple test cell of zinc in contact with Zn^{2+} ions can be used to see if your test setup in the laboratory is correct in regard to the signs of measured electrode potentials.) But suppose that the connections to the test and reference electrodes would be reversed. Then the sign of the measured physical electrode potential would also be reversed.

Thus, the sign of a physical electrode potential is a convention. This is in addition to the earlier convention that the forward direction of a half-cell reaction is taken to be the direction in which reduction occurs.

We want to relate a direction-sensitive quantity ΔG to a direction-insensitive observable E. This is the origin of all confusion regarding electrode potentials. Enlightening comments about this situation have been written by Anson [7] and deBethune [8].

Thus, in the American system of signs conventions, the sign of a thermodynamic half-cell reaction is sign bivariant (depending on whether the half-cell reaction is a reduction or an oxidation reaction), but the number of electrons transferred is always positive when using Eqs. (26), (27) or the Nernst equation, Eq. (36).

Mention should also be made of the European sign convention, used by Pourbaix [1]. According to this system

$$Zn^{2+}(a = 1) + 2e^{-} \rightarrow Zn(s) \quad E = -\,0.762\,V$$

and also

$$Zn(s) \rightarrow Zn^{2+}(a = 1) + 2e^{-} \quad E = -0.762\,V$$

Thus, in the European system of signs conventions, the sign of a thermodynamic half-cell reaction is invariant (i.e., the same for a reduction or an oxidation reaction). However, the number of electrons transferred is sign bivariant (negative when the half-cell reaction is a reduction reaction and positive when the half-cell reaction is anoxidation reaction). In this system, $\Delta G = nFE$ and the Nernst equation also has a slightly different form. See Table 4.2 for a comparison of these two conventions.

Confusing? Sort of, but not if one follows a given convention consistently. Throughout this text, we will be adhering to the American sign convention.

Factors Affecting Electrode Potentials

Actual physical electrode potentials are affected by several variables including (i) the nature of the metal, (ii) the chemical nature of the aqueous solution, (iii) the presence of oxide films on the metal surface, (iv) the presence of adsorbed gases on the metal surface, and (v) the presence of mechanical stress on the metal [9]. See Table 4.3, which lists the magnitude of these effects.

Table 4.2 A comparison of American and European sign conventions for thermodynamic systems

American convention	European convention
$Zn^{2+}(a = 1) + 2e^- \rightarrow Zn(s)$ $E^o = -0.762$ V	$Zn^{2+}(a = 1) + 2e^- \rightarrow Zn(s)$ $E^o = -0.762$ V
$Zn(s) \rightarrow Zn^{2+}(a = 1) + 2e^-$ $E^o = +0.762$ V	$Zn(s) \rightarrow Zn^{2+}(a = 1) + 2e^-$ $E^o = -0.762$ V
$Cu^{2+}(a = 1) + 2e^- \rightarrow Cu(s)$ $E^o = +0.342$ V	$Cu^{2+}(a = 1) + 2e^- \rightarrow Cu(s)$ $E^o = +0.342$ V
$Cu(s) \rightarrow Cu^{2+}(a = 1) + 2e^-$ $E^o = -0.342$ V	$Cu(s) \rightarrow Cu^{2+}(a = 1) + 2e^-$ $E^o = +0.342$ V
$\Delta G^o = -nFE^o$ E^o is sign bivariant n is sign invariant (always positive)	$\Delta G^o = nFE^o$ E^o is sign invariant n is sign bivariant (negative for reduction, positive for oxidation)
Nernst equation $E = E^o - \frac{2.303RT}{nF} \log \frac{a_C^c\, a_D^d}{a_A^a\, a_B^b}$ E^o is sign bivariant n is sign invariant (always positive)	Nernst equation $E = E^o + \frac{2.303RT}{nF} \log \frac{a_C^c\, a_D^d}{a_A^a\, a_B^b}$ E^o is sign invariant n is sign bivariant (negative for reduction, positive for oxidation

Table 4.3 Effect of various factors on the electrode potential [9]

Effect	Magnitude of effect on electrode potential
Nature of the metal	Whole volts
Chemical nature of the aqueous solution	Tenths to whole volts
Surface state (oxide films)	Tenths of a volt
Adsorbed gases	Hundreds to tenths of volts
Mechanical stress	Thousandths to hundreds of volts

Problems

1. Show that $\mu^o(H^+(aq)) = 0$ is a consequence of defining $E^o = 0.000$ as the standard electrode potential for reduction of the hydrogen ion:

$$2H^+(aq) + 2e^- \rightarrow H_2\ (g)$$

2. (a) Calculate $2.303RT$ in joules per mole and calories per mole at 25°C.
   (b) Show that $2.303RT/F$ is equal to 0.0591 V equiv/mol at 25°C.
3. Calculate the electrode potential at 25°C for the Peterson–Groover Ag/AgCl reference electrode when it is used in natural seawater, for which the Cl^- concentration is 0.6 M. (b) How does this electrode potential compare with that for the saturated calomel electrode?

4. Dissolved lead can enter the public drinking supply through the use of lead piping systems and lead solder. The present standard for allowable dissolved Pb^{2+} ions is 15 μg/L. Calculate the reduction potential for lead metal in contact with this concentration of dissolved Pb^{2+} ions:

$$Pb^{2+}(aq) + 2e^{-} \rightarrow Pb(s)$$

5. For health reasons the concentration of dissolved copper ions in drinking water should be less than 2 mg/L of Cu^{2+}[10]. Calculate the reduction potential of copper metal in contact with a solution containing 2 mg/L of dissolved Cu^{2+} at 25°C.
6. Calculate whether copper will spontaneously corrode at 25°C by the following half-cell reaction:

$$2Cu(s) + H_2O(l) + O_2(g) \rightarrow 2CuO(s) + H^{+}(aq) + OH^{-}(aq)$$

given the following standard chemical potentials μ^{o} (in cal/mol):

$$CuO\ (s) = -30{,}400;\ OH^{-}\ (aq) = -37,\ 595;\ H_2O\ (l) = -56{,}690.$$

7. Research has been conducted on molybdates as replacements for chromates in the surface treatment of various metals in order to reduce corrosion by the formation of a surface film of MoO_2. The overall reaction with iron is

$$2Fe(s) + 3MoO_4^{2-}(aq) + 6H^{+}(aq) \rightarrow Fe_2O_3(s) + 3MoO_2(s) + 3H_2O(l)$$

(a) Given the following standard chemical potentials (μ^{o}) in calories per mole at 25°C: Fe_2O_3 (s) = –177,100, MoO_2 (s) = –120,000, H_2O (l) = –56,690, MoO_4^{-2} (aq) = –205,420:

(a) Is the above overall reaction spontaneous under standard conditions?
(b) Calculate the standard electrode potential for the above overall reaction.
(c) Write an expression for the electrode potential as a function of molybdate ion concentration and pH.
(d) Is the overall reaction spontaneous for a molybdate ion concentration of 0.001 M at pH 7.0?
(e) Based solely on thermodynamic considerations, what do the results in (a) and (d) suggest about the possibility of molybdates as chromate replacements?

8. Calculate the standard chemical potential of Sn^{2+} ions, $\mu^{o}(Sn^{2+}\ (aq))$, from the appropriate standard electrode potential in Table 3.1.
9. Magnesium corrodes in an aqueous solution of pH 9.0 to produce precipitated $Mg(OH)_2$. What is the reduction potential of magnesium in this solution at 25°C for the reaction

$$Mg^{2+}(aq) + 2e^{-} \rightarrow Mg(s) \quad E^{o} = -2.372\ V$$

The solubility product of $Mg(OH)_2$ is $K_{sp} = 1.8 \times 10^{-11}$.
10. Waste acid of pH 3.0 is stored in a lead-lined vessel, as shown in the figure below. If the walls of the container undergo corrosion by the following reaction:

$$Pb(s) + 2H^{+}(aq) \rightarrow Pb^{2+}(aq) + H_2(g)$$

what is the equilibrium concentration of dissolved Pb^{2+} at 25°C if the pressure of hydrogen gas in the closed vessel is 0.5 atm? The standard chemical potential for Pb^{2+} (aq) is $\mu^o =$ –5,810 cal/mol.

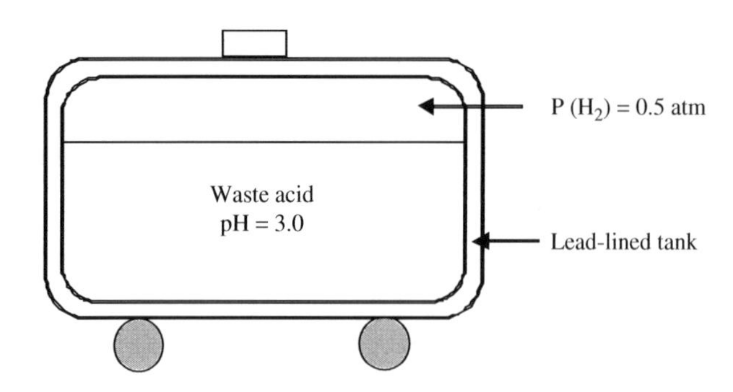

11. The following two half-cell reactions have the standard electrode potentials as given:

$$Fe^{3+}(aq) + e^- \rightarrow Fe^{2+}(aq) \quad E^o = +0.771\text{ V}$$
$$Fe^{2+}(aq) + 2e^- \rightarrow Fe(s) \quad E^o = -0.447\text{ V}$$

Calculate E^o for the following reaction:

$$Fe^{3+}(aq) + 3e^- \rightarrow Fe(s)$$

Note that if we add the first two half-cell reactions, we get the desired half-cell reaction. Can we add the corresponding two values of E^o to obtain E^o for the desired reaction? Explain your answer.

12. What qualitative statement can be made about the change in entropy when a metal atom in a crystalline lattice passes into solution to become a metal cation. Explain your answer.

References

1. M. Pourbaix, "Atlas of Electrochemical Equilibria in Aqueous Solutions", National Association of Corrosion Engineers, Houston, TX (1974).
2. M. W. Chase, Jr., Ed. "NIST-JANAF Thermochemical Tables", American Chemical Society and American Institute of Physics, Woodbury, NY (1998).
3. A. J. Bard, R. Parsons, and J. Jordan, Eds., "Standard Potentials in Aqueous Solutions", Marcel Dekker, New York, NY (1985).
4. W. M. Latimer, "The Oxidation States of the Elements and Their Properties in Aqueous Solutions", Prentice-Hall, New York NY (1952).
5. G. N. Lewis, M. Randall, K. S. Pitzer, and L. Brewer, "Thermodynamics", McGraw-Hill, New York, NY (1961).
6. "Standard Practice for Conventions Applicable to Electrochemical Measurements in Corrosion Testing", ASTM G3-89, "Annual Book of ASTM Standards", Vol. 3.02, p. 36, ASTM, West Conshohocken, PA (1998).
7. F. C. Anson, *J. Chem. Ed.*, *36*, 395 (1959).
8. A. J. deBethune, *J. Electrochem. Soc.*, *102*, 288C (1955).
9. G. V. Akimov, *Corrosion*, *11*, 477 t (1955).
10. D. J. Fitzgerald, *Am. J. Clinical Nutrition*, *67* (5), 1098S, (Supplement S) (1998).

Chapter 5
Thermodynamics of Corrosion: Electrochemical Cells and Galvanic Corrosion

Introduction

The previous chapter has dealt largely with the corrosion thermodynamics of half-cell reactions for single metals. This chapter considers the effect of coupling two half-cells to form an electrochemical cell. The galvanic corrosion of coupled dissimilar metals is also treated in this chapter.

Electrochemical Cells

Consider the two half-cells shown in Fig. 5.1(a). In the compartment on the left, zinc ions are in equilibrium with the solid zinc electrode. This means that the rate at which Zn^{2+} ions pass into solution is equal to the rate at which they are reduced to solid zinc. The standard reduction potential is

$$Zn^{2+}\,(aq) + 2e^{-} \rightarrow Zn(s) \quad E^{o} = -0.762\,V$$

In the compartment on the right, copper ions are in equilibrium with the solid copper electrode. Thus, the rate at which Cu^{2+} ions pass into solution is equal to the rate at which they are reduced to solid copper. The standard reduction potential is

$$Cu^{2+}(aq) + 2e^{-} \rightarrow Cu\,(s) \quad E^{o} = +0.342\,V$$

When the two unconnected half-cells are coupled by closing the switch in the external circuit, as in Fig. 5.1(b), an electrochemical cell is formed. One of the metals will be the anode and the other will be the cathode, although we cannot yet tell which is which. First suppose that the zinc electrode is the anode. Then

At the cathode

$$Cu^{2+}(aq) + 2e^{-} \rightarrow Cu\,(s) \quad E^{o} = +0.342\,V$$

At the anode

$$Zn(s) \rightarrow Zn^{2+}(aq) + 2e^{-} \qquad -E^{o} = -(-0.762\,V)$$

E. McCafferty, *Introduction to Corrosion Science*, DOI 10.1007/978-1-4419-0455-3_5,

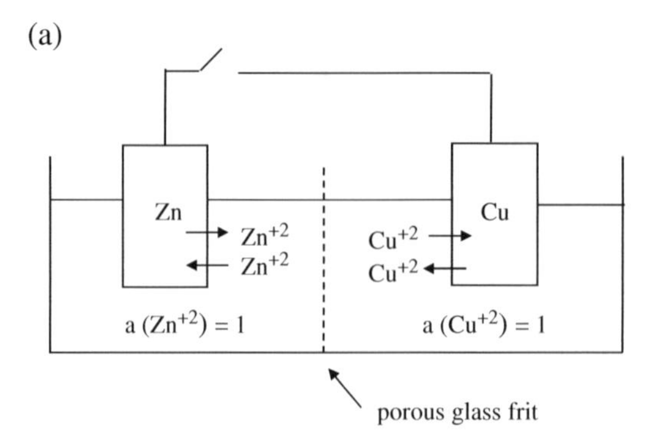

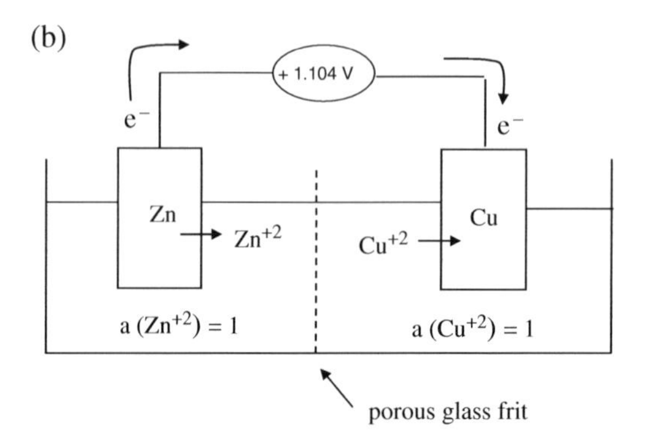

Fig. 5.1 An electrochemical galvanic cell

Adding the two half-cell reactions gives the overall chemical reaction:

$$Cu^{2+}(aq) + Zn(s) \rightarrow Cu(s) + Zn^{2+}(aq) \tag{1}$$

and adding the half-cell potentials gives

$$E^{o}_{cell} = E^{o}_{cathode} - E^{o}_{anode} \tag{2}$$

or $E^{o}_{cell} = +0.342\text{V} - (-0.762\,\text{V}) = +1.104\,\text{V}$. From

$$\Delta G_{cell} = -nFE_{cell} \tag{3}$$

the free energy change is negative because the cell potential is positive. Thus, the reaction proceeds spontaneously as written above in Eq. (1), and we have correctly identified the anode and the cathode in the coupled cell.

If we had assumed initially that the copper electrode was the anode and the zinc electrode the cathode, then the calculated cell potential would be $E^{\circ}_{cell} = -1.104$ V. But then the calculated change in free energy would have been positive so that the assumed reaction would not be spontaneous. Thus, we would have reached the same conclusion as above.

The following general conclusions can be drawn:

(1) In an electrochemical cell, the metal with the lower electrode potential (more negative reduction potential) in the emf series is the anode; the metal with the higher electrode potential (more positive reduction potential) is the cathode.
(2) The cell potential (also called the cell voltage or cell emf) is given by

$$E_{cell} = E_{cathode} - E_{anode} \tag{4}$$

where both $E_{cathode}$ and E_{anode} are reduction potentials.

If the concentration of dissolved ions is not unit activity, the half-cell potentials can be calculated using the Nernst equation, and the procedure is then continued as described above.

Electrochemical Cells on the Same Surface

An electrochemical cell can also exist on different portions of the same metal surface, as shown in the following example.

Example 5.1: Calculate whether copper can corrode at 25°C in an acid solution of pH 2.0 to produce a solution containing 0.10 M Cu^{2+} ions and 0.5 atm hydrogen gas.
Solution: The overall chemical reaction is

$$Cu(s) + 2H^+(aq) \rightarrow Cu^{2+}(aq) + H_2\,(g)$$

At the local cathodes

$$2H^+\,(aq) + 2e^- \rightarrow H_2(g) \quad E_{cathode}$$

At the local anodes

$$Cu\,(s) \rightarrow Cu^{2+}\,(aq) + 2e^- \quad E_{anode}$$

We calculate the electrode potentials for each half-cell reaction *written as a reduction reaction.* Using the Nernst equation

$$E_{cathode} = E^{o}_{H^+/H_2} - \frac{0.0591}{2}\log\frac{P(H_2)}{[H^+]^2}$$

$$E_{cathode} = 0.000 - \frac{0.0591}{2}\log\frac{0.5}{[1.0x10^{-2}]^2}$$

or

$$E_{cathode} = -0.109\text{ V}$$

Then

$$E_{anode} = E^{O}_{Cu^{+2}/Cu} - \frac{0.0591}{2}\log\frac{1}{[Cu^{+2}]}$$

$$E_{\text{anode}} = +\,0.342 \;-\; \frac{0.0591}{2}\log\frac{1}{0.10}$$

or

$$E_{\text{anode}} = +0.312$$

Thus

$$E_{\text{cell}} = E_{\text{cathode}} - E_{\text{anode}}$$
$$E_{\text{cell}} = -0.109 - (+0.312) = -0.421\ \text{V}$$

Because the cell voltage is negative, $\Delta G_{\text{cell}} = -\,nFE_{\text{cell}}$ is positive, and thus the reaction will not proceed spontaneously to produce the conditions which were described.

Galvanic Corrosion

Galvanic corrosion occurs when two dissimilar metals are in physical (and electrical) contact in an aqueous electrolyte, as shown in Fig. 5.2. Alternately (and less usual), the two metals may be connected by an external metal path. Galvanic coupling of metals in equilibrium with their own ions rarely occurs, but instead each of the two individual metals is usually immersed in a common electrolyte. Some examples of galvanic corrosion include the following:

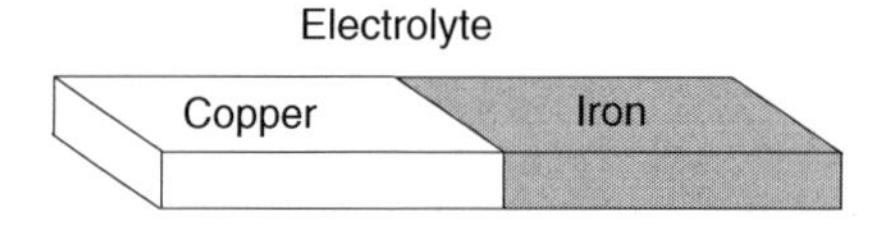

Fig. 5.2 Two dissimilar metals in a galvanic couple

(1) Copper piping connected to steel tanks.
(2) Boats having a nickel alloy hull and steel rivets.
(3) Zinc-coated screws in a sheet of copper.
(4) Tin-plated electrical connector pins mated with gold-plated sockets.
(5) A stainless steel screw in contact with a cadmium-plated steel washer.

The galvanic series described below is very useful in indicating whether combined pairs of coupled metals will be prone to galvanic corrosion.

Galvanic Series

As noted in Chapter 3, the emf series has limited practical use. The standard electrode potentials are for half-cell reactions for metals in solutions of their own ions at unit activity. This set of conditions is important in the development of the concept of electrode potentials, but these conditions are quite restrictive and are not those found in most corrosion applications. The concentration of dissolved metal ions in equilibrium with a given metal depends on the environment and on the details of the system, such as the corrosion rate, the degree of mass transfer, and the volume of the solution.

The restriction of unit activity in the emf series can be removed by using the Nernst equation to calculate the potential for a half-cell reaction at some concentration other than unit activity. However, there is still the restriction in the emf series that the solution contains cations of only the metal of interest. In addition, the emf series applies only to pure metals and not to alloys.

Any metal or alloy placed in a corrosive environment has its own electrode potential, called the corrosion potential E_{corr}. The galvanic series (in seawater) is an ordered listing of experimentally measured corrosion potentials in natural seawater for both pure metals and alloys. See Fig. 5.3[1]. Metals and alloys with more positive potentials (like platinum) are called noble metals, and metals with more negative potentials (like magnesium) are called active metals. Note that the electrode potentials in the galvanic series in Fig. 5.3 are measured relative to a saturated calomel electrode (although any suitable reference electrode can be used), whereas standard half-cell electrode potentials are always referred to as the standard hydrogen electrode. Note also that all the alloys given in Fig. 5.3 exhibit a range of electrode potentials.

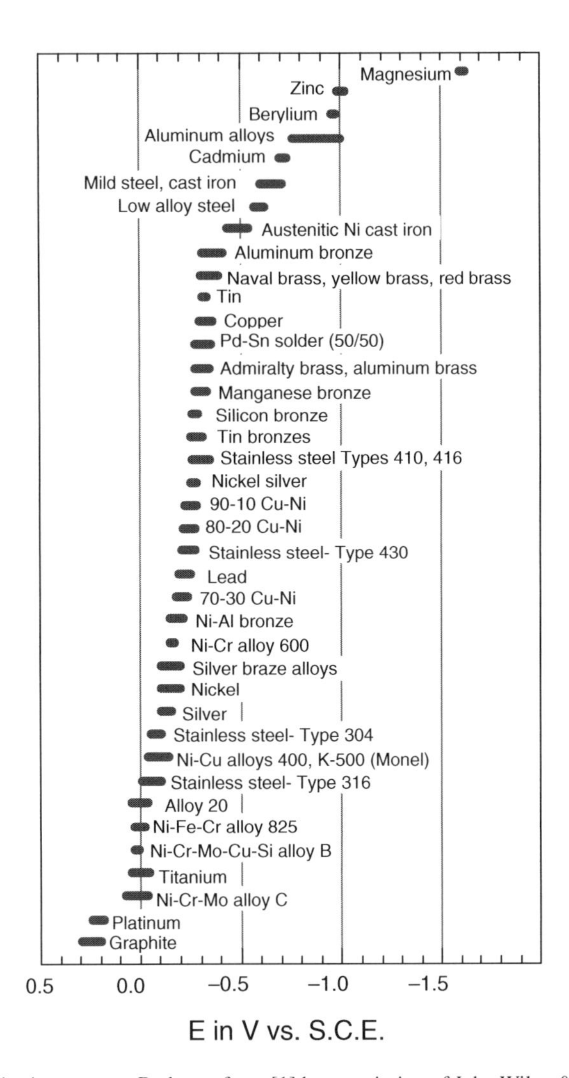

Fig. 5.3 The galvanic series in seawater. Redrawn from [1] by permission of John Wiley & Sons, Inc

The utility of the galvanic series is as follows:

In a galvanic couple, the metal or the alloy with the more negative electrode potential is the anode. The metal or the alloy with the more positive electrode potential is the cathode.

This conclusion follows from the analysis of electrochemical cells using the emf series.

Example 5.2: Composite materials consisting of magnesium containing graphite-reinforcing fibers have been developed to increase the mechanical strength of the light metal magnesium. Will the introduction of graphite into magnesium pose a problem if this composite material is used in seawater?

Solution: From Fig. 5.3, the electrode potential of graphite in seawater is approximately +0.25 V vs. SCE. The electrode potential of magnesium is –1.6 V vs. SCE. Thus, in a magnesium–graphite couple, magnesium will be the anode. Because of the large difference in the two electrode potentials, there will be a large galvanic effect between magnesium and graphite. This couple should not be used in seawater unless the composite material is provided corrosion protection, such as a barrier of paint or some other organic coating.

The order of a series of metals in any galvanic series may not be the same as their order in the emf series. Thus, the emf series cannot be used reliably to predict the corrosion tendencies of coupled metals in other environments.

Similarly, the galvanic series for seawater should not be used to predict corrosion tendencies in solutions which are very much different than seawater. For example, the galvanic series in seawater may be applied (with some caution) to the behavior of metals and alloys used for joint replacements in the human body. This is because a simulated physiological fluid known as Ringer's solution contains a mixture of dissolved NaCl, KCl, and $CaCl_2$, which is 0.16 M in total chlorides. Thus, body fluids and Ringer's solution can be considered to be dilute seawater so that the galvanic series for seawater can be used as a first approximation, although data in Ringer's solution itself should be used when available.

While the use of the galvanic series in seawater is possible for similar aqueous solutions, its extension to additional solutions, such as hydrochloric acid, for example, should be considered with much reservation.

Another limitation of any galvanic series is that the series gives no information about rates of corrosion, but only about corrosion tendencies. Actual corrosion rates must be determined in separate experiments or tests, to be discussed in Chapter 7.

Although it is not possible to predict corrosion rates from the use of the galvanic series, the anodic metal will have a higher corrosion rate in the couple than in the freely corroding (uncoupled) condition. LaQue [1] has tabulated seawater corrosion data for aluminum coupled to various metals located above it in the galvanic series in seawater. In each case, aluminum was the anode in the galvanic couple, and its corrosion rate when coupled was accelerated relative to its corrosion rate in its freely corroding (uncoupled) condition. Figure 5.4 shows corrosion data in natural seawater for aluminum coupled to various metals after 30 days of immersion in seawater flowing at 7.8 ft/s. In each case, the electrode potential of uncoupled aluminum was more negative than the electrode potential of the uncoupled second metal, as may be seen by referring to the galvanic series in Fig. 5.3. Thus, aluminum is the anode in each metal couple. As seen in Fig. 5.4, the corrosion rate in seawater in each galvanic couple was greater than the corrosion rate of freely corroding aluminum. The galvanic effect in each case is the difference in the corrosion rate of aluminum in each couple and the corrosion rate of aluminum coupled to itself.

For each couple in Fig. 5.4, the anodic to cathodic area ratios were held constant. The effect of the ratio of anodic and cathodic areas on galvanic corrosion will be considered in Chapter 7.

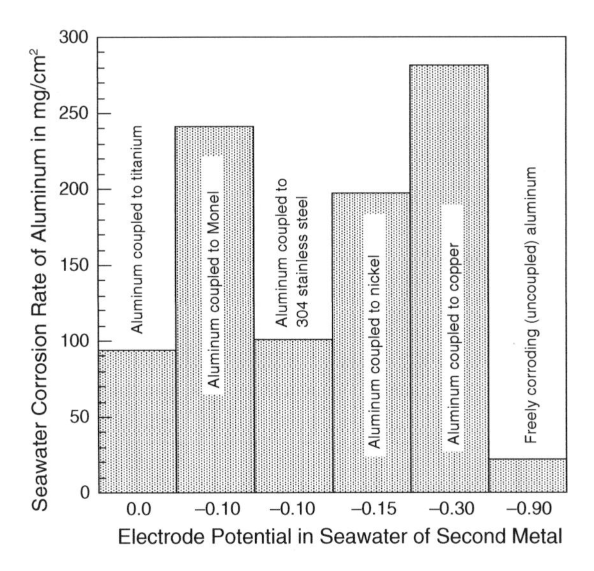

Fig. 5.4 Galvanic corrosion data for various metals coupled to aluminum in seawater [1]

To date, the galvanic series in seawater is the most extensive set of corrosion potentials developed for any corrosive environment. However, galvanic series exist for metals in flowing hot water [2], in phthalate buffers [3], in solutions of $NaHCO_3$ or Cl^--containing $NaHCO_3$ solutions [4].

Cathodic Protection

Galvanic corrosion is usually not a desired occurrence. However, by an appropriate selection of metal pairs, galvanic corrosion can be employed as a method of corrosion protection. Zinc metal plates fastened to the steel hulls of ships are used to protect the ship hulls by sacrificially corroding, rather than having the hull itself corrode. Such a practice is called cathodic protection, and its principle is illustrated in Fig. 5.5.

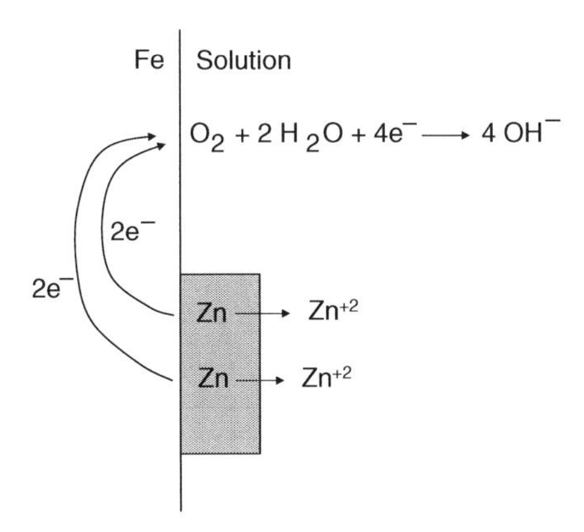

Fig. 5.5 Schematic diagram of the cathodic protection of iron by zinc

As shown in Fig. 5.3, the electrode potential of low alloy steel in seawater is approximately –0.6 V vs. SCE. The electrode potential of zinc is –1.0 V. Thus, when zinc is coupled to steel, zinc will be the anode and steel will be the cathode, as illustrated in Fig. 5.5. Hence the term "cathodic protection."

In actual practice, the exterior hull of the ship will most likely be painted. But when the layer of paint protection breaks down locally ("holidays" in the paint film), then cathodic protection takes over. Cathodic protection by sacrificial zinc anodes is used to protect bare steel in interior holding tanks of commercial and military vessels, in hot water heaters, water storage tanks, buried pipelines, and steel reinforcing bars in concrete structures located in marine environments, to name a few examples.

In addition to the use of *sacrificial anodes*, as described above, cathodic protection can also be provided through the use of external *impressed current systems*. In this latter approach, a source of direct current is applied using a power supply and an external anode located some distance away but having an electrical connection to the protected structure. See Fig. 5.6. Various anodes have been used, including high silicon cast iron, graphite, and platinum claddings.

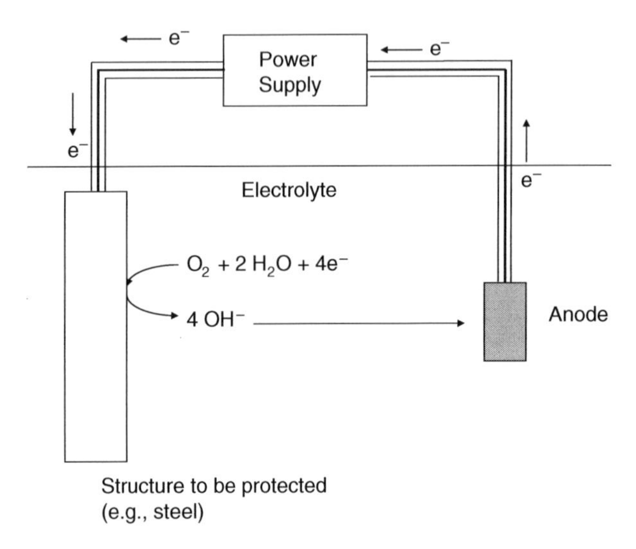

Fig. 5.6 Schematic diagram of cathodic protection by an impressed current device

Two Types of Metallic Coatings

There are two types of metallic coatings used to protect underlying metal substrates. These are (i) sacrificial and (ii) noble metal coatings.

The first of these types, sacrificial coatings, functions by cathodic protection of the substrate. The most common example is the use of an outer layer of zinc on a steel substrate ("galvanized" steel). Zinc has a more negative electrode potential than does steel in most environments, so zinc is again the anode when it is coupled to steel. When defects such as a pinhole or a crack develop in the outermost zinc coating, the underlying steel is protected by the sacrificial corrosion of zinc, as shown schematically in Fig. 5.7.

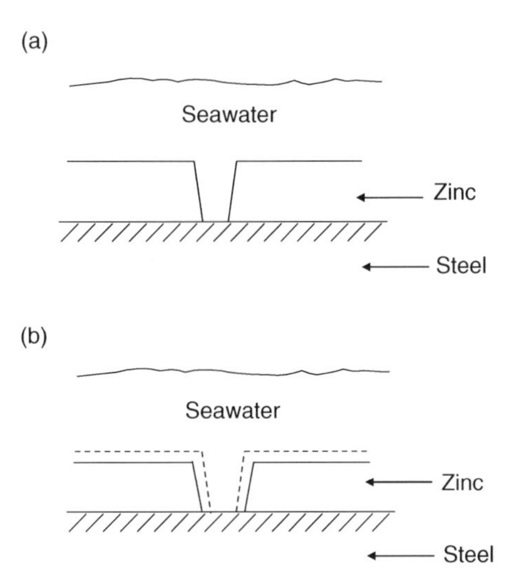

Fig. 5.7 A sacrificial protective coating. When a break or an imperfection develops in the coating and extends down to the substrate, as shown in (**a**), the zinc coating will galvanically corrode to protect the underlying substrate, as shown in (**b**)

Fig. 5.8 The corrosion of a galvanized roof in the natural atmosphere. When the protective zinc coating is breached, the underlying steel eventually suffers corrosion

Galvanized steel sheet is a common roofing material for both industrial and agricultural buildings. The thickness of the zinc coating is usually of the order of 0.001 in. (0.025 mm) [5], and the lifetime of galvanized roofs in outdoor atmospheric exposures is about 20 years or more, depending on the severity of the corrosive environment [5, 6]. When the entire zinc coating has been consumed due to corrosion, the effect of cathodic protection is lost, and the underlying steel substrate corrodes to display the familiar red color often seen in the roofs of deteriorating structures. See Fig. 5.8.

An alloy coating of zinc and aluminum has also been used in outdoor environments as a sacrificial coating. This zinc–aluminum coating provides cathodic protection while lasting longer than conventional zinc galvanized coatings [6].

The second type of metallic coating, the noble metal coating, is illustrated by a coating of nickel on steel, as shown in Fig. 5.9. In most aqueous environments, the electrode potential of nickel is more positive than the electrode potential of steel. Thus, steel is the anode when coupled to nickel.

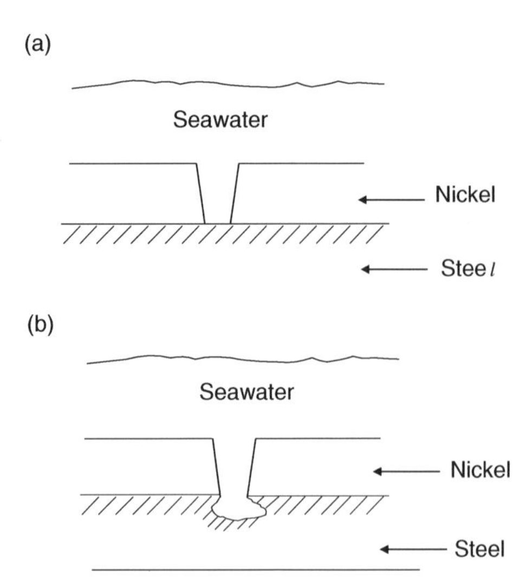

Fig. 5.9 A noble metal protective coating. When a break or an imperfection develops in the nickel coating and extends down to the substrate, as shown in (**a**), the underlying steel substrate will galvanically corrode, as shown in (**b**)

The protective effect of the nickel coating lies in its greater corrosion resistance in the corrosive environment. However, when defects arise in the noble metal coating, then the steel substrate undergoes corrosion, and the attack is concentrated in the steel substrate at the base of the defect. This is not a desired situation, as the attack can be deep and localized. This is in contrast to sacrificial coatings where the substrate is protected when a breach develops in the coating, and the attack is distributed to some extent along the coating.

In general, the situation involving a small anode area coupled to a large cathode area is unfavorable, because corrosion will be concentrated at the anode and will usually lead to intense localized attack.

Titanium Coatings on Steel: A Research Study

Plasma spraying is a useful technique for applying protective metal coatings to engineering alloys. However, plasma-sprayed coatings generally exhibit porosity because the molten droplets deposited by the plasma cool so rapidly that they cannot flow and completely wet the surface of the metal substrate. Thus small voids are trapped as the coating builds up droplet by droplet. As discussed above, voids or defects in metallic coatings bring on the onset of electrochemical cells between the coating and the substrate.

In order to improve the corrosion behavior of a porous titanium coating on steel (a noble metal coating), Ayers and co-workers [7] have employed the use of a high-power laser to melt partially through the porous plasma-sprayed titanium layer so as to consolidate the pores in the coating. (More about laser processing for improved corrosion resistance is given in Chapter 16).

Figure 5.10 shows electrode potentials measured in natural seawater for various samples. After 15 days of immersion, the electrode potential of pure titanium was approximately –0.2 V vs. SCE, whereas the electrode potential of the mild steel substrate was approximately –0.7 V vs. SCE. Thus, the titanium coating is a noble coating, and if coupled to steel, the steel member of the couple would be the anode.

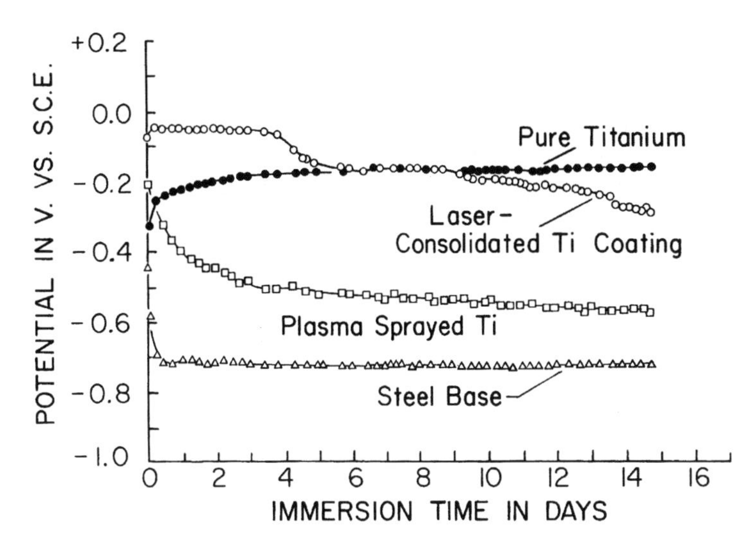

Fig. 5.10 Electrode potentials of titanium, steel, and titanium coatings in seawater [7]. Reprinted with permission of © NACE International 1981

Figure 5.10 shows that the electrode potential of the laser-consolidated coating was similar to that of pure titanium. During the immersion period this laser-processed coated sample remained unattacked and had the same general appearance of pure titanium. In contrast, the plasma-sprayed titanium coating had an electrode potential closer to that of steel rather than titanium. This sample showed visible rust staining during the test period.

Figure 5.11 shows cross sections through the two types of titanium coatings after the 15-day immersion period. As seen in the figure, corrosion pitting occurred at the steel substrate beneath the porous coating, and the attack was localized as deep pits, as illustrated schematically in Fig. 5.9 for a noble metal coating. Figure 5.11 also shows that laser remelt consolidation prevented the occurrence of such pitting by successfully eliminating pores in the laser-melted region.

Protection Against Galvanic Corrosion

Practices for minimizing galvanic corrosion are as follows [8]:

(1) Select combinations of metals as near to each other as possible in the galvanic series.
(2) Insulate the contact between dissimilar metals whenever possible.
(3) Apply organic coatings, but coat both members of the couple or coat only the cathode. Do not coat only the anode, because if a defect (holiday) develops in the organic coating, an accelerated attack will occur because of the unfavorable effect of a small anode area and a large cathode area, as shown in Fig. 5.9.
(4) Avoid the unfavorable area effect of having a small anode coupled to a large cathode.
(5) Install a third metal which is anodic to both metals in the galvanic couple.

Several of the measures listed above are illustrated schematically in Fig. 5.12. Area effects in galvanic corrosion will be revisited in Chapter 7 when the kinetic aspects of corrosion are considered.

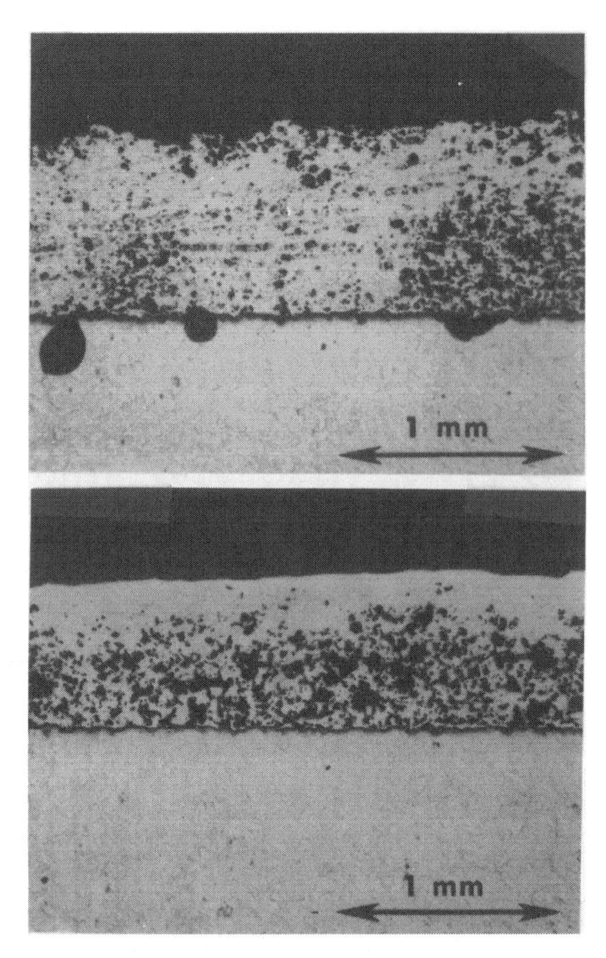

Fig. 5.11 Cross-sectional views through the coated specimen after 15 days of immersion in flowing seawater. *Top*: The unconsolidated sample shows pitting attack of the steel at the coating/substrate interface. *Bottom*: The sample was consolidated by laser melting approximately one-third of the coating thickness and showed no evidence of corrosion [7]. Reprinted with permission of © NACE International 1981

Differential Concentration Cells

As seen above, a difference in electrode potentials will exist for two different metals located in the same solution. A difference in electrode potentials will also exist for the same metal placed in two different concentrations of the same solution. Such a cell consisting of two half-cells of the same metal but with differing ionic or molecular concentrations is called a *differential concentration cell*.

There are two main types of differential concentration cells: (i) metal ion concentration cells and (ii) oxygen concentration cells. Either can cause corrosion to occur.

Metal Ion Concentration Cells

Consider two isolated (i.e., initially unconnected) copper electrodes in contact with different concentrations of dissolved cupric ions (Cu^{2+}), as in Fig. 5.13.

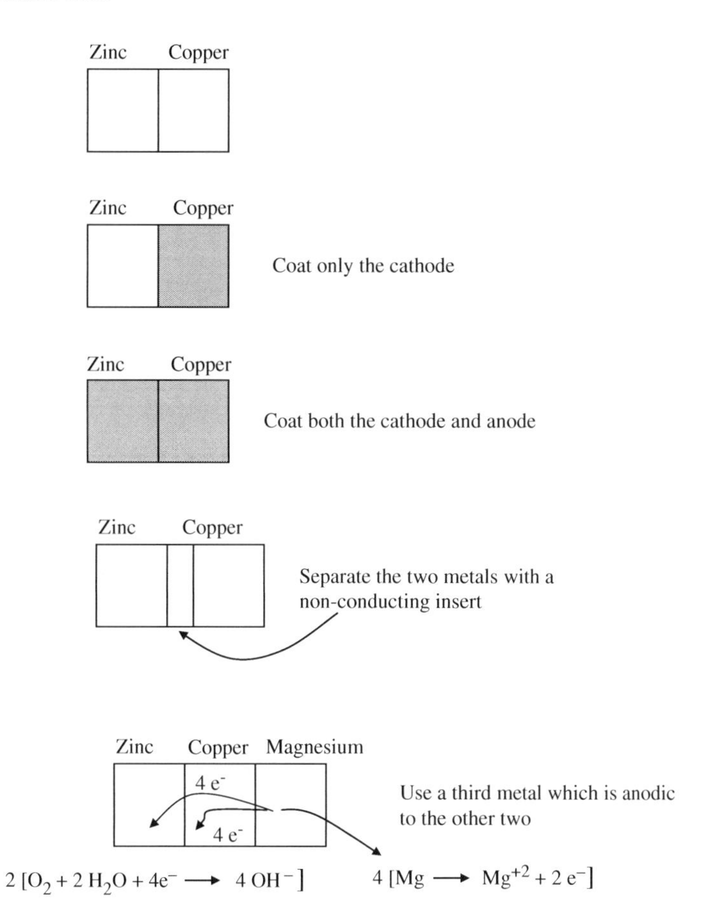

Fig. 5.12 Protective measures against galvanic corrosion

For the half-cell reaction

$$Cu^{2+}(aq) + 2e^{-} \rightarrow Cu(s)$$

application of the Nernst equation at 25°C gives

$$E = E^{O} - \frac{0.0591}{2} \log \frac{1}{[Cu^{2+}]} \qquad (5)$$

With $E^{\circ} = +\,0.342$ V vs. SHE, the electrode potential for the copper metal in contact with the 0.01 M Cu^{2+} solution is calculated to be $E = +\,0.283$. The electrode potential for the copper metal in contact with the 0.5 M Cu^{2+} solution is calculated to be $E = +\,0.333$. Thus, the copper electrode in contact with the more dilute Cu^{2+} solution has more negative (i.e., less positive) electrode potential when the two electrodes are coupled. The cell voltage is given in the usual manner as

$$E_{\text{cell}} = E_{\text{cathode}} - E_{\text{anode}}$$

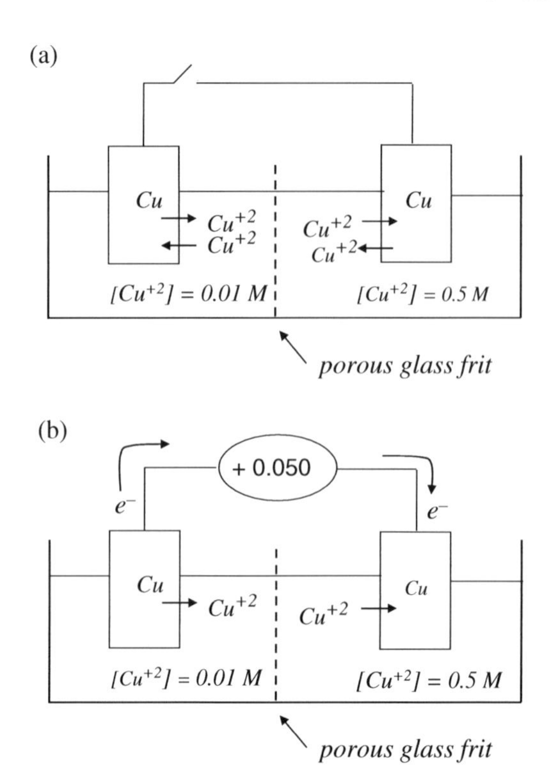

Fig. 5.13 A copper ion concentration cell

$$E_{\text{cell}} = 0.333\text{V} - 0.283 = 0.050\ \text{V}$$

Thus, there is a difference in electrode potential between the two half-cells, and a galvanic cell results due to the differences in metal ion concentrations. The copper specimen in the more dilute solution of Cu^{2+} will continue to corrode until the metal ion concentrations in the two half-cells become equal.

In general, areas on a surface where the electrolyte contains a lower quantity of the metal's ions will be anodic compared to locations where the metal ion concentration is higher.

For example, a copper pipe in contact with copper ion solutions of different concentrations will corrode at the part in contact with the more dilute solution of Cu^{2+}. Differences in concentration can be set up when the copper pipes carry moving water. The faster flowing portions of solutions will contain less Cu^{2+} as the dissolved ion is swept away, while the more stagnant areas will accumulate Cu^{2+}. The result is that the parts of the copper surface in contact with the faster moving fluid will experience corrosion due to the operation of a differential concentration cell.

Oxygen Concentration Cells

U. R. Evans has reported on a simple but elegant experiment performed to show the existence of oxygen concentration cells (also called differential oxygen cells or differential aeration cells) [9]. As shown in Fig. 5.14, two separate electrodes of the same metal were placed in a two-compartment

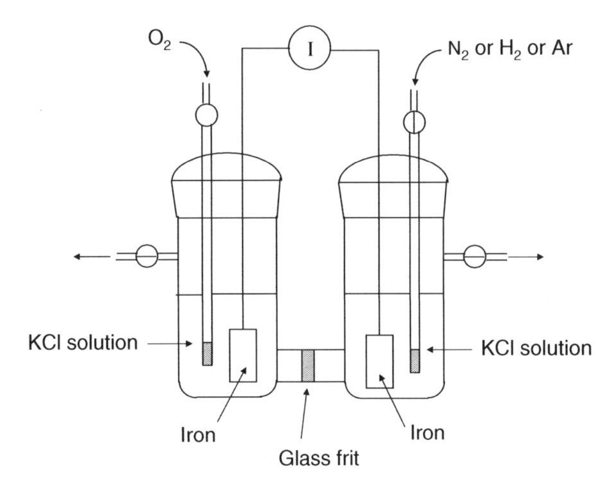

Fig. 5.14 An oxygen concentration cell [9]

cell, with each compartment containing the same solution but separated by a porous diaphragm. Oxygen (or air) was bubbled through one compartment, and when the two electrodes were coupled, a small current was observed to flow between them. It was established that the metal exposed to the lower concentration of oxygen was the anode. The direction of the cell current could be reversed by bubbling oxygen through the other compartment instead of the one originally used.

Modifications of this setup have also been used in which oxygen is bubbled through one compartment of the cell, and the other compartment is deaerated with hydrogen or nitrogen. The effect is the same as above.

The Nernst equation can be used again to explain a difference in electrode potentials. Suppose that two portions of the same metal are immersed into solutions which are identical except that each solution contains a different concentration of dissolved oxygen. The two solutions are again in contact through porous partition, as in Fig. 5.14. For the oxygen reduction reaction

$$O_2+2H_2O + 4e^- \rightarrow 4OH^-$$

the Nernst equation at 25°C gives

$$E_1 = E^o - \frac{0.0591}{4}\log\frac{[OH^-]^4}{P(O_2)_1} \tag{6}$$

and

$$E_2 = E^o - \frac{0.0591}{4}\log\frac{[OH^-]^4}{P(O_2)_2} \tag{7}$$

where E_1 and E_2 are the half-cell potentials for electrodes 1 and 2 (chemically the same metal) having oxygen pressures $P(O_2)_1$ and $P(O_2)_2$, respectively. The pH is assumed to be the same for both half-cells so that $[OH^-]$ is the same in both cases. Then Eqs. (6) and (7) combine to give

$$E_1 - E_2 = 0.0148 \log \frac{P(O_2)_1}{P(O_2)_2} \tag{8}$$

If $P(O_2)_1 < P(O_2)_2$, then it follows that $E_1 < E_2$. Or, the half-cell potential of specimen 1 is less than the half-cell potential of specimen 2. Thus, specimen 1 (in contact with the solution of lower oxygen content) is anodic (i.e. more negative) to specimen 2 (in contact with the solution of higher oxygen content).

In general, areas on a surface in contact with an electrolyte having a low oxygen concentration will be anodic relative to those areas where more oxygen is present.

The Evans Water Drop Experiment

U. R. Evans also conducted a well-known water drop experiment to illustrate the operation of an oxygen concentration cell [10]. Evans placed a drop of dilute NaCl on a horizontal steel surface and then followed the development of anodic and cathodic areas under the drop. Evans used small amounts of phenolphthalein and ferricyanide indicators, as in Problem 2.14. Cathodic areas turned red due to the formation of OH^- ions, and anodic areas turned blue due to the formation of Fe^{2+} ions.

Areas near the periphery of the drop had ready access to oxygen from the air and functioned as cathodes. Areas under the center of the drop had less access to oxygen. This is because dissolved oxygen needed to diffuse through the volume of the drop to replenish the oxygen initially consumed at the steel surface beneath the center of the drop. Thus, areas under the center of the drop became low in oxygen and functioned as anodes. These events are depicted schematically in Fig. 5.15. With time, the central anodic areas migrated outward to meet the cathodic areas, and rings of rust were deposited at their juncture.

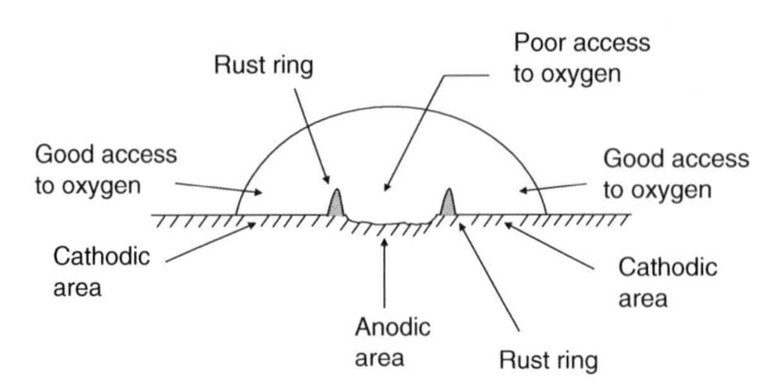

Fig. 5.15 The Evans water drop experiment. Redrawn from [10] by permission of Edward Arnold (Publishers) Ltd

Waterline Corrosion

A related phenomenon is the corrosion of partially immersed metals at a location just below the waterline, i.e., the level of submersion [10]. The electrode areas near the water surface again have easy access to oxygen, whereas electrode areas below the waterline have less access to oxygen, as shown in Fig. 5.16. Thus an oxygen concentration cell is established, with the upper part of the metal acting as a cathode and the nearby submerged areas acting as an anode.

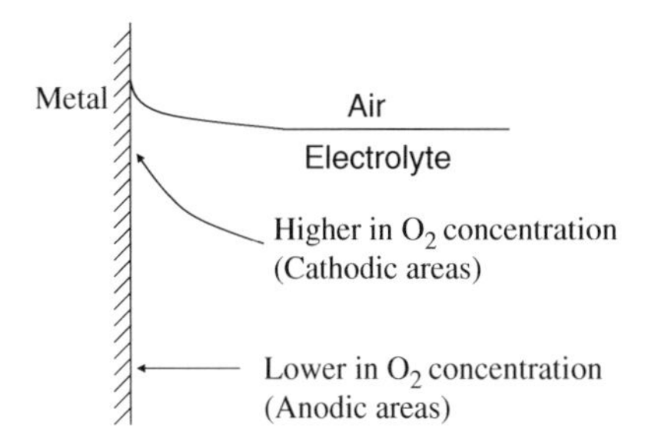

Fig. 5.16 Schematic diagram of waterline corrosion [10]

Waterline corrosion can be a problem in partially filled tanks and in partially submerged structures.

Crevice Corrosion: A Preview

The most common occurrence of differential oxygen cells in corrosion processes is in the phenomenon of crevice corrosion. A metal confined within a narrow clearance experiences a reduced concentration of dissolved oxygen relative to the metal outside the crevice, which is exposed to bulk electrolyte. This differential oxygen cell makes the metal inside the crevice anodic to the metal outside the crevice. The stage is thus set for more action. Crevice corrosion is treated further in Chapter 10.

Problems

1. An electrochemical cell consists of copper immersed in a solution of 0.01 M Cu^{2+} ions and nickel immersed in a solution of 0.1 M Ni^{2+} ions at 25°C. (a) Which electrode will corrode when the cell is short-circuited? (b) Calculate the cell emf. (c) Write the overall cell reaction.
2. An electrochemical cell consists of nickel immersed in a solution of 0.10 M Ni^{2+} ions and aluminum immersed in a solution of 0.15 M Al^{3+} ions at 25°C. (a) Which electrode will corrode when the cell is short-circuited? (b) Calculate the cell emf. (c) Write the overall cell reaction.
3. An electrochemical cell consists of cadmium immersed in a solution of $CdSO_4$ and iron immersed in a solution of $FeSO_4$. What concentration ratio $[Cd^{2+}]/[Fe^{2+}]$ is required at 25°C to prevent corrosion in the short-circuited cell?
4. Lead piping sometimes undergoes concentration cell corrosion. (a) Which electrode is the anode at 25°C in a concentration cell consisting of one lead electrode immersed in 0.01 M $PbSO_4$ and a second lead electrode immersed in 0.5 M $PbSO_4$. (b) Calculate the cell emf at 25°C
5. Compare the relative location of the following pairs of metals in the emf series and in the galvanic series for seawater.

   (a) Zinc and copper
   (b) Copper and tin
   (c) Nickel and silver
   (d) Titanium and iron (use low alloy steel for iron in the galvanic series for seawater)

What does the relative position of these various pairs of metals tell you about the use of the emf series to predict possible galvanic corrosion in seawater?

6. (a) Can magnesium be used instead of zinc to cathodically protect steel in seawater? (b) Can aluminum protect steel? (c) Refer to the galvanic series for seawater and select one additional metal or alloy which can be used to cathodically protect steel.
7. The electrode potential of tin in grapefruit juice is –0.740 V vs. a calomel electrode. The electrode potential of steel in grapefruit juice is –0.650 V vs. the same reference electrode [11]. If a tin can (a coating of tin on steel) develop a break in the tin coating which extends down to the steel substrate, which member of the iron/tin couple will galvanically corrode in the grapefruit juice electrolyte?
8. Refer to the galvanic series for seawater and predict the possibility of a galvanic effect in the following systems when they are used in seawater. If there is a galvanic effect, which metal in the couple is attacked?

(a) Aluminum/graphite composite

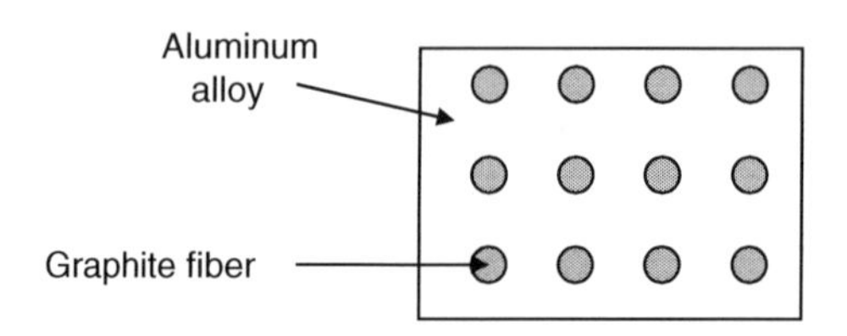

(b) Aluminum/alumina composite

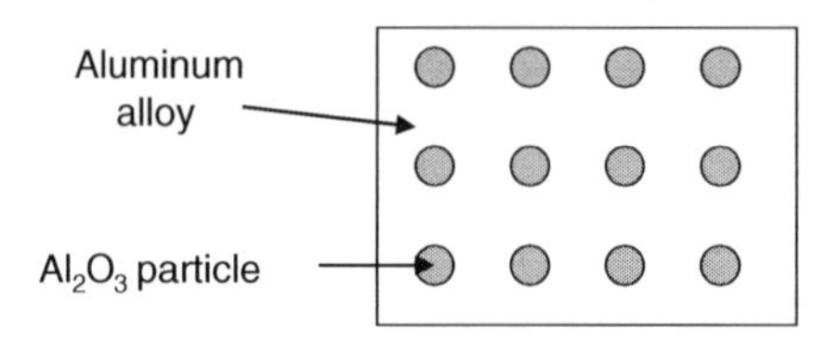

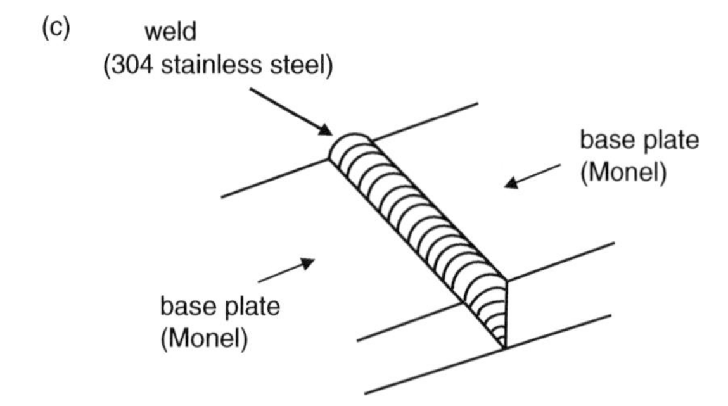

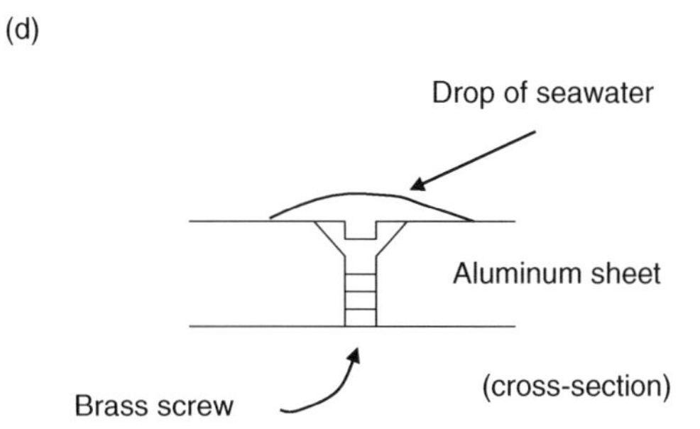

9. The commercial aluminum alloy 2024 contains approximately 4% copper, 1.5% magnesium, and 0.5% each of manganese and iron. These alloying elements produce second-phase intermetallic compounds in the aluminum matrix, as shown below.

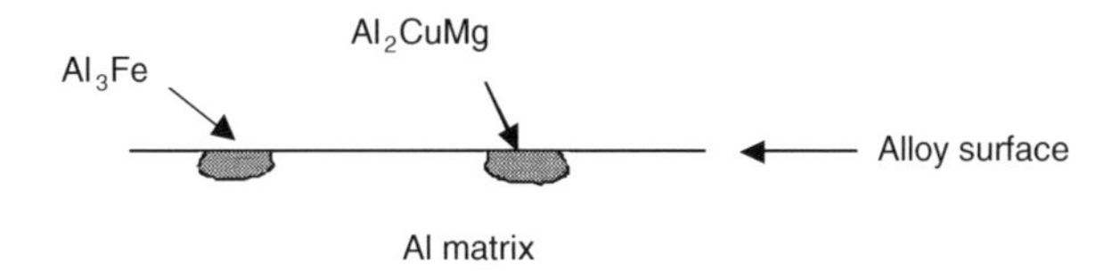

The following electrode potentials in a test solution of 53 g/l NaCl plus 3 g/l H_2O_2 were reported for bulk uncoupled samples of the aluminum matrix and for the second-phase particles Al_2CuMg and Al_3Fe [12]:

Al matrix –0.81 V vs. SCE
Al_2CuMg –0.91 V vs. SCE
Al_3Fe –0.47 V vs. SCE

In the commercial 2024 alloy, what is the galvanic effect between the second-phase particles of Al_2CuMg and the aluminum matrix and between the second-phase particles of Al_3Fe and the aluminum matrix? In each case, state whether the aluminum matrix or the second-phase particle preferentially corrodes in the test solution.

10. Cadmium coatings are often used to protect steel substrates in marine environments. (a) Classify this type of metallic coating. (b) Which metal will corrode if a break develops in the cadmium coating and extends down to the steel substrate? (For the electrode potential of steel, use the electrode potential of low alloy steel in the galvanic series.)

11. (a) For a metallic coating of type 430 stainless steel on mild steel, what happens if an imperfection or a break develops in the coating, as shown below?

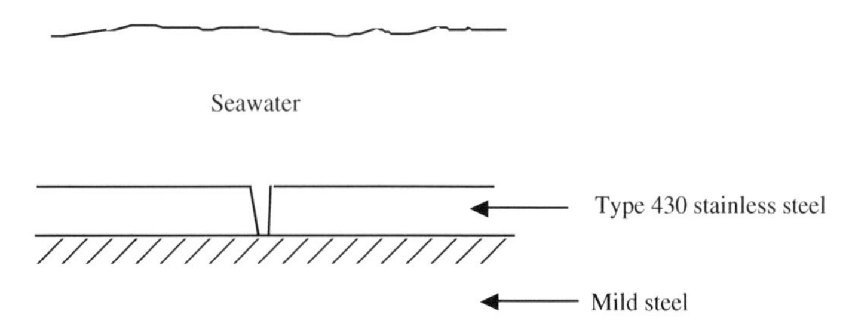

12. (a) For the duplex coating shown below, what happens if an imperfection or a break develops in the outermost nickel coating and extends down to the chromium inner coating? (b) If the

imperfection continues down to the steel substrate, what will happen? The electrode potentials in seawater are –0.20 V for nickel, –0.40 V for chromium, and –0.61 V for carbon steel, all measured vs. SCE [13, 14].

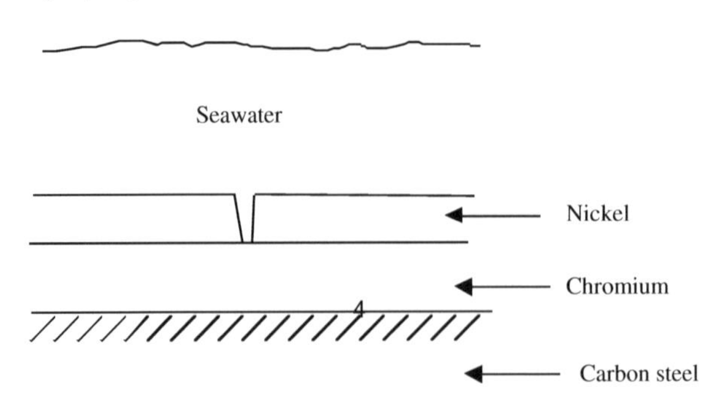

13. In Case A below, two steel plates are joined with a Monel bolt. In Case B, the reverse is done. Both systems are to be immersed in seawater, for which

    E (steel) = –0.61 V
    E (Monel) = –0.08 V.

    Thus, there is a galvanic effect in which the steel member of the couple will be attacked. In Case A, the steel plates will be galvanically attacked. In Case B, the steel bolt which keeps them together will be galvanically attacked. One case is much worse than the other. Which case is it? Why?

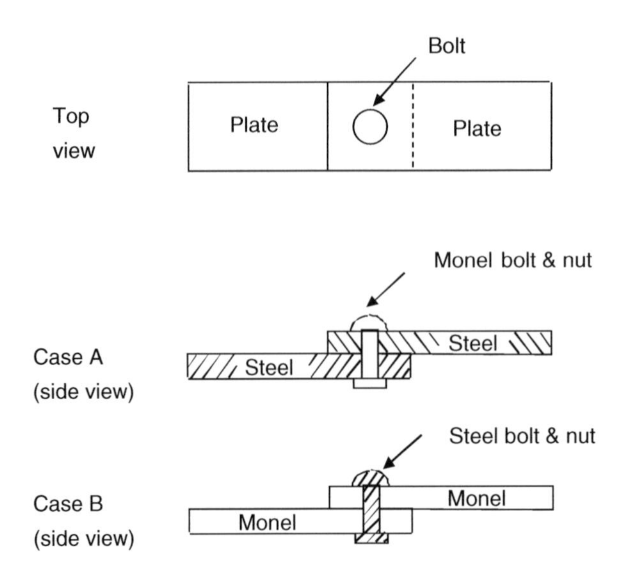

    Note: Assume that all crevices are sealed so that crevice corrosion is not a problem.

14. The electrode potential of aluminum in 0.1 M NaCl which has been de-aerated with argon gas is –1.4 V vs. SCE. The electrode potential of aluminum in the same solution but with the solution open to the air is –0.6 V vs. SCE. Would waterline corrosion of aluminum be expected to be possible for aluminum samples partially immersed in a saltwater estuary? Explain your answer.

References

1. F. L. LaQue, "Marine Corrosion", Chapter 6, John Wiley, New York (1975).
2. G. Butler and H. C. K. Ison, "Corrosion and Its Prevention in Waters", p. 104, Leonard Hill, London (1966).
3. H. Kaesche, "Metallic Corrosion", p. 293, National Association of Corrosion Engineers, Houston, TX (1985).
4. G. Butler, P. E. Francis, and A. S. McKie, *Corros. Sci.*, *9*, 715 (1969).
5. H. E. Townsend, L. Allegra, R. J. Dutton, and S. A. Kriner, *Mater. Perform.*, *25* (8), 36 (1986).
6. H. E. Townsend and A. R. Borzillo, *Mater. Perform.*, *35* (4), 30 (1996).
7. J. D. Ayers, R. J. Schaeffer, F. D. Bogar, and E. McCafferty, *Corrosion*, *37*, 55 (1981).
8. M. G. Fontana and N. D. Greene, "Corrosion Engineering", p. 37, McGraw-Hill, New York (1978).
9. U. R. Evans, "The Corrosion and Oxidation of Metals", p. 128, Edward Arnold, London (1971).
10. U. R. Evans, "An Introduction to Metallic Corrosion", pp. 35–37, 197–199, Edward Arnold, London (1982).
11. G. G. Kamm and A. R. Wiley, "First International Congress on Metallic Corrosion", p. 493, Butterworths, London (1962).
12. R. G. Buchheit, *J. Electrochem. Soc.*, *142*, 3994 (1995).
13. F. W. Fink and W. K. Boyd, "The Corrosion of Metals in Marine Environments", p. 7, Defense Metals Information Center, Columbus, OH (1970).
14. V. L'Hostis, C. Dagbert, and D. Feron, *Electrochim. Acta*, *48*, 1451 (2003).

Chapter 6
Thermodynamics of Corrosion: Pourbaix Diagrams

Introduction

Marcel Pourbaix has developed a unique and concise method of summarizing the corrosion thermodynamic information for a given metal in a useful potential–pH diagram. These diagrams indicate certain regions of potential and pH where the metal undergoes corrosion and other regions of potential and pH where the metal is protected from corrosion. Such diagrams are usually called "Pourbaix diagrams" but are sometimes called "equilibrium diagrams" because these diagrams apply to conditions where the metal is in equilibrium with its environment. Pourbaix diagrams are available for over 70 different metals [1].

An example of a Pourbaix diagram is given in Fig. 6.1, which shows the Pourbaix diagram for aluminum. The abscissa in the diagram is the pH of the aqueous solution, which is a measure of the chemical environment. The ordinate is the electrode potential E, which is a measure of the electrochemical environment. In a Pourbaix diagram, there are three possible types of straight lines:

(1) Horizontal lines, which are for reactions involving only the electrode potential E (but not the pH),
(2) Vertical lines, which are for reactions involving only the pH (but not the electrode potential E),
(3) Slanted lines, which pertain to reactions involving both the electrode potential E and the pH.

Pourbaix diagrams also contain regions or fields between the various lines where specific chemical compounds or species are thermodynamically stable. The Pourbaix diagram for aluminum in Fig. 6.1 identifies the various regions where the species Al (solid), Al_2O_3 (solid), Al^{3+} ions, and AlO_2^- ions are each stable. When the stable species is a dissolved ion, the region on the Pourbaix diagram is labeled as a region of "corrosion." When the stable species is either a solid oxide or a solid hydroxide, the region on the Pourbaix diagram is labeled as a region of "passivity," in which the metal is protected by a surface film of an oxide or a hydroxide. When the stable species is the unreacted metal species itself, the region is labeled as a region of "immunity". Kruger has described the Pourbaix diagram as being a "map of the possible" [2].

Example 6.1: After immersion in natural seawater (pH 7.6) for 70 days, the electrode potential of pure aluminum was observed to be –1.47 V vs. SCE [3]. What behavior is expected in terms of corrosion, passivity, or immunity?

Solution: First convert the electrode potential to the standard hydrogen scale. From Table 3.2

E. McCafferty, *Introduction to Corrosion Science*, DOI 10.1007/978-1-4419-0455-3_6,

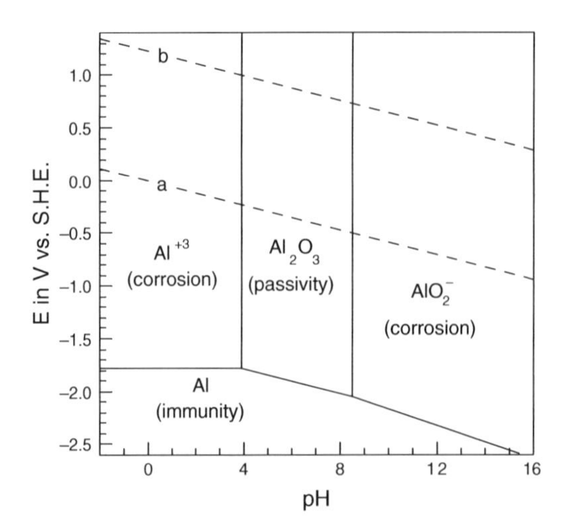

Fig. 6.1 Pourbaix diagram for aluminum at 25°C. Drawn from data in [1]

$$E \text{ vs. SCE} = E \text{ vs. SHE} - 0.242$$
$$-1.47\text{ V} = E \text{ vs. SHE} - 0.242\text{ V}$$
$$\therefore E \text{ vs. SHE} = -1.47\text{ V} + 0.242\text{ V} = -1.23\text{ V}$$

By referring to the Pourbaix diagram for pure aluminum in Fig. 6.1, it is seen that this electrode potential at a pH of 7.6 corresponds to a region of *passivity.*

Example 6.2: An Al–1% Mn alloy immersed in 0.1 M sodium citrate solution (pH 3.5) for 24 h displayed an electrode potential of – 1.25 V vs. SCE [4]. What behavior is expected in terms of corrosion, passivity, or immunity? What assumptions are you making in your analysis?

Solution: First convert the electrode potential to the standard hydrogen scale

$$E \text{ vs. SHE} = -1.25\text{ V} + 0.242\text{ V} = -1.01\text{ V}$$

From the Pourbaix diagram for pure aluminum in Fig. 6.1, it is seen that this electrode potential at a pH of 3.5 corresponds to a region of *corrosion*. The assumptions made are that (1) the Pourbaix diagram for pure aluminum also holds for the dilute Al alloy (1% Mn) and (2) the system is in equilibrium after the relatively short immersion period of 24 h.

Pourbaix Diagram for Aluminum

Construction of the Pourbaix Diagram for Aluminum

Pourbaix diagrams are constructed from the first principles of thermodynamics. The Pourbaix diagram for aluminum serves as a simple example as to how these diagrams are developed. It is first necessary to assemble the appropriate chemical information about the specific metal being considered. Aluminum undergoes dissolution in acid solutions as Al^{3+} ions and in basic solutions as aluminate ions (AlO_2^-). In neutral or nearly neutral solutions, aluminum is covered with a protective oxide film.

Thus, in the case of the Al/H_2O system, the various chemical entities and their chemical potentials are as follows [1]:

Dissolved substances	μ^o(cal/mol)
Al^{3+}	– 115,000
AlO_2^-	– 200,710
H^+	0
Solid substances	
Al	0
Al_2O_3	– 384,530
Liquid substances	
H_2O	– 56,690

The chemical reactions involving these species are [1]

$$Al \longrightarrow Al^{3+} + 3e^- \tag{1}$$

$$2Al^{3+} + 3H_2O \longrightarrow Al_2O_3 + 6H^+ \tag{2}$$

$$2Al + 3H_2O \longrightarrow Al_2O_3 + 6H^+ + 6e^- \tag{3}$$

$$Al_2O_3 + H_2O \longrightarrow 2AlO_2^- + 2H^+ \tag{4}$$

$$Al + 2H_2O \longrightarrow AlO_2^- + 4H^+ + 3e^- \tag{5}$$

The dissolution of aluminum as Al^{3+}ions by Eq. (1) is a reaction involving only the electrode potential E (but not the pH). We first rewrite Eq. (1) as a reduction reaction. Then, for

$$Al^{3+} + 3e^- \longrightarrow Al \tag{6}$$

the Nernst equation gives

$$E = E^0 - \frac{2.303RT}{nF} \log \frac{1}{[Al^{3+}]} \tag{7}$$

From Table 6.1, $E^o = -1.663$ V vs. SHE, and with $2.303RT/F = 0.0591$ V and $n = 3$, Eq. (7) becomes

$$E = -1.663 + 0.0197 \log[Al^{3+}] \tag{8}$$

The numerical value of the electrode potential E for the dissolution of Al as Al^{3+} ions depends on the concentration of the dissolved ion. Figure 6.2 shows the plot of Eq. (8) for two different values of $[Al^{3+}]$. As seen in Fig. 6.2, the electrode potential for the Al/Al^{3+} reaction becomes more positive with increasing $[Al^{3+}]$ concentration. Thus, for a given concentration of Al^{3+}, say 10^{-6} M, the oxidized form (Al^{3+}) is stable for potentials above the appropriate straight line at concentrations equal to or greater than 10^{-6} M. Below the given straight line, the oxidized species does not exist at

Table 6.1 Standard (reduction) potentials for various electrochemical reactions involving aluminum and/or water [1]

Equation number in text	Reaction (oxidation)	E^o (for reduction) (V vs. SHE)
(1)	$Al \longrightarrow Al^{3+} + 3e^-$	−1.662
(3)	$2Al + 3H_2O \longrightarrow Al_2O_3 + 6H^+ + 6e^-$	−1.550
(5)	$Al + 2H_2O \longrightarrow AlO_2^- + 4H^+ + 3e^-$	−1.262
(22)	$2H_2O + 2e^- \longrightarrow H_2 + 2OH^-$	−0.828

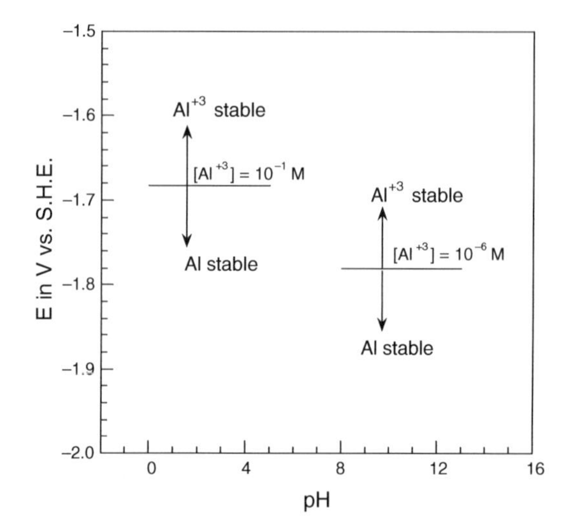

Fig. 6.2 Partial Pourbaix diagram for aluminum at 25°C

the concentration being considered. That is, below the straight line, the reduced species (Al atoms) are stable.

A clearer way to see this is to consider an electrode potential which is considerably below the straight line for $[Al^{3+}] = 10^{-6}$ M, say $E = -2.5$ V vs. SHE. Substituting this value of E into Eq. (8) gives the concentration of $[Al^{3+}]$ to be 2.9 x 10^{-43} M. This is a negligibly small value, so the oxidized species (Al^{3+}) is not stable, and thus the reduced species (solid Al) is stable. In general, for any electrochemical reaction represented on a Pourbaix diagram, the oxidized species of the couple is stable above the straight line of the Nernst equation for the reaction, and the reduced species is stable below the straight line.

By convention, Pourbaix diagrams are usually constructed under the condition that corrosion is considered to have occurred when the minimum concentration of the dissolved ion is 1.0×10^{-6} M. That convention is followed in this text.

Equation (2) is a chemical rather than an electrochemical reaction in that there is no electron transfer involved. The reaction depends on the pH but not on the electrode potential. Thus, from Chapter 4

$$\Delta G^o = \mu^o(Al_2O_3(s)) + 6\mu^o(H^+(aq)) - [2\mu^o(Al^3(aq) + 3\mu^o(H_2O(l))] \qquad (9)$$

Using the various values of μ^ogiven earlier gives $\Delta G^o = +15{,}540$ cal/mol Al_2O_3 for Eq. (2). Also from Chapter 4

$$\Delta G^{o} = -2.303RT \log K \tag{10}$$

where the equilibrium constant K for Eq. (2) is

$$K = \frac{[H^{+}]^{6}}{[Al^{3+}]} \tag{11}$$

Equation (10) yields

$$+15{,}540 \frac{\text{cal}}{\text{mol}} = -2.303 \left(1.98 \frac{\text{cal}}{\text{mol. K}}\right) (298\ \text{K}) \log K \tag{12}$$

or

$$\log K = -11.436 \tag{13}$$

Combining Eqs. (11) and (13) gives

$$3\,\text{pH} + \log [Al^{3+}] = 5.718 \tag{14}$$

When $[Al^{3+}] = 1.0 \times 10^{-6}$ M, then Eq. (14) gives the result pH = 3.91. This is the pH which results when dissolved Al^{3+} ions of concentration 1.0×10^{-6} M react with water according to Eq. (2). See Fig. 6.3, which summarizes the thermodynamic results so far. It can be shown easily that Al^{3+} ions are stable to the left of the line pH 3.91 and $Al_2O_3(s)$ is stable to the right of this line. For example, when the pH is 7.0, the concentration of Al^{3+} calculated from Eq. (14) is $[Al^{3+}] = 5.2 \times 10^{-16}$M. That is, dissolved Al^{3+} ions are not stable for pH values greater than 3.91, but instead solid Al_2O_3 is stable in that region.

Continuing with the construction of the Pourbaix diagram for aluminum, we next turn to Eq. (3), which is a reaction depending on both the electrode potential and the pH. After Eq. (3) is first recast as a reduction reaction, writing the corresponding Nernst equation gives

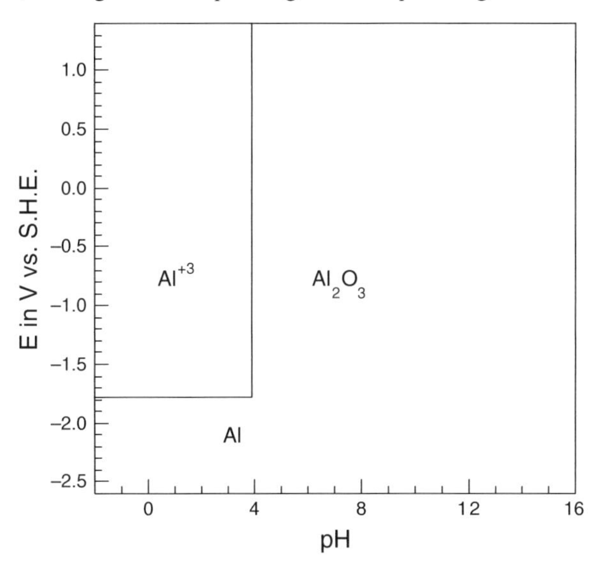

Fig. 6.3 Partial Pourbaix diagram for aluminum at 25°C

$$E = E^0 - \frac{2.303\,RT}{nF}\log\frac{1}{[H^+]^6} \tag{15}$$

The value of E^o for Eq. (3) is $E^o = -1.550$ V vs. SHE [1]. See Table 6.1. Alternately, E^o can be calculated from the free energy change for Eq. (3) using the chemical potentials. This approach is left as an exercise in Problem 6.6. With $E^o = -1.550$ V and $n = 6$, Eq. (15) gives

$$E = -1.550 - 0.0591\text{pH} \tag{16}$$

See Fig. 6.4, which adds this line to the thermodynamic data accumulated so far and also summarizes the regions of stability for the various species as has been computed so far.

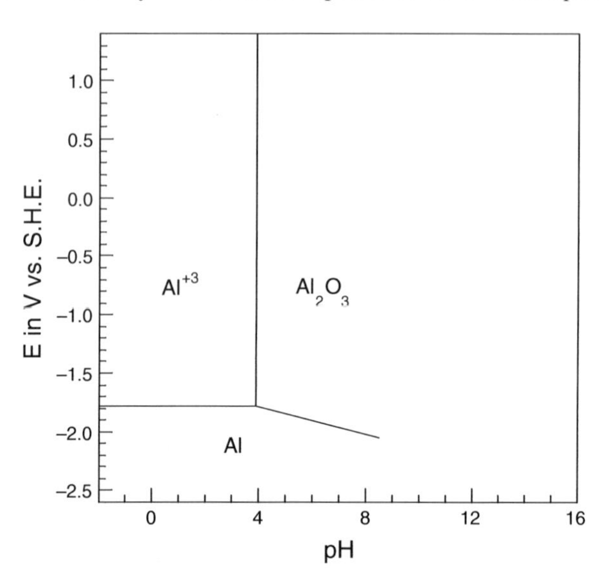

Fig. 6.4 Partial Pourbaix diagram for aluminum at 25°C

Equation (4) in the set of possible reactions for the Al/H_2O system can be treated in the same fashion as Eq. (2) because both are chemical (i.e., non-electrochemical reactions). The result for Eq. (4) is

$$-\text{pH} + \log[AlO_2^-] = -14.644 \tag{17}$$

This result is left as an exercise in Problem 6.7. When the concentration of the dissolved ion, in this case AlO_2^-, is 1.0×10^{-6} M, Eq. (17) gives the result

$$\text{pH} = 8.64 \tag{18}$$

The Nernst equation for the electrochemical reaction in Eq. (5) gives the following result

$$E = -1.262 + 0.0197\,\log[AlO_2^-] - 0.0788\,\text{pH} \tag{19}$$

See Problem 6.8. When $[AlO_2^-] = 1.0 \times 10^{-6}$ M, Eq. (19) gives

$$E = -1.380 - 0.0788\text{pH} \tag{20}$$

Addition of the straight lines for Eqs. (18) and (20) to Fig. 6.4 gives the completed Pourbaix diagram for aluminum shown in Fig. 6.1. The lines labeled "a" and "b" are discussed below.

When the diagram indentifies only the regions of corrosion, passivity, and immunity (rather than citing the individual stable species), the diagram is said to be a "simplified Pourbaix diagram."

Comparison of Thermodynamic and Kinetic Data for Aluminum

According to the Pourbaix diagram in Fig. 6.1, aluminum corrodes only at low pH and at high pH values and does not corrode at pH values between 3.9 and 8.6. This thermodynamic information is in excellent agreement with corrosion rate data for aluminum given earlier in Fig. 2.11 [5]. By referring back to Fig. 2.11, it can be seen that aluminum in aerated 1 M NaCl adjusted to various pH values has its lowest corrosion rates between pH 4 and pH 8.

Figure 6.5 shows additional corrosion data for pure aluminum and an aluminum alloy in aqueous solutions of various pH values [6]. Again, the lowest corrosion rates for aluminum and the aluminum alloy were at intermediate pH values. Thus, while Pourbaix diagrams do not provide actual corrosion rates, the thermodynamic trends given by the Pourbaix diagram are expected to be consistent with the results determined separately for experimental corrosion rates.

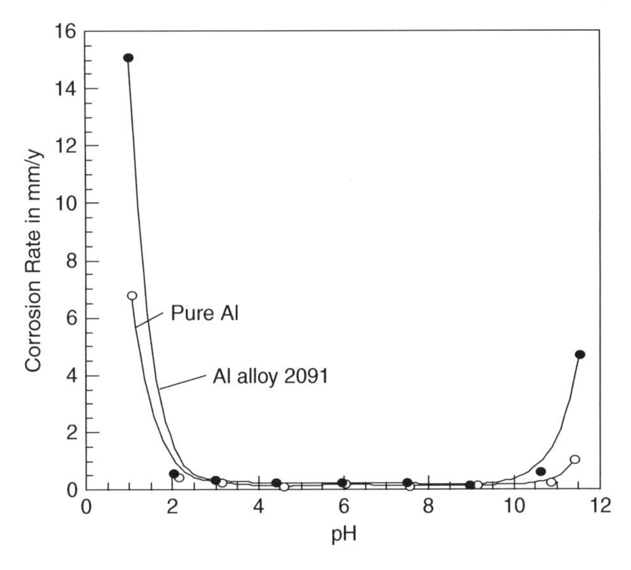

Fig. 6.5 Corrosion rates of pure aluminum and aluminum alloy 2091 (minor additions of Cu, Li, and Mg and others) in aqueous solutions of various pH values. Redrawn from [6] with the permission of Elsevier, Ltd

Pourbaix Diagram for Water

The line labeled "a" in Fig. 6.1 is for the cathodic evolution of hydrogen. In acid solutions, the cathodic reaction is

$$2H^+ + 2e^- \longrightarrow H_2 \tag{21}$$

and in basic solutions, the cathodic reaction is

$$2H_2O + 2e^- \longrightarrow H_2 + 2OH^- \tag{22}$$

For both Eqs. (21) and (22), the Nernst expression at 25°C is

$$E = 0.000 - 0.0591\text{pH} \tag{23}$$

which is plotted as the "a" line in Fig. 6.1. As shown in Fig. 6.6, the reduced species (H_2) in Eqs. (21) or (22) is stable below the "a" line. Above the "a line," the oxidized species H^+ is stable in acid solutions, and OH^- is stable in basic solutions.

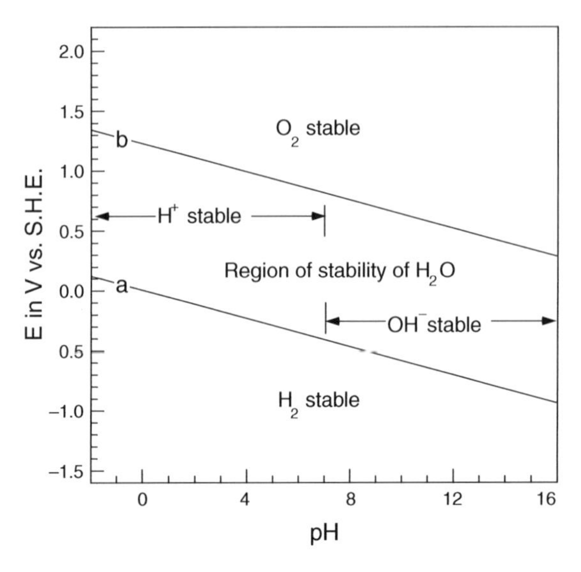

Fig. 6.6 Pourbaix diagram for water at 25°C [1]

That Eq. (23) also applies to the cathodic reaction in Eq. (22) is posed as Problem 6.9.

The line labeled "b" in Fig. 6.1 is for the anodic evolution of oxygen. At sufficiently high electrode potentials, water molecules dissociate to liberate oxygen by the following anodic reaction:

$$2H_2O \longrightarrow O_2 + 4H^+ + 4e^- \tag{24}$$

The Nernst expression for this reaction at 25°C and 1 atm of O_2 is

$$E = 1.228 - 0.0591\,\text{pH} \tag{25}$$

Equation (25) is plotted as the "b" line in Figs. 6.1 and 6.6. Below the "b" line the reduced species (H_2O) is stable, and above the "b" line the oxidized species (O_2) is stable. Between the "a" and "b" lines, water is thermodynamically stable. As stated earlier, below the "a" line, H_2 evolution is possible; and above the "b" line, O_2 evolution is possible.

Figure 6.6 is referred to as the Pourbaix diagram for water because this diagram shows region of stability for H_2O (l), H^+(aq), OH^-(aq), H_2(g), and O_2(g). The "a" and "b" lines are usually superimposed on the Pourbaix diagrams for metals. The "a" line is of particular interest because it shows conditions where hydrogen evolution is possible. As discussed in Chapter 2, hydrogen atoms are formed by the reduction of H^+ ions, with two hydrogen atoms combining to form one molecule of

H_2 gas. If, however, some hydrogen atoms migrate into the interior of the metal rather than combining to form H_2, then the presence of H atoms in stressed regions can promote stress-corrosion cracking by the process of hydrogen embrittlement. Hydrogen embrittlement is a serious form of failure in various metals and alloys, as will be discussed in Chapter 11.

Example 6.3: In experiments on the stress-corrosion cracking of 4340 steel (predominately iron with minor amounts of Cr, Mn, and Ni), the local pH of the electrolyte contained within the stress-corrosion crack attained values of 3.0–4.0, even though the pH of the bulk solution was approximately 7. The local electrode potentials corresponding to the pH values of 3.0 and 4.0 were –0.52 V and –0.58 V vs. SCE, respectively [7]. Is hydrogen embrittlement a possible mechanism of stress-corrosion cracking for this alloy?
Solution: First convert the electrode potentials to the standard hydrogen scale. For pH 3.0

$$E \text{ vs. SCE} = \text{E vs. SHE} - 0.242$$
$$-0.52\text{ V} = \text{E vs. SHE} - 0.242\text{ V}$$
$$E \text{ vs. SHE} = -0.52\text{ V} + 0.242\text{ V} = -0.28\text{ V}$$

Similarly, at pH 4.0, the electrode potential is

$$E \text{ vs. SHE} = -0.58\text{ V} + 0.242\text{ V} = -0.34\text{ V}$$

Each of the points (3.0, – 0.28) and (4.0, –0.34) when superimposed onto the Pourbaix diagram for water (Fig. 6.6) or for iron (Fig. 6.8) lie below the "a" line for hydrogen evolution. Thus, hydrogen evolution is thermodynamically possible. Accordingly, H atoms can be formed, and hydrogen embrittlement is a possible mechanism of stress-corrosion cracking for this alloy.

Pourbaix Diagrams for Other Metals

Pourbaix Diagram for Zinc

Figure 6.7 shows the Pourbaix diagram for zinc. (Throughout this chapter, Pourbaix diagrams are calculated using thermodynamic data and electrode potentials given by Pourbaix [1].) The diagram for zinc is similar to that for aluminum because zinc, like aluminum, undergoes dissolution in acid solutions (as Zn^{2+} ions) and in basic solutions (as zincate ions, ZnO_2^{2-}). This thermodynamic information is in agreement with kinetic data in which zinc has a high corrosion rate at both low and at high pH values, and a lower corrosion rate at intermediate pH values [8].

Pourbaix Diagram for Iron

The Pourbaix diagram for iron is shown in Fig. 6.8. This diagram is of considerable interest because of the widespread use of iron and its alloys as structural materials. Iron can undergo corrosion in acid or neutral solutions in two different oxidation states, i.e., Fe^{2+} or Fe^{3+}. Passivity is provided by oxide films of Fe_3O_4 or Fe_2O_3. Corrosion in alkaline solutions occurs as the complex anion $HFeO_2^-$, which is analogous to the dissolved ions AlO_2^- and ZnO_2^{2-} for aluminum and zinc, respectively, in alkaline solutions.

The Pourbaix diagram for iron shown in Fig. 6.8 is constructed from the set of equilibrium reactions and expressions given in Table 6.2. The simplified Pourbaix diagram for iron shown in Fig. 6.9

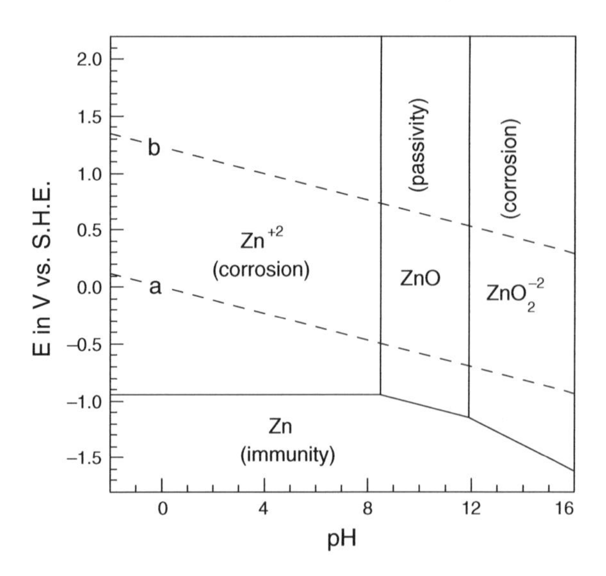

Fig. 6.7 Pourbaix diagram for zinc at 25°C. Drawn from data in [1]

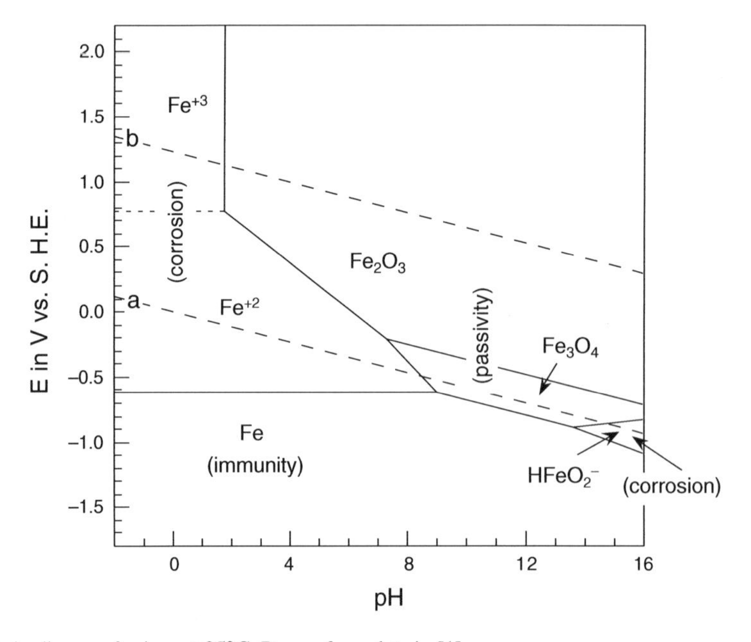

Fig. 6.8 Pourbaix diagram for iron at 25°C. Drawn from data in [1]

suggests three different means of corrosion protection so as to remove iron from the region of corrosion. For instance, a pH of 6.0 and an electrode potential of –0.4 V vs. SHE corresponds to a region of corrosion as Fe^{2+} ions. The three corrosion control measures are as follows:

(1) If the electrode potential is changed in the negative direction to a value below –0.7 V SHE, the iron electrode is forced into a region of immunity. (This process is called *cathodic protection*, which has been discussed earlier in Chapter 5).

Table 6.2 Equilibrium reactions and thermodynamic expressions for the Fe/H_2O system [1]

$Fe(s) \longrightarrow Fe^{2+}(aq) + 2e^-$
$E = -0.440 - 0.0295 \log [Fe^{2+}]$

$3Fe(s) + 4H_2O(l) \longrightarrow Fe_3O_4(s) + 8H^+(aq) + 8e^-$
$E = -0.085 - 0.0591pH$

$3Fe^{2+}(aq) + 4H_2O(l) \longrightarrow Fe_3O_4(s) + 8H^+(aq) + 2e^-$
$E = 0.980 - 0.2364pH - 0.0886\log[Fe^{2+}]$

$2Fe^{2+}(aq) + 3H_2O(l) \longrightarrow Fe_2O_3(s) + 6H^+(aq) + 2e^-$
$E = 0.728 - 0.1773pH - 0.0591\log[Fe^{2+}]$

$2Fe_3O_4(s) + 4H_2O(l) \longrightarrow 2Fe_2O_3(s) + 2H^+(aq) + 2e^-$
$E = 0.221 - 0.0591pH$

$Fe(s) + 2H_2O(l) \longrightarrow HFeO_2^-(aq) + 3H^+(aq) + 2e^-$
$E = 0.493 - 0.0886pH + 0.0295 \log [HFeO_2^-]$

$3HFeO_2^-(aq) + H^+(aq) \longrightarrow Fe_3O_4(s) + 2H_2O(l) + 2e^-$
$E = -1.819 + 0.0295pH - 0.0886 \log [HFeO_2^-]$

$Fe^{2+}(aq) \longrightarrow Fe^{3+} + (aq) + e^-$
$E = 0.771 + 0.0591 \log ([Fe^{3+}]/[Fe^{2+}])$

$2Fe^{3+}(aq) + 3H_2O(l) \longrightarrow Fe_2O_3(s) + 6H^+$
$\log[Fe^{3+}] = -0.72 - 3pH$

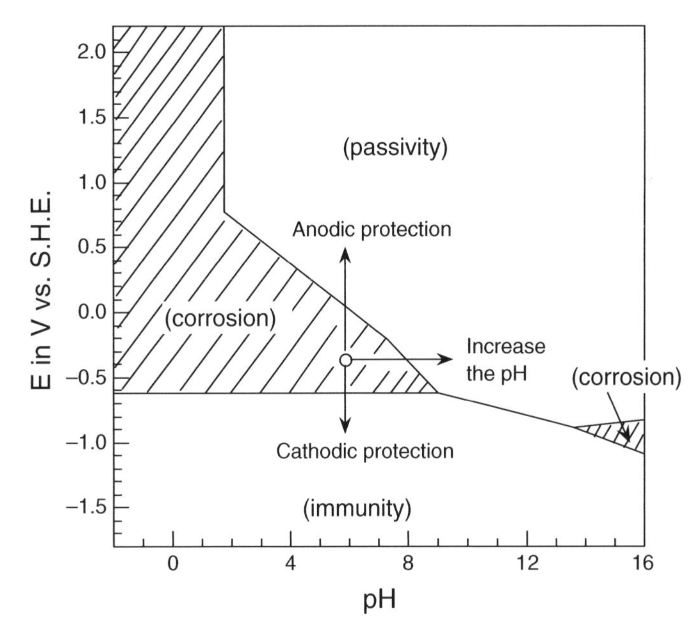

Fig. 6.9 Simplified Pourbaix diagram for iron at 25°C

(2) If instead, the electrode potential is changed in the positive direction to values above approximately 0.0 V vs. SHE, the iron electrode is forced into a region of passivity. This process is called *anodic protection* and will be taken up further in Chapter 9.

(3) The third method of protection is to change the pH of the aqueous solution. If the pH is increased to approximately 8 or higher, the iron electrode will then also reside in a region of passivity.

Figure 6.10 shows a drawing (after Marcel Pourbaix) which illustrates in an amusing but striking manner the differences between immunity, passivity, and corrosion for iron.

Fig. 6.10 A humorous look at the Pourbaix diagram for iron (after a colleague of M. Pourbaix). Figure kindly provided by A. Pourbaix of CEBELCOR (Centre Belge d'Etude de la Corrosion)

Pourbaix Diagrams for Additional Metals

The Pourbaix diagram for chromium is shown in Fig. 6.11. A distinguishing feature of this diagram is that the passivity of chromium can be destroyed by increasing the electrode potential (at a given pH). For instance, at pH 7 and an electrode potential of –0.4 vs. SHE, the chromium electrode resides in a region of passivity. But by increasing the electrode potential, the electrode can be shifted into a region of corrosion (as chromate ions). This phenomenon in which corrosion occurs at high electrode potentials (beyond the existence of passivity) is called *transpassive dissolution*.

Figure 6.12 shows the Pourbaix diagram for copper. An interesting feature of the diagram is that the oxidation reactions occur at electrode potentials above the "a" line for hydrogen evolution. Thus, the specific reactions

$$Cu + 2H^+ \longrightarrow Cu^{2+} + H_2 \tag{26}$$

or

$$2Cu + H_2O \longrightarrow Cu_2O + H_2 \tag{27}$$

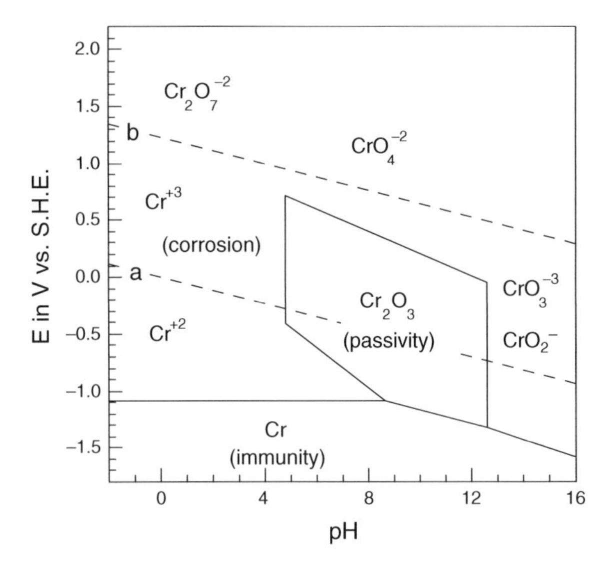

Fig. 6.11 Pourbaix diagram for chromium at 25°C. Drawn from data in [1]

are not thermodynamically favored. That is, copper cannot be oxidized by hydrogen ions or by water molecules to produce H_2 gas (because the product H_2 is not stable in regions where Cu^{2+} or Cu_2O exists). Instead of Eq. (27), the following reaction is favored:

$$2Cu + H_2O \longrightarrow Cu_2O + 2H^+ + 2e^- \tag{28}$$

Compare this feature of the copper diagram with the Pourbaix diagrams for aluminum, zinc, and, iron, for example.

Figures 6.13 and 6.14 show Pourbaix diagrams for two noble metals silver and palladium, respectively, each of which are used as electronic materials, as dental alloys, and in jewelry. In

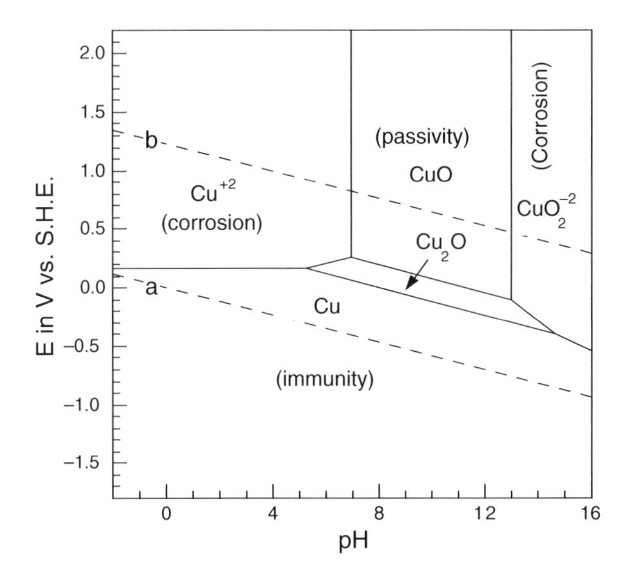

Fig. 6.12 Pourbaix diagram for copper at 25°C. Drawn from data in [1]

addition, palladium finds an important application in catalytic converters used to reduce emissions of hydrocarbons from gasoline-powered vehicles.

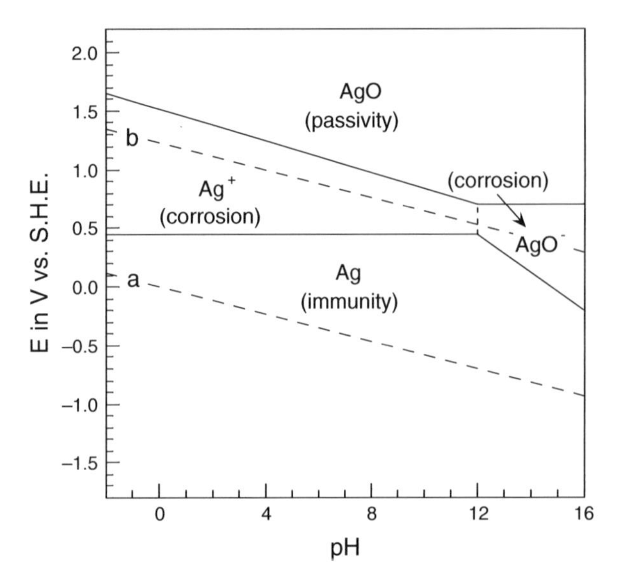

Fig. 6.13 Pourbaix diagram for silver at 25°C (assuming passivity is due to AgO). Drawn from data in [1]

Applications of Pourbaix Diagrams to Corrosion

Several applications of Pourbaix diagrams to corrosion have already been considered. Various applications can be catalogued as follows [1, 2]:

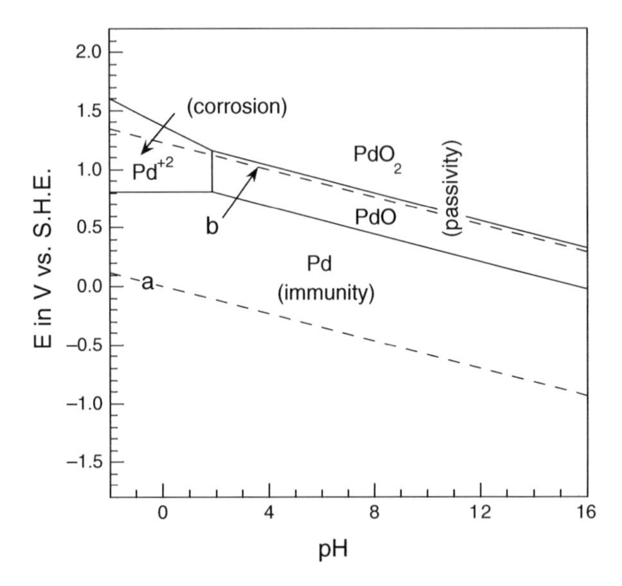

Fig. 6.14 Pourbaix diagram for palladium at 25°C. Drawn from data in [1]

(1) The resistance of metals to uniform corrosion in aqueous solutions.
(2) The basis for establishing which metals can be expected to have passivity over a wide range of conditions of pH and potential.
(3) Evaluation of the possible use of oxidizing inhibitors.
(4) Identification of the set of internal conditions within a localized corrosion cell.

The first application listed above, i.e., the corrosion resistance of metals in aqueous solutions has already been treated here in some detail. This method is based on establishing pH–potential domains of corrosion, passivity, and immunity.

The basis for establishing which metals can be passive over a wide range of conditions of pH and potential lies in examining the Pourbaix diagrams for various metals. For instance, the Pourbaix diagram for titanium, given in Fig. 6.15, shows that titanium exhibits passivity over a wide range of pH values due to its stable oxide film, although titanium is susceptible to corrosion in acid solutions. The Pourbaix diagram for tantalum, shown in Fig. 6.16, is even more appealing. As seen in Fig. 6.16, tantalum is either immune to corrosion or forms a passive oxide over the entire potential–pH range. Tantalum is known to be one of the most corrosion-resistant metals available and is used in a variety of applications. However, as pointed out by Macdonald et al. [9], tantalum is subject to corrosion in solutions where the complex ion tantalate (TaO_3^-) can be formed.

An example involving Pourbaix diagrams and corrosion inhibitors is provided by the use of chromates to inhibit the corrosion of iron in nearly neutral solutions. The chromate ion containing Cr^{6+} is reduced to Cr^{3+} by the following reaction:

$$2CrO_4^{2-}(aq) + 10H^+(aq) + 6e^- \longrightarrow Cr_2O_3(s) + 5H_2O(l) \tag{29}$$

(Compare this reaction with the transpassive dissolution of chromium.) By superimposing the region of stability for Cr_2O_3 onto the Pourbaix diagram for iron, as in Fig. 6.17, it can be seen that there is a region of overlap where the oxides Fe_2O_3 and Cr_2O_3 are both stable (the cross-hatched region in Fig. 6.17). In addition, Fig. 6.17 shows that it is possible to prevent the corrosion of Fe

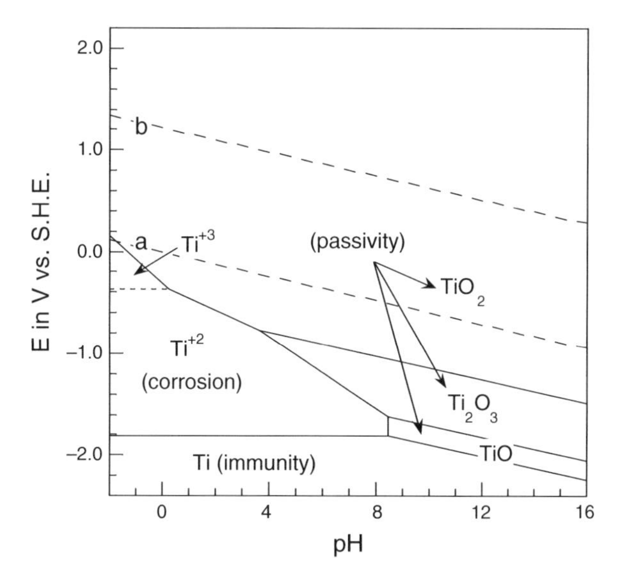

Fig. 6.15 Pourbaix diagram for titanium at 25°C. Drawn from data in [1]

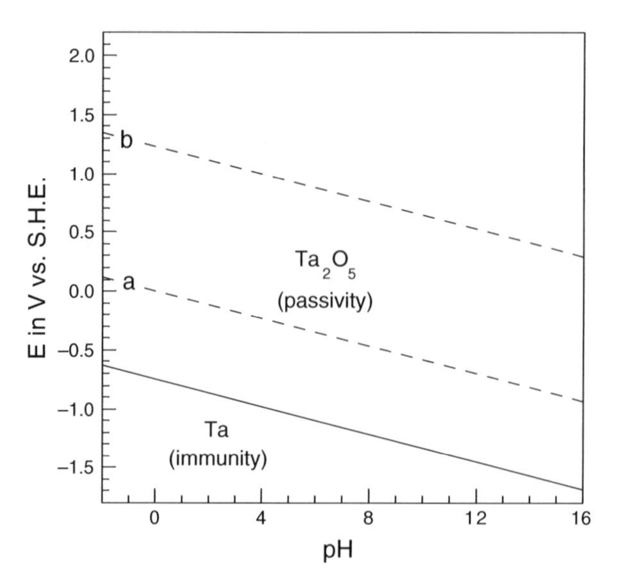

Fig. 6.16 Pourbaix diagram for tantalum at 25°C. Drawn from data in [1]

as Fe^{2+} at approximately neutral pH values through the formation of a Cr_2O_3 passive film. The surface analysis of oxide films formed on iron in chromate solutions has shown that the passive film is a bilayer consisting of an outer layer of Cr_2O_3 and an inner layer of Fe_2O_3 [10]. Thus, films of Cr_2O_3 are responsible for corrosion inhibition by chromates, as predicted on the basis of corrosion thermodynamics.

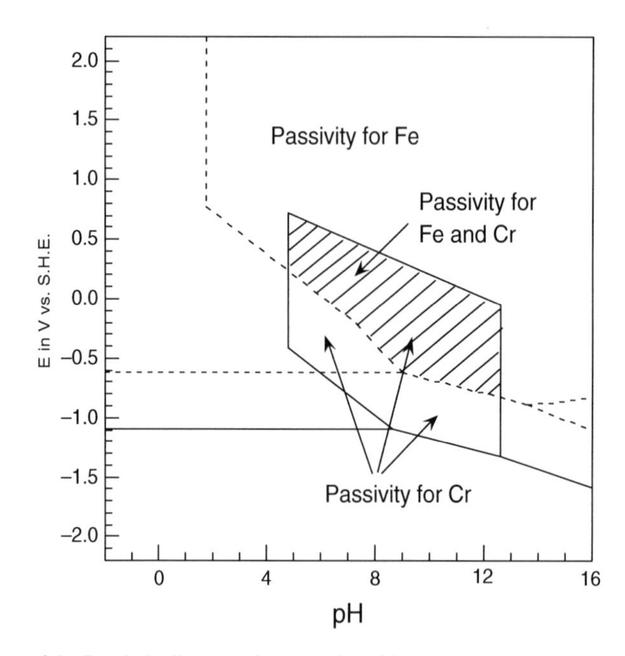

Fig. 6.17 Superposition of the Pourbaix diagrams for iron (*dotted lines*) and chromium (*solid lines*) at 25°C showing regions of passivity

Pourbaix diagrams can also be applied to localized corrosion processes. Pourbaix diagrams are useful in identifying potential–pH conditions which exist within a localized corrosion cell (crevice, corrosion pit, or stress-corrosion crack), provided that the values of E and pH pertain to the *local* electrode potential and to the *internal* pH, as in Example 6.3. Further applications to localized corrosion are taken up in Chapter 10 and Chapter 11.

Limitations of Pourbaix Diagrams

Pourbaix diagrams provide a first guide as to the corrosion behavior of a given metal. Despite their usefulness, as seen above, however, Pourbaix diagrams are subject to several important limitations. These limitations are as follows:

(1) Equilibrium is assumed. (But in practical cases, the actual conditions may be far from equilibrium.)
(2) Pourbaix diagrams give no information on actual corrosion *rates*.
(3) Pourbaix diagrams apply to single elemental metals only and not to alloys. (For a solid solution binary alloy, the Pourbaix diagrams of the two constituents may be superimposed as a first estimate, as will be shown later. For engineering alloys, experimental Pourbaix diagrams may be developed, as will be seen in Chapter 10).
(4) Passivation is ascribed to all oxides or hydroxides, regardless of their actual protective properties. (Corrosion may sometimes proceed by diffusion of ions through oxide films, a process which is ignored in the construction of the diagrams).
(5) Pourbaix diagrams do not consider localized corrosion by chloride ions. (Special experimental diagrams must be constructed, as will be seen in Chapter 10.)
(6) Conventional Pourbaix diagrams apply to a temperature of 25°C. (Pourbaix diagrams exist for elevated temperatures, and examples are given later.)

Pourbaix Diagrams for Alloys

As mentioned above, one of the limitations of Pourbaix diagrams is that the diagrams can be calculated rigorously only for single elemental metals. However, Pourbaix diagrams can extended to metal alloys by three approaches:

(1) For binary solid solution alloys, the two individual Pourbaix diagrams may be superimposed onto each other.
(2) In the case of a mixed binary oxide, the corresponding Pourbaix diagram can be calculated if all appropriate thermodynamic data are available.
(3) For more complex alloys, including engineering alloys, experimental Pourbaix diagrams can be determined from electrochemical polarization measurements, as will be seen in Chapter 10.

Copper and nickel form solid solution alloys over their entire composition range; so as a first guide as to corrosion trends for Cu-Ni alloys, the Pourbaix diagrams for copper and nickel can be superimposed, as shown in Fig. 6.18. This superposition implies that each of the two metals behaves independently of the other. This assumption is reasonable for solid solution alloys but not for heterogeneous alloys where galvanic effects may arise from the existence of second phases. Figure 6.18 shows that there are regions of potential vs. pH where the preferential dissolution of copper or nickel occurs, as well as regions where both metals undergo dissolution. Films of either

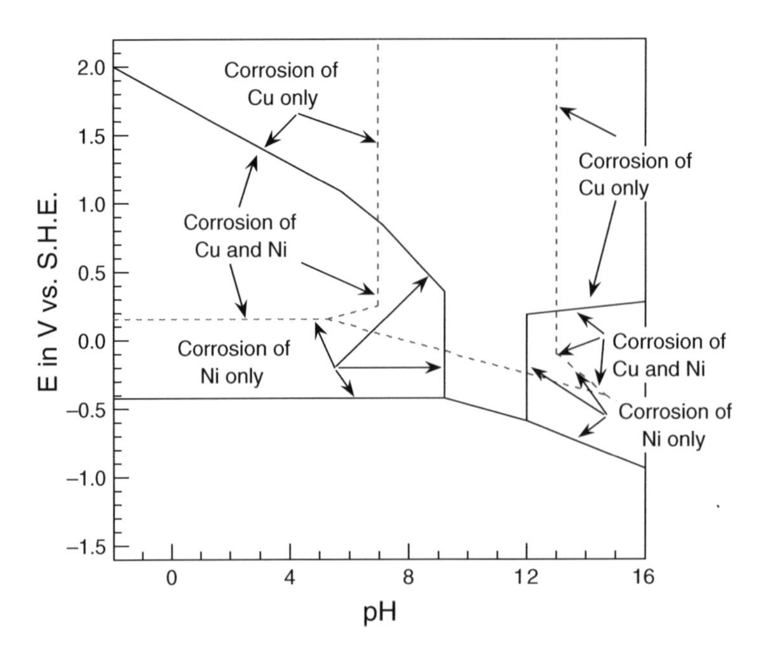

Fig. 6.18 Superposition of the Pourbaix diagrams for copper (*dotted lines*) and nickel (*solid lines*) at 25°C

Cu_2O, CuO, or NiO provide passivity at neutral to mildly alkaline pHs. In practice, alloys of Cu–10% Ni and Cu–30% Ni find wide usage in various freshwater and seawater applications.

When thermodynamic information exists for mixed metal oxides, the Pourbaix diagrams can be calculated using such information. Examples include the Fe-Cu-H_2O system for the mixed oxide $FeCuO_2$ [11] and the Fe-Cr-H_2O system for the mixed oxide $FeCr_2O_4$ [12]. Pourbaix diagrams have been computed also for other binary alloys [12] although these diagrams tend to be unduly complex, given their limitations.

Pourbaix Diagrams at Elevated Temperatures

The construction of Pourbaix diagrams at elevated temperatures is similar to their construction at 25°C, except that it is necessary to add a temperature correction for the free energy change of each individual reaction. The temperature affects:

(1) the electrode potential E of electrochemical reactions through the relationship $\Delta G = -nFE$, in which the free energy change ΔG is a function of temperature,
(2) the Nernst equation, through the factor $(2.303RT/F)$,
(3) the equilibrium constant K in non-electrochemical reactions through the relationship $\Delta G = -RT \ln K$, where ΔG and (of course) T are temperature dependent.

Consider the thermodynamic cycle shown in Fig. 6.19, which relates a chemical reaction at 25°C (29)8 K) to the same reaction at an elevated temperature T. As shown in Appendix B, the standard free energy change ΔG^{o}_{T} at temperature T is related to the standard free energy change ΔG^{o}_{298} at 25°C by the following expression:

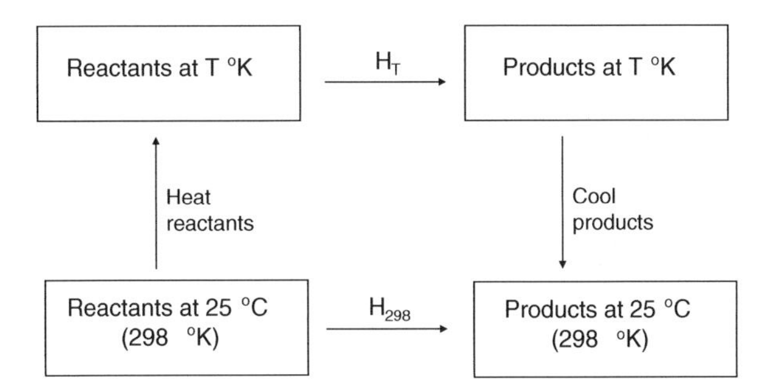

Fig. 6.19 Thermodynamic cycle for a chemical reaction at two different temperatures

$$\Delta G^o_T = \Delta G^0_{298} + \int_{298}^{T} \Delta C^o_p dT - T\int_{298}^{T} \Delta C^o_p d\ln T - (\Delta T)\Delta S^o_{298} \tag{30}$$

where $\Delta C_p{}^o$ is a difference in heat capacities given by $\Delta C_p{}^o = C_p{}^o$ (products) – $C_p{}^o$ (reactants), $\Delta T = T - 298$, and $\Delta S_{298}{}^o$ is the standard entropy change at 25°C.

The heat capacities required in Eq. (30) are either measured or estimated. Ionic entropies are usually estimated by the empirical correlation method of Criss and Cobble [13, 14] which relates entropies at elevated temperatures to entropies at 298 K.

One of the first Pourbaix diagrams for elevated temperatures was constructed by Townsend [15], who calculated the diagram for iron at temperatures up to 200°C. The 25°C and 200°C diagrams are compared in Fig. 6.20. The two diagrams are similar, with the main difference being that the regime of stability for the complex ion $HFeO_2{}^-$ increases with increasing temperature. In addition, the reaction

$$2HFeO_2^- \longrightarrow Fe_2O_3 + H_2O + 2e^- \tag{39}$$

was considered to occur at high temperatures but not at 25°C. In the Pourbaix diagram for iron at 200°C, the regions between the complex ion $HFeO_2{}^-$ and the oxides Fe_3O_4 and Fe_2O_3 are

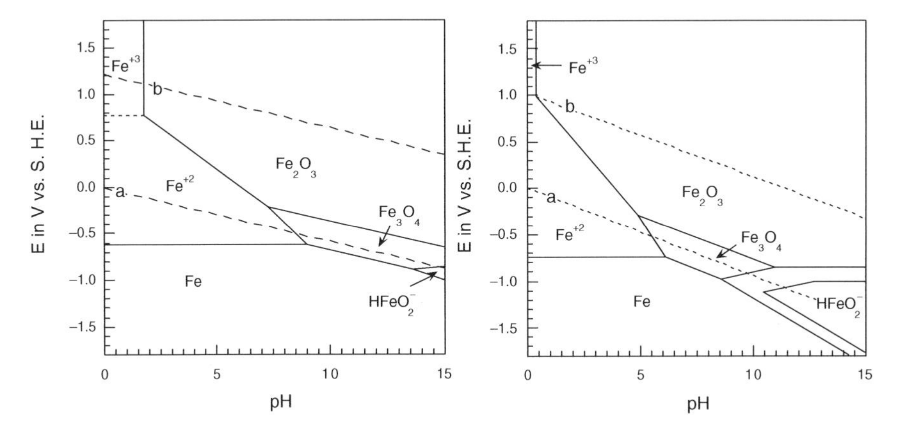

Fig. 6.20 Pourbaix diagrams for Fe at 25°C (*top*) and at 200°C (*bottom*). The bottom figure is redrawn from [15] with the permission of Elsevier, Ltd

shown as bands rather than lines due to uncertainties in the calculation of ΔG^o as a function of temperature [15].

Pourbaix diagrams have been constructed by Cowan and Staehle [16] for the Ni-H_2O system in the temperature range of 25–200°C. Analogous to the case for iron, the range of stability of a complex ion $HNiO_2^-$ increases with increasing temperature, but for the most part, the Ni–H_2O diagram at elevated temperatures retains the same general features as the diagram at 25°C.

Pourbaix diagrams at elevated temperatures are available for a variety of single metals and alloys [9, 15–19].

Problems

1. The following electrode potentials were observed after a 1-year immersion period in natural seawater of pH 8.0 [20]:

Mild steel	–0.65 V vs. SCE
Zinc	–1.02 V vs. SCE
Mild steel/zinc couple	– 0.95 V vs. SCE

   (a) Refer to the appropriate Pourbaix diagrams and indicate what behavior is expected for each of the two uncoupled electrodes. (b) Based on the Pourbaix diagrams, what behavior is expected for the mild steel and zinc components in the short-circuited couple? (c) Is the behavior in part (b) consistent with that predicted based on the galvanic couple approach in Chapter 5?
2. Suppose that an acid flux is inadvertently left on the surface of a silver contact and that in the presence of a thin film of condensed moisture, the acid residue has a pH of 5.0. Under what range of electrode potentials could the corrosion of silver occur in the thin-film electrolyte?
3. Palladium (like gold and platinum) is a noble metal. However, the Pourbaix diagram for palladium (Fig. 6.14) contains a region where palladium is subject to corrosion. Can palladium corrode by the following reaction:

$$Pd(s) + 2H^+(aq) \longrightarrow Pd^{2+}(aq) + H_2(g)$$

   Explain your answer.
4. The electrode potential of titanium immersed in a sulfuric acid solution of pH approximately 1 was measured to be –0.70 V vs. SCE [21]. (a) Confirm that titanium undergoes active corrosion under these conditions and (b) consult the Pourbaix diagram to suggest three different means to provide corrosion protection.
5. In the interpretation of Pourbaix diagrams, corrosion is considered to occur when the concentration of dissolved metal ions attains a minimum concentration of 1.0×10^{-6} M. Suppose that a 0.50-cm^2 sample of aluminum immersed in 1.0 L of an aqueous solution corrodes at a constant current density of 100 $\mu A/cm^2$. How long will it take to produce a concentration of 1.0×10^{-6} M Al^{3+}?
6. Calculate the standard reduction potential E^o for the reaction

$$Al_2O_3(s) + 6H^+(aq) + 6e^- \longrightarrow 2Al(s) + 3H_2O(l)$$

given the following chemical potentials (in calories per mole): $\mu^o(Al_2O_3(s)) = -384{,}530$, $\mu^o(H_2O(l)) = -56{,}690$.

7. For the chemical reaction

$$Al_2O_3(s) + H_2O(l) \longrightarrow 2AlO_2^-(aq) + 2H^+(aq)$$

show the result given in Eq. (17), i.e.,

$$-pH + \log[AlO_2^-] = -14.644$$

8. For the electrochemical reaction

$$Al(s) + 2H_2O(l) \longrightarrow AlO_2^-(aq) + 4H^+(aq) + 3e^-$$

show the result given in Eq. (19), i.e.,

$$E = -1.262 + 0.0197 \log [AlO_2^-] - 0.0788\,pH$$

9. For the cathodic evolution of hydrogen from neutral or basic solutions

$$2H_2O(l) + 2e^- \longrightarrow H_2(g) + 2OH^-(aq)$$

for which the standard electrode potential is $E^o = -0.828$ V vs. SHE, show that

$$E = 0.000 - 0.0591pH$$

10. Refer to the Pourbaix diagram for chromium in Fig. 6.11. The equilibrium reaction for the line separating the regions of passivation by Cr_2O_3 and transpassive dissolution as chromate ions (CrO_4^{2-}) corresponds to the reaction

$$2CrO_4^{2-}(aq) + 10H^+(aq) + 6e^- \longrightarrow Cr_2O_3(s) + 5H_2O(l)$$

Show that the reduction potential for this half-cell reaction depends on both the pH and the concentration of dissolved chromate ions. The standard electrode potential for the reduction reaction is $E^o = 1.311$ V vs. SHE.

11. Refer to Fig. 6.15, which is the Pourbaix diagram for titanium.

   (a) Write the electrochemical half-cell reactions for the step-wise oxidation of titanium by the following reactions

      Ti (s) to TiO (s)
      TiO (s) to Ti_2O_3 (s)
      Ti_2O_3 (s) to TiO_2 (s)

   Note: H_2O is a reactant in each of the above reactions.

   (b) What is the slope dE/dpH for each of these reactions?

12. The following thermodynamic data are for the Mg/H_2O system [1].

Dissolved substances	μ^o (cal/mol)
Mg^{2+}	– 108,990
H^+	0
Solid substances	
Mg	0

$Mg(OH)_2$ – 199,270

Liquid substances

H_2O – 56,690

The chemical reactions involving these species are

$$Mg \longrightarrow Mg^{2+} + 2e^- \quad E^o \text{ (reduction) vs. SHE} = -2.363V$$
$$Mg + 2H_2O \longrightarrow Mg(OH)_2 + 2H^+ + 2e^- \quad E^o \text{ (reduction) vs. SHE} = -1.862V$$
$$Mg^2 + 2H_2O \longrightarrow Mg(OH)_2 + 2H^+$$

Calculate and construct the Pourbaix diagram for magnesium, assuming that the passivity of magnesium is provided by a layer of solid $Mg(OH)_2$. Label all regions of corrosion, passivity, and immunity.

13. Construct the Pourbaix diagram for iron from the set of electrochemical and chemical equilibria given in Table 6.2. Label all regions of corrosion, passivity, and immunity.
14. If iron is immersed in a solution of pH 3.0, what electrode potential must be maintained to prevent the evolution of hydrogen gas?
15. In a study on the corrosion of iron in model crevices or pits, Pickering and Frankenthal [22] observed the egress of gas bubbles from the surface of iron contained in the model crevice. For a solution of 0.5 M sodium acetate/0.5 M acetic acid (pH 5.5) contained in the crevice, the electrode potential within the crevice was measured to be –0.5 V vs. SHE [23]. Based on thermodynamic considerations, what gas was evolved?
16. Consult the Pourbaix Atlas [1] or other suitable source for the Pourbaix diagram for niobium. How does this diagram compare with that for tantalum? Based on the location of niobium and tantalum in the periodic table of the elements, would you expect the two Pourbaix diagrams to be similar?

References

1. M. Pourbaix, "Atlas of Electrochemical Equilibria in Aqueous Solutions", National Association of Corrosion Engineers, Houston, TX (1974).
2. J. Kruger in "Equilibrium Diagrams: Localized Corrosion", R. P. Frankenthal J. Kruger, Eds., p. 45, The Electrochemical Society, Pennington, NJ (1984).
3. R. E. Groover, J. A. Smith, and T. J. Lennox, Jr., *Corrosion*, *28*, 101 (1972).
4. E. McCafferty, P. G. Moore, and G. T. Peace, *J. Electrochem. Soc.*, *129*, 9 (1982).
5. M. J. Pyror and D. S. Keir, *J. Electrochem. Soc.*, *105*, 629 (1958).
6. R. Ambat and E. S. Dwarakadasa, *Corros. Sci.*, *33*, 681 (1992).
7. J. A. Smith, M. H. Peterson, and B. F. Brown, *Corrosion*, *26*, 539 (1970).
8. T. K. Christman, J. Payer, G. W. Kurr, J. B. Doe, D. S. Carr, and A. L. Ponikvar, "Zinc: Its Corrosion Resistance", pp. 4, 114, International Lead and Zinc Research Institute, New York, NY (1983).
9. H. S. Betrabet, W. B. Johnson, D. D. Macdonald, and W. A. T. Clark in "Equilibrium Diagrams: Localized Corrosion", R. P. Frankenthal and J. Kruger, Eds., p. 83, The Electrochemical Society, Pennington, NJ (1984).
10. E. McCafferty, M. K. Bernett, and J. S. Murday, *Corros. Sci.*, *28*, 559 (1988).
11. D. Cubicciotti, *Corrosion*, *44*, 875 (1988).
12. D. Cubicciotti, *J. Nucl. Mater.*, *201*, 176 (1993).
13. C. M. Criss and J. W. Cobble., *J. Am. Chem. Soc.*, *86*, 5385 (1964).
14. J. W. Cobble, *J. Am. Chem. Soc.*, *86*, 5394 (1964).
15. H. E. Townsend, Jr., *Corros. Sci.*, *10*, 343 (1970).
16. R. L. Cowan and R. W. Staehle, *J. Electrochem. Soc.*, *118*, 557 (1971).
17. J. B. Lee, *Corrosion*, *37*, 467 (1981).
18. B. Beverskog and I. Puigdomenech, *J. Electrochem. Soc.*, *144*, 3476 (1997).
19. R. J. Lemire and G. A. McRae, *J. Nuclear Mater.*, *294*, 141 (2001).

20. M. H. Peterson and T. J. Lennox, Jr., *Mater. Perform.*, *23* (3), 15 (1984).
21. M. Stern and H. Wissenberg, *J. Electrochem. Soc.*, *106*, 755 (1959).
22. H. W. Pickering and R. P. Frankenthal, *J. Electrochem. Soc.*, *119*, 1297 (1972).
23. H. W. Pickering in "Equilibrium Diagrams: Localized Corrosion", R. P. Frankenthal and J. Kruger, Eds., p. 535, The Electrochemical Society, Pennington, NJ (1984).

Chapter 7
Kinetics of Corrosion

Introduction

As seen in the previous chapter, potential–pH diagrams (Pourbaix diagrams) are a useful first guide as to the corrosion behavior of many different systems. However, Pourbaix diagrams do not give any information as to corrosion rates. Although a given reaction may be spontaneous, it does not necessarily proceed at a fast rate. The reaction may in fact proceed "slowly" rather than "quickly", but we cannot determine this difference from thermodynamics alone.

This chapter deals with the rates of corrosion reactions. The corrosion rate of a given metal or alloy in its environment is a crucial factor in determining the lifetime of both structural and electronic materials.

Units for Corrosion Rates

Various units have been used to express corrosion rates. These include weight loss per unit area per unit time, penetration rates, and electrochemical rates. There is no standard unit to express corrosion rates, and many different units have been used, as given in Table 7.1.

Methods of Determining Corrosion Rates

Corrosion rates for metals undergoing uniform corrosion can be determined by any of the following methods:

(a) Weight loss
(b) Weight gain
(c) Chemical analysis of solution
(d) Gasometric techniques (when one of the reaction products is a gas)
(e) Thickness measurements
(f) Electrical resistance probes
(g) Inert marker method
(h) Electrochemical techniques

Each of these methods is discussed briefly below. Most of this chapter, however, concerns the measurement of corrosion rates by means of the electrochemical polarization method.

E. McCafferty, *Introduction to Corrosion Science*, DOI 10.1007/978-1-4419-0455-3_7,

Table 7.1 Some units commonly used to express corrosion rates

Weight loss
g/cm^2 h
g/cm^2 day
g/m^2 h
g/m^2 day
mg/m^2 s
mdd (mg/dm^2 day)
Penetration
ipy (inches per year)
mpy (mils per year)[a]
mm/year
μm/year
Corrosion current density
$\mu A/cm^2$
mA/cm^2
A/cm^2
A/m^2

[a] 1 mil = 0.001 in.

Weight Loss Method

In this method, previously weighed metal samples are removed from the solution or the environment at various timed intervals, and the loss in weight due to metallic corrosion is determined (per unit area of the sample). For this method to be accurate, any solid corrosion products must be removed from the metal surface.

If the corrosion products are loose and non-adherent, they can be removed by mechanical means. Tight and adherent corrosion products can be removed by chemical or electrochemical methods, which are usually specific to the metal or the alloy being tested [1].

Figure 7.1 shows an example of weight loss data for the uniform corrosion of aluminum alloy 7075 in 0.5 M hydrochloric acid [2]. After an initial incubation period of about 2 h., the rate of weight loss becomes constant with time. The corrosion rate is simply the slope of the straight line, as shown in the figure.

Figure 7.2 shows weight loss data taken for a much longer period of time for the atmospheric corrosion of various steels exposed in the natural atmosphere of Kearney, New Jersey [3, 4]. Note that the time axis is in years of exposure and that the corrosion rate is not constant over the entire exposure period. The instantaneous corrosion rate at any given time is the slope of the tangent line drawn at that given time. (The trends shown in Fig. 7.2 hold for up to 20 years [3, 4].)

Weight Gain Method

This method is generally not useful in aqueous corrosion studies but is more suited to dry oxidation where oxide films or products are more adherent than is in the case of aqueous corrosion. Laboratory weight gain measurements in dry environments have the useful feature that the gain in weight can be measured continuously and without removal of the specimen from the environment.

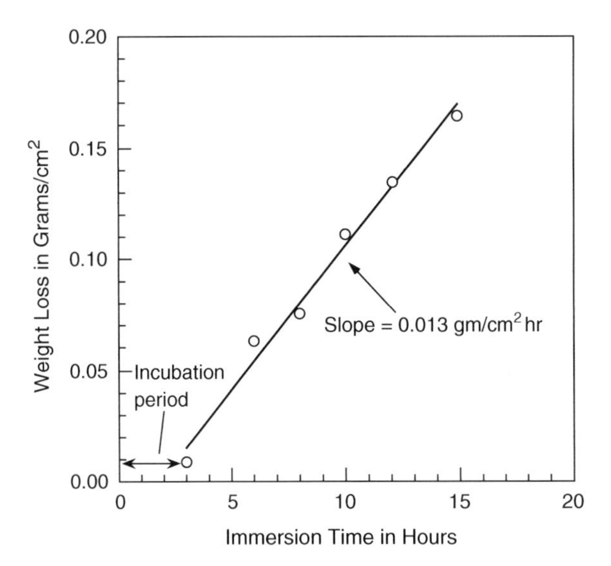

Fig. 7.1 Weight loss vs. time for Al alloy 7075 in 0.5 M HCl [2]. (Al alloy 7075 contains approximately 5% Zn, 1% Cu, and 2% Mg)

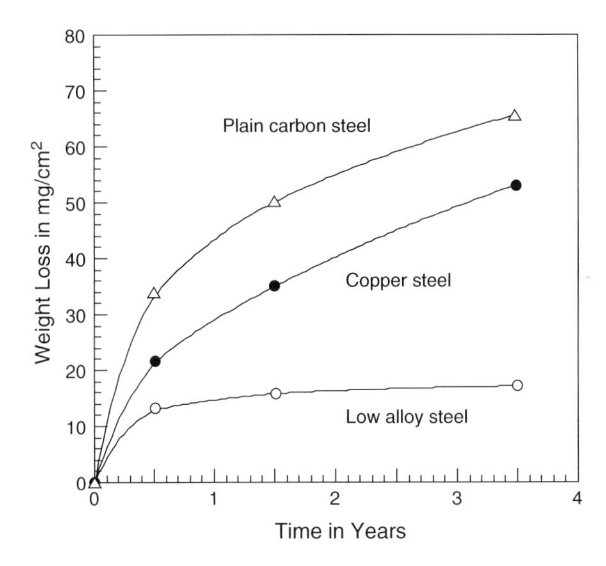

Fig. 7.2 Corrosion–time curves for the atmospheric corrosion of various steels exposed in the natural atmosphere of Kearney, New Jersey. Redrawn from [3] by permission of ECS – The Electrochemical Society

Chemical Analysis of Solution

When a metal undergoes corrosion in an electrolyte of a fixed volume, cations of the corroding metal will accumulate in the solution. Accordingly, the solution becomes more concentrated in the dissolved cation with the progression of time. Thus, chemical analysis of withdrawn aliquots of the solution as a function of time allows determination of the corrosion rate.

Figure 7.3 shows chemical analysis data for the corrosion of aluminum alloy 7075 in 0.5 M HCl corresponding to the weight loss data in Fig. 7.1. From Fig. 7.3, the corrosion rate is 4.80×10^{-3} mol

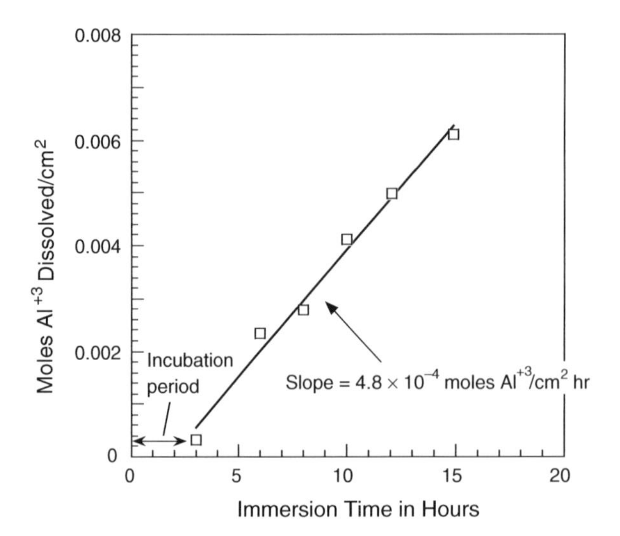

Fig. 7.3 Amount of dissolved Al^{3+} in 0.5 M HCl as a function of time corresponding to the weight loss data in Fig. 7.1. (The amount of dissolved Al^{3+} is calculated assuming that the anodic reaction is $Al \rightarrow Al^{3+} + 3e^-$)

Al^{3+}/cm^2 day. This corrosion rate is converted to the rate in terms of weight loss by the following simple calculation:

$$\left(\frac{4.80\ \times\ 10^{-4}\text{molesAl}^{+3}}{\text{cm}^2\text{hr}}\right)\left(\frac{1\text{ mole Al}}{1\text{ mole Al}^{+3}}\right)\left(\frac{27.0\text{ gm Al}}{1\text{ mole Al}}\right)=\frac{0.0130\text{ gm Al}}{\text{cm}^2\text{hr}}$$

The use of standard colorimetric techniques in the visible range is a common method for chemical analysis of solutions. This method depends on the formation of a colored complex, the intensity of which is proportional to the concentration of the complex formed and correspondingly to the concentration of the dissolved cation in solution. The intensity of the complex is determined by measuring the percent transmittance at a given wavelength of light and by comparison with a previously prepared standard plot.

The specific colorimetric technique to be used depends on the particular cation to be analyzed, and standard reference texts should be consulted for more details [5]. The corrosion rate of iron, aluminum, or titanium in different environments has been determined using various specific colorimetric procedures [6–10].

Another useful method for chemical analysis of solutions is atomic absorption spectrophotometry. The sample size required for this method is very small (typically about 10 mg). The sample is accurately weighed and then sprayed into the flame of the instrument and atomized. Again, light of a suitable wavelength for a particular element is directed through the flame, and some of this light is absorbed by the atoms of the sample. The amount of light absorbed is proportional to the concentration of the element in the solution. The method is sensitive to trace elements down to the parts per million level [11].

Gasometric Techniques

When aluminum (or a dilute alloy of aluminum) corrodes in acid solutions, the overall ionic reaction is

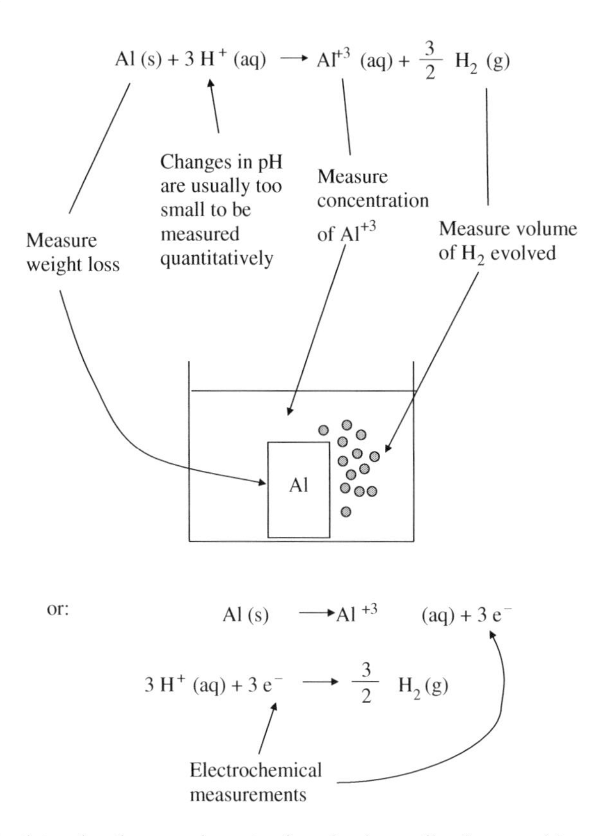

Fig. 7.4 Various ways to determine the corrosion rate of an aluminum alloy immersed in an acid solution

$$\mathrm{Al(s) + 3H^+(aq)Al^{3+}(aq) + \frac{3}{2}H_2(g)}$$

As depicted in Fig. 7.4, the corrosion rate may be determined by measuring the change in amounts of reactant Al or of products Al^{3+} or H_2 as the reaction proceeds.

Figure 7.5 shows data for the volume of hydrogen gas evolved for corrosion of aluminum alloy 7075 in 0.5 M HCl corresponding to the weight loss data in Fig. 7.1. From Fig. 7.5, the rate of hydrogen evolution is 16.1 ml H_2 (STP)/cm^2 h, where STP refers to conditions of standard temperature (0°C) and pressure (1 atm).

Example 7.1: Convert the rate of hydrogen evolution in Fig. 7.5 to the corresponding weight loss of metallic aluminum in terms of grams per square meter hour.
Solution: We use the fact that under conditions of STP, 1 mol of a gas occupies 22.4 L. Thus

$$\left(\frac{16.1\,\mathrm{mL\,H_2(STP)}}{\mathrm{cm^2h}}\right)\left(\frac{1\mathrm{L}}{1000\,\mathrm{mL}}\right)\left(\frac{1\,\mathrm{mol\,H_2}}{22.4\,\mathrm{L\,H_2(STP)}}\right)$$
$$\times\left(\frac{1\,\mathrm{mol\,Al}}{\frac{3}{2}\mathrm{mol\,H_2}}\right)\times\left(\frac{27.0\,\mathrm{g\,Al}}{1\,\mathrm{mol\,Al}}\right)=\frac{0.0129\,\mathrm{g\,Al}}{\mathrm{cm^2h}}$$

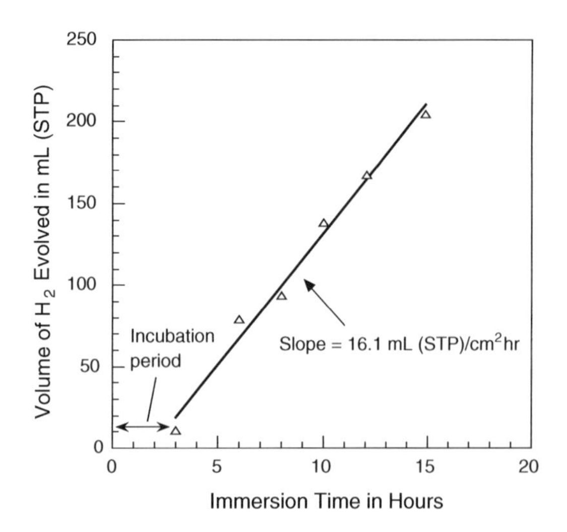

Fig. 7.5 Volume of hydrogen gas evolved as a function of time when an aluminum alloy is immersed in 0.5 M HCl. The amount of H_2 gas evolved is calculated from the weight loss data in Fig. 7.1 and the reaction $Al(s) + 3H^+ (aq) \rightarrow Al^{3+} (aq) + \frac{3}{2} H_2 (g)$

Thickness Measurements

Measurement of the thickness of a metal as a function of time is a method to monitor the corrosion behavior of a system [12]. This method is used to survey the practical corrosion behavior of a system in service, rather than as a laboratory technique, and has been used to track the progress of corrosion in railroad hopper cars, storage tanks, and pipelines. Thickness measurements are based on the response of a specimen to an ultrasonic or a magnetic signal.

Electrical Resistance Method

Electrical resistance methods depend on an increase in the electrical resistance as a test metallic film or wire is thinned due to corrosion [12, 13]. For example, the resistance of a metal wire of length l and area A is given by $R = \rho l/A$, where ρ is the resistivity. As the area A of the metal wire decreases due to corrosion, the resistance R increases. These types of instruments are commercially available and are generally used to monitor the progression of corrosion rather than to determine the actual corrosion rate. One drawback of this method is that conductive deposits on the sensor elements will distort the resistance reading. However, this method is useful for solutions of low conductivity or for non-aqueous solutions.

Inert Marker Method

This is a novel experimental method which is suited to the laboratory and requires a sophisticated ion implantation system. In this method, an inert marker, such as ion-implanted Xe (xenon), is used in conjunction with the surface analysis technique of Rutherford backscattering (RBS). Ion implantation is a technique in which any ion can be placed in the near-surface region of any substrate (including metals), by allowing a high-energy beam of a selected ion to bombard the metal surface.

More information about ion implantation is given in Chapter 16. The RBS method is a surface analysis method which uses He atoms (α particles) as collider particles to probe the composition of a surface. This method is described in detail elsewhere [14].

Figure 7.6 shows RBS data for the ion-implanted Xe in pure titanium before immersion into boiling 1 M H_2SO_4 and for two different brief immersion times [10]. After immersion, the Xe signal is diminished as the titanium sample corrodes.

Corrosion rates were determined in the following manner. It is assumed that the shape of the Xe profiles in Fig. 7.6 is changed only by the removal of Xe atoms from the surface as the surface moves inward. The shift in energy of the remaining Xe profile with respect to the initial profile (before corrosion) is converted into a thickness of material removed by means of a well-known energy loss rate d(Energy)/dx for He ions in titanium. For example, 2.5 MeV He ions incident on the sample that are scattered by a Xe atom and are subsequently detected have an effective d(Energy)/dx = 0.9 keV nm^{-1} (1 nm = 1.0 $\times$ 10^{-9} m). If an energy shift of 10 keV is measured in the Xe profile, the thickness of the material removed t is 10 keV/0.9 keV nm^{-1} = 11.1 nm.

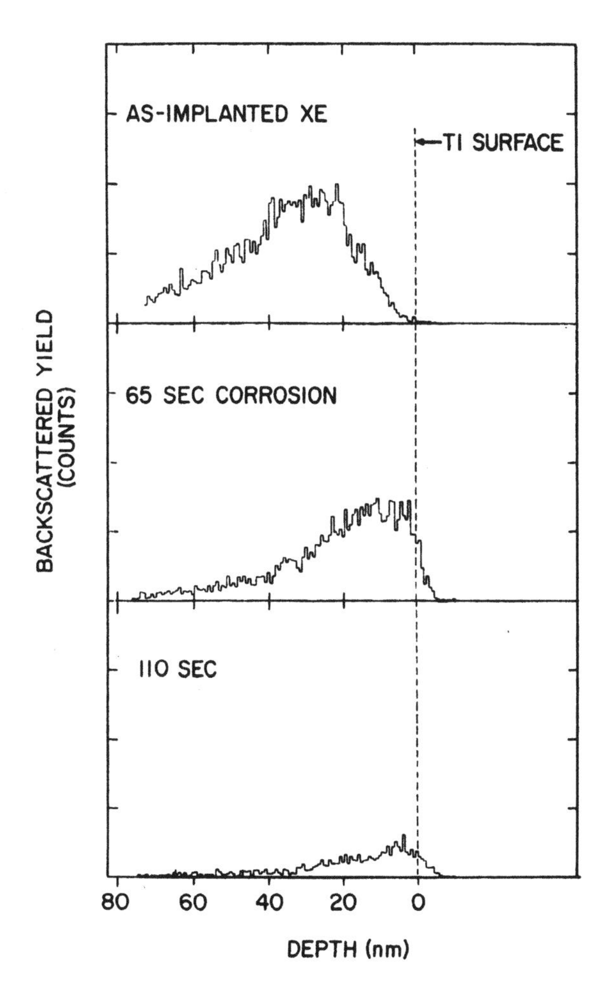

Fig. 7.6 Rutherford backscattering (RBS) profiles for Xe implanted into titanium before and after immersion in boiling 1 M H_2SO_4 [10]. Reproduced from [10] with the permission of Elsevier, Ltd

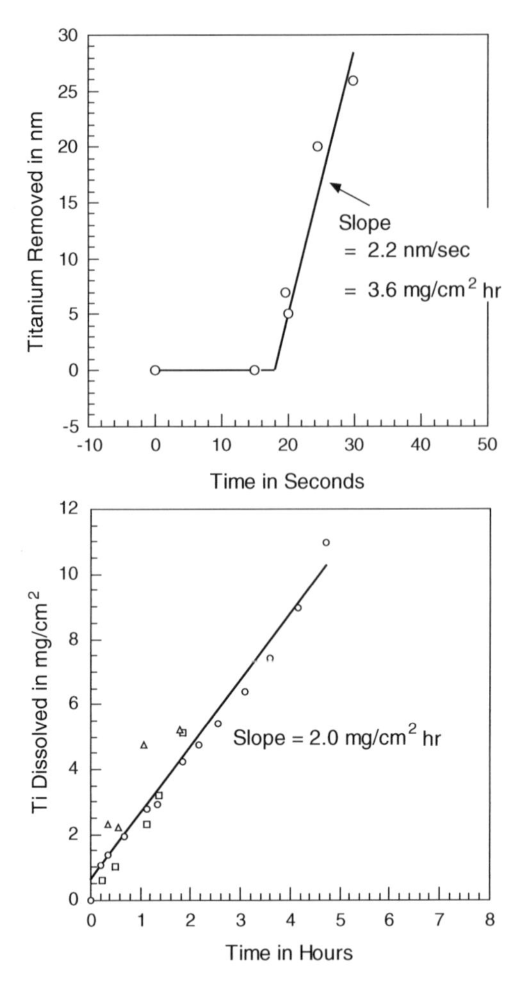

Fig. 7.7 *Top*: Titanium thickness removed as a function of immersion time in boiling 1 M H_2SO_4 as determined from RBS profiles. *Bottom*: Colorimetric analysis of dissolved titanium in boiling 1 M H_2SO_4 assuming the anodic reaction is $Ti \rightarrow Ti^{2+} + 2e^-$. Reproduced from [10] with the permission of Elsevier, Ltd

In this manner, Fig. 7.7(top), which is a plot of titanium thickness removed as a function of time, was determined. After an initial incubation time of approximately 15 s to remove the initial air-formed oxide film, the rate of titanium dissolution was constant. From Fig. 7.7(top), the corrosion rate is 2.2 nm/s. The use of the density of titanium (4.51 g/cm^3) gives the corrosion rate to be 3.6 mg/cm^2 h, in approximate agreement with the rate of 2.2 mg/cm^2 h determined by colorimetric analysis of the solution for dissolved titanium. See Fig. 7.7 (bottom).

Electrochemical Techniques

Each of the methods described above has an application in either the laboratory or the field. But none of the methods above can provide insights into the underlying mechanism of corrosion. Electrochemical techniques have this capability and in addition offer the possibility of mitigating corrosion by controlling the electrode potential. There are various methods of determining the corrosion rate by electrochemical techniques, but this chapter will consider only the direct current (DC) polarization method and its offshoots. Alternating current (AC) techniques, which are very powerful in determining corrosion rates and in monitoring the progress of corrosion, are treated later in a separate chapter.

Electrochemical Polarization

Electrochemical polarization (usually referred to simply as "polarization") is the change in electrode potential due to the flow of a current. There are three types of polarization:

(1) *Activation polarization* is polarization caused by a slow electrode reaction.
(2) *Concentration polarization* is polarization caused by concentration changes in reactants or products near an electrode surface.
(3) *Ohmic polarization* is polarization caused by IR drops in solution or across surface films, such as oxides (or salts).

The degree of polarization is defined as the *overvoltage* (or overpotential) η given by the following equation:

$$\eta = E - E_o \tag{1}$$

where E is the electrode potential for some condition of current flow and E_o is the electrode potential for zero current flow (also called the open-circuit potential, corrosion potential, or rest potential). Note that the electrode potential of zero current flow E_o should not be confused with the standard electrode potential E^o, which plays a prominent role in corrosion thermodynamics.

Anodic and Cathodic Polarization

Either an anode or a cathode can be polarized:

Anodic polarization is the displacement of the electrode potential in the positive direction so that the electrode acts more anodic.
Cathodic polarization is the displacement of the electrode potential in the negative direction so that the electrode acts more cathodic.

These processes are represented schematically in Fig. 7.8.

Visualization of Cathodic Polarization

Consider the hydrogen evolution reaction occurring at a metal surface. First, hydrogen atoms are produced by the reduction reaction

$$2H^+ + 2e^- \rightarrow 2H_{ads,} \tag{2}$$

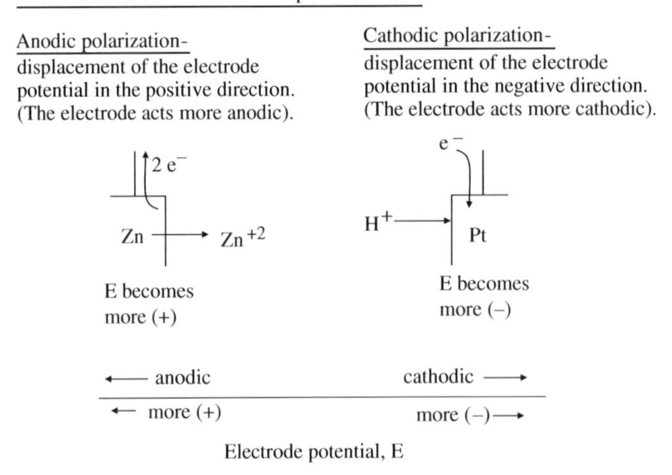

Fig. 7.8 Schematic representation of anodic and cathodic polarization

followed by their combination:

$$2H_{ads} \rightarrow H_2 \tag{3}$$

The process of activation polarization involves a slow step in the electrode reaction. Suppose that electrons are supplied to the metal electrode faster than they can react to form H atoms, as represented in Fig. 7.9. Then the concentration of electrons is increased at the metal side of the interface. The result is that the electrode potential E becomes more negative, due to *activation polarization* [15].

Suppose instead that there are concentration effects near the electrode surface for the hydrogen reduction reaction. If reactant hydrogen ions H^+ are slow to diffuse to the electrode surface, as illustrated in Fig. 7.10, then electrons again can accumulate at the metal side of the interface. The result is that the electrode potential E again becomes more negative, but this time due to *concentration polarization* [15].

Visualization of Anodic Polarization

Consider the following anodic reaction:

$$Fe \rightarrow Fe^{2+} + 2e^{-} \tag{4}$$

Suppose that the oxidation of Fe atoms to Fe^{2+} ions is slow. Then electrons exit the electrode faster than Fe atoms leave the metal matrix, as illustrated in Fig. 7.11. This means that the electron concentration is decreased at the metal side of the interface. The electrode potential E thus becomes more positive due to *activation polarization* [15].

Suppose next that the products of the anodic reaction, i.e., Fe^{2+} ions, are slow to diffuse away from the metal surface, as represented in Fig. 7.12. Then the surface becomes more positively charged due to the accumulation of Fe^{2+} ions. The electrode potential E again becomes more positive, but this time due to *concentration polarization* [15].

Cathodic polarization (activation)

Suppose the reaction is slow.

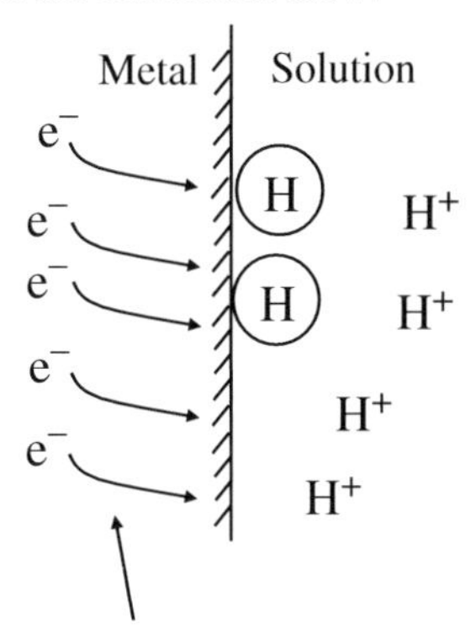

Electrons are supplied to the metal faster than they can react.

Thus, the concentration of e^- is increased at the metal side of the interface.

E becomes more (–) due to activation polarization .

Fig. 7.9 Schematic representation of activation polarization for a cathode

Cathodic polarization (concentration)

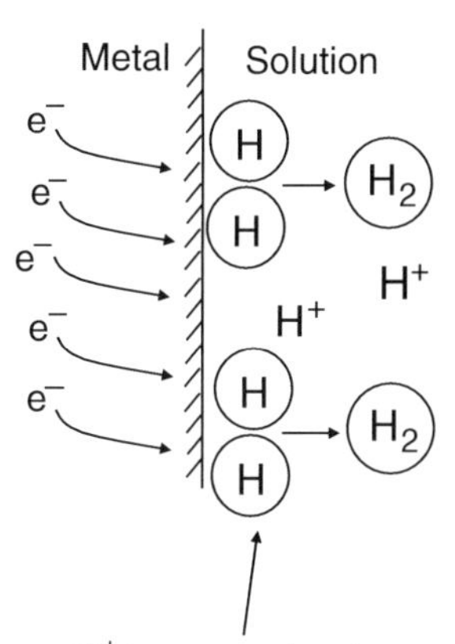

Suppose H^+ ions are slow to move into the electrical double layer.

Then, electrons flow into the surface faster than they can be consumed.

The concentration of e^- is again increased at the metal side of the interface.

E becomes more (–) due to concentration polarization.

Fig. 7.10 Schematic representation of concentration polarization for a cathode

Anodic polarization (activation)

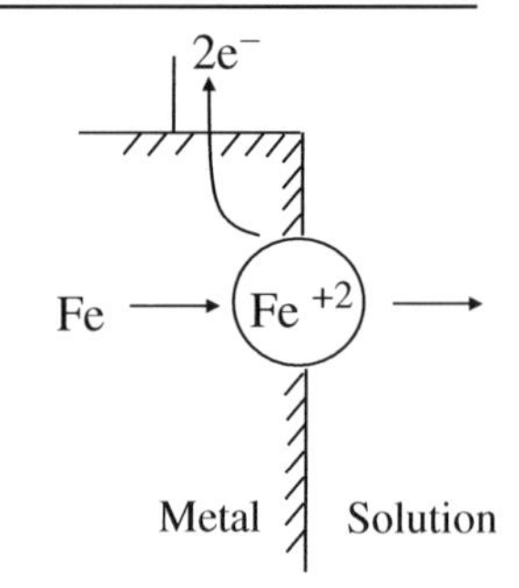

Suppose the reaction is slow.

Then, electrons exit from the surface faster than Fe atoms leave the matrix.

The concentration of e^- is decreased at the metal side of the interface.

E becomes more (+) due to activation polarization.

Fig. 7.11 Schematic representation of activation polarization for an anode

Anodic polarization (concentration)

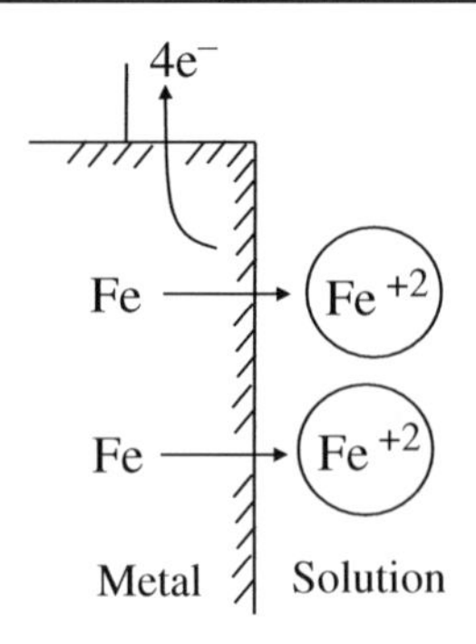

Suppose the products are slow to diffuse away from the interface.

Again, the surface becomes more positively charged.

E becomes more (+) due to concentration polarization.

Fig. 7.12 Schematic representation of concentration polarization for an anode

Ohmic Polarization

Ohmic polarization within a solution is due to the inability to place the reference electrode directly at the metal surface under investigation, as illustrated in Fig. 7.13. As mentioned in Chapter 3, the use of a Luggin–Haber capillary minimizes the error due to IR drops [16–18]. Barnartt [16] found that the current distribution around an electrode surface is unaffected if the distance between the electrode and the capillary is twice the capillary outer diameter. The term $IR_{solution}$ is usually

Ohmic polarization

Ohmic polarization is caused by the IR drop due to the inability to place the reference electrode directly at the metal surface under study

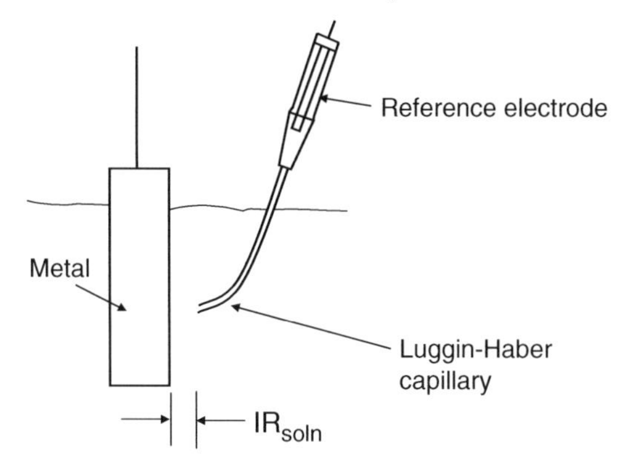

IR_{soln} is negligible for high conductivity solutions.

IR_{soln} is high for low conductivity solutions.
(organic media, some solis)

Fig. 7.13 Schematic representation of ohmic polarization in an aqueous solution

negligible for high-conductivity solutions, such as most aqueous solutions, but can be a problem for low-conductivity solutions, such as organic media, and for some soils.

If the IR drop is due to the existence of a surface film on a metal, such as an oxide, hydroxide, oxyhydroxide, or salt film, then such films are usually understood to be part of the system and the electrode potential is considered to be that for the total system: metal/film/solution.

Electrode Kinetics for Activation Polarization

The remainder of this chapter deals with corrosion kinetics when both the anodic and the cathodic processes are under activation polarization. Situations in which concentration polarization occurs are treated in Chapter 8.

Absolute Reaction Rate Theory

The electrode kinetics of activation polarization is treated using the absolute reaction rate theory of Glasstone, Laidler, and Eyring [19]. According to the absolute reaction rate theory, a reaction proceeds along a reaction co-ordinate which measures the extent of the reaction. For reactants to be converted into products, a transitory state is first reached, in which an "activated complex" is formed. Formation of such a complex requires overcoming a free energy barrier of height $\Delta G^{\neq}$, as shown in Fig. 7.14. For the general reaction

$$A + B \rightarrow [AB]^{\neq} \rightarrow \text{products} \tag{5}$$

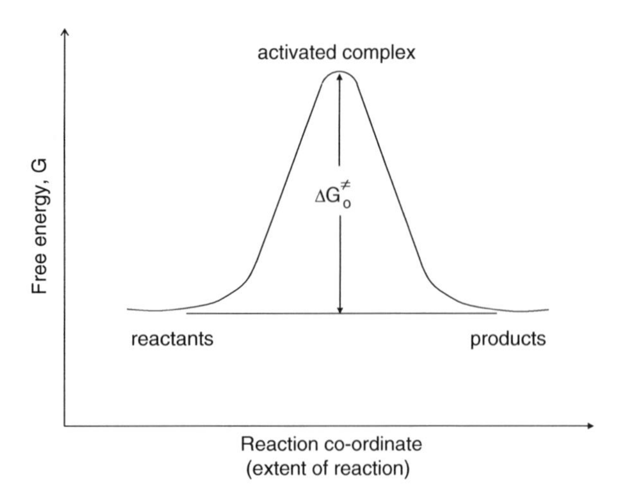

Fig. 7.14 Diagram of a free energy barrier to a chemical reaction

where $[AB]^{\neq}$ is the activated complex. The rate of the reaction depends on the concentration of activated complexes and their rate of passage over the energy barrier. A result of absolute reaction rate theory is that the rate constant is given by the following equation:

$$\text{rate constant} = \frac{kT}{h} e^{-\Delta G^{\neq}/RT} \tag{6}$$

The derivation of this equation is given in Appendix C. According to Eq. (6), the larger the free energy barrier $\Delta G^{\neq}$, the smaller the rate constant (and thus the lesser the rate of the reaction). This expression will be used in the sections to follow.

Electrode Kinetics for Non-Corroding Metals

Consider a substance Z in equilibrium with its ions Z^{n+}. Some examples are the following:

(1) Cu in equilibrium with Cu^+ ions (see Fig. 7.15)

$$Cu(s) \rightleftharpoons Cu^+(aq) + e^- \tag{7}$$

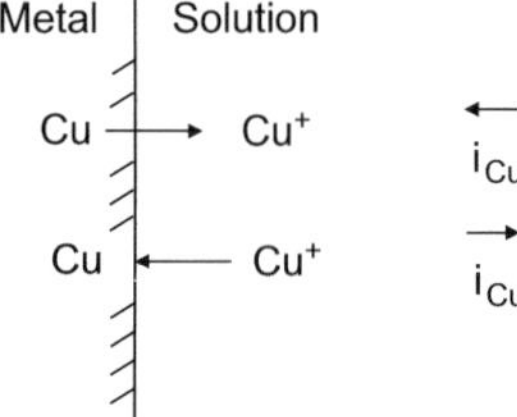

Fig. 7.15 The equilibrium between a solid copper metal and its dissolved ions. At equilibrium the rate of Cu^+ dissolution is equal to the rate of Cu deposition

(2) Pb in equilibrium with Pb^{2+} ions

$$Pb(s) \rightleftharpoons Pb^{2+}(aq) + 2e^- \tag{8}$$

(3) Pt supporting the reaction

$$Fe^{3+}(aq) + e^- \rightleftharpoons Fe^{2+} \tag{9}$$

(4) Pt supporting the reaction

$$2H^+(aq) + e^- \rightleftharpoons H_2(g) \tag{10}$$

At equilibrium, the rate of oxidation of Z is equal to the rate of reduction of Z^{n+}. That is

$$|\vec{i}_Z| = \overleftarrow{i}_Z = i_o \tag{11}$$

where $|\vec{i}_Z|$ is the cathodic current density for the reduction reaction:

$$Z^{n+}(aq) + ne^- \rightarrow Z(s) \tag{12}$$

and $\overleftarrow{i}_Z$ is the anodic current density for the oxidation reaction

$$Z(s) \rightarrow Z^{n+}(aq) + ne^- \tag{13}$$

and i_o, the rate of the reaction in either direction at the equilibrium open-circuit potential E_o, is called the *exchange current density*. The currents $\vec{i}_z$ and $\overleftarrow{i}_z$ flow in opposite directions. The sign of the cathodic current $\vec{i}_z$ is negative, and the sign of the anodic current $\overleftarrow{i}_z$ is positive.

At the open-circuit potential, the *net* rate of the reaction is zero. For the net cathodic rate at open-circuit potential E_o,

$$|\vec{i}_{net}| = |\vec{i}_Z| - \overleftarrow{i}_Z = 0 \tag{14}$$

At the open-circuit potential E_o, the net rate of the anodic reaction is also zero (15):

$$\overleftarrow{i}_{net} = \overleftarrow{i}_Z - |\vec{i}_Z| = 0 \tag{15}$$

As described by Bockris and Reddy [20], there is a “two-way traffic” of electrons in and out of the electrode at equilibrium. The net rate of traffic is zero, but there is traffic flow in either direction.

It is possible to directly measure E_o at the open-circuit potential, but it is not possible to directly measure i_o at the open-circuit potential. To measure i_o, it is necessary to perturb the system from equilibrium. When the reaction is activation controlled, i.e., when metal atoms pass into solution by transcending an energy barrier, then Fig. 7.14 applies.

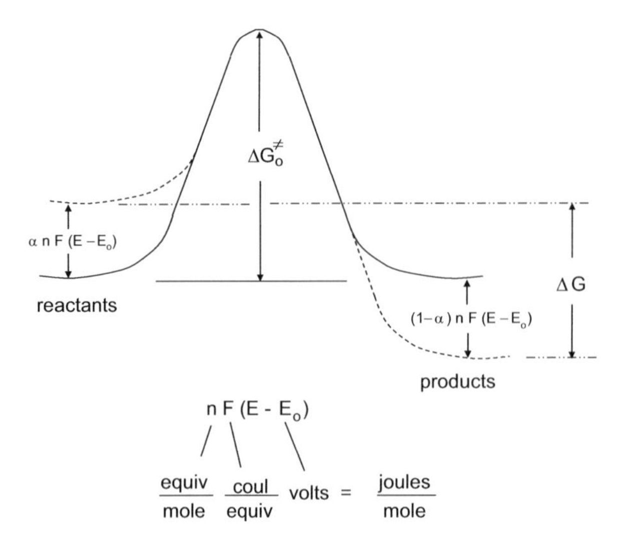

Fig. 7.16 Effect of a change in electrode potential on the height of the free energy barrier. When the potential is changed from E_o to some new electrode potential E, the shape of the free energy curve is changed from the solid line to the dotted line

Consider the oxidation of Z(s) to Z^{n+} (aq). If the electrode potential is changed from E_o to some other value E, the rate of oxidation is either raised or lowered to some new value i depending on whether the free energy barrier $\Delta G^{\neq}$ is raised or lowered. If the free energy barrier is lowered, as in Fig. 7.16, then

$$\Delta G^{\neq} = \Delta G_o^{\neq} - \alpha nF(E - E_o) \tag{16}$$

where $\Delta G_o^{\neq}$ is the free energy barrier at E_o, and α is a dimensionless parameter (between 0 and 1, but usually 0.5). The parameter α is a measure of the symmetry of the free energy barrier as modified by the new applied potential E. When $\alpha = 0.5$, the degree of decrease in the free energy barrier in the forward direction is the same as its degree of increase in the reverse direction.

Substitution of Eq. (16) in Eq. (6) gives

$$\text{rate constant} = \frac{kT}{h} \mathrm{e}^{-[\Delta G_o^{\neq} - \alpha nF(E - E_o)]/RT} \tag{17}$$

or

$$\text{rate constant} = \frac{kT}{h} \mathrm{e}^{-\Delta G_o^{\neq}/RT} \mathrm{e}^{\alpha nF(E - E_o)/RT} \tag{18}$$

From Fig. 7.17, the total current i across a surface of area A is

$$I = (\text{rate constant}) C^{\text{surf}} nFA \tag{19}$$

where C^{surf} is the concentration of the surface reactant. (In this case the surface concentration of the species Z.). Substitution of Eq. (18) in Eq. (19) and setting $I/A = i$ gives the anodic current density to be

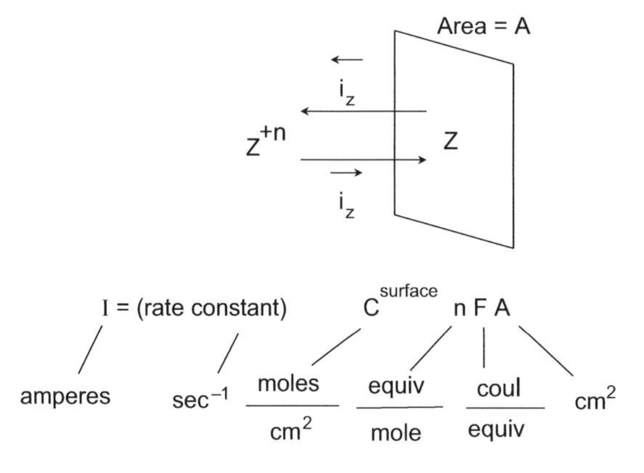

Fig. 7.17 Current flow across the surface of an electrode

$$\overleftarrow{i_Z} = \frac{I}{A} = \frac{kT}{h} e^{-\Delta G_o^{\neq}/RT} \times e^{\alpha nF(E-E_O)/RT} C_Z^{surf} nF \tag{20}$$

When $E = E_o$, $\overleftarrow{i_Z} = i_o$ and Eq. (20) becomes

$$i_o = \frac{kT}{h} e^{-\left[\Delta G_o^{\neq}/RT\right.} C_Z^{surf} nF \tag{21}$$

Use of Eq. (21) in Eq. (20) gives

$$\overleftarrow{i_Z} = i_o e^{\alpha nF(E-E_o)/RT} \tag{22}$$

Similarly, the rate of the back reaction (reduction reaction) is

$$|\overrightarrow{i_Z}| = i_o e^{-(1-\alpha)nF(E-E_o)/RT} \tag{23}$$

The net anodic reaction is thus

$$\overleftarrow{i_{net}} = \overleftarrow{i_Z} - |\overrightarrow{i_Z}| \tag{24}$$

or

$$\overleftarrow{i_{net}} = i_o \left[e^{\alpha nF(E-E_o)/RT} - e^{-(1-\alpha)nF(E-E_O)/RT} \right] \tag{25}$$

Equation (25) is usually called the Butler–Volmer equation (19) (but is sometimes called the Erdey–Gruz and Volmer equation) and is an important expression in electrode kinetics for both corroding and non-corroding metals. Equation (25) clearly shows that the rate of an electrochemical reaction depends on the electrode potential E.

An expression similar to Eq. (25) holds for the net cathodic reaction. That is

$$|\overrightarrow{i_{net}}| = i_o \left[e^{-(1-\alpha)nF(E-E_o)/RT} - e^{\alpha nF(E-E_o)/RT} \right] \tag{26}$$

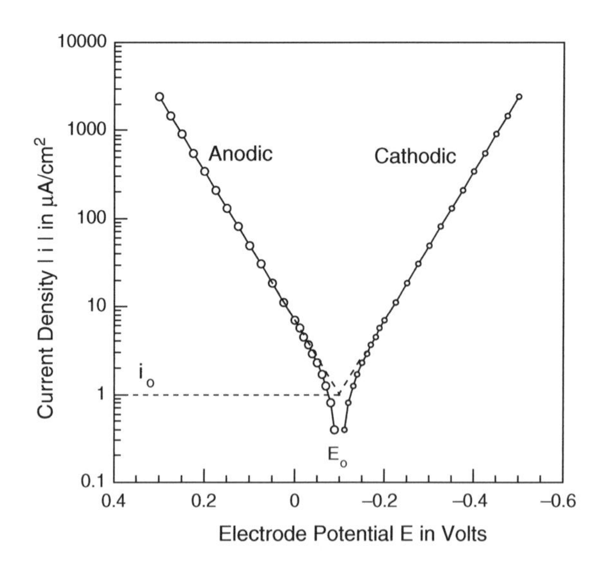

Fig. 7.18 Plot of the Butler–Volmer equation for the following set of electrode kinetic parameters: $E_o = -0.100$ V vs. SHE, $i_o = 1\ \mu A/cm^2$, $b_a = +0.100$ V/decade, and $b_c = -0.100$ V/decade

Figure 7.18 shows a plot of the Butler–Volmer equation for $E_o = -0.100$ V and Tafel slopes of dE/dlog $|i|$ ± 0.100 V/decade. The logarithm of the absolute value of the current density is plotted vs. the overvoltage ($E - E_o$), or vs. the potential E. (Plotting the logarithm of the absolute value of the current density is a convenience in making both current scales positive, in that cathodic currents have a negative sign, as mentioned earlier). As seen in Fig. 7.18, at sufficiently high overvoltages, both the cathodic and the anodic polarization curves display linear regions in the plot of log $|i|$ vs. E, as is discussed later.

Plots of log $|i|$ vs. E or vs. ($E - E_o$) are called *polarization curves*. The polarization curve is the basic kinetic law for any electrochemical reaction (16).

How to Plot Polarization Curves?

Polarization curves have been plotted in various ways by various workers. Very often, polarization curves are plotted as in Fig. 7.19(a), in which log $|i|$ is given along the abscissa, even though it is the electrode potential E and *not* the current which is the independent variable. This practice dates back to the very early days of corrosion science when polarization curves were determined galvanostatically, i.e., by applying a constant current and then observing the resultant electrode potential. Before the development of electronic potentiostats (i.e., constant potential devices), galvanostats were easier to construct and to operate than were potentiostats. Today, however, most polarization curves are determined potentiostatically so that the electrode potential is the independent experimental variable. In addition, according to absolute reaction rate theory, it is by changing the electrode potential that the free energy barrier is either lowered or raised so that a concomitant current flow is observed. Thus, the electrode potential is properly the independent variable and should be plotted on the abscissa in a polarization curve.

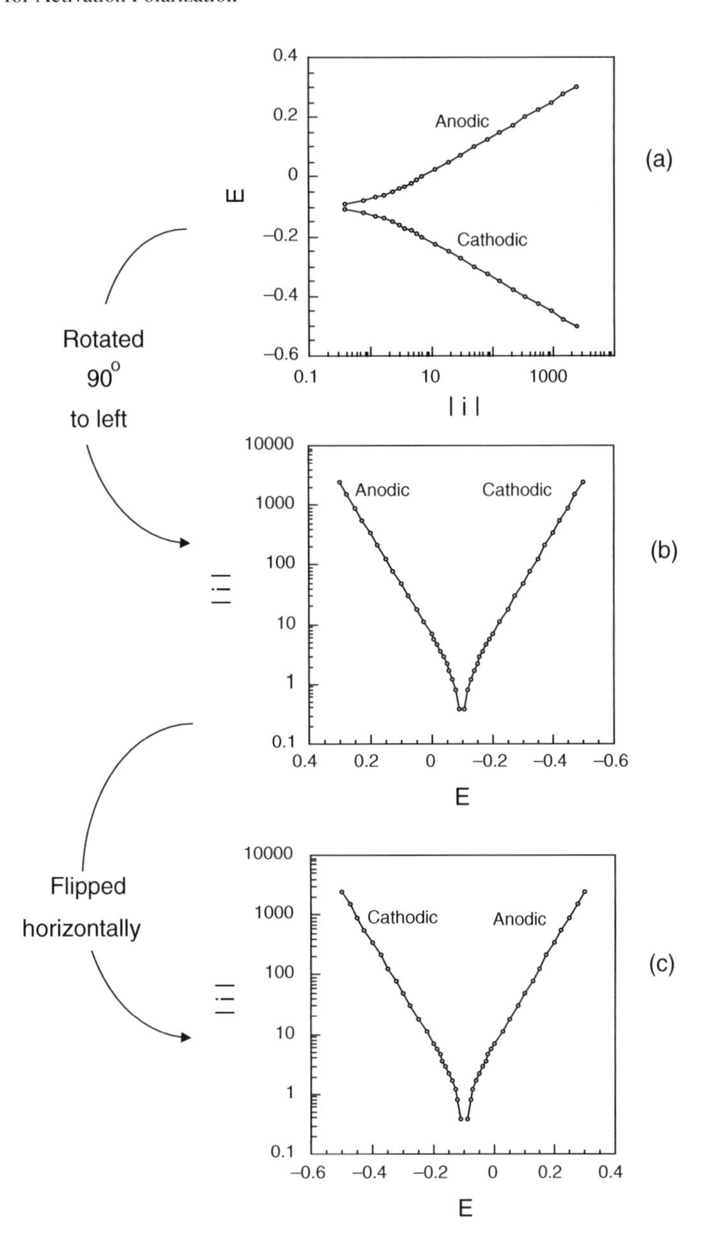

Fig. 7.19 Various methods of presenting polarization curves

By rotating Fig. 7.19(a) 90° to the left, the independent variable E is placed along the abscissa, as shown in Fig. 7.19(b). The "disadvantage" of Fig. 7.19(b) is that electrode potentials increase in the negative direction along the X-axis. However, this is a common convention in electroanalytical chemistry [21].

Plots as in Fig. 7.19(b) are used in this text when both anodic and cathodic polarization curves are shown. If only the anodic or the cathodic curve is given, then the curve is drawn from left to right and the sense of the X-axis is clearly labeled.

Increasingly positive potentials can appear to the right along the X-axis if Fig. 7.19(b) is flipped horizontally, as in Fig. 7.19(c). But such curves are then two operations removed from the familiar plots of Fig. 7.19(a).

The Tafel Equation

The semi-logarithmic plot in Fig. 7.18 is not linear near the open-circuit potential (i.e., near zero overvoltage). This is because the other half-cell reaction is still appreciable and contributes to the total current. However, at sufficiently high overvoltages, the contribution of the reverse reaction becomes negligible. In this linear "Tafel" region, the straight lines shown in Fig. 7.18 can be extrapolated back to zero overvoltage (i.e., back to E_o) to give the open-circuit exchange current density i_o, as shown below.

The well-known Tafel equation follows from the Butler–Volmer equation as follows. At sufficiently high overvoltages, the rate of the reverse reaction becomes negligible so that Eq. (25) can be written as

$$\overleftarrow{i}_{\text{net}} = i_o\, e^{\alpha n F(E-E_o)/RT} \tag{27}$$

or simply

$$i = i_o\, e^{\alpha n F(E-E_o)/RT} \tag{28}$$

Taking logarithms in Eq. (28)

$$\log i = \log i_o + \frac{\alpha n F}{2.303\, RT}(E - E_o) \tag{29}$$

Thus, a plot of log $|i|$ vs. the overvoltage (E–E_o) (or of log $|i|$ vs. the electrode potential E) gives a straight line, as shown in Fig. 7.18. From Eq. (29), when $E = E_o$, $i = i_o$. Thus, the Tafel region can be extrapolated back to $E = E_o$ to give the exchange current density i_o, as shown in Fig. 7.18.

Equation (29) can be rewritten as

$$\eta_a = b_a \log \frac{i}{i_o} \tag{30}$$

which is one form of the Tafel equation, where η_a is the anodic overvoltage and b_a is the anodic Tafel slope given by

$$b_a = \frac{dE}{d \log i} = \frac{2.303\, RT}{\alpha\, n\, F} \tag{31}$$

Equation (30) can also be written as

$$\eta_a = a + b_a \log i \tag{32}$$

which is another form of the Tafel equation, where a is a constant given by

$$a = -\frac{2.303\, RT}{\alpha n F} \log i_o \tag{33}$$

Similar considerations also hold for the cathodic branch of the polarization curve. For the cathodic direction, when the back reaction (now the anodic reaction) becomes negligible, then Eq. (26) produces a cathodic Tafel region which can also be extrapolated back to $E = E_o$ to give the exchange current density i_o, as also shown in Fig. 7.18. Moreover, Eq. (26) leads to

$$\eta_c = b_c \log \frac{i}{i_o} \tag{34}$$

and

$$\eta_c = a' + b_c \log |i| \tag{35}$$

where η_c is the cathodic overvoltage and b_c is the cathodic Tafel slope

$$b_c = \frac{dE}{d \log |i|} = -\frac{2.303\,RT}{(1-\alpha)nF} \tag{36}$$

and a' is

$$a' = \frac{2.303\,RT}{(1-\alpha)nF} \log i_o \tag{37}$$

From Fig. 7.18, it can be easily seen that cathodic Tafel slopes have negative signs and anodic Tafel slopes have positive signs. Tafel slopes have the units of volts or millivolts per decade of current density.

It should be noted that the Tafel slope (b_a or b_c) is the geometric slope of the linear portion of the semi-logarithmic plot in Fig. 7.19(a), but *not* the geometric slope in Fig. 7.19(b, c). The Tafel slope is always defined as $dE/d \log |i|$.

Reversible and Irreversible Potentials

So far, we have been dealing with reversible electrode potentials. That is, the metal in question exists in solutions of its own ions, with the rate of the dissolution reaction being equal to the rate of the reverse deposition reaction. As discussed above, we have the general condition that the reaction $Z^{n+1}(aq) + ne^- \rightleftharpoons Z(s)$ is in equilibrium. If the dissolved ions are present at unit activity in solution, then the electrode potential is the standard electrode potential E^o for the couple Z^{n+} (aq)/Z (s). If the concentration of dissolved ions Z^{n+} is not at unit activity, the electrode potential is still a reversible one and can be related to E^o through the Nernst equation.

However, in most cases, the following circumstances hold:

(1) The solution does not initially contain ions of the metal.
(2) The solution contains foreign ions, e.g., Cl^-, SO_4^{2-}, CO_3^{2-}, PO_4^{3-}, H^+.
(3) The solution contains cations which are different than those of the corroding metal, e.g., copper immersed in a solution of ferric chloride.
(4) Metal atoms pass into solution continuously and irreversibly.

In these cases the electrode potential is said to be an *irreversible potential* and is not given by the Nernst equation. Examples of irreversible potentials have been given in Chapter 2 (although not identified there as such). For example, the dissolution of iron in acid solutions

$$\mathrm{Fe\,(s) + 2H^+\,(aq) \rightarrow Fe^{2+}\,(aq) + H_2\,(g)} \tag{38}$$

proceeds under an irreversible potential which involves two separate oxidation–reduction reactions:

$$\mathrm{Fe\,(s) \rightarrow Fe^{2+}\,(aq) + 2}e^- \tag{39}$$

and

$$\mathrm{2H^+\,(aq) + 2}e^- \rightarrow \mathrm{H_2\,(g)} \tag{40}$$

See Fig. 7.20. Neither half-cell reaction (39) nor (40) operates reversibly, and the electrode potential under which Eq. (38) proceeds is an irreversible one, determined by mixed potential kinetics, as described below.

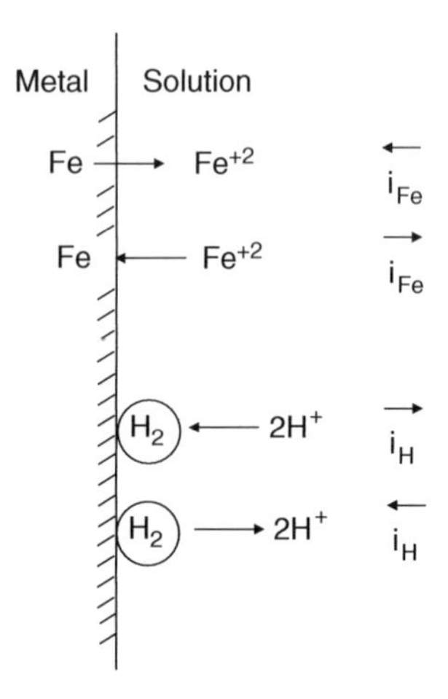

Fig. 7.20 Various processes occurring on an iron electrode immersed in an acid solution and the corresponding current vectors

The fact that Eqs. (39) and (40) can exist at different sites on the same metal surface is due to the physical and chemical heterogeneity of the metal surface, as has been discussed in Chapter 2.

Mixed Potential Theory (Wagner and Traud)

According to the Wagner and Traud mixed potential theory of corrosion [22]:

(1) Any electrochemical reaction can be divided into two or more partial oxidation and reduction reactions.
(2) At equilibrium, the total cathodic rate is equal to the total anodic rate (by the conservation of charge). Thus, for the dissolution of Fe in acidic solutions, as in Eq. (38)

$$|\overrightarrow{i}_{\mathrm{H}}| + |\overrightarrow{i}_{\mathrm{Fe}}| = \overleftarrow{i}_{\mathrm{H}} + \overleftarrow{i}_{\mathrm{Fe}} \tag{41}$$

where $|\vec{i}_H|$ is the current density for the reduction reaction

$$2H^+(aq) + 2e^- \rightarrow H_2(g)$$

$|\vec{i}_{Fe}|$ is the current density for the reduction reaction

$$Fe^{2+}(aq) + 2e^- \rightarrow Fe(s)$$

$\overleftarrow{i}_H$ is the current density for the oxidation reaction

$$H_2(g) \rightarrow 2H^+(aq) + 2e^-$$

and $\overleftarrow{i}_{Fe}$ is the current density for the oxidation reaction

$$Fe(s) \rightarrow Fe^{2+}(aq) + 2e^-$$

(3) The electrode potential for the steady-state, freely corroding condition given by Eq. (41) is called the corrosion potential E_{corr}.
(4) The corrosion potential is not related to the standard E^o values for either half-cell reaction by the Nernst equation.
(5) The corrosion potential E_{corr} lies between the electrode potentials for the two half-cell reactions and is called a *mixed potential.*

From Eq. (41), at equilibrium

$$\underbrace{\overleftarrow{i}_{Fe} - |\vec{i}_{Fe}|}_{\substack{\text{net rate of}\\ \text{iron dissolution}}} = \underbrace{|\vec{i}_H| - \overleftarrow{i}_H}_{\substack{\text{net rate of}\\ \text{hydrogen evolution}}} = i_{corr} \quad (42)$$

where i_{corr} is the corrosion rate of the freely corroding metal. At equilibrium, there is no net current flow, as per Eq. (41), but there is a loss of metal, i.e., corrosion, as per Eq. (42). The electrode potential for the steady-state, freely corroding condition in Eq. (42) is called the *corrosion potential* E_{corr} (also called the *open-circuit potential* or the *rest potential*).

We can directly measure the corrosion potential E_{corr}, but we cannot measure i_{corr} without polarizing the electrode away from the corrosion potential. That is the individual oxidation rates $\left(\overleftarrow{i}_{Fe} \text{ or } \overleftarrow{i}_H\right)$ or individual reduction rates ($|\vec{i}_{Fe}|$ or $|\vec{i}_H|$) cannot be measured.

Instead, it is necessary to make the entire corroding metal either the anode or the cathode in an electrolytic cell and to deduce the steady-state condition in Eq. (42) from the polarized condition. For the anodic direction, the Tafel equation gives

$$\eta_a = b_a \log \frac{\overleftarrow{i}_H + \overleftarrow{i}_{Fe}}{i_{corr}} \tag{43}$$

where $\left(\overleftarrow{i}_H + \overleftarrow{i}_{Fe}\right)$ is the total anodic current density. At any potential, the net anodic current density is

$$\overleftarrow{i}_{net} = \left(\overleftarrow{i}_H + \overleftarrow{i}_{Fe}\right) - \left(|\overrightarrow{i}_H| + |\overrightarrow{i}_{Fe}|\right) \tag{44}$$

Substitution of Eq. (44) into Eq. (43) gives

$$\eta_a = b_a \log \frac{\overleftarrow{i}_{net} + (|\overrightarrow{i}_H| + |\overrightarrow{i}_{Fe}|)}{i_{corr}} \tag{45}$$

But the cathodic current densities $|\overrightarrow{i}_H|$ and $|\overrightarrow{i}_{Fe}|$ are negligible at sufficiently high anodic overvoltages so that Eq. (45) reduces to

$$\eta_a = b_a \log \frac{\overleftarrow{i}_{net}}{i_{corr}} \tag{46}$$

or simply

$$\eta_a = b_a \log \frac{i}{i_{corr}} \tag{47}$$

An expression similar to Eq. (47) holds for the cathodic branch. That is

$$\eta_c = b_c \log \frac{i}{i_{corr}} \tag{48}$$

According to either Eq. (47) or Eq. (48), $i = i_{corr}$ when $\eta = 0$. Thus, anodic or cathodic Tafel lines can be extrapolated back to the corrosion potential E_{corr} to give the corrosion rate i_{corr}.

Thus, the Wagner–Traud theory of mixed potentials places the electrochemistry of local action cells on a firm theoretical basis (described qualitatively in Chapter 2). A consequence of the Wagner–Traud theory is that the Butler–Volmer equation can be written as follows for a corroding metal:

$$\overleftarrow{i}_{net} = i_{corr}\left[e^{\alpha nF(E-E_{corr})/RT} - e^{-(1-\alpha)nF(E-E_{corr})/RT}\right] \tag{49}$$

Comparison of Eqs. (49) and (25) for a single electrochemical reaction shows that for a corroding metal, i_{corr} replaces i_o, and E_{corr} replaces E_{corr}. However, unlike E_o, the corrosion potential E_{corr} does not have any thermodynamic significance attached to it and E_{corr} is determined by the kinetics of the system.

Example 7.2: Suppose that a metal M has an oxidation–reduction exchange current density of 0.1 μA/cm^2 at its reversible potential of –0.160 V. The anodic Tafel slope is + 0.060 V and the cathodic Tafel slope is –0.060 V for the system $M^+ + e^- \rightarrow M$. For hydrogen evolution on the surface of M, the anodic and the cathodic Tafel slopes are + 0.100 and – 0.100 V, respectively. The exchange current density for hydrogen evolution on the metal M is 1.0 μA/cm^2 at the reversible potential of 0.00 V.

(a) Construct the individual polarization curves for the systems $M^+ + e^- \rightleftharpoons M$ and $2H^+ + 2e^- \rightleftharpoons H_2$
(b) What is the corrosion potential?
(c) What is the corrosion rate?
(d) Plot the experimentally observed anodic and cathodic polarization curves.

Solution: The schematic polarization curves for the electrochemical reactions $M^+ + e^- \rightleftharpoons M$ and $2H^+ + 2e^- \rightleftharpoons H_2$ are drawn graphically, as shown in Fig. 7.21. The corrosion potential and the corrosion rate are indicated on the figure. The experimentally observed anodic and cathodic polarization curves are determined by first setting up Table 7.2, in which the net cathodic and anodic current densities are determined graphically from Fig. 7.21. The net current densities given in Table 7.2 are then plotted as open circles in Fig. 7.21 to give the experimental polarization curves.

Schematic polarization diagrams as in Fig. 7.21 are called *Evans diagrams*, after the noted corrosion scientist U.R. Evans (1889–1980), who was instrumental in their development and interpretation.

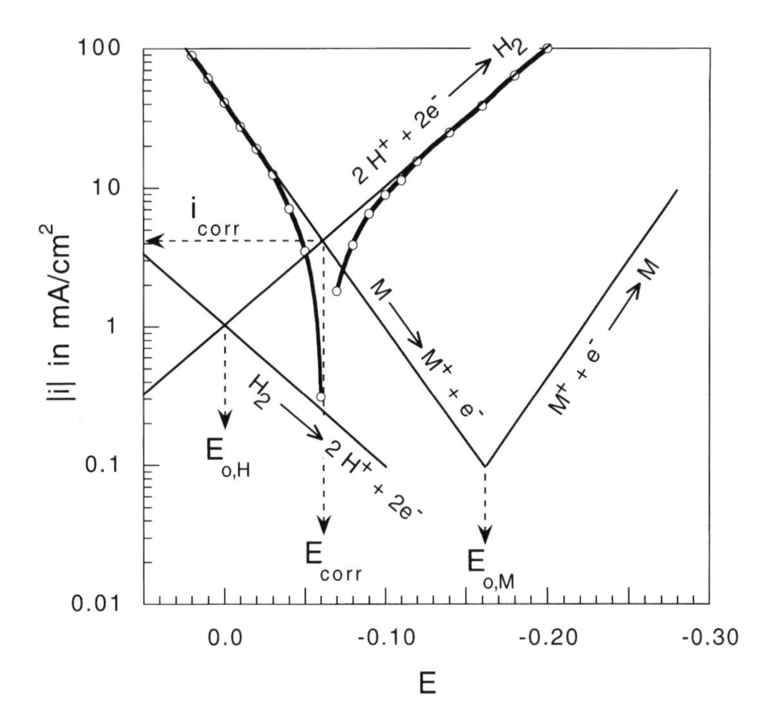

Fig. 7.21 Evans diagrams for the half cell reactions $M^+ + e^- \leftrightharpoons M$ and $2H^+ + 2e^- \leftrightharpoons H_2$, as described by the electrode kinetic parameters given in Example 7.2. Reproduced by permission of ECS – The Electrochemical Society

Table 7.2 Determination of the experimental polarization curve for Example 7.2

Cathodic: $|\vec{i}_{net}| = |\vec{i}_H| + |\vec{i}_M| - (\overleftarrow{i}_H + \overleftarrow{i}_M)$

E	$\vec{\lvert i_H \rvert}$ H^+reduction	$\vec{\lvert i_M \rvert}$ M deposition	$\overleftarrow{i}_H$ H_2 oxidation	$\overleftarrow{i}_M$ M dissolution	$\vec{\lvert i_{net} \rvert}$ Cathodic
–0.070	5.0	–	0.2	3.0	1.8
–0.080	6.2	–	0.16	2.1	3.9
–0.090	8.0	–	0.13	1.4	6.5
–0.100	10	–	0.10	1.0	8.9
–0.110	12	–	–	0.68	11.3
–0.120	16	–	–	0.45	15.5
–0.140	25	–	–	0.2	24.8
–0.160	39	–	–	0.1	38.9
–0.180	64	–	–	–	64
–0.200	100	–	–	–	100

Anodic: $\overleftarrow{i}_{net} = \overleftarrow{i}_H + \overleftarrow{i}_M - \left(|\vec{i}_H| + |\vec{i}_M|\right)$

E	$\overleftarrow{i}_H$ H_2 oxidation	$\overleftarrow{i}_M$ M dissolution	$\vec{\lvert i_H \rvert}$ H^+ reduction	$\vec{\lvert i_M \rvert}$ M deposition	$\overleftarrow{i}_{net}$ Anodic
–0.060	0.26	4.4	4.2	–	0.46
–0.050	0.32	6.4	3.2	–	3.5
–0.040	0.4	9.2	2.5	–	7.1
–0.030	0.5	14	2.0	–	12.5
–0.020	0.7	20	1.6	–	19.1
–0.010	0.8	28	1.3	–	27.5
0.000	1.0	41	1.0	–	41
+0.010	1.3	60	0.8	–	61
+0.020	1.6	88	0.6	–	89

Electrode Kinetic Parameters

Table 7.3 lists exchange current densities for the hydrogen evolution reaction on various metals as compiled by Trasatti [23]. The value of the exchange current density for a given metal depends on various experimental factors, such as the particular electrolyte used, its concentration, the purity of the metal, its surface cleanliness, the time of immersion in solution, and whether equilibrium conditions had been attained.

The exchange current density for hydrogen evolution increases with the work function of the metal, as shown in Fig. 7.22. The work function is a measure of the ability of the metal to expel (i.e., donate) electrons. Thus, the greater the electron-donating ability of the metal, the greater the rate of the electron-accepting reaction:

$$2\,H^+ + 2e^- \rightarrow H_2$$

The relationships in Fig. 7.22 were first observed by Conway and Bockris [24].

Table 7.3 Exchange current densities $i_{o,H}$ for the hydrogen evolution reaction [23]

Metal	$i_{o,H}$ (A/cm^2)	Metal	$i_{o,H}$ (A/cm^2)
Ag	1.3×10^{-8}	Ni	5.6×10^{-6}
Al	1.0×10^{-8}	Os	7.9×10^{-5}
Au	3.2×10^{-7}	Pb	4.0×10^{-12}
Bi	1.6×10^{-8}	Pd	7.9×10^{-4}
Cd	2.5×10^{-12}	Pt	1.0×10^{-3}
Co	5.0×10^{-6}	Re	1.0×10^{-3}
Cr	1.0×10^{-7}	Rh	3.2×10^{-4}
Cu	1.6×10^{-8}	Ru	6.3×10^{-5}
Fe	2.5×10^{-6}	Sb	7.9×10^{-6}
Ga	4.0×10^{-9}	Sn	1.6×10^{-8}
In	3.2×10^{-10}	Ta	3.0×10^{-9}
Ir	2.5×10^{-4}	Ti	5.0×10^{-9}
Mn	1.3×10^{-11}	Tl	2.5×10^{-10}
Mo	5.0×10^{-8}	W	4.0×10^{-7}
Nb	4.0×10^{-9}	Zn	3.2×10^{-11}

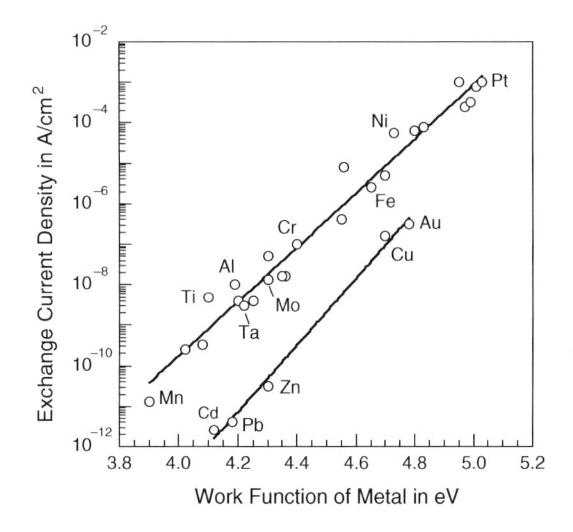

Fig. 7.22 The exchange current density for hydrogen evolution as a function of the work function of the metal. Most metals follow the trend shown in the upper set of data, but several metals follow the trend shown in the lower set of data. Plotted from data compiled by Trasatti [23]

Exchange current densities for the dissolution (or deposition) of several metals are listed in Table 7.4. In most cases, the numerical value of the exchange current density depends on the nature of the solution, the concentration of the dissolved cation, and the pH of the solution.

Cathodic Tafel slopes for the hydrogen evolution reaction have been usually observed to be near the theoretical value of –0.120 V/decade. This value is obtained from Eq. (36) with $\alpha = 0.5$ and $n = 1$ assuming a single electron transfer step of H^+ to H_{ads}, followed by the combination of two H_{ads} atoms to H_2, or by the following reaction [25]:

$$H_{ads} + H^+ + e^- \rightarrow H_2$$

Table 7.4 Exchange current densities $i_{o,M}$ for metal dissolution/metal deposition half-cell reactions

Metal	Solution	Exchange current density (A/cm^2)	Source
Zn	1 M $ZnSO_4$	2×10^{-5}	Bockris [25]
	$Zn(ClO_4)$ 0.057–0.46 M Zn^{2+}, 3 M ClO_4^-	3.5×10^{-4}	Clark and Hampson [26]
Cu	1 M $CuSO_4$	2×10^{-5}	Bockris [25]
	0.05–0.5 M $CuSO_4$ + 0.5 M H_2SO_4	7.0×10^{-3}	Despic [27]
Fe	1 M $FeSO_4$	10^{-8}	Bockris [25]
	1 M HCl	4.0×10^{-8} to 1.0×10^{-7}	Kaesche and Hackerman [28]
	4% NaCl (pH 1.5)	4.1×10^{-8}	Stern and Roth [29]
	4% NaCl (acidified)	1.0×10^{-7}	Stern [30]
	0.1 M citric acid	9.3×10^{-8}	Stern [30]
	0.1 malic acid	1.5×10^{-8}	Stern [30]
	$FeSO_4 + Na_2SO_4$ (pH = 3.1)	2.2×10^{-6}	Bockris and Drazic [31]
Ni	1 M $NiSO_4$	2.0×10^{-9}	Bockris [25]
Cd	0.049 M Cd^{2+}	2.3×10^{-3}	Hampson et al. [32]
	0.452 M Cd^{2+}	2.5×10^{-2}	
Pb	0.0027 M Pb^{2+}	2.62×10^{-2}	Hampson and Larkin [33]
	0.5 M Pb^{2+}	7.14×10^{-2}	
Ag	0.001 M Ag^+	0.15	Vetter [34]
	0.1 M Ag^+	4.5	

Experimental anodic Tafel slopes for the dissolution of various metals usually range from 0.040 to 0.080 V/decade. Chapter 16 deals with electrode kinetics and reaction mechanisms in more detail. See Problem 7.6 at the end of this chapter.

Applications of Mixed Potential Theory

Metals in Acid Solutions

One of the first applications of mixed potential theory was the dissolution of metals in acid solutions. Figure 7.23 considers the case of zinc immersed in hydrochloric acid. The overall chemical reaction

$$Zn + 2H^+ \rightarrow Zn^{2+} + H_2 \tag{50}$$

can be separated into its two half-cell reactions:

$$Zn \rightarrow Zn^{2+} + 2e^- \tag{51}$$

and

$$2H^+ + 2e^- \rightarrow H_2 \tag{52}$$

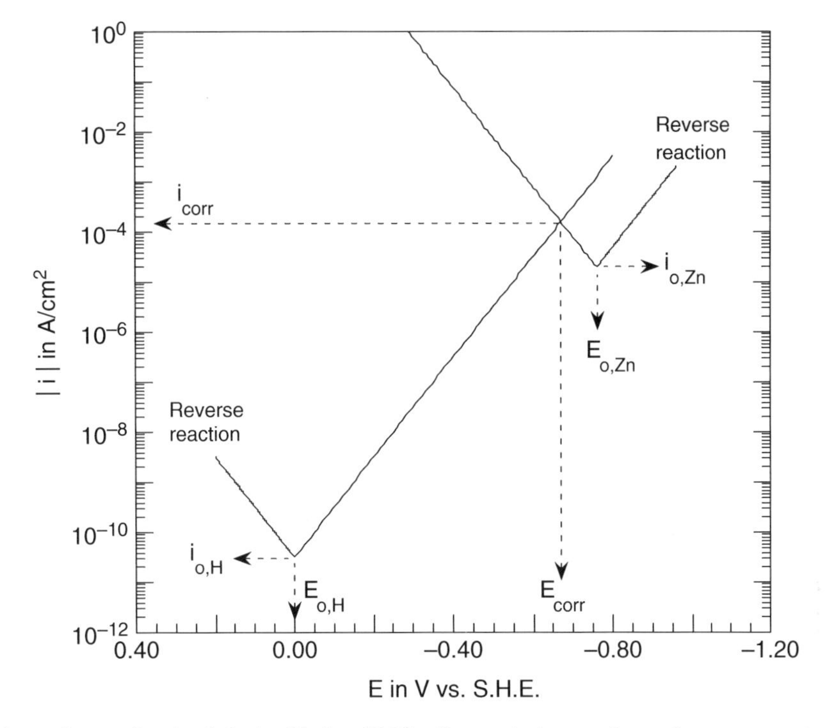

Fig. 7.23 Evans diagram for zinc in hydrochloric acid. The diagram is drawn using exchange current densities from Tables 7.3 and 7.4. Tafel slopes are taken to be ± 100 mV/decade for H_2 evolution and deposition and ± 60 mV/decade for zinc dissolution and deposition

Each half-cell reaction proceeds under its own set of polarization curves with its own individual open-circuit potential E_o and its own exchange current density i_o. For the half-cell reaction in Eq. (52), i_o is $i_{o,H}$, as shown in Fig. 7.23, and is taken from Table 7.3 to be 3.2×10^{-11} A/cm^2 for hydrogen evolution on zinc [23]. The open-circuit equilibrium potential for Eq. (52) is exactly the standard electrode potential E^o only if a(H^+) = 1.0 and H_2 gas is present at 1.0 atm pressure. Tafel slopes are taken to be ±100 mV/decade for H_2 evolution and deposition.

Similarly, the half-cell reaction in Eq. (51) for zinc dissolution/deposition has its own exchange current density $i_{o,Zn}$, taken in Fig. 7.23 to be 2.0×10^{-5} A/cm^2 (see Table 7.4). Again, the equilibrium open-circuit potential E_o is exactly the standard electrode potential E^o for zinc if a(Zn^{2+}) = 1.0. Tafel slopes are taken to be ±60 mV/decade for metal dissolution and deposition. As in Example 7.2, the polarization curve for the total system is given by the superposition of the polarization curves for the two half-cell reactions.

From Fig. 7.23 it can be seen that the total cathodic current density at any electrode potential is essentially that due to the reaction

$$2H^+ + 2e^- \rightarrow H_2 \qquad (53)$$

(On a logarithmic scale, the contributions to the total cathodic density due to the reaction Zn^{2+} + $2e^-$ → Zn are negligible.) Similarly, from Fig. 7.23, the total anodic current density is essentially that due to the reaction

$$Zn \rightarrow Zn^{2+} + 2e^- \quad (54)$$

(Again, on a logarithmic scale, the contributions to the total anodic density due to the reaction $H_2 \rightarrow 2H^+ + 2e^-$ are negligible.)

At the intersection of the polarization curves for reactions (53) and (54), the total anodic current density is equal to the total cathodic current density. Thus, this intersection defines the corrosion potential E_{corr} and the corrosion current density i_{corr}.

The experimentally determined polarization curves are given by the net anodic and cathodic polarization curves, as in Example 7.2.

Tafel Extrapolation

The Tafel extrapolation method can be used to determine the corrosion rate of a metal when metallic dissolution is under activation control. The most common application is for metals immersed in de-aerated acid solutions for which the anodic reaction is

$$M \rightarrow M^{n+} + ne^-$$

and the cathodic reaction is

$$2\,H^+ + 2e^- \rightarrow H_2$$

De-aeration of the solution restricts the cathodic reaction to hydrogen evolution alone, rather than also including the cathodic reduction of oxygen. Also, in de-aerated acid solutions, oxide films initially present on the metal surface are dissolved by the acid solution en route to attainment of the steady-state, open-circuit potential. Thus, the sole anodic reaction is the dissolution of the bare metal surface.

Figure 7.24 shows experimental anodic and cathodic polarization curves for iron immersed in HCl solutions of various concentrations [7]. These curves were obtained after a steady-state, open-circuit potential had been first obtained. If a well-defined Tafel region exists, as in Fig. 7.24, the anodic and cathodic Tafel regions can be extrapolated back to zero overvoltage. The intersection of the anodic and cathodic Tafel slopes gives the corrosion potential E_{corr} and the corrosion current density i_{corr}, as indicated in Fig. 7.24.

The basis for the Tafel extrapolation method stems from Eq. (49). When $E = E_{corr}$, then $i = i_{corr}$. The contribution of the back reaction to the forward reaction is usually negligible for overvoltages ranging from 60 to 120 mV. (See Problem 7.9.) That is, linear Tafel regions are usually observed for overvoltages 59–120 mV away from the open-circuit potential.

The Tafel extrapolation method is valid if the following conditions apply:

(1) Both the anodic and the cathodic branches of the polarization curves are under activation control. (Concentration polarization may occur toward the tail end of the Tafel regions, but this is a minor effect. For instance, for high anodic dissolution rates, the accumulation of dissolved cations near the electrode surface may cause a concentration effect, which is manifested as a slight deviation from the anodic Tafel region at high anodic overvoltages. Stirring the solution can minimize concentration polarization effects.)
(2) Well-defined anodic and cathodic Tafel regions exist (over at least one decade of current).

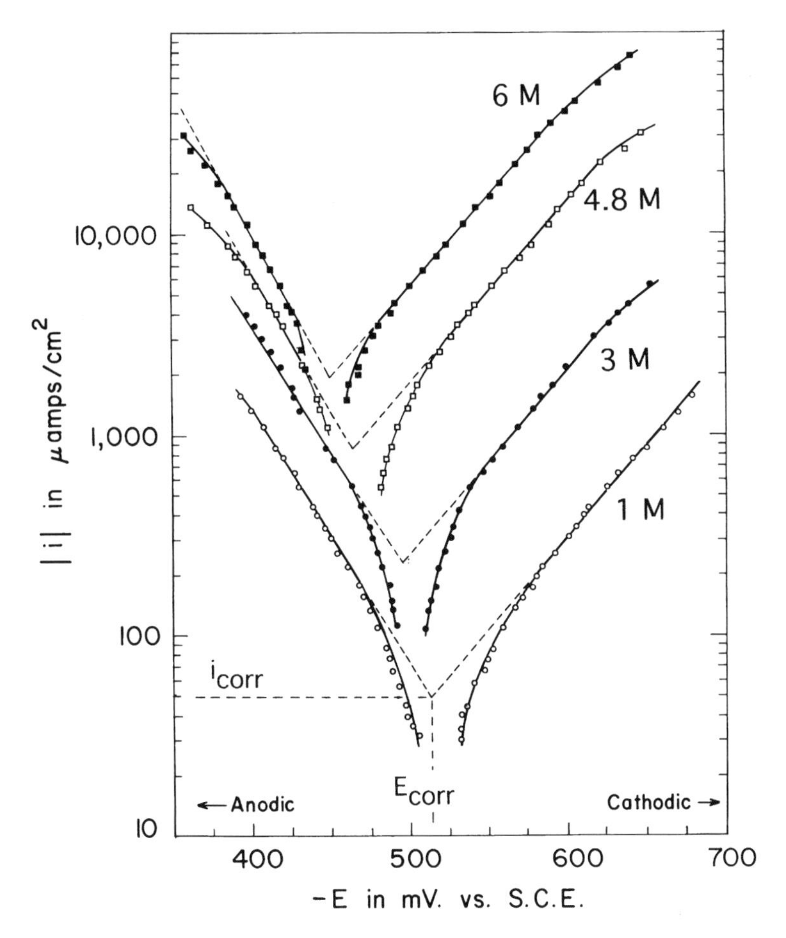

Fig. 7.24 Experimental polarization curves for iron immersed in hydrochloric acid of various concentrations (which are indicated on the figure) [7]. The corrosion potential E_{corr} and the corrosion rate i_{corr} are indicated for iron in 1 M HCl. Reproduced by permission of ECS – The Electrochemical Society

(3) The anodic and cathodic reactions which occur at the corrosion potential are also the only reactions which occur during determination of the polarization curves. That is, changes in electrode potential should not induce additional electrochemical reactions in either the anodic or the cathodic direction. (This also means that corrosion product films are not formed at anodic potentials.)

(4) Corrosion is general (i.e., uniform) in nature, and localized corrosion does not occur. (For instance, rigorous preferential attack of the metal along grain boundaries may cause individual grains of the metal to become dislodged from the metal surface. These dislodged grains would continue to produce dissolved metal ions in solution but are disconnected from the electrochemical circuit so as not to contribute to the corrosion rate measured by the Tafel method. This phenomenon is called the "chunk effect" [35].)

The corrosion rate can also be determined by Tafel extrapolation of *either* the cathodic or the anodic polarization curve alone. If only one polarization curve alone is used, it is generally the cathodic curve which usually produces a longer and better defined Tafel region. Anodic polarization may sometimes produce concentration effects, as noted above, as well as roughening of the surface which can lead to deviations from Tafel behavior. Thus, extrapolation of the cathodic Tafel region back to zero overvoltage gives the net rate of the cathodic reaction at the corrosion potential; but from Eq. (42), this is also the net rate of the anodic reaction at the corrosion potential. Use of both the anodic and the cathodic Tafel regions is preferred, of course, over the use of just one Tafel region.

Verification of Corrosion Rates Obtained by Tafel Extrapolation

It is now widely understood and accepted that the rate of general corrosion can be determined by the Tafel extrapolation method (subject to the restrictions listed earlier). However, this acceptance was not forthcoming until the theory of polarization curves for corroding metals had been placed on a firm theoretical basis, as was done by the work of many individual investigators [15, 17, 22, 25, 28–30, 36–40]. Final acceptance of the polarization method was gained upon an experimental validation of the corrosion rates determined electrochemically. This validation was provided by comparison of corrosion rates determined from polarization curves with corrosion rates measured by an independent non-electrochemical method, such as weight loss measurements, colorimetric analysis of the solution containing the dissolved metal, or determination of the amount of hydrogen gas evolved by the corrosion reaction (for metals in acid solutions).

Table 7.5 compares corrosion rates determined electrochemically by the Tafel extrapolation method for iron in various solutions with corrosion rates determined separately by a second non-electrochemical method. Data are also given for the corrosion of other metals in various solutions. Weight loss measurements or solution analyses are converted into equivalent current densities using Faraday's law. For gas evolution measurements, the equivalent current density follows from the molar volume of a gas, 22,400 ml at standard temperature and pressure (0°C and 760 Torr), and Faraday′s law. In each case in Table 7.5, there is a good agreement between the results for electrochemical and non-electrochemical measurements.

Cathodic Protection of Iron in Acids

Chapter 5 considered the cathodic protection of iron by zinc from a thermodynamic point of view. In seawater, for example, the electrode potential of zinc is more negative than that of iron. Thus, when zinc and iron are coupled, zinc is the anodic member of the couple and the iron electrode in the couple is cathodically protected.

This phenomenon of cathodic protection also applies in acid solutions and can be examined by considering the polarization curves for the zinc/acid and iron/acid systems. Figure 7.25 compares schematic Evans diagrams for iron and zinc in hydrochloric acid. The Evans diagram for iron is constructed using $i_{o,H} = 2.5 \times 10^{-6}$ A/cm^2 (Table 7.3) and taking $i_{o,M}$ to be 1.0×10^{-7} A/cm^2 (Table 7.4).

It can be seen that the corrosion rate of iron and of zinc is approximately the same, even though the electrode potential of zinc is more active (i.e., more negative) than that of iron. The exchange current density for hydrogen evolution on iron is higher than that on zinc. However, the exchange current density for zinc dissolution $i_{o,Zn}$ is higher than that for iron dissolution $i_{o,Fe}$ (see Table 7.4). These differences combine to produce corrosion rates which are similar for iron and zinc in this particular acid solution. This observation illustrates the statement made in Chapter 5 that the emf series or galvanic series cannot accurately predict the relative order of corrosion rates but can only identify the anodic and cathodic members of a galvanic couple.

Table 7.5 Comparison of corrosion rates determined by the Tafel extrapolation method with corrosion rates determined by a second method [41]

	Environment	Immersion time	i_{corr} Tafel method	i_{corr} Second method	Type of second method	Reference
Iron	1 M HBr	120 h	23 μA/cm^2	16 μA/cm^2	Colorimetric	McCafferty [6]
	1 M HI	110 h	10 μA/cm^2	7 μA/cm^2	Colorimetric	
Iron	1 M HCl	24 h	30 μA/cm^2	36 μA/cm^2	Colorimetric	McCafferty [8]
Iron	1 M HCl	8 h	50 μA/cm^2	50 μA/cm^2	Colorimetric	Kaesche [28]
Iron	O_2-free 4% NaCl, pH 1	100 h	10.5 μA/cm^2	11.1 μA/cm^2	Weight loss	Stern [40]
Fe + 0.11% Cu	O_2-free 4% NaCl, pH 1	100 h	154 μA/cm^2	156 μA/cm^2	weight loss	Stern [40]
Fe + 0.08% Si	O_2-free 4% NaCl, pH 1	100 h	270 μA/cm^2	290 μA/cm^2	Weight loss	Stern [40]
Fe + 0.017% P	O_2-free 4% NaCl, pH 1	100 h	390 μA/cm^2	400 μA/cm^2	Weight loss	Stern [40]
Zinc	10% NH_4Cl	–	100 μA/cm^2	100 μA/cm^2	AAS of solution[a]	Maja et al. [42]
Titanium	Boiling 1 M H_2SO_4	2.5	1.9 mA/cm^2	2.2 mA/cm^2	Colorimetric	McCafferty [41]
				4.0 mA/cm^2	Inert marker	
Al alloy 3003	Acetate buffer, pH 4.6	1 week	8.2 μA/cm^2	9.1 μA/cm^2	H_2 evolution	Evans and Koehler [43]
	Citrate buffer, pH 2.5		1.2 μA/cm^2	1.3 μA/cm^2	H_2 evolution	
	Prune juice		1.4 μA/cm^2	1.5 μA/cm^2	H_2 evolution	
Al alloy 3003	0.1 M HCl	24 hrs.	31 μA/cm^2	31 μA/cm^2	Colorimetric	McCafferty, Moore, and Peace [9]
	1.0 M HCl	4 hrs.	7.5 mA/cm^2	7.3 mA/cm^2	Colorimetric	
Copper	0.5 M H_2SO_4 25 °C	72 hrs	25.7 μA/cm^2	22.5 μA/cm^2	Weight loss	Quartarone [44]
	55 °C	72 hrs.	54.6 μA/cm^2	39.3 mA/cm^2	Weight loss	
Fe-14 Cr	1 N H_2SO_4 30 °C	20 min.	0.438 mA/cm^2	0.479 mA/cm^2 [b]	Weight loss	Yau and Streicher [45]

[a] Atomic absorption spectrophotometry.

[b] Assumes divalent dissolution of each metal and stoichiometric dissolution of each component of the binary alloy on an atomic fraction basis.

When iron and zinc are coupled in hydrochloric acid, the situation is shown in Fig. 7.26. The total cathodic current density for the Fe/Zn couple at any electrode potential is essentially that for hydrogen evolution on iron, because contributions from other reduction reactions are negligible on a logarithmic basis. Similarly, the total anodic current density for the Fe/Zn couple at any electrode potential is essentially that for zinc dissolution, because contributions from other oxidation reactions are negligible on a logarithmic scale. Thus, the corrosion potential for the Fe/Zn couple in acid solutions and the corresponding corrosion current for zinc in the couple are given by the intersection of the polarization curves for the reactions

$$2\,H^+ + 2e^- \rightarrow H_2\,(\text{on iron})$$

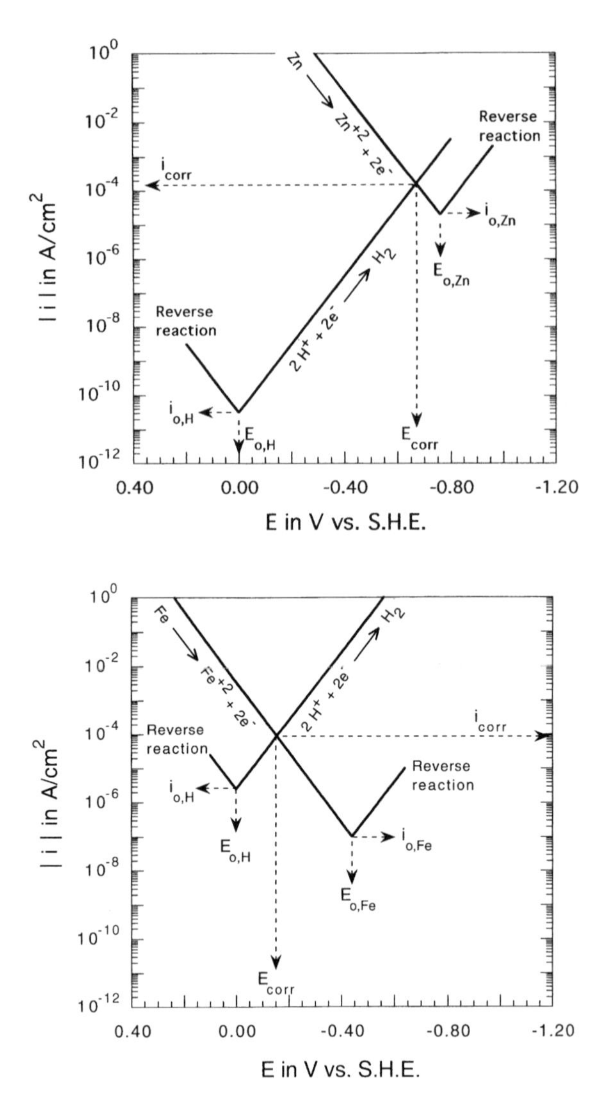

Fig. 7.25 Comparison of Evans diagrams for zinc and iron in hydrochloric acid. The diagrams are drawn using exchange current densities from Tables 7.3 and 7.4. Tafel slopes are taken to be ± 100 mV/decade for H_2 evolution and deposition and ± 60 mV/decade for metal dissolution and deposition

and

$$Zn \rightarrow Zn^{2+} + 2e^-$$

The intersection point is indicated in Fig. 7.26.

Figure 7.26 also shows the corrosion potential and the corrosion current density for iron alone (uncoupled) and for zinc alone (uncoupled).

Figure 7.26 shows the following:

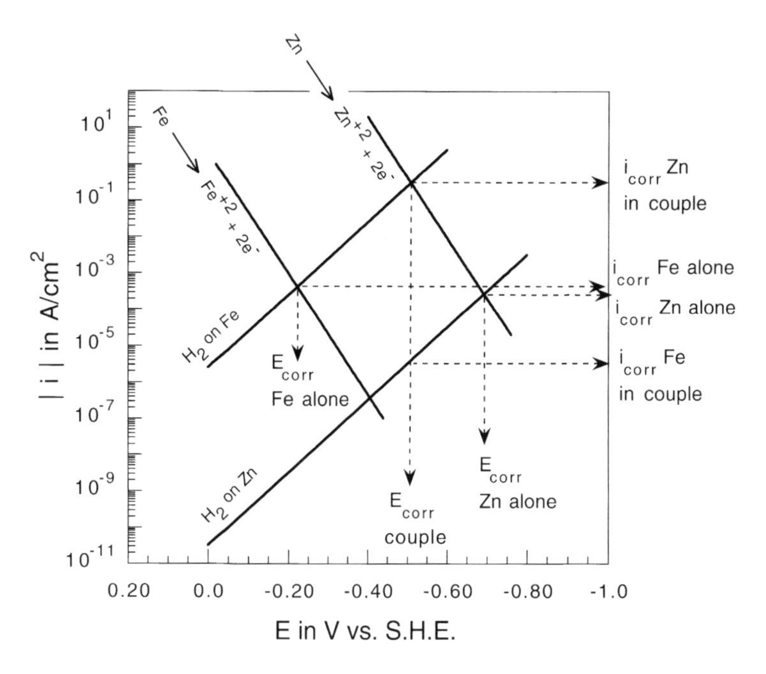

Fig. 7.26 Kinetic explanation of the cathodic protection of iron by zinc in an acid solution

(1) The corrosion potential for the Fe/Zn couple is intermediate between the corrosion potentials for each of the two uncoupled electrodes.
(2) The corrosion current density for zinc in the Fe/Zn couple is *greater* than the corrosion current density of uncoupled zinc.
(3) The corrosion current density for iron in the Fe/Zn couple is *less* than the corrosion current density of uncoupled iron.

Thus, mixed potential theory shows that the coupling of iron to zinc in acid solutions results in the cathodic protection of iron by zinc. See Fig. 7.27 for a schematic representation and compare this figure with Fig. 5.5.

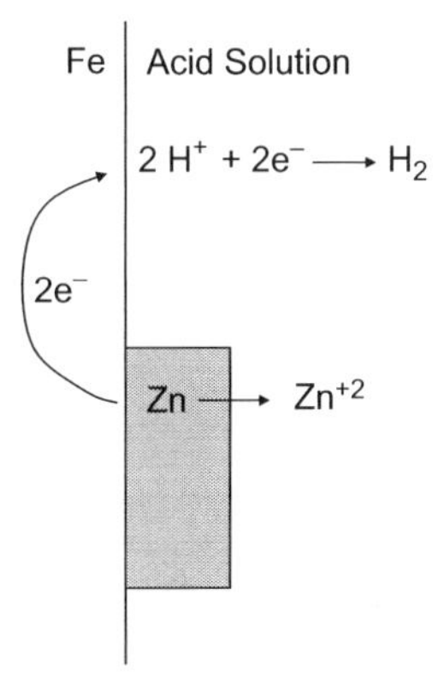

Fig. 7.27 Schematic illustration of cathodic protection of iron by zinc in an acid solution showing the principal electrode reactions

Effect of the Cathodic Reaction

The previous section has shown that the corrosion rate of zinc in the Zn/Fe couple is greater than that of uncoupled zinc because the exchange current density $i_{o,H}$ for the cathodic reaction (hydrogen evolution) is greater on iron than on zinc. As seen in Table 7.3, the exchange current density for hydrogen evolution on Pt, Fe, and Zn follows the following order:

$$i_{o,H}(pt) > i_{o,H}(Fe) > i_{o,H}(Zn)$$

Accordingly, the Evans diagrams shown in Fig. 7.28 show that the corrosion rates follow the following order:

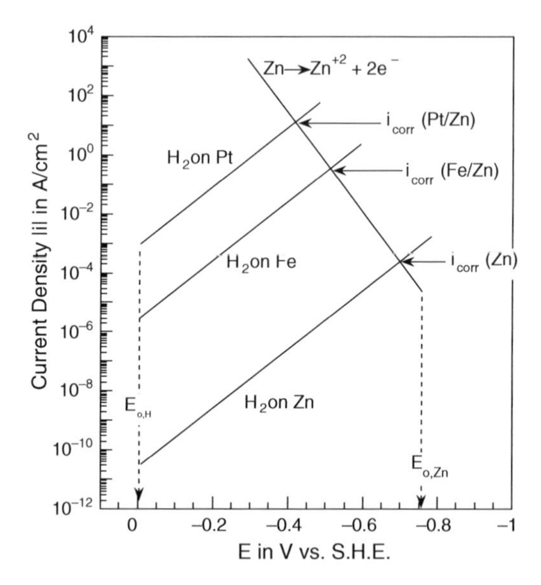

Fig. 7.28 Effect of the cathode metal on the corrosion rate of zinc in a galvanic couple in an acid solution. (The cathodic Tafel slope is taken to be –100 mV/decade for H_2 evolution and the anode slope is +60 mV/decade for zinc dissolution)

$$i_{corr}(\text{Zn coupled to pt}) > i_{corr}(\text{Zn coupled to Fe}) > i_{corr}(\text{Zn alone})$$

Thus, the greater the rate of the hydrogen evolution reaction on the cathode, the greater the corrosion rate.

Effect of Cathode Area on Galvanic Corrosion

The relative areas of cathode and anode are important in galvanic behavior. Figure 7.29 illustrates this effect on a galvanic couple of iron and zinc. As noted previously, when iron and zinc are coupled, the iron member of the couple is the cathode and zinc is the anode.

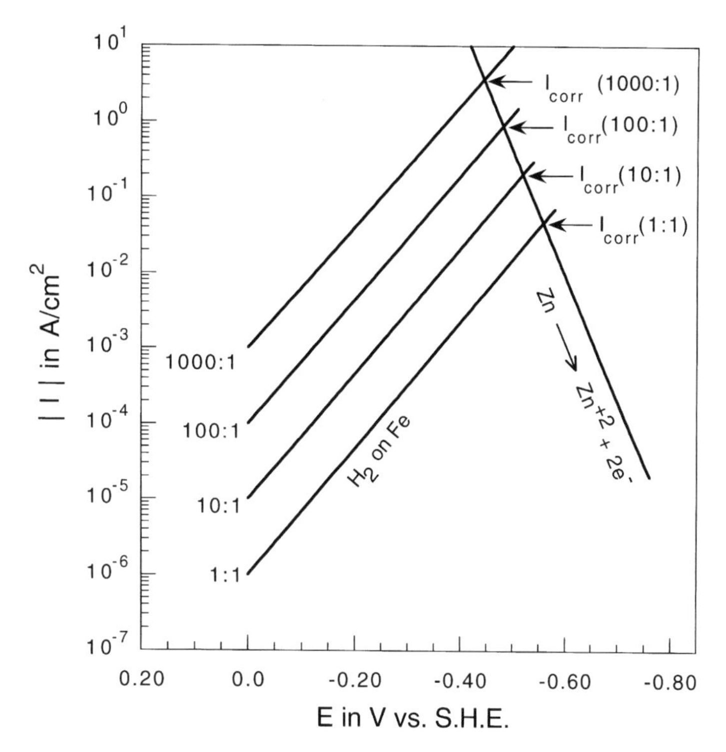

Fig. 7.29 Effect of the area ratios of cathode to anode on the corrosion rate of zinc in an iron/zinc couple in an acid solution (The cathodic Tafel slope is again taken to be –100 mV/decade for H_2 evolution and the anode slope is again +60 mV/decade for zinc dissolution)

The cathodic Tafel slope is again taken to be –100 mV/decade for H_2 evolution on iron, and the anode slope for zinc dissolution is again +60 mV/decade. The exchange current density for hydrogen evolution on iron is 1.0×10^{-6} A/cm^2 (see Table 7.3) but the exchange current density for hydrogen on zinc is only 3.2×10^{-11} A/cm^2 (also Table 7.3). Thus, the total current due to hydrogen evolution is due to that on the iron member (the cathode) of the couple. The exchange current density for zinc dissolution/deposition is 2.0×10^{-5} A/cm^2 (Table 7.4). Total current rather than current density is used in Fig. 7.29, but if 1.0 cm^2 of zinc is considered, the total current and the current density for the zinc member of the couple are numerically identical.

As seen in Fig. 7.29, when the cathode/anode area ratio is 1:1, the current density for the corroding iron member is 4.0×10^{-2} A/cm^2. If the cathode/anode area ratio is increased to 10:1, the total current for the couple is also increased, and the current density on the corroding zinc member is 1.8×10^{-1} A/cm^2. Figure 7.29 shows that the corrosion current density on the zinc continues to increase with increasing area of the iron cathode in the couple.

Figure 7.30 shows that the logarithm of the corrosion current density increases linearly with the logarithm of the area ratio of cathode to anode. Verification of this relationship is left as an exercise in Problem 7.10.

The situation involving a small anode and a large cathode is unfavorable in the case of a noble coating on an active substrate, e.g., for a copper coating on steel, as shown in Fig. 7.31. When a break or a pore develops in the noble metal coating, then a small anode (steel) is exposed to a larger area cathode (copper), and the anodic current density increases with the area of the cathode, as shown above and mentioned previously in Chapter 5.

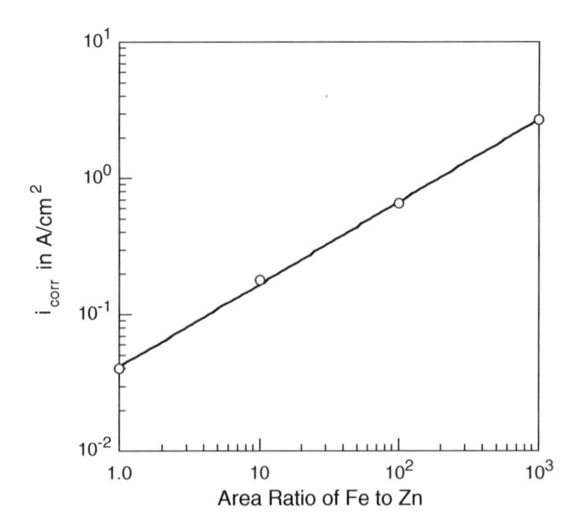

Fig. 7.30 The corrosion current density in an acid solution for 1 cm^2 of zinc coupled to various surface areas of iron

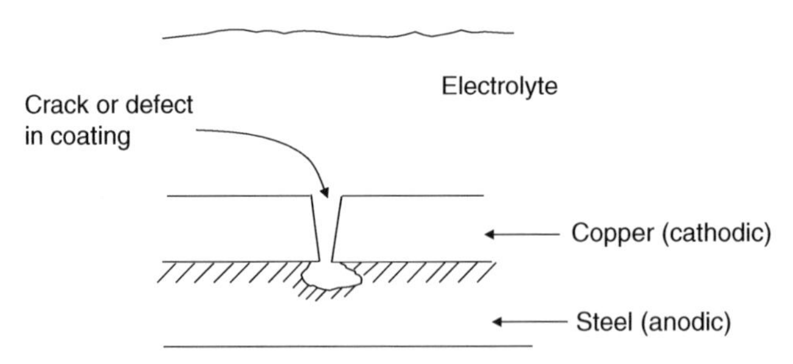

Fig. 7.31 Schematic diagram of a crack or a defect in a noble metal coating showing accelerated attack at the anodic substrate

Multiple Oxidation–Reduction Reactions

Systems with more than two oxidation–reduction reactions can also be treated by the Wagner–Traud theory of mixed potentials. The example given in Fig. 7.32 is taken from Stern [39] for a system of three oxidation–reduction reactions involving species H^+/H_2, M^{2+}/M, and Y^{2+}/Y. The electrode kinetic parameters describing this system of three oxidation–reduction reactions are given in Table 7.6. It can be seen in Fig. 7.32 that the total anodic current density is essentially that due to the reaction

$$M \rightarrow M^{2+} + 2e^-$$

(because the contributions to the total current density from the other two anodic half-cell reactions are negligible). The total cathodic current density indicated by the dotted line in Fig. 7.32 is the sum of the current densities due to the following two cathodic reactions:

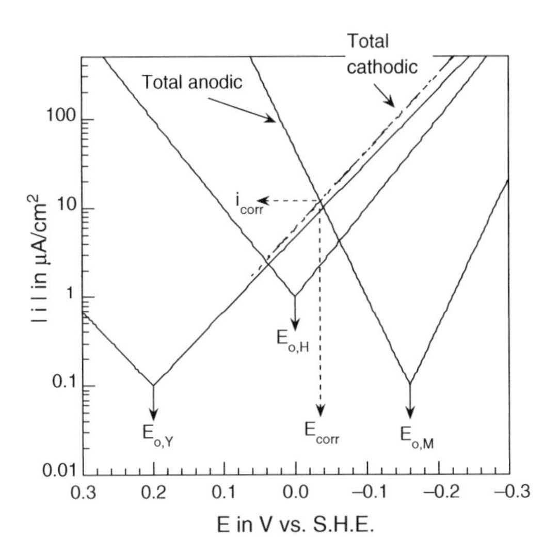

Fig. 7.32 Evans diagrams for a system of three oxidation–reduction reactions having the electrode kinetic parameters given in Table 7.6. The total anodic curve is essentially that for the reaction $M \rightarrow M^{+2} + 2e^-$, but the total cathodic curve (dotted line) is given by the sum of the curves for the two half-cell reactions $Y^{2+} + 2e^- \rightarrow Y$ and $2H^+ + 2e^- \rightarrow H_2$. The intersection of the total anodic and total cathodic curves defines the corrosion potential E_{corr} and the corrosion rate i_{corr}, as shown. Redrawn from [39] by permission of ECS – The Electrochemical Society

Table 7.6 Electrode kinetic parameters for the three oxidation–reduction reactions considered in Fig. 7.32 [39]

System	E_o V vs. SHE	i_o (μA/cm^2)	b_a (V/decade)	b_c (V/decade)
$2H^+ + 2e^- \rightleftharpoons H_2$	0.000	1.0	+0.100	–0.100
$M^{2+} + 2e^- \rightleftharpoons M$	–0.160	0.1	+0.060	–0.060
$Y^{2+} + 2e^- \rightleftharpoons Y$	+0.200	0.1	+0.120	–0.120

$$2H^+ + 2e^- \rightarrow H_2$$

and

$$Y^{2+} + 2e^- \rightarrow Y$$

(The contribution to the total cathodic current density from the M^{2+}/M system is negligible). The corrosion potential and the corrosion rate are then given by the intersection of the total anodic and the total cathodic curves, as shown in Fig. 7.32.

Figure 7.33 shows the experimentally observed polarization curve corresponding to the Evans diagrams in Fig. 7.32. The experimentally observed polarization curve is determined in the usual way. That is, the net anodic rate i_a at any anodic overvoltage is again the sum of all the anodic current densities minus all the cathodic current densities:

$$i_a = i_{net} = (\overleftarrow{i}_H + \overleftarrow{i}_M + \overleftarrow{i}_Y) - (|\overrightarrow{i}_H| + |\overrightarrow{i}_M| + |\overrightarrow{i}_Y|) \tag{55}$$

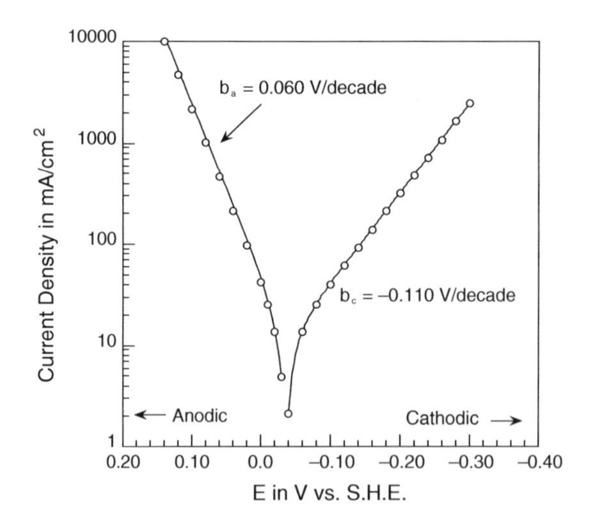

Fig. 7.33 Experimental polarization curve corresponding to the Evans diagrams in Fig. 7.32. Note that the experimental cathodic Tafel slope is intermediate between the values for b_H and b_Y in the Evans diagrams

Similarly, the net cathodic rate i_c at any cathodic overvoltage is the sum of all the cathodic current densities minus all the anodic current densities:

$$i_c = \vec{i}_{net} = \left(| \vec{i}_H | + | \vec{i}_M | + | \vec{i}_Y | \right) - \left(\overleftarrow{i}_H + \overleftarrow{i}_M + \overleftarrow{i}_Y \right) \tag{56}$$

A Tafel region is observed for both anodic and cathodic plots, but a special feature of this case is that the cathodic Tafel slope is intermediate between the values for b_H and b_Y.

An example of a system with three simultaneous oxidation-reduction reactions is the corrosion of stainless steels in acid solutions containing Fe^{3+} ions [39], for which the additional oxidation–reduction reaction is

$$Fe^{3+}(aq) + e^- \rightarrow Fe^{2+}(aq)$$

Industrial acids may contain dissolved ions such as Fe^{3+} due to corrosion or contamination.

Anodic or Cathodic Control

Corrosion reactions can be classified as being under anodic, cathodic, or mixed control. When polarization occurs mostly at the local anodes, the corrosion reaction is said to be under *anodic control.* This situation is illustrated in Fig. 7.34(a), in which small changes in the current density result in larger changes in the anode potential rather than in the cathode potential. An experimental example is passivation of various metals by chromate inhibitors.

When polarization occurs mostly at the local cathodes, as shown in Fig. 7.34(b), the corrosion reaction is under *cathodic control.* An example is the corrosion of iron in natural waters, where the cathodic reaction given by

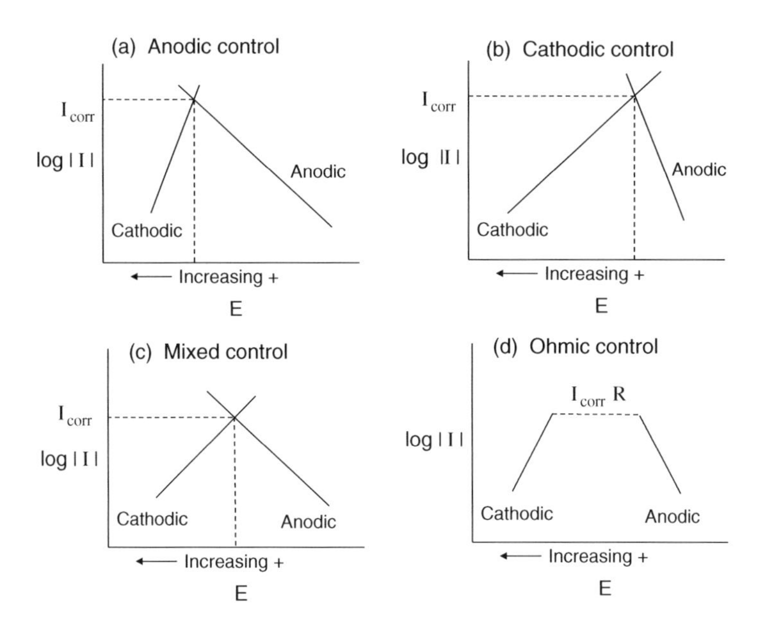

Fig. 7.34 (**a**) Anodic control, (**b**) cathodic control, (**c**) mixed control, and (**d**) ohmic control in a corrosion process

$$O_2 + 2H_2O + 4e^- \rightarrow 4OH^-$$

occurs under diffusion control, as discussed in Chapter 8.

It is common for polarization to occur to some degree at both local anodes and cathodes, in which case the corrosion process is under *mixed control*, as in Fig. 7.34(c). If there is a large IR drop between local anodes and local cathodes, as in Fig. 7.34(d), then the corrosion reaction is under *resistance* or *ohmic control*. Examples include the corrosion of metals in organic media, or where there is a porous insulating coating covering a metal surface [46], or in narrow crevices or stress-corrosion cracks, within which IR drops can exist.

Most corrosion reactions are under cathodic or mixed control, and anodic control is less common.

The Linear Polarization Method (Stern and Geary)

One of the problems with the Tafel extrapolation method is that the surface of the sample is often changed considerably after extensive polarization. The cathodic branch of the polarization curve should be measured first, because cathodic polarization is usually non-destructive. However, if there is considerable uptake of hydrogen atoms by the metal during hydrogen evolution, the steady-state corrosion potential may not be regained after the polarizing potential has been turned off. More serious is the possibility that after extensive anodic polarization, the metal surface may become etched and roughened, sometimes displaying its grain structure. These changes do not lend the Tafel extrapolation method to re-measurement of the corrosion rate of a given individual specimen at some later time of immersion.

These problems are overcome by the linear polarization method of Stern and Geary [37]. This method depends on the fact that the observed current density is a linear function of the applied

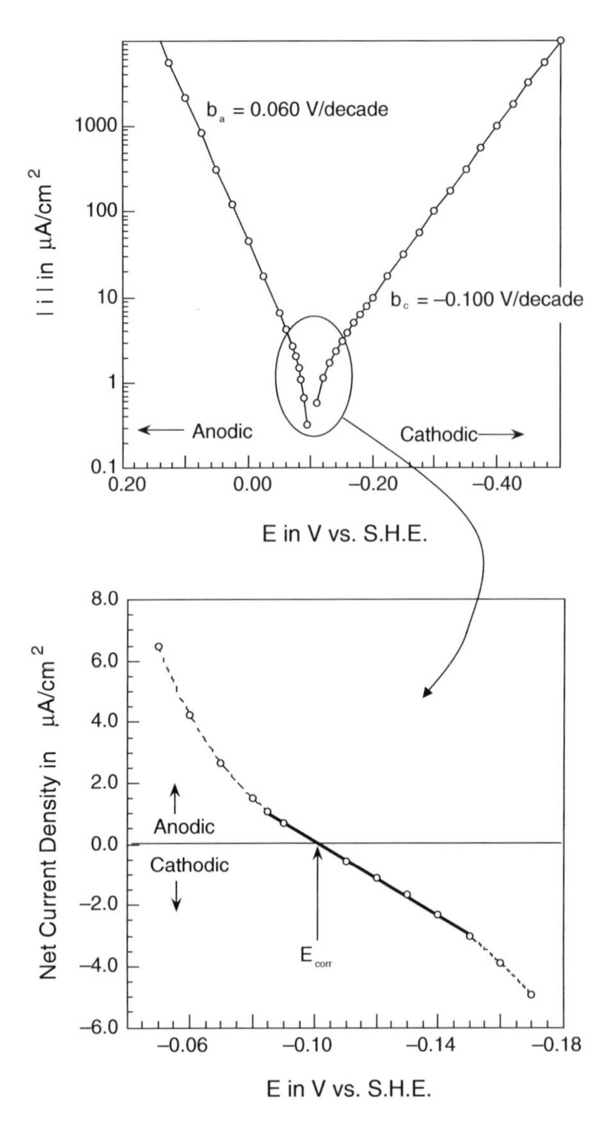

Fig. 7.35 *Top*: An experimental polarization curve for which E_{corr} = –0.100 V vs. SHE, i_{corr} = 1 μA/cm^2, b_a = +0.060 V/decade, and b_c = –0.100 V/decade. *Bottom*: Plot of the same polarization curve near the corrosion potential. Note the linear region near E_{corr}

potential near the corrosion potential. Fig. 7.35 (top) shows the usual semi-logarithmic polarization curves for a metal having the following parameters: E_{corr} = –0.100 V, i_{corr} = 1 μA/cm^2, b_a = 0.060 V/decade, and b_c = –0.100 V/decade. Figure 7.35 (bottom) shows linear plots near the corrosion potential. It can be seen that there is a limited range over which the current density is a linear function of the electrode potential (50 mV in the cathodic direction and only 15 mV in the anodic direction for this example). The extent of the linear region depends on the values of b_a and b_c and is symmetric about E_{corr} if the anodic and the cathodic Tafel slopes are numerically equal, as in Fig. 7.18.

Simmons [47] and Skold and Larson [48] were among the first to observe linear regions in a polarization curve near the corrosion potential. For a study of steel and cast iron in waters containing NaCl and $NaHCO_3$, Skold and Larson [48] also showed empirically that the slope $\Delta E/\Delta i$ of the linear polarization region was correlated to the corrosion rate determined by weight loss measurements. The greater the parameter $\Delta E/\Delta i$, the less the corrosion rate.

A theoretical analysis has been provided by Stern and Geary [37], and the linear polarization method is based on the following derivation. For small overvoltages, the exponential terms in the Butler–Volmer equation for a corroding metal can be expanded using a MacLaurin series.

For any small variable x

$$\begin{aligned} e^{x} &= 1 + x + \tfrac{x^2}{2!} + \ldots \\ e^{-x} &= 1 - x + \tfrac{x^2}{2!} + \ldots \end{aligned} \tag{57}$$

The use of Eq. (57) in Eq. (49) and dropping second-order terms gives

$$\overleftarrow{i}_{\mathrm{net}} = i_{\mathrm{corr}} \left\{ 1 + \frac{\alpha\, n\, F}{RT}\eta - \left[1 - \frac{(l-\alpha)\, n\, F}{RT}\eta \right] \right\} \tag{58}$$

or

$$\overleftarrow{i}_{\mathrm{net}} = i_{\mathrm{corr}} \left[\frac{\alpha\, n\, F}{RT} + \frac{(l-\alpha)\, n\, F}{RT} \right] \eta \tag{59}$$

so that

$$\overleftarrow{i}_{\mathrm{net}} = 2.303\, i_{\mathrm{corr}} \left[\frac{1}{b_{\mathrm{a}}} + \frac{1}{-b_{\mathrm{C}}} \right] \eta \tag{60}$$

Solving for η and taking differentials gives

$$\left(\frac{\mathrm{d}\,\eta}{\mathrm{d}\,\overleftarrow{i}_{\mathrm{net}}} \right)_{\eta \to 0} = \frac{1}{2.303\, i_{\mathrm{corr}}\left(\frac{1}{b_{\mathrm{a}}} + \frac{1}{|b_{\mathrm{c}}|}\right)} \tag{61}$$

(Recall that b_{c} has a negative sign.) Solving for i_{corr} gives

$$i_{corr} = \frac{1}{2.303 \left(\frac{d\eta}{d\,\overleftarrow{i}_{\mathrm{net}}} \right)_{\eta \to 0} \left(\frac{1}{b_{\mathrm{a}}} + \frac{1}{|b_{\mathrm{c}}|}\right)} \tag{62}$$

The differential in Eq. (61) is called the polarization resistance R_{p}, because it has the units of resistance

$$\left(\frac{\mathrm{d}\,\eta}{\mathrm{d}\,\overleftarrow{i}_{\mathrm{net}}} \right)_{\eta \to 0} \equiv R_{\mathrm{p}} \tag{63}$$

so that Eq. (62) can be written as

$$i_{\mathrm{corr}} = \frac{1}{2.303\, R_{\mathrm{p}} \left(\frac{1}{b_{\mathrm{a}}} + \frac{1}{|b_{\mathrm{c}}|}\right)} \tag{64}$$

which is the *Stern–Geary equation* . An expression similar to Eq. (64) holds for linear polarization in the cathodic direction except that the term $d\eta/di$ (i.e., R_p) is evaluated for a net cathodic current near the corrosion potential.

Thus, the corrosion rate can be determined from Eq. (64) by polarization measurements close to the corrosion potential. The required Tafel slopes are usually measured in separate experiments, although methods [49] have been developed to estimate b_a and b_c from polarization measurements made near the corrosion potential.

Example 7.3: Titanium alloys are finding recent use as dental implant materials. In a study [50] on the electrochemical behavior of a commercial titanium alloy, the polarization resistance R_p was observed to have the value 45,400 Ω cm^2 in an electrolyte containing Cl^- and HPO_4^- ions (among other constituents) intended to approximate the electrolyte found in the human mouth. What is the corrosion rate of the titanium alloy in this electrolyte if the anodic and cathodic Tafel slopes are 0.100 and –0.100 V, respectively?
Solution: The experimental data can be entered directly into the Stern–Geary equation to give

$$i_{corr} = \frac{1}{2.303\,(45{,}400\,\Omega\,\mathrm{cm}^2)\,\left(\frac{1}{0.100\,\mathrm{V}} + \frac{1}{0.100\,\mathrm{V}}\right)}$$

The result is $i_{corr} = 0.48\ \mu A/cm^2$. Thus, the commercial alloy has a low corrosion rate in the electrolyte of interest.

The form of Eq. (64) is also useful in applications where the cathodic reaction is under diffusion control. As will be seen in Chapter 8, the effective Tafel slope is infinity for a diffusion-controlled cathodic process which reaches a limiting current density. This value for b_c can be conveniently entered into Eq. (64) because $1/\infty = 0$.

Advantages and Possible Errors for the Linear Polarization Technique

An advantage of the linear polarization method is that the linearly polarized sample can be re-used repetitively to give corrosion rates as a function of immersion time. This is because the extent of polarization is at most about 50 mV removed from E_{corr}, and any changes in the electrode surface due to polarization are minimal (if any). Thus, the linear polarization method can be used to monitor continuously the corrosion rate of test coupons as a function of time.

Possible errors in the linear polarization method include the following:

(1) Application of the Stern–Geary equation under conditions where the Butler–Volmer equation does not apply (inclusion of the case when the cathodic reaction is under concentration polarization has been mentioned above).
(2) Application of the Stern–Geary equation under non-steady-state conditions.
(3) Errors in achieving linearity near the corrosion potential.
(4) Changes in Tafel slopes with time.
(5) Errors because the equilibrium potentials of anodic or cathodic half-cell reactions are close to the corrosion potential. This limitation has been treated in detail by various authors [51, 52].

In brief, when the corrosion potential E_{corr} is close to the equilibrium potentials of either half-cell reaction, it cannot be assumed that the total anodic or cathodic current density in the mixed couple is carried by a single half-cell reaction, as is assumed in the derivation of the Stern–Geary equation. While important from a theoretical standpoint, this information is not always tractable, especially for complex systems.

Applications of the Linear Polarization Technique

This method has been used in a variety of applications [12]. Some examples include the evaluation of corrosion inhibitors [53, 54], the behavior of protective zinc coatings [55], atmospheric corrosion monitors [56]), the oil industry [57, 58], pipelines [59, 60], chemical process industries [61], in pulp and paper plants [62], in water treatment applications [63], the corrosion of steel-reinforcing bars in chloride-contaminated concrete [64, 65], the corrosion behavior of biomedical implant alloys in simulated body fluids [50, 66], and the corrosion of aluminum alloys in beverages or liquid foods [43], or of aluminum alloys in crevices [67].

Two studies on reinforced steels in concrete systems serve to illustrate the utility of the linear polarization method. For reinforcing steel in a saturated $Ca(OH)_2$ solution, Andrate and Gonzalez [64] have observed an excellent correlation between weight losses calculated from the linear polarization method and weight losses determined from integration of current–time curves. For steel embedded in chloride-contaminated concrete, Lambert et al. [65] have observed an excellent agreement between weight losses calculated from the linear polarization method and actual gravimetric weight losses.

Figure 7.36 shows how the polarization resistance of an in-situ inaccessible sample can be related to the instantaneous corrosion rate [67]. Plots such as in Fig. 7.36 can be used to monitor continuously the durability of corrosion test coupons.

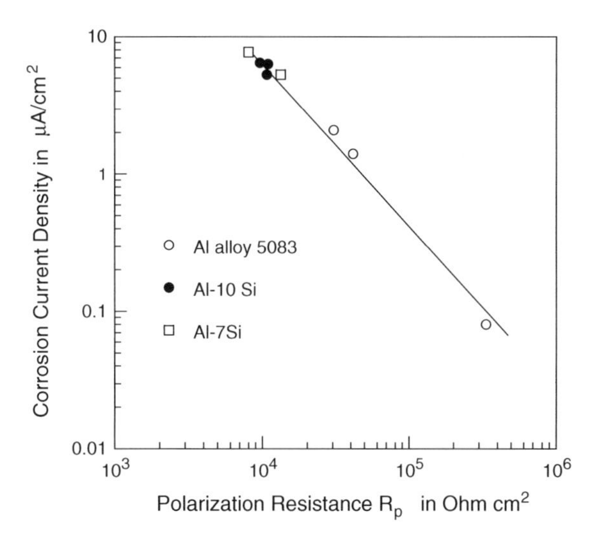

Fig. 7.36 Relationship between current density calculated from weight loss measurements and polarization resistance of three different aluminum alloys in crevices in air-saturated 5% NaCl. Plotted from data in [67] by permission of NACE International 1981

Table 7.7 Comparison of corrosion rates determined by the linear polarization technique with results obtained by a second method

Metal	Solution	Corrosion rate by linear polarization	Corrosion rate by a second method	Second method	Reference
Iron	1 M HCl	46 μA/cm^2	42 μA/cm^2	Tafel	McCafferty [6]
	1 M HBr	21 μA/cm^2	23 μA/cm^2	Tafel	
			16 μA/cm^2	Colorimetric	
	1 M HI	10 μA/cm^2	10 μA/cm^2	Tafel	
			7 μA/cm^2	Colorimetric	
Plain carbon steel	0.5 M H_2SO_4	157 μA/cm^2	165 μA/cm^2	Tafel	Driver and Meakins [54]
	0.5 M H_2SO_4 + 10^{-4} M inhibitor	12 μA/cm^2	14 μA/cm^2	Tafel	
Low alloy steel	Hydrocarbon/ water mixture 85°C	0.11 mm/year	0.08 mm/year	Weight loss	Jasinski and Efird [58]
Fe–18Cr	Molten carbonate	4.7 A/m^2	5.6 A/m^2	Tafel	Zhu et al. [68]
Al alloy 3003	Acetate buffer, pH 4.6	8.1 μA/cm^2	8.2 μA/cm^2	Tafel	Evans and Koehler [43]
			9.1 μA/cm^2	Hydrogen evolution	
Al brass	0.6 M NaCl (flowing)	0.068 mm/year	0.072 mm/year	Weight loss	Tamba [69]
Cu–Ni (90–10)	Seawater (flowing)				Syrett and Macdonald [70]
	0.045 mg/dm^3 O_2	0.49 mg/day	0.42 mg/day	Weight loss	
	26.3 mg/dm^3 O_2	0.35 mg/day	0.39 mg/day	Weight loss	

Various investigators have compared corrosion rates determined by the linear polarization method with the corrosion rate determined by an independent method. A partial list of examples for various metals is given in Table 7.7. Additional examples have been compiled by Callow et al. (52). A sufficient number of suitable comparisons have been obtained over the years so that the linear polarization method has achieved widespread acceptance (subject, of course, to its limitations).

The use of alternating current techniques to measure the polarization resistance R_p is discussed in Chapter 14.

Small-Amplitude Cyclic Voltammetry

As noted earlier, the linear polarization technique should be applied under steady-state conditions. That is, when the electrode potential is changed from E_{corr} to some new value near E_{corr}, sufficient time must be allowed so that the resulting current reaches a steady-state value before the potential is changed again. Very often, the potential is ramped with time so that the resulting current may be a function of the rate of potential sweep. Macdonald [71] has studied the response of copper–nickel alloys to a small-amplitude triangular potential excitation and has observed that there was significant

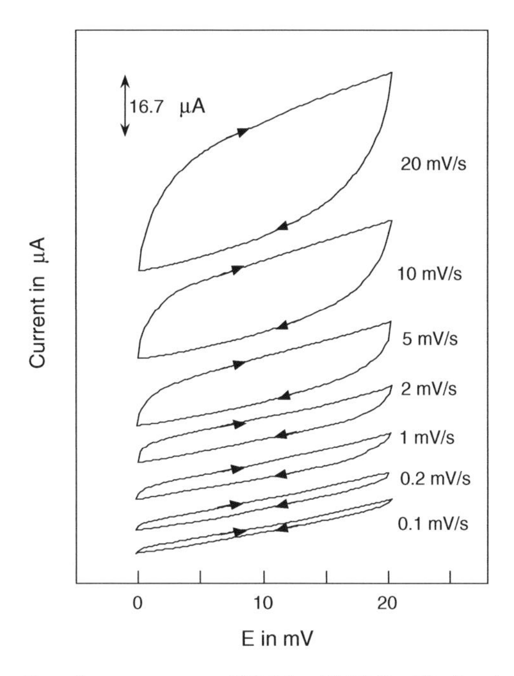

Fig. 7.37 Small-amplitude cyclic voltammograms at 25°C for 90:10 Cu–Ni alloy in flowing seawater containing dissolved oxygen. Specimen area = 11.05 cm^2, exposure time = 50 h. Redrawn from [71] by permission of ECS – The Electrochemical Society

hysteresis in the current–potential curves for high sweep rates. Small-amplitude cyclic voltammetry (SACV) plots are shown in Fig. 7.37 for 90:10 Cu–Ni alloys in flowing seawater. Macdonald concluded that the proper data to be used in the Stern–Geary equation were from SACV plots which exhibited minimum hysteresis.

Experimental Techniques for Determination of Polarization Curves

This section is intended to present only a brief description of experimental techniques used to determine polarization curves in the laboratory.

Electrode Samples

For polarization studies in the laboratory, several different types of metal electrodes can be used, including cylindrical rods, flat specimens, thin sheets of metal, or wires. The typical size of a specimen for laboratory studies is 1–10 cm^2 in area. Sample preparation to clean the surface prior to immersion of the electrode into the electrolyte may vary from mechanical polishing only to mechanical polishing followed by a chemical etch or electropolishing. The electrode surface is usually given a final cleaning in a mild solvent, such as methanol, and is then allowed to air-dry.

Electrode Holders

Specimen holders are needed to introduce the metal sample into solution in such a manner that only the metal surface of interest contacts the electrolyte. All extraneous metal surfaces (such as connecting rods or wires) must be isolated from the electrolyte. For cylindrical metal specimens or sufficiently thick planar samples, a Stern–Makrides electrode holder is often used [72, 73]. See Fig. 7.38. Electrical contact to the metal sample is made by means of a stainless steel threaded rod. This threaded rod is placed within a heavy-walled glass tube and the electrode sample is forced firmly against a compression Teflon® piece by means of pressure hand applied to a knurled knob located at the top of the electrode holder. When the holder is properly constructed, the electrolyte does not leak into the interior of the heavy-walled glass tube.

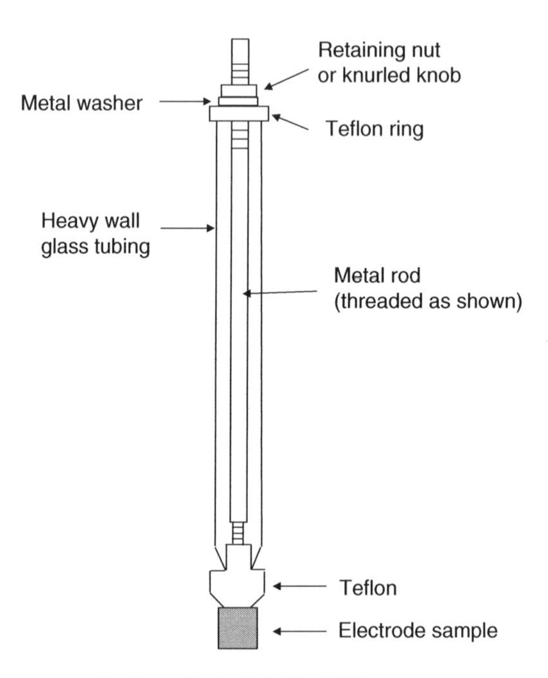

Fig. 7.38 A Stern–Makrides compression-type electrode holder [72]

Electrical contact to flat samples can be made by a screw inserted into a hole drilled and tapped in the metal sample, as shown in Fig. 7.39. Extraneous surfaces including the edges of the sample can be covered with an insulating varnish or lacquer which has sufficient adhesion so as to preclude undercutting of the varnish or the lacquer by the aqueous solution. Following Wilde et al. [74], the author has found a commercial alkyd resin to be effective. Flat samples can also be mounted in non-metallic materials, such as epoxy mounts. This arrangement makes handling of the sample easier for polishing prior to immersion, but if localized corrosion inadvertently occurs in a crevice formed between the mount and the metal sample itself, then the experimental results are suspect.

Thin or brittle samples can also be used in epoxy mounts with an electrical contact provided to the backside of the metal, as described by Jones [75].

Wire or thin rod metal specimens can be encased in heat-shrinkable Teflon® or polymer tubing so as to expose only a certain fixed area of the sample to the solution.

Various types of samples and electrode holders are illustrated in Fig. 7.39.

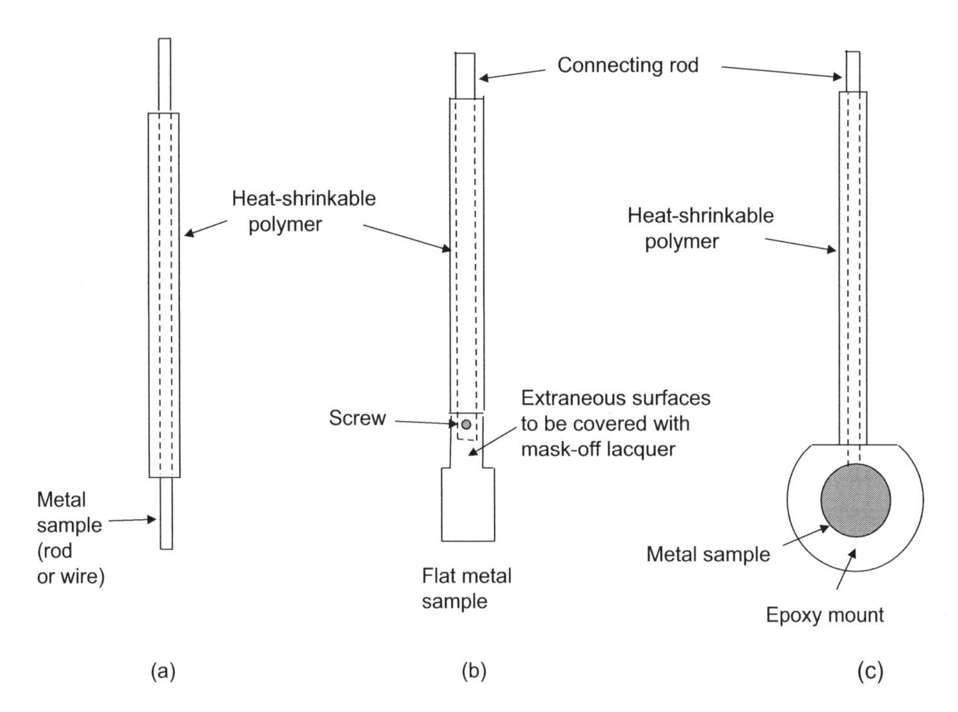

Fig. 7.39 Various types of metal samples and electrode holders

Electrochemical Cells

Figure 7.40 shows the design of an electrochemical cell used for polarization measurements. The cell has ports for the introduction of the working (test) electrode, an auxiliary electrode (usually platinum) which is needed to complete the electrical circuit, and the reference electrode which is housed in a Luggin–Haber capillary compartment. The cell also has gas inlets and outlets, which can be used to purge the electrolyte and the cell interior of oxygen by using an inert gas such as argon. The electrolyte can also be bubbled with oxygen if it is desired to conduct experiments in oxygen-saturated solutions. Bubbling with a gas also stirs the solution and minimizes concentration polarization effects. In addition, experiments can be conducted in quiescent solutions open to the air by simply keeping the gas ports open to the atmosphere.

In this type of cell, a split auxiliary electrode should be used, in which identical auxiliary electrodes are placed on each side of the working electrode. The two parts of the auxiliary electrode are connected externally to the cell. The purpose of this arrangement is to minimize problems in current distribution which might otherwise exist between the working and the auxiliary electrodes. Cells similar to that in Fig. 7.40 have been described elsewhere [73, 75].

A two-compartment cell can also be used [6, 17] in which the working electrode and counter electrode are contained in separate compartments joined by a porous glass frit. With this arrangement, any changes in electrolyte composition which may result from electrochemical reactions at the counter electrode are not experienced by the solution in the compartment which houses the working electrode.

In either type of cell, graphite auxiliary electrodes can be used if platinum is not available.

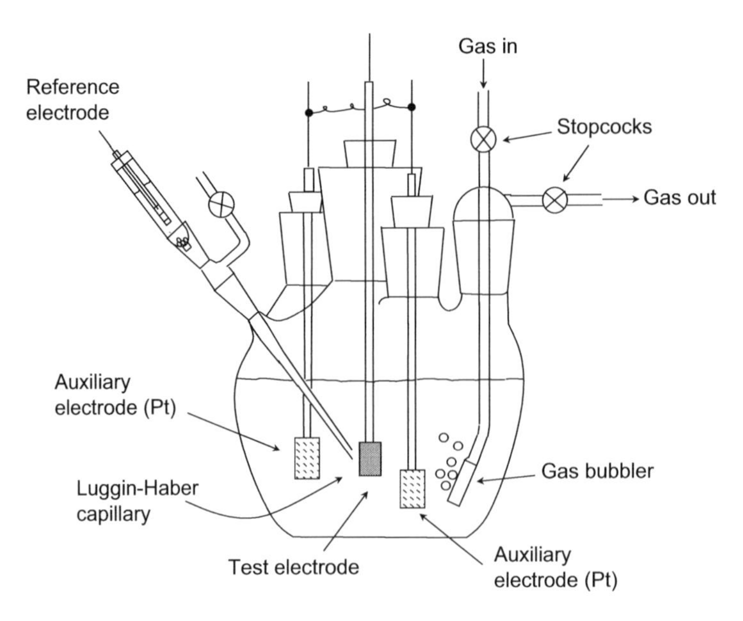

Fig. 7.40 Schematic diagram of a glass cell commonly used for making polarization measurements

Instrumentation and Procedures

Modern potentiostats (constant potential devices) are commercially available having computer control and a wide variety of software packages. Potentiostats are based on operational amplifiers, and other sources should be consulted as to their theory and design [76, 77]. These devices can be operated either in the potentiostatic or the galvanostatic mode. In the potentiostatic mode, the electrode potential is held constant and the resulting current is recorded. In the galvanostatic mode, the current is held constant and the resulting electrode potential is recorded. Either mode can be used unless there is an active–passive transition, as shown schematically in Fig. 7.41. In such cases, the galvanostatic mode, which determines E as a function of i, will not detect the "nose" in the anodic polarization curve. (The subject of passivity is treated in Chapter 9.)

It should be noted that a simple and inexpensive galvanostat can be constructed using a 45 V battery in series with a bank of variable resistors and with the corrosion cell itself [7]. Because the electrolytic resistance of the solution in the corrosion cell is much less than that of the variable resistors, a constant current is essentially defined by the setting of the variable resistors.

Polarization curves can be determined by the potential step or the potential sweep method. In the potential step method, the electrode potential is changed abruptly from one discrete value to another, and the resulting current is then recorded as a function of time until the current reaches a steady-state value which either is invariant with time or changes only minimally with time. The process is then repeated, potential by potential, until the entire polarization curve is determined.

The time required to attain a steady-state current for a given potential depends on the nature of the system under investigation. For example, in a study on the corrosion of iron in 1 M HCl with and without organic inhibitors, potential steps of 10–15 mV were used, and the currents were observed to be constant after approximately 20 min [8]. The potential step method is sometimes tedious but ensures that steady-state polarization curves are obtained.

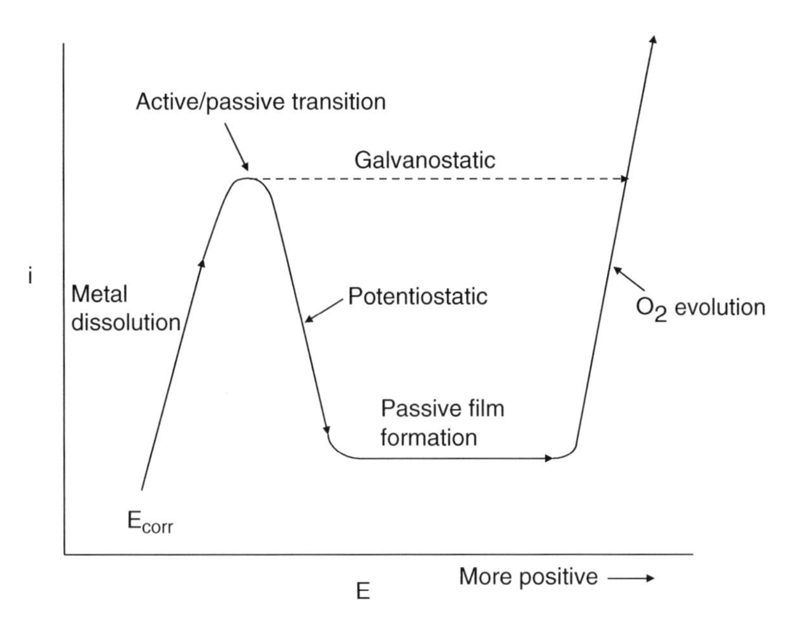

Fig. 7.41 An anodic polarization curve showing a transition from a region of active corrosion to a region of passivity and the response to this transition for galvanostatic vs. potentiostatic measurements

In the potential sweep method, the electrode potential is ramped continuously with time, and the resulting current is simultaneously recorded along with the applied potential. Care must be taken that the sweep rate is sufficiently slow so that the steady state is approximated by the sweep. Sweep rates of 10–100 mV/min are commonly used.

Polarization curves should be determined only after the open-circuit electrode potential of the test metal has attained a steady-state value. The length of time required to reach the steady state depends markedly upon the nature of the system. For the well-defined set of laboratory conditions for iron in de-aerated 1 M HCl referred to above [8], only 24 h was required to reach a steady-state electrode potential.

However, for the decidedly more complex environment of natural seawater open to the air, much longer times are required for immersed metals to reach stable open-circuit potentials. For example, the open-circuit electrode potential of zinc was stable after 75 days of immersion, whereas mild steel required 250 days [78]. Electrode potentials of copper immersed in laboratory cells containing seawater or in a seawater flume were stable after 50 days of immersion, but similar copper specimens immersed in seawater under a pier exhibited unstable potentials which were highly variable even after 1 year of immersion [78].

The next chapter will consider in detail the corrosion behavior of metals immersed in electrolytes open to the air where the principal cathodic reaction is the reduction of dissolved oxygen.

Problems

1. The following weight loss data were observed for the corrosion of Al alloy 7075 in 1 M HCl [2]:

Time (h)	Weight loss (g)
1.0	0.0628
2.0	0.3955
3.0	1.3097
4.0	1.8796
6.0	3.3372
7.6	3.8788

The sample size was 15.0 cm^2 in each case.

(a) What is the corrosion rate in grams per square centimeter hour?
(b) What is the equivalent current density in milliamperes per square centimeter assuming that the anodic half-cell reaction is

$$Al \rightarrow Al^{3+} + 3e^-$$

2. The following weight loss data were observed for the atmospheric corrosion of plain carbon steel in an industrial atmosphere [79]:

Time (years)	Weight loss (mg/cm^2)
0.0	0.0
0.5	51.5
1.0	82.7
2.0	149.3
3.0	189.9
4.0	226.9
5.0	257.8
7.0	295.2

(a) What is the instantaneous corrosion rate in milligrams per square centimeter per year at 1 year exposure?
(b) What is the trend in instantaneous corrosion rates with time?
(c) What does the trend in (b) indicate about the protective nature of the rust layer formed on the carbon steel in this particular environment?

3. Suppose that a 2.0 cm^2 sample of iron is immersed in 500 ml of acid solution and the corrosion rate is determined by Tafel extrapolation to be 10.0 mA/cm^2. Assume that the iron corrodes to form Fe^{2+} ions with the evolution of hydrogen. After a 10 h immersion period:

(a) What is the total concentration of dissolved Fe^{2+} in solution?
(b) What is the weight loss of the iron specimen per unit area of sample?
(c) What volume of hydrogen gas in ml (STP) has been evolved per unit area of sample?

4. Convert the corrosion rate for titanium given in Fig. 7.7 in terms of nanometer per square centimeter per second to the equivalent weight loss of titanium metal in milligrams per square centimeter per hour. The density of titanium is 4.51 g/cm^3.
5. Alloy C-276 is used in various applications where enhanced corrosion resistance is required. The alloy has the following composition by weight of the major components: 54.6% Ni, 16.2% Cr, 15.3% Mo, and 5.8% Fe. In a study [80] on the accelerated corrosion of alloy C-276, the alloy was anodically polarized in a small volume of electrolyte (300 μL) as shown in the figure below:

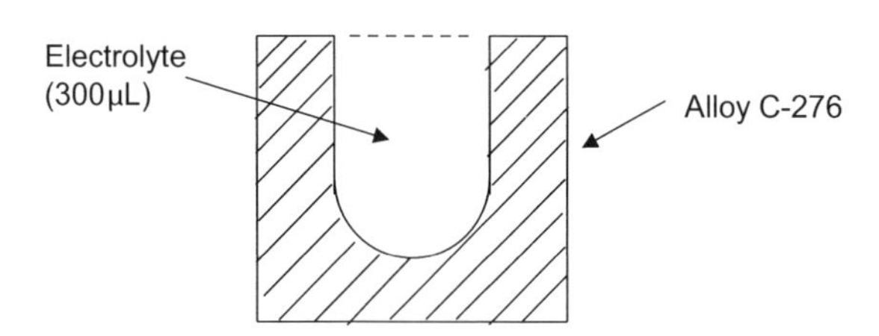

After anodic polarization, the ionic composition of the electrolyte was chemically analyzed using atomic absorption spectrophotometry. The results of one trial were as follows:

Fe	Cr	Mo	Ni
0.93	1.62	2.10	11.51

where the numbers refer to the total milligrams of each dissolved metal in the electrolyte volume. Under the conditions of this study, does the alloy corrode stoichiometrically in the electrolyte? That is, does each alloying element in the alloy corrode in proportion to its amount in the bulk alloy? If not, which elements, if any, are preferentially corroded?
6. Calculate the theoretical Tafel slope for the anodic reaction

$$Fe(s) \rightarrow Fe^{2+}(aq) + 2e^-$$

when the reaction is under activation control. Assume that the transfer coefficient $\alpha = 0.5$ and that the anodic reaction occurs in a single-step two-electron transfer.
7. Sketch the experimentally observed anodic and cathodic polarization curves for zinc which result from the Evans diagrams shown in Fig. 7.23.
8. The following data were taken for iron immersed in 1 M HCl [8]

(a) What is the corrosion potential?
(b) What is the corrosion rate?

Cathodic polarization		Anodic polarization	
E vs. SCE	*i* (μA/cm^2)	*E* vs. SCE	*i* (μA/cm^2)
–0.465	8.0	–0.443	5.7
–0.475	19.6	–0.440	30.3
–0.485	33.9	–0.435	71.4
–0.495	51.8	–0.425	178
–0.505	71.4	–0.415	339
–0.515	94.6	–0.405	607
–0.525	127	–0.395	1000
–0.535	171	–0.385	1964
–0.545	223		
–0.555	294		
–0.570	446		
–0.590	803		
–0.600	892		
–0.610	1363		
–0.620	1690		

9. For a corrosion reaction under activation control,

   (a) At what overvoltage does the rate of the reverse reaction become 10% of the rate of the forward reaction?

   (b) At what overvoltage does the rate of the reverse reaction become 1% of the rate of the forward reaction?

10. Given the following Evans diagram, show that the logarithm of corrosion current density i_{corr} in a galvanic cell is a linear function of the logarithm of the cathode to anode ratio (A_c/A_c). That is, show that i_{corr} vs. A_c/A_c obeys a linear log–log relationship. Assume that $b_a = -b_c = b$.

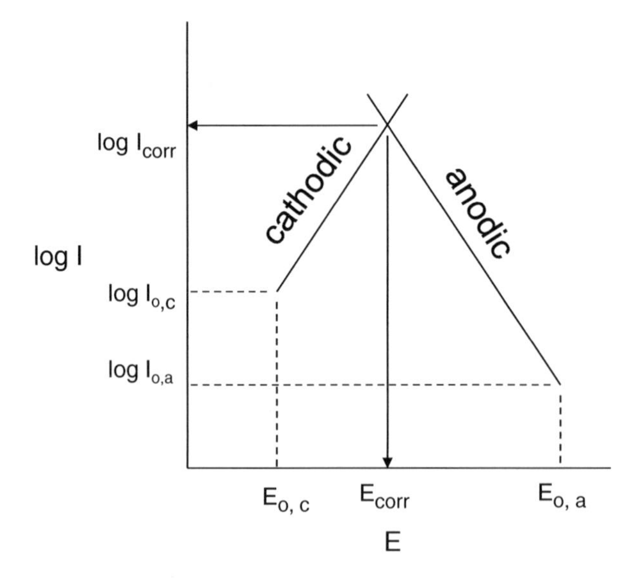

11. (a) Determine the corrosion rate of nickel in 1 M HCl graphically using the data in Tables 7.3, 7.4, and 3.1. Assume that $b_a = 0.060$ V and $b_c = -0.100$ V. (b) What is the corrosion potential?
12. Evans and Koehler [43] determined the corrosion rate of Al alloy 3003 in apple juice by the linear polarization method. If the measured polarization resistance is 0.0136 MΩ cm^2, and the anodic and cathodic Tafel slopes are +0.100 and –0.100 V, respectively, what is the corrosion rate in microamperes per square centimeter of the aluminum alloy in apple juice?
13. The following data were obtained for the cathodic linear polarization of iron in 1 M HCl [6]:

Overvoltage η (V)	Cathodic current density (μA/cm^2)
–0.010	–25
–0.020	–47
–0.030	–69
–0.040	–77

From the full cathodic polarization curve, the cathodic Tafel slope was determined to be –0.100 V, and the corrosion rate was 43.7 μA/cm^2. What is the expected value of the anodic Tafel slope?

14. By retaining quadratic terms in the MacLaurin expansion show that the Butler–Volmer equation for a corrosion metal can be written as

$$\overleftarrow{i}_{\text{net}} = 2.303\, i_{\text{corr}} \left[\left(\frac{1}{b_a} + \frac{1}{|b_c|} \right) \eta + 1.152 \left(\frac{1}{b_a^2} - \frac{1}{b_c^2} \right) \eta^2 \right]$$

(a) How could the corrosion rate be determined from this equation?
(b) Does this approach offer any advantage over the linear polarization method?

15. Consider a set of tests (e.g., the addition of various corrosion inhibitors) in which the anodic and cathodic Tafel slopes do not vary from sample to sample. Show that under these conditions, a plot of log i_{corr} vs. log R_p should have a slope of –1.

References

1. M. G. Fontana and N. D. Greene, "Corrosion Engineering", p. 129, McGraw-Hill, New York, NY (1978).
2. E. McCafferty, *Corrosion*, *54*, 862 (1998).
3. C. P. Larrabee, *Trans. Electrochem. Soc.*, *85*, 297 (1944).
4. C. P. Larrabee and S. K. Coburn, "First International Congress on Metallic Corrosion", p. 276, Butterworths, London, (1962).
5. E. B. Sandell, "Colorimetric Determination of Traces of Metals", Interscience Publishers., New York (1959).
6. E. McCafferty and A. C. Zettlemoyer, *J. Phys. Chem.*, *71*, 2444 (1967).
7. E. McCafferty and N. Hackerman, *J. Electrochem. Soc.*, *119*, 999 (1972).
8. E. McCafferty and J. V. McArdle, *J. Electrochem. Soc.*, *142*, 1447 (1995).
9. E. McCafferty, P. G. Moore, and G. T. Peace, *J. Electrochem. Soc.*, *129*, 9 (1982).
10. G. K. Hubler and E. McCafferty, *Corros. Sci.*, *20*, 103 (1980).
11. R. J. Williams and D. E. Bause in "A Guide to Materials Characterization and Chemical Analyses", 2nd edition, J. P. Sibilia, Ed., p. 115, VCH Publishers, New York, NY (1996).
12. C. F. Jaske, J. A. Beavers, and N. G. Thompson, *Corros. Prevention & Control*, *49* (1), 3 (2002).

13. S. D. Cramer and B. S. Covino, Jr., Eds., "Corrosion, Fundamentals, Testing, and Protection", ASM Handbook, Vol. 13A, pp. 82, 419, 514, 536, 544, 884, 889, ASM International, Materials Park, OH (2003).
14. L. C. Feldman in "Ion Spectroscopies for Surface Analysis", A. W. Czanderna and D. M. Hercules, Eds., p. 311, Plenum Press, New York (1991).
15. G. V. Akimov, *Corrosion*, *11*, 515t (1955).
16. S. Barnartt, *J. Electrochem. Soc.*, *99*, 549 (1952).
17. A. C. Makrides, *Corrosion*, *18*, 338 (1962).
18. D. T. Sawyer and J. L. Roberts, "Experimental Electrochemistry for Chemists", p. 118, John Wiley, New York (1974).
19. H. Eyring, S. Glasstone, and K. J. Laidler, *J. Chem. Phys.*, *7*, 1053 (1939).
20. J. O'M. Bockris and A. K. N. Reddy, "Modern Electrochemistry", Vol. 2, Chapter 8, Plenum Press, New York (1977).
21. A. J. Bard and L. R. Faulkner, "Electrochemical Methods", Wiley, New York (2001).
22. C. Wagner and W. Traud, *Z. Elektrochem.*, *44*, 391 (1938).
23. S. Trasatti, *J. Electroanal. Chem.*, *39*, 163 (1972).
24. B. E. Conway and J. O'M. Bockris, *J. Chem. Phys.*, *26*, 532 (1957).
25. J. O'M. Bockris in "Modern Aspects of Electrochemistry", J. O′M. Bockris and B. E. Conway, Eds., Vol. 1, p. 180, Academic Press, New York (1954).
26. J. T. Clark and N. A. Hampson, *J. Electroanal. Chem.*, *26*, 307 (1970).
27. A. R. Despic in "Comprehensive Treatise of Electrochemistry", Vol. 7, B. E. Conway, J. O′M. Bockris, E. Yeager, S. U. Khan, and R. E. White, Eds., p. 451, Plenum Press, New York (1983).
28. H. Kaesche and N. Hackerman, *J. Electrochem. Soc.*, *105*, 191 (1958).
29. M. Stern and R. M. Roth, *J. Electrochem. Soc.*, *104*, 390 (1957).
30. M. Stern, *J. Electrochem. Soc.*, *102*, 609 (1955).
31. J. O'M. Bockris and D. Drazic, *Electrochim. Acta*, *7*, 293 (1962).
32. N. A. Hampson, R. J. Latham, and D. Larkin, *J. Electroanal. Chem.*, *23*, 211 (1969).
33. N. A. Hampson and D. Larkin, *Trans. Faraday Soc.*, *65*, 1660 (1969)
34. K. J. Vetter, "Electrochemical Kinetics", p. 668, Academic Press, New York (1967).
35. G. A. Marsh and E. Schaschl, *J. Electrochem. Soc.*, *107*, 960 (1960).
36. H. H. Uhlig, *Proc. New York Acad. Sci.*, *40*, 276 (1954).
37. M. Stern and A. L. Geary, *J. Electrochem. Soc.*, *104*, 56 (1957).
38. M. Stern, *J. Electrochem. Soc.*, *104*, 559 (1957).
39. M. Stern, *J. Electrochem. Soc.*, *104*, 645 (1957).
40. M. Stern, *J. Electrochem. Soc.*, *102*, 663 (1955).
41. E. McCafferty, *Corros. Sci.*, *47*, 3202 (2005).
42. M. Maja, N. Penazzi, G. Farnia, and G. Sandona, *Electrochim. Acta*, *38*, 1453 (1993).
43. S. Evans and E. L. Koehler, *J. Electrochem. Soc.*, *108*, 509 (1961).
44. G. Quartarone, T. Bellomi, and A. Zingales, *Corros. Sci.*, *45*, 715 (2003).
45. Y.-H. Yau and M. A. Streicher, *Corrosion*, *47*, 352 (1991).
46. H. H. Uhlig and W. R. Revie, "Corrosion and Corrosion Control", p. 51, John Wiley and Sons, New York (1985).
47. E. J. Simmons, *Corrosion*, *11*, 255t (1955).
48. R. V. Skold and T. E. Larson, *Corrosion*, *13*, 139t (1957).
49. F. Mansfeld, *J. Electrochem. Soc.*, *120*, 515 (1973).
50. H.-H. Huang, *Biomaterials*, *24*, 275 (2003).
51. F. Mansfeld and K. B. Oldham, *Corros. Sci.*, *11*, 787 (1971).
52. L. M. Callow, J. A. Richardson, and J. L. Dawson, *Br. Corros. J.*, *11*, 123, 132 (1976).
53. T. J. Butler, R. M. Hudson, and C. J. Warning, *Electrochem. Technol.*, *6*, 227 (1968).
54. R. Driver and R. J. Meakins, *Br. Corros. J.*, *9*, 227 (1974).
55. G. W. Walter, *Corros. Sci.*, *15*, 47 (1975).
56. F. Mansfeld, S. L. Jeanjaquet, M. W. Kendig, and D. K. Doe, *Atmos. Environ.*, *20*, 1179 (1986).
57. C. F. Britton, *Corros. Prevention & Control*, *27* (2), 10 (1980).
58. R. J. Jasinski and K. D. Efird, *Corrosion*, *41*, 658 (1988).
59. R. G. Asperger and P. G. Hewitt, *Mater. Perform.*, *25* (8), 47 (1986).
60. K. van Gelder, L. van Bodegom, and A. Vissar, *Mater. Perform.*, *27* (4), 17 (1988).
61. C. G. Arnold and D. R. Hixon, *Mater. Perform.*, *21* (4), 25 (1982).
62. D. C. Crowe and Y. A. Yeske, *Mater. Perform.*, *25* (6), 18 (1986).

63. R. Rizzi, S. Bonato, and L. Forte, "Proc. 6th European Symposium on Corrosion Inhibitors", p. 1145, Ferrara, Italy (1985).
64. C. Andrate and J. A. Gonzalez, *Werkstoffe Korros.*, *29*, 515 (1978).
65. P. Lambert, C. L. Page, and P. R. W. Vassie, *Mater. Struct.*, *24*, 351 (1991).
66. G. Lewis and K. Daigle, *J. Appl. Biomater.*, *4*, 47 (1993).
67. M. Suzuki and Y. Sato, *Corrosion*, *37*, 123 (1981).
68. B. Zhu, G. Lindbergh, and D. Simonsson, *Corros. Sci.*, *41*, 1497 (1999).
69. A. Tamba, *Br. Corros. J.*, *17*, 29 (1982).
70. B. C. Syrett and D. D. Macdonald, *Corrosion*, *35*, 505 (1979).
71. D. D. Macdonald, *J. Electrochem. Soc.*, *125*, 1443 (1978).
72. M. Stern and A. C. Makrides, *J. Electrochem. Soc.*, *107*, 782 (1960).
73. "Standard Reference Test Method for Making Potentiostatic and Potentiodynamic Anodic Polarization measurements", ASTM Test Method G 5, "1998 Annual Book of ASTM Standards", Section 3, p. 54, ASTM, West Conshohocken, PA (1998).
74. N. D. Greene, W. D. France, Jr., and B. E. Wilde, *Corrosion*, *21*, 275 (1965).
75. D. A. Jones, "Principles and Prevention of Corrosion", pp. 104–105, Prentice Hall, Upper Saddle River, NJ (1996).
76. D. T. Sawyer and J. L. Roberts, "Experimental Electrochemistry for Chemists", p. 236, John Wiley, New York (1974).
77. A. J. Bard and L. R. Faulkner, "Electrochemical Methods", p. 632, John Wiley & Sons, New York, NY (2001).
78. M. H. Peterson and T. J. Lennox, Jr., *Mater. Perform.*, *23* (3), 15 (1984).
79. R. Todoroki, S. Kado, and A. Teramae, "Proceedings of the Fifth International Congress on Metallic Corrosion", p. 764, National Association of Corrosion Engineers, Houston, TX (1974).
80. M. A. Cavanaugh, J. A. Kargol, J. Nickerson, and N. F. Fiore, *Corrosion*, *39*, 144 (1983).

Chapter 8
Concentration Polarization and Diffusion

Introduction

The previous chapter has considered the electrode kinetics of corrosion processes under the control of activation polarization. Most of the examples in Chapter 7 were for metals corroding in acid solutions, in which the cathodic reaction is the evolution of hydrogen, a reaction which usually proceeds under activation control. As was also discussed in Chapter 7, a second major type of polarization process is *concentration polarization*, in which the electrode is polarized due to a concentration effect that occurs at the electrode surface. For cathodic processes, the most important concentration effect is the diffusion of dissolved oxygen *to the metal surface* and the subsequent reduction of oxygen by the half-cell reaction

$$O_2(g) + 2H_2O(l) + 4e^- \rightarrow 4OH^-(aq) \tag{1}$$

For anodic processes, concentration polarization occurs due to the slow diffusion *away from the metal surface* of the metal cations which are produced by anodic dissolution. For example, the anodic reaction

$$Fe(s) \rightarrow Fe^{2+}(aq) + 2e^- \tag{2}$$

will produce a concentration effect if Fe^{2+} ions are slow to diffuse away from the metal surface.

Concentration polarization effects can also occur in the corrosion of metals in acid solutions (for which the cathodic reaction is hydrogen evolution rather than oxygen reduction). But these instances occur far less frequently than does concentration polarization due to oxygen reduction. Discussion of concentration polarization for the case of hydrogen reduction will be given later in this chapter.

Where Oxygen Reduction Occurs

Oxygen reduction is the predominant cathodic reaction for any solution of neutral to alkaline pH and which is open to the atmosphere. This is because the electrolyte involved will contain a certain concentration of dissolved O_2 from the atmosphere. Thus, oxygen reduction occurs during the corrosion of metals in a wide variety of instances, including corrosion in

- natural waters, such as lakes, rivers, or streams
- seawater and salt spray environments
- structural metals in the natural outdoor atmosphere

E. McCafferty, *Introduction to Corrosion Science*, DOI 10.1007/978-1-4419-0455-3_8,

- electronic components in indoor environments
- piping systems which contain dissolved O_2, such as copper pipes carrying drinking water
- industrial cooling systems
- waste water systems
- hot water heaters and steam generators
- moist soils
- O_2-containing body fluids near joints or in the oral cavity
- systems with standing water, such as rain gutters, drains, and nozzles
- oil-well and geothermal brines.

Oxygen reduction can also occur in acidic solutions open to the air, and some examples will be considered later in this chapter.

Concentration Polarization in Current Density–Potential Plots

Figure 8.1 shows schematically a typical current density–potential curve displaying activation polarization for both the anodic and the cathodic branches. This curve is typical of the corrosion of metals in oxygen-free acid solutions and is the type of polarization curve considered in much detail in the previous chapter.

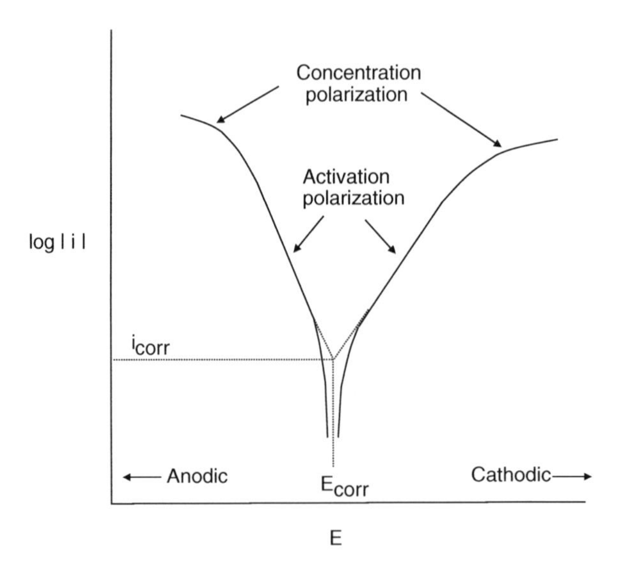

Fig. 8.1 Schematic diagram showing concentration polarization at the tail end of an activation polarization curve

Concentration polarization may also occur in such cases but usually as a secondary effect at the tail end of the Tafel regions, as shown in Fig. 8.1. On the anodic side of the curve, increased metal dissolution at large anodic overvoltages may lead to the accumulation of dissolved cations near the electrode surface. If these accumulated dissolved cations diffuse away slowly, then concentration polarization will result. On the cathodic side of the curve, increased hydrogen evolution at large cathodic overvoltages may cause a depletion of reacting hydrogen ions near the electrode surface. In addition, the increased number of hydrogen gas molecules which are produced may be slow to diffuse away from the electrode surface. Either event will result in concentration polarization for the cathodic reaction. However, as seen in Fig. 8.1, such concentration polarization effects are minor compared to the effect of activation polarization, and the corrosion rate can be easily determined by Tafel extrapolation.

Figure 8.2 shows an experimentally determined polarization curve for iron in 0.6 M NaCl solution open to the air [1]. The anodic curve is under activation control and is similar to the anodic polarization curve for iron in acid solutions shown in Fig. 7.24. However, the cathodic curve for oxygen reduction in Fig. 8.2 is quite different than the cathodic curves for hydrogen evolution in Fig. 7.24. The cathodic polarization curve in Fig. 8.2 "flattens out." In the example shown, for electrode potentials more negative than –800 mV vs. Ag/AgCl, the current density essentially becomes independent of the electrode potential. This is a typical polarization curve under concentration control and is a characteristic curve for the diffusion-limited reduction of oxygen.

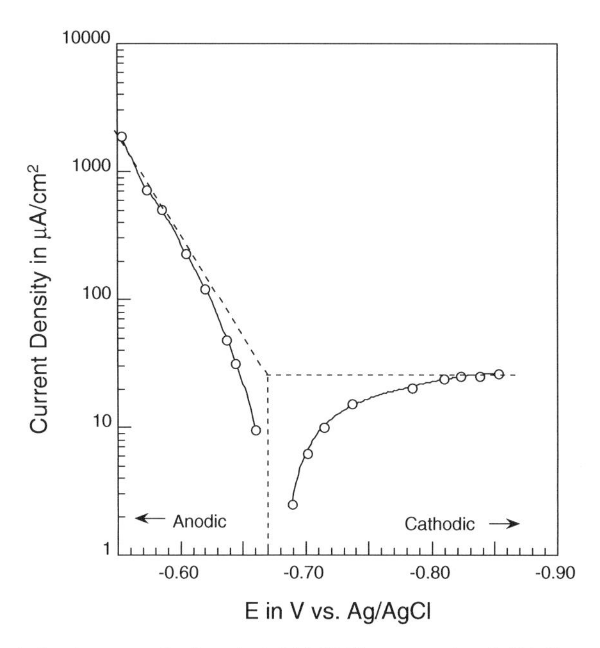

Fig. 8.2 Experimental polarization curve for iron in 0.6 M NaCl open to the air [1]. Reproduced by permission of ECS – The Electrochemical Society

Solubility and Diffusion

Solubility of Oxygen in Aqueous Solutions

Any solution open to the air is in contact with oxygen molecules in the gaseous phase. Upon initial contact of the solution with the air, there is a dynamic exchange between O_2 molecules in the gas phase and O_2 molecules which enter the liquid phase. Some of the dissolved O_2 molecules in the near-surface region escape back out into the gaseous phase, but other O_2 molecules venture further into the interior of the liquid phase by the process of diffusion. See Fig. 8.3. Diffusion is the random movement of molecules (or ions) arising from the translation due to thermal energy. By the process of diffusion, the entire liquid phase eventually gets saturated in O_2 and becomes uniform throughout in its composition of dissolved O_2.

Oxygen molecules can dissolve in water or in an aqueous solution due to intermolecular forces which exist between O_2 and H_2O molecules. The solubility of dissolved oxygen in an electrolyte

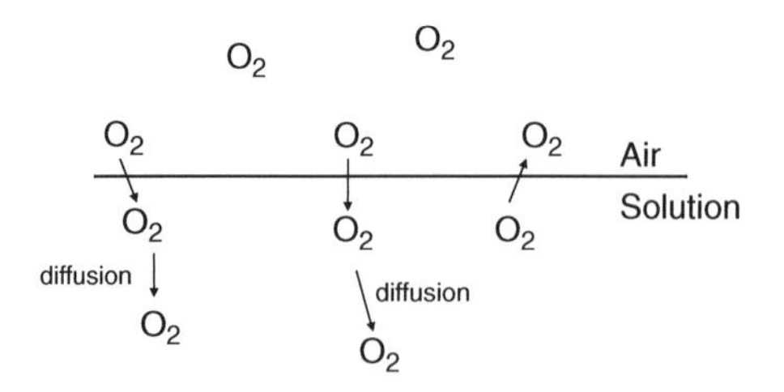

Fig. 8.3 Schematic diagram showing the dynamic equilibrium between H_2O molecules in the gas phase and the liquid surface. Also represented is the process of diffusion by which H_2O molecules enter the bulk of the liquid

depends on three factors: (i) the temperature of the solution, (ii) the pressure of the external gas phase, and (iii) the nature and concentration of the electrolyte.

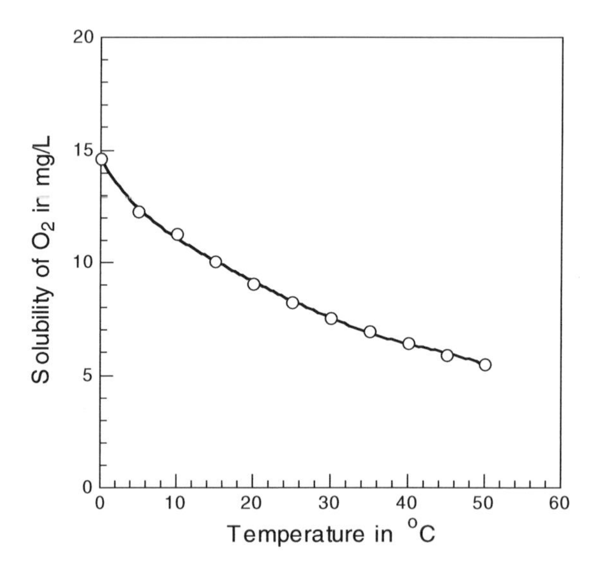

Fig. 8.4 Solubility of O_2 in water as a function of temperature at 1 atm pressure [2]

The solubility of dissolved oxygen decreases with increasing temperature of the liquid, as shown in Fig. 8.4, which gives the solubility of dissolved O_2 in water in equilibrium with 1 atm of air. With increasing temperature, the translational motion of molecules in the solution becomes more rapid so that dissolved O_2 molecules are more likely to enter the interface and escape back into the gaseous phase. However, with increasing external partial pressure of oxygen, more O_2 molecules from the gas phase contact the surface of the liquid so that there is a greater likelihood that gaseous O_2 molecules enter the liquid phase and are dissolved. If the atmosphere of air is replaced by one of pure oxygen, the solubility of O_2 in the liquid will increase according to Henry's law

$$[O_2] = K_H P_{O_2} \tag{3}$$

where $[O_2]$ is the concentration of dissolved oxygen in the liquid, K_H is the Henry's law constant, and P_{O2} is the partial pressure of O_2 in the gas. (The partial pressure of oxygen in air is $P_{O2} =$ 0.21 bar). More information on units of pressure is given in Table 8.1.

Table 8.1 Units of pressure

1 Torr is the pressure produced by a column of mercury 1 mm high
1 Pascal (Pa) equals a force of 1 Newton per square meter (N/m^2) (a Newton is the force required to give a mass of 1 kg an acceleration of 1 m/s)
1 bar is equal to 100,000 Pa = 100 kPa

Units of atmospheric pressure:
1 atmosphere (atm) = 760 mm Hg
= 760 Torr
= 101,325 Pa
= 1.01325 bar

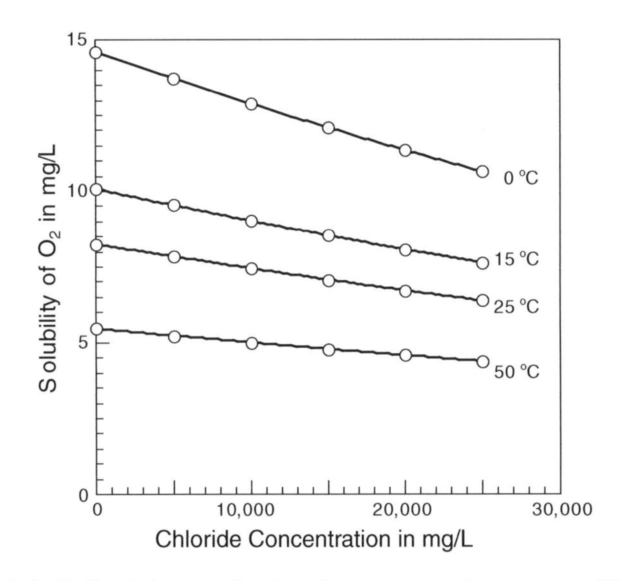

Fig. 8.5 Solubility of O_2 in NaCl solutions as a function of temperature at 1 atm pressure [2]

Figure 8.5 shows the solubility of oxygen in NaCl solutions open to the air. It can be seen that the concentration of dissolved O_2 decreases with increasing salt concentration (for a given temperature).

The solubility of O_2 in an electrolyte also depends on the identity of the electrolyte. Figure 8.6 shows solubility data for several different electrolytes [3].

Fick's First Law of Diffusion

The flux *J* due to diffusion past a given plane is given by Fick's first law:

$$\begin{array}{ccc} J = & -\mathrm{D} & \dfrac{\mathrm{d}C}{\mathrm{d}x} \\ \dfrac{\mathrm{mol}}{\mathrm{cm^2\,s}} & \dfrac{\mathrm{cm^2}}{\mathrm{s}} & \dfrac{\mathrm{mol/cm^3}}{\mathrm{cm}} \end{array} \tag{4}$$

where the units are as indicated above. The proportionality constant *D* is called the *diffusion coefficient* and has the value of approximately 10^{-5} cm^2/s for most ions and small molecules in water

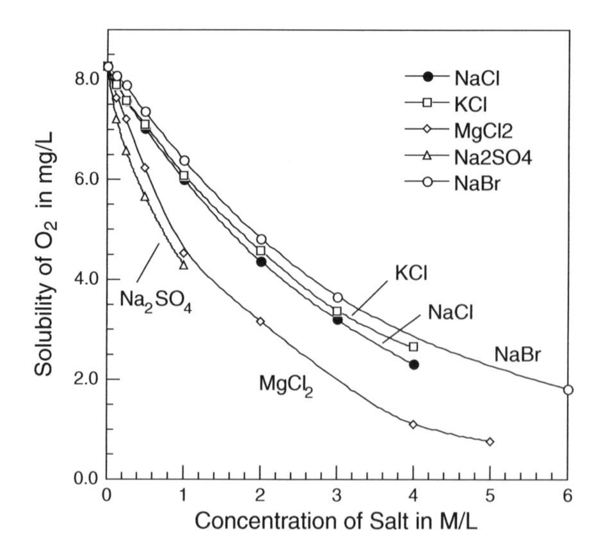

Fig. 8.6 Solubility of O_2 in various salt solutions at 25°C [3]

[4, 5]. The quantity dC/dx in Eq. (4) is the concentration gradient. The minus sign in Eq. (4) arises because the diffusion flux J and the concentration gradient dC/dx act in opposite directions, as shown schematically in Fig. 8.7.

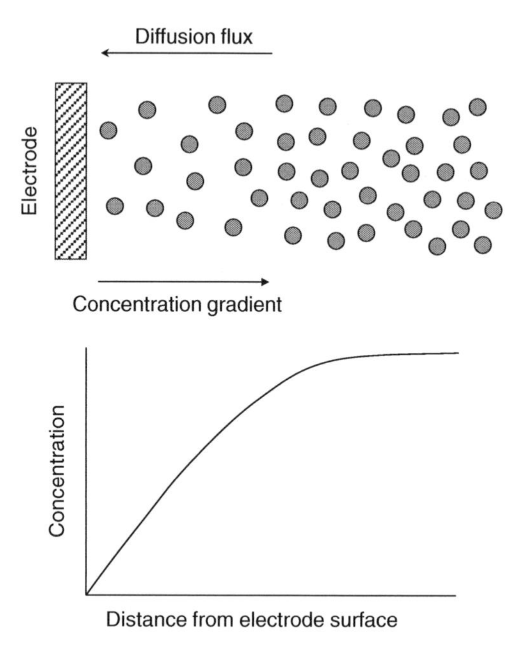

Fig. 8.7 Diffusion of a molecule or an ion toward an electrode surface

Diffusion and Random Walks

The driving force for the diffusion of a gas molecule (or an ion) in a solution is a difference in concentrations of the molecule (or ion) in different regions of the solution. On a microscopic scale, diffusion can be considered to be a random walk of molecules or ions. For the diffusion of O_2 molecules in water, for example, an O_2 molecule will travel only a short distance before it encounters a solvent H_2O molecule. As a result, the O_2 molecule wanders about to find a "space" between H_2O molecules and in doing so executes a random walk. (The concept of spaces or holes between molecules in the liquid phase arises from the theory of viscosity of liquids, in which liquid molecules flow by movement into vacant sites in a liquid lattice-like structure [6].)

In the classic one-dimensional random walk, a diffusing molecule can take a random step from the origin either to the right or to the left along the *X*-axis (each step having equal probability). The question is to determine how far the molecule can travel in some given time *t*. The distance traversed is referred to as a root mean square distance $\sqrt{<d^2>}$, because the actual net distance finally traveled may have either a positive sign (to the right of the origin) or a negative sign (to the left of the origin). In the random walk, each diffusing molecule is assumed to move independently of all the others, and each new step is independent of the previous step. For the diffusion of O_2 in H_2O, for example, the O_2–H_2O intermolecular forces which exist after a given step do not restrict the movement of the O_2 molecule in the next step.

The random motions of a diffusing molecule, such as O_2, actually occur in three dimensions. Thus one-dimensional space is not realistic, but three-dimensional space is more difficult to visualize, so instead let us consider two-dimensional random walks. Let the diffusing molecule be able to jump 1 distance unit in the directions N, NE, E, SE, S, SW, W, and NW, each occurring with equal probability 1/8. In this example, the jump distance is restricted to be 1 distance unit because it is unlikely that an O_2 molecule in a liquid would take large jumps without encountering water molecules.

Figure 8.8 shows several 20-step random walks which resulted from the use of a random number generator, with the integers 1–8 being used to designate the directions N, NE, E, etc. In each case, the root mean square distance $<d>$ is indicated in Fig. 8.8. The 20-step process shown in Fig. 8.8 was carried out 300 times using a simple computer program which inputs a sequence of 20 random

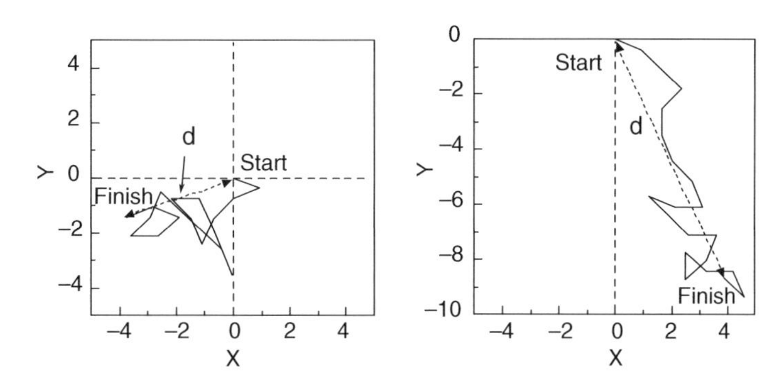

Fig. 8.8 Examples of a two-dimensional random walk showing the net distance (*d*) traveled in each case. Each random walk begins at the origin (0, 0)

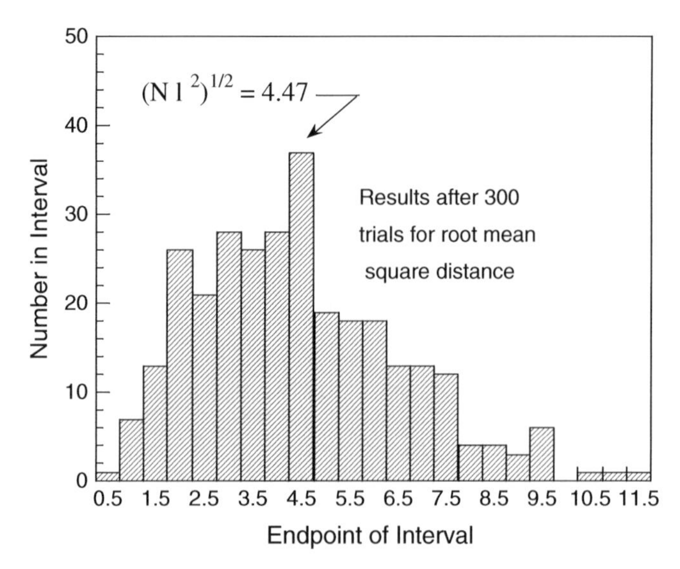

Fig. 8.9 Histogram of the root mean square distance traveled for 300 two-dimensional random walks having 20 jumps each of one distance unit

numbers varying from 1 to 8 and then tracks the successive change in co-ordinates as the particle in question is translated from the origin. The resulting net distance traveled d is obtained by calculating the distance between the origin and the final co-ordinates of the point. Figure 8.9 shows that after 300 trials, the calculated values of $<d>$ are distributed around a central range of 4.0–4.5 distance units.

For a random walk (in one, two, or three dimensions), the expected root mean square distance $<d^2>$ is given by [7]

$$<d^2> = Nl^2 \tag{5}$$

where N is the number of jumps and l is the average distance of a single jump. This equation is well known for a one-dimensional random walk [5–7]; the derivation of this equation for a two-dimensional walk is given in Appendix D. For the random walks constructed here, the expected root mean square distance is $<d> = (20 \times 1^2)^{1/2} = 4.47$ distance units. Thus, the observed histogram in Fig. 8.9 shows a distribution around the expected value of $<d>$. The distribution is not symmetrical around the mean expected value $<d>$. This is because distances less than $<d>$ are constrained by the lower limit of zero, but large net distances traveled, e.g., $<d> \approx 10$ distance units, are possible, though infrequent, as seen in the histogram. (It should be noted that the two-dimensional walk modeled here is not entirely random in that movement has been restricted to certain discrete directions and the jump distance has been confined to a unit length.)

An important result from random walk theory is that the root mean square distance from the random walk approach is related to the diffusion coefficient D. The relationship is [4, 5]

$$<d> = \sqrt{ADt} \tag{6}$$

where A is a constant and t is the time required to travel the distance $<d>$. The constant $A = 2$ for a one-dimensional random walk, $A = 4$ for a two-dimensional random walk, and $A = 6$ for a three-dimensional random walk [4, 5]. Equation (6) is called the Einstein–Smoluchowski equation.

Example 8.1: If a thin-layer electrolyte 0.3 mm in thickness [8] is condensed from the atmosphere onto a metal roof of a building in a seacoast location, how long will it take for oxygen gas to diffuse throughout the thickness of the condensed electrolyte?

Solution: Use the Einstein–Smoluchowski equation for three dimensions:

$$<d> = \sqrt{6Dt}$$

with $<d>$ being the thickness of the electrolyte. Because the condensed electrolyte will contain chloride ions, we should use a value of D characteristic of chloride solutions. We use $D = 2.0 \times 10^{-5}$ cm^2/s. See Table 8.2. then

$$0.03\text{cm} = \sqrt{6\left(\frac{2.0 \times 10^{-5}\text{cm}^2}{\text{s}}\right)(t\text{s})}$$

Solving for t gives: $t = 7.5$ s. Thus, diffusion of oxygen into the thin-film electrolyte occurs rapidly.

Table 8.2 Data for O_2 in aqueous solutions

Standard electrode potentials

Electrolyte	Cathodic reaction	E^o vs. SHE [33]
Neutral or alkaline solutions	$O_2 + 2H_2O + 4e^- \longrightarrow 4OH^-$	+0.401 V
Acidic solutions	$O_2 + 4H^+ + 4e^- \longrightarrow 2H_2O$	+1.229 V

Diffusion coefficients D for O_2 dissolved in water

Temperature (°C)	D (cm^2 s^{-1})	Source
15	1.67×10^{-5}	CRC Handbook [34]
20	2.01×10^{-5}	
25	2.42×10^{-5}	
25	1.9 to 2.6×10^{-5}	Compiled by Lorbeer and Lorenz [19]

Diffusion coefficients D for O_2 dissolved in electrolytes at 25°C

Electrolyte	Concentration	D (cm^2 s^{-1})	Source
NaCl	0.08–0.85 M	2.2×10^{-5}	van Stroe and Janssen [35]
KCl	0.05	2.08×10^{-5}	Ikeuchi et al. [36]
	1.0	2.00×10^{-5}	
	3.0	1.75×10^{-5}	
LiCl	0.05	2.18×10^{-5}	Ikeuchi et al. [36]
	1.0	1.91×10^{-5}	
	5.0	1.09×10^{-5}	
$MgCl_2$	0.05	1.61×10^{-5}	Ikeuchi et al. [36]
	1.0	1.34×10^{-5}	
	3.0	0.61×10^{-5}	
Na_2SO_4	0.5 M	1.57×10^{-5}	Lorbeer and Lorenz [19]

Electrode Kinetics for Concentration Polarization

Concentration Profile Near an Electrode Surface

When an electrolyte is open to the air, gaseous O_2 molecules first dissolve in the electrolyte and then diffuse throughout the liquid to become uniformly distributed therein, as described above. If a metal electrode is present in solution, then O_2 molecules which diffuse to the electrode surface can undergo reduction by the half-cell reaction

$$O_2 + 2H_2O + 4e^- \rightarrow 4OH^- \tag{1}$$

Thus, the concentration of dissolved O_2 near the electrode surface will decrease as O_2 molecules are converted to OH^- ions by the reaction above. The concentration profiles for dissolved O_2 molecules near the electrode surface are depicted schematically in Fig. 8.10. If all the dissolved O_2 near the metal surface are reduced to hydroxyl ions, the O_2 concentration at the metal surface will go to zero.

The actual concentration profile near a metal surface for a diffusing species is given by [9]

$$C = C^{\mathrm{o}} - \frac{1}{\sqrt{D}}\left[\frac{2\sqrt{t}}{\sqrt{\pi}}e^{-\frac{x^2}{4Dt}} - \frac{x}{\sqrt{D}}\mathrm{erfc}\left(\frac{x^2}{4Dt}\right)^{1/2}\right] \tag{7}$$

where C is the concentration of the diffusing species at a distance x from the electrode surface, C^{o} is the concentration in the bulk of the solution, D is the diffusion coefficient, and t is the elapsed time. The error function complement (erfc) has the form

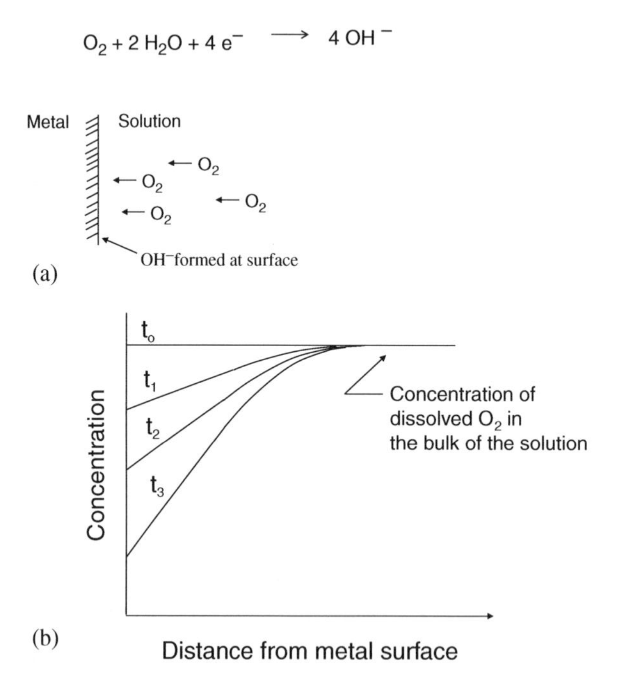

Fig. 8.10 (**a**) Oxygen depletion and OH^- production near a metal surface due to the oxygen reduction reaction. (**b**) Schematic diagram of oxygen concentration profiles near a metal surface ($t_o < t_1 < t_2 < t_3$)

$$\text{erfc}(w) = 1 - \frac{2}{\sqrt{\pi}} \int_0^w e^{-U^2} du \tag{8}$$

As will be seen in the next section, rather than using the expression in Eq. (7), the expression for the concentration profile near an electrode surface can be simplified considerably.

Limiting Diffusion Current Density

When a diffusing species participates in an electrochemical reaction, the flux (J) is related to the current density (i) by

$$J = -\frac{|i|}{nF} \tag{9}$$

where n is the number of electrons transferred and F is the Faraday. The minus sign results from the fact that the diffusion flux and the electron flux operate in opposite directions, as illustrated in Fig. 8.11. Using Fick's first law of diffusion, Eq. (4), in Eq. (9) and solving for the current density gives

$$|i| = nFD\frac{dC}{dx} \tag{10}$$

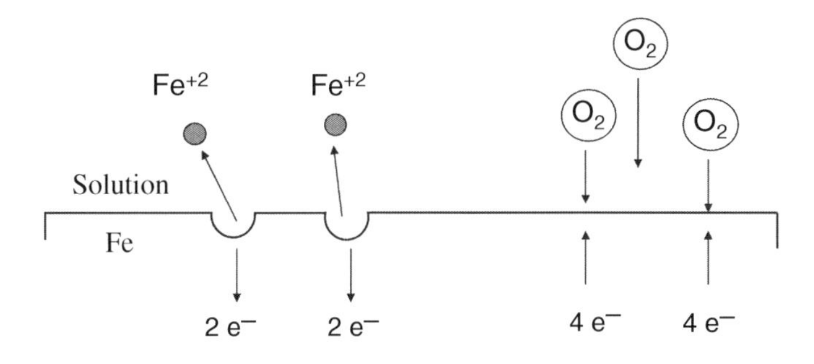

At anodic sites: $2[Fe \longrightarrow Fe^{+2} + 2e^-]$

At cathodic sites: $O_2 + 2\,H_2O + 4e^- \longrightarrow 4\,OH^-$

Overall reaction: $2\,Fe + O_2 + 2\,H_2O \longrightarrow 2\,Fe^{+2} + 4\,OH^-$

Fig. 8.11 Schematic diagram to show that the oxygen diffusion flux and the electron flux at a corroding metal have different directions. (So too do the electron flux and the diffusion flux due to metal cations which are generated)

The concentration gradient dC/dx can be obtained from Eq. (7), but instead the concentration profile can be simplified by using a linear approximation near the metal surface, as shown in Fig. 8.12. If we linearize the concentration gradient near the surface, then

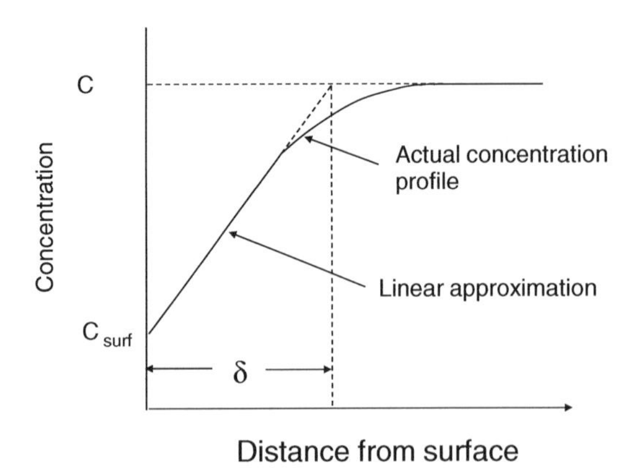

Fig. 8.12 The linear approximation to an O_2 concentration profile near a metal surface. The parameter δ is the thickness of the oxygen diffusion layer

$$|i| = nFD\,\frac{C - C_{\text{surf}}}{\delta} \tag{11}$$

The maximum current density is given by the condition that $C_{\text{surf}} = 0$. In the case of oxygen reduction, for example, this means that all oxygen molecules near the surface are reduced to OH^- ions, as per Eq. (1). Then Eq. (11) becomes

$$|i_{\text{L}}| = \frac{nFDC}{\delta} \tag{12}$$

where i_{L} is called the limiting diffusion current, C is the concentration of the diffusing species in the bulk electrolyte (in moles per cubic centimeter), and δ is the thickness of the diffusion layer.

Example 8.2: Calculate the thickness of the oxygen diffusion layer on iron in 0.6 M NaCl using the data in Fig. 8.2.

Solution: From Fig. 8.2, the limiting diffusion current density (i.e., the plateau value) for the cathodic reaction at 25°C is 25 μA/cm^2 = 2.5×10^{-5} A/cm^2. From Table 8.2, we use 2.0×10^{-5} cm^2/s for the diffusion coefficient of O_2 in NaCl. The concentration of Cl^- ion in 0.6 M NaCl is given by

$$\left(\frac{0.6\,\text{mol Cl}^-}{\text{L}}\right)\left(\frac{35.5\,\text{g Cl}^-}{\text{mol Cl}^-}\right)\left(\frac{1000\,\text{mg}}{\text{g}}\right) = \frac{21300\,\text{mg Cl}^-}{\text{L}}$$

Then, from tabulated data [2] or from Fig. 8.5, the concentration of dissolved O_2 is 6.7 mg/L. Then use

$$|i_{\text{L}}| = \frac{nFDC}{\delta}$$

$$2.5 \times 10^{-5}\,\frac{\text{A}}{\text{cm}^2} = \frac{\left(\frac{4\,\text{equiv}}{\text{mol}}\right)\left(\frac{96{,}500\,\text{coul}}{\text{equiv}}\right)\left(\frac{2.5\times10^{-5}\,\text{cm}^2}{\text{s}}\right)\left(\frac{6.7\times10^{-3}\,\text{g O}_2}{1000\,\text{cm}^3}\right)\left(\frac{1\,\text{mol}}{32\,\text{g O}_2}\right)}{\delta}$$

Solving for δ gives $\delta = 0.065$ cm.

Diffusion Layer vs. The Diffuse Layer

The diffusion layer is the thickness over which a concentration gradient exists for a diffusing electroactive species (such as oxygen). As seen in Example 8.2, the diffusion layer for the reduction of oxygen on iron in 0.06 M NaCl has a thickness of 0.065 cm or 6.5×10^6 Å.

Recall that the diffuse layer, i.e., the diffuse part of the electrical double layer (edl), is that outer portion of the edl which balances the charge on the inner part of the edl. As mentioned in Chapter 3, the thickness of the diffuse part of the edl is about 100 Å. Thus, the diffusion layer which arises from concentration gradients near an electrode surface is much greater in thickness than the diffuse part of the edl. In fact, for this example, the diffusion layer is about 65,000 times as thick as the diffuse layer. Thus, the entire edl including its diffuse (Gouy–Chapman) component fits easily inside the oxygen diffusion layer on iron in 0.06 M NaCl.

The water molecule can be used as a metric to size the thickness of the O_2 diffusion layer. From the density and the molecular weight of water, the diameter of a water molecule is easily calculated to be 3.8 Å Thus, the O_2 diffusion layer would embrace 1,600,000 water molecules laid end to end. In contrast, the diffuse part of the edl would contain 26 H_2O molecules laid end to end.

Current–Potential Relationship for Concentration Polarization

The most important instance of concentration polarization in corrosion applications is the diffusion-limited reduction of oxygen:

$$O_2 + 2H_2O + 4e^- \rightarrow 4OH^- \tag{1}$$

For this reaction to occur, O_2 molecules first diffuse up to the electrode surface, are next reduced to OH^- ions at cathodic sites on the metal surface, and finally the OH^- ions which are formed diffuse away from the electrode out toward the bulk of the solution. Because diffusion is a slow process, this sequence of events occurs under conditions of equilibrium or near-equilibrium. Thus, the Nernst equation can be used to describe the half-cell reaction. The Nernst equation for Eq. (1) is

$$E = E^o - \frac{2.303\,RT}{4F} \log \frac{[OH^-]^4}{[O_2]_{surf}} \tag{13}$$

where E^o is the standard electrode potential for Eq. (1), $[OH^-]$ is the concentration of the hydroxyl ions which are formed, and $[O_2]_{surf}$ is the concentration of dissolved oxygen at the electrode surface. The quantity $[O_2]_{surf}$ can be obtained by dividing Eq. (11) by Eq. (12) to give

$$\frac{|i|}{|i_L|} = 1 - \frac{[O_2]_{surf}}{[O_2]} \tag{14}$$

where $[O_2]$ is the concentration of dissolved oxygen in the bulk of the electrolyte. Thus,

$$[O_2]_{surf} = [O_2]\left(1 - \frac{|i|}{|i_L|}\right) \tag{15}$$

Use of Eq. (15) in Eq. (13) gives

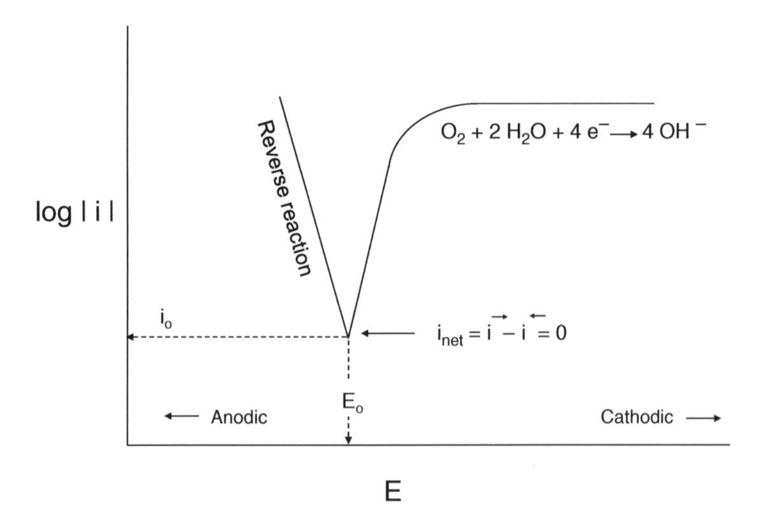

Fig. 8.13 Schematic Evans diagrams for O_2 reduction and OH^- oxidation

$$E_o = E^o - \frac{2.303\,RT}{4F}\log\frac{[OH^-]^4}{[O_2]} + \frac{2.303\,RT}{4F}\log\left(1 - \frac{|i|}{|i_L|}\right) \tag{16}$$

When there is no net current flow, $i = 0$ and $E = E_o$. See Fig. 8.13. Note the difference between E_o (the potential corresponding to zero net current flow) and E^o (the standard electrode potential). Under the condition of zero net current, Eq. (16) becomes

$$E_o = E^o - \frac{2.303\,RT}{4F}\log\frac{[OH^-]^4}{[O_2]} \tag{17}$$

Equation (17) describes how the thermodynamic standard electrode potential E^o is modified by the particular pH of the solution and the given concentration of dissolved O_2 to yield the potential E_o of zero net current flow for the oxygen half-cell reaction. Use of Eq. (17) in Eq. (16) gives

$$E = E_o - \frac{2.303\,RT}{4F}\log\left(1 - \frac{|i|}{|i_L|}\right) \tag{18}$$

With $E - E_o = \eta_{conc}$, the overvoltage due to concentration polarization, Eq. (18), becomes

$$\eta_{conc} = -\frac{2.303\,RT}{4F}\log\left(1 - \frac{|i|}{|i_L|}\right) \tag{19}$$

A plot of Eq. (19) is shown in Fig. 8.14. Figure 8.14 shows clearly that the cathodic current density due to concentration polarization reaches a limiting value and that the limiting value is reached in this instance after a cathodic overvoltage of approximately –0.040 V. The plot in Fig. 8.14 is similar to the experimental cathodic curve in Fig. 8.2, where the limiting current is attained after a slightly larger overvoltage (due to the existence of the back anodic reaction due to iron dissolution).

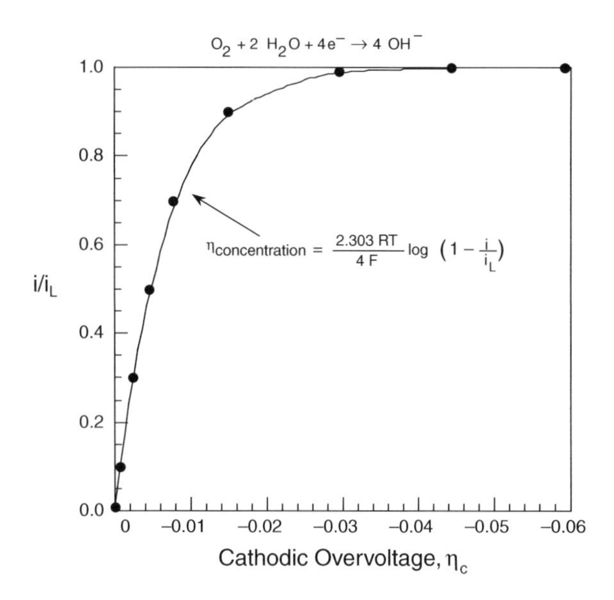

Fig. 8.14 Plot of Eq. (19) showing the attainment of a diffusion-controlled limiting cathodic current density

Wagner–Traud Theory for Concentration Polarization

In the previous chapter, it has been shown that the experimentally determined polarization curve for a corroding metal under activation control is given by the addition of the individual polarization curves for the two oxidation–reduction reactions. The same approach holds if one of the individual oxidation–reduction reactions is under diffusion control (concentration polarization). This will be illustrated for the corrosion of iron in quiescent (unstirred) solutions open to the air.

The anodic reaction is the usual

$$\mathrm{Fe} \rightarrow \mathrm{Fe}^{2+} + 2e^{-} \tag{2}$$

and the cathodic reaction is the reduction of O_2 according to the following reaction:

$$\mathrm{O_2} + 2\mathrm{H_2O} + 4e^{-} \rightarrow 4\mathrm{OH}^{-} \tag{1}$$

Schematic Evans diagrams are shown in Fig. 8.15. The anodic dissolution of Fe is usually under activation control, as shown in Fig. 8.15, but the cathodic polarization curve in the unstirred air-saturated solution is under diffusion control, as also shown in the figure. Both partial polarization curves are drawn using the standard electrode potentials given in Tables 3.1 and 8.2, although each of these standard electrode potentials would be modified by use of the appropriate Nernst equation if Eqs. (1) and (2) proceeded under non-standard conditions.

The corrosion potential and the corrosion rate are given by conditions at the intersection of the total anodic and total cathodic polarization curves. It can be seen that the corrosion rate is in fact equal to the limiting current density for oxygen reduction. That is, the corrosion rate for iron is controlled by the rate of the oxygen reduction reaction. This is because at the corrosion potential

$$\overleftarrow{i_{\mathrm{Fe}}} + \overleftarrow{i_{\mathrm{O_2}}} = |\overrightarrow{i_{\mathrm{Fe}}}| + |\overrightarrow{i_{\mathrm{O_2}}}| \tag{20}$$

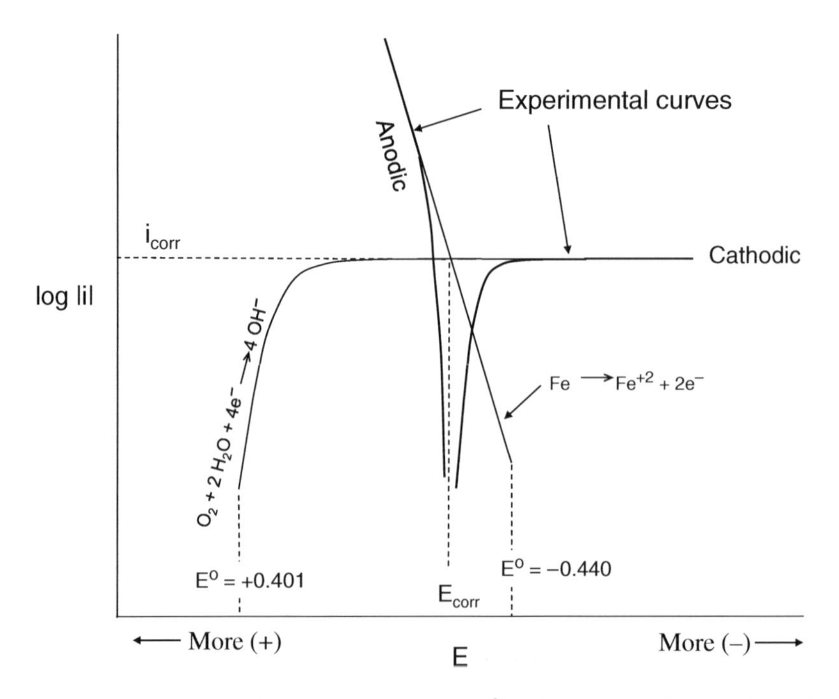

Fig. 8.15 Schematic Evans diagrams for the O_2/OH^- and Fe/Fe^{2+} oxidation–reduction couples and the resulting experimental polarization curves

where the notation is the same as used in the previous chapter.

Thus

$$\overleftarrow{i_{Fe}} - |\overrightarrow{i_{Fe}}| = |\overrightarrow{i_{O_2}}| - \overleftarrow{i_{O_2}} \tag{21}$$

That is, the net rate of iron corrosion is equal to the net rate of oxygen reduction.

The experimental polarization curves as derived from the Evans diagrams are also shown schematically in Fig. 8.15. As before, the net anodic current density at any potential is the total anodic current density minus the total cathodic current density. An analogous expression holds for the cathodic branch of the curve. The schematic experimental polarization curves are similar to the actual experimental curves observed in Fig. 8.2.

Thus, extrapolating the anodic Tafel slope for iron dissolution back to E_{corr} gives the corrosion rate, which is the same as the limiting diffusion current density for oxygen reduction. This is a classic case in which the corrosion reaction is under cathodic control.

Effect of Environmental Factors on Concentration Polarization and Corrosion

As was just shown, the corrosion rate of a metal is determined by the rate of oxygen reduction when oxygen reduction is under diffusion control. Thus, environmental factors which affect the limiting diffusion current for oxygen reduction will, in turn, affect the corrosion rate. The effect of several different environmental factors is considered below.

Effect of Oxygen Concentration

Figure 8.16 shows schematically the effect of increasing the concentration of dissolved oxygen on the limiting current density for oxygen reduction. Increasing the concentration of dissolved O_2increases the limiting current density i_L. This effect stems directly from Eq. (12), in which it is seen that i_L is directly proportional to C, the concentration of dissolved O_2.

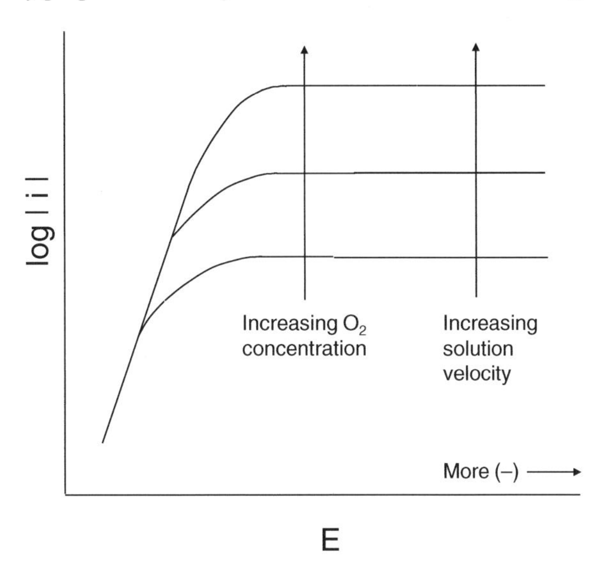

Fig. 8.16 Schematic diagram showing the effect of O_2 concentration and solution velocity on the limiting diffusion current for O_2 reduction

Various experimental studies have shown that the corrosion rate increases with increasing O_2 content of the electrolyte [10, 11]. For example, Uhlig et al. [10, 11] have observed that for the corrosion of mild steel in water containing small amounts of $CaCl_2$, the corrosion rate increases with increasing O_2 concentration, as shown in Fig. 8.17. This effect can be explained by the fact that the anodic polarization curve for iron dissolution intersects the cathodic polarization curve for oxygen reduction at increased values of i_L, which result for increased O_2 concentrations, as shown in Fig. 8.18.

However, Uhlig et al. [10, 11] also observed that additional increases in O_2 concentration beyond the range shown in Fig. 8.17 produced a decrease in the corrosion rate, as shown in Fig. 8.19. This decrease is due to the onset of a new anodic reaction, i.e., the passivation of the iron surface as it received an additional supply of oxygen. This increased supply of O_2 serves to form a protective oxide film on the surface, which acts to lower the corrosion rate. (More details on passivity are given in Chapter 9).

In systems where the corrosion rate is under the control of the oxygen reduction reaction, the addition of certain chemical compounds can reduce the corrosion rate by reducing the concentration of dissolved O_2. These chemicals are called "oxygen scavengers," and two which are used frequently are sodium sulfite (Na_2SO_3) and hydrazine (N_2H_4). Sodium sulfite reacts with O_2 as follows:

$$2Na_2SO_3 + O_2 \rightarrow 2Na_2SO_4$$

and thus serves to reduce the concentration of dissolved oxygen so as to lower the limiting diffusion current density, as per Eq. (12). The corrosion rate is then lowered, as in Fig. 8.18.

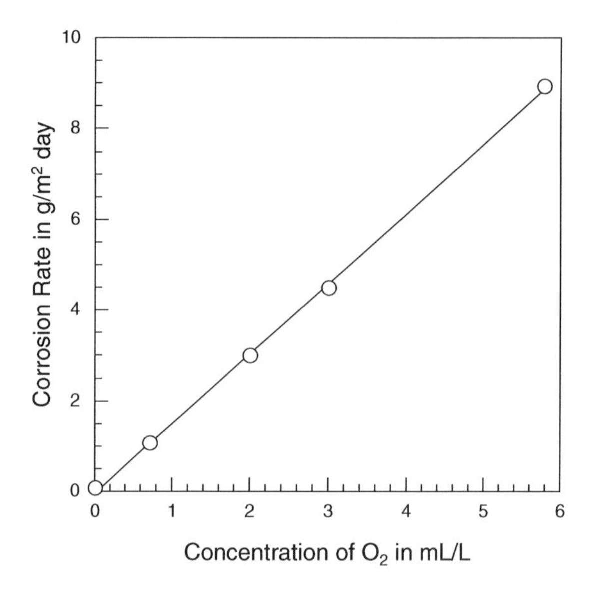

Fig. 8.17 Effect of oxygen concentration on the corrosion of mild steel in water containing 165 ppm $CaCl_2$. Redrawn from Uhlig and Revie [10] by permission of John Wiley & Sons, Inc

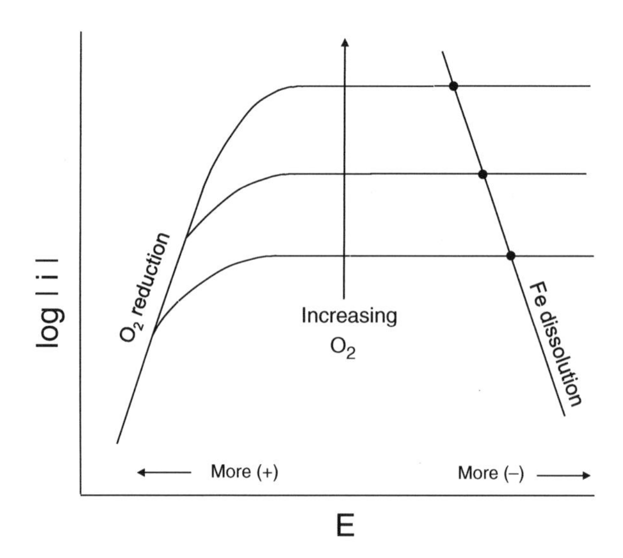

Fig. 8.18 Schematic Evans diagrams for Fe in aqueous solutions where the oxygen reduction reaction is diffusion controlled. The intersection of the oxygen reduction and iron dissolution curves gives the corrosion potential and the corrosion rate

Effect of Solution Velocity

Figure 8.16 also shows the effect of increasing velocity on the limiting diffusion current for oxygen reduction. Increasing the velocity of the solution brings reactants up to the electrode surface more easily, thus reducing δ, the thickness of the oxygen diffusion layer. As δ decreases, the value of the limiting diffusion current i_L will increase, as per Eq. (12).

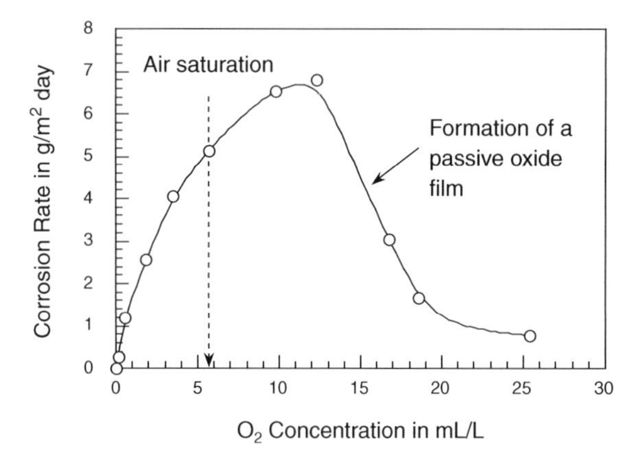

Fig. 8.19 Effect of oxygen concentration on the corrosion of mild steel in slowly moving distilled water. Redrawn from Uhlig and Revie [10] by permission of John Wiley & Sons, Inc

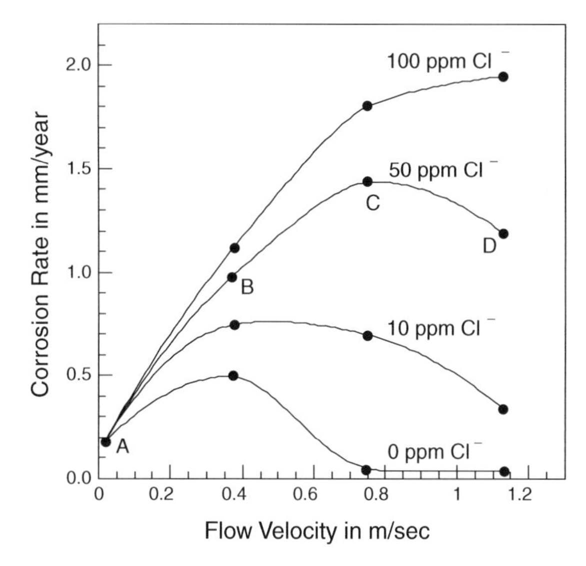

Fig. 8.20 Effect of solution velocity on the corrosion of carbon steel in water containing various amounts of dissolved Cl^-. Redrawn from Matsushima [12] by permission of John Wiley & Sons, Inc

Figure 8.20 shows the effect of solution velocity on the corrosion of a carbon steel in water containing various amounts of dissolved Cl^- (12). For the highest concentration of Cl^- (100 ppm), the corrosion rate increases with increasing velocity, as described above. However, for a lower concentration of dissolved Cl^- (e.g., 50 ppm), the corrosion rate first increases with increasing velocity up to point C in Fig. 8.20 and then decreases with increasing velocity (point D). This decrease is again due to an increased supply of O_2 and to the concomitant passivation of the metal surface, as shown schematically in Fig. 8.21.

In a more aggressive environment, such as natural seawater, where passivation does not occur, the corrosion rate of steel increases with increasing velocity [13].

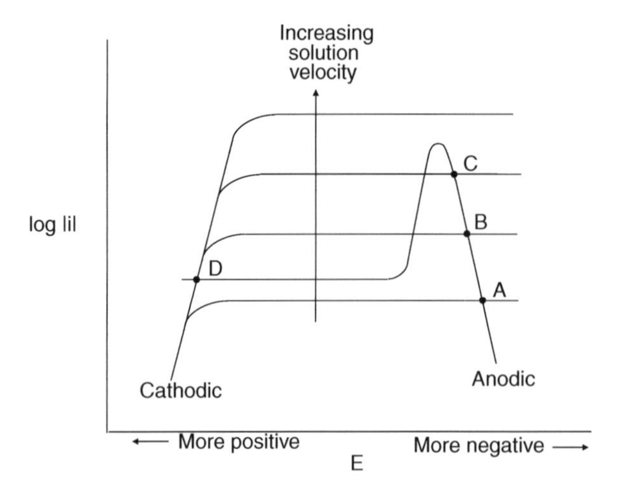

Fig. 8.21 Schematic Evans diagrams showing the effect of anodic passivation on the corrosion behavior when the O_2 reduction reaction is under diffusion control. The corrosion rates are given by the points A through D, which correspond to similar points in Fig. 8.20

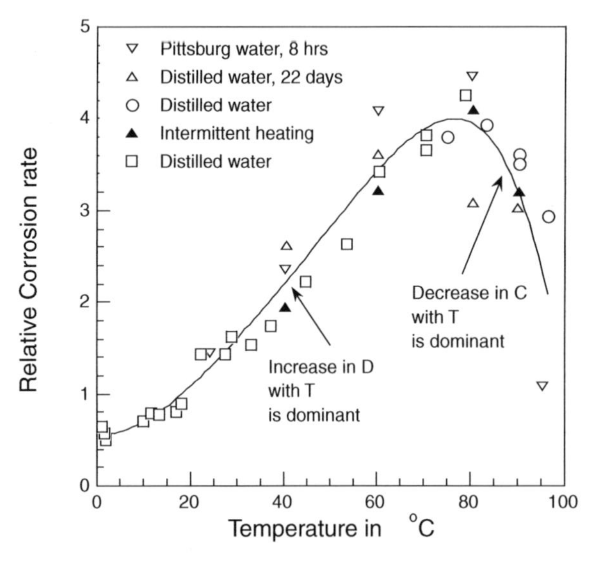

Fig. 8.22 Corrosion rate as a function of temperature for iron in water open to the atmosphere. Redrawn from Butler and Ison [14]

Effect of Temperature

Figure 8.22 shows the effect of temperature on the corrosion rate of iron in water open to the atmosphere and containing dissolved oxygen [14]. The corrosion rate first increases with increasing temperature up to about 80°C and then decreases with further increases in temperature.

There are two competing effects at work here. First, the diffusion coefficient D increases with temperature, and the effect is that D increases about 2% per degree centigrade (see Table 8.2). With an increase in D, the limiting diffusion current density i_L also increases, as per Eq. (12), and thus

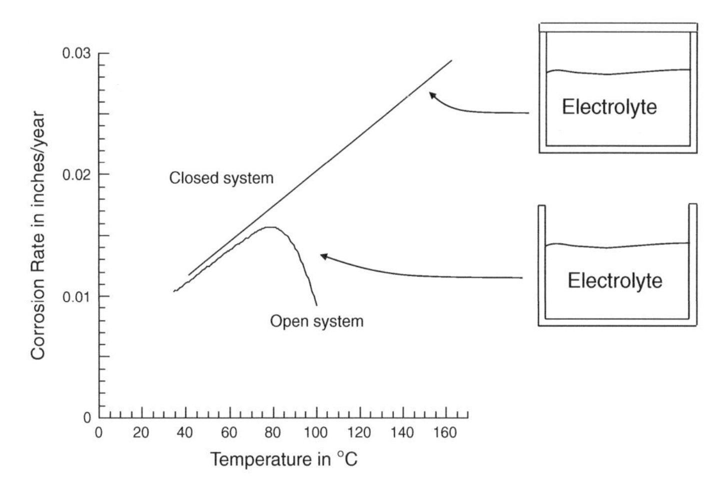

Fig. 8.23 Schematic diagram showing the effect of temperature on the corrosion rate for a metal immersed in an aqueous solution in an open vs. a closed system. Redrawn from Uhlig and Revie [10] by permission of John Wiley & Sons, Inc

so will the corrosion rate. This effect is predominant for the portion of the curve in Fig. 8.22 over which the corrosion rate increases with temperature.

The second temperature effect is that the concentration of dissolved O_2 decreases with increasing temperature, as discussed earlier. Thus, as the O_2 concentration decreases, so does i_L, as per Eq. (12), and accordingly, so does the corrosion rate. This effect is predominant for the portion of the curve in Fig. 8.22 over which the corrosion rate decreases with increasing temperature.

For a closed system, the effect of increasing temperature in an oxygen-containing electrolyte is somewhat different and is shown schematically in Fig. 8.23 [10]. In the closed system, dissolved O_2 molecules which enter the vapor phase cannot escape out into the open atmosphere, but instead O_2 molecules first accumulate in the vapor phase above the liquid. Then, the pressure of O_2 above the liquid increases, which in turn produces an increase in the concentration of dissolved O_2 in the electrolyte by Henry's law, Eq. (3). This effect predominates so that the corrosion rate increases monotonically with increasing temperature for a closed system.

Further Applications of Concentration Polarization Curves

Cathodic Protection

The corrosion rate of iron in seawater is 15 mpy (3.3×10^{-5} A/cm^2), and the corrosion rate of zinc in seawater is 2 mpy (3.4×10^{-6} A/cm^2) [15]. For both metals in seawater, the corrosion rate is determined by the rate of oxygen reduction, which is under diffusion control. Thus, we can construct the Evans diagrams and the experimental polarization curves for both metals, as shown in Fig. 8.24.

The open-circuit corrosion potential for uncoupled zinc is more negative than that for iron, so when the two metals are coupled, the zinc electrode is the anode of the pair. This is shown in Fig. 8.25, which displays several important features:

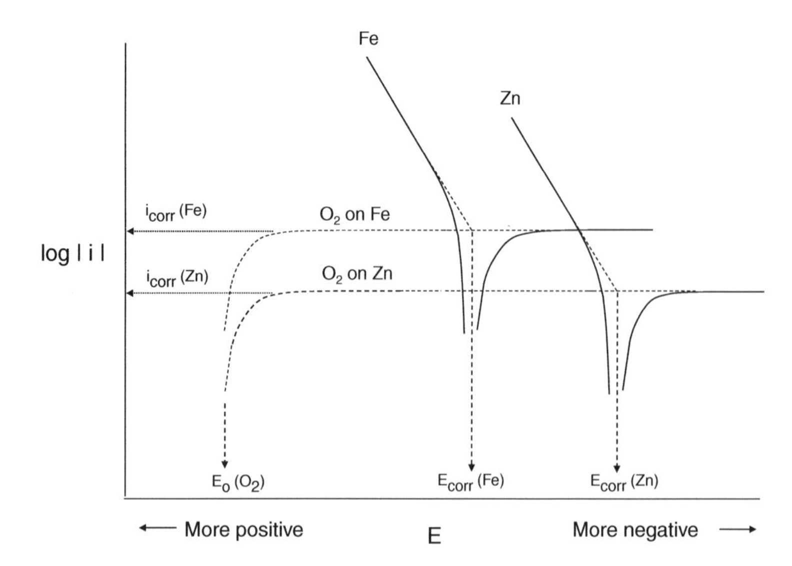

Fig. 8.24 Schematic Evans diagrams for iron and zinc immersed in an aqueous solution where the cathodic reaction is O_2 reduction under diffusion control. The experimentally observed polarization curves are given for each metal by the solid lines

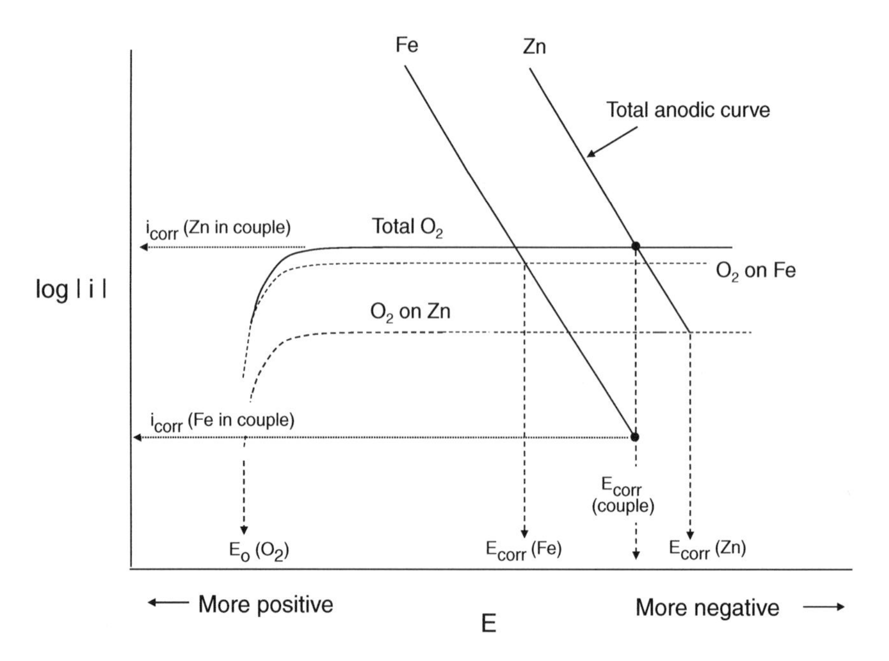

Fig. 8.25 Effect of galvanic coupling on the Fe/Zn system shown in Fig. 8.24

(1) The total cathodic curve for the Fe/Zn couple intersects the total anodic curve for the couple at an electrode potential which is slightly different than the corrosion potential of the uncoupled zinc. (This observation is often described as iron being polarized to the potential of zinc.)
(2) The corrosion rate of zinc in the Fe/Zn couple is higher than the corrosion rate of uncoupled zinc.

(3) The corrosion rate of iron in the Fe/Zn couple is much lower than the corrosion rate of uncoupled iron.

Figure 8.25 forms the kinetic basis for cathodic protection in neutral electrolytes where the corrosion rate of the individual metals is determined by the diffusion-limited oxygen reduction reaction. Compare Fig. 8.25 with Fig. 5.5, which presents a thermodynamic view of cathodic protection.

Area Effects in Galvanic Corrosion

When copper and iron are coupled in seawater, iron is the anode and copper is the cathode, as may be seen by referring to the galvanic series in Fig. 5.3. The effect of increasing the area of the copper cathode is shown schematically in Fig. 8.26 (a). Increasing the cathodic area results in an increase in the value of the limiting diffusion current (i.e., the *total current*, not the current density). Thus the total cathodic curve intersects the total anodic curve (for a fixed area of the iron anode) at increasingly higher values. This intersection defines the corrosion rate so that the corrosion rate increases with the area of the cathode.

Actual data are given in Fig. 8.26(b) for copper and iron coupled in seawater [16]. This effect is similar to that seen in Chapter 7 for galvanic couples in acid solutions.

Linear Polarization

The linear polarization technique can also be used to determine the corrosion rate if the cathodic reaction is under concentration polarization. From Chapter 7, the Stern–Geary equation for linear polarization near the corrosion potential is

$$i_{\mathrm{corr}} = \frac{1}{2.303\left(\frac{\mathrm{d}\eta}{\mathrm{d}i}\right)_{\eta\to 0}\left(\frac{1}{b_{\mathrm{a}}}+\frac{1}{|b_{\mathrm{c}}|}\right)} \tag{22}$$

This equation follows from the Butler–Volmer equation and, strictly speaking, is applicable only for the situation where both the anodic and cathodic polarization curves are under activation control. However, if the cathodic process is controlled by concentration polarization, the "effective" cathodic Tafel slope $b_{\mathrm{c}} = \mathrm{d}E/\mathrm{d}\log|i|$ can be taken to be infinity for the region where a limiting diffusion current density is observed. This can be seen by referring to Fig. 8.2. Thus, with $|b_{\mathrm{c}}| = \infty$, Eq. (22) reduces to

$$i_{\mathrm{corr}} = \frac{1}{2.303\left(\frac{\mathrm{d}\eta}{\mathrm{d}i}\right)_{\eta\to 0}\left(\frac{1}{b_{\mathrm{a}}}\right)} \tag{23}$$

for a system in which the cathodic process is under diffusion control.

Example 8.3: In a study on the corrosion of mild steel in an alkaline lithium bromide solution [17], the cathodic process was observed to be under the control of diffusion-limited concentration polarization, and the anodic process was under activation control with an anodic Tafel slope of +0.062 V. Linear polarization measurements in the cathodic region near the corrosion potential gave a value of $\mathrm{d}\eta/\mathrm{d}i = 0.089\ \mathrm{V\ cm^2/\mu a}$. What is the corrosion rate of mild steel in this solution?
Solution: Use the Stern–Geary equation with $|b_{\mathrm{c}}| = \infty$.

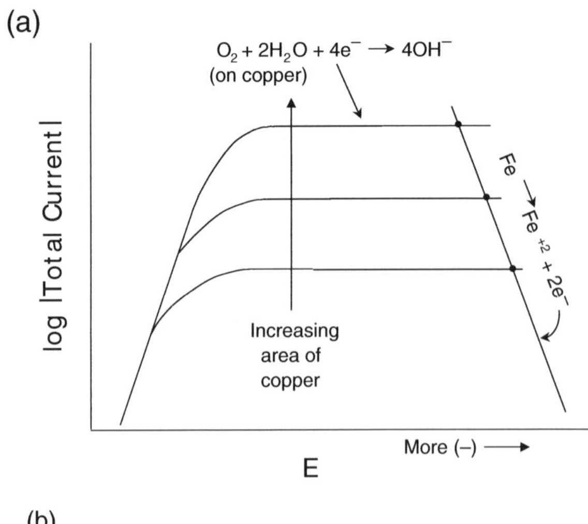

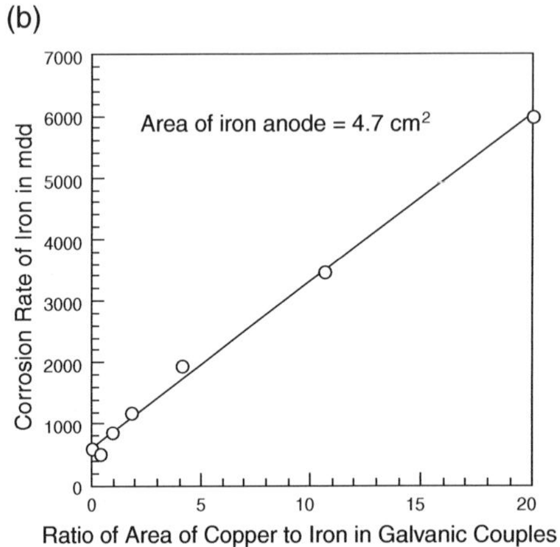

Fig. 8.26 (**a**) Schematic Evans diagrams showing the effect of increasing cathode area on the corrosion rate of a copper/iron couple when the O_2 reduction reaction is under diffusion control. The intersection of the anodic and cathodic polarization curves defines the corrosion potential and the corrosion rate. (**b**) Effect of cathode area on the experimental corrosion rates for a copper/iron couple in seawater. Reproduced by permission of ECS – The Electrochemical Society

$$i_{\text{corr}} = \frac{1}{2.303\left(\frac{0.089\ \text{V cm2}}{\mu\text{A}}\right)\left(\frac{1}{0.062\ \text{V}} + \frac{1}{\infty}\right)}$$

Then, with $1/\infty = 0$, we can solve for i_{corr} to get $i_{\text{corr}} = 0.30\ \mu\text{A/cm}^2$.

Concentration Polarization in Acid Solutions

In the previous chapter, we considered that hydrogen reduction in acid solutions was under activation control. That is, a cathodic Tafel region was observed experimentally. The half-cell reaction is the usual

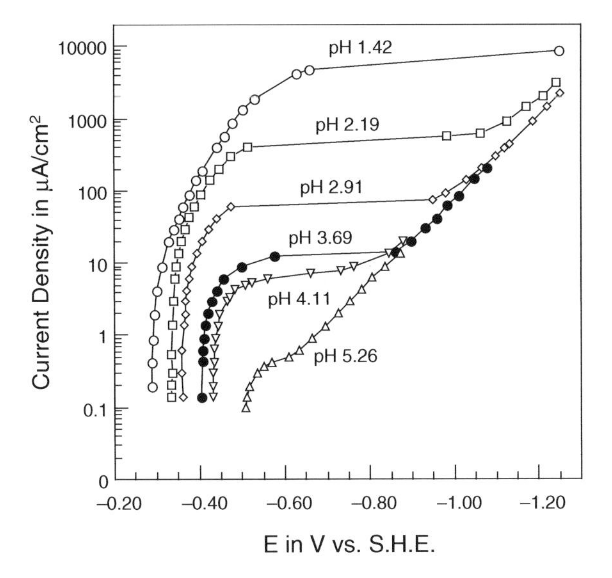

Fig. 8.27 Experimental cathodic polarization curves for iron in unstirred, H_2-saturated acid solutions [18]. Reproduced by permission of ECS – The Electrochemical Society

$$2H^+ + 2e^- \rightarrow H_2$$

Stern [18] has shown that the hydrogen reduction reaction in unstirred air-free, H_2-saturated solutions can undergo concentration polarization. Stern's results are shown in Fig. 8.27, in which it is seen that the limiting diffusion current density for hydrogen reduction increases with decreasing pH (i.e., with increasing acidity). This result follows by writing Eq. (12) for the limiting diffusion current density for hydrogen reduction:

$$i_L = \frac{nFD\,(10^{-3})\,a(H^+)}{\delta} \tag{24}$$

where $a(H^+)$ is the hydrogen ion activity in moles per liter. As $a(H^+)$ increases, (pH decreases), i_L increases, as per Eq. (24), as was observed by Stern. Stern took D, the diffusion coefficient for H^+, as 7.39×10^{-5} cm^2/s, $\delta = 0.05$ cm, and $n = 1$ (see Chapter 7, which briefly discusses two different mechanisms for the hydrogen evolution reaction shown above). The overall number of electrons transferred is two, but either reaction mechanism proceeds in two consecutive *one-electron* transfer steps. Use of the parameters listed above in Eq. (24) gives

$$\log i_L = -0.845 - \text{pH} \tag{25}$$

Stern's results for the limiting diffusion current density for hydrogen reduction as a function of pH are given in Fig. 8.28. The observed experimental values for i_L are in good agreement with values calculated using Eq. (25).

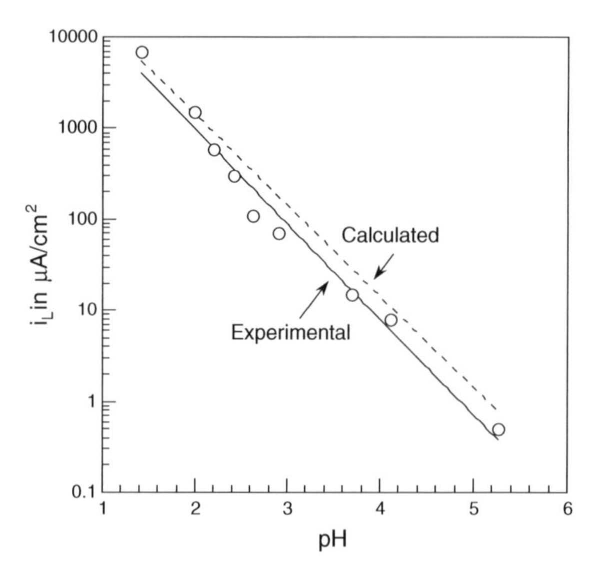

Fig. 8.28 Comparison of experimental and calculated limiting diffusion current densities for hydrogen reduction on iron [18] in Fig. 8.27. Calculated values were obtained using Eq. (24). Reproduced by permission of ECS – The Electrochemical Society

Combined Activation and Concentration Polarization

Activation polarization and concentration polarization can occur simultaneously for the same cathodic reaction. This situation is illustrated in Fig. 8.29, in which it is seen that the total cathodic polarization curve is the sum of the two individual processes. These are the types of curves observed by Stern [18] in Fig. 8.27 for hydrogen reduction in unstirred acid solutions.

It is also possible for O_2 reduction and H_2 evolution to occur in the same solution. This may take place in acid solutions open to the air. In this case, the O_2 reduction reaction can be under diffusion control, and the H^+ reduction reaction can be under activation control. Then, both a limiting diffusion current density for O_2 reduction (at lower cathodic overvoltages) and a Tafel region for H^+ reduction (at higher cathodic overvoltages) are observed, as illustrated in Fig. 8.30. Such polarization curves have been observed for iron [19, 20] and cadmium [21] in acidified solutions open to the air and containing dissolved oxygen.

The Rotating Disc Electrode

Throughout this chapter we have referred to solutions as being quiescent (unstirred) or stirred. Thus, flow conditions have not been stated rigorously, although it has been noted that the limiting diffusion current density increases if the solution velocity increases. Reproducible control of flow conditions can be achieved using a rotating disc electrode, as shown in Fig. 8.31.

If a planar disc rotates at a high speed in the liquid, the liquid layer close to the electrode moves under centrifugal force toward the perimeter of the disc, as shown in Fig. 8.31 [22]. Liquid from the bulk of the electrolyte will flow toward the center of the disc.

Under these conditions, the limiting diffusion current density is given by the Levich equation [22, 23]:

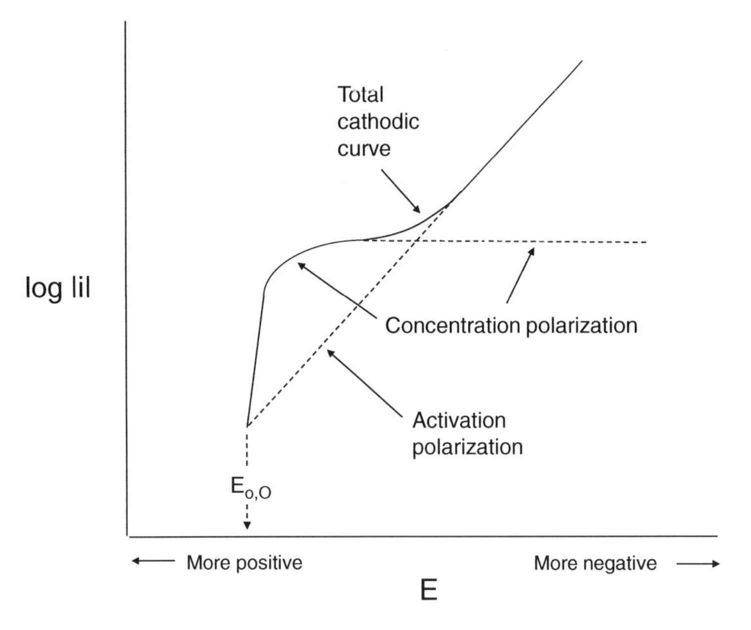

Fig. 8.29 Schematic representation of co-existing activation polarization and concentration polarization for a cathodic reaction

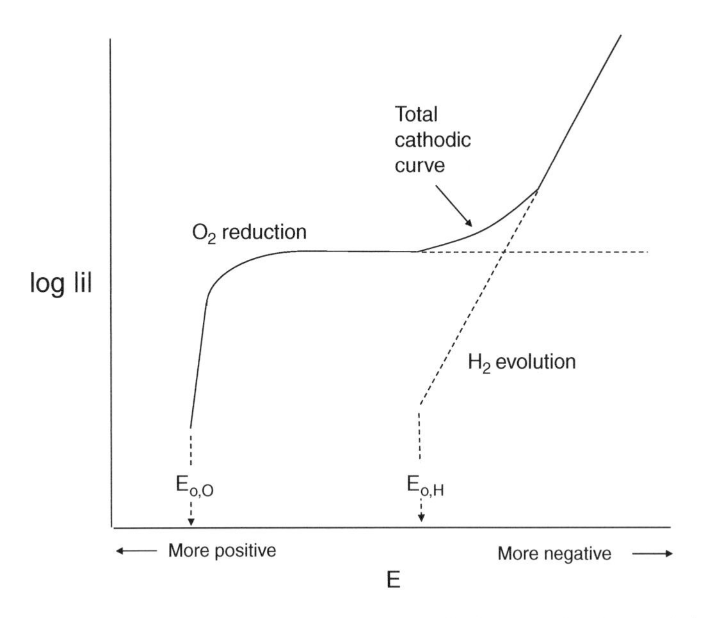

Fig. 8.30 Two concurrent cathodic reactions – O_2 reduction under diffusion control and H_2 evolution under activation control

$$|i_L| = 0.62\, n\, FD^{\frac{2}{3}}\, \omega^{\frac{1}{2}}\, \nu^{-\frac{1}{6}} C^{\text{o}} \tag{26}$$

where i_L is the limiting diffusion current density, n is the number of electrons transferred, D is the diffusion coefficient of the reacting species of bulk concentration C^{o} (moles per cubic centimeter),

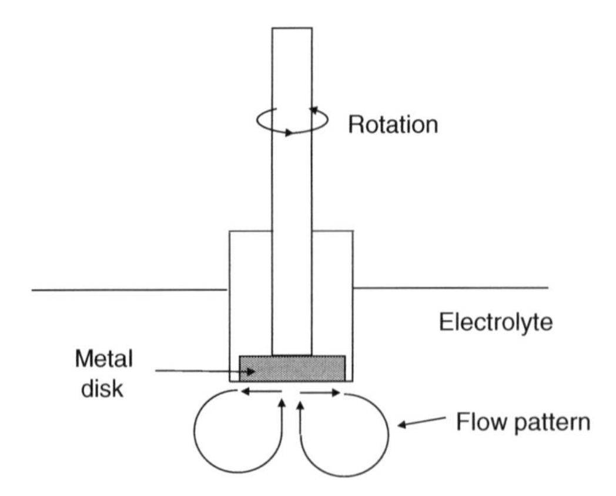

Fig. 8.31 Liquid flow patterns in the vicinity of a rapidly rotating disc electrode [22]

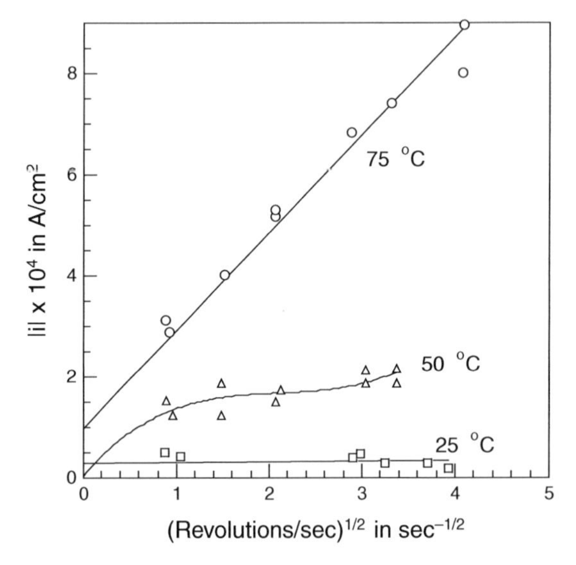

Fig. 8.32 Effect of stirring on the corrosion rate of copper in 0.1 M H_2SO_4 solutions at various temperatures [24, 25]. Reproduced by permission of Elsevier Ltd

ω is the angular frequency of rotation of the disc (2π times the rotation frequency), and ν is the kinematic viscosity (in square centimeter per second).

Rotating disc electrodes have found a large number of applications in mechanistic studies of corrosion processes (as well as in electrodeposition, electrocatalysis, and electrochemical machining). Only two examples will be given here.

A rotating disc electrode was used by Zembura [24, 25] to follow the corrosion of copper as the reaction changed from activation control to diffusion control with increasing temperature. Figure 8.32 shows the corrosion current density of copper in air-saturated 0.1 M H_2SO_4 as a function of the square root of the rotation velocity [24, 25]. At 25°C, the reaction is under activation control as the corrosion current density is independent of rotation rate. At 75°C, the corrosion current density obeys the Levich equation, and the reaction is under diffusion control. At an intermediate temperature of 50°C, the reaction is transitioning from activation control to diffusion control.

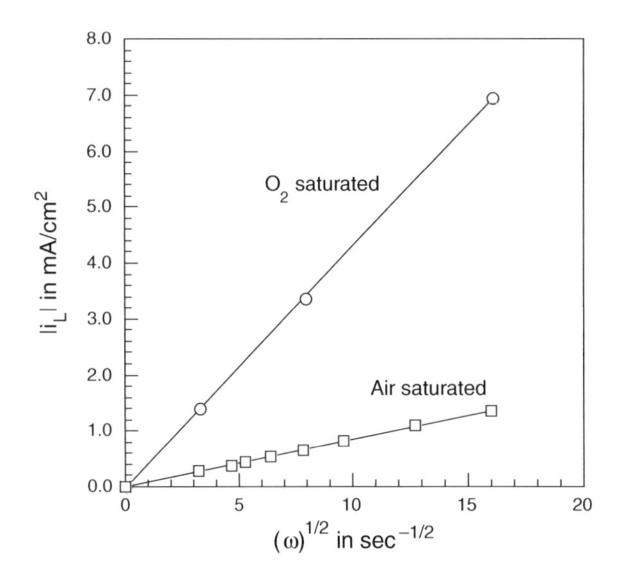

Fig. 8.33 Levich plots for iron in 0.1 M NaCl solutions at pH 7 [26]. Reproduced by permission of Elsevier Ltd

In a study on the oxygen reduction reaction on rust-free iron surfaces in 0.1 M NaCl solutions, Miyata and Asakura [26] reported the Levich plots shown in Fig. 8.33. From the slope of the Levich plots and published values for the O_2 concentration, diffusion coefficient, and kinematic viscosity of the solution, Miyata and Asakura calculated the number of electrons transferred to be $n = 3.9$ and 3.8 for aerated and O_2-saturated solutions, respectively. This result is consistent with the following overall reduction reaction which has been assumed throughout this chapter:

$$O_2 + 2H_2O + 4e^- \rightarrow 4OH^-$$

However, it is unlikely that all four electrons are transferred simultaneously in a single step. Various mechanisms have been proposed for the cathodic reduction of oxygen [27], but the exact mechanism is not known.

A variation of the rotating disc electrode is the rotating ring disc electrode (RRDE). In the RRDE, a rotating disc is surrounded by a concentric co-planar ring electrode, with the ring electrode being electrically isolated from the disc. The ring electrode does not disturb the flow pattern along the disc so that cations produced by corrosion at the disc flow outward toward the ring. Moreover, the potentials of the ring and the disc electrodes can be varied independently so that cations produced at the disc can be reduced at the ring electrode and identified by their reduction potentials. Thus, this technique can be used to study the mechanism of corrosion of metal alloys [28].

Problems

1. Nitrate ions (NO_3^-) are often used to inhibit the corrosion of iron in aqueous solutions. Suppose a nitrate ion takes 10,000 random three-dimensional jumps in water, with the distance of each jump being the diameter of a NO_3^- ion (3.0 Å).

(a) What is the expected root mean square distance traveled by the NO_3^- ion after 10,000 jumps?
(b) If the diffusion coefficient of the NO_3^- ion is 1.0×10^{-5} cm²/s, what is the jump frequency (number of jumps per second)?

2. If an amine inhibitor is delivered into the solution just outside the opening of a crevice which is 2.0 cm in depth, as shown below, how long will it take for the amine molecule to diffuse into the deepest part of the crevice if the diffusion coefficient of the amine is 5.0×10^{-5} cm²/s?

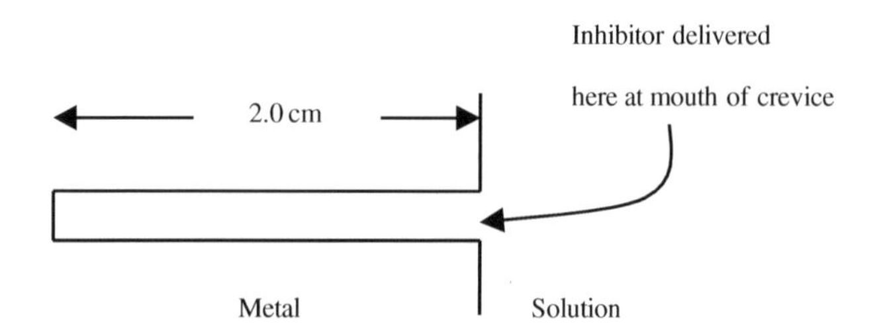

3. Suppose that the crevice in Problem 8.2 is extremely tight so that the diffusion of the amine into the crevice is restricted to a two-dimensional random walk. How long will it then take for the amine molecule to diffuse into the deepest part of the crevice, assuming the same diffusion coefficient as in Problem 8.2?
4. How long will it take for an O_2 molecule to diffuse through the oxygen diffusion layer formed on iron in quiescent 0.6 M NaCl open to the air?
5. Oxygen diffusion layers in quiet electrolytes with natural convection are usually between 0.05 and 0.10 cm in thickness [29]. For a 0.1 M NaCl solution at 25°C, calculate the corresponding range of the limiting diffusion current for oxygen reduction when the electrolyte is in contact with one atmosphere of air.
6. A piping system for cooling water of pH 7 is constructed of copper. The open-circuit corrosion potential for copper in this cooling water is $E_{corr} = +0.5$ V vs. SHE. Suppose that the cathodic half-cell reaction involving O_2 reduction is under diffusion control and that the equilibrium potential for O_2 reduction is $E^o = +1.3$ V vs. SHE. The anodic half-cell reaction involving copper oxidation is under activation control with an equilibrium potential of $E^o = +0.34$ V.

   (a) Sketch theoretical polarization curves (Evans diagrams) for the cathodic and anodic half-cell reactions. Label the corrosion potential and corrosion rate on your diagram.
   (b) How does an increase in the flow rate of the cooling water affect the corrosion rate? Why? Indicate this effect in your diagram.
   (c) Hydrazine (N_2H_4) added to the cooling water reacts as follows:

$$N_2H_4 + O_2 \rightarrow N_2 + 2H_2O$$

   What is the effect of the addition of hydrazine to the corrosion rate? Why? Indicate this effect on the polarization diagram.

7. Under what conditions is the corrosion rate of a metal or an alloy equal to the limiting current density for oxygen reduction?

8. The corrosion rate of 99.9% nickel in seawater after immersion in seawater near the Panama Canal Zone for 16 years was observed to be 1.2 mpy (mils per year) [30]. (The corresponding corrosion current density is 2.8×10^{-6} A/cm^2.)

   (a) Draw the hypothetical Evans diagram for this system when corrosion is under the control of the diffusion-limited oxygen reduction reaction. Assume that $E_{o,O2} = +0.40$ V vs. SHE and that $i_{o,O2} = 1.0 \times 10^{-9}$ A/cm^2. Assume that the anodic reaction for nickel dissolution is under activation control with an anodic Tafel slope of +0.060 V. Use values for $E_{o,Ni}$ and $i_{o,Ni}$ from Tables 3.1 and 7.4, respectively.
   (b) Based on the Evans diagram, what is the corrosion potential for nickel in seawater?
   (c) How does the corrosion potential from part (b) compare with the electrode potential for nickel given in the galvanic series for seawater?
   (d) Sketch the anodic and cathodic experimental polarization curves which result from the Evans diagram in part (a).

9. The polarization resistance of a 90Cu–10Ni alloy in flowing seawater containing 6.6 mg O_2/L was observed to be $R_P = 2.7$ kΩ after an immersion time of 40 h [31]. The sample area was 11.05 cm^2. The cathodic process was under O_2 diffusion control, and the anodic process was under activation control with an anodic Tafel slope of + 0.070 V.

   (a) What was the corrosion rate of the alloy in microamperes per square centimeter at 40 h immersion?
   (b) What was the corresponding weight loss in grams per square meter day assuming that corrosion of the alloy occurs mainly as:

$$Cu \rightarrow Cu^{2+} + 2e^-.$$

10. The following occurrence of corrosion as shown in the figure below can occur in heat exchangers where there are temperature differences along the metal [32]. Explain why corrosion occurs preferentially at the hot end of the metal.

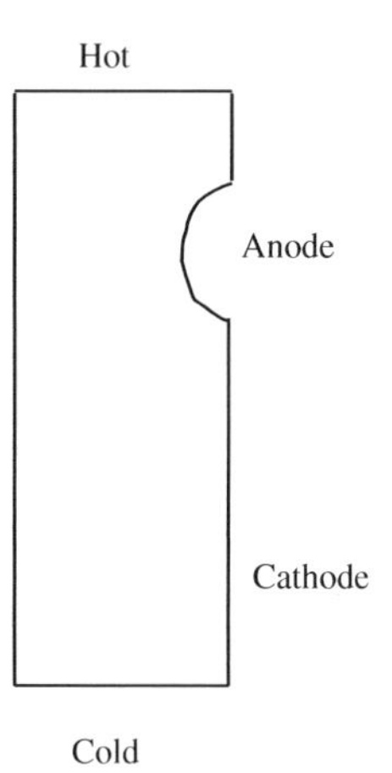

11. Suppose that the corrosion rate of a metal is 50.0 μA/cm^2 when the metal is immersed in a solution open to the air and that the corrosion rate is under O_2 diffusion control. What is the corrosion rate when the air above the solution is replaced with pure O_2 at 1 atm pressure?

(Assume that passivation does not occur when the atmosphere is pure O_2.) *HINT*: Use Henry's law and the fact that the partial pressure of O_2 in air is 0.21 atm.

References

1. E. McCafferty, *J. Electrochem. Soc.*, *121*, 1007 (1974).
2. R. L. Droste, "Theory and Practice of Water and Wastewater Treatment", p. 776, John Wiley & Sons, New York, NY (1997).
3. R. Battino, Ed., "Solubility Data Series, Vol. 7, Oxygen and Ozone", Pergamon Press, Oxford (1981).
4. H. C. Berg, "Random Walks in Biology", Chapter 1, Princeton University Press, Princeton, NJ (1993).
5. W. Jost, "Diffusion in Solids, Liquids, and Gases", pp. 25–30, Academic Press, New York, NY (1960).
6. J. O. Hirschfelder, C. F. Curtis, and B. R. Bird, "Molecular Theory of Gases and Liquids", p. 625, John Wiley & Sons, New York, NY (1954).
7. J. O'M. Bockris and A. K .N. Reddy, "Modern Electrochemistry", Second Edition, p. 411, Plenum Press, New York, NY (1998).
8. I. L. Rozenfeld, "Atmospheric Corrosion of Metals", p. 65, NACE, Houston, TX (1972).
9. J. O'M. Bockris and A. K .N. Reddy, "Modern Electrochemistry", Second Edition, p. 395, Plenum Press, New York, NY (1998).
10. H. H. Uhlig and R. W. Revie, "Corrosion and Corrosion Control", pp. 91–106, John Wiley and Sons, New York, NY (1985).
11. H. H. Uhlig, D. N. Triadis, and M. Stern, *J. Electrochem. Soc.*, *102*, 59, (1955).
12. I. Matsushima in "Uhlig's Corrosion Handbook", Second Edition, W. R. Revie, Ed., p. 529, John Wiley & Sons, Inc, New York, NY (2000).
13. F. L. La Que, "Marine Corrosion: Causes and Prevention", p. 143, John Wiley & Sons, Inc, New York, NY (1975).
14. G. Butler and H. C. K. Ison, "Corrosion and Its Prevention in Waters", p. 138, Leonard-Hill, London (1966).
15. S. C. Dexter, "Handbook of Oceanographic Engineering Materials", pp. 38, 167, John Wiley & Sons, New York, NY (1979).
16. F. L. La Que, "Marine Corrosion: Causes and Prevention", p. 182, John Wiley & Sons, Inc, New York, NY (1975).
17. A. Cohen and R. V. Jelinek, *Corrosion*, *22*, 39 (1966).
18. M. Stern, *J. Electrochem. Soc.*, *102*, 609 (1955).
19. P. Lorbeer and W. J. Lorenz, *Electrochim. Acta*, 25, 375 (1980).
20. Z. A. Iofa and M. A. Makhuba, *Prot. Metals USSR*, *3*, 329 (1967).
21. W. A. Badawy, F. M. Al-Kharafi, and E. Y. Al-Hassan, *Corros. Prevention Control*, *45* (4), 95 (1998).
22. J. O'M. Bockris and A. K. N. Reddy, "Modern Electrochemistry" Volume 2, p. 1070, Plenum Press, New York, NY (1970).
23. A. J. Bard and L. R. Faulkner"Electrochemical Methods", p. 338, John Wiley & Sons, New York, NY (2001)
24. Z. Zembura, *Corros. Sci.*, *8*, 703 (1968).
25. Yu. V. Pleskov and V. Yu Filinovski, "The Rotating Disc Electrode", p. 266, Consultants Bureau, New York, NY (1976).
26. Y. Miyata and S. Asakura, *Corros. Sci.*, *44*, 589 (2002).
27. V. Jovancicevic and J. O'M. Bockris, *J. Electrochem. Soc.*, *133*, 1797 (1986).
28. S. L. F. A. da Costa, S. M. L. Agostinho, and K. Nobe, *J. Electrochem. Soc.*, *140*, 3483 (1993).
29. N. D. Tomashov, "Theory of Corrosion and Protection of Metals", p. 186, The MacMillan Company, New York, NY (1966).
30. F. W. Fink and W. K. Boyd, "The Corrosion of Metals in Marine Environments", p. 34, Defense Metals Information Center, Battelle Memorial institute, Columbus, OH (1970).
31. B. C. Syrett and D. D. Macdonald, *Corrosion*, *35*, 505 (1979).
32. G. Butler and H. C. K. Ison, "Corrosion and Its Prevention in Waters", p. 11, Leonard-Hill, London (1966).
33. P. Vanysek in "CRC Handbook of Chemistry and Physics", D. R. Lide, Ed., p. 8–21, CRC Press, Boca Raton, FL (2001).
34. D. R. Lide, Ed., "CRC Handbook of Chemistry and Physics", 82nd Edition, p. 6–194, CRC Press, Boca Raton, FL (2001).
35. A. J. van Stroe and L. J. J. Janssen, *Anal. Chim. Acta*, *279*, 213 (1993).
36. H. Ikeuchi, M. Hayafuji, Y. Aketagawa, J. Taki, and G. Sato, *J. Electroanal. Chem.*, *396*, 553 (1995).

Chapter 9
Passivity

Introduction

So far this text has considered (or has mentioned) several different methods of corrosion control. These methods have included cathodic protection, corrosion protection by organic coatings, the use of corrosion inhibitors, and the use of additives which can change the pH of the aqueous solution. One of the most effective means of corrosion protection is the use of metals or alloys which have inherently low corrosion rates in a solution due to the existence of a *passive* oxide film.

As noted by Macdonald [1], metals such as iron, nickel, chromium, and aluminum are all inherently reactive, as evidenced by the fact that they occur in nature as their ores rather than in elemental form. Yet these metals are used in industry because they react with water and/or oxygen to form stable passive oxide films. Macdonald states that "passivity is the key to our metals-based civilization."

This chapter discusses the phenomenon of passivity and the nature of passive films.

Aluminum: An Example

Aluminum and its alloys are important structural metals because of their light weight, high strength (when alloyed), and excellent corrosion resistance. However, consider the reaction of aluminum with water, as in conditions of immersion or exposure in the atmosphere:

$$4Al(s) + 12H_2O\,(l) + 3O_2(g) \rightarrow 2Al_2O_3(s) + 12H^+(aq) + 12\,OH^-\ (aq) \qquad (1)$$

The free energy ΔG^o of the reaction can be determined from the thermodynamic principles considered in Chapters 4 and 6. Thus

$$\Delta G^\circ = 2\mu^\circ(Al_2O_3(s)) + 12\mu^o(OH^-(aq)) - 12\mu^o(H_2O(l)) \qquad (2)$$

where μ^o refers to the chemical potential. Using values tabulated by Pourbaix [2]:

$$\Delta G^o = 2(-\,376{,}000\text{ cal}) + 12(-\,37{,}595\text{ cal}) - 12(-\,56{,}690\text{ cal}) = -\,522{,}860\text{ cal}$$

This large negative value of the free energy change for Eq. (1) indicates that the reaction of aluminum with water is spontaneous. Thus, the surface of aluminum reacts with water to form an oxide film, as depicted in Fig. 9.1. However, the reaction essentially stops after the aluminum surface is covered with an oxide film, which may be only hundreds to thousands of angstroms in thickness.

E. McCafferty, *Introduction to Corrosion Science*, DOI 10.1007/978-1-4419-0455-3_9,

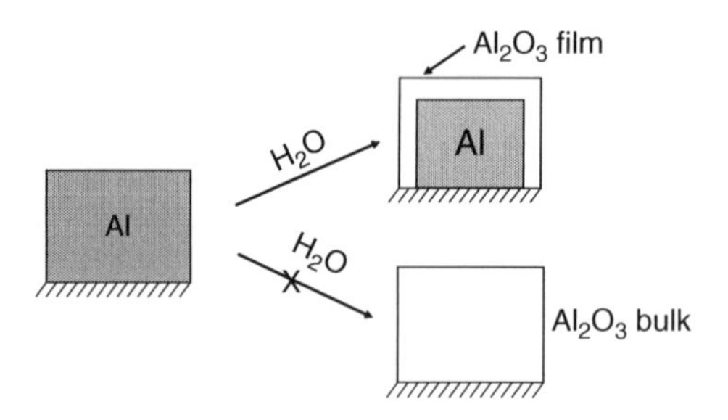

Fig. 9.1 Formation of a protective oxide film on aluminum (schematic)

In view of the spontaneous nature of Eq. (1), the question arises as to why the reaction does not continue until the entire thickness of the aluminum sample is converted to Al_2O_3. The answer to this question is that the reaction stops because the aluminum metal has formed a *passive oxide film* which protects the underlying metal from further attack. Thus, the passive oxide film on aluminum and its alloys is what is responsible for the excellent corrosion resistance of aluminum in aqueous environments. In general, the phenomenon of passivity is the single most important factor which imparts chemical stability to a wide variety of metals and alloys for their use as structural or electronic materials.

What is Passivity?

Passivity is defined as the reduction in chemical or electrochemical activity of a metal due to the reaction of the metal with its environment so as to form a protective film on the metal surface. This definition does not indicate what properties constitute a "protective" film nor does this definition tell us anything about the nature of the "film" itself. Today's studies on passivity continue to be directed toward gaining a better understanding of the chemical and physical properties of passive films.

Early History of Passivity

Observations on passivity date back to over 200 years. Uhlig [3] has traced the first use of the word "passivity" to C. Schonbein in 1836, although earlier observations on passivity had been made by M. Lomonosov in Russia in 1738, C. Wenzel in Germany in 1782, J. Keir in England in 1790, and subsequently by Schonbein and Michael Faraday. These early investigators established that the dissolution of iron in certain acid solutions ceased after first undergoing vigorous active dissolution. Early investigators also observed that the passive state could be destroyed by mechanical scratching, by the presence of chlorides, or by electrochemical reduction of the passive film.

Thickness of Passive Oxide Films

Oxide films on metals are often (but not always) very thin and are not visible to the eye. These films must be studied by special surface analytical techniques, which will be addressed later in this chapter. The transition metals (e.g., Fe, Cr, Co, Ni, Mo) and their alloys (e.g., the Fe–Cr stainless steels) tend to have thin passive films, which are tens to hundreds of angstroms (Å) in thickness. The air-formed oxide film on titanium is 30–80 Å (3–8 nm) in thickness [4–6] and increases to 250 Å (25 nm) after 4 years of exposure to air [7].

The non-transition metals (e.g., Zn, Cd, Cu, Mg, Pb) tend to have much thicker passive films, which can be thousands to tens of thousands of angstroms in thickness. For instance, the passive film on copper piping in domestic water systems consists largely of Cu_2O and is typically about 5,000 Å (500 nm) in thickness [8]. In the open atmosphere, copper forms the familiar green patina that is clearly visible on copper roofs, sheathing, and statues. In its most stable form, the patina consists of basic copper sulfate, $CuSO_4 \cdot 3Cu(OH)_2$, although the patina may also contain chlorides if formed in marine atmospheres or carbonates if formed in industrial atmospheres [8]. The presence of this patina protects the underlying copper from further atmospheric corrosion.

Aluminum can have either thin or thick passive films. The natural air-formed oxide film on aluminum which provides some measure of protection is only 30–40 Å thick (3–4 nm) [4, 5], whereas anodizing techniques can provide much thicker passive films. For example, anodization of an aluminum alloy in phosphoric acid produced an oxide film approximately 4,000 Å (400 nm) in thickness [9].

Purpose of This Chapter

The passivity of metals and alloys is a subject of much current interest to corrosion scientists and engineers. The first international symposium on passivity was convened in 1957, and such meetings continue to be hotbeds of ideas, presentations, and discussions. For example, the Eighth International Symposium on Passivity of Metals and Semiconductors held in Jasper Park, Canada, in 1999 attracted over 100 participants and featured over 100 oral presentations or poster papers. The conference proceedings for this one symposium alone amount to over 900 pages. Thus, studies on the nature of passivity continue to be a topic of ongoing interest, so this brief chapter cannot summarize adequately the latest research progress. Instead, an introduction to passivity is given and several current research ideas are presented, although no attempt at completeness is claimed.

The breakdown of passive films by the localized corrosion processes of pitting, crevice corrosion, and stress-corrosion cracking is treated in Chapters 10 and 11.

Electrochemical Basis for Passivity

The electrochemical basis for passivity is found in the anodic polarization curve illustrated in Fig. 9.2. The shape of this curve is typical of various metals (e.g., iron) which undergo an active–passive transition in acid solutions (such as sulfuric acid).

Beginning with the open-circuit corrosion potential and moving in the anodic direction, we first pass through a region of active corrosion in which the open-circuit corrosion rate can be determined as usual by the Tafel extrapolation method. But at a certain potential called the *Flade potential*, further increases in potential cause a decrease in the anodic current density. This is because a passive film is being formed on the metal surface, and the metal is said to be undergoing an *active/passive transition*. The current is observed to drop precipitously as the passive oxide is formed and as the metal enters the *passive region* of the anodic polarization curve. The current density at the Flade potential is called the *critical current density for passivation*, and the current density in the passive region is called the *passive current density*. With further increases in anodic potential, the current

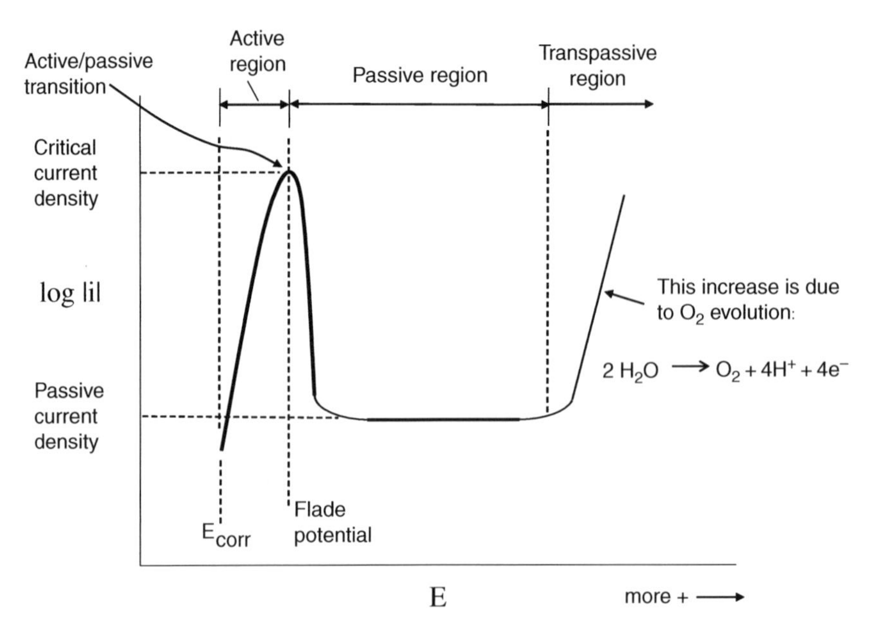

Fig. 9.2 Anodic polarization curve (schematic) showing the formation of a passive film after an active/passive transition

again increases, but this increase is not due to metallic corrosion but rather is due to the evolution of oxygen by the breakdown of water in the electrolyte:

$$2H_2O(l) \rightarrow O_2(g) + 4H^+(aq) + 4e^- \tag{3}$$

The region beyond the passive region in which the current density again increases with increasing potential is called the *transpassive region.*

Figure 9.3 shows the effect of pH on the anodic polarization curves for iron in phosphate solutions [10]. It can be seen that the Flade potential varies with pH, becoming more negative as the pH increases. The variation in Flade potential with pH is shown in Fig. 9.4, which also contains data for iron in additional aqueous solutions [11]. For these solutions, the Flade potential E_F follows the empirical relationship:

$$E_F = A - B\text{pH} \tag{4}$$

where A and B are constants, which differ numerically for the two aqueous solutions.

The critical current density for passivation also depends on the pH of the solution. As seen in Fig. 9.3, the critical current density for passivation increases with increasing acidity (decreasing pH). Note that for the most alkaline phosphate solution (pH 11.50), iron achieves passivity without first undergoing extensive anodic dissolution.

In fact, it is not necessary for a metal to undergo an active/passive transition in order for it to enter the passive state. For example, oxidizing inhibitors such as chromate solutions (of neutral or basic pH) can confer passivity without considerable anodic dissolution, as shown schematically in

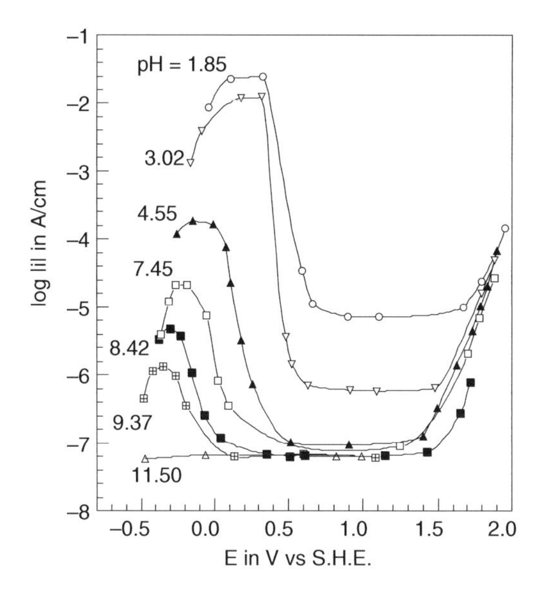

Fig. 9.3 Anodic polarization curves for iron in phosphate solutions at various pH values [10]. Reproduced by permission of ECS – The Electrochemical Society

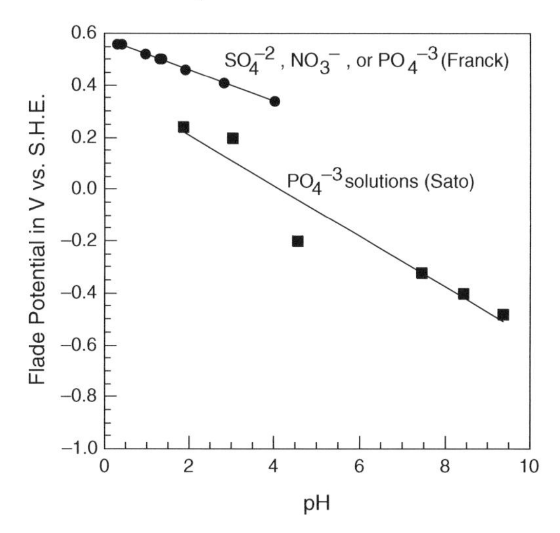

Fig. 9.4 Variation of the Flade potential of iron with pH [10, 11]

Fig. 9.5. The chromate inhibitor passivates the iron surface through the formation of a mixed oxide of Fe_2O_3 and Cr_2O_3 [12] by the following overall reaction:

$$2Fe + 2CrO_4^{2-} + 4H^+ \rightarrow Fe_2O_3 + Cr_2O_3 + 2H_2O \tag{5}$$

In addition to the evolution of oxygen in the transpassive region, it is possible for either of two additional reactions to occur, depending on the specific nature of the system. These reactions occur between the Flade potential and the potential for O_2 evolution and hence will take place before the potential for O_2 evolution is reached. The first of these two possible reactions involves dissolution of

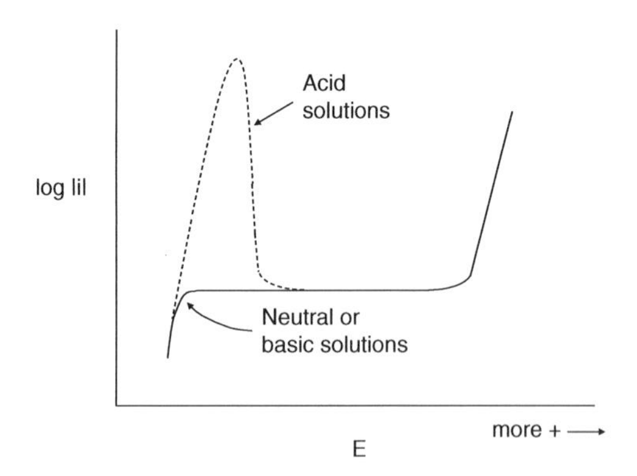

Fig. 9.5 Anodic polarization curves (schematic) showing the formation of a passive film in acid, neutral, or basic solutions

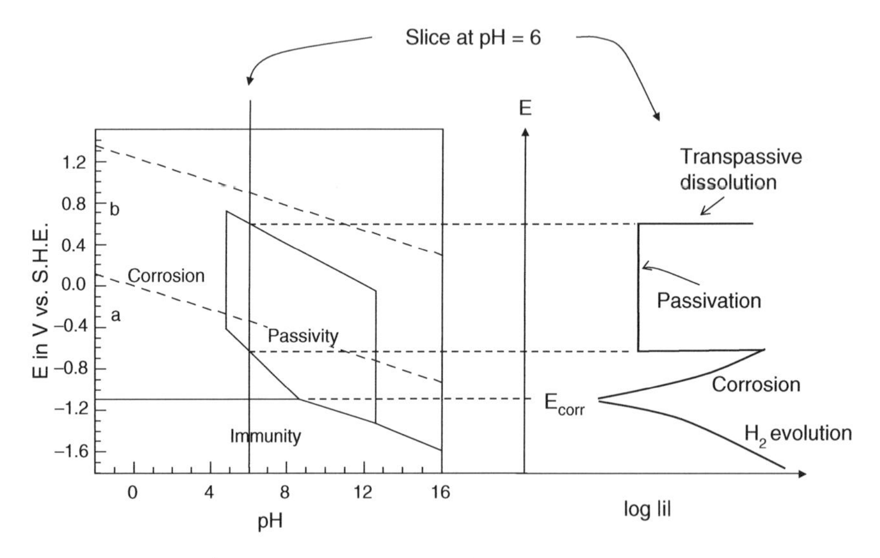

Fig. 9.6 Correlation between the Pourbaix diagram for chromium and its anodic polarization curve at pH 6

the oxide film and will occur if the oxide film is not stable over the entire passive region of the anodic polarization curve. An example is provided by chromium. The Pourbaix diagram for chromium given in Fig. 9.6 (and also in Fig. 6.11) shows that Cr_2O_3 is dissolved to form CrO_4^{2-} ions at an electrode potential below that for O_2 evolution. Figure 9.6 also schematically shows the anodic polarization curve for Cr in solutions of pH 6, where the polarization behavior is linked to the thermodynamic behavior in the Pourbaix diagram.

A second type of anodic reaction which can occur in the passive region prior to the electrode potential for O_2 evolution is due to pitting corrosion in the presence of aggressive anions, such as chloride ions. Chloride ions can locally attack the oxide film so that the electrode potential of oxygen evolution for Eq. (3) will not be reached. Instead the anodic current density will increase at a potential before that for oxygen evolution, with the rise in current density due to active dissolution within developing corrosion pits. This situation is shown in Fig. 9.7 for the pitting of 18Cr–8Ni stainless steel in chloride solutions [13]. Pitting corrosion is considered in Chapter 10.

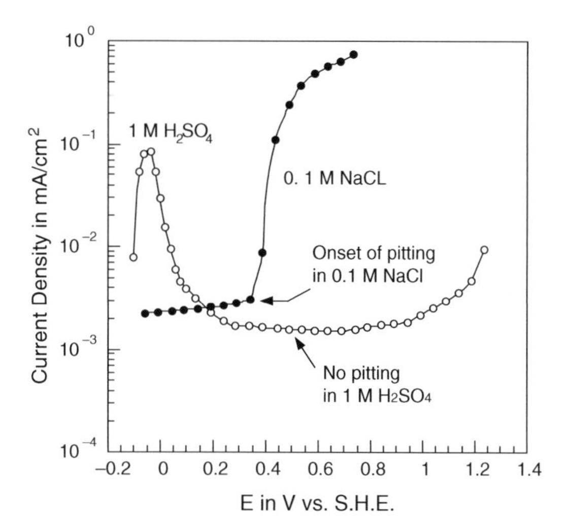

Fig. 9.7 Anodic polarization curves for 18Cr–8Ni stainless steel in 1 M H_2SO_4 or 0.1 M NaCl [13]. Reproduced by permission of ECS – The Electrochemical Society

The above description of passivity is based on the kinetic point of view. We have already considered passivity from a simple thermodynamic viewpoint in the treatment of Pourbaix diagrams. Recall that any oxide film that was formed was considered to be a passive film. This chapter will address in more detail the characteristics and properties of passive films.

Theories of Passivity

There are three main theories of passivity. These are (i) the adsorption theory [14–16], (ii) the oxide film theory [17], and (iii) the film sequence theory [18]. The film sequence theory actually combines the main ideas of the first two theories and attempts to bridge the differences in the first two points of view.

Adsorption Theory

According to this view of passivity, a chemically adsorbed (chemisorbed) monolayer of oxygen (i.e., one molecular layer in thickness) reduces the reactivity of surface metal atoms and thus provides protection against further attack. The adsorbed monomolecular film, of course, continues to grow in thickness; but proponents of this theory maintain that the initial act of chemisorption is the primary cause of passivity [14–16].

Support for the adsorption theory lies in the following observations:

(1) Electrochemical measurements yield the number of coulombs required to achieve passivity to correspond to one layer of oxygen.
(2) The transition metals (e.g., Fe, Cr, Ni) have high heats of adsorption of O_2, corresponding to chemisorption or surface bond formation.

(3) The observed Flade potential for iron is consistent with the formation of a chemisorbed monolayer, as will be seen below.
(4) The critical alloy composition for the passivity of binary alloys can be explained by chemisorption effects which respond to the electron configurations of the alloying metals. Passivity in binary alloys is treated later in this chapter.

In regard to the first point listed above, coulometric measurements have determined that the amount of charge required to form a passive film on iron in NaOH solutions [19], 0.1 M Na_2SO_4 [20], or in a borate buffer [21] corresponds to a monolayer of oxide. See Problem 9.1.

Uhlig [22] was able to explain the Flade potential of iron on the basis that a chemisorbed monolayer is formed. The Flade potential for iron in sulfuric acid at pH 0 is +0.58 V vs. SHE, as observed by Franck [11] and given in Fig. 9.4. However, this value for the Flade potential cannot be explained on the basis of a simple electrochemical reaction which occurs between iron and the solution to produce a stoichiometric oxide. This can be seen from Table 9.1, in which none of the standard electrochemical reactions involving iron or its bulk oxides can be related to the observed Flade potential for iron at pH 0.

Table 9.1 Thermodynamic relationships for iron and its oxides in regard to the Flade potential E_F

Reaction	Nernst equation [2]	E at pH 0 (V vs. SHE)
$Fe + H_2O \rightleftharpoons FeO + 2H^+ + 2e^-$	$E = -0.0417 - 0.0591$ pH	-0.0417
$3Fe + 4H_2O \rightleftharpoons Fe_3O_4 + 8H^+ + 8e^-$	$E = -0.085 - 0.0591$ pH	-0.085
$2Fe + 3H_2O \rightleftharpoons Fe_2O_3 + 6H^+ + 6e^-$	$E = -0.051 - 0.0591$ pH	-0.051
$2Fe_3O_4 + H_2O \rightleftharpoons 3Fe_2O_3 + 2H^+ + 2e^-$	$E = 0.221 - 0.0591$ pH	0.221
Fe in 1 M H_2SO_4 (experimental)	–	$E_F = +0.58$ V [22]

Uhlig assumed that the following surface reaction was responsible for passivation:

$$Fe(s) + 3H_2O(l) \rightarrow Fe(O_2 \cdot O)_{ads} + 6H^+(aq) + 6e^- \quad (6)$$

where $Fe(O_2.O)_{ads}$ refers to the chemisorbed monolayer with a second layer of adsorbed O_2 molecules. The approach was to calculate the standard free energy change ΔG^o for Eq. (6) and then to obtain E^o from $\Delta G^o = -nFE^o$. Details of these calculations are given in Appendix E, but in brief, Uhlig obtained $\Delta G^o = 78{,}222$ cal per mole of $Fe(O_2.O)_{ads}$ for Eq. (6) and $E^o = -0.57$ V vs. SHE. But this value of the electrode potential is for the oxidation reaction, and the standard electrode potential is for the reduction reaction so that $E^o = +0.57$ V vs. SHE. This value is in good agreement with the Flade potential of + 0.58 V for Franck's data for iron in sulfuric acid.

Oxide Film Theory

According to this theory, which is derived from the early work of U. R. Evans and co-workers, a thin three-dimensional oxide film separates the metal from its environment. This oxide films acts as a barrier to the passage of the corrosive environment into the film and the passage of metal cations from the substrate out through the film.

Support for the oxide film theory lies in the following observations:

(1) Oxide films have been detached from the metal surface using chemicals such as alcoholic bromine, and the films have been analyzed by techniques such as electron microscopy and electron diffraction [17, 23, 24].
(2) There have been numerous studies on the chemical composition of passive films using modern surface analytical techniques, such as X-ray photoelectron spectroscopy (XPS).
(3) The presence of certain beneficial alloying elements (such as Cr, Ni, and Mo) in passive films has established that the chemical composition of a passive film on an alloy is an important factor in imparting passivity to the alloy.

Early evidence for three-dimensional oxide films was provided in the 1920s by U. R. Evans [17], who isolated the passive film on iron by stripping the film from the underlying metal by using a solution of bromine or iodine in methanol, as depicted in Fig. 9.8. Various workers have isolated oxide films from iron, stainless steels, and aluminum and studied their properties by chemical analysis, optical and electron microscopy, and electron diffraction [23–27]. For example, early studies on passive films isolated from stainless steels showed that the films were about 30 Å in thickness, were crystalline or semi-crystalline, and that both chromium and iron were present in the passive film [25].

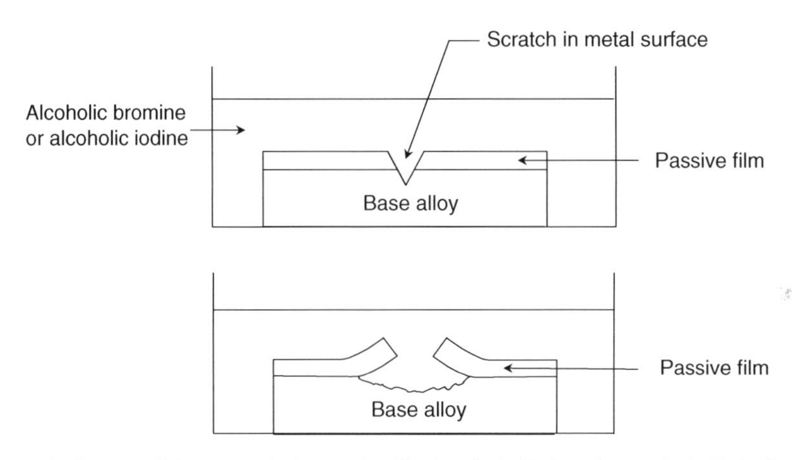

Fig. 9.8 Schematic diagram of the removal of a passive film by alcoholic bromine or alcoholic iodine. The bottom portion of the figure shows a cross section after a brief immersion in the solution

Proponents of the adsorption theory of passivity did not question the existence of such oxide films but claimed that these films were not the primary cause of passivity but rather were its end result.

With the advent of modern surface analytical techniques, much information became available about the composition of passive films on a wide range of alloys. For instance, for a series of Fe–Cr alloys passivated in sulfuric acid, X-ray photoelectron spectroscopy (XPS) was used to determine the chemical composition of oxide films as a function of the alloy composition [28]. Results are shown in Fig. 9.9, in which it is striking that the oxide film becomes enriched in Cr^{3+} at approximately 13 at. % Cr in the alloy, which is the critical alloy composition in Fe–Cr alloys for the onset of passivity by alloying.

More information about surface analysis techniques and the passivity of binary alloys is given later in this chapter.

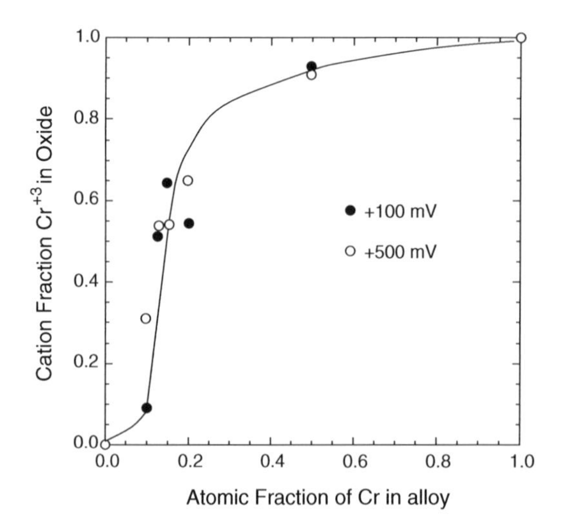

Fig. 9.9 Chromium content in the passive film on Fe–Cr binary alloys as a function of the chromium content in the alloy after immersion in 1 M H_2SO_4 at two different electrode potentials [28]. Reproduced by permission of Elsevier Ltd

Film Sequence Theory

The film sequence theory proposed by Hackerman [18] attempts to reconcile the differences between the adsorption theory and the oxide film theory. Hackerman noted that an adsorbed film can cause large potential changes associated with passivity, but that it is unlikely that an adsorbed monolayer could provide long-term protection against an aggressive environment. Thus, Hackerman proposed the film sequence theory to take into account both adsorptive and oxide film characteristics of passivity.

According to the film sequence theory, a passive film is formed in a sequence of steps, which involve: (i) chemisorption of O_2, (ii) splitting of the adsorbed O_2 molecule to form two adsorbed oxygen atoms O_{ads}, (iii) formation of a charged surface species O_{ads}^-, (iv) intrusion of metal ions from the lattice into the adsorbed layer, and (v) growth of a three-dimensional oxide. In addition, the film must be able to regenerate itself when damaged so that steps (i) though (v) are repeated when the oxide film is breached. Thus, the film sequence theory draws on the properties of adsorbed films or three-dimensional oxides as needed.

A link between the adsorption theory and the oxide film theory of passivity is provided by Frankenthal's study of the passive film on an Fe–24Cr alloy [29]. Coulometric measurements showed that the primary film responsible for passivity was of the order of one monomolecular layer of oxide ions, but that a secondary and more stable thicker film was formed at electrode potentials higher than the potential of primary passivation.

Surface Analysis Techniques for the Examination of Passive Films

Today it is generally accepted and understood that passivity is provided by a thin but three-dimensional oxide film. Prior to approximately 1970, passive films were examined mostly by electrochemical methods, with some use of electron diffraction techniques, either in the transmission mode for films stripped from metal surfaces or in the reflection mode for films intact but removed from solution.

Starting in the 1970s and continuing today, there has been an explosion of new surface analytical techniques that have opened new approaches to corrosion science, especially in the area of passivity. These surface analytical techniques, which have been used to study the chemical composition or the structure of passive films, include X-ray photoelectron spectroscopy (XPS), scattering ion mass spectrometry (SIMS), Mössbauer spectroscopy, and X-ray absorption spectroscopy (XAS), which includes X-ray absorption near-edge spectroscopy (XANES) and extended X-ray absorption fine structure (EXAFS). These and other techniques which have been applied to the study of passive films are summarized in Table 9.2.

Table 9.2 Various surface analytical techniques which may be applied to the study of passive films [30–32]

Technique	Incident radiation	Detected radiation	Information obtained
Auger electron spectroscopy (AES)	Electrons	Electrons	Elemental analysis Depth profiles
X-ray photoelectron spectroscopy (AES)	X-rays (photons)	Electrons	Elemental analysis Oxidation states Depth profiles
Secondary ion mass spectrometry (SIMS)	Ions	Ions	Elemental analysis Depth profiles
Reflection high-energy electron diffraction (RHEED)	Electrons	Electrons	Surface structure Crystallinity
Transmission electron microscopy (TEM)	Electrons	Electrons	Morphology Microstructure
Scanning electron microscopy (SEM)	Electrons	Electrons	Morphology Microstructure
Energy dispersive analysis by X-rays (EDAX)	Electrons	X-rays	Chemical analysis
Mössbauer spectroscopy	γ-rays	Electrons	Fe-containing phase identification (also Co- and Sn-)
Infrared spectroscopy	Infrared radiation (photons)	Photons	Functional groups Chemical structure
X-ray absorption near-edge spectroscopy (XANES)	X-rays (photons)	Photons	Chemical composition
Extended X-ray absorption fine structure (EXAFS)	X-rays (photons)	Photons	Bond distances
Scanning tunneling microscopy (STM)	Electrons	Electrons	Physical topography Atomistic resolution
Rutherford backscattering (RBS)	H^+ ions (α-particles)	α-particles	Chemical composition

Throughout this chapter, we will draw on the results of various surface analytical studies on passivity, without going into details of the various techniques themselves. However, it is instructive to first provide a brief description of three techniques which have been very useful in the study of passivity. These are X-ray photoelectron spectroscopy (XPS), X-ray absorption spectroscopy (XAS), and scanning tunneling microscopy (STM).

X-ray Photoelectron Spectroscopy (XPS)

The "workhorse" of these surface techniques has proved to be X-ray photoelectron spectroscopy. In XPS, the surface to be analyzed is placed inside a vacuum system and then irradiated with a beam of X-rays of known energy. Transfer of the energy of the X-ray photons to atoms in the surface of the sample causes core electrons (photoelectrons) to be ejected, as shown schematically in Fig. 9.10. These photoelectrons are collected and analyzed by the XPS spectrometer, which measures the kinetic energy of the photoelectrons E_K. The kinetic energy of the photoelectrons, however, depends on the energy $h\nu$ of the incident X-rays. The intrinsic material property is the binding energy of the electron E_B given by

$$E_B = h\nu - E_K - \phi_{sp} \tag{7}$$

where ϕ_{sp} is the work function of the spectrometer, which can be measured. Thus, the binding energy E_B can be readily determined from the three quantities on the right-hand side of Eq. (7). The binding energy is a characteristic quantity for the various elements and their different oxidation states, and extensive tables of binding energies have been compiled for identification of various peaks which are obtained in an XPS spectrum.

An example of a typical XPS photopeak is given in Fig. 9.11, which shows the O 1s photopeak for the air-formed oxide film on titanium [4, 5]. (XPS photopeaks are described in terms of their quantum numbers.) The O 1s peak in Fig. 9.11 has been resolved into three constituent subpeaks, showing that there are three oxygen-containing species in the oxide film: O^{2-} oxide ions,

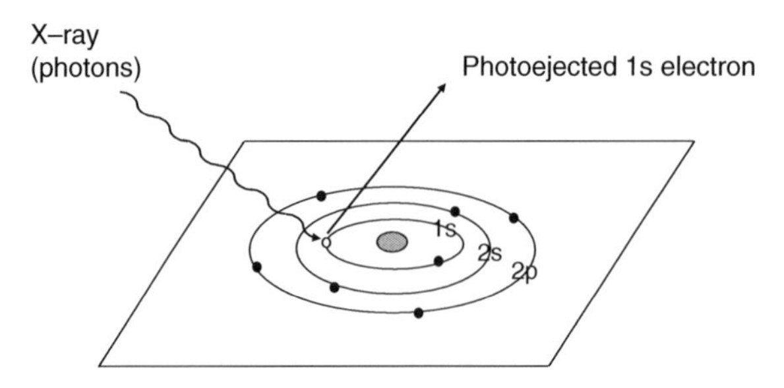

Fig. 9.10 Schematic diagram of the ejection of photoelectrons as used in X-ray photoelectron spectroscopy (XPS)

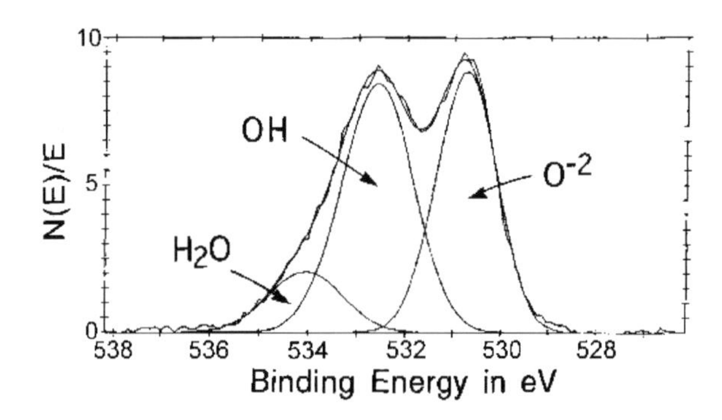

Fig. 9.11 The XPS O 1s photopeak for the air-formed oxide film on titanium [4, 5]. Reproduced by permission of ECS – The Electrochemical Society

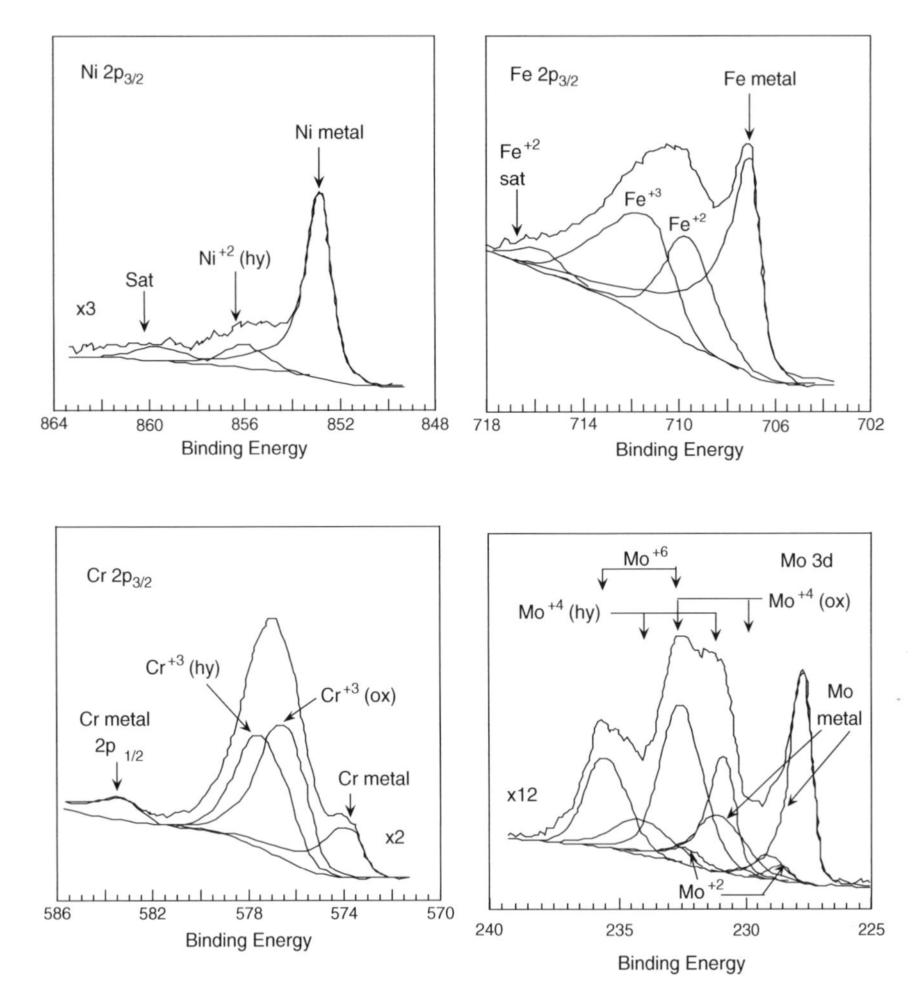

Fig. 9.12 XPS spectra for Fe–20Cr–20Ni–6Mo after polarization for 10 min in 0.1 M HCl + 0.4 M NaCl [33]. "Hy" refers to hydroxide and "ox" to oxide. Reproduced by permission of ECS – The Electrochemical Society

chemisorbed OH groups which result from the interaction of the oxide film with water vapor in the ambient atmosphere, and adsorbed water molecules.

Figure 9.12 gives an example of the application of XPS in a study on the passive film formed on an Fe–20Cr–20Ni-6 Mo stainless steel in a chloride solution [33]. As seen in Fig. 9.12, each of the alloying elements is also found in the passive film. The Cr peak consists of three constituents, representing Cr^{3+} bound as an oxide and a hydroxide, as well as a constituent from the underlying metal substrate. Molybdenum in the passive film is found as Mo^{2+}, Mo^{4+}, and Mo^{6+} species.

Thus, XPS provides the elemental composition as well as the oxidation state of each element detected. XPS is surface sensitive to approximately the first 5–30 Å of the sample (depending on the kinetic energy of the emitted electrons) and can detect less than monolayer quantities. Information about the depth of detected constituents can be obtained by sputtering off the surface by using ions such as Ar^+ or by varying the angle which the incident X-ray beam makes with the sample surface.

A disadvantage of XPS is that the sample must be removed from the aqueous electrolyte and placed into a high-vacuum system for surface analysis, thus raising the question as to whether withdrawal from the electrolyte or exposure to a high vacuum alters the character of the passive film. This problem can be minimized, however, through the use of special chambers which allow the sample to be removed from the electrolyte with its liquid film intact and to be transferred under a blanket of inert gas into the spectrometer. However, the effect of exposure of the film to the vacuum still remains open to question.

X-ray Absorption Spectroscopy

X-ray absorption spectra for various iron oxides and hydroxides have shown that the spectrum for the passive film does not match that of any of the individual oxides or hydroxides [34, 35]. Figure 9.13 shows XPS data for several oxides and for the passive film on iron grown in a borate buffer solution [35]. (Models for the passive film on iron are considered later in this chapter.)

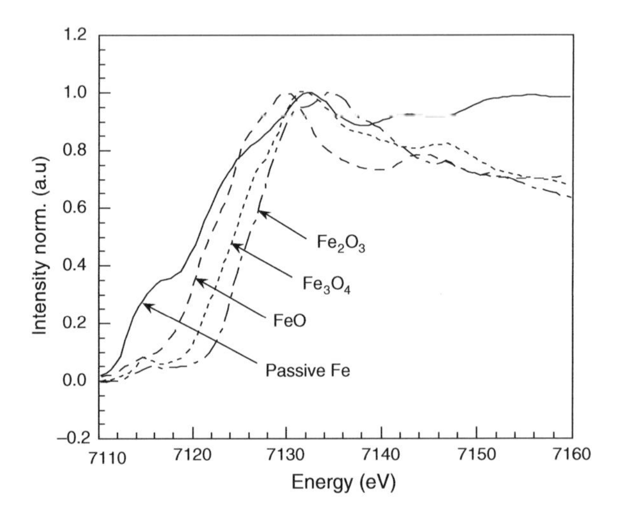

Fig. 9.13 X-ray absorption spectra for several iron oxides and for the passive film on iron in a pH 8.4 buffer solution. Redrawn from Schmuki et al. [35] by permission of ECS – the electrochemical society

Each spectrum in Fig. 9.13 consists of two parts: an initial sharp increase in absorbance (an X-ray edge) and a region of oscillations following the edge. The observation and analysis of the edge portion of the spectrum is called X-ray absorption near-edge spectroscopy (XANES), while the information contained in the subsequent oscillation is termed X-ray absorption fine structure (EXAFS).

Absorption edges occur when the energy of the incident X-ray excites a core electron in the sample to an empty valence level or beyond. The positions (i.e., energies) of these edges are characteristic of the absorbing atom and are designated according to the core electron level which has been excited. Thus, a K edge arises from the 1s level and an L edge from a 2s or a 2p level.

The oscillations which arise in the EXAFS portion of the spectrum are due to interactions of the photoelectron wave with its nearest neighbor atoms. Thus, this region contains information on atom–atom bond distances. For details, more advanced texts should be consulted [32, 36]. In addition, the pre-edge region can be expanded to yield more additional information as to similarities or differences between various samples being investigated.

X-ray absorption spectroscopy requires a monochromatic high-intensity X-ray beam which can be tuned through a wide range of energies. Such experiments are carried out using synchrotron X-ray sources [36].

Scanning Tunneling Microscopy

The scanning tunneling microscope (STM) is capable of characterizing surface topography on an atomistic scale. In the STM, a voltage is applied between the surface to be imaged and a sharp probe whose tip can be narrowed to a single atom. As the tip is moved to within a few atomic diameters of the surface, a current flows between the sample and the probe. This is due to the overlap between the electronic wave functions of the atoms in the sample and the tip. The current falls off exponentially with the distance above the surface and is thus a measure of the height of the tip above the surface. The tip can be scanned across the surface in the *x*- and *y*-directions, as shown in Fig. 9.14.

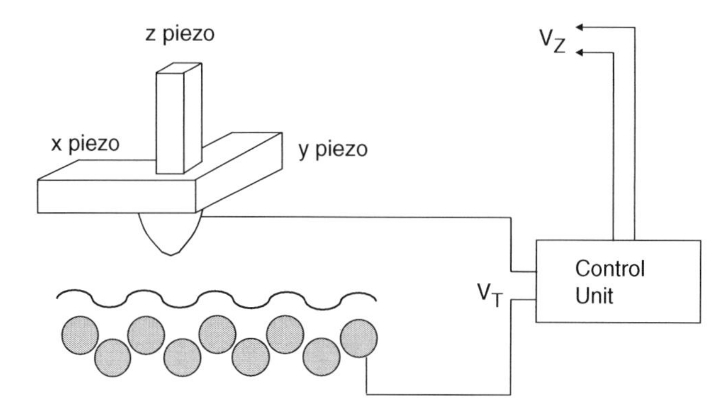

Fig. 9.14 Schematic diagram of a scanning tunneling microscope (STM), as adapted from Vickerman [32]. V_T is the sample bias voltage and V_Z is the voltage applied to the z-piezo to maintain constant tunneling current. Reproduced by permission of John Wiley & Sons, Inc

The STM can operate in two modes: constant current or constant height. In the constant current mode, the distance of the tip above the surface adjusts to maintain the constant current and thus responds to the topography of the surface. In the constant height mode, the current output changes as the distance of the tip above the surface changes with respect to the atomic surface topography.

An example of the use of the STM in studies on passivity is given in Fig. 9.15, which shows an STM image of the passive film on an Fe–14 at.% Cr alloy [37]. The STM image shows that the passive film is crystalline and has a slightly distorted close-packed hexagonal lattice.

A technique related to the STM is the atomic force microscope (AFM), which consists of a cantilever beam having a sharp tip which can sense forces between the tip and the sample. As the interaction force between the tip and the sample varies, deflections are produced in the cantilever, and these are used to construct a topographical map of the surface.

Spacing of corrugations ~ 3.1 Å

Fig. 9.15 In situ STM image of the passive film on Fe–14Cr in 0.01 M H_2SO_4 at +0.400 V vs. SCE [37]. Reproduced by permission of ECS – The Electrochemical Society

Models for the Passive Oxide Film on Iron

The nature of the passive film on iron has been studied in great detail by many investigators who have used a variety of techniques. However, even with this intense study and even with the remarkable capabilities of modern surface analytical techniques, the exact nature of the passive film on elemental iron remains elusive. This is due in part to the fact that the passive film on iron is very thin (of the order of 1.5–5.0 nm, i.e., 15–50 Å). In addition, the nature of the passive film may vary with the details of its formation and with the electrolyte in which the film exists.

Based on various experimental investigations, there are four principal models for the passive film on iron. These are (i) the bilayer model, (ii) the hydrous oxide model, (iii) the bipolar-fixed charge model, and (iv) the spinel/defect model. Each of these models is discussed briefly below.

Bilayer Model

The bilayer model for the passive film on iron was developed by M. Cohen and co-workers at the National Research Council of Canada [38–40]. According to this model, which is illustrated

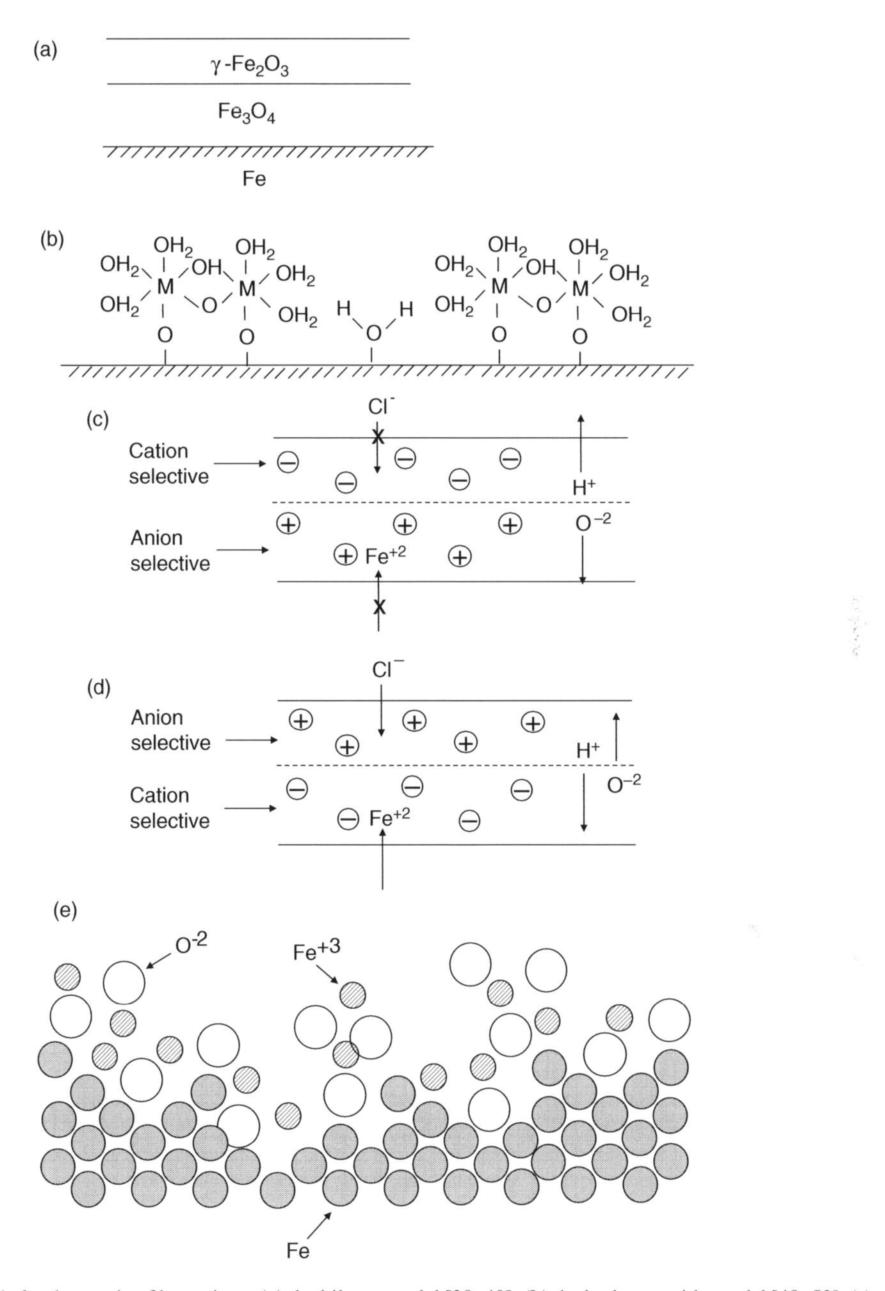

Fig. 9.16 Models for the passive film on iron: (**a**) the bilayer model [38–40], (**b**) the hydrous oxide model [48–53], (**c**) and (**d**) the bipolar model [10, 59], (**e**) the spinel/defect model [34, 61, 62]. Figure 9.16b reproduced with permission from Elsevier Ltd. Figure 9.16c, d reproduced with the permission of ECS – The Electrochemical Society

schematically in Fig. 9.16(a), the passive film on iron consists of an inner layer of Fe_3O_4 adjacent to the metal and an outer layer of γ-Fe_2O_3. (In addition, there is an outermost thin layer of hydroxyl groups which exists on every oxide film and which usually consists of only one to several monolayers [4, 5, 41] and so is much thinner than the oxide film itself.)

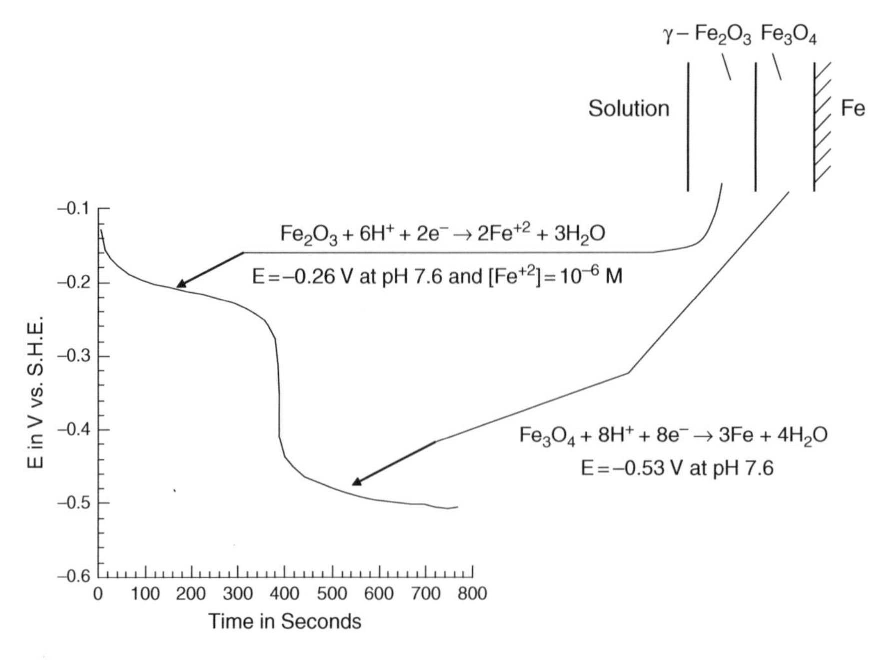

Fig. 9.17 Cathodic reduction of a thin film of γ-Fe_2O_3 on Fe_3O_4 on iron [38]

Evidence that γ-Fe_2O_3 is in the outer layer was obtained from electron diffraction studies of the passive film. The presence of a bilayer was determined from cathodic reduction experiments conducted in nearly neutral solutions, as in Fig. 9.17, which shows two plateaus in the reduction of a system consisting of layers of γ-Fe_2O_3 on Fe_3O_4 on iron. The first plateau is due to the reduction of the outer layer of γ-Fe_2O_3 by the reaction

$$Fe_2O_3 + 6H^+ + 2e^- \rightarrow 2Fe^{2+} + 3H_2O \tag{8}$$

The thermodynamic electrode potential for this half-cell reaction is given by

$$E = 0.728 - 0.1773\text{pH} - 0.0591\log[Fe^{2+}] \tag{9}$$

The second plateau is due to the reduction of the inner layer of the oxide by the half-cell reaction

$$Fe_3O_4 + 8H^+ + 8e^- \rightarrow 3Fe + 4H_2O \tag{10}$$

for which

$$E = -0.085 - 0.0591\text{pH} \tag{11}$$

Cathodic reduction of the passive films gave reduction profiles similar to Fig. 9.17, except that the first plateau frequently appeared as an inflection point and the second plateau did not always correspond to the thermodynamic potential given above. However, this may be because there is no sharp phase boundary between the two thin bilayers but instead a varying concentration from the metal/oxide interface to the oxide/solution interface. See Problem 9.2 at the end of this chapter.

Support for this bilayer model has been provided from research involving surface analytical techniques, including AES, RHEED, and SIMS [30, 42, 43], which have been used to analyze the composition of the passive film. In particular, the results of SIMS data showed that the outer layer was "dry" γ-Fe_2O_3 rather than "wet" γ-FeOOH. The yield of the FeO^+ fragment for the passive film was higher than for the fragment $FeOH^+$, in agreement with the results for an Fe_2O_3 standard but not for FeOOH, as shown in Fig. 9.18.

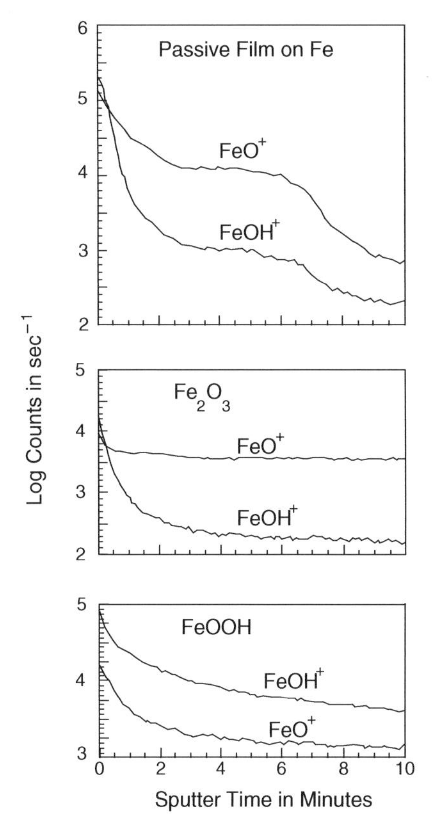

Fig. 9.18 SIMS profiles for the passive film on iron, for Fe_2O_3, and for FeOOH [30]. Reproduced by permission of Elsevier Ltd

Modifications of the basic bilayer structure discussed above have also been proposed in which the outer of the two layers contains hydrogen atoms [44], is γ-FeOOH rather than γ-Fe_2O_3 [45, 46], or is a mixture of Fe_3O_4, γ-Fe_2O_3, and an unknown ferric oxide, hydroxide, or oxyhydroxide [47].

Hydrous Oxide Model

In contrast to the above model (of an anhydrous passive film), the main feature of the hydrous oxide model is that the passive film on the iron surface contains water, the exact form of which remains unclear. Possibilities include molecular H_2O, H atoms, and OH groups.

There are considerable experimental indications that water (in some form) exists in passive oxide films. The earliest evidence is due to Rhodin [48], who found bound water in the passive film of stainless steel by chemical analysis of the film stripped from the stainless steel surface.

Okamato and Shibata [49, 50] have used radiotracer techniques with stainless steel passivated in a tritiated acid solution. The passivated surface was removed from the tritiated solution, washed thoroughly with ordinary water, dried, and then immersed in a solution containing non-tritiated water. The desorption of bound tritiated water from the passive film was detected by an increase in the counting rate upon immersion, as shown in Fig. 9.19. Similar experiments using radiotracer techniques have also been carried out on iron [51, 52] and have shown the existence of a hydrous species in the passive film on iron.

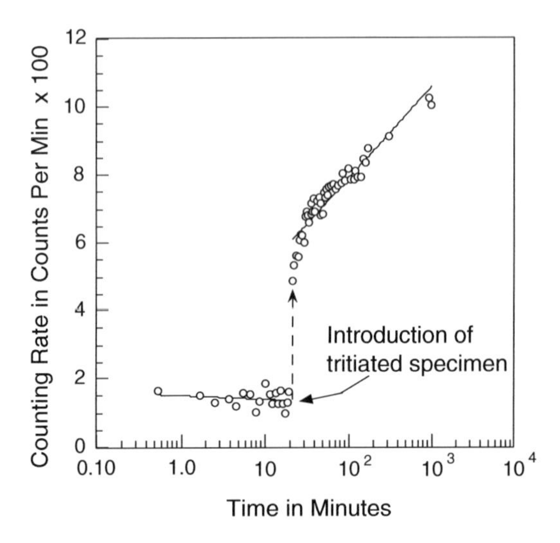

Fig. 9.19 Change in counting rate with time during the immersion into a scintillating solution for 304 stainless steel which had been previously passivated in a tritiated acid solution [49]. Reproduced by permission from Macmillan Publishers: Nature [49], copyright (1965)

Surface analytical measurements with SIMS [53] and Mössbauer spectroscopy [54–56] have also suggested the presence of water in the passive film on iron. Using SIMS, Murphy et al. [53] observed a constant ratio of OH^-/O^- throughout the passive film formed on iron in a borate buffer and concluded that water is contained throughout the thickness of the passive film. (This conclusion is directly opposite to that reached by Mitchell and Graham [30] in their SIMS work.) The results of Mössbauer investigations have led to various interpretations, but all involve the presence of water in the passive film. These interpretations are that the passive film contains a non-stoichiometric Fe^{3+} species [54], contains γ-FeOOH [55], is ferric hydroxide with excess water [56], and consists of polymeric chains of iron atoms bonded by dioxy and dihydroxy bridging bonds and linked together by water molecules [57].

The hydrous oxide model is represented schematically in Fig. 9.16(b) [58].

Bipolar-Fixed Charge Model

The bipolar-fixed charge model was proposed by Sakashita and Sato [10, 59] and is based on the transport properties of hydrous oxide films in regard to anions or cations.

If the outer part of the passive oxide film is cation selective and the inner part of the film is anion selective, as in Fig. 9.16(c), then the inward passage of aggressive anions (such as Cl^-) through the film is not favored. In addition, the outward passage of cations produced by metallic dissolution at the iron/oxide interface is similarly not favored. The passive film also becomes dehydrated in that H^+ ions move outward through the cation-selective layer into solution, and OH^- ions move inward through the anion-selective layer to reinforce passivity at the metal/oxide interface. The joint action of the selectivity properties of the film and dehydration of the film is proposed to produce a protective oxide film.

Reversal of the bipolarity of the film, as in Fig. 9.16(d), favors passage of aggressive anions into the film, passage of cations produced by corrosion out through the film, as well as continued hydration of the film. Thus, this set of conditions produces a non-protective oxide film.

At present, it is not clear how the passive oxide film on an elemental metal (e.g., Fe) can produce a bipolar ion-selective film. This model may be more better suited to the passivity of alloys where the oxide film may contain various ionic species. In fact, the bipolar model of passivity has been extended by Clayton and Lu [60] to Fe–Cr–Ni–Mo stainless steels. A detailed surface analysis of the passive film formed anodically in 0.1 M HCl showed the presence of Mo (as MoO_4^{2-}) in the outer part of the oxide film and of Cr^{3+} (as $Cr(OH)_3$) in the inner regions of the passive film. Clayton and Lu have suggested that MoO_4^{2-} is the basis for a cation-selective outer layer and $Cr(OH)_3$ is the basis for an anion-selective inner layer.

Spinel/Defect Model

Based on in situ X-ray studies of the passive film formed on iron in a buffer solution [34, 35, 61, 62], a model of the passive film has been proposed which suggests that the film is a single-phase nanocrystalline spinel oxide which is related to but different than γ-Fe_2O_3 or Fe_3O_4.

The oxides Fe_3O_4 and γ-Fe_2O_3 are closely related, consisting of a unit cell of 32 cubic close-packed oxygen ions containing Fe cations as follows [63]:

Fe_3O_4:	8 Fe^{3+} in tetrahedral sites, 8 Fe^{2+} plus 8 Fe^{3+} in octahedral sites.
γ-Fe_2O_3:	21 and 1/3 Fe^{3+} statistically scattered in the above 24 positions.

(See Fig. 9.20 for more about tetrahedral and octahedral sites.) The passive film on iron, however, was observed to have a distribution of Fe^{3+} or Fe^{2+} ions which does not follow either of the above patterns for Fe_3O_4 or γ-Fe_2O_3.

The groundwork for this model was laid by earlier work of Long et al. [64], who using surface EXAFS showed that passive films formed on iron in nitrate or chromate solutions had Fe–Fe bond distances different from that of the known iron oxides or hydroxides.

In addition, the passive film was observed to contain point defects, such as vacancies and interstitials, and also extended defects, such as finite crystallite size and stacking faults. The crystalline nature of the passive film on iron was confirmed by scanning tunneling microscopy studies [65].

In addition, Schmuki et al. [35] showed that it is possible to have a two-stage reduction process (as observed by Cohen and co-workers) when there is only a single-phase Fe_2O_3 oxide film (so that the observation of a two-stage reduction does not necessarily imply a bilayer structure).

This model of the passive film is represented by Fig. 9.16(e).

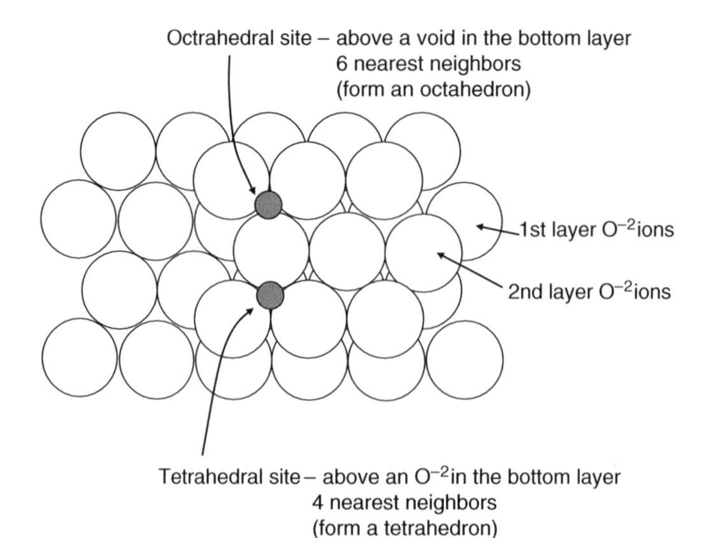

Fig. 9.20 Octahedral and tetrahedral sites in iron oxides. The small gray circles refer to Fe^{3+} ions and the larger white circles refer to O^{2-} ions

What Do These Various Models Mean?

Elucidation of the nature of the passive film on iron is primarily a matter of scientific concern. From the practical point of view, the corrosion engineer needs to know how to achieve passivity but does not have to understand in detail its precise mechanism. However, there is much interest in understanding the nature of the passive film on iron because the results for iron provide a starting point for understanding passivity when alloying elements are added to iron (to produce stainless steels, for example). In addition, an improved understanding of mechanisms of passivity may lead to the design of new and improved corrosion-resistant bulk or surface alloys.

Passive Oxide Films on Aluminum

Aluminum reacts readily with water or aqueous solutions to form an oxide film, as shown in the "Introduction to this Chapter." The composition, structure, and thickness of the oxide films vary greatly, depending upon details of their formation.

The aluminum–water system is much simpler than the iron–water system because the aluminum cation exists in only a single oxidation state (3+), unlike the case for iron. The possible reaction products are aluminum hydroxide, $Al(OH)_3$, and aluminum oxide, Al_2O_3, although the latter may exist in various degrees of hydration and in several different crystal structures [66]. See Table 9.3. The presence of water in oxide films on aluminum has been determined through the use of infrared spectroscopy and Rutherford back scattering spectrometry [66–68].

Table 9.3 Major reaction products formed between aluminum and water [66]

$Al(OH)_3$	
γ-Al_2O_3	
γ-$Al_2O_3 \cdot H_2O$ (γ-AlOOH)	(boehmite)
β-$Al_2O_3 \cdot 3H_2O$	(Bayerite)

Pseudoboehmite is a poorly crystalline modification of boehmite.

Air-Formed Oxide Films

The air-formed oxide film on aluminum is of the order of 30–40 Å in thickness and consists entirely of Al_2O_3 [4, 5]. The outermost layer consists of surface hydroxyls which result from the interaction of the oxide film with water vapor in the atmosphere, but this hydroxylated region is only 8 Å in thickness and does not persist into the inner portion of the oxide film [4, 5]. Using a simple weight gain technique, Godard [69] found that the oxide film thickness on pure aluminum was approximately 200 Å after 5 years of exposure to 52% relative humidity and 1,700 Å after 5 years of exposure to 100% relative humidity.

Films Formed in Aqueous Solutions

Aluminum reacts readily with water to form a hydrous oxide film. Growth of the film in water occurs in two stages [70, 71]: a pseudoboehmite film is produced initially and is then covered by a layer of bayerite crystals, as shown in Fig. 9.21. The thickness of the film in Fig. 9.21 was 2.8 μm (28,000 Å) after 4 days of immersion in distilled water at 40°C [71]. Using weight gain measurements, Hart [70] observed that film growth was complete after about 12 days of immersion in water at 20°C. From the measured weight gain and the density of bayerite, the limiting film thickness is calculated to be 50,000 Å.

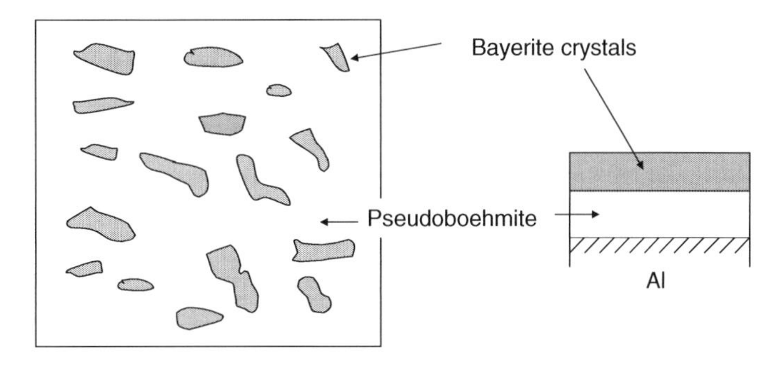

Fig. 9.21 Idealized schematic representation of the growth of an oxide film on aluminum in water at 40°C [70, 71]

Oxide films on aluminum can be increased in thickness by the process of anodization, in which oxide films are grown in aqueous solutions under conditions of controlled anodic potential. Two types of anodized films can be formed: barrier oxides and porous oxides. Ammonium tartrate, boric acid, and ammonium pentaborate are electrolytes commonly used to produce barrier-type oxides, which are compact films that can be varied in thickness between approximately 100 and 2,000 Å [72, 73].

The second type of anodic oxide film, the porous oxide film, typically has a total oxide thickness of 3–35 μm (30,000–350,000 Å) [74]. This type of film is commonly formed by anodization in sulfuric or phosphoric acids and consists of an inner compact layer and an outer porous layer. The outer porous layer is characterized by a "scalloped" structure of hexagonal cells, each containing a central pore. See Fig. 9.22, which is a transmission electron micrograph of a cross section of a phosphoric acid anodized film [75]. A schematic representation of a phosphoric acid anodized film is shown in Fig. 9.23. The fingers or protrusions of oxide shown in the figure are believed to improve the adhesion of organic coatings or adhesives by interlocking with the organic overlayer [9, 76].

The open cells in the porous structure are usually sealed in boiling water to improve the corrosion resistance [77]. Organic dyes or pigments can be incorporated into the porous oxide to produce a colored anodic coating [78].

Properties of Passive Oxide Films

Passive oxide films have the following properties [79–81]:

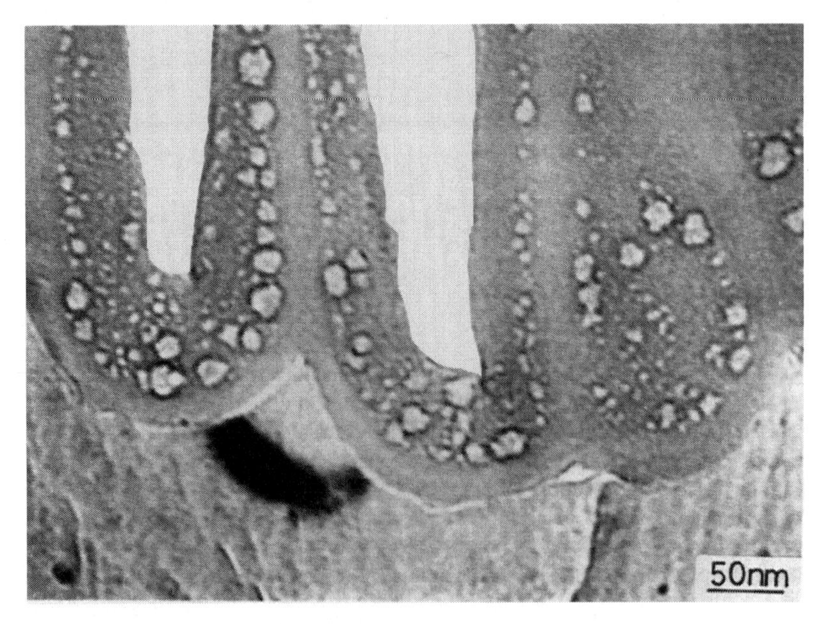

Fig. 9.22 Transmission electron micrograph of the cross section of an anodized film formed on aluminum in phosphoric acid [75]. Reproduced by permission of Taylor and Francis Ltd and www.informaworld.com

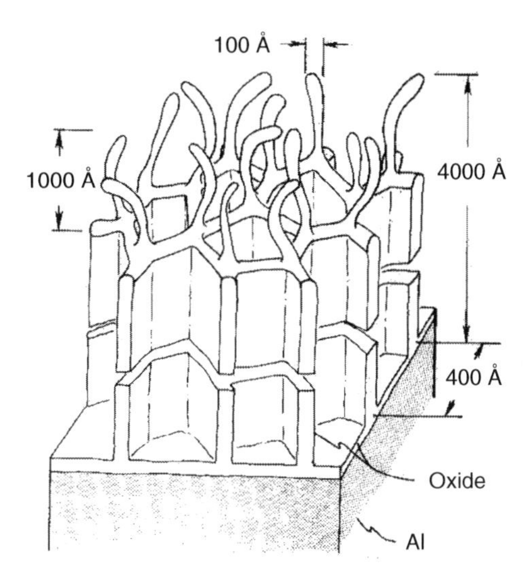

Fig. 9.23 Schematic drawing of the oxide structure on an aluminum alloy after anodizing in phosphoric acid [9, 76]. Reproduced by permission of Elsevier Ltd

Thickness

As discussed earlier, the thickness of passive films varies from tens to hundreds of thousands of angstroms (essentially from nanometers to micrometers), depending on the metal, the electrolyte, and the conditions under which the film was formed.

Electronic and Ionic Conductivity

Passive layers on the transition metals (e.g., Fe, Cr, Ni) are generally thin (less than 100 Å, i.e., 10 nm) and possess good electronic conductivity. This is evidenced by the fact that passive films on iron, for example, can support electron transfer reactions such as the $Fe(CN)_6^{4-}$/ $Fe(CN)_6^{3-}$ exchange [82]. There is considerable evidence, for example, that passive films on Fe, Ti, or Sn are n-type semiconductors, while the passive film on Cu or Ni is a p-type semiconductor [81]. In contrast, barrier passive films on metals like Al, Ta, and Pb are thicker and generally possess poor electronic conductivity.

The ionic conductivity of all passive films should be low to prevent the passage of aggressive ions such as chlorides inward though the film and the transfer of dissolved cations outward through the film.

Chemical Stability

The oxide film should have low chemical solubility and should be stable under a wide range of electrode potentials.

The range of stability for a given a passive film on a given metal is conveniently provided by the Pourbaix diagram of the metal. For example, Fig. 9.24 compares Pourbaix diagrams for Al, Cr, and

Nb. The passive film on Cr is stable over a wider range of pH values than is the passive film on Al. However, the passive film on Nb is stable over an even wider range of both pH values and electrode potentials.

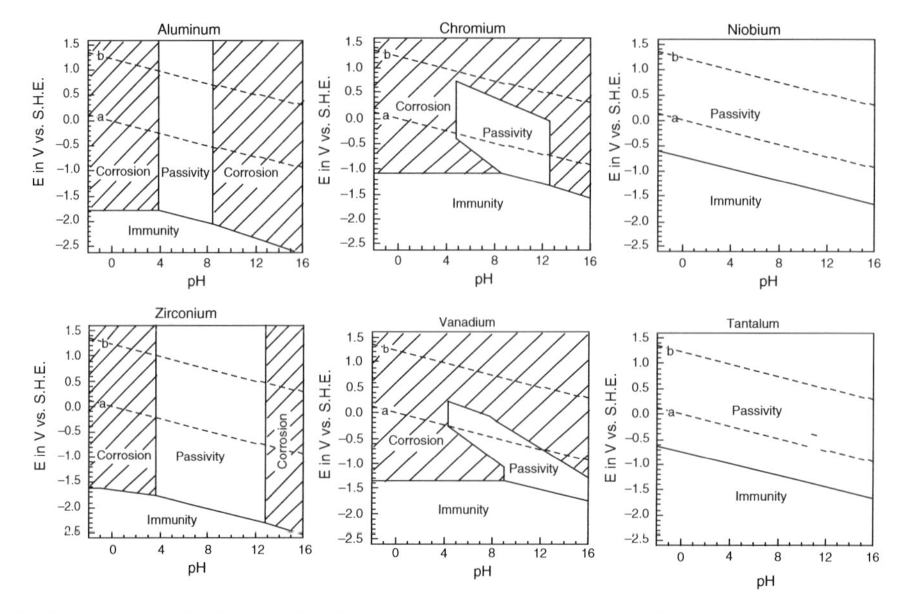

Fig. 9.24 *Top row*: Pourbaix diagrams for aluminum, chromium, and niobium. *Bottom row*: Pourbaix diagrams for zirconium, vanadium, and tantalum

Figure 9.24 also compares the Pourbaix diagrams for Zr, V, and Ta. See Problem 9.3 at the end of this chapter.

Mechanical Properties

The mechanical properties of passive oxide films are important because the metal may undergo scratching, abrasion, or deformation. The oxide film should have a high compressive strength and should have good adhesion to the underlying metal. However, relatively little is known about the mechanical properties of passive oxides, due in large part to the fact that they are thin films.

Stresses have been measured in the passive films on various metals, including 304 stainless steel, nickel, tantalum, niobium, aluminum, zirconium, titanium, and tungsten [83–86]. Using a straining technique, Bubar and Vermilyea found that oxide films on tantalum exhibited great ductility and could be deformed as much as 50% before fracture [87]. Oxide films on zirconium, 304 stainless steel, and iron also had appreciable ductility, but oxide films on aluminum were more brittle [87, 88].

Lateral stresses in passive films can arise due to the epitaxial misfit between the oxide and the underlying metal. A measure of this misfit is given by the Pilling–Bedworth ratio [89], defined as

$$R = \frac{\text{Molecular volume of oxide}}{\text{Molecular volume of metal}} \tag{12}$$

or

$$R = \frac{\dfrac{M_{\text{oxide}}}{d_{\text{oxide}}}}{n\dfrac{A_{\text{metal}}}{d_{\text{metal}}}} \tag{13}$$

where M_{oxide} is the molecular weight of the oxide, A_{metal} is the atomic weight of the metal, d_{oxide} and d_{metal} refer to densities, and n is the number of metal atoms per molecule of oxide (e.g., $n = 2$ for Fe_2O_3). If the Pilling–Bedworth ratio R is of the order of 1, then a given volume of metal produces the same volume of oxide, there is a good spatial match between the two, and the oxide is protective. A value of $R < 1$ means that the metal produces insufficient oxide to cover the metal, and the oxide is unprotective. A value of R slightly greater than 1 means that there is a favorable compressive strength in the oxide, but values much greater than 1 introduce large compressive strengths which can lead to cracking and fracture of the oxide.

Table 9.4 lists Pilling–Bedworth ratios R calculated from Eq. (13). It can be seen that each of the non-passive alkali metals and alkaline earths has values of $R < 1$ (with the exception of beryllium). Metals which are normally passive or can be passivated have values of R between 1 and 2. However, the Pilling–Bedworth ratio is greater than 2 for some oxides which provide stable passive films. For example, the Pilling–Bedworth ratios for Fe_2O_3 and Fe_3O_4 are 2.14 and 2.10, respectively, although these oxides are components of the passive film on iron. In addition, the Pilling–Bedworth ratio for Ta_2O_5 is 2.44, although the oxide film on tantalum is extremely resistant to localized attack by pitting corrosion.

These breakdowns in comparison are due to two causes. First, the passive oxide films are very thin and may be non-stoichiometric so that their densities are not necessarily those of the bulk oxides which have been used to calculate the Pilling–Bedworth ratio. Second, comparison of molar volumes is only one aspect in assessing the mechanical properties of oxide films, and other factors are important, such as the inherent stress–strain behavior of the oxide film. Stresses may arise in passive films due to external mechanical loading, intrusion of chloride ions into the passive film, or by hydration/dehydration events. Pilling–Bedworth ratios are more useful in the high-temperature oxidation of metals, where the thicker oxide scales which form are more closely related to bulk oxide properties.

There has been much recent interest in the study of mechanical properties of passive films using new approaches, such as microindentation techniques [90] and the atomic force microscope [32].

If a passive film undergoes fracture so as to expose the underlying metal to the environment, then the continuing corrosion resistance of the metal depends upon the ability of the metal to repassivate faster than active corrosion can ensue. The topic of repassivation will be discussed in Chapter 10, which considers the localized breakdown of passive films.

Structure of Passive Films

Passive films may be amorphous, nanocrystalline, or crystalline. For example, the air-formed oxide film on pure aluminum is amorphous in that it does not give an electron diffraction pattern characteristic of a crystalline solid [70]. However, when immersed in water, the oxide film which develops is crystalline [66, 70, 71]. An example of a nanocrystalline passive film is provided by work on iron which has shown that the passive film is crystalline, but with a fine grain structure [65, 91].

Table 9.4 Pilling–Bedworth ratios R for various metal/metal oxides[a]

$R < 1$		$1 < R < 2$		$R > 2$	
Li_2O	0.57	Ag_2O	1.57	Co_2O_3	2.41
Na_2O	0.58	AgO	1.61	Fe_2O_3	2.14
K_2O	0.46	Al_2O_3	1.29	Fe_3O_4	2.10
Rb_2O	0.42	BeO	1.67	MoO_2	2.10
MgO	0.81	CdO	1.22	MoO_3	3.25
CaO	0.65	Ce_2O_3	1.28	Mn_2O_3	2.10
SrO	0.61	CeO_2	1.09	MnO_2	2.27
BaO	0.71	CoO	1.75	Nb_2O_5	2.66
		Co_3O_4	1.98	PtO_2	2.11
		Cr_2O_3	2.00	SiO_2	2.13
		Cu_2O	1.68	Ta_2O_5	2.44
		CuO	1.78	Sb_2O_5	2.35
		FeO	1.69	U_3O_8	2.68
		MnO	1.75	UO_3	3.14
		NbO	1.38	VO_2	2.25
		NbO_2	1.95	V_2O_5	3.20
		NiO	1.68	WO_2	2.10
		PbO	1.30	WO_3	3.38
		Pb_2O_3	1.25		
		Pb_3O_4	1.40		
		PtO	1.64		
		Sb_2O_3	1.43		
		SnO	1.28		
		SnO_2	1.35		
		TaO_2	1.93		
		TiO	1.21		
		Ti_2O_3	1.51		
		TiO_2	1.77		
		UO_2	1.98		
		U_4O_9	1.96		
		VO	1.37		
		V_2O_3	1.81		
		ZnO	1.58		
		ZrO_2	1.55		

[a]Metal and oxide densities are from "Handbook of Chemistry and Physics," 82nd Ed. (2001–2002).

Amorphous metals generally have reduced corrosion rates compared to their crystalline counterparts. For example, Fig. 9.25 shows polarization curves for an 85Co–15Nb alloy in 0.5 M NaCl [92]. The crystalline alloy readily undergoes anodic dissolution at anodic potentials, but the amorphous alloy exhibits an anodic current density which is several orders of magnitude lower over a range of approximately 1.0 V up to the localized breakdown of the passive film, which occurred above +0.4 V.

In a study on the effect of progressive addition of Cr to iron, McBee and Kruger [93] showed by electron diffraction that the passive films go from polycrystalline to amorphous with the addition of increased amounts of Cr to the alloy. In view of observations like this and with the information available on the improved corrosion behavior of amorphous metals [94], it has been proposed that an important factor in the passivity of metals is the structure of the passive film [94, 95]. An amorphous

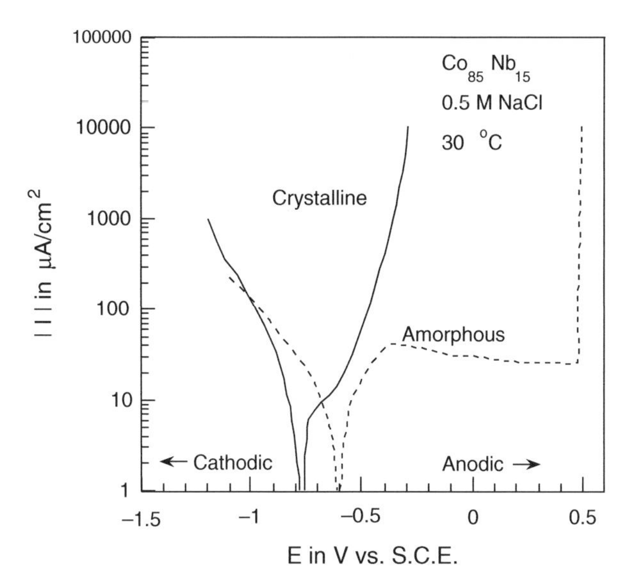

Fig. 9.25 Anodic polarization curves for crystalline and amorphous 85Co–15Nb in 0.5 M NaCl [92]. Reproduced by permission of ECS – The Electrochemical Society

oxide is believed to provide a more protective passive film due to the lack of grain boundaries and other defects which would otherwise assist the outward migration of cations though the film.

Passivity in Binary Alloys

Chromium is naturally passive when exposed to the outdoor atmosphere and remains bright for years, unlike iron which corrodes in a short time. In order to achieve passivity with iron, it is necessary to produce a passive film electrochemically (by anodic polarization) or chemically (by the action of inhibitors such as chromates).

Another approach toward attaining passivity in iron is by alloying. Alloying of iron with chromium, for example, facilitates the passivation process, as shown in Fig. 9.26. The addition of 18% Cr to iron produces a sharper active/passive transition, a decreased critical current density for passivation, and a decreased passive current density. Each of these effects increases the ease of passivation.

It is well known that the introduction of 13 at.% Cr into iron passivates the surface of the binary iron–chromium alloy. Figure 9.27 shows the effect of progressive Cr additions on the Flade potential and on the passive current density in Fe−Cr binary alloys [16]. It can be seen that Fe−Cr alloys become passive at a critical alloy composition of approximately 13 at.% Cr. This fact forms the basis for the protection provided by type 400 stainless steels, which are essentially Fe−Cr alloys.

A number of other binary alloys also exhibit a critical alloying composition, in which introduction of a certain minimum amount of the alloying element into the host metal causes the resulting binary alloy to become passive or to have reduced corrosion rates. For example, copper–nickel alloys display a reduction in corrosion rates in sodium chloride solutions for the addition of 30–40 wt%

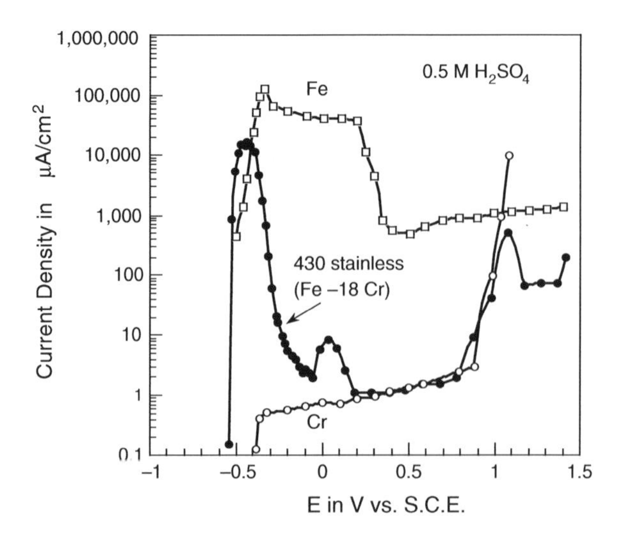

Fig. 9.26 Anodic polarization curves for Fe, Cr, and Fe–18Cr (430 stainless steel) in 1 N H_2SO_4. Redrawn from [96] with permission of ECS – The Electrochemical Society and [97] with © NACE International (1962)

(32–43 at.%) nickel to copper [98], as shown in Fig. 9.28. In another example, the passive current densities for aluminum–zirconium alloys in 1 M HCl decrease dramatically for a critical alloy composition of 24 at.% Zr [99], as shown in Fig. 9.29.

The Fe−Cr binary alloy system is the most important from an industrial point of view and has also attracted the most scientific interest. Several theories have been advanced to explain the occurrence of a critical alloying composition in Fe−Cr binary alloys [93, 100–105], the most recent of which has also been extended to other binary alloys [104–109]. These theories are (i) the electron configuration theory, (ii) oxide film effects, (iii) the percolation theory, and (iv) the graph theory model. The last three theories are not mutually exclusive, and in fact, the last two provide different explanations of oxide film effects. Each of these theories is discussed briefly below.

Electron Configuration Theory

The electron configuration theory was proposed by Uhlig [100] and is an extension of the adsorption theory of passivity. Uhlig noted that a number of transition metals (which have unfilled d-shells) become passive at certain critical compositions when alloyed with a second metal. According to the electron configuration theory, the passivity of a binary alloy depends on the tendency of unfilled d-bands of transition metals to become filled with electrons from alloying elements capable of donating electrons in such a way that chemisorption of oxygen is favored. For Fe−Cr alloys, the electron configuration of the two alloying elements are the following:

Fe [Ar] $3d^6\ 4s^2$
Cr [Ar] $3d^5\ 4s^1$

where [Ar] denotes that the inner shells of the atom have the same electron configuration as in the argon atom, and $3d^6$, for example, denotes that there are six electrons in the 3d level. Uhlig's idea was that a passive binary alloy would have an electron configuration as close as possible to that of

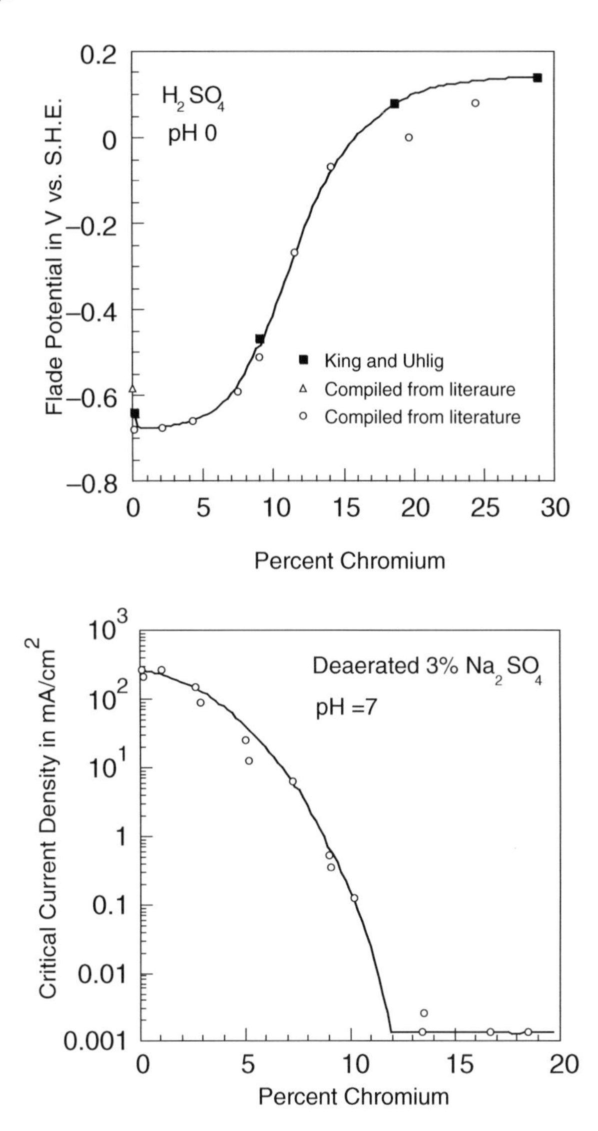

Fig. 9.27 *Top*: the Flade potential of Fe–Cr alloys in H_2SO_4 (pH 0) as a function of the alloy composition. *Bottom*: The passive current density as a function of alloy composition [16]. Reproduced by permission of the American Chemical Society

an inert noble gas atom. In the Fe−Cr pair, Cr is the passive metal so that its d-bands become filled with 4s electrons donated by Fe atoms. Electron configurations in a condensed (solid) metal are different than in the isolated metal atom, as given in Table 9.5. The number of d-vacancies for the metal surface atom in Table 9.5 is the number of vacancies in the metal plus one due to the donation of one electron from each metal surface atom to a chemisorbed oxygen atom to form the O^- surface species, as shown in Fig. 9.30.

Thus, a Cr atom in the metal surface contains 4.0 d-electron vacancies. Iron atoms in the surface of the alloy can each donate (1.8–1.0) = 0.8 electrons to Cr. Thus, the number of Fe atoms required

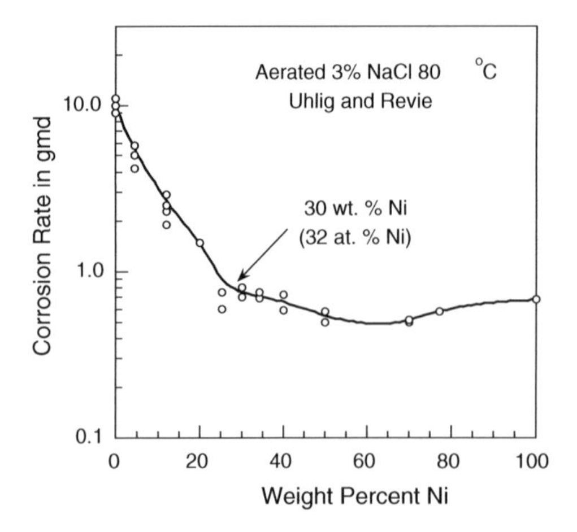

Fig. 9.28 Corrosion rates of copper–nickel binary alloys in 3% NaCl [98]. Reproduced by permission of John Wiley & Sons, Inc

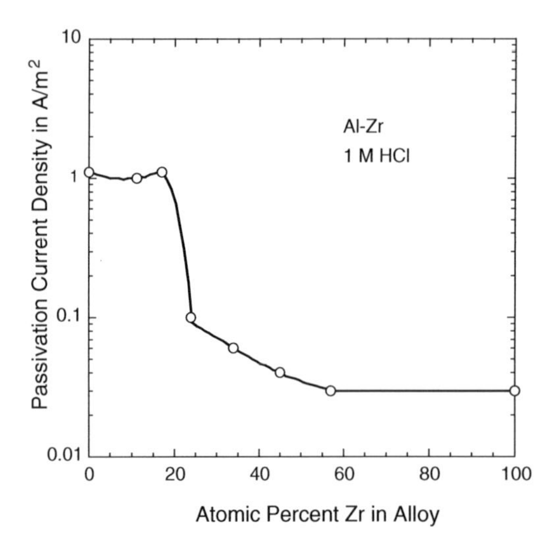

Fig. 9.29 Passive current densities for aluminum–zirconium binary alloys in 1 M HCl [99]

to donate electrons to fill the vacant d-bands of a surface Cr atom is 4.0/0.8 = 5.0 Fe atoms; the atomic percent of Cr is given by Cr/(Cr + Fe) = 1.0/(1.0 + 5.0) = 16.7 at.%, in nominal agreement with the experimental value of 13 at.% where Fe–Cr alloys attain passivity.

This approach has also been used by Uhlig to explain critical alloy compositions in Cu–Ni, Fe–Ni, Fe–Co, Ni–Mo, Ni–Cr, and Co–Cr binary alloys, as shown in Table 9.6. In each case, there is a good agreement between predicted and observed critical alloy compositions. However, the usual criticism of the electron configuration theory is that it fails to take into account the properties of the oxide film.

Table 9.5 Electron configurations for various transition metals and for copper [100]

Metal	Atomic configuration	d-Vacancies in atom	d-Vacancies in metal[a]	d-Vacancies in metal surface	4s electrons demoted to d-band
Cr	[Ar] $3d^5$ 4s	5	–	4.0[b]	2.0
Fe	[Ar] $3d^6$ $4s^2$	4	2.2	3.2	1.8
Co	[Ar] $3d^7$ $4s^2$	3	1.7	2.7	1.3
Ni	[Ar] $3d^8$ $4s^2$	2	0.6	1.6	1.4
Cu	[Ar] $3d^{10}$ 4s	0	0.0	0.0	0.0
Mo	[Kr] $4d^6$ 5s	4	3.0	4.0	1.0

[a] From magnetic measurements.
[b] Extrapolated value.

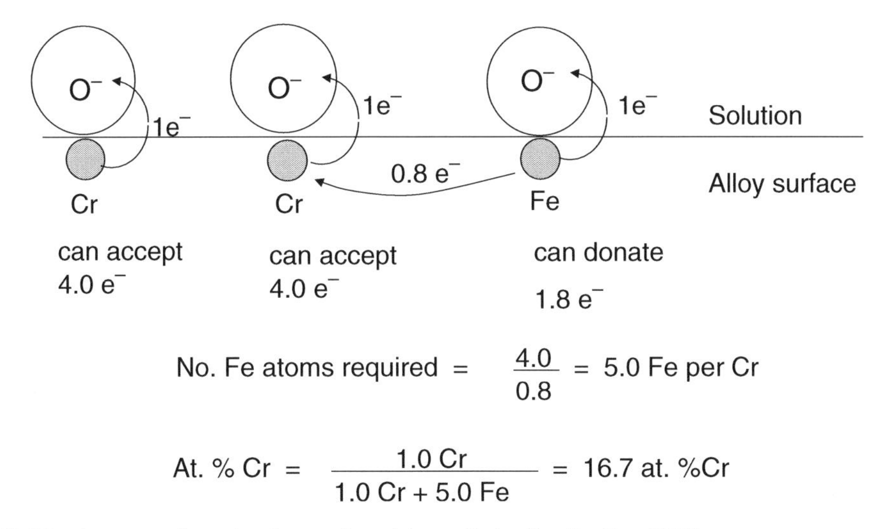

Fig. 9.30 The electron configuration theory of passivity applied to Fe–Cr alloys [100]

Oxide Film Properties

There is considerable evidence that the properties of the oxide film are responsible for the onset of passivity in Fe-Cr binary alloys to occur at 13 at.% Cr. Various investigators have found that Cr is enriched in the passive film on Fe-Cr alloys [28, 101, 102]. The results of Asami et al. [28] have been given in Fig. 9.9, which shows that the cation fraction of Cr^{3+} in the oxide film increases dramatically at 10–13 at.% of Cr in the binary alloy. Recent in situ results of Oblonsky et al. [102] using XANES also show an enrichment of Cr^{3+} in the passive films in a series of Fe–Cr alloys.

McBee and Kruger [93] showed the importance of the structure of the passive film on Fe–Cr alloys in a study using electron diffraction. As Cr was added to Fe in a series of Fe–Cr alloys, the passive films passed from crystalline to amorphous, as shown in Table 9.7. The authors attributed the enhanced corrosion resistance of the higher content Cr alloys to the existence of an amorphous oxide film, with its concomitant improved properties, as discussed earlier.

Table 9.6 Electron configurations and critical alloy compositions [100]

Binary alloy	Acceptor or donor?	d-Vacancies in metal surface[a]	Number of donor electrons	Predicted critical composition		Observed critical composition (wt%)[b]
				At.%	Wt%	
Ni	A	1.6				
Cu	D		1.0	38.5% Ni	36.6% Ni	30–40% Ni
Cr	A	4.0				
Fe	D		0.8	16.7% Cr	15.7 % Cr	12% Cr
Ni	A	1.6				
Fe	D		0.8	33.3% Ni	34% Ni	40% Ni
Co	A	2.7				
Fe	D		0.8	22.8% Co	23.8% Co	21–25% Co
Mo	A	4.0				
Ni	D		0.4	9.1% Mo	14.7% Mo	10–15% Mo
Cr	A	4.0				
Ni	D		0.4	9.1% Cr	8.2% Cr	7–9% Cr
Cr	A	4.0				
Co	D		0.3	7.0% Cr	6.2 % Cr	10% Cr

[a] After donation of one electron to O^- surface species.
[b] For specific aqueous environments, see [100].

Table 9.7 Structural characteristics for the passive oxide films on Fe–Cr binary alloys [93]

Percentage Cr in alloy	Structure of passive film
0	Well-oriented spinel
5	Well-oriented spinel
12	Poorly oriented spinel
19	Mainly amorphous
24	Completely amorphous

Percolation Theory

Sieradzki and Newman [103] applied percolation theory to explain the critical Cr composition for passivity in Fe–Cr alloys. These authors considered that Cr atoms must come close enough to each other so that adsorbed oxygen anions can bridge what is to become oxidized Cr. Based on atomic distances and the concept of a cluster of Cr atoms in the body centered cubic lattice, percolation theory was able to explain the threshold composition for passivity in Fe–Cr alloys.

As shown in Fig. 9.31, when an oxygen ion (O^{2-}) bridges two Cr^{3+} ions, the maximum separation between the two Cr^{3+} ions is equal to twice the radius (0.69 Å) of the Cr^{3+} plus twice the radius (1.40 Å) of the O^{2-} ion, or 4.18 Å. The distances between various nearest neighbors in the bcc Fe–Cr lattice having a lattice parameter $a = 2.87$ Å are given in Table 9.8.

Note that the distance of 4.18 Å required to bridge two Cr atoms is approximately the distance between third nearest neighbors in the bcc Fe–Cr lattice. Percolation processes show a sharp threshold p_c, which represents the concentration of occupied sites at which a connected cluster suddenly appears [110]. The p_c for interactions up to third nearest neighbors is 10%. The p_c for interactions up to fourth nearest neighbors is 12%. These values correspond to Cr concentrations producing passivity in Fe–Cr alloys. As the Cr atom concentration increases above p_c, the infinite cluster grows rapidly absorbing smaller clusters [110].

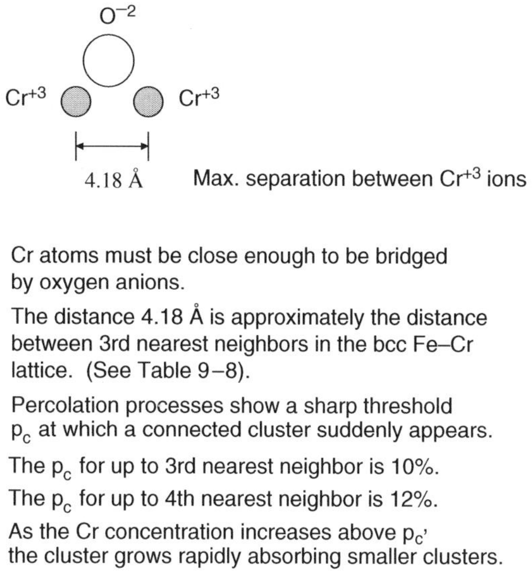

Fig. 9.31 Percolation theory and the passivity of Fe–Cr alloys [103]

Table 9.8 Distance between nearest neighbors in the Fe–Cr body centered cubic lattice [103]

Nearest neighbors			
1st	2nd	3rd	4th
$\frac{\sqrt{3}}{2}a$ 2.49 Å	a 2.87 Å	$\sqrt{2}a$ 4.06 Å	$\sqrt{3}a$ 4.97 Å

a is the lattice parameter.

Graph Theory Model

Another approach [104–109] has recently been taken using graph theory and the concept of a connected network of –M–O–M– bridges in the passive oxide film, where M is the component of the binary alloy giving rise to passivity. For Fe–Cr alloys, passivity is imparted by the addition of Cr to Fe so that a connected network of –Cr–O–Cr– would lead to passivity.

A graph is a collection of points (vertices) connected by lines called edges. For example, Fig. 9.32 is a graph of a floorplan in a residential home. The individual rooms are represented by points (vertices), and rooms which are physically connected in the house are connected on the graph by edges. Graphs have been used to represent and analyze various types of relationships in biology, the social sciences, and in systems research [111].

In regard to passivity, an oxide may be represented by a mathematical graph in which vertices and edges are used to designate atoms and their bonds. For Fe–Cr binary alloys, a network of uninterrupted –Cr–O–Cr– bridges in the oxide film is first considered, and then random graph theory is used to insert Fe^{3+} ions in a stochastic manner into the –Cr–O–Cr– network to form an oxide of composition $xFe_2O_3{\cdot}(1{-}x)Cr_2O_3$. The critical concentration of Fe^{3+} (and accordingly also of Cr^{3+}) where the –Cr–O–Cr– network loses its connectivity is thus calculated. Or equivalently, the critical concentration of Cr^{3+} in the oxide film (and also of Fe^{3+}) is determined where the –Cr–O–Cr–

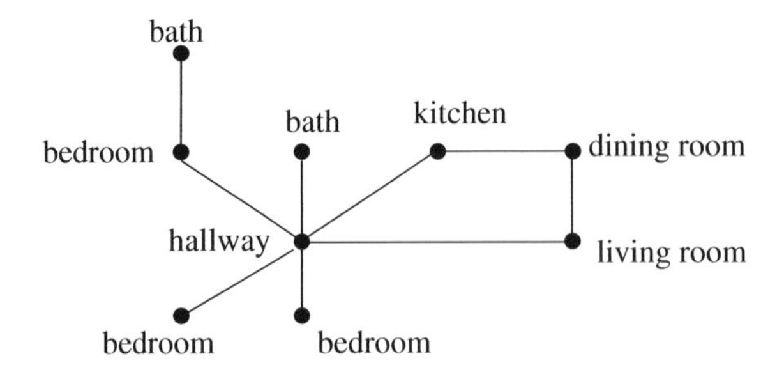

Fig. 9.32 The mathematical graph of the floorplan of a residence

network becomes connected. The composition of the oxide film is then related to the composition of the alloy substrate by means of empirical surface analysis information.

Based on the STM studies of Ryan et al. [65], as shown in Fig. 9.15, the structure of the passive oxide film on an Fe–25 at.% Cr binary alloy can be represented by an array of irregular distorted hexagons. The corresponding *mathematical graph* is given in Fig. 9.33, where the vertices of the graph represent Cr^{3+} ions and the edges represent O^{2-} ions. If the total number of Cr^{3+} ions in the network is given by N, then the total number of edges (O^{2-} ions) is simply $A(G_o) = (3/2)N$.

The –Cr–O–Cr– network, as represented by its graph G_o, is characterized by two properties: its size and configuration. The size of G_o is given by the number of edges $A(G_o)$, and the configuration is given by a quantity called the Randic index $X(G_o)$ [112]:

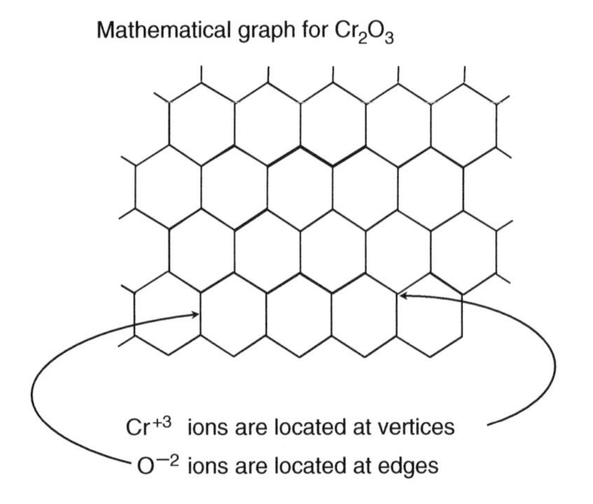

Fig. 9.33 The mathematical graph for Cr_2O_3 [104–106]. The vertices refer to Cr^{3+} ions and the edges connecting them refer to O^{2-} ions

$$X(G_o) = \sum_{\text{edges}} (ij)^{-0.5} \tag{14}$$

where i and j are the degrees of the vertices connecting the edge in question. (A vertex has a degree 2, for example, if two edges meet at that vertex.) The Randic index has been used to characterize the degree of branching of organic compounds, as shown in Fig. 9.34.

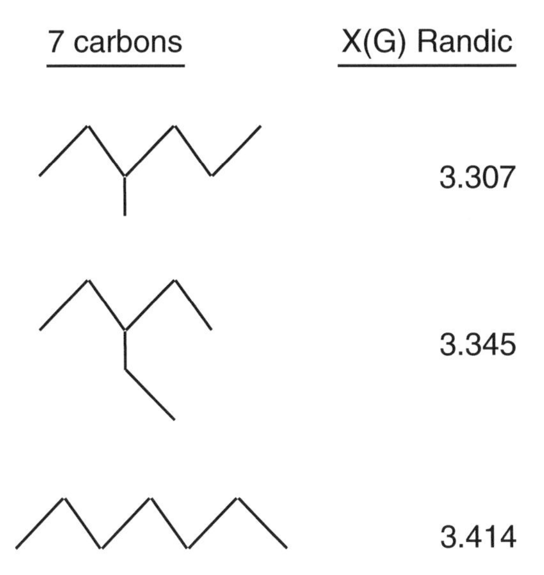

Fig. 9.34 Hydrocarbon isomers and the Randic index

Next, the composition of the oxide network is changed conceptually by insertion of Fe^{3+} ions to produce a new graph G for the mixed oxide having new properties $A(G)$ and $X(G)$. The relative change in the connectivity of the network is given by

$$\delta = \frac{A(G)X(G) - A(G_o)X(G_o)}{A(G_o)X(G_o)} \tag{15}$$

Large values of δ indicate that the connectivity of the new graph G (which contains the disrupting cation) is very much different than that for the original graph G_o, whereas small values of δ mean that the connectivity of the two graphs is similar. When the connectivities of the two graphs are identical, then $\delta = 0$, as can be seen by letting $G = G_o$ in Eq. (15).

The Randic index $X(G_o)$ for the $-$Cr$-$O$-$Cr$-$ network is obtained as follows. Each vertex in Fig. 9.33 has a degree 3 so that each edge contributes one-third to the Randic index:

$$(ij)^{-\frac{1}{2}} \equiv \left(\frac{1}{3 \times 3}\right)^{\frac{1}{2}} = \frac{1}{3} \tag{16}$$

But there are (3/2)N total edges so that

$$X(G_o) = \frac{3}{2}N \times \frac{1}{3} = \frac{N}{2} \tag{17}$$

See Table 9.9, which collects the various input quantities necessary to calculate the connectivity δ.

Table 9.9 Values of various quantities used in graph theory approach to passivity in Fe–Cr binary alloys[a]

*Original graph G_o for -Cr-O-Cr_*
Number of edges
$A(G_O) = \frac{3}{2}N$
Randic index
$X(G_O) = \frac{N}{2}$
New graph G containing Fe^{3+}
Number of edges
$A(G) = N\left[\frac{3-5x}{2(1-x)}\right]$
$X(G) = \frac{\left(\frac{1}{5}\right)^5 N}{(1-x)^5}(3-5x)\left\{24x^4 + 24\sqrt{2}x^3(3-5x) + 4(\sqrt{3}+3)x^2(3-5x)^2 + 2\sqrt{6}x(3-5x)^3 + \frac{1}{2}(3-5x)^4\right\}$
Randic index

[a] N is the number of Cr^{3+} ions in the original graph G_o for Cr_2O_3.
x is the number of moles of Fe_2O_3 in the mixed oxide: $xFe_2O_3{\cdot}(1{-}x)\,Cr_2O_3$.

Next, Fe^{3+} ions are introduced into the graph to produce the new graph G. See Fig. 9.35. The composition of the oxide network after insertion of Fe^{3+} ions is $xFe_2O_3{\cdot}(1{-}x)Cr_2O_3$. If D is the number of edges deleted in G_o to form the new graph G and one Fe^{3+} is inserted per edge deletion (to preserve charge neutrality) then

$$\frac{Fe^{3+}}{Cr^{3+}} = \frac{D}{N} = \frac{x \text{ moles } Fe_2O_3\left(\frac{2 \text{ moles } Fe^{3+}}{\text{mole } Fe_2O_3}\right)}{(1-x) \text{ moles } Cr_2O_3\left(\frac{2 \text{ moles } Cr^{3+}}{\text{mole } Cr_2O_3}\right)} \tag{18}$$

Then, $A(G) = (3/2)N - D$ or

$$A(G) = N\left[\frac{3-5x}{2(1-x)}\right] \tag{19}$$

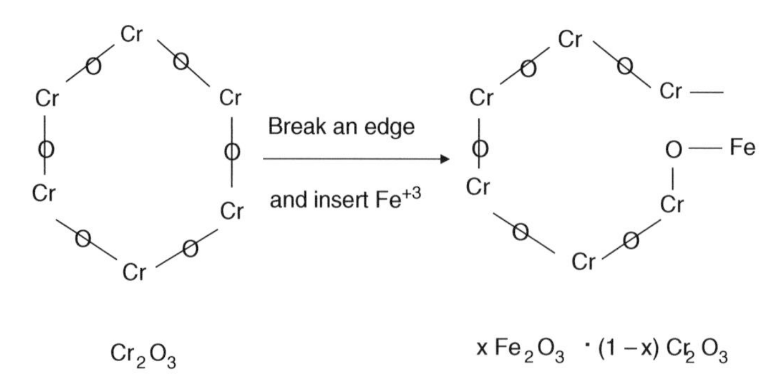

Fig. 9.35 Insertion of Fe^{3+} ions into the graph for Cr_2O_3

Unlike $X(G_o)$, the term $X(G)$ cannot be calculated directly because it is not known which particular edges have been deleted from G_o to form G. A stochastic calculation is employed, the details of which are given in Appendix F. The result is

$$X(G) = \frac{\left(\frac{1}{3}\right)^5 N}{(1-x)^5}(3-5x)\left\{24x^4 + 24\sqrt{2}x^3(3-5x) + 4(\sqrt{3}+3)x^2(3-5x)^2 + 2\sqrt{6}x(3-5x)^3 + \frac{1}{2}(3-5x)^4\right\} \quad (20)$$

Insertion of expressions for $A(G)$, $X(G)$, $A(G_o)$, and $X(G_o)$ (as listed in Table 9.9) into Eq. (15) gives the following result:

$$\delta = \frac{2\left(\frac{1}{3}\right)^6 (3-5x)^2}{(1-x)^6}\left\{24x^4 + 24\sqrt{2}x^3(3-5x) + 4(\sqrt{3+3})x^2(3-5x)^2 + 2\sqrt{6}x(3-5x)^3 + \frac{1}{2}(3-5x)^4\right\} - 1 \quad (21)$$

Figure 9.36 shows the behavior of δ as a function of the composition of the oxide film. ($\delta \approx 0$ when $G \approx G_o$.) It can be seen that an atomic fraction of Fe^{3+} in the oxide film of 0.70 suffices to break down the –Cr–O–Cr– network. That is, an atomic fraction of 0.30 of Cr^{3+} ions in the oxide film provides a continuous –Cr–O–Cr– network, for which $\delta \approx 0$. (The calculated results for δ pass through 0 for $(1-x) \approx 0.30$ but are plotted as absolute values in Fig. 9.36). Figure 9.36 clearly shows that there is a critical concentration of Cr^{3+} ions *in the oxide film* which is needed to provide a continuous network of –Cr–O–Cr– linkages *in the oxide film.*

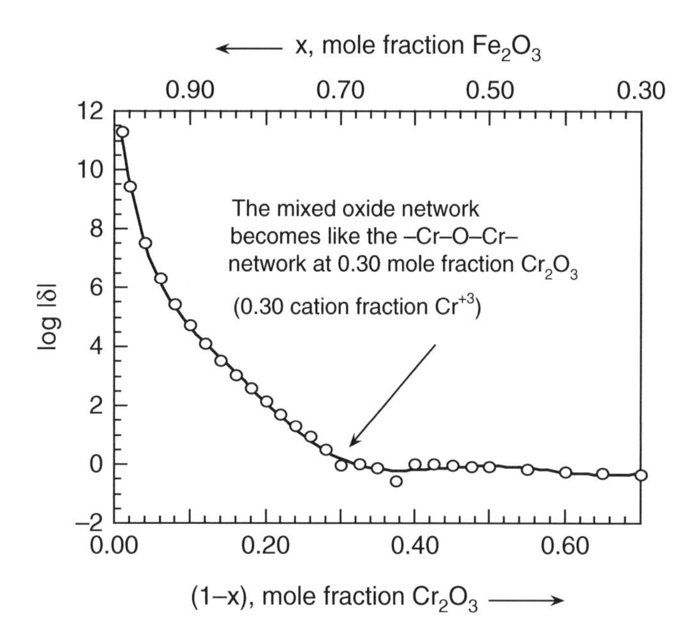

Fig. 9.36 The behavior of the connectivity function δ as a function of the oxide composition for Fe_2O_3–Cr_2O_3 mixed oxides [104–106]. Reproduced by permission of Elsevier Ltd

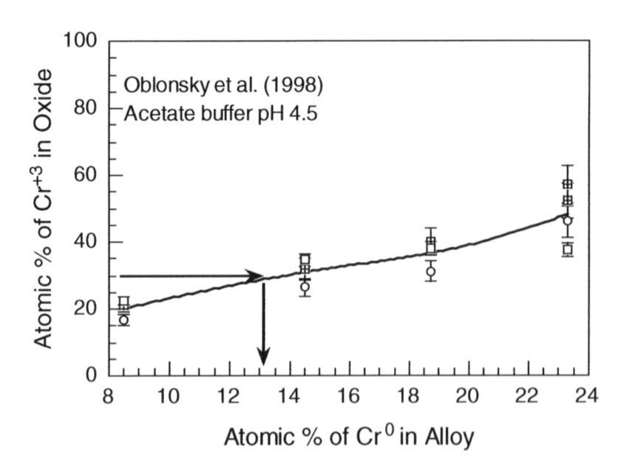

Fig. 9.37 Chromium content in the passive film on Fe−Cr binary alloys as a function of the chromium content in the alloy after immersion in an acetate buffer. Redrawn from [102]. Reproduced by permission of ECS – The Electrochemical Society

From the XPS data of Asami et al. [28] for Fe−Cr alloys shown in Fig. 9.9, a cation fraction of 0.30 Cr^{3+} in the oxide film corresponds to a metal alloy composition of approximately 13 at.% Cr. More recent in situ XANES data of Oblonsky et al. [102] show that an atomic fraction of 0.30 Cr^{3+} in the oxide film corresponds to an atomic fraction of 0.13 Cr in the Fe−Cr alloy. See Fig. 9.37. This result is in excellent agreement with the well-known observation that 13 at.% chromium induces passivity in Fe−Cr alloys.

The graph theory approach has successfully explained the occurrence of critical alloy compositions in 18 different binary alloys [104–109]. Some results are given in Table 9.10.

The exact mechanism of protection by a continuous network of the passivating oxide is not known with certainty. One possibility is that the network of −M−O−M− chains is more stable in the aqueous solution than is the second component of the oxide film. A second possibility (related to the first) is that the tangled network of −M−O−M− chains restricts the transport of ions and defects through the passive oxide film.

Table 9.10 Results from application of graph theory to the passivity of various binary alloys [104–109]

Alloy system	Critical mole fraction of oxide obtained from graph theory δ	Calculated ionic mole fraction in oxide	Corresponding alloy composition from surface analysis data	Observed critical alloy composition from corrosion data
Fe−Cr	0.31 Cr_2O_3	0.31 Cr^{3+}	13 at.% Cr	13 at.% Cr
Ni−Cr	0.13 Cr_2O_3	0.23Cr^{3+}	7 at.% Cr	8–10 at.% Cr
Cu−Ni	0.45 NiO	0.45 Ni^{2+}	~30 at.% Ni	30–40 at.% Ni
Fe−Si	0.36 SiO_2	0.22 Si^{4+}	30 at.% Si	26–30 at.% Si
Co−Cr	0.13 Cr_2O_3	0.23 Cr^{3+}	6 at.% Cr	9–11 at.% Cr
Al−Cr	0.31 Cr_2O_3	0.31 Cr^{3+}	~50 at.% Cr	40–50 at.% Cr
Cr−Mo	0.28 MoO_2	0.16 Mo^{4+}	21 at.% Mo	20 at.% Mo
Cr−Zr	0.28 ZrO_2	0.16 Zr^{4+}	16 at.% Zr	10 at.% Zr
Mo−Ti	0.21 TiO_2	0.21 Ti^{4+}	58 at.% Ti	˜60 at.% Ti

Passivity in Stainless Steels

Stainless steels are commercial alloys containing a minimum of approximately 11 wt% chromium. This amount of chromium prevents the formation of rust in the natural atmosphere, and this observation is the origin of the term "stainless." See Fig. 9.38. The stainless steel "family tree" is depicted in Fig. 9.39, and the compositions of several common stainless steels are given in Table 9.11. Type 304 stainless steel is commonly used for applications requiring good corrosion resistance and formability. However, this alloy is subject to pitting and crevice corrosion, and type 316 is frequently used in its stead when localized corrosion poses a problem. Higher alloys, such as

Fig. 9.38 Not all steels are created equally. From the Washington Post, January 24, 2004 and by permission of Dave Coverly and Creators Syndicate, Inc

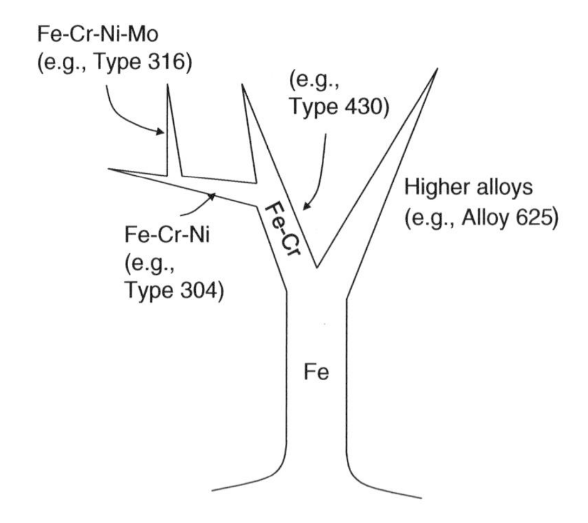

Fig. 9.39 The stainless steel family tree

Table 9.11 Composition (by wt%) of selected stainless steels and higher alloys [135]

UNS[a]	AISI[b]	Cr	Ni	Mo	Fe	C	Mn	Other
S40300	403	11.5–13			Balance	0.15	1.0	Si 0.5
S43000	430	16–18			Balance	0.12	1.0	Si 1.0
S30400	304	18–20	8–10		Balance	0.08	2.0	Si 1.0
S30403	304L	18–20	8–12		Balance	0.03	2.0	Si 1.0
S31600	316	16–18	10–14	2–3	Balance	0.08	2.0	Si 1.0
N06625	625	20–23	Balance (~ 60)	8–10	5	0.1	0.5	Nb 4.15
N10276	C-276	14.5–16.5	Balance (~55)	15–17	7	0.02	1.0	Co 2.5 W 4.5

[a] Unified Numbering System.
[b] American Iron and Steel Institute Designation.

alloy 625 and alloy C-276, which are Cr–Ni–Mo alloys (rather than steels), provide increased corrosion resistance in aggressive environments; but these two alloys can also suffer crevice corrosion in some instances.

The corrosion resistance of each of these alloys depends on the existence of a passive oxide film.

Electrochemical Aspects

Figure 9.40 compares anodic polarization curves in for pure iron, type 430 stainless steel, and type 304 stainless steel in 0.5 M H_2SO_4 [113]. As seen in Table 9.11, type 430 stainless steel contains chromium as an alloyed element, and type 304 contains both chromium and nickel. The critical current density for passivation follows the order: Fe > 430 stainless steel > 304 stainless steel. Thus, it is easier to passivate type 304 stainless steel than either type 430 stainless or pure iron. (The passive current densities for type 430 stainless steel and type 304 are similar but are each considerably lower than that for iron).

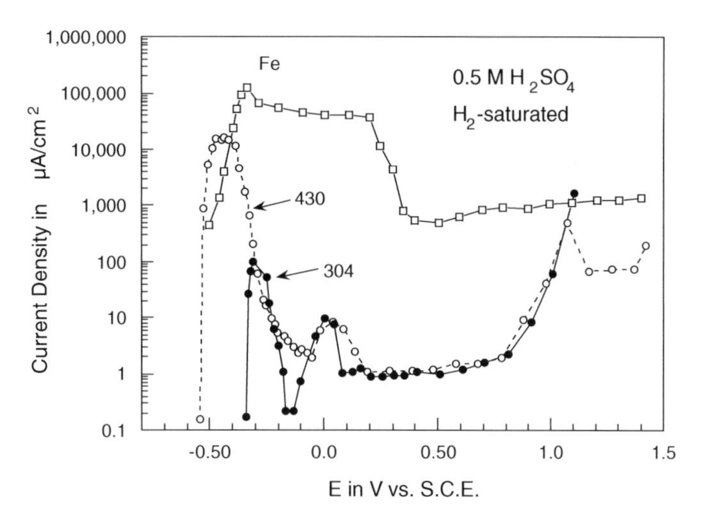

Fig. 9.40 Anodic polarization curves for iron, 430 stainless steel (Fe–18Cr), and 304 stainless steel (Fe–18Cr–8Ni) in 0.5 M H_2SO_4 [113]. Reproduced with permission from The McGraw-Hill Companies

The effect of various alloying elements on electrochemical passivation has been determined by studies in which one element is systematically added to one or two others. For example, Fig. 9.41 shows the effect of Mo additions on the anodic polarization curves of an Fe−Cr alloy in 0.5 M H_2SO_4 [114]. As seen in Fig. 9.41, progressive additions of Mo to the Fe−Cr alloy decrease the critical current density for passivation i_{crit} and also the passive current density i_{pass}. By collecting observations similar to this, a set of "rules" can be formulated as to whether a given alloying element is beneficial to passivation. (An element will be considered to be beneficial if its addition reduces the critical current density for passivation i_{crit} and/or the passive current density i_{pass}).

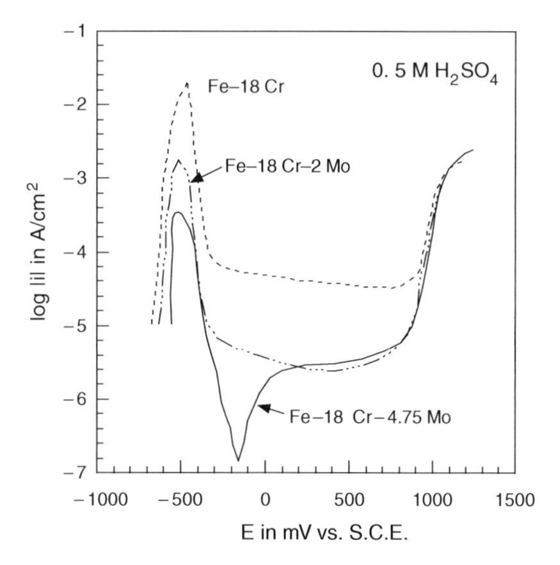

Fig. 9.41 Effect of Mo additions on the anodic behavior of Fe–18Cr alloys in 0.5 M H_2SO_4 [114]. Reproduced by permission of ECS – The Electrochemical Society

Thus, the following effects have been observed in passivation:

Cr is beneficial to Fe [15, 16, 115].
Cr is beneficial to Ni [116].
Cr is beneficial to Fe–Ni [117].
Cr is beneficial to Fe–Ni–Mo [118].
Cr is not beneficial to Mo [119, 120].

Mo is not beneficial to Fe [118, 121, 122].
Mo is not beneficial to Ni [123].
Mo is beneficial to Cr [119, 120].
Mo is beneficial to Fe–Cr [114, 124].
Mo is beneficial to Fe–Cr–Ni [118].

Hence, Mo is beneficial only when Cr is present.

Ni is beneficial to Fe [125].
Ni is beneficial to Mo [123].
Ni is not beneficial to Cr [116].

From these "rules", the effect of Cr on the passivation of Fe, Ni, or Mo is illustrated in Fig. 9.42, where the arrow indicates the direction of beneficial alloying. The effect of alloying by the other three elements is also shown in Fig. 9.42. While the relationships in Fig. 9.42 hold for binary alloys, the trends may break down for ternary alloys.

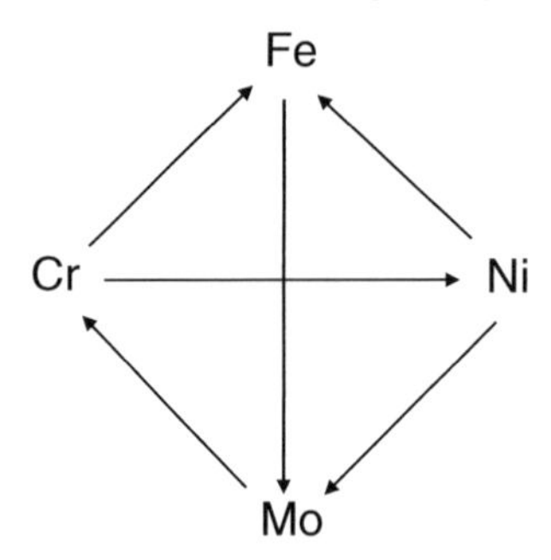

Fig. 9.42 The effect of various metallic elements on passivity in binary alloys. The arrow indicates that the effect is a beneficial one. For example, additions of Cr to Fe are beneficial, and additions of Mo to Fe are not beneficial

Polarization curves can provide interesting data on the effects of alloying on passivation but give no information on the nature of the passive films which are formed on the alloy. Instead, it is necessary to employ surface analytical techniques to experimentally obtain the composition or the structure of the passive films.

Composition of Passive Films on Stainless Steels

The nature of the passive film on stainless steels is complex but is a mixture of metal oxides and hydroxides or oxyhydroxides, with the films possibly containing bound water. The results of one of the earliest analyses of passive films using XPS are shown in Fig. 9.43 for type 304 stainless steel

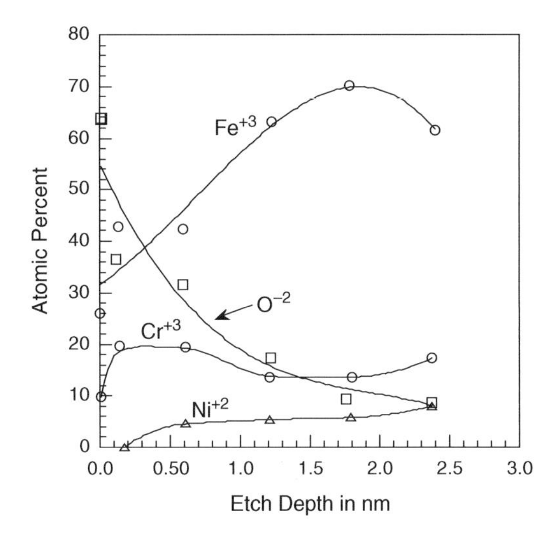

Fig. 9.43 XPS sputter profiles of the passive film formed on type 304 stainless steel in flowing deoxygenated water at 80°C [126]. Reproduced by permission of Elsevier Ltd

[126]. It can be seen that the passive film contains oxides of both Cr and Ni, as well as of Fe. The outer region of the film contains Cr^{3+}, while the inner and outer regions contain both Cr^{3+} and Ni^{2+}.

The results of a more recent XPS investigation on the passive film on a Mo-containing stainless steel were given earlier in Fig. 9.12. When the stainless steel contains molybdenum as an alloying element, Mo^{4+} is usually incorporated into the inner region of the passive film, whereas Mo^{6+} is present in the outer layer [60, 127]. It has been suggested that molybdate ions may act as cation-selective species in the outer layer producing a bipolar passive film, with the inner layer which contains Cr_2O_3 being anion selective, as discussed earlier. Thus, the inward passage of aggressive anions (such as Cl^-) through the film is not favored. In addition, the outward passage of cations produced by metallic dissolution at the metal/oxide interface is similarly not favored.

This model of the passive oxide film on Mo-containing stainless steels is given in Fig. 9.44 [60, 127].

The surface analysis of passive films on stainless steels continues to be a subject of active experimental interest.

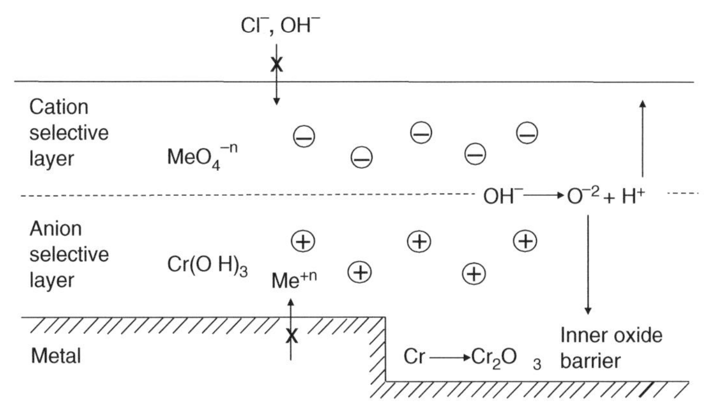

Fig. 9.44 The bipolar model of passivity applied to Mo-containing stainless steels [60, 127]. Reproduced by permission of ECS – The Electrochemical Society

Passivity by Alloying with Noble Metals

Metals which undergo an active/passive transition can be made passive by alloying with noble metals such as palladium or platinum. This idea [128, 129] is illustrated in Fig. 9.45 for Ti-Pd alloys in hot acid solutions. Titanium can be passivated in boiling sulfuric acid solutions, but only after undergoing an active/passive transition following extensive anodic dissolution. The cathodic polarization curve for titanium in Fig. 9.45 is drawn using the exchange current density for hydrogen evolution on titanium given in Table 7.3 and with a cathodic Tafel slope of –0.100 V/decade. As seen in Fig. 9.45, the cathodic and anodic polarization curves for titanium intersect in the active region of the anodic curve so that at open-circuit potential, titanium freely undergoes active dissolution.

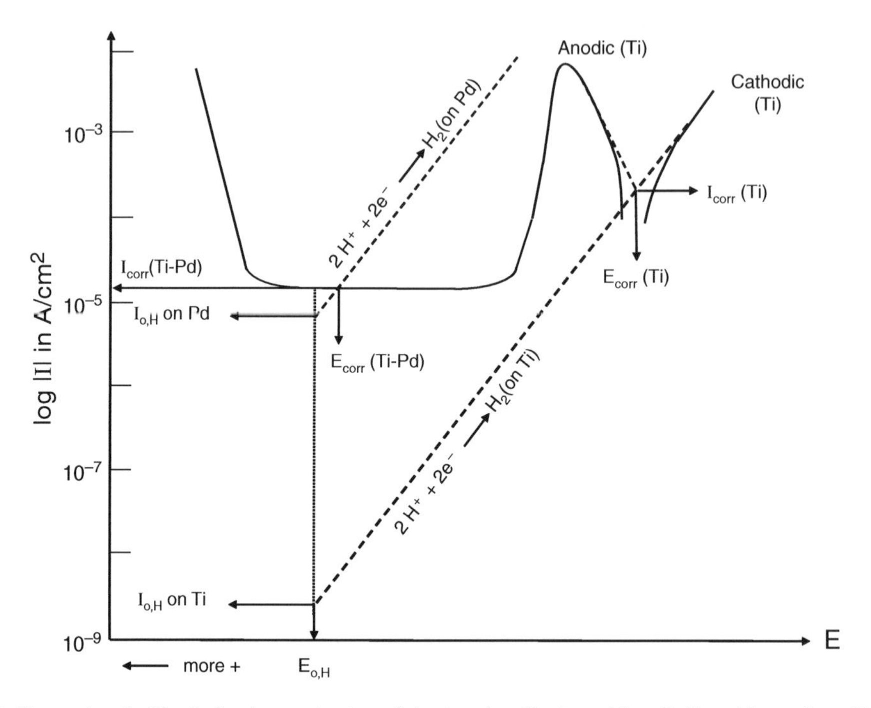

Fig. 9.45 Electrochemical basis for the passivation of titanium by alloying with palladium (drawn for a Ti–2% Pd alloy)

However, when titanium is alloyed with a noble metal, such as palladium, the hydrogen evolution reaction occurs both on Ti surface sites and on Pd surface sites, as depicted in Fig. 9.46. The exchange current density for hydrogen evolution on Pd is five orders of magnitude higher than on Ti so that the total rate of hydrogen evolution at any electrode potential is dominated by the reaction on Pd sites, even for a dilute alloy, as shown in Fig. 9.45, which is constructed for a Ti–2 at.%

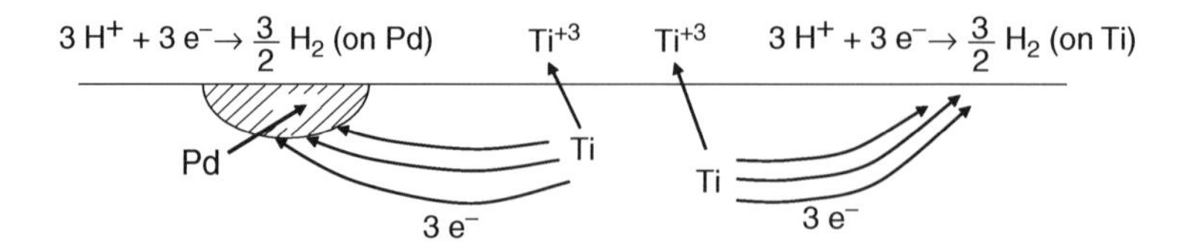

Fig. 9.46 Types of surface sites on Ti-Pd alloys

Pd alloy. The resulting effect is that the total cathodic polarization curve intersects the total anodic polarization curve in the passive region of titanium. (The anodic dissolution of Pd contributes only negligibly to the total anodic polarization curve.) Thus, alloying of titanium with Pd self-passivates the Ti–Pd alloy.

Corrosion rates for Ti–Pd or Ti–Pt alloys as determined by Stern and Wissenberg [128] are given in Fig. 9.47. It can be seen that Pd or Pt additions reduce the corrosion rate of unalloyed titanium by a factor of approximately 1/1,000.

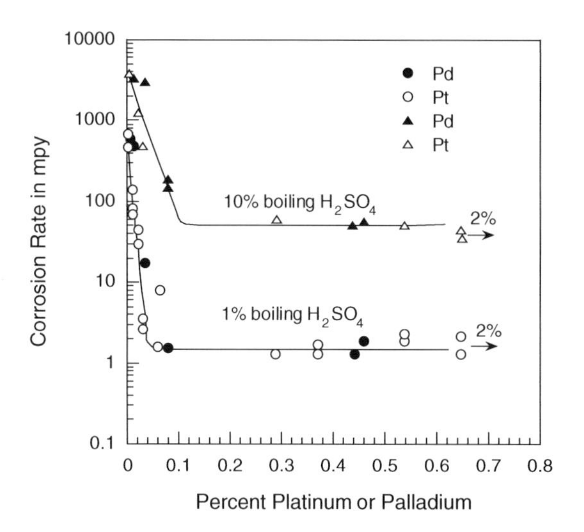

Fig. 9.47 Corrosion rates of Ti–Pd or Ti–Pt alloys in boiling sulfuric acid [128]. Reproduced by permission of ECS – The Electrochemical Society

Similar results have been obtained for Cr-containing steels [130] and for Ti–Pd–Mo and Ti–Pd–Cr ternary alloys in acid solutions [129]. In each case, the mechanism of passivation was a shift of the corrosion potential into the passive region of the host metal by an increase in the rate of hydrogen evolution on the noble metal constituent of the alloy. The use of additional noble metals into various base metals and their specific applications have been reviewed [131, 132].

Anodic Protection

Anodic protection is a technique of corrosion protection in which the electrode potential is maintained in the passive region. This technique is applicable only to metals or alloys which can be passivated, and the method is based on the fact that the corrosion rate is very low in the passive region and is given by the passive current density. Figure 9.48 illustrates suitable applied potentials for anodic protection and for cathodic protection for a metal which undergoes an active/passive transition.

Various techniques can be used to maintain the metal surface in the passive region. These include the following:

(1) Oxidizing inhibitors, such as chromates, molybdates, or nitrates,
(2) Alloying to produce passive films, i.e., the use of stainless steels,

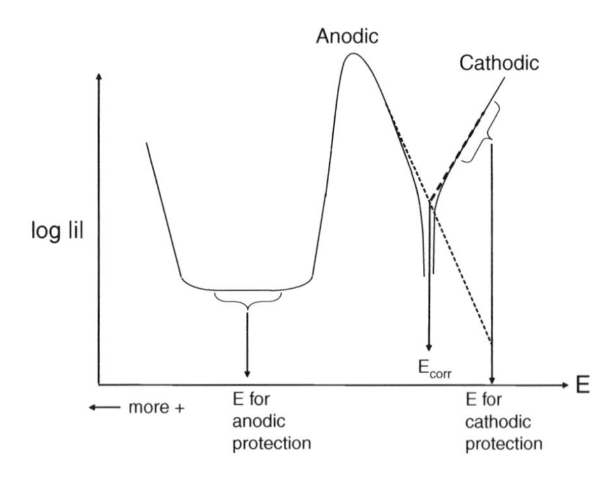

Fig. 9.48 Comparison of the electrochemical basis for anodic protection and for cathodic protection

(3) Noble metal alloying, e.g., Pd–Ti alloys,
(4) Impressed potential systems.

However, the use of inhibitors can result in contamination of the liquid contained in the system, and the use of passive alloys may be costly, so impressed voltage systems have been employed. Figure 9.49 illustrates an anodic protection system for a storage tank [133]. The cathode must be able to withstand the corrosive environment, and cathode materials which have been used include platinum-clad brass, copper, and type 304 stainless steel. Various reference electrodes have been used, including platinum and metal/metal oxide electrodes [133].

Anodic protection has been used most extensively to protect metals used in the manufacture, transport, and storage of sulfuric acid. Anodic protection has also been used in the pulp and paper industry and in the storage of chemical fertilizers and alkaline liquids [133, 134].

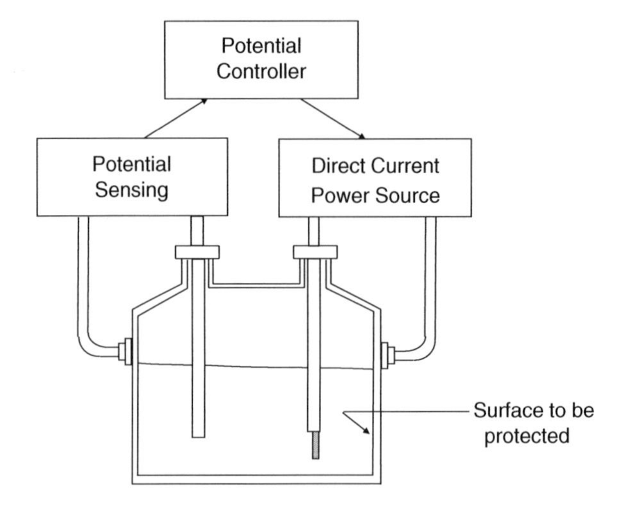

Fig. 9.49 Schematic diagram of an anodic protection setup [133]

Table 9.12 compares some of the features of anodic and cathodic protection. Each method of corrosion control has its own advantages and disadvantages.

Table 9.12 Comparison of cathodic and anodic protection

	Cathodic protection	Anodic protection
Applicability	All metals	Active/passive metals only
Environments	Weak to moderate in corrosivity	Weak to aggressive
Corrosion rate of protected member	Can be very small	Small but finite

Problems

1. The charge required to form a passive film on iron in a borate buffer solution was determined from coulometric measurements to range from 4.8×10^{-4} to 9.8×10^{-4} C/cm^2, depending on the electrode potential at which the film was formed [21]. How many layers of oxide ion O^{2-} are contained in the passive layer assuming that O^{2-} is formed by the following reaction:

$$O_2 + 4e^- \rightarrow 2O^{2-}$$

   (The radius of the O^{2-} ion is 1.40 Å.)
2. In the cathodic reduction of the passive film on iron, Cohen and co-workers estimated the thickness of the outer layer consisting of γ- Fe_2O_3 to be 32 Å [38–40]. The density of γ–Fe_2O_3 is 5.2 g/cm^3 and its molecular weight is 159.69 g/mol. The pH of the solution used in the cathodic reduction experiments was 7.6 and its approximate volume was 150 ml. Based on this information, calculate the thermodynamic electrode potentials for the reduction of γ–Fe_2O_3 to Fe^{2+} and for Fe_3O_4 to Fe. How do these values compare with the two plateaus in Fig. 9.17?
3. If zirconium, vanadium, or tantalum could be selected for an application where the metal is placed in a corrosive environment subject to fluctuations in pH or in electrode potential, which *one* metal would you choose based on corrosion behavior alone?
4. The Flade potential for Cr in 0.5 M H_2SO_4 (pH 0.3) is –0.350 V vs. SHE. [16]. How does the Flade potential compare with the equilibrium electrode potential for the reaction of Cr with water to form Cr_2O_3?

$$2Cr + 3H_2O \rightarrow Cr_2O_3 + 6H^+ + 6e^-$$

   The standard electrode potential for this reaction is $E^o = -0.654$ V vs. SHE.
5. Sketch an anodic polarization curve for a metal having the following characteristics:
   (a) The corrosion potential is –0.50 V vs. SHE.
   (b) The corrosion rate for active dissolution is 10 μA/cm^2.
   (c) The Flade potential is –0.10 V vs. SHE.
   (d) The critical current density for passivation is 100 μA/cm^2.
   (e) The passive current density is 1 μA/cm^2.
   (f) Oxygen evolution begins at + 1.0 V vs. SHE.
6. The corrosion behavior of Co–Ni binary alloys has not been determined. However, use the electron configuration theory to predict the critical alloy composition for Co–Ni alloys, assuming

that Ni is the acceptor in the alloy pair. What reservations do you have in using the electron configuration theory?

7. From film stripping studies have been conducted [23] on 304 stainless steel having the composition by weight of 18% Cr, 8% Ni, and the balance Fe. The oxide film on this stainless steel was found to contain 32 wt% Fe, 13.4 wt% Cr, and 9.4 wt% Ni (the balance is due to oxygen either in the oxide or contained in bound water). The oxide film thickness was 30 Å, and the typical weight gain due to oxide film formation was 2.0 $\mu g/cm^2$. The density of 304 stainless steel is 7.0 g/cm^3. Assume that the oxide film is a mixture of Fe_2O_3, Cr_2O_3, and NiO. Calculate the Pilling–Bedworth ratio for this oxide and use this ratio to predict whether the oxide film would be protective. To calculate the Pilling–Bedworth ratio, it is first necessary to calculate the effective molecular weight of the oxide, its effective value of n, and the effective atomic weight of the metal.
8. Use the alloying trends in Fig. 9.42 to predict the effect of Mo additions to Fe–Ni alloys on the passivity in the resulting ternary alloy. Keep a constant nickel content in the ternary alloy.
9. A steel tank car is used to transport a fertilizer solution. This solution is corrosive to steel, so it is necessary to provide some form of protection to the internal surfaces of the tank car. The tank car is equipped with a device which can control the electrode potential of the internal steel surface when the tank car is filled. The polarization curve for steel in the fertilizer solution is shown below.

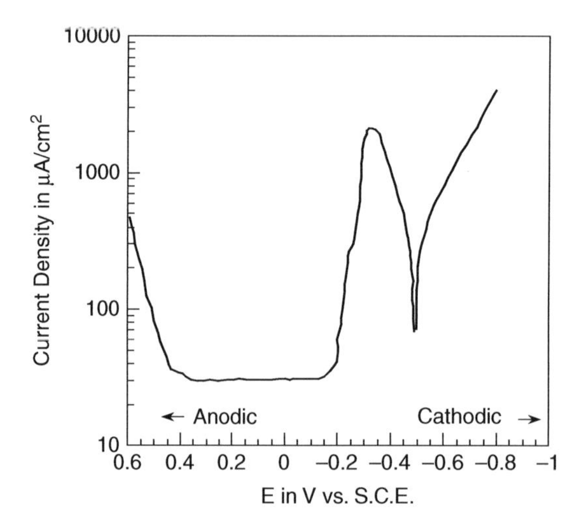

   (a) What is the corrosion current density of the steel in the freely corroding condition?
   (b) What potential can be safely used in providing anodic protection?
   (c) What is the corrosion current density at this applied anodic potential?
   (d) Under the condition in Part (c), what is the loss of thickness of the tank car wall after a 10-day service period? Assume that the corrosion reaction is $Fe \rightarrow Fe^{2+} + 2e^-$. The atomic weight of Fe is 55.85 and its density is 7.87 g/cm^3.
   (e) What potential can be used if cathodic protection is applied to give the same corrosion rate as in Part (c)?

10. Sketch the experimentally observed anodic and cathodic polarization curves which result from each of the following sets of theoretical curves. Indicate the corrosion potential on each of the experimental curves and indicate the anodic and cathodic portions of each experimental curve. What can you say about the stability of the system for curve (c) below?

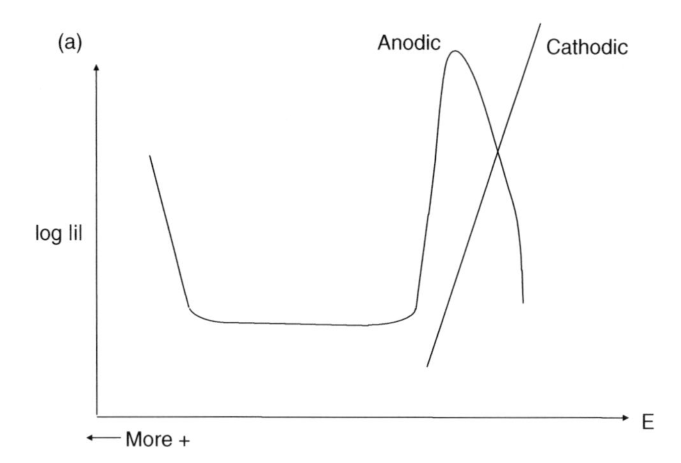
(a)
Anodic
Cathodic
log lil
E
More +

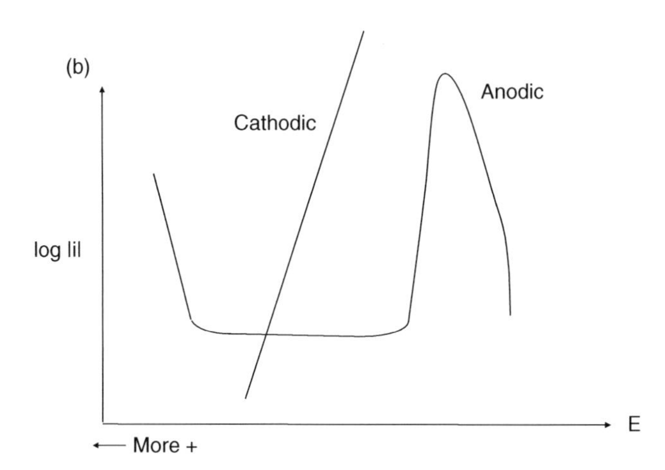
(b)
Cathodic
Anodic
log lil
E
More +

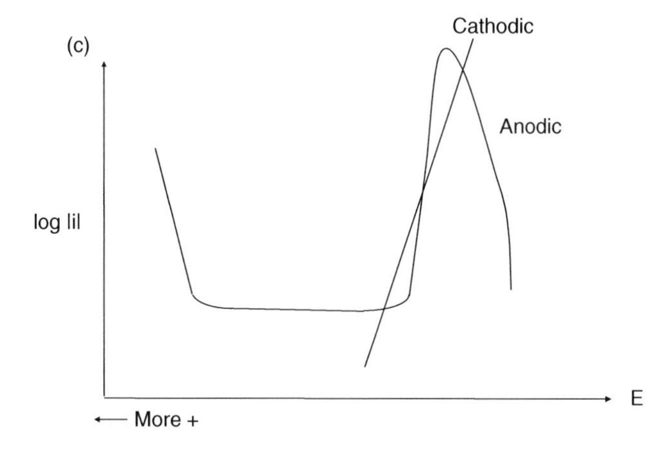
(c)
Cathodic
Anodic
log lil
E
More +

References

1. D. D. Macdonald, *Pure Appl. Chem., 71*, 951 (1999).
2. M. Pourbaix, "Atlas of Electrochemical Equilibria in Aqueous Solutions", National Association of Corrosion Engineers, Houston, TX (1974).
3. H. H. Uhlig in "Passivity of Metals", R. P. Frankenthal and J. Kruger, Eds., p. 1, The Electrochemical Society, Princeton, NJ (1978).
4. E. McCafferty and J, P. Wightman, *Surf. Interface Anal., 26*, 549 (1998).
5. E. McCafferty, J. P. Wightman, and T. F. Cromer, *J. Electrochem. Soc., 146*, 2849 (1999).
6. Z. Tun, J. J. Noël, and D. W. Shoesmith, *J. Electrochem. Soc., 146*, 988 (1999).
7. V. V. Andreeva, *Corrosion, 20*, 35t (1964).
8. C. A. Sequeira in "Uhlig's Corrosion Handbook", R. W. Revie, Ed., p. 729, John Wiley and Sons, New York (2000).
9. J. D. Venables, D. K. McNamara, J. M. Chen, T. S. Sun, and R. L. Hopping, *Appl. Surf. Sci., 3*, 88 (1979).
10. N. Sato in "Passivity of Metals", R. P. Frankenthal and J. Kruger, Eds., p. 29, The Electrochemical Society, Princeton, NJ (1978).
11. U. F. Franck, *Z. Naturforsch, 4a*, 378 (1949).
12. E. McCafferty, M. K. Bernett, and J. S. Murday, *Corros. Sci, 28*, 559 (1988).
13. H. P. Leckie and H. H. Uhlig, *J. Electrochem. Soc., 113*, 1262 (1966).
14. H. H. Uhlig, *Official Digest*, p. 650, October (1952).
15. Y. M. Kolotrykin, "First International Congress on Metallic Corrosion", p. 10, Butterworths, London (1962).
16. P. F. King and H.H. Uhlig, *J. Phys. Chem., 63*, 2026 (1959).
17. U. R. Evans, *J. Chem. Soc. (London), 127*, 1020 (1927).
18. N. Hackerman, *Z. Elektrochem., 62*, 632 (1958).
19. B. Kabanov, R. Burstein, and A. Frumkin, *Discussions Faraday Soc., 1*, 259 (1947)
20. E. S. Snavely, Jr. and N. Hackerman, *Can. J. Chem., 37*, 268 (1959).
21. R. P. Frankenthal, *Electrochim. Acta, 16*, 1845 (1971).
22. H. H. Uhlig, *Z. Elektrochem., 62*, 626 (1958).
23. N. A. Nielsen and T. N. Rhodin, Jr., *Z. Elektrochem., 62*, 707 (1958).
24. T. N. Rhodin, Jr., *Corrosion., 12*, 123t (1956).
25. E. M. Mahla and N. A. Nielsen, *J. Electrochem. Soc., 93*, 1 (1948).
26. C. L. Foley, J. Kruger, and C. J. Bechtoldt, *J. Electrochem. Soc., 114*, 994 (1967).
27. M. J. Pryor and D. S. Keir, *J. Electrochem. Soc., 102*, 370 (1955).
28. K. Asami, K. Hashimoto, and S. Shimodaira, *Corros. Sci., 18*, 151 (1978).
29. R. P. Frankenthal, *J. Electrochem. Soc., 114*, 542 (1967).
30. M. J. Graham, *Corros. Sci., 37*, 1377 (1995).
31. J. F. Watts, "An Introduction to Surface Analysis by Electron Spectroscopy", p. 67, Oxford University Press, Oxford, England (1990).
32. J. C. Vickerman, Ed., "Surface Analysis- The Principal Techniques", John Wiley & Sons, New York, NY (1997).
33. F. Falkenberg and I. Olejford in "Passivity of Metals and Semiconductors", M. B. Ives, J. L. Luo, and J. R. Rodda, Eds., p. 570, The Electrochemical Society, Pennington, NJ (2001).
34. A. J. Davenport and M. Sansome, *J. Electrochem. Soc., 142*, 725 (1995).
35. P. Schmuki, S. Virtanen, A. J. Davenport, and C. M. Vitus, *J. Electrochem. Soc., 143*, 574 (1996).
36. D. C. Koningsberger and R. Prins, Eds., "X-Ray Absorption", John Wiley & Sons, New York, NY (1988).
37. M. Ryan, R. C. Newman, and G. E. Thompson, *J. Electrochem. Soc., 141*, L164 (1994).
38. C. D. Stockbridge, P. B. Sewell, and M. Cohen, *J. Electrochem. Soc., 108*, 928 (1961).
39. P. B. Sewell, C. D. Stockbridge, and M. Cohen, *J. Electrochem. Soc., 108*, 933 (1961).
40. M. Nagayama and M. Cohen, *J. Electrochem. Soc., 109*, 781 (1962).
41. E. McCafferty and A. C. Zettlemoyer, *Discuss. Faraday Soc., 52*, 239 (1971).
42. M. Seo, J. B. Lumsden, and R. W. Staehle, *Surf. Sci., 42*, 337 (1974).
43. M. Cohen, D. F. Mitchell and K. Hashimoto, *J. Electrochem. Soc., 126*, 442 (1979).
44. M. C. Bloom and L. Goldenberg, *Corros. Sci., 5*, 623 (1965).
45. M. Cohen in "Passivity of Metals", R. P. Frankenthal and J. Kruger, Eds., p. 521, The Electrochemical Society, Princeton, NJ (1978).
46. S. C. Tjong and E. Yeager, *J. Electrochem. Soc., 128*, 2251 (1981).
47. V. Schroder and T. Devine, *J. Electrochem. Soc.*, 146, 406 (1999).
48. T. N. Rhodin, Jr., *Annals N.Y. Acad. Sci., 58*, 855 (1954).

49. G. Okamota and T. Shibata, *Nature*, *206*, 1350 (1965).
50. G. Okamota and T. Shibata, "Proc. 3rd International Congress on Metallic Corrosion", Moscow (1966), Vol. 1, p. 396 (1969).
51. G. Okamota, and T. Shibata, *Corros. Sci.*, *10*, 371 (1970).
52. H. T. Yolken, J. Kruger, and J. P. Calvert, *Corros. Sci.*, *8*, 103 (1968).
53. O. J. Murphy, J. O'M. Bockris, T. E. Pou, D. L. Cocke, and G. Sparrow, *J. Electrochem. Soc.*, *129*, 2149 (1982).
54. U. Stumm, W. Meisel, and P. Gütlich, *Hyperfine Inter.*, *28*, 923 (1986).
55. C. Vertes, M. Lakatos-Varsanyi, A. Vertes, E. Kuzmann, W. Meisel, and P. Gütlich, *Hyperfine Inter.*, *69*, 731 (1991).
56. J. Eldridge and R. W. Hoffman, *J. Electrochem. Soc.*, *136*, 955 (1989).
57. W. E. O'Grady, *J. Electrochem. Soc.*, *127*, 555 (1980).
58. G. Okamoto and T. Shibata, in Passivity of Metals", R. P. Frankenthal and J. Kruger, Eds., p. 646, The Electrochemical Society, Princeton, NJ (1978).
59. M. Sakashita and N. Sato in Passivity of Metals", R. P. Frankenthal and J. Kruger, Eds., p. 479, The Electrochemical Society, Princeton, NJ (1978).
60. C. R. Clayton and Y. C. Lu, *J. Electrochem. Soc.*, *133*, 2465 (1986).
61. L. J. Oblonsky, A. J. Davenport, M. P. Ryan, H. S. Isaacs, and R. C. Newman, *J. Electrochem. Soc.*, *144*, 2388 (1997).
62. L. J. Oblonsky, A. J. Davenport, M. P. Ryan, and M. F. Toney, in "Passivity of Metals and Semiconductors", M. B. Ives, J. L. Luo, and J. R. Rodda, Eds., p. 173, 962, The Electrochemical Society, Pennington, NJ (2001).
63. A. F. Wells, "Structural Inorganic Chemistry", p. 552, Clarendon Press, Oxford (1984).
64. G. G. Long, J. Kruger, D. R. Black, and M. Kuriyama, *J. Electroanal. Chem.*, *150*, 603 (1983).
65. M. P. Ryan, R. C. Newman, and G. E. Thompson, *J. Electrochem. Soc.*, *142*, L177 (1995).
66. R. S. Alwitt in "Oxides and Oxide Films", J. W. Diggle and A. K. Vijh, Eds., Vol. 4. p. 169, Marcel Dekker, New York (1976).
67. W. Vedder and D. A. Vermilyea, *Trans Faraday Soc.*, *65*, 561 (1969).
68. T. Hurlen, A. T. Haug, and G. A. Salomonsen, *Electrochim. Acta*, *29*, 1161 (1984).
69. H. P. Godard, *J. Electrochem. Soc.*, *114*, 354 (1967).
70. R. K. Hart, *Trans. Faraday Soc.*, *52*, 1020 (1957).
71. R. S. Alwitt and L. C. Archibald, *Corrosion Sci.*, *13*, 687 (1973).
72. S. Wernick, R. Pinner, and P. G. Sheasby, "The Surface Treatment and Finishing of Aluminum and its Alloys", Vol. 1, p. 471, ASM International, Metals Park, OH (1987).
73. G. R. T. Scheuller, S. R. Taylor, and E. E. Hajcsar, *J. Electrochem. Soc.*, *139*, 2799 (1992).
74. S. Wernick, R. Pinner, and P. G. Sheasby, "The Surface Treatment and Finishing of Aluminum and its Alloys", Vol. 1, p. 292, ASM International, Metals Park, OH (1987).
75. G. E. Thompson, Y. Xu, P. Skeldon, K. Shimizu, S. H. Han, and G. C. Wood, *Phil. Mag. B*, *55*, 651 (1987).
76. S. Wernick, R. Pinner, and P. G. Sheasby, "The Surface Treatment and Finishing of Aluminum and its Alloys", Vol. 1, p. 453, ASM International, Metals Park, OH (1987).
77. S. Wernick, R. Pinner, and P. G. Sheasby, "The Surface Treatment and Finishing of Aluminum and its Alloys", Vol. 2, p. 773, ASM International, Metals Park, OH (1987).
78. S. Wernick, R. Pinner, and P. G. Sheasby, "The Surface Treatment and Finishing of Aluminum and its Alloys", Vol. 2, p. 729, ASM International, Metals Park, OH (1987).
79. V. Brusic in "Oxides and Oxide Films", Vol. 1, p. 1, J. W. Diggle, Ed., Marcel Dekker, New York (1972).
80. J. Kruger, *Inter. Mater. Revs.*, *33*, 113 (1988).
81. J. W. Schultze and M. M. Lohrengel, *Electrochim. Acta.*, *45*, 2499 (2000).
82. F. M. Delnick and N. Hackerman in "Passivity of Metals", R. P. Frankenthal and J. Kruger, Eds., p. 116, The Electrochemical Society, Princeton, NJ (1978).
83. D. Rodriguez-Marek, M. Pang, and D. F. Bahr, *Met. Mater. Trans. A.*, *34*, 1291 (2003).
84. J.-D. Kim and M. Seo, *J. Electrochem. Soc.*, *150*, B193 (2003).
85. L. C. Archibald and J. S. L. Leach, *Electrochim. Acta*, *22*, 15 (1971).
86. D. A. Vermilyea, *J. Electrochem. Soc.*, *110*, 345 (1963).
87. S. F. Bubar and D. A. Vermilyea, *J. Electrochem. Soc.*, *113*, 892 (1966).
88. S. F. Bubar and D. A. Vermilyea, *J. Electrochem. Soc.*, *114*, 882 (1967).
89. A. S. Khanna, "Introduction to High Temperature Oxidation and Corrosion", ASM International, Materials Park, OH (2002).
90. M. Seo, M. Chiba. and K. Suzuki, *J. Electroanal. Chem.*, *473*, 49 (1999)
91. A. J. Davenport, L. J. Oblonsky, M. P. Ryan, and M. F. Toney, *J. Electrochem. Soc.*, *147*, 2162 (2000).

92. H. Kobayashi, T. Yashiro, A. Kawashima, K. Asami, K. Hashimoto, and H. Fujimora in "Proceedings of the Symposium on Corrosion, Electrochemistry, and Catalysis of Metallic Glasses", p. 254, R. B. Diegle and K. Hashimoto, Eds, The Electrochemical Society, Pennington, NJ (1988).
93. C. L. McBee and J. Kruger, *Electrochim. Acta*, *17*, 1337 (1972).
94. A. G. Revescz and J. Kruger in "Passivity of Metals", R. P. Frankenthal and J. Kruger, Eds., p. 137, The Electrochemical Society, Princeton, NJ (1978).
95. J. P. Fehlner in "Passivity of Metals", R. P. Frankenthal and J. Kruger, Eds., p. 181, The Electrochemical Society, Princeton, NJ (1978).
96. R. F. Steigerwald and N. D. Greene, *J. Electrochem. Soc.*, *109*, 1026 (1962).
97. N. D. Greene, *Corrosion*, *18*, 136t (1962).
98. H. H. Uhlig and R. W. Revie, "Corrosion and Corrosion Control", p. 82, John Wiley and Sons, New York, NY (1985).
99. H. Yoshioka, H. Habazaki, A. Kawashima, K. Asami, and K. Hashimoto, *Corros. Sci.*, *33*, 425 (1992).
100. H. H. Uhlig, *Z. Elektrochem.*, *62*, 700 (1958).
101. J. E. Holliday and R. P. Frankenthal, *J. Electrochem. Soc.*, *119*, 1190 (1972).
102. L. J. Oblonsky, M. P. Ryan, and H. S. Isaacs, *J. Electrochem. Soc.*, *145*, 1922 (1998), and references therein.
103. K. Sieradzki and R. C. Newman, *J. Electrochem. Soc.*, *133*, 1979 (1986).
104. E. McCafferty, *Electrochem. Solid State Lett.*, *3*, 28 (2000).
105. E. McCafferty, *Corros. Sci.*, *42*, 1993 (2000).
106. E. McCafferty, *Corros. Sci.*, *44*, 1393 (2002).
107. E. McCafferty, *J. Electrochem. Soc.*, *149*, B333 (2002).
108. E. McCafferty, *J. Electrochem. Soc.*, 150, B238 (2003).
109. E. McCafferty, *Corros. Sci.* *47*, 1765 (2005).
110. D. Stauffer and A. Ahorony, "Introduction to Percolation Theory", Taylor and Francis, Washington, DC (1992).
111. R. J. Wilson and J. J. Watkins, "Graphs: An Introductory Approach", John Wiley & Sons, New York, NY (1990).
112. M. Randic, *J. Am. Chem. Soc.*, *97*, 6609 (1975).
113. M. G. Fontana and N. D. Greene, "Corrosion Engineering", p. 336, McGraw-Hill, New York, NY (1978).
114. J. B. Lumsden in "Passivity of Metals", R. P. Frankenthal and J. Kruger, Eds., p. 730, The Electrochemical Society, Princeton, NJ (1978).
115. H. H. Uhlig and G. E. Woodside, *J. Phys. Chem.*, *57*, 280 (1953).
116. A. P. Bond and H. H. Uhlig, *J. Electrochem. Soc.*, *107*, 488 (1960).
117. K. Osazawa and H.-J. Engell, *Corros. Sci.*, *6*, 389 (1966).
118. K. Sugimoto and Y. Sawada, *Corros. Sci.*, *17*, 425 (1977).
119. M. Klimmeck, *Electrochim. Acta*, *25*, 1375 (1980).
120. P. Y. Park, E. Akiyama, A. Kawashima, K. Asami, and K. Hashimoto, *Corros. Sci*, *37*, 1843 (1995).
121. D. A. Stout, J. B. Lumsden, and R. W. Staehle, *Corrosion*, *35*, 141 (1979).
122. V. V. Plaskeev and V. M. Knyazheva, *Prot. Metals*, *31*, 35 (1995).
123. M. B. Ives, V. Mitrovic-Scepanovic, and M. Moriyama in "Passivity of Metals and Semiconductors", M. Froment, Ed., p. 175, Elsevier, New York, NY (1983).
124. M. B. Rockel, *Corrosion*, *29*, 393 (1973).
125. K. Shiobara, Y. Sawada, and S. Morioka, *Trans. J. I. M.*, *6*, 97 (1965).
126. J. E. Castle and C. R. Clayton, *Corros. Sci.*, *17*, 7 (1977).
127. C. R. Clayton and I. Olejford in "Corrosion Mechanisms in Theory and Practice", P. Marcus and J. Oudar, Eds, p. 175, Marcel Dekker, New York, NY (1995).
128. M. Stern and H. Wissenberg, *J. Electrochem. Soc.*, *106*, 759 (1959).
129. N. D. Tomashov, R. M. Altovsky, and G. P. Chernova, *J. Electrochem. Soc.*, *108*, 113 (1961).
130. N. D. Tomashov, *Corrosion*, *14*, 229t (1958).
131. J. H. Potgieter, *J. Appl. Electrochem.*, *21*, 471 (1991).
132. R. W. Schutz, *Corrosion*, *59*, 1043 (2003).
133. O. L. Riggs and C. E. Locke, "Anodic Protection", Plenum Press, New York (1981).
134. J. I. Munro and W. W. Shim, *Mater. Perform.*, *40* (5), 22 (2001).
135. A. J. Sedriks, "Corrosion of Stainless Steels", Wiley-Interscience, New York, NY (1996).

Chapter 10
Crevice Corrosion and Pitting

Introduction

Passive films, which were studied in the previous chapter, are remarkable in their ability to provide corrosion protection to a wide variety of metals and alloys. However, passive films are not perfect and often suffer localized breakdown at certain specific areas. Although the remainder of the passive film may remain intact and continue to offer protection against general corrosion, the occurrence of localized corrosion is dangerous because it is concentrated at a fixed area and can lead to catastrophic failure of the metal piece. In addition, the presence of localized corrosion is usually more difficult to detect than general corrosion.

The most common forms of localized attack are crevice corrosion, pitting, and stress-corrosion cracking. The first two of these forms are considered in this chapter. Crevice corrosion is localized corrosion that occurs within narrow clearances or under shielded metal surfaces. Pitting is the localized breakdown of passive films, usually by chloride ions, and the subsequent attack of the underlying metal at certain fixed specific sites. As will be seen, crevice corrosion and pitting differ in their mechanisms of initiation, but the mechanism of propagation is similar for these two forms of localized corrosion.

The research literature on pitting corrosion alone (like that on passivity) is voluminous. This chapter deals only with the most important aspects of pitting and does so in an introductory manner. No attempt has been made to present a comprehensive research review.

Crevice Corrosion

Crevice corrosion can occur in geometrical clearances, such as

- under gaskets or seals
- under bolt heads
- between overlapping metal sheets
- within screw threads
- within strands of wire rope

or crevice corrosion can occur under deposits, such as

- corrosion products
- dust particles
- barnacles (in seawater)

E. McCafferty, *Introduction to Corrosion Science*, DOI 10.1007/978-1-4419-0455-3_10,

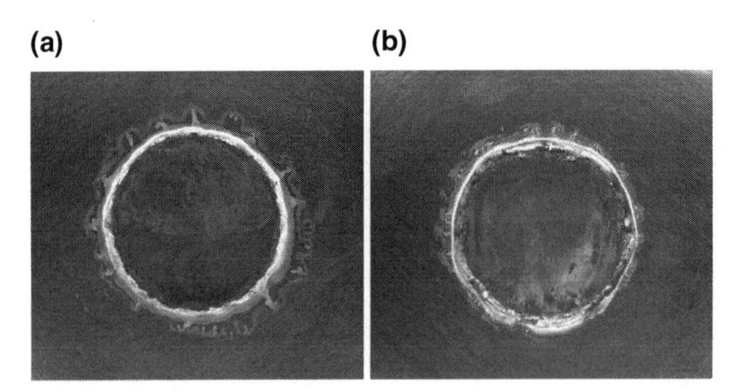

Fig. 10.1 Alloy 625 after crevice corrosion testing at 65°C in natural seawater [1]. (*Left*) +200 mV vs. Ag/AgCl for 24 h, (*Right*) +400 mV vs. Ag/AgCl for 24 h. The crevice (removed in the photographs) was provided with a compressed gasket. Reused with permission from F. J. Martin, K. E. Lucas, and E. A. Hogan, Copyright 2002, American Institute of Physics

At first glance it would appear that metals located within a crevice would be protected from the electrolyte because the metal is sheltered from the bulk of the corrosive environment. But this is not the case. For example, Fig. 10.1 shows crevice attack on an alloy 625 specimen after crevice corrosion testing in natural seawater [1]. It can be seen that localized attack is quite severe beneath the region where a crevice had been formed with a compressed gasket.

There are two distinct stages of crevice corrosion: (i) initiation and (ii) propagation [2].

Initiation of Crevice Corrosion

Crevice corrosion initiates due to the operation of a differential oxygen cell. Oxygen reduction occurs both on the metal surface which is exposed to the bulk electrolyte and also on the portion of the metal surface which is contained within the crevice, as depicted in Fig. 10.2. The cathodic reaction is the same on both internal and external metal surfaces:

$$O_2 + 2H_2O + 4e^- \rightarrow 4OH^- \tag{1}$$

However, the metal exposed to the bulk electrolyte is in contact with an open supply of oxygen from the atmosphere, so as O_2 is consumed near the external metal surface by Eq. (1), additional O_2 molecules diffuse to the metal surface and a steady-state concentration of O_2 is maintained near the surface of the external metal. However, when O_2 molecules are consumed within the narrow clearance of the crevice, they are not easily replaced due to the long narrow diffusion path formed by the crevice. Thus, oxygen becomes depleted within the crevice, as shown in Fig. 10.3, which gives experimental measurements for stainless steel within crevices in seawater [3].

Hence, an oxygen concentration cell is formed between the metal surface outside the crevice and the metal surface within the crevice. This difference in O_2 concentration between the bulk solution and the sheltered crevice has two effects.

First, as was seen in Chapter 5, the metal exposed to the lower concentration of oxygen has a more negative potential (i.e., less positive) for oxygen reduction than does the metal exposed to a higher concentration of oxygen.

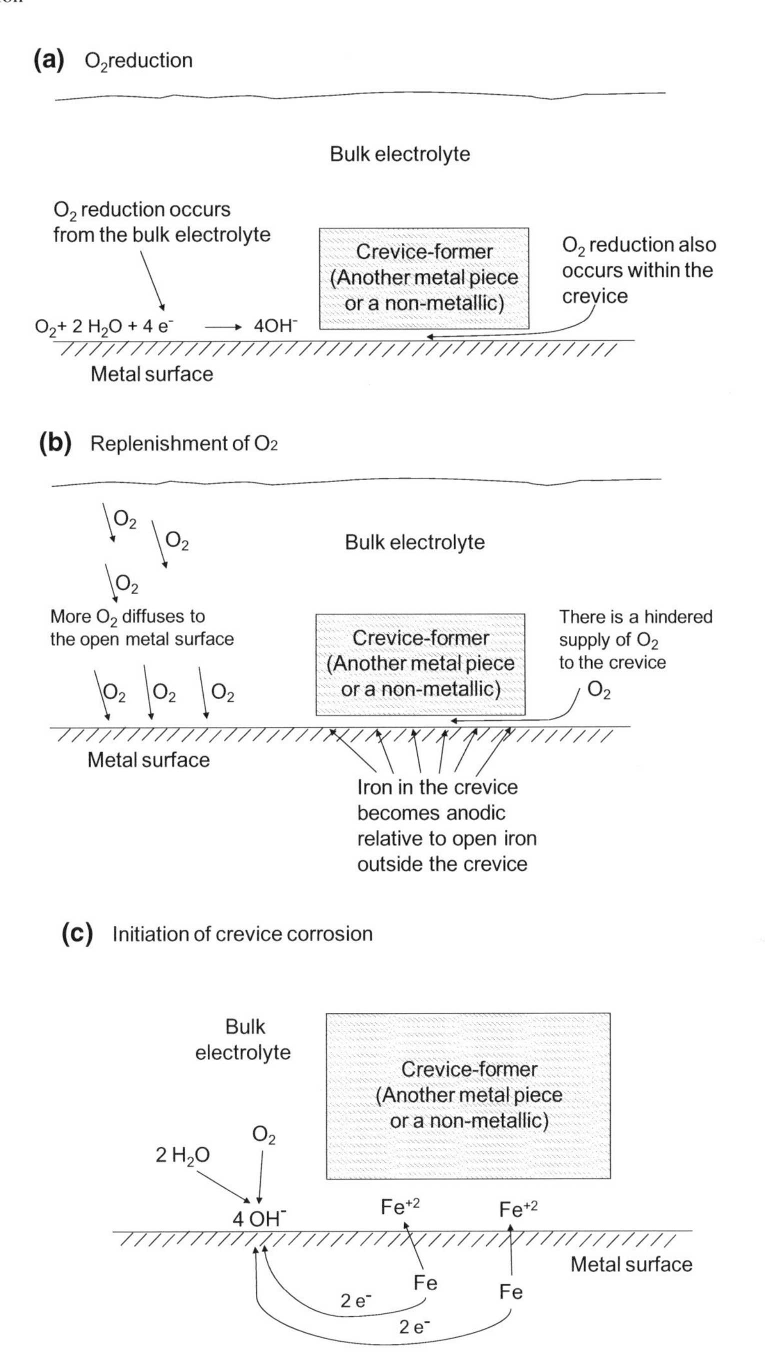

Fig. 10.2 Schematic illustration of the initiation stage of crevice corrosion: (**c**) is an expanded view of (**a**) and (**b**)

This can be seen easily by application of the Nernst equation to Eq. (1), for which $E^0 = 1.228$ V vs. SHE. At a pH of 7.0, when the concentration of dissolved O_2 is 15 mg/L (often the case for bulk solutions), then $E(O_2) = 1.596$ V vs. SHE. When the concentration of dissolved O_2 within a crevice is 1 mg/L (see Fig. 10.3), then $E(O_2) = 1.579$ V vs. SHE.

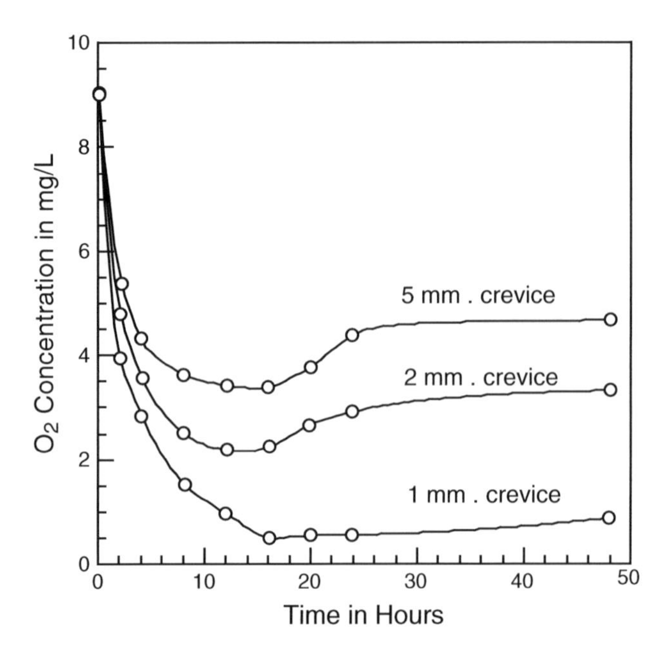

Fig. 10.3 Variation of the oxygen concentration within crevices on a Cr-containing stainless steel. Redrawn from [3]

The second effect (which is more important) is that the limiting current density i_L for O_2 reduction within the crevice is decreased relative to that for the bulk solution. This is because the concentration C of dissolved O_2 decreases within the crevice, so that i_L decreases according to

$$|i_L| = \frac{nFDC}{\delta} \tag{2}$$

Figure 10.4 shows schematic Evans polarization curves which illustrate these effects. The corrosion potential for creviced metal becomes more positive with time as oxygen is depleted within the crevice. The differences in electrode potential between open and crevice metal may amount to only tens of millivolts. Figure 10.5 shows that there is only a 25 mV difference in the electrode potential between open iron and iron in crevices in 0.6 M NaCl [4]. But this difference is enough to initiate corrosion within the crevice.

Figure 10.5 also shows that the electrode potential for metal within the crevice becomes more negative with time before leveling off, as depicted in Fig. 10.4.

Figure 10.6 shows experimental cathodic polarization curves obtained in 0.06 M NaCl for iron in crevices as well as for iron open to the bulk electrolyte [4]. The polarization curves for the iron in crevices were obtained using the corrosion cell shown in Fig. 10.7 which featured a glass disc which could be adjusted to provide a desired crevice height. Figure 10.6 shows that the limiting cathodic density is less for the creviced iron than for open iron. In fact, the limiting cathodic rate is suppressed to a constant value for crevice heights between 0.13 and 0.50 mm. The crevice height of 0.50 mm is significant because it is approximately the thickness of the oxygen diffusion layer near the metal surface. The thickness δ of the diffusion layer is given by Eq. (2) and was calculated in Chapter 8 to be 0.65 mm for iron in NaCl solutions. Thus, when the crevice height is comparable to the thickness

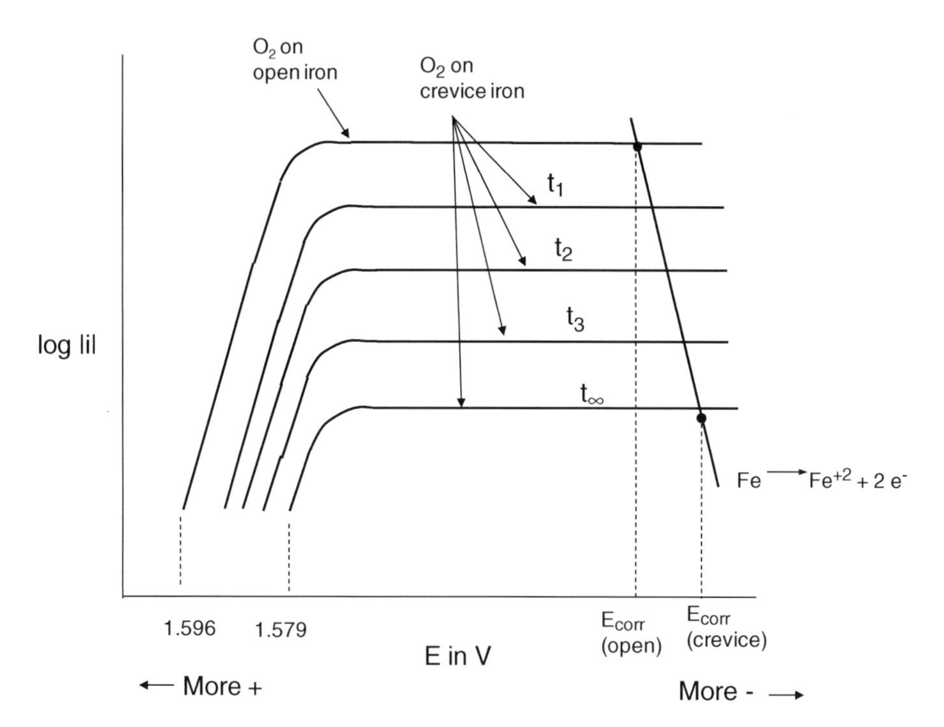

Fig. 10.4 Evans diagrams for electrode reactions on an open or creviced metal (the *X*-axis is not to scale)

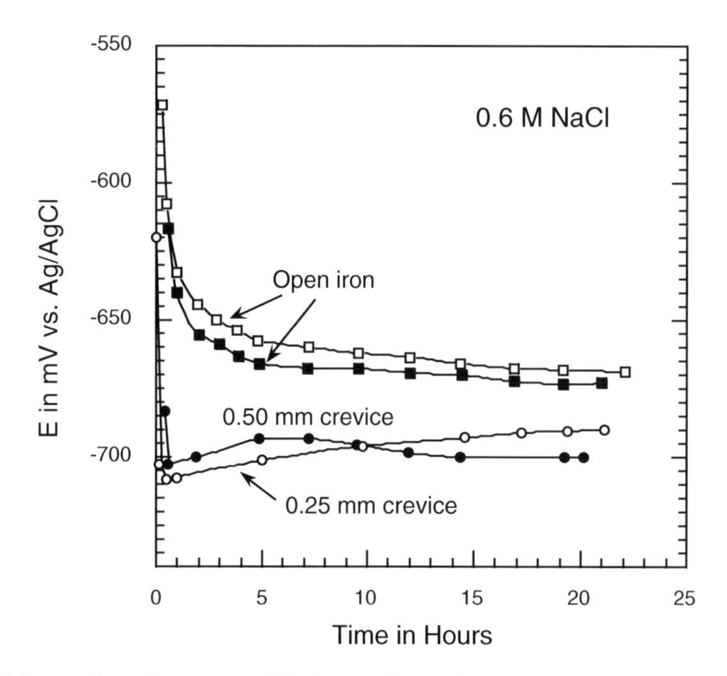

Fig. 10.5 Establishment of steady-state potentials for open iron and iron within crevices in 0.6 M NaCl [4]. (Note that the electrode potential for creviced iron is more negative than that for open iron.) Reproduced by permission of ECS – The Electrochemical Society

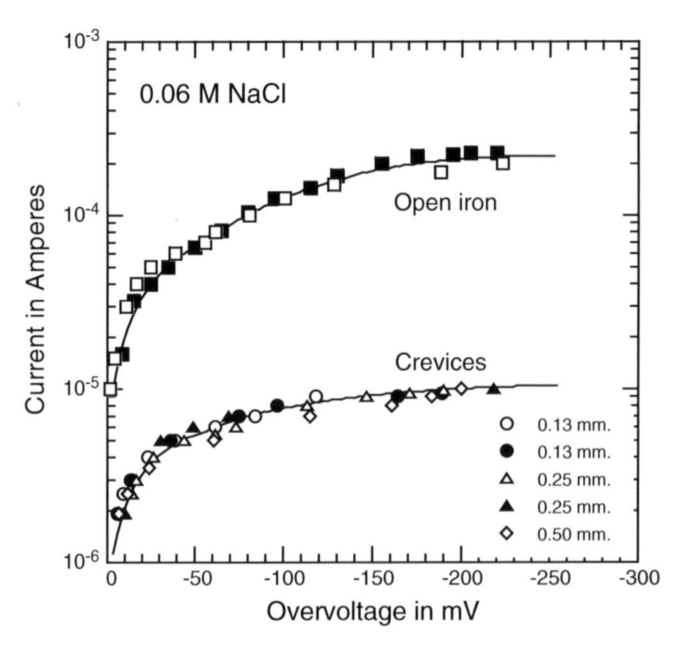

Fig. 10.6 Cathodic polarization curves for open iron and iron in crevices in 0.06 M NaCl [4]. (The numbers in the figure refer to crevice heights.) Reproduced by permission of ECS – The Electrochemical Society

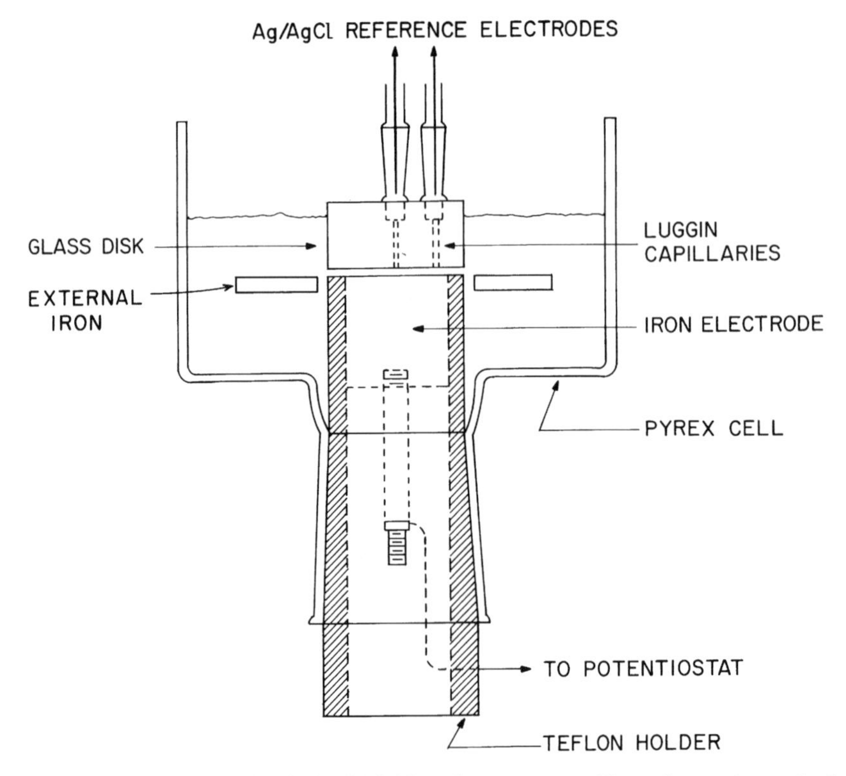

Fig. 10.7 A crevice corrosion electrochemical cell [4]. The micrometer assembly used to set the crevice height is not shown. Reproduced by permission of ECS – The Electrochemical Society

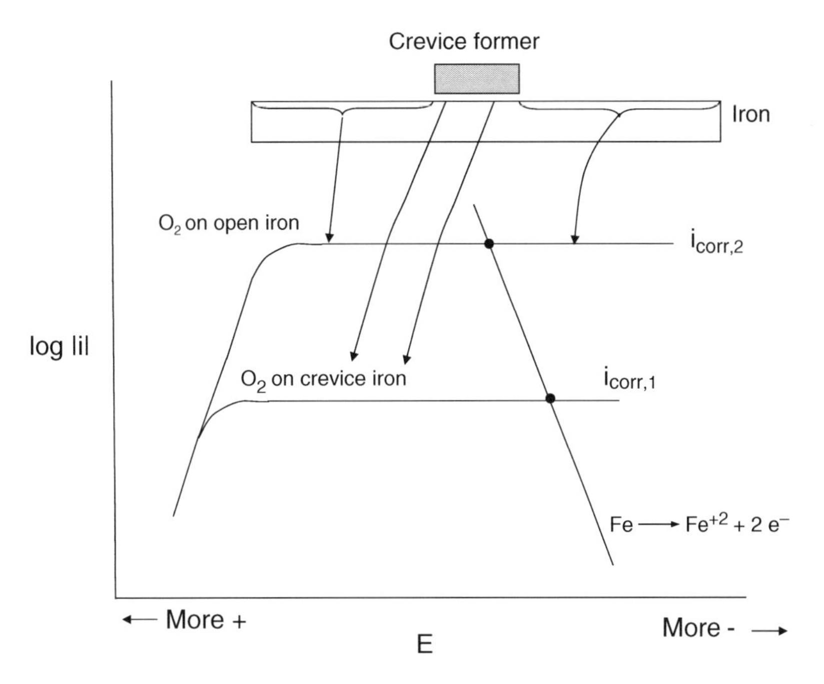

Fig. 10.8 Evans diagrams for the crevice corrosion of iron

of the oxygen diffusion layer, diffusion of oxygen into the crevice is impeded, and the rate of oxygen reduction is decreased.

Figure 10.8 shows hypothetical Evans diagrams for iron in a crevice coupled to an open iron electrode. As seen in Fig. 10.8, if the iron within the crevice were not coupled to external open iron (an "isolated crevice") then the corrosion rate would be given by $i_{corr,1}$ (the value for the limiting cathodic current density for oxygen reduction within the crevice). However, when the crevice iron is coupled to external iron, the initial corrosion rate is given by the larger value $i_{corr,2}$ (the value for the limiting cathodic current density for oxygen reduction on open external iron). And thus, corrosion is initiated in the crevice.

Propagation of Crevice Corrosion

Crevice corrosion propagates by changes in the electrolyte composition within the crevice. In particular, the crevice electrolyte will become *acidic* in nature and will also contain concentrated amounts of cations discharged from the metal or alloy. In chloride solutions, the internal electrolyte within the crevice will also become concentrated in chloride ions. This internal electrolyte is sufficiently aggressive to break down the passive film on the metal.

These changes in the composition of the crevice electrolyte occur because of the narrow geometrical character of the crevice, which allows only restricted exchange between the crevice and bulk electrolytes. For instance, Fe^{2+} ions which are formed within a crevice on iron do not easily diffuse out of the crevice due to the crevice geometry. Thus, these ions accumulate within the crevice and hydrolyze (react with water) to produce hydrogen ions, as follows:

$$Fe \rightarrow Fe^{2+} + 2e^- \tag{3}$$

$$Fe^{2+} + H_2O \rightarrow FeOH^+ + H^+ \tag{4}$$

$$Fe^{2+} + 2H_2O \rightarrow Fe(OH)_2 + 2H^+ \tag{5}$$

The degree of acidity depends on the identity of the dissolved cation. For Eq. (4)

$$K_h = \frac{[H^+]^2}{[Fe^{2+}]} \tag{6}$$

where K_h is the hydrolysis constant. From thermodynamic data compiled by Peterson [5], Eq. (6) can be written as follows:

$$pH = 4.75 - \frac{1}{2} \log [Fe^{2+}] \tag{7}$$

so if the concentration of dissolved Fe^{2+} ions within the crevice is 1 M, the internal crevice electrolyte is 4.75; see Table 10.1. Thus, a local internal pocket of acidity develops within the active crevice even though the external electrolyte has a neutral pH of 7.

Table 10.1 Expressions for the pH of various crevice hydrolysis reactions [5, 8]

Reaction	Equilibrium pH	pH of 1 M solution
$Fe^{2+} + H_2O \rightarrow FeOH^+ + H^+$	$pH = 4.75 - 0.500 \log [Fe^{2+}]$	4.75
$Fe^{2+} + 2H_2O \rightarrow Fe(OH)_2 + 2H^+$	$pH = 6.64 - 0.500 \log [Fe^{2+}]$	6.64
$Fe^{3+} + 3H_2O \rightarrow Fe(OH)_3 + 3H^+$	$pH = 1.61 - 0.333 \log [Fe^{3+}]$	1.61
$Cr^{2+} + 2H_2O \rightarrow Cr(OH)_2 + 2H^+$	$pH = 5.50 - 0.500 \log [Cr^{2+}]$	5.50
$Cr^{3+} + 3H_2O \rightarrow Cr(OH)_3 + 3H^+$	$pH = 1.60 - 0.333 \log [Cr^{3+}]$	1.60
$Ni^{2+} + 2H_2O \rightarrow Ni(OH)_2 + 2H^+$	$pH = 6.09 - 0.500 \log [Ni^{2+}]$	6.09
$Al^{3+} + H_2O \rightarrow Al(OH)^{2+} + H^+$	$pH = 2.43 - 0.500 \log [Al^{3+}]$	2.43

Experimentally measured values for the pH of crevice electrolytes on iron have ranged from 3.1–4.7 to 5.5 [4, 6, 7], in nominal agreement with the value of 4.75 calculated from Eq. (7).

As shown in Table 10.1, an even lower pH is expected within a crevice on chromium. This has been verified experimentally by the work of Bogar and Fujii [7], who showed that the pH within crevices on Fe–Cr alloys decreased with increasing Cr content until a constant pH of approximately 2.0 was attained, as shown in Fig. 10.9. Peterson et al. [5] measured the internal pH to be 1.2–2.0 on Type 304 stainless steel in crevices.

Example 10.1: Chromium contained within a crevice of 0.5 cm^2 area and 0.2 mm in height corrodes at the rate of 1 $\mu A/cm^2$. What is the pH within the crevice after 10 days if Cr corrodes as Cr^{3+}, and Cr^{3+} ions hydrolyze to form $Cr(OH)_3$? Assume that all Cr^{3+} ions produced by crevice corrosion remain within the crevice.

Solution: The amount of Cr^{3+} ions produced is given by

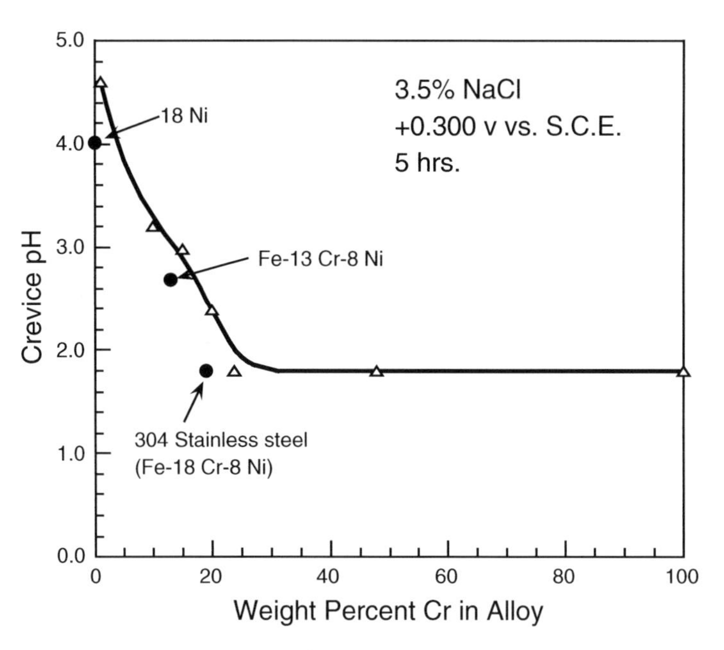

Fig. 10.9 Crevice pH of Fe–Cr binary alloys as a function of the chromium content of the alloy [7]. Data are also shown for two Fe–Cr–Ni alloys

$$\left(\frac{1.0\times10^{-6}\,\mathrm{A}}{\mathrm{cm}^2}\right)(0.5\,\mathrm{cm}^2)(10\,\mathrm{day})\left(\frac{24\mathrm{h}}{\mathrm{day}}\right)\left(\frac{3600\,\mathrm{s}}{\mathrm{h}}\right)=0.432\,\mathrm{C}$$

$$(0.432\,\mathrm{C})\left(\frac{1\,\mathrm{equiv}}{96{,}500\,\mathrm{C}}\right)\left(\frac{1\,\mathrm{mol\,Cr}^{3+}}{3\,\mathrm{equiv}}\right)=1.49\times10^{-6}\,\mathrm{mol\,Cr}^{3+}$$

The volume of the crevice is

$$(0.5\,\mathrm{cm}^2)(2\times10^{-2}\,\mathrm{cm})=1.0\times10^{-2}\mathrm{cm}^2$$

The concentration of Cr^{3+} within the crevice is

$$[\mathrm{Cr}^{3+}]=\left(\frac{1.49\times10^{-6}\,\mathrm{mol}}{1.0\times10^{-2}\,\mathrm{cm}^3}\right)\left(\frac{1000\,\mathrm{cm}^3}{\mathrm{L}}\right)=0.149\,\mathrm{M}$$

Then, from Table 10.1, the pH due to hydrolysis to $Cr(OH)_3$ is

$$\mathrm{pH}=1.60-0.333\log\,[\mathrm{Cl}^{3+}]$$

or

$$\mathrm{pH}=1.60-0.333\log\,(0.149)=1.9$$

With the accumulation of H^+ ions and metallic cations within an active crevice, Cl^- ions then migrate from the bulk electrolyte to the crevice electrolyte in order to maintain charge neutrality within the

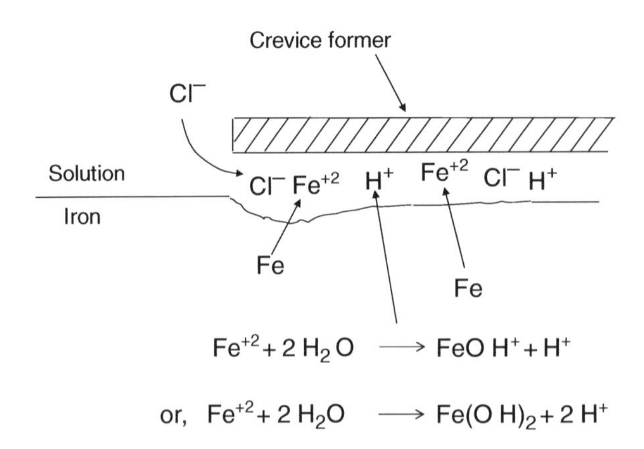

Fig. 10.10 Schematic illustration of the propagation stage of crevice corrosion of iron

crevice solution. The presence of increased chloride levels (3–12 M) within corroding crevices has been determined experimentally for artificial crevices on various stainless steel alloys [9, 10].

Thus, the propagation stage of crevice corrosion involves the formation of a highly corrosive internal electrolyte which is acidic and also concentrated in chloride ions and dissolved cations of the metal or alloy, as shown in Fig. 10.10 [11]. The initiation stage of crevice corrosion can be quite prolonged (months to years), but propagation may proceed rapidly due to the highly corrosive crevice environment which is formed.

Crevice Corrosion Testing

Crevice corrosion can be studied using a variety of experimental test setups. To screen or test various alloys for their susceptibility to crevice corrosion, metal samples can be situated beneath flat crevice formers made of plastic or fluorocarbon. The crevice can also be formed under a compressed rubber gasket. Metals in the form of pipes or tubing can be equipped with tight-fitting polymer sleeves to form a crevice. A crevice test assembly has been used in which a serrated fluorocarbon washer containing multiple grooves is compressed against a metal surface to provide multiple crevices on the same metal specimen [12]. Martin and co-workers [1] have developed a crevice test assembly consisting of a Viton™ gasket and an acrylic test assembly which is capable of providing a reproducible crevice tightness by controlling the compression of the gasket against the metal surface.

In these types of testing, there is often a variability between replicate samples as to the occurrence of crevice corrosion on a given alloy. In multiple crevice assemblies, for instance, not all crevice locations may show crevice corrosion, but the number of sites showing attack in a given time is an indication of the susceptibility (or resistance) to initiation, and the depth of attack gives the rate of propagation.

The tests described above are useful for alloy testing and development, but provide no information on the mechanism of crevice corrosion. In order to apply electrochemical techniques, a split electrode must be used in which the metal in the crevice is physically separated (but electrically connected) to the open metal outside the crevice. A typical arrangement, sometimes called a "remote crevice assembly," has been given in Fig. 10.7. By short circuiting the internal (crevice) electrode to the

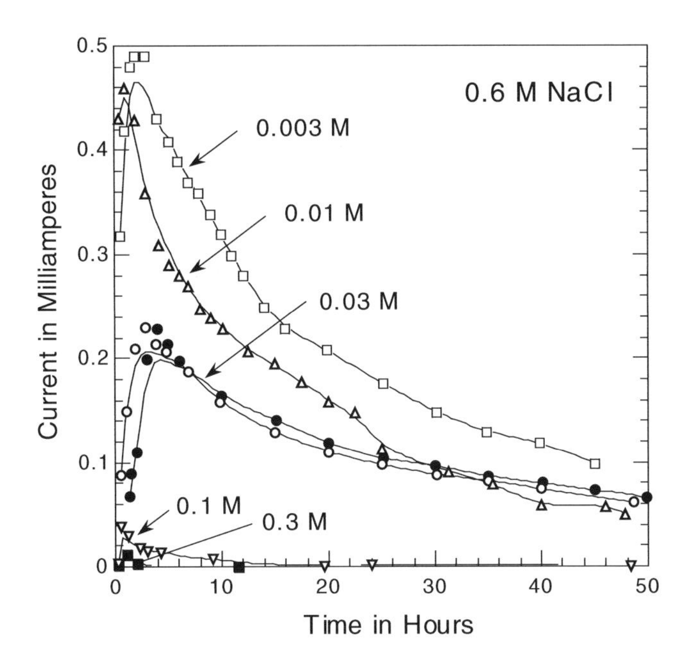

Fig. 10.11 Effect of chromate concentration on the crevice corrosion of iron in 0.6 M NaCl [13]. Chromate concentrations are indicated on the figure. Crevice height = 0.25 mm, area of crevice iron = 7.9 cm^2, area of outer iron electrode = 50 cm^2. Reproduced by permission of ECS – The Electrochemical Society

external (open) electrode, the crevice corrosion current can be measured as a function of time for various parameters, such as crevice dimensions, alloy composition, and electrolyte concentration.

As an example, Fig. 10.11 shows the behavior of iron in a crevice in an aqueous solution of 0.6 M NaCl containing various amounts of chromate inhibitor [13]. The crevice was formed with the crevice electrolyte in place so as to eliminate the problem of diffusion of chromate inhibitor from the bulk solution into the crevice electrolyte. For a fixed chloride concentration in the electrolyte, increasing concentrations of chromate decrease the crevice corrosion rate, as shown in Fig. 10.11. For a given chloride concentration, crevice corrosion can be inhibited by using an appropriate minimum amount of $CrO_4{}^{-2}$ inhibitor, as shown in Fig. 10.11 and Table 10.2.

The propagation stage of crevice corrosion can be studied by using electrolytes which approximate the concentrated acid chloride solutions found within corroding crevices. For example, the

Table 10.2 Minimum concentration of chromate required to inhibit crevice corrosion of iron in chloride solutions [13]. Area of internal crevice iron = 7.9 cm^2, area of external open iron ≈ 50 cm^2, crevice height = 0.25 mm

Concentration of Cl^- (M)	Minimum concentration of $CrO_4{}^{2-}$ required (M)
2.0×10^{-4}	1.0×10^{-3}
1.0×10^{-3}	3.0×10^{-3}
0.01	0.01
0.1	0.03
0.6	0.1

ASTM Test Method G48 for crevice corrosion consists of forming crevices on specimens with a fluorocarbon block and measuring the metal weight loss in a 6 wt% ferric chloride ($FeCl_3$) solution after 72 h immersion [14]. Although $FeCl_3$ solutions provide an extremely aggressive electrolyte, it should be noted that active crevices on iron have been found to contain Fe^{2+} ions rather than Fe^{3+} ions [5].

The ferric chloride solution, however, simulates (perhaps incorrectly) the crevice electrolytes found within propagating crevices only on iron or plain carbon steels. For Fe–Cr stainless steels, for example, the test electrolyte should contain both $FeCl_3$ and $CrCl_3$. For studying the propagation of crevice corrosion on aluminum or its alloys, the test solution should be a concentrated solution of $AlCl_3$. (The pH of a 0.5 M solution of $AlCl_3$ is 2.5 [15]). In general, the test solution for the propagation of crevice corrosion for a given alloy should contain chlorides of each of the cations expected to be present within the corroding crevice.

Table 10.3 shows the chemical analysis of internal electrolytes formed within crevices on alloy 625 or alloy C-276 after immersion in natural seawater for 160–170 days [16]. (The crevice electrolytes were analyzed by thin-layer chromatography.) As shown in Table 10.3, as a result of crevice corrosion, the crevice electrolyte contained dissolved cations of each of the major alloying elements for each of the alloys. Experimental anodic polarization curves determined for each alloy in synthetic crevice solutions similar to those in 10.3 have shown extensive active dissolution, thus confirming that the alloys are subject to crevice corrosion propagation [16, 17].

Table 10.3 Composition of Cr–Ni–Mo alloys and of crevice electrolytes formed in natural seawater after immersion for 160–170 days [16]

Nominal composition of alloy in weight percent					
	Cr	Ni	Mo	Fe	
Alloy 625	22	60	8–10	5	
Alloy C-276	15	57	16	6	
Composition of electrolyte in crevices in natural seawater [16]					
	Cr^{3+}	Ni^{2+}	Mo^{3+}	Fe^{2+}	pH
Alloy 625	0.5 M	0.6–1.2 M	0.008 M	0.06–0.12 M	~–1.0[a]
Alloy C-276	1.5 M	5.6 M	0.002 M	0.5 M	~–1.0[a]

[a]Measured in separate experiments using synthetic crevice electrolytes.

Area Effects in Crevice Corrosion

The rate of crevice corrosion increases with an increase in the area of open metal outside the crevice. An example is given in Fig. 10.12 for crevices on a 17% Cr stainless steel immersed in seawater [18]. As the area of the external (open) metal increases, the total current for oxygen reduction also increases. Thus, the polarization curve for the limiting cathodic current density for oxygen reduction intersects the anodic polarization curve of the creviced metal (in the propagating electrolyte) at increasingly higher values of current, as shown in Fig. 10.13. With the area of the internal creviced metal being constant from case to case, the corrosion current density for the metal within the crevice increases linearly, as shown in Fig. 10.12.

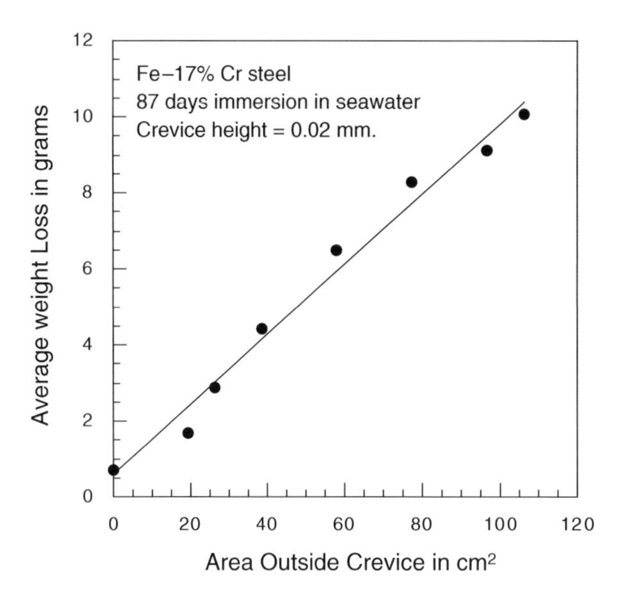

Fig. 10.12 Effect of the area outside the crevice on the weight loss of an Fe-17% Cr steel inside a crevice in natural seawater. Redrawn from [18] by permission of © NACE International (1951)

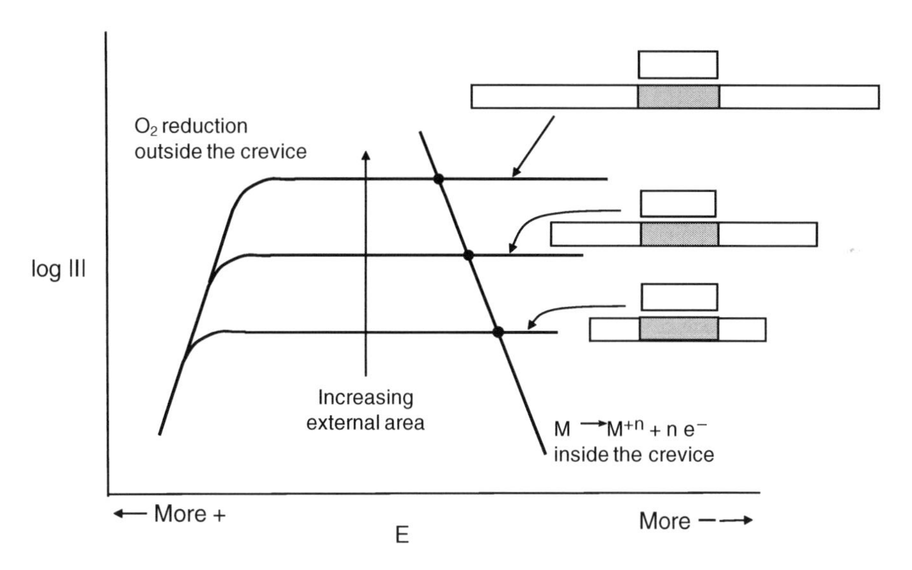

Fig. 10.13 Schematic Evans diagrams showing the effect of external area on the rate of crevice corrosion

Protection Against Crevice Corrosion

Passive metals or alloys are particularly susceptible to crevice corrosion, because these metals depend on their passive oxide films for protection, but these passive films can be destroyed locally by the acidic conditions which develop within propagating crevices.

Crevice corrosion can be prevented or minimized by several protective measures. These include the following:

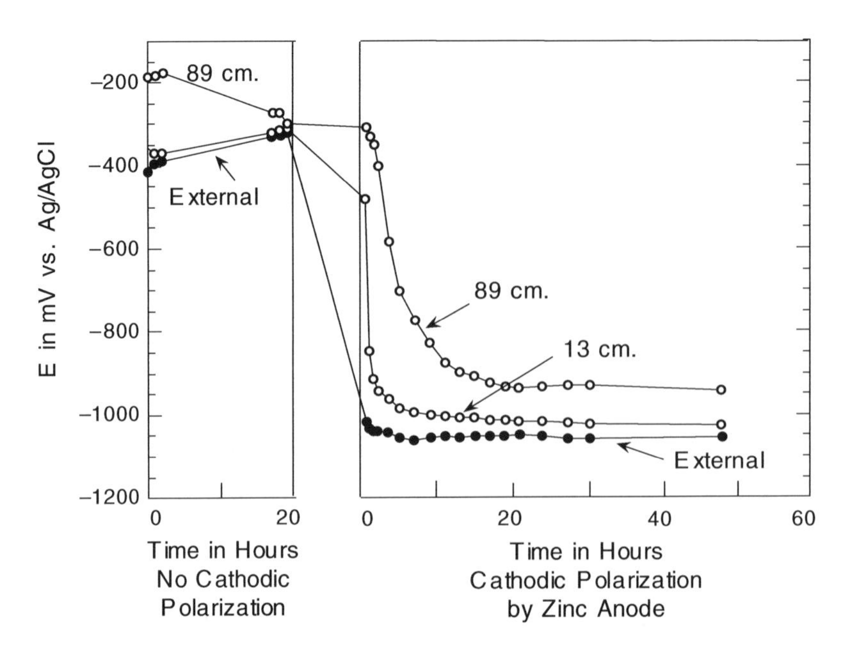

Fig. 10.14 Cathodic polarization of type 304 stainless steel in 0.05 mm crevices in 0.6 M NaCl [19]. The numbers in the figure indicate the distance (in cm) from the crevice opening. Reproduced by permission of © NACE International (1974)

(1) Use of corrosion inhibitors, such as chromates, dichromates, nitrites, or phosphates.
(2) Cathodic protection (using anodes located outside the crevice). Lennox and Peterson [19] have shown that it is possible to cathodically protect deep crevices on stainless steel or copper. Figure 10.14 shows experimental data for stainless steel in 0.6 M NaCl. After 50–70 h of cathodic protection, the electrode potential and pH for stainless steel at a location 35 in. (13.8 cm) inside the crevice were –948 mV vs. SCE and 11.4, respectively, showing that deep crevices can be protected. Similar results were obtained when the initial pH within the crevice was 1.9, showing that cathodic protection is effective even after crevice corrosion has been initiated. Similar results were obtained for copper in crevices.
(3) Design considerations to minimize the existence of crevices. For example, Fig. 10.15 shows that the use of continuous welds is preferred over intermittent welds in order to prevent the occurrence of multiple crevices [20]. Figure 10.16 shows that a discharge valve can be re-designed to prevent crevice corrosion under deposits [8].
(4) Materials selection: titanium and its alloys and Mo-containing alloys such as alloy 625 or alloy C-276 are more resistant to crevice corrosion (at ambient temperatures) than conventional stainless steels. The presence of nitrogen is also beneficial to crevice corrosion resistance in alloys which contain molybdenum [21].

Example 10.2: Why does the internal pH within a crevice increase within a crevice when the crevice is being cathodically protected, as in the work of Lennox and Peterson [19]?

Solution: The principal reaction within the cathodically protected crevice is the reduction of oxygen:

$$O_2 + 2\,H_2O + 4e^- \rightarrow 4\,OH^-$$

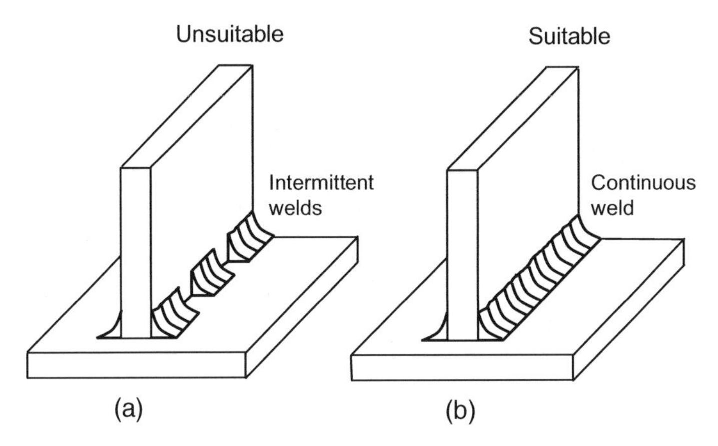

Fig. 10.15 Continuous welds in (**b**) are preferable to intermittent welds in (**a**) because intermittent welds introduce additional crevices [20].

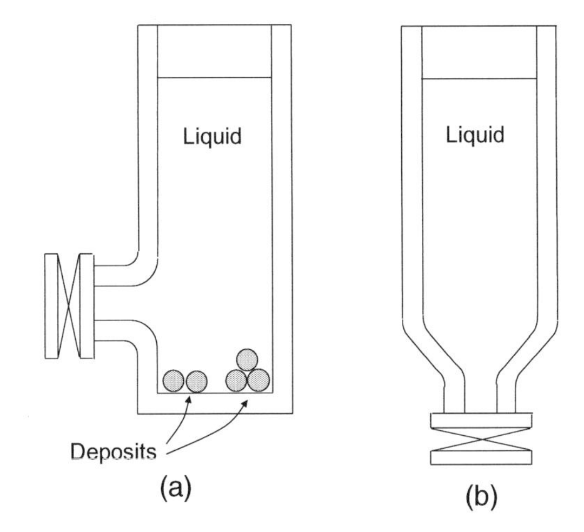

Fig. 10.16 Discharge valves can be re-designed as in (**b**) to prevent crevice corrosion under deposits [8]. Reproduced by permission of © NACE International (1974)

The product of this reaction, hydroxyl ions (OH^-), accumulates within the narrow confines of the crevice, so that the crevice electrolyte becomes more alkaline. Thus, the pH increases.

Pitting

Pitting is a form of localized corrosion in which the attack is confined to a small fixed area of the metal surface. Pitting occurs due to localized breakdown of a passive film, usually by chloride ions. Figure 10.17 shows a cross section of a corrosion pit formed on Al alloy 6061 (Al-1% Mg) in 0.1 M NaCl by anodic polarization [22].

Pitting is a dangerous form of corrosion attack for several reasons. Pits can result in the perforation of a metal component while the rest of the metal piece remains unattacked. In the presence of an applied stress, pits can serve as sites to initiate stress-corrosion cracking, another catastrophic form

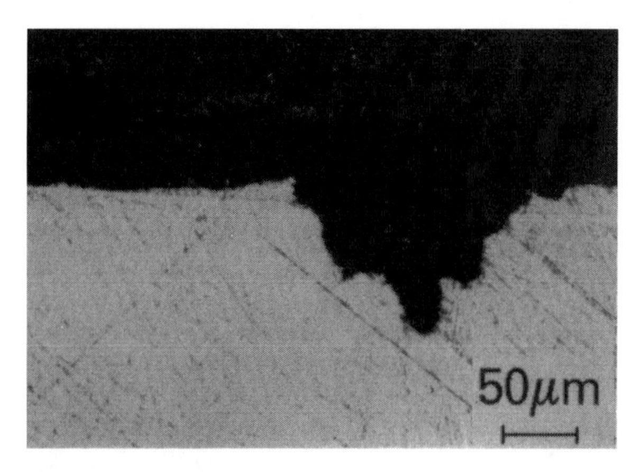

Fig. 10.17 Cross-sectional view of a corrosion pit on Al 6061 formed by anodic polarization in 0.1 M NaCl [22]. Reproduced by permission of © NACE International (1990)

of corrosion attack. Finally, the presence of pits can be difficult to detect if they are covered with corrosion products.

Pits can require a long time to appear in actual service (months to years), so that their absence in the short term is not a certainty that the metal or alloy is immune to pitting. All passive metals are subject to pitting corrosion, although most experimental studies have involved stainless steels (and their alloying components), aluminum, and copper. Pitting is caused by the presence of an "aggressive" anion in the electrolyte, usually Cl^- ions, but pitting of various metals or alloys has also occurred in the presence of other anions, including Br^-, I^-, SO_4^{2-}, or NO_3^-. A more complete list of ions which cause pitting attack has been given elsewhere [23].

The chloride ion has a special importance in piitting corrosion for several reasons. First of all, Cl^- ions are ubiquitous, being constituents of seawater, brackish waters, de-icing salts, and airborne salts. In addition, chlorides are found in the human body, so they can cause the pitting of biomedical implants. Chlorides are also common contaminants in various electronic systems due to handling and processing. The chloride ion is a strong Lewis base (electron donor) and tends to interact with Lewis acids (electron acceptors), such as metal cations. In addition, the chloride ion is a relatively small anion and has a high diffusivity.

Critical Pitting Potential

The tendency of a metal or alloy to undergo pitting is characterized by its critical pitting potential, as illustrated in Fig. 10.18. In the absence of chloride ions, the metal retains its passivity up to the electrode potential of oxygen evolution. However, in the presence of chloride ions, the passive film suffers localized attack, and pitting initiates at a well-defined potential called the *critical pitting potential* (or, usually just the *pitting potential*). (The critical pitting potential is also sometimes called the *breakthrough potential*, *breakdown potential*, or *rupture potential*). Once corrosion pits initiate, they usually propagate rapidly, as shown by the sharp rise in current density at electrode potentials just beyond the critical pitting potential.

An experimentally determined polarization curve for the pitting of type 304 stainless steel is shown in Fig. 10.19 [24]. In the absence of chlorides, the stainless steel remains passive up to the

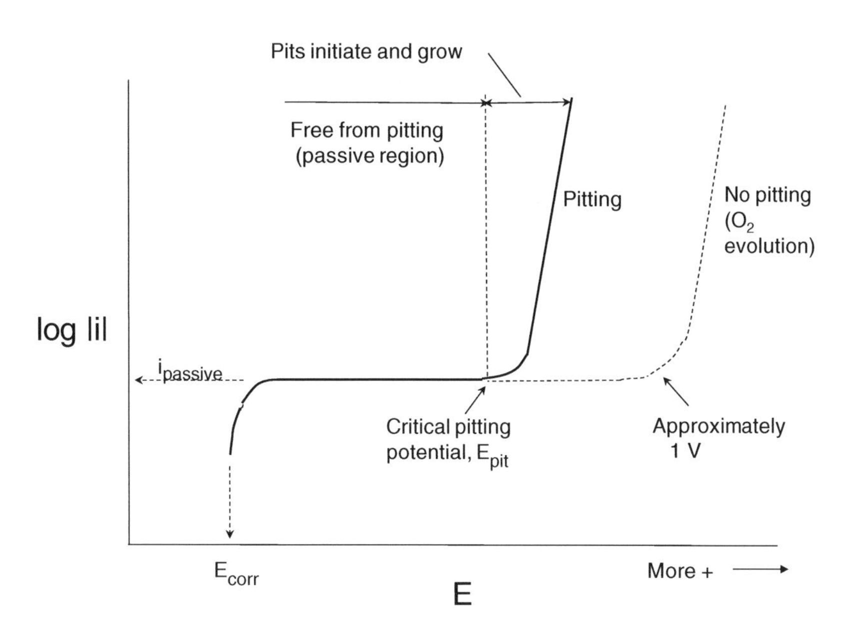

Fig. 10.18 Schematic anodic polarization curve showing the critical pitting potential (for a passive metal)

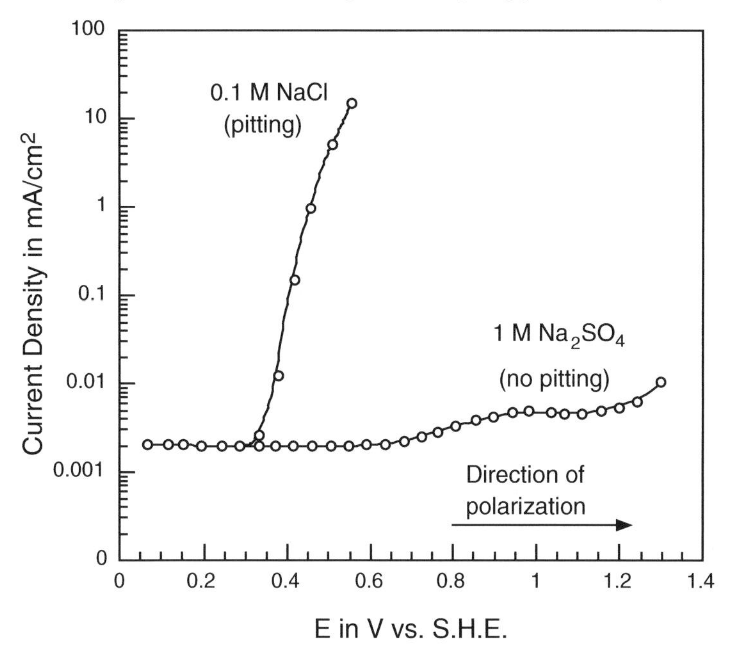

Fig. 10.19 Experimental anodic polarization curves for type 304 stainless steel in 0.1 M NaCl or in 1 M Na_2SO_4. Redrawn from [24] by permission of ECS – The Electrochemical Society

potential of oxygen evolution (about +1.2 V vs. SHE.) but in the 0.1 M NaCl solution, pitting occurs at +0.35 V vs. SHE (the pitting potential).

The critical pitting potential, E_{pit}, is a characteristic property of a given metal or alloy, although the value of E_{pit} for a given metal depends on the chloride concentration. For a given chloride concentration, the more positive the critical pitting potential, the more resistant the metal or alloy to

Table 10.4 Critical pitting potentials, E_{pit}, in 0.1 M NaCl for various metals and alloys

Metal or alloy[a]	E_{pit} in V vs. SCE	Source
Zinc	−1.02	Alvarez and Galvele [94]
Aluminum	−0.70	Natishan et al. [95]
Al alloy 5656 (Al-5 Mg)	−0.68	Trzaskoma et al. [96]
Iron	−0.41	Strehblow and Titze [97], Alvarez and Galvele [98]
M-50 steel (Fe-4 Cr-5 Mo-1 V)	−0.23	Wang et al. [99]
Copper	−0.04	Thomas and Tiller [100]
Molybdenum	+0.055	Wang et al. [99]
Nickel	+0.08	Strehblow and Titze [97]
Chromium	+0.125	Wang et al. [99]
Zirconium	+0.22	Cragnolino and Galvele [101]
304 stainless steel	+0.30	McCafferty and Moore [62]
316 stainless steel	+0.50	McCafferty and Moore [62]
Titanium	$> +1.0$[b]	Dugdale and Cotton [102]

[a]The resistance to pit initiation increases down this table.
[b]This value refers to dielectric breakdown of the oxide film rather than to pitting.

pit initiation. Table 10.4 lists critical pitting potentials for various metals and alloys in 0.1 M NaCl. In each case, the significance of the pitting potential is that

- below E_{pit}, pitting does not occur,
- above E_{pit}, corrosion pits initiate and propagate.

The resistance to pit initiation decreases down Table 10.4. However, the characteristic pitting potential is not the only parameter which should be used to assess the resistance to pitting attack. Other important parameters are the density of pits (number per unit area) and their propagation rate, which will determine the depths of pits and, accordingly, the lifetime of the structure.

Practical support for the existence of a critical pitting potential has been provided by long-term immersion tests in natural seawater of a series of 19 different aluminum alloys (having various heat treatments as well) [25]. The electrode potential of each alloy was monitored over a 1-year immersion period, after which the alloys were removed from seawater, the corrosion products removed, and the depth of attack measured. As can be seen in Table 10.5, aluminum alloys which maintained electrode potentials below (more negative) than −0.89 V vs. Ag/AgCl displayed little or no pitting attack. Thus, the value of −0.89 V vs. Ag/AgCl can be taken to be the pitting potential of aluminum alloys in natural seawater.

Experimental Determination of Pitting Potentials

Critical pitting potentials are usually determined from steady-state anodic polarization curves. The most reliable approach is to use a potentiostatic technique in which a constant potential is applied, and the current is recorded as a function of time, as shown schematically in Fig. 10.20. At electrode potentials below the pitting potential, the current decays to a constant value. But at the pitting potential and above, the current increases with time as corrosion pits initiate and grow. At potentials sufficiently beyond the pitting potential, the anodic current sometimes increases to a limiting value.

Table 10.5 Corrosion behavior of various aluminum alloys in seawater after 1-year immersion [25]

Alloy designation[a]	E vs. Ag/AgCl	Depth of attack in mils	
5257-H25	−1.24	<1	↑
5154-H38	−1.05	1–4	
X7005-T63	−1.01 to −1.02	<1	
Alc. X7002-T6		1–4	
5052-H34		1–4	
5052-H32		<1	
5456-H321	−0.96 to −0.98	<1	
5086-H32		<1	No pitting
5086-H34		<1	
Alc. 7178-T6		1–4	
7106-T63	−0.92 to −0.94	<1	
6061-T651		<1	
5086-H112		<1	
5050-H34		1–4	
1100-H14		1–4	
5083-O		<1	↓
1100-F	−0.88 to −0.89	5–10	← E_{pit}
3000-H14		11–20	↑
6061-T6		5–10	
7178-T6	−0.73 to −0.79	25–40	
X7002-T6		5–10	Pitting
7075-T7351		11–20	
7079-T6		11–20	
2014-T6		5–10	
2024-T351	−0.69 to −0.70	25–40	
2219-T87			↓

[a] Alloy 1100 is commercially pure aluminum. The other alloys have the following principal additions: 2000 series, Cu; 3000 series, Mn; 5000 series, Mg; 6000 series, Mg and Si; 7000 series, Zn. The amount of the alloy addition is generally 3–5 wt.%.

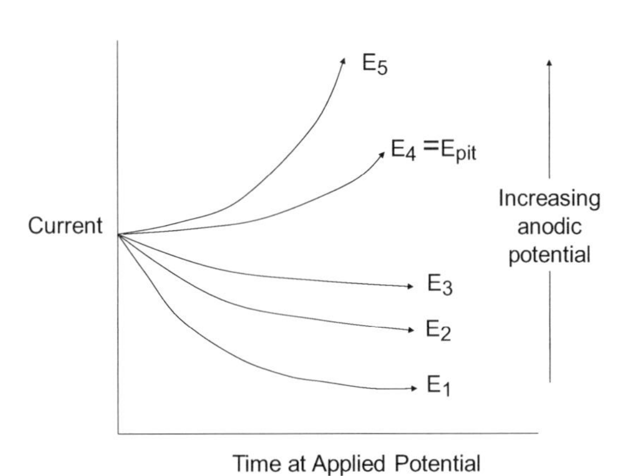

Fig. 10.20 Schematic illustration of experimental current–time curves obtained by the potential step method in the measurement of the critical pitting potential. Below the pitting potential, the current decays with time in response to a potential step. But at and above the pitting potential, when pits initiate and grow, the current increases with time

This is because the current within the propagating pit becomes diffusion-limited due to the formation of a concentrated solution within the pit, as is discussed later.

Pitting potentials have also been determined by anodic sweep (potentiodynamic) measurements, but the measured value for E_{pit} usually depends on the sweep rate. This is caused by not allowing sufficient time at each electrode potential for pit initiation events to transpire.

In addition to the sweep rate, the critical pitting potential also depends on the surface finish of the metal. (Rough surfaces present more initiation sites and generally have lower values of E_{pit} than for smooth surfaces.) The pitting potential does not seem to depend on the oxide film thickness [26].

Effect of Chloride Ions on the Pitting Potential

Chloride ions (among others) cause pitting to occur. The critical pitting potential decreases (is less positive) as the chloride concentration increases. Said differently, for a dilute chloride solution, it is necessary to drive the electrode potential farther in the positive direction to cause pitting to occur. Figure 10.21 shows the effect of chloride concentration on the pitting potential of type 304 stainless steel [24] and aluminum [27]. In each case, the critical pitting potential, E_{pit}, is a linear function of the logarithm of the chloride ion concentration:

$$E_{pit} = a + b \log [\mathrm{Cl}^-] \tag{8}$$

where a and b are constants.

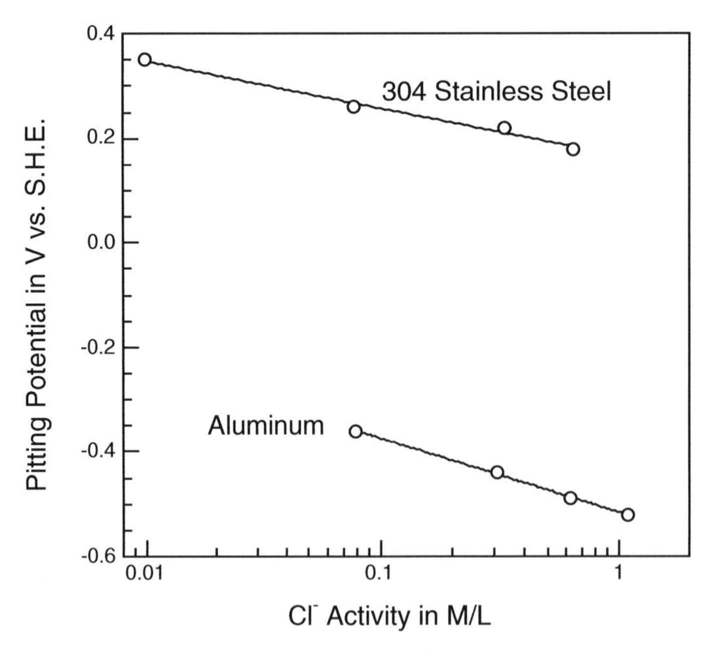

Fig. 10.21 Effect of chloride concentration (activity) on the pitting potential of aluminum and type 304 stainless steel [24, 27]. Reproduced by permission of ECS – The Electrochemical Society

Effect of Inhibitors on the Pitting Potential

The addition of a corrosion inhibitor to a solution increases the pitting potential. Figure 10.22 shows that sulfate additions to a 0.1 M NaCl solution raise the pitting potential of type 304 stainless steel [24]. More about the action of corrosion inhibitors is given in Chapter 12.

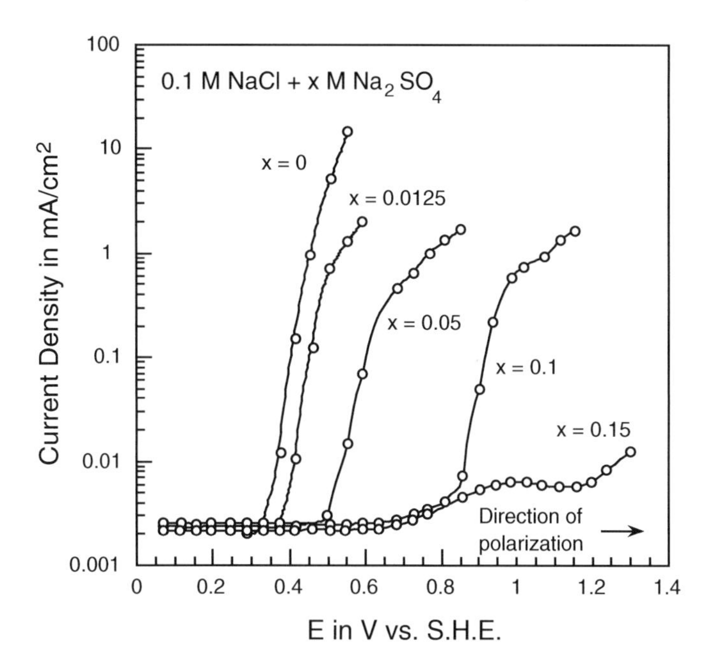

Fig. 10.22 Effect of sulfate additions on the pitting potential of type 304 stainless steel in 0.1 M NaCl solutions. Redrawn from [24] by permission of ECS – The Electrochemical Society

Mechanism of Pit Initiation

Like crevice corrosion, pitting corrosion may be divided into initiation and propagation stages. We will consider the initiation stage to include both the breakdown of the passive film and the onset of an anodic current at the metal surface. The exact mechanism of pit initiation is not known with great certainty, but it is generally understood [23, 28–30] that three main mechanisms are possible. These are (i) the penetration mechanism, (ii) the film thinning mechanism (also referred to as the adsorption mechanism), and (iii) the film rupture mechanism.

In the *penetration mechanism*, aggressive anions are transported through the oxide film to the underlying metal surface where they participate in localized dissolution at the metal/oxide interface. This mechanism is depicted schematically in Fig. 10.23. There is recent evidence from X-ray photoelectron spectroscopy and X-ray absorption spectroscopy that Cl^- ions penetrate passive films on both stainless steel and aluminum [31, 32]. More details on chloride penetration are given in a later section on the pitting of aluminum. The mechanism by which Cl^- ions penetrate the oxide film is not completely understood, but the radius of the Cl^- ion is only slightly larger than that of an oxide ion (1.81 vs. 1.40 Å, respectively), so that Cl^- migration through oxygen vacancies is a possible mechanism of chloride entry into the passive film [33].

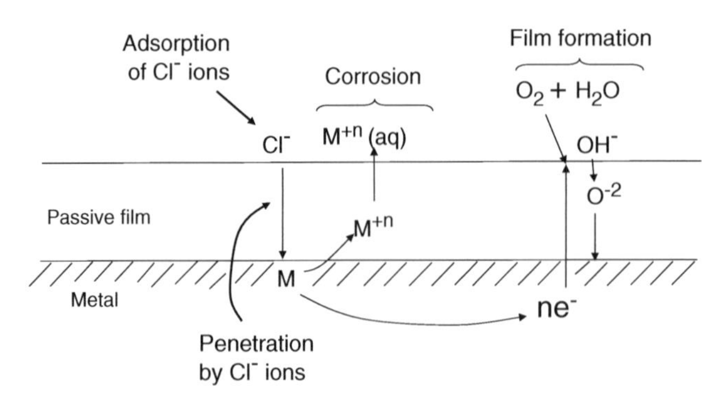

Fig. 10.23 The penetration mechanism of pitting showing the competing processes of film rupture and film formation. Redrawn from [29]. Copyright Wiley VCH Verlag Gmbh & Co. Reproduced with permission

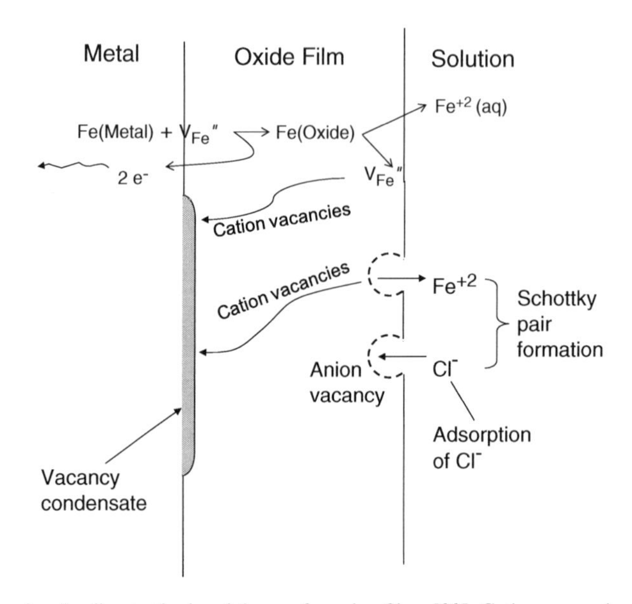

Fig. 10.24 One mechanism leading to the breakdown of passive films [33]. Cation vacancies are transported inward and chloride ions adsorb by occupying anion (oxygen ion) vacancies. V_{Fe}'' represents a cation vacancy of effective charge –2, Fe(oxide) represents a metal cation in a cation site. Reproduced by permission of ECS – The Electrochemical Society

Macdonald and co-workers [33] have developed a point defect model for the breakdown of passivity involving the action of vacancies within the passive film. It is assumed that cation vacancies migrate from the oxide/electrolyte interface inward (as shown in Fig. 10.24), which is equivalent to the migration of cations outward from the metal to the solution. If the cation vacancies accumulate at the metal/oxide interface, the resulting voids which form due to vacancy condensation lead to stresses within the passive film and to pit initiation. In addition, Cl^- ions are carried inward by migration through oxygen vacancies so as to assist in localized dissolution at the metal substrate.

An alternate mechanism of Cl^- transport has been proposed by Bockris and co-workers [34, 35] who have suggested that chloride ions migrate through localized water channels in the hydrous oxide film.

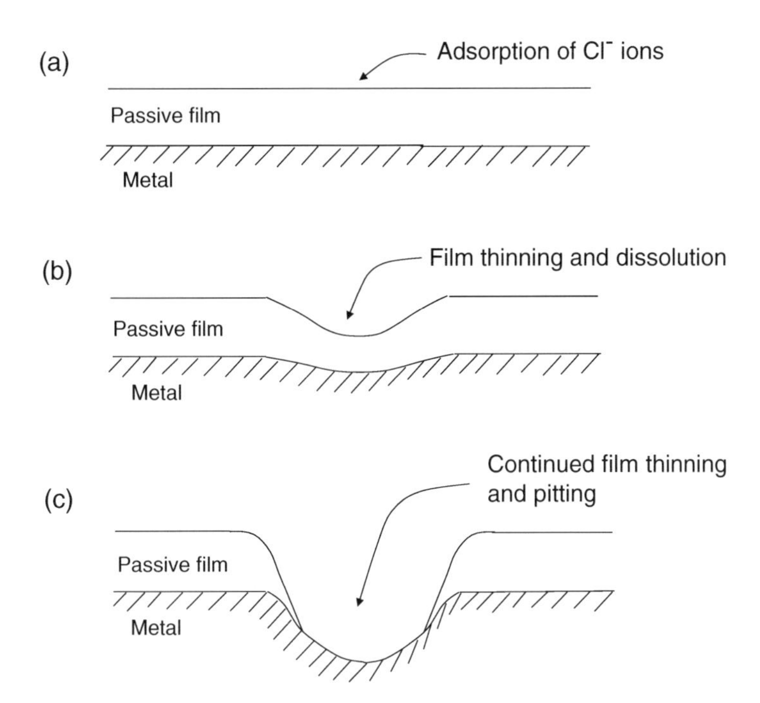

Fig. 10.25 The film thinning mechanism of pitting in which chloride adsorption (**a**) initiates the process of film thinning (**b**) leading to pitting (**c**) [29]. Copyright Wiley VCH Verlag Gmbh & Co. Reproduced with permission

Because pitting can also be caused by ions which are larger in size than the Cl^- ion (e.g., Br^- or I^-), additional mechanisms of initiation must be considered which would apply not only to Cl^- ions but also to other aggressive anions as well. In the *film thinning mechanism*, the aggressive ions (chloride or others) first adsorb on the oxide surface (perhaps in clusters) and then form surface complexes with the oxide film which lead to local dissolution and thinning of the passive film, as shown schematically in Fig. 10.25. Nguyen and Foley [36] have shown that aluminum oxide dissolves to some extent in the presence of sodium chloride solutions. Yu et al. [37] using XPS recently showed that the oxide film thickness on aluminum in 0.1 M NaCl remained essentially invariant at electrode potentials leading up to the pitting potential, but then the passive film displayed thinning at −0.750 V (i.e., 0.050 V before the pitting potential). XPS measurements have shown that F^-, Cl^-, Br^-, and I^-, ions all cause thinning of the passive film on iron [38].

In the *film rupture mechanism* (Fig. 10.26), chloride ions penetrate the oxide through cracks or flaws in the film. In addition to pre-existing defects in the oxide film, flaws may further develop by hydration/dehydration events in the oxide film and by the intrusion of Cl^- ions into the film. According to Sato [39], the presence of a high electric field in the oxide can lead to an electromechanical breakdown of the passive film. It should be noted that pre-existing flaws cannot extend down to the underlying metal substrate because the metal would quickly react with water molecules in the electrolyte and would re-oxidize. Then, chloride ions must penetrate the re-formed oxide film by mechanisms discussed above.

Passive films can also be ruptured or disrupted due to metallurgical variables, such as grain boundaries, impurity atoms, and inclusions. The role of sulfide inclusions in the pitting of stainless steels is discussed later in this chapter.

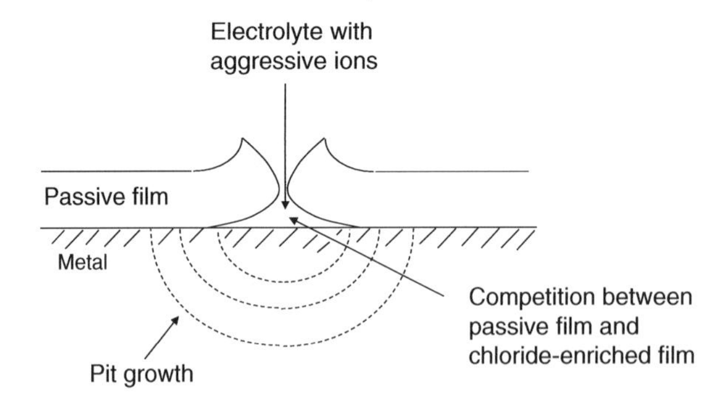

Fig. 10.26 The film rupture mechanism of pitting [29]. Copyright Wiley VCH Verlag Gmbh & Co. Reproduced with permission

These three mechanisms of pit initiation (chloride penetration, film thinning, and film rupture) are not necessarily mutually exclusive. For example, when Cl^- ions are transported through oxygen vacancies, there may also be concomitant oxide film dissolution because atoms surrounding vacancy defects are expected to be less bonded to the bulk oxide [40]. Yu et al. [37] have shown evidence for both chloride penetration and film thinning for the pitting of aluminum in chloride solutions. More details on these three mechanisms of pit initiation are given elsewhere [29].

Mechanism of Pit Propagation

The mechanism of pit propagation is similar to that for the propagation of crevice corrosion, as illustrated in Fig. 10.27 [41]. When the corrosion pit has been initiated, the corresponding local current density is very high because the current is confined to a small active geometrical area, with

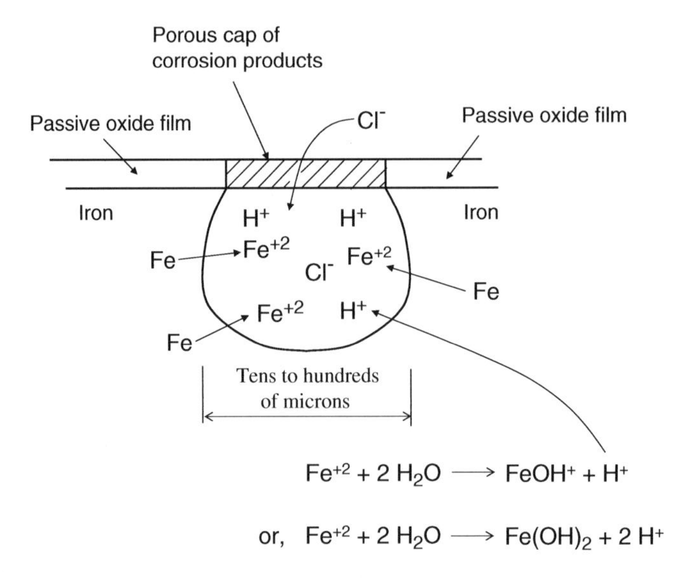

Fig. 10.27 Schematic representation of the propagation stage of pitting. Note the similarity to Fig. 10.10

the oxide film adjacent to the pit remaining passive and unattacked. As the pit grows, its volume increases, but dissolved metal cations are confined within the pit and do not diffuse out into the bulk electrolyte due to the confinement of a restricted geometry or a cap of porous corrosion products, which exists in some cases. As a result, accumulated metal cations undergo hydrolysis, as in the case of crevice corrosion, and a local acidity develops within the pit. Finally, the accumulation of H^+ ions and cations within an active pit, Cl^- ions migrate from the bulk electrolyte to the crevice electrolyte in order to maintain charge neutrality within the pit solution.

One of the earliest pieces of experimental evidence which showed that corrosion pits are acidified was provided by Butler et al. [42], who scanned across a corrosion pit on iron using small diameter (5–25 μm) antimony/antimony oxide pH electrodes. Their results given in Fig. 10.28 show that the interior of a corrosion pit is acidified and that the pH adjacent to the pit becomes alkaline because the cathodic reaction (reduction of O_2 to OH^-) occurs on the passive surface outside the pit. Butler et al. also measured electrode potentials across the same pit using a Ag/AgCl micro-electrode. Results given in Fig. 10.28 also show that the electrode potential above the pit is more active (more negative) than above the areas adjacent to the pit.

Thus, the propagation stage of pitting, like that of crevice corrosion, involves the formation of a highly corrosive internal electrolyte which is acidic and concentrated in chloride ions and in dissolved cations of the metal or alloy. When the corrosive pit electrolyte has been formed, pitting

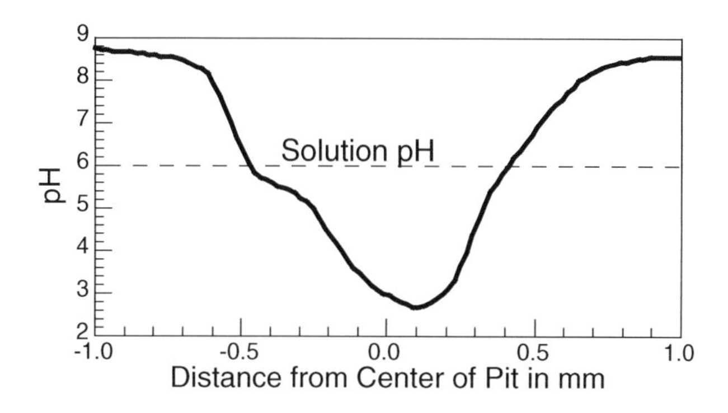

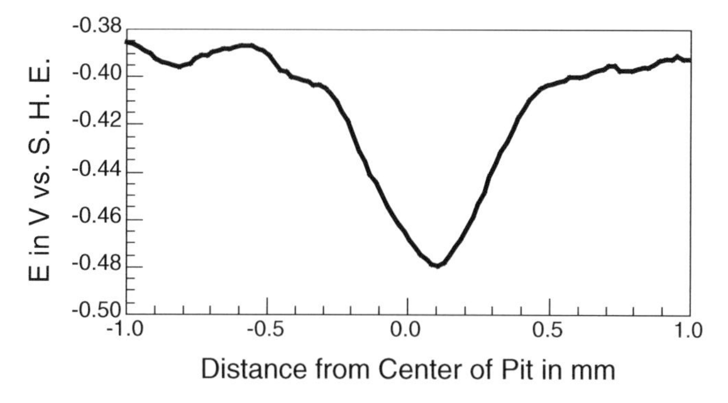

Fig. 10.28 Variations in pH (*top*) and electrode potential (*bottom*) across a growing pit on iron in a dilute chloride solution. Redrawn from [42] by permission of Maney Publishing

is considered to be autocatalytic in nature. That is, the local conditions which have developed are capable of sustaining further pit growth.

Protection Potential

The concept of a protection potential against pitting arose from cyclic anodic polarization curves in which the scan direction was changed at anodic potentials beyond the pitting potential. As shown in Fig. 10.29, when the scan direction is changed, the current density decreases with decreasing electrode potential until the reverse curve crosses the forward curve in the region of passivity. The electrode potential at this intersection is called the protection potential, E_{prot}. The significance attached to the protection potential is that at this potential, the growth of active pits is diminished or possibly stopped (because the passive current density has been regained).

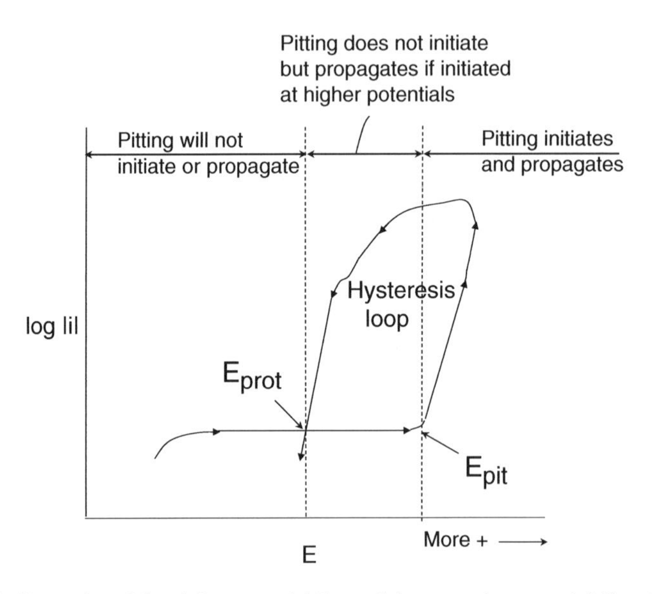

Fig. 10.29 Schematic illustration of the pitting potential E_{pit} and the protection potential E_{prot}. The *arrows* show the direction of polarization

Wilde and Williams [43] investigated this concept for various alloys which had been immersed under crevices in natural seawater for 4.25 years. Cyclic anodic polarization curves were also determined in 3.5% NaCl solutions for open samples cut from the same specimens previously immersed in seawater for the 4.25 years of exposure period. Figure 10.30 lists the conditions used in measurement of the electrochemical polarization curves for the open samples. For each alloy, Wilde and Williams [43] determined the pitting potential, E_{pit}, from the forward sweep of the anodic polarization curve and the protection potential, E_{prot}, from the reverse sweep of the anodic polarization curve.

As shown in Fig. 10.30, the amount of hysteresis for the open samples, as measured by the difference ($E_{pit} - E_{prot}$), correlates with the amount of crevice corrosion (the "first cousin" to pitting corrosion). That is the less the hysteresis in the cyclic polarization curve, the less the amount of

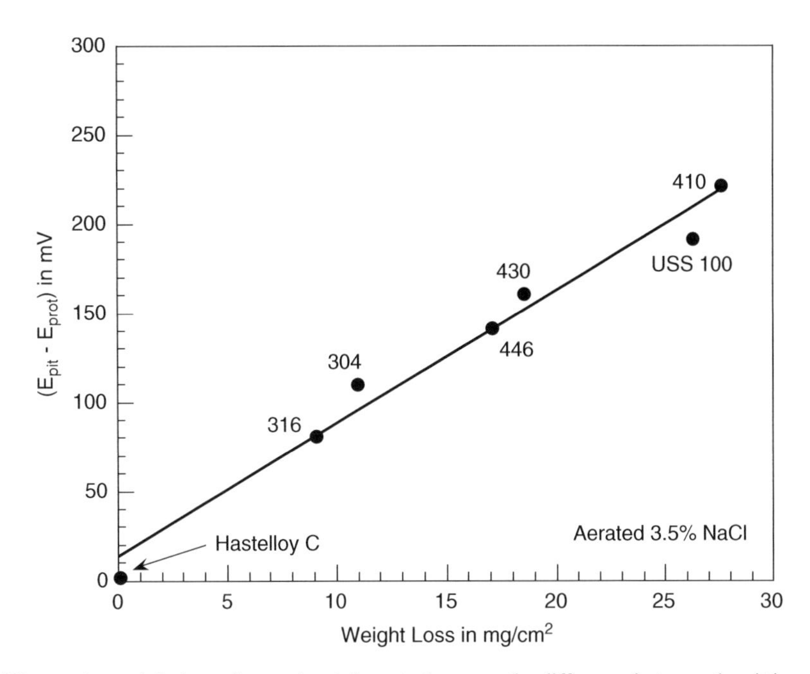

Fig. 10.30 The crevice weight loss of several stainless steels versus the difference between the pitting potential E_{pit} and the protection potential E_{prot} for open samples [43]. The pitting potential was determined in aerated 3.5% NaCl at 25°C at a potential sweep rate of 10 mV/min, and the protection potential determined after reversing the sweep at a current density of 200 $\mu A/cm^2$. The sample area was 1220 cm^2 and the crevice area approximately 20 cm^2. Alloy types 4xx and USS 100 are Fe–Cr alloys, alloy types 3xx are Fe–Cr–Ni alloys, and Hastelloy C (alloy C-276) is primarily a Ni–Cr–Mo alloy; see Table 9.11. Reproduced by permission of Elsevier Ltd

crevice corrosion observed in the actual seawater immersion test. At first glance, then, the use of the protection potential is useful as an indicator of corrosion behavior.

However, Wilde and Williams [43] also established that the measured protection potential depends on the experimental conditions used in its determination. That is, once the electrode potential has exceeded E_{pit}, corrosion pits initiate and then propagate with concomitant acid hydrolysis, buildup of corrosion products, and increase in chloride ion concentration within the pit. When the sweep is reversed, the propagation processes which are potential dependent decrease until the rate of propagation becomes zero at the passive current density. Thus, the electrode potential for this occurrence, i.e., E_{prot}, should depend on the amount of propagation which has taken place within the pit. This was found to be the case for type 430 and 304 stainless steels, as shown in Fig. 10.31. Thus, the protection potential should not be regarded as a material property, although it can be used (with caution) in ranking various alloys as to their pitting behavior.

Pessall and Liu [44] introduced a technique to measure pitting potentials by scribing through the passive film and then observing its ability to re-form following mechanical rupture of the oxide film. The electrode potentials for the repair of oxide films on various Fe-Cr alloys, as determined by this scratch test, were approximately 100 mV more negative than the critical pitting potential determined by the conventional anodic polarization technique. These "scratch potentials" are now considered to be associated with repassivation of the oxide film rather than with its initial chemical breakdown and are a measure of the protection potential.

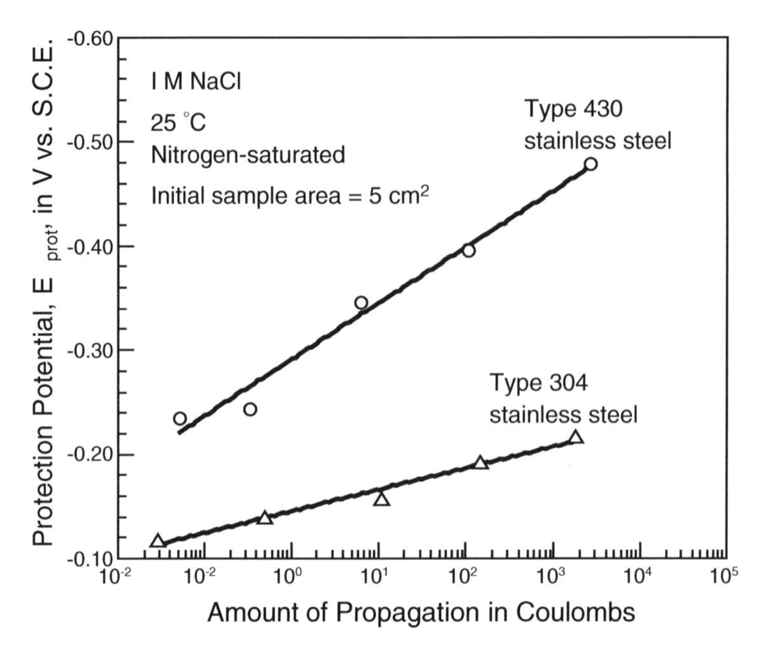

Fig. 10.31 Effect of extent of propagation on the value of the protection potential [43]. Reproduced by permission of Elsevier Ltd

Metastable Pits and Repassivation

The current–time traces shown schematically in Fig. 10.18 are idealized curves. At electrode potentials below the pitting potential, the current–time curves often contain transient excursions as shown in Fig. 10.32 [45]. These transients are due to metastable pits, i.e., pits that grow for a limited time but are repassivated and stop growing. (The pitting potential marks the initiation of stable pits which can grow and propagate.) Metastable pits have a limited lifetime because the concentrated acidic chloride solution which promotes pit propagation has not yet developed within the metastable pit, so that the metastable pit can be repassivated. Metastable pits are generally of micron size and have a lifetime of the order of seconds or less [30].

The mechanism by which metastable pits repassivate is not known with certainty. It has been suggested that salt films are formed at the base of metastable pits because the solubility product of the metal salt is easily exceeded, and that these salt films provide protection from further pit growth by limiting diffusion of dissolved cations out of the metastable pit [46]. The anodic behavior of metals in the presence of concentrated electrolytes, including salt films, depends on the chloride ion concentration, pH, and electrode potential. Because of the small size of corrosion pits, repassivation processes have usually been studied using scaled-up artificial pits.

The importance of the internal pH has been shown by Jones and Wilde [47], who examined the anodic behavior of type 304 stainless steel in concentrated chloride solutions. In 1 N $CrCl_3$, the stainless steel exhibited an "active/passive transition" similar to those discussed in Chapter 9, except that the value of the current density at the nose of the anodic curve was of the order of 10 mA/cm^2 and was much higher than values normally associated with passivity. In more concentrated $CrCl_3$ solutions (which also were more acidic), the "active/passive" transition was absent and only active

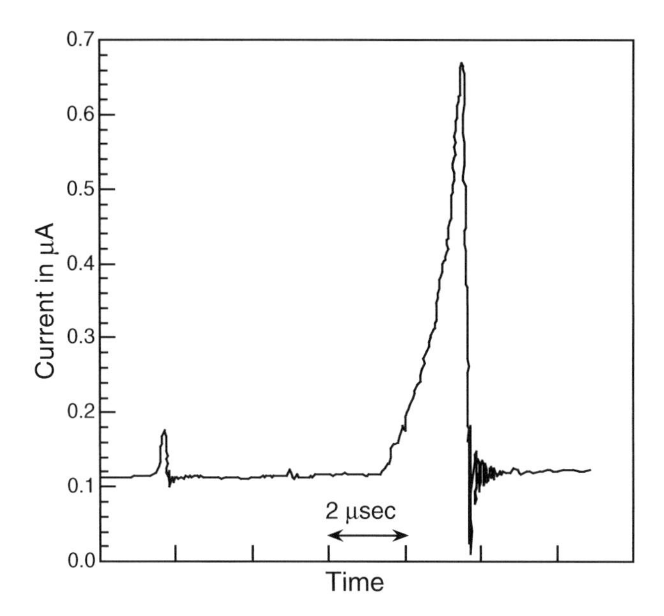

Fig. 10.32 Metastable pitting current transients observed for type 302 stainless steel in 0.1 M NaCl at +0.420 V vs. SCE. Redrawn from [45] by permission of © NACE International (1987)

dissolution was observed. Similar results were observed for type 304 stainless steel in passing from 1 N NaCl (pH 1) (active/passive transition) to 1 M NaCl (pH 0) (active dissolution only).

Using artificial pits, type 304 stainless steel and Fe-Cr-Ni alloys were observed [48, 49] to undergo active dissolution in simulated pit electrolytes. These simulated pit electrolytes were solutions concentrated in H^+ ions (i.e., were acidic) and were also concentrated in Cl^- ions and cations of the dissolving alloy.

The corrosion rate of pure (open) iron in concentrated acidic chloride solutions depends on both the chloride ion and hydrogen ion concentrations [50], as shown in Fig. 10.33. For 1.0 M chloride solutions, the open-circuit corrosion rate varied only slightly with increasing H^+ concentration. But the open-circuit corrosion rate increased as the chloride concentration increased (at a fixed H^+ concentration). For a 6.0 M chloride solution, the open-circuit corrosion rate increased with increasing acidity. This result shows the synergistic effect of H^+ and Cl^- ions in promoting corrosion in concentrated chloride solutions and suggests that pit growth in occluded concentrated electrolytes follows a similar synergy.

In recent years, there has been increased interest in understanding the repassivation of metastable pits, and this subject is an area of much current research.

Experimental Pourbaix Diagrams for Pitting

As discussed in Chapter 6, conventional Pourbaix diagrams do not consider localized corrosion by chloride ions, and to do so, special experimental diagrams must be constructed. This construction can now be done using the concepts of the pitting potential and the protection potential.

Figure 10.34 shows anodic polarization curves for iron in solutions of different pH but containing 10^{-2} M Cl^- [51, 52]. The pitting potential and protection potential were first determined for

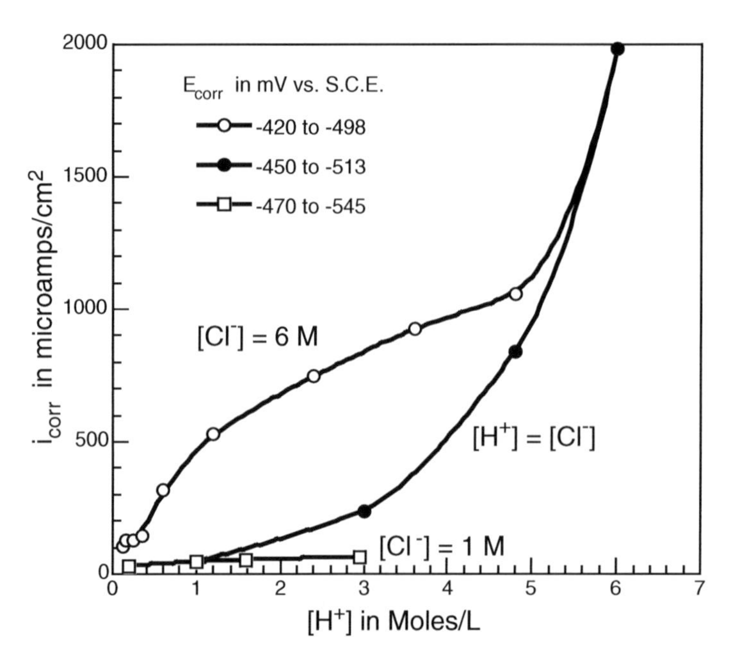

Fig. 10.33 Effect of the concentration of H^+ and Cl^- on the corrosion of open iron in concentrated solutions [50]

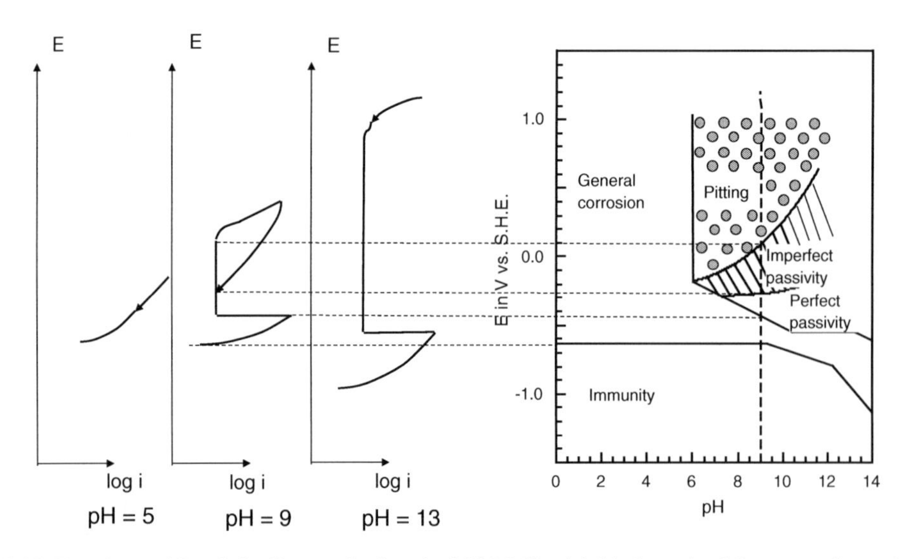

Fig. 10.34 Experimental Pourbaix diagram for iron in 0.01 M Cl^- (*right*) determined from experimental anodic polarization curves, as on the *left*. Redrawn from [51, 52]. The *arrows* show the effect of a reverse potential sweep. The polarization curve furthest to the *left* shows general corrosion only; the polarization curve in the *middle* displays a pitting potential and a protection potential; the polarization curve on the *right* shows general corrosion or passivity, but not pitting. Reproduced by permission of Elsevier Ltd

each solution in which pitting occurs, and then this information was transferred onto a potential–pH diagram to establish various regions of corrosion behavior. These regions are regimes of immunity, general corrosion, perfect passivity, imperfect passivity, and pitting and are defined in Fig. 10.34.

The condition of *perfect passivity* exists in the passive range below the protection potential where pits do not nucleate or grow. The region of *imperfect passivity* lies between the protection potential and the pitting potential and is a region where previously formed pits can grow before they are repassivated at potentials at or below E_{prot}. The region of *pitting* exists above the pitting potential.

Thus, the experimental Pourbaix diagram for iron in Fig. 10.34 takes into account pitting corrosion with regions now defined for pit initiation, growth, or repassivation. Such a diagram is more useful for engineering alloys rather than for pure metals because the general and localized corrosion behavior of complex alloys can be summarized in a convenient experimental Pourbaix diagram (for a fixed concentration of chloride ions). Experimental Pourbaix diagrams taking into account pitting corrosion have been determined for various alloys [53–56] and are similar to Fig. 10.34.

Effect of Molybdenum on the Pitting of Stainless Steels

The addition of molybdenum to Fe-Cr alloys increases the pitting potential [57], as shown in Fig. 10.35. The addition of Mo to alloys which contain Cr has already been noted to have a beneficial effect on passivity, as noted in Chapter 9. The mechanism by which Mo promotes resistance to localized breakdown of the passive film has been studied by various authors, and several mechanisms of increased resistance to pitting have been proposed [58]. These proposed mechanisms of protection include the following:

(1) Active sites are covered with molybdenum oxyhydroxide or molybdate salts, thereby inhibiting localized attack.

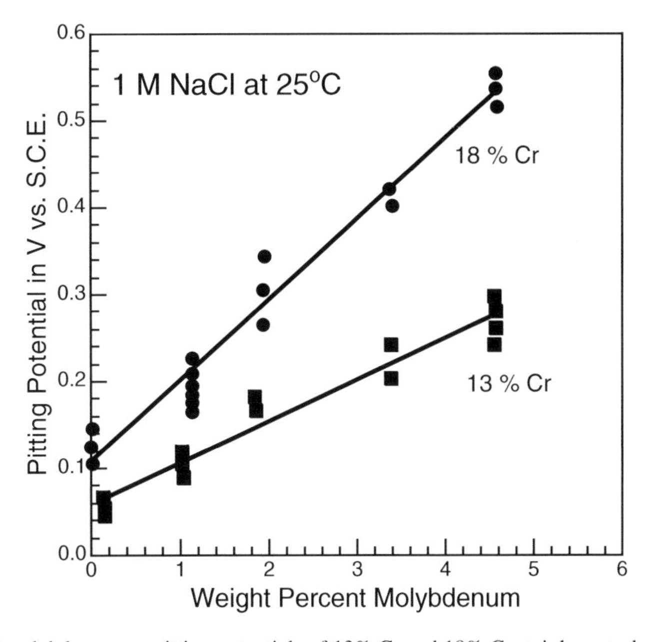

Fig. 10.35 Effect of molybdenum on pitting potentials of 13% Cr and 18% Cr stainless steels in 1 M NaCl at 25°C [57]. Reproduced by permission of John Wiley & Sons, Inc

(2) Dissolution of Mo in the alloy produces molybdate ions, which then act as a corrosion inhibitor near the metal surface.
(3) Mo^{6+} species in the passive film interact with cation vacancies of the opposite sign and reduce the flux of cation vacancies to the metal substrate (depicted in Fig. 10.24).
(4) Molybdate ions act as a cation-selective outer layer in the passive film, as shown schematically in Fig. 9.44, and thus hinder the transport of Cl^- ions through the passive film.
(5) Mo interferes with the kinetics of active dissolution at the base of the developing pit.

The role of molybdenum on the improved pitting resistance of stainless steels is a topic of continuing research.

Effect of Sulfide Inclusions on the Pitting of Stainless Steels

The integrity of the passive film on a metal or alloy can be affected by metallurgical variables in the metal. These include compositional heterogeneities at grain boundaries and disruption of the passive film by impurity atoms or inclusions. Sulfide inclusions, especially manganese sulfide (MnS), are known to be pit initiation sites on stainless steels [59–61], although pitting also occurs on the metal matrix itself and thus does not require the existence of sulfides.

Sulfur is present in stainless steels as an impurity (0.03% maximum) or as an intentional additive (approximately 0.3%) to aid in the machining of the metal [61]. The most common sulfide which is formed in stainless steels is MnS, although other sulfide inclusions are also known to exist. Figure 10.36 shows a scanning electron micrograph of a typical sulfide inclusion on type 304 stainless steel [62]. The approximate size of the inclusion is 10 μm in length.

It has long been postulated that such anodic dissolution of sulfide inclusions leads to an aggressive electrolyte composition near the inclusion, so as to disrupt the passive film. Recent studies using a scanning vibrating electrode (SVE) or nanometer-scale secondary ion mass spectrometry (SIMS) have verified that pits initiate at the edge of the sulfide inclusion and that anodic zones exist around the inclusion [63, 64]. Ryan et al. attribute the anodic zones to be due to an area around the MnS particle which is depleted in chromium, and thus susceptible to localized attack [64]; see Fig. 10.36. The MnS inclusion itself is unstable in the Cl^--rich acidic local environment which is established in the propagating pit(s).

The pitting resistance of type 304 stainless steel can be improved by laser surface melting (see Chapter 16). The effect of treating the surface with a high-power laser is to remove large-scale sulfide inclusions during laser processing. The sulfides are melted and redistributed upon solidification into a larger number of smaller particles [62] which are less disruptive to the passive film.

Effect of Temperature

The pitting potentials of type 304 and 316 stainless steels in an aqueous solution containing 200 mg/L Cl^- are given as a function of temperature in Fig. 10.37 [65]. It can be seen that at all temperatures, the pitting potential of the Mo-containing 316 stainless steels is higher than that of type 304 stainless steel. In addition, the pitting potential of each alloy decreases with increasing temperature.

Brigham [66] has developed the concept of a critical pitting temperature as a criterion to rate the susceptibility of a series of alloys to pitting corrosion. A series of Mo-containing alloys was immersed in 10% $FeCl_3$ solutions held at a constant temperature, and the alloys were examined

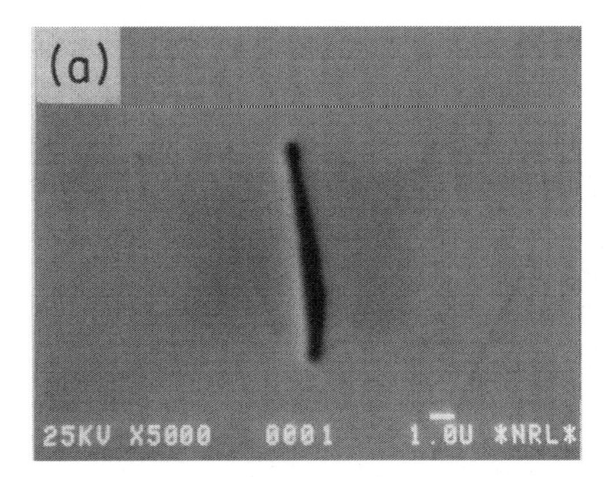

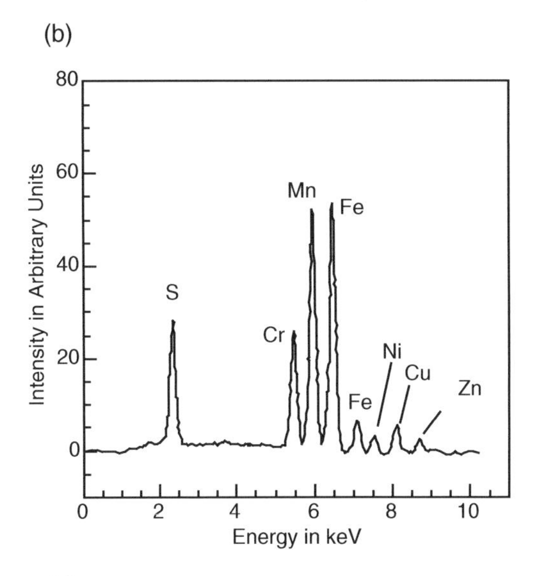

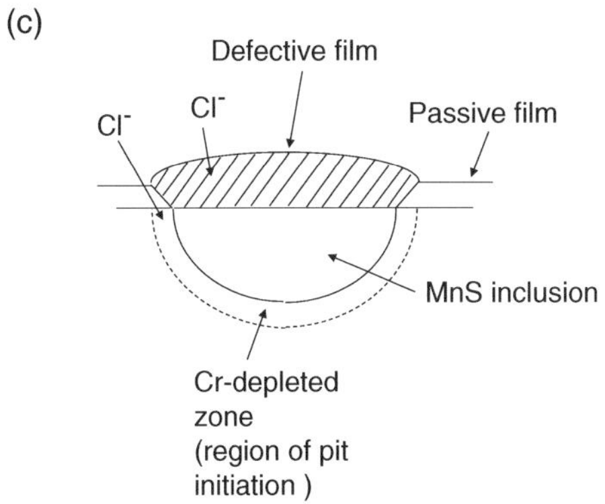

Fig. 10.36 (**a**) A top view scanning electron micrograph of a sulfide inclusion in type 304 stainless steel [62]. (**b**) EDAX (energy dispersive analysis by X-rays) showed that the sulfide inclusion contained Mn, Fe, Cr, and S. (**c**) Cross-sectional illustration of a MnS particle as a pit initiation site. Figure 10.36a–b adapted from [62] by permission of ECS – The Electrochemical Society. Figure 10.36c reprinted with permission from Macmillan Publishers Ltd: Nature, copyright (2002) [64] (as shown)

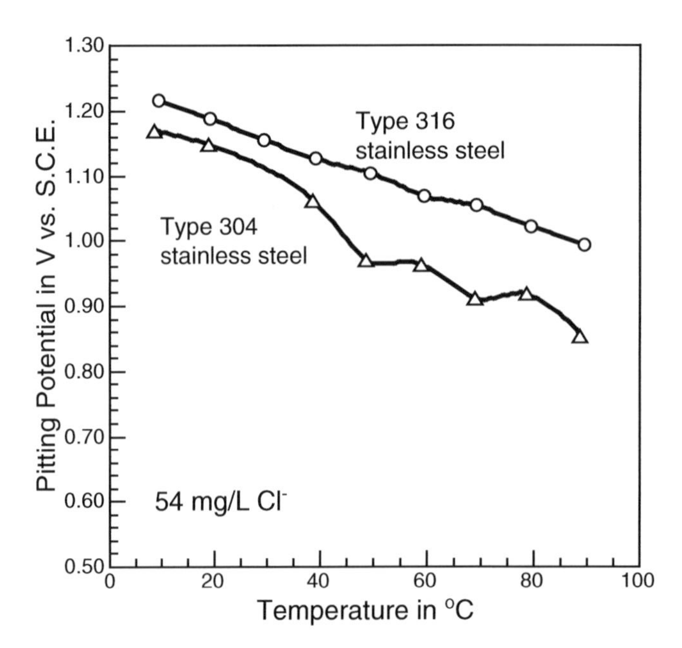

Fig. 10.37 The effect of temperature on the pitting potential of type 304 and type 316 stainless steels in a dilute chloride solution. Redrawn from [65] by permission of © NACE International (1979)

for signs of pitting after a 24 h exposure period. If no pitting was observed, the temperature was increased, and this procedure was repeated until the first signs of pitting occurred at the "critical pitting temperature" for that alloy. The higher the critical pitting temperature, the more resistant the alloy to pitting Fig. 10.38 shows that the pitting resistance increases with the Mo content of the alloy for a series of alloys having the nominal base composition Fe-18 Cr-20 Ni.

A similar set of tests was conducted in 10% $FeCl_3$ solutions for specimens having crevices, and a "critical crevice temperature" was determined, as also shown in Fig. 10.38. Again, increasing amounts of Mo were beneficial. Figure 10.38 also shows that the critical crevice temperature is lower than the critical pitting temperature. This is because severe conditions already exist within the occluded crevice in the $FeCl_3$ solution, but must develop within a corrosion pit on an open surface.

Protection Against Pitting

Corrosion control measures to prevent or minimize pitting include the following:

(1) Maintain the electrode potential below (more negative, i.e., less positive than) the critical pitting potential.
(2) Add inhibitors to raise the critical pitting potential.
(3) Materials selection: metals and alloys which are resistant to crevice corrosion are also usually resistant to pitting. The resistance to pitting corrosion increases in the order:

Fe-12% Cr < 304 stainless steel < 316 stainless steel < alloy C-276 < titanium.

The order of this series is due to progressive alloying changes, as shown in Fig. 10.39.

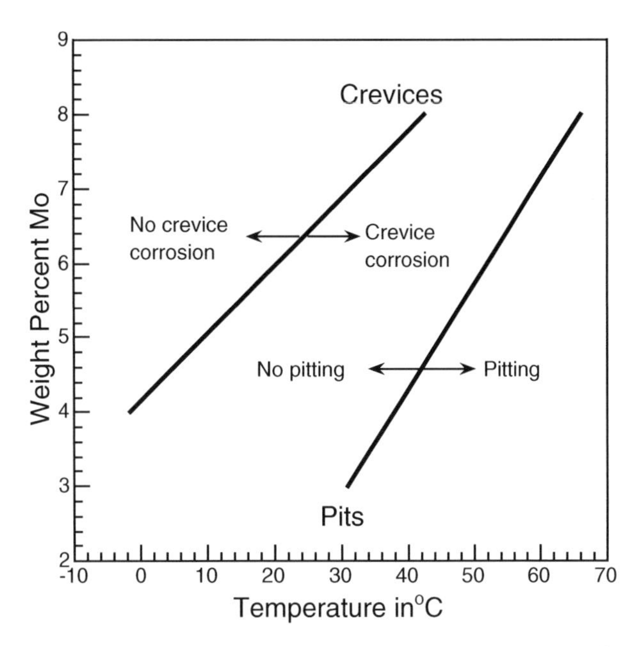

Fig. 10.38 Critical pitting temperature and critical crevice temperature as a function of molybdenum content for several different stainless steels having the nominal composition Fe-18 Cr-20 Ni-x Mo. Plotted from data in [66] by permission of © NACE International (1974)

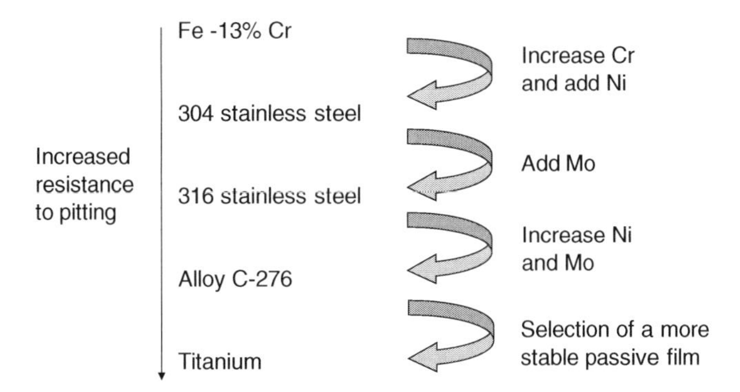

Fig. 10.39 Effect of alloy selection on the pitting potential in solutions of fixed chloride concentration

Pitting of Aluminum

The purpose of this section is to give more details on the pitting corrosion of one particular metal. Aluminum is subject to general corrosion in acidic and alkaline environments but is passive in solutions of intermediate pH, as shown by the Pourbaix diagram in Fig. 6.1. However, in the presence of chloride ions, the passive film on aluminum suffers localized breakdown, and pitting ensues. In practice, aluminum is usually alloyed with various elements to improve its mechanical strength, and these alloying additions result in the formation of intermetallic second-phase particles which can also disrupt the passive film on the alloy surface. However, the pitting corrosion of pure aluminum is

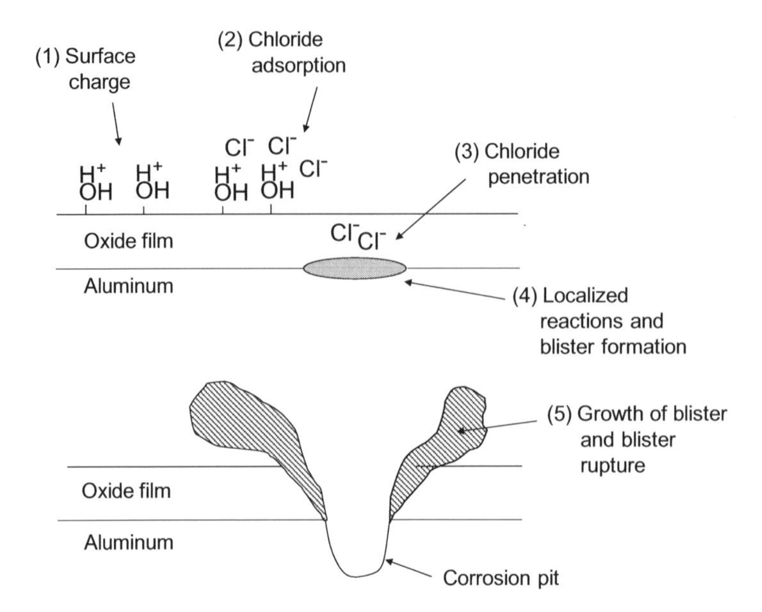

Fig. 10.40 Sequence of steps in the pitting of aluminum [67]

discussed here in order to illustrate the sequence of events that occurs from passive film breakdown through pit initiation to pit propagation.

The steps involved in the pit initiation process involve (i) adsorption of chloride ions at the oxide surface, (ii) penetration of the oxide film by chloride ions, and (iii) chloride-assisted dissolution which occurs beneath the oxide film at the metal/oxide interface [67]; see Fig. 10.40. The propagation stage of pitting on aluminum produces blisters beneath the oxide film [68], due to continuing localized reactions within the pit which lead to a localized acidic environment. The blisters subsequently rupture due to the formation of hydrogen gas in the corrosion pit.

The first step in the pitting process is adsorption of chloride ions onto the oxide-covered surface. It is well known that the outermost surface of an oxide or an oxide film is covered with a layer of hydroxyl groups, which results from the interaction of the outermost layer of the oxide film with the aqueous solution. In the case of aluminum oxide films in nearly neutral solutions, these hydroxyl groups interact with protons in solution to produce a positively charged surface, as is discussed further in Chapter 16:

$$-\mathrm{AlOH_{surf}} + \mathrm{H^+_{(aq)}} \rightleftharpoons \mathrm{AlOH_2^+}_{\,\mathrm{surf}} \tag{9}$$

Thus, the adsorption of chloride anions on the oxide film occurs through the operation of attractive coulombic forces between the positively charged oxide surface and the negatively charged Cl^- ion. There is considerable evidence that Cl^- ions adsorb on the oxide film on aluminum [31, 34, 35, 37, 67], and additional work using a scanning chloride micro-electrode suggests that adsorption occurs in clusters of Cl^- ions on the oxide-covered 304 stainless steel surface [69].

Chloride ions next penetrate the oxide film, as has been shown by X-ray photoelectron spectroscopy (XPS). Figure 10.41 shows an XPS Cl 2p spectrum for an aluminum surface held below the pitting potential [31]. Each Cl 2p spectrum is actually a doublet, and there are two distinct sets of doublets. The lower binding energy doublet was removed by sputtering, indicating that this doublet was characteristic of surface chloride. The higher binding energy doublet persisted after sputtering,

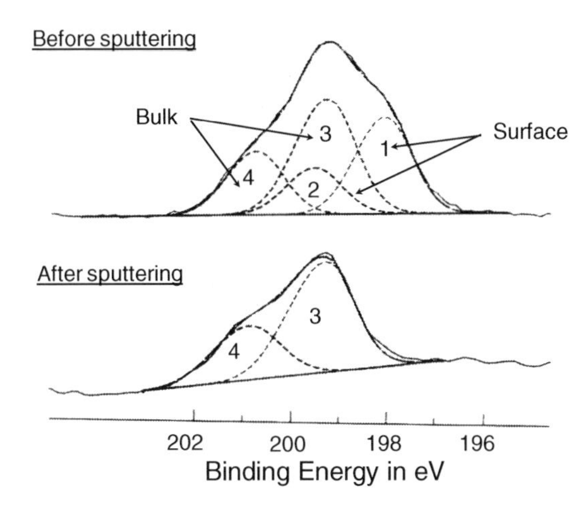

Fig. 10.41 XPS spectra of chloride on aluminum for a sample polarized at −0.750 V vs. SCE [31]. (*Top*) Before sputtering. (*Bottom*) After sputtering. Peaks 1 and 2 are $2p_{3/2}-2p_{1/2}$ doublets, as are peaks 3 and 4. The fact that the doublets 1 and 2 can be removed by sputtering shows that this doublet is due to surface chloride, while the remaining doublet 3 and 4 is due to chloride in the bulk of the oxide film. Reproduced by permission of ECS – The Electrochemical Society

so that the higher energy species is related to chloride contained within the interior of the oxide film. Additional evidence for chloride entry into the oxide film on aluminum has been obtained using X-ray absorption spectroscopy (XAS) [31, 37].

In addition, there is evidence from several studies using XPS or radiotracer techniques that the total amount of chloride (surface plus bulk) increases with increasing electrode potential en route to the pitting potential [31, 37, 67, 70]. Figure 10.42 shows that the uptake of chloride ion by the passive film on aluminum increases up to the pitting potential [31]. At potentials above (more positive than) the pitting potential, the amount of Cl^- in the passive film decreases as chloride ions are expelled from the ruptured pit [37, 70].

Thus, it is clear that chloride ions penetrate the oxide film on aluminum, although the exact mechanism by which this occurs is not known with certainty. The various possibilities, which have been mentioned earlier, include transport of chloride ions through the oxide film by way of oxygen vacancies, transport of chloride ions through water channels in the film, localized film dissolution or thinning, or penetration through cracks or flaws in the film.

The propagation of corrosion pits on aluminum occurs with the formation and rupture of blisters at the aluminum/oxide interface [68, 71–73]. Figure 10.43 shows such blisters in several stages of development. Figure 10.43(a) shows a blister in an early stage formed beneath the oxide film due to electrochemical reactions which are occurring at the oxide/metal interface. Figure 10.43(b) shows a blister with a surface crack. Figure 10.43(c) shows a blister in which corrosion has occurred along its periphery due to changes in the local solution chemistry which occur due to acid hydrolysis reactions. This blister also contains surface cracks due to the mounting pressure of hydrogen within its interior. As propagation proceeds, and when the hydrogen pressure within the blister becomes sufficiently large, the blister ruptures as shown in Fig. 10.43(d); and the corrosion pit is opened to the bulk electrolyte.

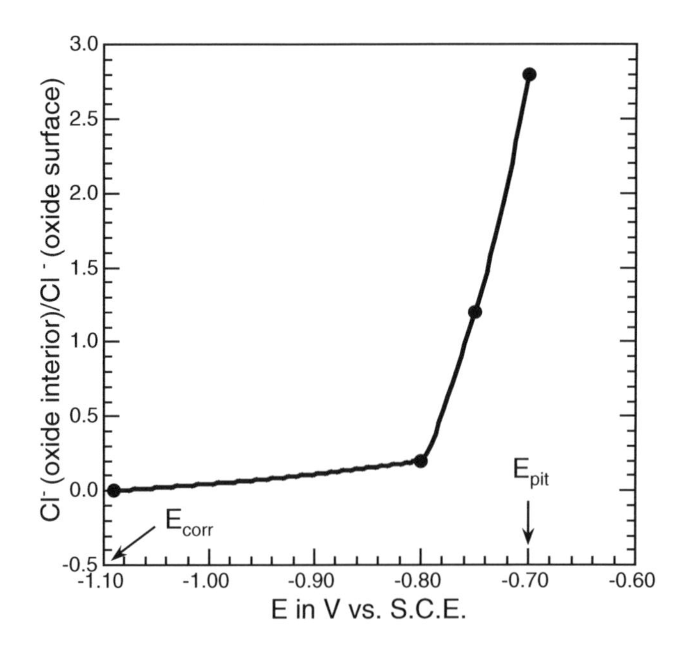

Fig. 10.42 XPS measurement of the uptake of Cl^- by the passive film on aluminum as a function of electrode potential [31]

Such blisters were first observed on aluminum by Bargeron and Givens [71, 72], who also found the presence of chloride ions within blisters. Similar blisters have also been found on pits formed on various aluminum alloys immersed in natural seawater [73].

A number of investigators have observed the evolution of gas from such localized corrosion cells on aluminum and have determined that hydrogen is the gas which is evolved [74, 75]. This is consistent with observations that the interior of corrosion pits on aluminum are locally acidic, as is discussed below.

Occluded Corrosion Cells

Active crevices, corrosion pits, and stress-corrosion cracks each develop local internal acidities even when the bulk electrolyte is neutral or alkaline. For all three forms of localized corrosion, a special restrictive geometry "seals off" an active local corrosion cell by limiting the exchange of local and bulk electrolytes. In pitting corrosion, a cap of porous corrosion products acts as the barrier; within crevices and stress-corrosion cracks, the long narrow diffusion path limits the access to bulk electrolyte; see Fig. 10.44.

Thus, the common features of restrictive geometry and local acidity have led to the point of view that all three forms of localized corrosion are different geometric manifestations of the same general phenomenon of "occluded cell corrosion" [76]; that is, corrosion pits are "microcrevices" or crevices are "macropits". (It should be remembered, however, that the mechanisms of initiation are different for pitting and crevice corrosion.)

Brown [76] thus suggested that information gained in studying one form of localized corrosion can be validly transferred to other forms of localized attack, and that in addition, scaled-up

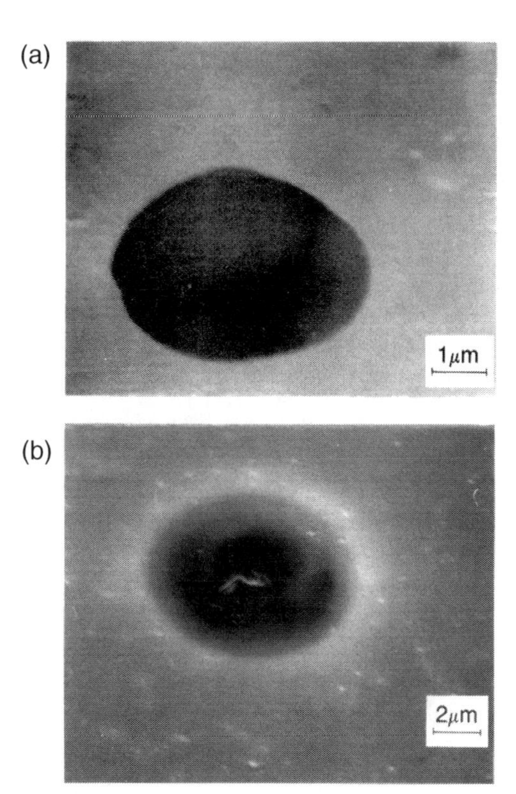

Fig. 10.43 Progressive growth of oxide blisters on aluminum surface alloys prepared by ion implantation [68] (see Chapter 16 for more on ion implantation). (**a**) Blister formation with no visible signs of cracks on the blister. (**b**) Cracks are visible on the blister. (**c**) The blister is cracked and crystallographic etching is observed under the blister around its periphery. (**d**) The blister is ruptured and is opened to the bulk electrolyte. Photographs courtesy of P. M. Natishan and reproduced by permission of ECS – The Electrochemical Society

experimental models may be used to study the general phenomenon of occluded cell corrosion. Figure 10.45 shows an example of an experimental device used to study occluded corrosion cell [77].

Occluded Corrosion Cell (OCC) on Iron

M. Pourbaix [51, 52] has suggested that when pitting, crevice corrosion, or stress-corrosion cracking occurs on iron in chloride solutions, the solution within the active cavities becomes saturated with respect to ferrous chloride ($FeCl_2$) and also contains magnetite (Fe_3O_4). The aqueous solution would then exist in the presence of three solid phases-: $FeCl_2{\cdot}4H_2O$, Fe_3O_4, and Fe. Pourbaix's sketch of a corrosion pit or stress-corrosion crack on iron is shown in Fig. 10.46. Based on this model for the OCC on iron, the Pourbaix diagram for iron within an OCC is given in Fig. 10.47. The equilibrium E and pH within the OCC are given at the "triple point" where $FeCl_2.4H_2O$, Fe_3O_4, and Fe co-exist. The result is $E = -0.368$ V vs. SHE (-0.590 V vs. Ag/AgCl) and pH 4.8.

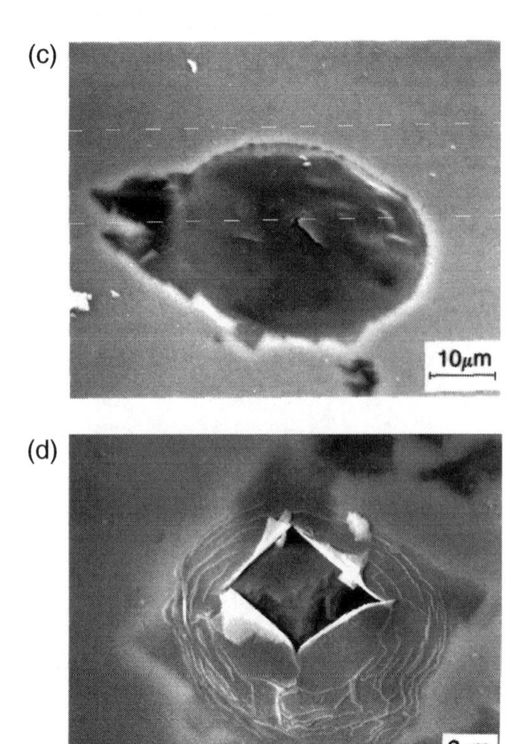

Fig. 10.43 (continued)

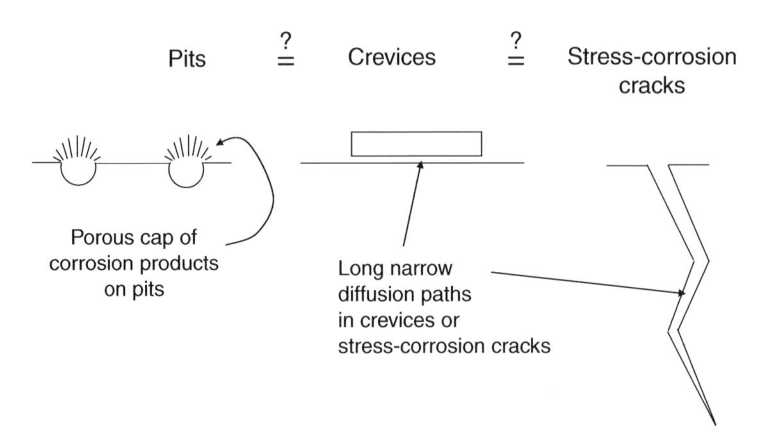

Fig. 10.44 Schematic illustration showing geometric similarities between pitting, crevice corrosion, and stress-corrosion cracking

These calculated results have been verified in various experimental studies, as shown in Table 10.6. For iron in crevices in various CrO_4^{2-}/Cl^- solutions, the internal electrode potential within an active crevice was always slightly more negative than −0.600 V vs. Ag/AgCl. Whenever there was inhibition of crevice corrosion of iron, the crevice potential was always more positive than

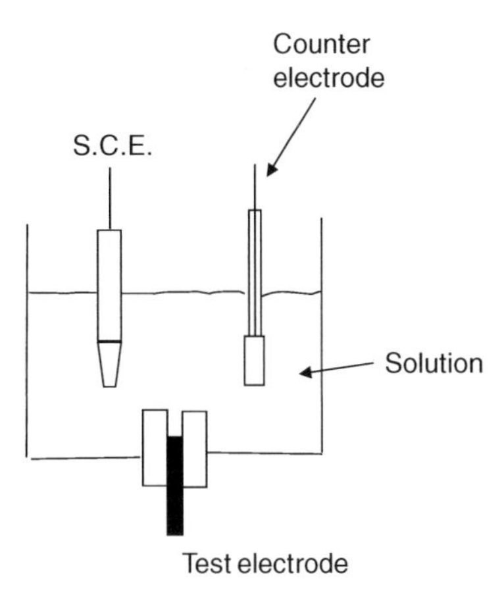

Fig. 10.45 One type of experimental electrochemical cell used to study artificial pits. Redrawn from [77]. The sample is a "lead-in-pencil" electrode prepared by mounting a 1 mm wire in a rod of epoxy to a depth of approximately 1 mm. Reproduced by permission of Elsevier Ltd

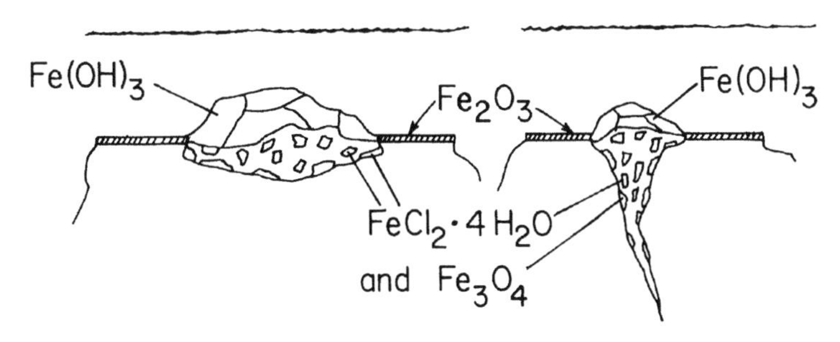

Fig. 10.46 Pourbaix's sketch of a corrosion pit or stress-corrosion crack on iron. Redrawn from [51, 52] by permission of © NACE International (1974)

−0.600 V vs. Ag/AgCl [13]. Thus, as seen in Table 10.6, experimentally measured electrode potentials within active crevices on iron agree with values calculated from thermodynamics or measured in scaled-up-occluded corrosion cells.

Occluded Corrosion Cells on Copper and Aluminum

Figure 10.48 is a schematic drawing of a corrosion pit formed on copper in cold tap water [51, 52]. This pit contains a layer of green malachite, $CuCO_3 \cdot Cu(OH)_2$, white crystals of cuprous chloride (CuCl), and a loose deposit of red cuprous oxide, Cu_2O. The solution at the bottom of the pit is in contact with Cu, Cu_2O, and CuCl. Pourbaix has calculated the thermodynamic conditions to be $E =$ +0.326 V vs. SHE and pH 2.45 [51, 52].

Using a scaled-up experimental OCC, the observed values for a pit interior were $E =$ +0.265 V vs. SHE and pH 3.5, in reasonable agreement with the thermodynamic model.

Descriptive models proposed for a corrosion pit on aluminum [51, 74, 78, 79] are similar to those for iron and copper in that the model consists of a porous cap of corrosion products over the pit

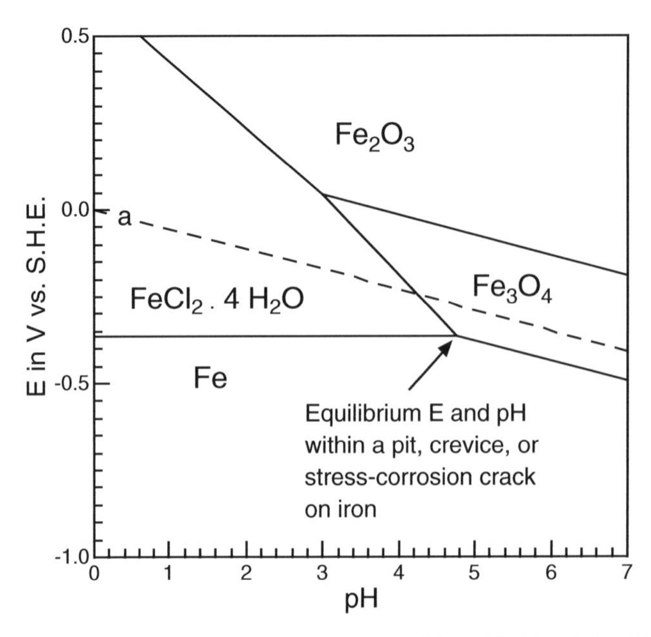

Fig. 10.47 The Pourbaix diagram for iron in a localized corrosion cell in a chloride solution [51, 52], as based on the model for a pit or stress-corrosion crack shown in Fig. 10.46. Reproduced by permission of © NACE International (1974)

Table 10.6 Calculated and experimental conditions within an occluded corrosion cell on iron

Source	*E* vs. Ag/AgCl	pH
M. Pourbaix [51, 52, 78, 79] Thermodynamic calculation for $Fe/Fe_3O_4/FeCl_2 . 4H_2O$	−0.59	4.8
Fujii [103] Iron in scaled-up-occluded corrosion cell in a slurry of Fe_3O_4 and saturated $FeCl_2 . 4H_2O$	−0.55 to −0.57	2.9−3.8
A. Pourbaix [104] Iron in scaled-up artificial crevice or pit in various chloride solutions	−0.57 to −0.60	3.1−4.3
Petersen et al. [105] Synthetic pit on steel in natural seawater	∼ −0.66	2.5−5.0
McCafferty [13] Real-sized crevices in various CrO_4^{2-}/Cl^- solutions	−0.62 to −0.66	4.2−5.5

and an acidic chloride solution within. The local pH and electrode potential have been measured for crevice corrosion [80, 81], stress-corrosion cracking [82, 83], and for a scaled-up model of occluded corrosion cells [84] (i.e., an artificial pit).

Experimental values for the internal *E* and pH in each case are superimposed on the conventional Pourbaix diagram for aluminum in Fig. 10.49. Each of the data sets for the various localized

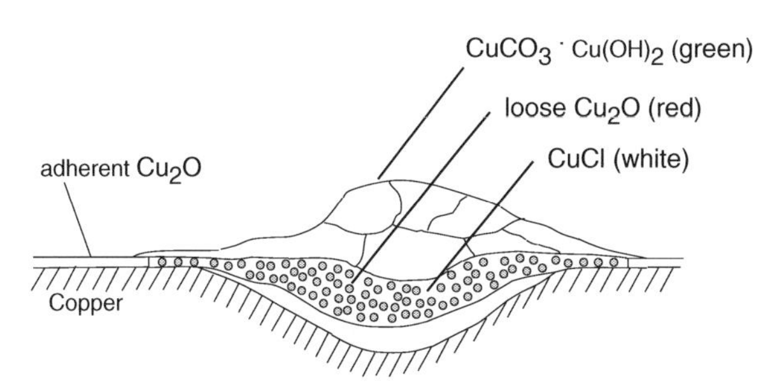

Fig. 10.48 Sketch of a copper pit in cold tap water. Redrawn from [51, 52]. Reproduced by permission of Elsevier Ltd

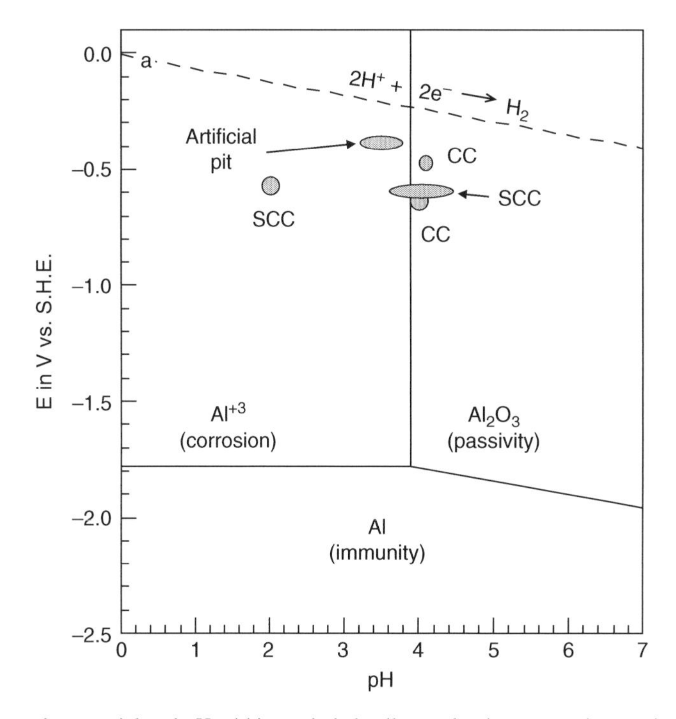

Fig. 10.49 The electrode potential and pH within occluded cells on aluminum superimposed on a partial Pourbaix diagram for aluminum. CC refers to crevice corrosion [80, 81] and SCC to stress-corrosion cracking [82, 83]. The results for an artificial pit on aluminum are due to Hagyard and Santhipillai [84]

corrosion cells either lie in the region of corrosion on the Pourbaix diagram for aluminum or else are very close to this region.

The experimental pH values within the localized corrosion cells range from 2 to 4.5 and experimental points for each of the various forms of localized attack lie below the "a" line for hydrogen evolution, showing that hydrogen evolution is thermodynamically possible in each of these forms of localized corrosion. Hydrogen bubbles have in fact been observed for aluminum in active crevices [80].

Differences Between Pitting and Crevice Corrosion

Despite similarities between pitting and crevice corrosion, there are significant differences between these two forms of localized attack. First, the electrode potential for crevice corrosion is more negative than the pitting potential, as shown for iron in Fig. 10.50. This is due in part to the fact that electrode potentials in Fig. 10.50 are measured within the crevice but outside the pit. Thus, the pitting potential is a mixed potential between events occurring on the passive surface and within the developing corrosion pit.

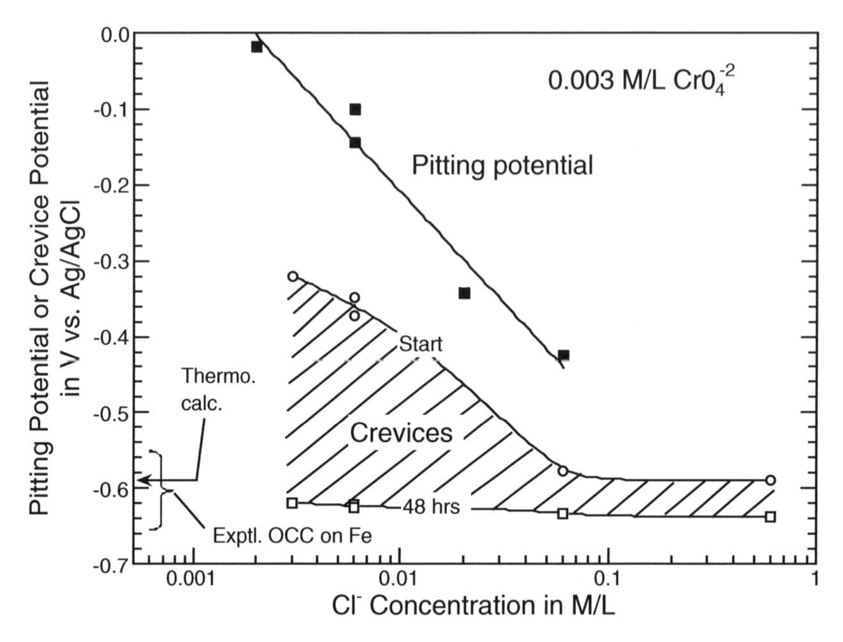

Fig. 10.50 Comparison of the pitting potential of open iron with the internal potential of iron within active crevices for various chloride concentrations in 0.003 M/L chromate

A second major difference is that the current density within a corrosion pit is much higher than that within a crevice. For example, the highest current densities for the iron crevice in Fig. 10.11 are about 0.06 mA/cm^2. By contrast, current densities within corrosion pits can be of the order of A/cm^2; see Problem 10.8.

Detection of Corrosion Pits

The ability to detect propagating corrosion pits on a metal surface has improved with advances in instrumentation. Figure 10.28 has shown that the distribution of pH and electrode potential above a propagating pit on iron, and similar scans were determined earlier by Rosenfeld and Danilov [85] for corrosion pits on an Fe-18 Cr-8 Ni steel. Scans such as these can be used to detect the physical location of corrosion pits on a metal surface, as discussed below.

From the potential field around a corrosion pit, Rosenfeld and Danilov were also able to determine the current distribution, because the current density, i, is proportional to the normal gradient of the potential E:

$$i = -\sigma \left(\frac{\delta E}{\delta z} \right) \tag{10}$$

where σ is the conductivity of the solution and z denotes the direction normal to the electrode surface; see Fig. 10.51. Isaacs [87] advanced the technique by scanning across a surface in the *XY* plane so as to produce traces of propagating pits. Later refinements [88] included a scanning vibrating electrode (SVE) which produces potential scans and also current scans by sampling the potential gradient normal to the surface. For example, Fig. 10.52 shows the distribution of current density obtained over active sites on an iron surface in a chloride solution [88].

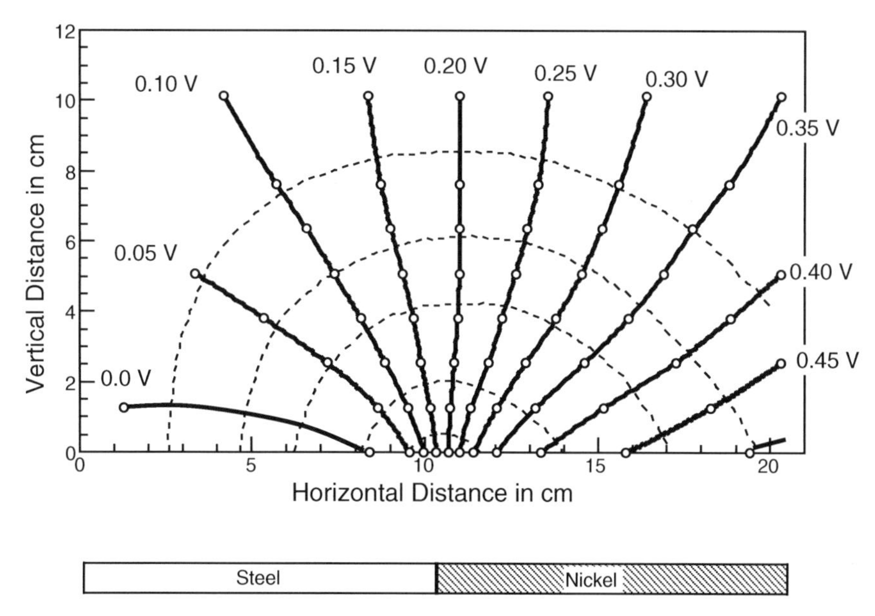

Fig. 10.51 Distribution of potential (*solid lines*) and current density (*dotted lines*) for a steel/nickel co-planar couple in tap water. (The lines of constant current density are at right angles to the equipotential lines.) Redrawn from [86] by permission of ECS – The Electrochemical Society

With advances in scanning probe techniques, increased attention has turned to the more difficult task of the detection of the initiation stage of pitting and of the imaging of sites which are pre cursors to pitting. In this regard, Smyrl and co-workers [89, 90] have successfully employed several new experimental scanning techniques. Smyrl et al. found that surface active sites corresponded to locations of pit formation (although not all active sites led to pits). Inclusions, second-phase particles, and other defects, such as grain boundaries, were among the sites of increased electrochemical activity.

Other advanced instrumental methods which have been used recently to study pit initiation include the use of the atomic force microscope (AFM) (discussed in Chapter 9) and the scanning Kelvin probe (SKP) (which measures changes in work function) [91, 92].

At present these newer scanning techniques are useful in laboratory research studies, and would be welcome inspection tools should they be transitioned into practice. At present there is a need for reliable technique for the detection of pit initiation sites on metal structures or electronic components.

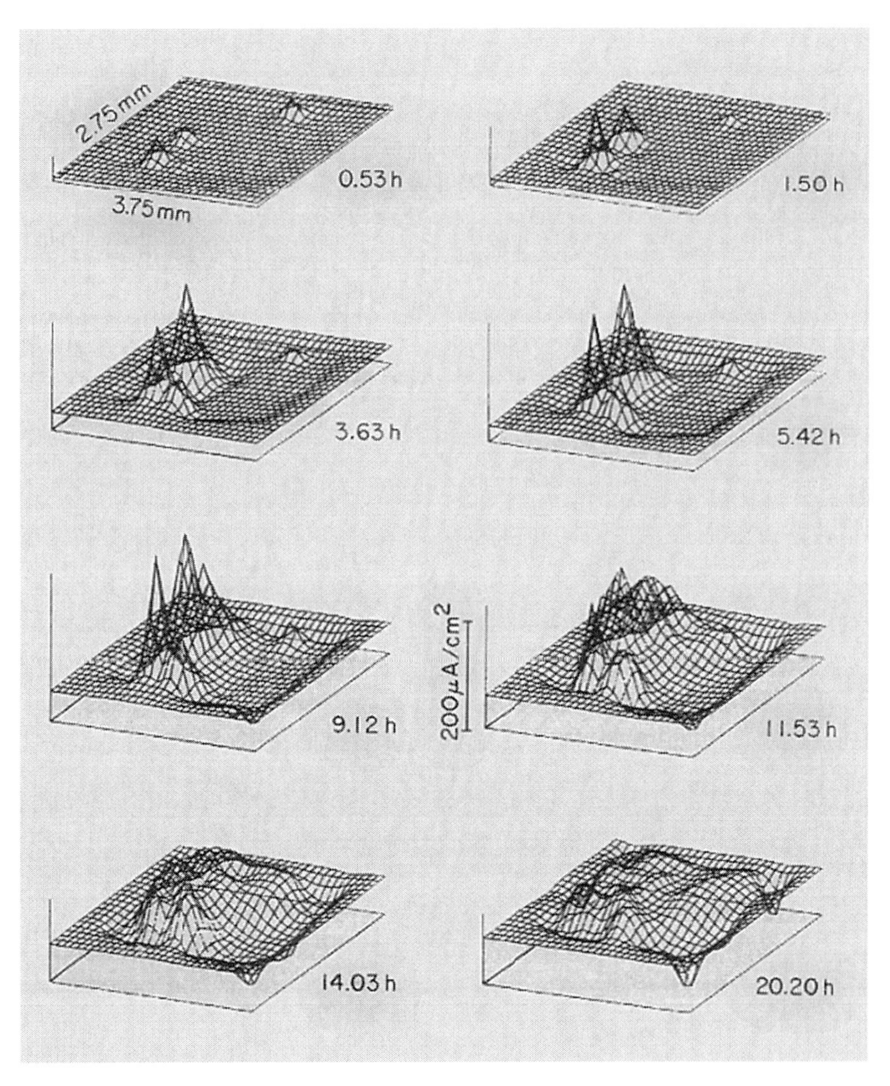

Fig. 10.52 The distribution of current density over an iron surface showing the location and progressive growth of a pit in a solution of 1 mM NaCl plus 1 mM Na_2SO_4, as determined using a vibrating probe electrode [88]. Reproduced by permission of © NACE International (1990)

Problems

1. In what ways are crevice corrosion and pitting similar? In what ways are they different?
2. (a) Name and discuss two different ways to test the resistance of an alloy to crevice corrosion.
   (b) Name and discuss two different ways to test the resistance to pitting corrosion.
3. (a) Sketch an Evans diagram showing the approximate shape of anodic and cathodic polarization curves for a low alloy steel which undergoes general corrosion in seawater. Assume that anodic dissolution is under activation control and that the cathodic half-cell reaction is controlled by oxygen diffusion.

(b) Show that the *total weight loss* for the low alloy steel in a crevice is the same for the following two cases:

Crevice area	External cathode area
1.0 cm^2	10.0 cm^2
2.0 cm^2	10.0 cm^2

(c) Which of the two cases above will have the greater weight loss *per unit area*?

4. Sketch the approximate shapes of cyclic anodic polarization curves to explain the variation in the protection potential, as shown in Fig. 10.31 for type 430 stainless steel. The corrosion potential is −0.50 V vs. SCE and the pitting potential is −0.100 V vs. SCE.
5. Make a sketch analogous to Fig. 10.46 for a corrosion pit on aluminum in a chloride solution.
6. The following laboratory data were taken for the anodic polarization of aluminum in 0.1 M NaCl. The sample size is 8.0 cm^2. Plot these data and determine the critical pitting potential.

E in V vs. SCE	Current in μA
−1.30	0.40
−1.20	0.58
−1.10	0.68
−1.00	0.70
−0.900	0.78
−0.800	0.71
−0.775	0.75
−0.750	0.76
−0.725	0.79
−0.700	1.20
−0.685	4.00
−0.670	104.
−0.655	750.
−0.640	2100.

7. What is the current density within a corrosion pit on aluminum if the pit size is given in Fig. 10.43(d)? Assume that the pit is hemispherical in shape and that the total current within the corroding pit is given by the maximum current in the data set above.
8. (a) The following data are for the pitting of pure aluminum in chloride solutions for a short-term immersion of 24 h [67]. What is the pitting potential for aluminum in 0.6 M NaCl (the chloride ion concentration in seawater)?

$[Cl^-]$	Pitting potential, E_{pit}, vs. SCE
0.01 M	−0.620
0.1 M	−0.685
0.5 M	−0.730
0.5 M	−0.725
1.0 M	−0.760
1.0 M	−0.745

(b) How does the value for pure aluminum in the short-term immersion test compare with the pitting potential of aluminum alloys as determined from the long-term practical immersion tests in Table 10.5?

(c) If there is a difference between the two values, explain why.

9. An Fe-12 Cr alloy was observed to have the following characteristic potentials in 0.1 M NaCl [55].

pH	E_{corr}	E_{Flade}	E_{pit}	E_{prot}
4.6	−0.580	−0.420	−0.040	−0.350
5.4	−0.610	−0.470	0.0100	−0.220
6.9	−0.700	−0.550	0.0100	−0.220
8.9	−0.740	−0.600	0.175	−0.440
10.7	−0.800	−0.445	0.210	−0.450

(a) Draw the corresponding experimental Pourbaix diagram and label regions of immunity, general corrosion, perfect passivity, imperfect passivity, and pitting. (b) Over what range of electrode potentials is it safe to use Fe-12 Cr in 0.1 M NaCl at pH 8?

10. Shown below are experimentally determined anodic polarization curves in 1 M H_2SO_4 and 1 M HCl for three alloys [93] (reproduced by permission of NACE International). These alloys are the conventional 304 stainless steel and two possible replacements, alloy A and alloy B:

304:	18% Cr, 8% Ni, balance Fe
Alloy A:	12% Cr, balance Fe
Alloy B:	12% Cr, 10% Ni, 1.5 % Si, 2% Mo, balance Fe

Based on these polarization curves alone, which alloy or alloys can be used instead of 304 stainless steel? Why?

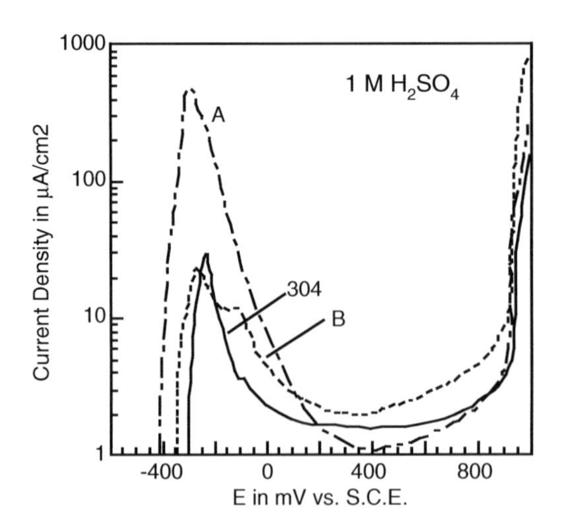

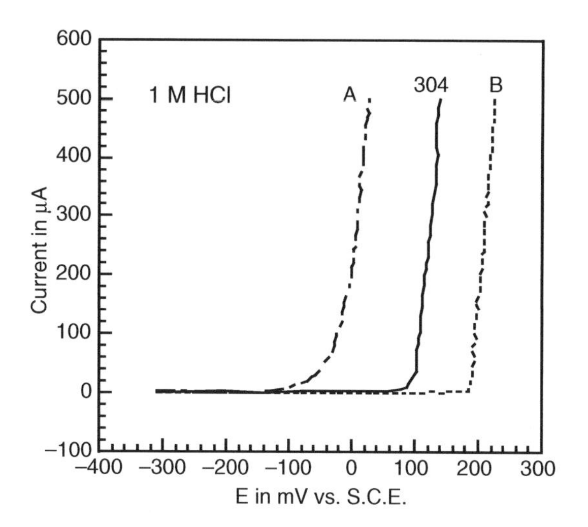

Reproduced by permission of © NACE International (1979).

11. Explain why the pitting potential decreases with increasing temperature.

References

1. F. J. Martin, K. E. Lucas, and E. A. Hogan, *Rev. Sci. Instruments*, *73*, 1273 (2002).
2. I. L. Rosenfeld and I. K. Marshakov, *Corrosion*, *20*, 115t (1964).
3. I. B. Ulanovski and Y. M. Korovin, *J. Appl. Chem. USSR*, *35*, 1683 (1962).
4. E. McCafferty, *J. Electrochem. Soc.*, *121*, 1007 (1974).
5. M. H. Peterson, T. J. Lennox, Jr., and R. E. Groover, *Mater. Protect. Perform.*, *9* (1), 23 (1970).
6. A. Turnbull, *Corros. Sci.*, *23*, 833 (1983).
7. F. D. Bogar and C. T. Fujii, "Solution Chemistry in Crevices on Fe-Cr Binary Alloys", NRL Report 7690, Naval Research Laboratory, Washington, DC, March 21, 1974
8. I. L. Rosenfeld in "Localized Corrosion", R. W. Staehle, B. F. Brown, J. Kruger, and A. Agrawal, Eds., p. 373, National Association of Corrosion Engineers, Houston, TX (1974).
9. T. Suzuki, M. Yamabe, and Y. Kitamura, *Corrosion*, *29*, 18 (1973).
10. J. Mankowski and Z. Szkarlarska-Smialowska, *Corros. Sci.*, *15*, 493 (1975).
11. M. G. Fontana and N. D. Greene, "Corrosion Engineering", p. 43, McGraw-Hill, New York (1978).
12. R. M. Kain, *J. Testing Evaluation*, *18*, 309 (1990).
13. E. McCafferty, *J. Electrochem. Soc.*, *126*, 385 (1979).
14. Standard Test Method G-48-97, "1998 Annual Book of ASTM Standards", ASTM, West Conshohocken, PA (1998).
15. K. Sotoudeh, T. H. Nguyen, R. T. Foley, and B. F. Brown, *Corrosion*, *37*, 358 (1981).
16. E. McCafferty, F. D. Bogar, E. D. Thomas, C. A. Creegan, K. E. Lucas, and A. I. Kaznoff, *Corrosion*, *53*, 755 (1997).
17. R. S. Lillard, M. P. Jurinski, and J. R. Scully, *Corrosion*, *50*, 251 (1994).
18. O. B. Ellis and F. L. LaQue, *Corrosion*, *7*, 362 (1951).
19. T. J. Lennox, Jr. and M. H. Peterson in "Localized Corrosion", R. W. Staehle et al., Eds., p. 173, National Association of Corrosion Engineers, Houston, TX (1974).
20. V. R. Pludek, "Design and Corrosion Control", p. 173, John Wiley and Sons, New York, NY (1977).
21. A. J. Sedriks, "Corrosion of Stainless Steels", p. 201, Wiley-Interscience, New York (1996).
22. E. McCafferty, P. P. Trazskoma, and P. M. Natishan in "Advances in Localized Corrosion", H. S. Isaacs, U. Bertocci, J. Kruger, and S. Smialowska, Eds., p. 181, National Association of Corrosion Engineers, Houston, TX (1990).

23. J. R. Galvele in "Passivity of Metals", R. P. Frankenthal and J. Kruger, Eds., p. 285, The Electrochemical Society, Princeton, NJ (1978).
24. H. P. Leckie and H. H. Uhlig, *J. Electrochem. Soc.*, *113*, 1262 (1966).
25. R. E. Groover, T. J. Lennox, Jr., and M. H. Peterson, *Mater. Protect.*, *8* (11), 25 (1969).
26. Z. A. Foroulis and M. J. Thubrikar, *J. Electrochem. Soc.*, *122*, 1296 (1975).
27. H. Böhni and H. H. Uhlig, *J. Electrochem. Soc.*, *116*, 906 (1969).
28. S. Szklarska-Smialowski in "Advances in Localized Corrosion", H. S. Isaacs, U. Bertocci, J. Kruger, and S. Smialowska, Eds., p. 41, National Association of Corrosion Engineers, Houston, TX (1990).
29. H.-H. Strehblow, *Werkstoffe u. Korros.*, *27*, 792 (1976).
30. G. S. Frankel, *J. Electrochem. Soc.*, *145*, 2186 (1998).
31. P. M. Natishan, W. E. O′Grady, E. McCafferty, D. E. Ramaker, K. Pandya, and A. Russell, *J. Electrochem. Soc.*, *146*, 1737 (1999).
32. L. A. Krebs and J. Kruger in "Passivity of Metals and Semiconductors", M. B. Ives, J. L. Luo, and J. R. Rodda, Eds., p. 561, The Electrochemical Society, Pennington, NJ (2001).
33. D. D. Macdonald, *J. Electrochem. Soc.*, *139*, 3434 (1992).
34. J. O′M. Bockris and L. V. Minevski, *J. Electroanal. Chem.*, *349*, 375 (1993).
35. J. O′M. Bockris and Y. K. Kang, *J. Solid State Electrochem.*, *1*, 17 (1997).
36. T. H. Nguyen and R. T. Foley, *J. Electrochem. Soc.*, *127*, 2563 (1980).
37. S. Yu, W. E. O′Grady, D. E. Ramaker, and P. M. Natishan, *J. Electrochem. Soc.*, *147*, 2952 (2000).
38. W. Khalil, S. Haupt, and H.-H. Strehblow, *Werkstoffe u. Korros.*, *36*, 16 (1985)
39. N. Sato, *Electrochim. Acta*, *16*, 1683 (1971).
40. W. A. Badawy, M. M. Ibrahim, M. M. Abou-Romia, and M. S. El-Bosiouny, *Corrosion*, *42*, 324 (1986).
41. I. L. Rosenfeld in "Proceedings of the Fifth International Congress on Metallic Corrosion", p. 53, NACE, Houston, TX (1974).
42. G. Butler, P. Stretton, and J. G. Beynon, *Br. Corros. J.*, *7*, 168 (1972).
43. B. E. Wilde and E. Williams, *Electrochim. Acta.*, *16*, 1971 (1971).
44. N. Pessall and C. Liu, *Electrochim. Acta*, *16*, 1987 (1971).
45. G. S. Frankel, L. Stockert, F. Hunkeler, and H. Boehni, *Corrosion*, *43*, 429 (1987).
46. I. L. Rosenfeld, I. S. Danilov, and R. N. Oranskaya, J. Electrochem. *Soc.*, *125*, 1729 (1978).
47. D. A. Jones and B. E. Wilde, *Corros. Sci.*, *18*, 631 (1978)
48. T. Hakkarainen in "Proceedings of the Eighth International Congress on Metallic Corrosion", Vol. 1, p. 157, DECHEMA, Frankfurt am Main, (1981).
49. R. C. Newman and H. S. Isaacs in "Passivity of Metals and Semiconductors", M. Froment, Ed., p. 269, Amsterdam (1983).
50. E. McCafferty and N. Hackerman, *J. Electrochem. Soc.* *119*, 999 (1972).
51. M. Pourbaix, *Corros. Sci.*, *12*, 161 (1970).
52. M. Pourbaix in "Localized Corrosion", R. W. Staehle, B. F. Brown, J. Kruger, and A. Agrawal, Eds., p. 12, National Association of Corrosion Engineers, Houston, TX (1974).
53. R. A. Bonewitz and E. D. Verinck, Jr., *Mater. Perform.*, *14* (8), 16 (1975).
54. K. K. Starr, E. D. Verinck, Jr., and M. Pourbaix, *Corrosion*, *32*, 47 (1976).
55. E. D. Verinck, Jr., and M. Pourbaix, *Corrosion*, *27*, 495 (1971).
56. E. D. Verinck, Jr., K. K. Starr, and J. M. Bowers, *Corrosion*, *32*, 60 (1976).
57. A. J. Sedriks in "Corrosion of Stainless Steels", p. 123, Wiley-Interscience, New York (1996).
58. C. R. Clayton and I. Olejford in "Corrosion Mechanisms in Theory and Practice", P. Marcus and J. Oudar, Eds, p. 175, Marcel Dekker, New York, NY (1995).
59. Z. Szklarska-Smialowska and E. Lunarska, *Werkstoffe u. Korros.*, *32*, 478 (1981).
60. A. J. Sedriks, *Intl. Met. Rev.*, *28*, 295 (1983).
61. A. J. Sedriks in A. J. Sedriks, "Corrosion of Stainless Steels", p. 24, Wiley-Interscience, New York (1996).
62. E. McCafferty and P. G. Moore, *J. Electrochem. Soc.*, *133*, 1090 (1986).
63. B. Vuillemin, X. Phillipe, R. Oltra, V. Vignal, L. Coudreuse, L. C. Dufour, and E. Finot, *Corros. Sci.*, *45*, 1143 (2003).
64. M. P. Ryan, D. E. Williams, R. J. Chater, B. M. Hutton, and D. S. McPhail, *Nature*, *415*, 770 (2002).
65. K. D. Efird and G. E. Moller, *Mater. Perform.*, *18* (7), 34 (1979).
66. R. J. Brigham, *Corrosion*, *30*, 396 (1974).
67. E. McCafferty, *Corros. Sci.*, *45,* 1421 (2003).
68. P. M. Natishan and E. McCafferty, *J. Electrochem. Soc.*, *136*, 53 (1989).
69. C. J. Lin, R. G. Du, and T. Nguyen, *Corrosion*, *56*, 41 (2000).

70. L. Tomcsanyi, K. Varga, I. Bartik, G. Horanyi, and E. Maleczki, *Electrochim. Acta*, *34*, 855 (1989).
71. C. B. Bargeron and R. B. Givens, *J. Electrochem. Soc.*, *124*, 1845 (1977).
72. C. B. Bargeron and R. B. Givens, *Corrosion*, *36*, 618 (1980).
73. J. Perkins, J. R. Cummings, and K. J. Graham, *J. Electrochem. Soc.*, *129*, 137 (1982).
74. W. Hübner and G. Wranglen in “Current Corrosion Research in Scandinavia”, p. 60, Kemian Keskusliitto, Helsinki, Finland (1965).
75. C. B. Bargeron and R. B. Givens, *J. Electrochem. Soc.*, *127*, 2528 (1980).
76. B. F. Brown, *Corrosion*, *26*, 249 (1970).
77. R. C. Newman, *Corros. Sci.*, *25*, 341 (1985).
78. M. Pourbaix in “The Theory of Stress Corrosion Cracking in Alloys”, J. C. Scully, Ed., NATO, Brussels, Belgium (1971).
79. M. Pourbaix in Localized Corrosion”, R. W. Staehle, B. F. Brown, J. Kruger, and A. Agrawal, Eds., p. 12, National Association of Corrosion Engineers, Houston, TX (1974).
80. D. W. Siitari and R. C. Alkire, *J. Electrochem. Soc.*, *129*, 481 (1982).
81. N. J. H. Holroyd, G. M. Scamans, and R. Hermann in “Corrosion Chemistry within Pits, Crevices and Cracks”, A. Turnbull, Ed., p. 495, Her Majesty′s Stationary Office, London (1987).
82. J. A. Davis in “Localized Corrosion”, R. W. Staehle, B. F. Brown, J. Kruger, and A. Agrawal, Eds., p. 168, National Association of Corrosion Engineers, Houston, TX (1974).
83. O. V. Kurov and R. K. Melekhov, *Protection Metals*, *15*, 249 (1979).
84. T. Hagyard and J. R. Santhipillai, *J. Appl. Chem.*, *9*, 323 (1959).
85. I. L. Rosenfeld and I. S. Danilov, *Corros. Sci.*, *7*, 129 (1967).
86. H. R. Copson, *Trans. Electrochem. Soc.*, *84*, 71 (1943).
87. H. S. Isaccs in “Localized Corrosion”, R. W. Staehle, B. F. Brown, J. Kruger, and A. Agrawal, Eds., p. 158, National Association of Corrosion Engineers, Houston, TX (1974).
88. H. S. Isaacs in “Advances in Localized Corrosion”, H. S. Isaacs, U. Bertocci, J. Kruger, and S. Smialowska, Eds., p. 221, National Association of Corrosion Engineers, Houston, TX (1990).
89. J. P. H. Sukamto, W. H. Smyrl, N. Casillas, M. Al-Odan, P. James, W. Jin, and L. Douglas, *Mater. Sci., Eng.*, *A198*, 177 (1995).
90. L. F. Garfias-Mesias and W. H. Smyrl, *J. Electrochem. Soc.*, *146*, 2495 (1999).
91. P. Leblanc and G. S. Frankel, *J. Electrochem. Soc.*, *149*, B239 (2002).
92. F. Andreatta, H. Terryn, and J. H. W. de Wit, *Electrochim. Acta*, *49*, 2851 (2004).
93. Y. C. Chen and J. R. Stephens, *Corrosion*, *35*, 443 (1979)
94. M. G. Alvarez and J. R. Galvele, *Corrosion*, *32*, 285 (1976).
95. P. M. Natishan, E. McCafferty, and G. K. Hubler, *J. Electrochem. Soc.*, *135*, 321 (1988).
96. P. P. Trzaskoma, E. McCafferty, and C. R. Crowe, *J. Electrochem. Soc.*, *130*, 1804 (1983).
97. H.-H. Strehblow and B. Titze, *Corros. Sci.*, *17*, 461 (1977).
98. M. G. Alvarez and J. R. Galvele, *Corros. Sci.*, *24*, 27 (1984).
99. Y. F. Wang , C. R. Clayton, G. K. Hubler, W. H. Lucke, and J. K. Hirvonen, *Thin Solid Films*, *63*, 11 (1978).
100. J. G. N. Thomas and A. K. Tiller, *Br. Corros. J.*, *7*, 256 91972).
101. G. Cragnolino and J. R. Galvele in “Passivity of Metals”, R. P. Frankenthal and J. Kruger, Eds, p. 1053, The Electrochemical Society, Princeton, NJ (1978).
102. I. Dugdale and J. B. Cotton, *Corros. Sci.*, *4*, 397 (1964).
103. C. T. Fujii, “Electrochemical and Chemical Aspects of Localized Corrosion”, Rapport Technique 213, CEBELCOR, Belgium (1974).
104. A. Pourbaix, *Corrosion*, *27*, 449 (1971).
105. C. W. Petersen, G. C. Soltz, and K. Mairs, *Corrosion*, *30*, 366 (1974).

Chapter 11
Mechanically Assisted Corrosion

Introduction

In the previous chapter we discussed two forms of localized corrosion – crevice corrosion and pitting. This chapter considers five more forms of localized corrosion, and these have the added common feature that they are assisted by mechanical processes. These five forms of mechanically assisted localized corrosion are stress-corrosion cracking, corrosion fatigue, cavitation corrosion, erosion corrosion, and fretting corrosion.

Stress-corrosion cracking is the cracking of a metal or alloy by the combined action of stress and the environment. Stress-corrosion cracking (SCC) can occur in stressed structures such as bridges and support cables, aircraft, pressure vessels, pipelines, and turbine blades, to name a few instances. Figure 11.1 shows the failure of a high strength rocket motor case due to stress-corrosion cracking in tap water [1].

During stress-corrosion cracking, most of the metal or alloy is virtually unattacked, as shown schematically in Fig.11.2 [2]. However, the presence of stress-corrosion cracks reduces the cross-section of the metal capable of carrying a load, as shown in Fig. 11.2. In the case of general corrosion, the same reduction in effective thickness due to the formation of rust layers may take years. Instances of stress-corrosion cracking are especially serious when they involve critical structures such as bridges or aircraft as described in Chapter 1, because they involve the loss of life or personal injury.

The study of stress-corrosion cracking (SCC) requires an interdisciplinary approach involving electrochemistry, materials science, and mechanics, as suggested in Fig. 11.3. Some of the variables which affect SCC within each of these disciplines are also given in Fig. 11.3.

Corrosion fatigue is the cracking of a metal or alloy due to the combined action of a repeated cyclic stress and a corrosive environment. Corrosion fatigue can occur in aircraft wings, in bridges, and in offshore platforms which are impacted by ocean waves, and in vibrating machinery. In addition, stainless steel and titanium alloys used as orthopedic implants (for knee and hip replacements) can also undergo corrosion fatigue under cyclic stresses in the electrolyte found within the human body.

Cavitation corrosion is the combined mechanical and corrosion attack caused by the collapse and impingement of vapor bubbles in a liquid near a metal surface. Cavitation corrosion can occur in ship propellers, within pumps, on turbine blades, on hydrofoils, and on other surfaces where there is a high-velocity fluid flow and where pressure changes are encountered. Figure 11.4 shows an example of cavitation corrosion.

Two additional forms of mechanically assisted corrosion are *erosion corrosion* and *fretting corrosion*. In erosion corrosion, the mechanical effect is provided by the movement of a corrosive liquid against the metal surface (without the need for cavitating bubbles). In fretting corrosion, the

E. McCafferty, *Introduction to Corrosion Science*, DOI 10.1007/978-1-4419-0455-3_11,

Fig. 11.1 Fracture of a high-strength steel rocket case due to stress-corrosion cracking [1]. The *arrows* indicate the location of the stress-corrosion crack

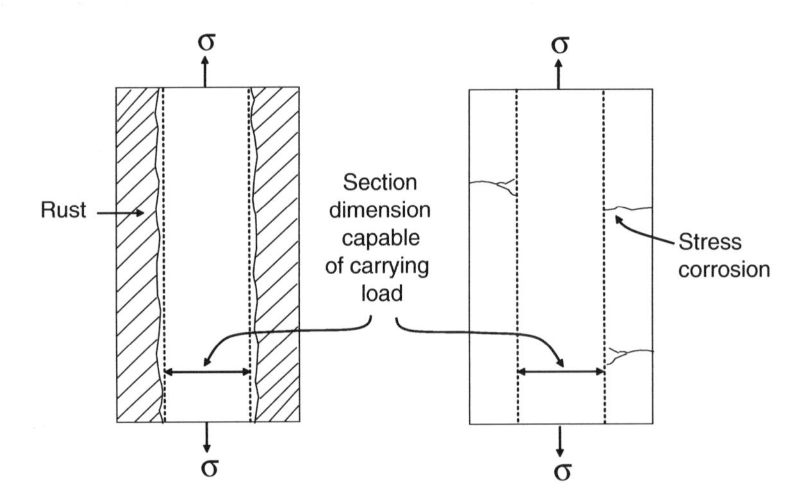

Fig. 11.2 Reduction in the effective cross section of a metal due to general corrosion or the presence of stress-corrosion cracks. The *arrows* show the direction of the applied stress, σ. Redrawn from Staehle [2] by permission of © NACE International 1969

mechanical effect is due to the abrasive wear of two metal surfaces sliding or vibrating past each other. Erosion corrosion can occur in various types of equipment exposed to fast-moving liquids. These include piping systems, bends, elbows, valves, and pumps. Polymer, glass, and metal surfaces on moving aircraft can be damaged by rainwater erosion [3]. Solids suspended in a liquid can also cause erosion corrosion, as in coal slurry pipelines. Fretting corrosion can occur in loaded interfaces which move past each other, such as in vibrating machinery, connecting rods, or springs. Fretting corrosion can also be a problem with orthopedic implants placed in the human body. Fretting corrosion is also called wear corrosion or tribocorrosion.

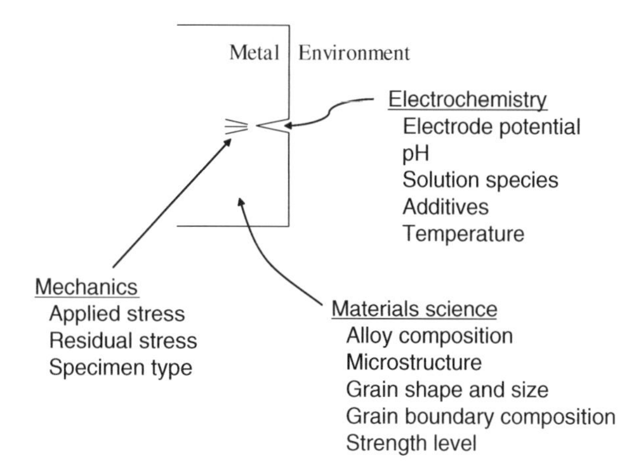

Fig. 11.3 Disciplines used in studying stress-corrosion cracking and variables which affect SCC within each discipline

Fig. 11.4 Cavitation damage on the outer edge of a propeller on a personal watercraft. Figure courtesy of Erik Axdahl, who took this photograph while he was an undergraduate research assistant at the University of Minnesota and who is now a Graduate Research Assistant at the Georgia Institute of Technology

Each of these types of mechanically assisted localized corrosion processes is discussed in this chapter, although most of this chapter deals with stress-corrosion cracking.

Stress-Corrosion Cracking

Mechanical Metallurgy

Before discussing stress-corrosion cracking, it is useful to briefly review the basic principles of mechanical metallurgy [4]. If a load *P* is applied perpendicular to a solid body of cross section *A*, as in Fig. 11.5, the *stress* σ is given by:

$$\sigma = \frac{P}{A} \tag{1}$$

and has English units of pounds per square inch or kilopounds per square inch (ksi). In metric units, the units of stress are pascals or megapascals (MPa); see Table 11.1. If the applied stress causes separation of material across the plane in which it acts, the stress is called a *tensile stress*. (The opposite effect is the compression of material due to a *compressive stress*.) When a tensile stress is applied, the solid body undergoes a small deformation, and the *strain* ε is defined as the change in length Δl divided by the original length l_o,

$$\varepsilon = \frac{\Delta l}{l_o} \tag{2}$$

The relationship between the stress and strain is shown in Fig. 11.5. At low applied stresses, the stress and strain are linearly related in a region of the stress–strain curve called the *elastic region*. In the elastic region,

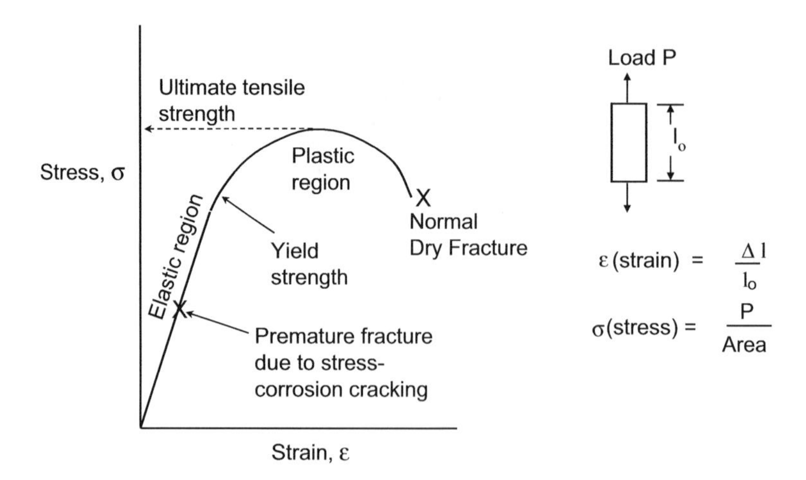

Fig. 11.5 Typical stress–strain curve for a metal

Table 11.1 Systems of units used in mechanics

Parameter	English units	Metric units
Force	Pounds[a] or kilopounds	$\frac{\text{kg m}}{\text{sec}^2}$ = netwtons (N)
Stress	Pounds/in.^{2} (psi) or kilopounds/in.^{2} (ksi)	$\frac{\text{N}}{\text{m}^2}$ = Pa
		$\frac{\text{MN}}{\text{m}^2}$ = MPa
Stress intensity factor, K_I	ksi $\sqrt{\text{in.}}$	MPa $\sqrt{\text{m}}$

[a]In English units, a pound is the force which imparts to a given mass called a "slug" an acceleration of 1 ft/s^2.
Useful conversion factors: 1 ksi = 6.89 MPa; 1 ksi $\sqrt{\text{in.}}$ = 1.10 MPa $\sqrt{\text{m}}$.

$$\sigma = E\varepsilon \tag{3}$$

where E is a proportionality constant called Young's modulus and is a property of the solid material. Equation (3) is called Hooke's law. In the elastic region the strains are reversible and if the load is removed, the material returns to its original length.

However, if material is loaded to higher stresses, the deformation becomes permanent or *plastic*, and Hooke's law no longer applies. The stress marking the beginning of plastic deformation is called the *yield stress* σ_Y (also called the *yield strength*), and the point where the material can support the highest load is called the *ultimate tensile stress* (also called the *ultimate tensile strength*). In Fig. 11.5, the material finally fractures at point X.

Characteristics of Stress-Corrosion Cracking

Figure 11.5 applies to the tensile testing of a solid metal in air or in the presence of an inert environment. The effect of an electrolyte is to cause the metal to fail prematurely at lower stresses due to stress-corrosion cracking, as indicated in Fig. 11.5.

The following are the most important characteristics of SCC [5]:

(1) *A tensile stress is required.* The tensile stress may be external, as with an applied load, or the stress may be a residual stress due to metal-working processes, such as machining or welding. The stress may also be self-generated, as may be caused by the wedging action of solid corrosion products within an emerging crack.
(2) *Metals or alloys which undergo SC usually have a good resistance to general corrosion.* For example, stainless steels, aluminum alloys, and titanium all undergo SCC but are usually resistant to general corrosion (in neutral solutions). One exception is that high-strength low-alloy steels may undergo both general corrosion and SCC at the same time.
(3) *Specific environments cause SCC in certain alloys.* For example, Cl^- solutions (but not NO_3^- solutions) cause SCC of type 304 stainless steels. On the other hand NO_3^- solutions (but not Cl^-solutions) cause SCC of mild steel. In addition, ammonia or amines are specific reagents for the SCC of brasses (Cu–Zn alloys). Some specific environments which cause SCC for various metals are listed in Table 11.2.

Table 11.2 Some examples of stress-corrosion cracking

Types of alloys	Environments
High-strength steels [33]	Seawater
	chloride solutions
	solutions containing H_2S
	NO_3^- solutions
Stainless steels [20]	Cl^-, Br^-, F^- solutions
	hydroxide solutions
	polythionic acids ($H_2S_xO_6$, $x = 3, 4, 5$)
	thiosulfate solutions
Aluminum alloys [24]	Chloride solutions
	most aqueous solutions
	organic liquids
Titanium alloys [1, 66, 37, 34]	Seawater
	Cl^-, Br^-, I^- solutions
	carbon tetrachloride and other organic solvents, including alcohols
	fuming nitric acid
Brass (Cu–Zn alloys) [41, 40]	Ammonia solutions
	amines
	citrate solutions
	tartrate solutions

(4) *The necessary corrodent species can be present in small concentrations.* For example, a few ppm of Cl^- in high-temperature water can cause SCC in certain stainless steels.
(5) *A threshold stress or stress intensity must be exceeded for SCC to occur.*
(6) *Stress-corrosion cracks are brittle in macroscopic appearance*, i.e., have little evidence of plastic deformation (but may show plastic flow on a smaller scale).

Stages of Stress-Corrosion Cracking

As with crevice corrosion and pitting, SCC also has two general stages – initiation and propagation. Measurement of the time to failure of a smooth specimen (without surface flaws) can be misleading because such measurements do not differentiate between the initiation and propagation stages. For instance, titanium alloys in seawater do not readily pit (often a precursor to SCC) but these alloys undergo fast fracture once a crack is initiated. Thus, there is a long time to failure (consisting largely of a long initiation time) which masks the susceptibility of titanium alloys to a much more rapid propagation stage.

Initiation of SCC

The initiation stage of SCC involves a series of events which are depicted in Fig. 11.6 [6, 7]. These events are as follows:

(1) localized breakdown of the passive oxide film due to mechanical damage or to chemical attack by aggressive ions, such as Cl^-,

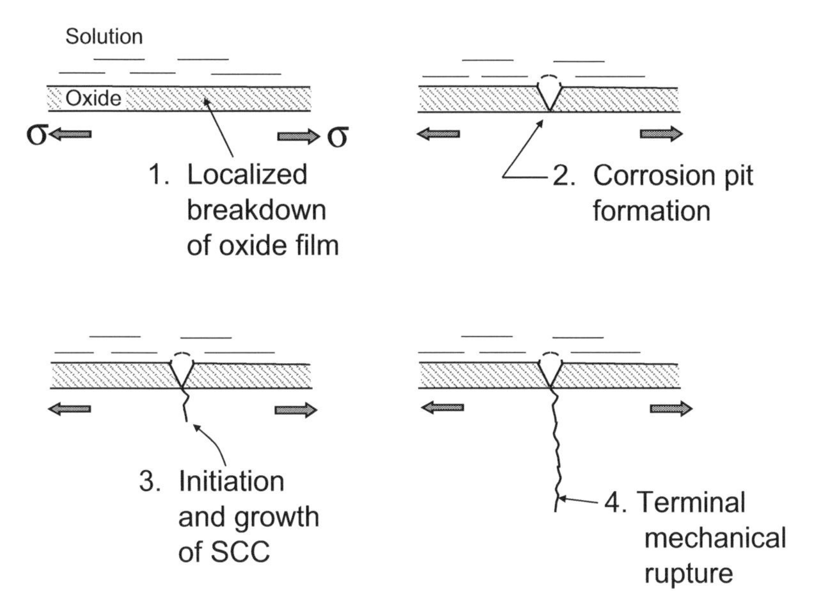

Fig. 11.6 Stages in the initiation of stress-corrosion cracks. (The fourth stage of mechanical rupture deals with propagation.) The *shaded arrows* denote the direction of applied stress, σ. Some systems such as titanium do not easily pit, but a stress-corrosion crack can initiate from a fatigue crack. After Brown [1, 6, 7]

(2) formation of a corrosion pit after the oxide film has been broken, and
(3) initiation and growth of a stress-corrosion crack.

The last event in Fig. 11.6 (mechanical rupture) deals with the propagation stage rather than with initiation.

The corrosion pit can generate a stress corrosion crack in two ways. First, the corrosion pit serves as a stress intensifier, so that the local stress due to an external load is magnified locally near the corrosion pit. (Stress intensity and the role of flaws in SCC are considered later in this chapter.) Second, the corrosion pit is a source of hydrogen ions, as discussed in Chapter 10. For instance, for stainless steels, the following reaction within the corrosion pit

$$Cr^{+3} + 3H_2O \rightarrow Cr(OH)_3 + 3\,H^+ \tag{4}$$

generates H^+ ions which can be reduced to hydrogen atoms at the tip of an emerging stress-corrosion crack. This reaction can also occur within propagating stress-corrosion cracks, as depicted in Fig. 11.7. This process of local acidification may lead to the phenomenon of hydrogen embrittlement, which assists crack growth, as discussed later in this chapter.

In some instances, stress-corrosion cracks may initiate without visible signs of pitting, as, for example, in the SCC of titanium alloys in seawater. In addition, stress-corrosion cracks may also initiate because of surface flaws, inherent microstructural defects (such as grain boundaries), or to mechanical damage (such as damage due to erosion).

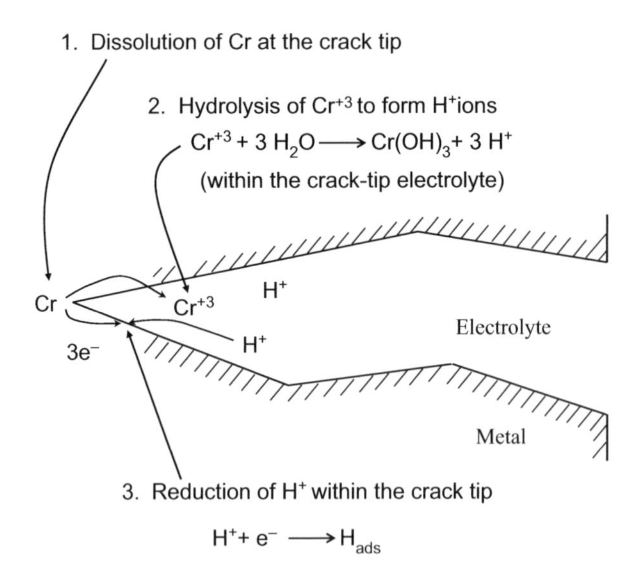

Fig. 11.7 Schematic illustration showing a propagating stress-corrosion cracking in a Cr-containing alloy. This figure shows the production of H^+ ions within a crack by the hydrolysis of dissolved Cr^{3+} ions and the reduction of H^+ ions near the crack tip

Propagation of SCC

As with crevice corrosion and pitting, the local electrolyte within a stress-corrosion crack becomes acidic in nature due to the hydrolysis of dissolved hydrogen metal cations, as, e.g., in Eq. (4). Thus, the stress-corrosion cracks propagate and grow due to the combined effects of the applied stress and the corrosive conditions within the crack arising from acidification of the crack electrolyte.

The acidification of local electrolytes within crack tips was shown by Brown and co-workers [7–10] in a series of experiments in which a slot was cut into a metal, a wedge (of the same metal) was pressed into the slot to initiate a crack, and the sample was then placed into an electrolyte. After several hours of crack growth, the specimen was removed and placed into liquid nitrogen to freeze the corrodent within the crack. The specimen was broken open, and the pH of the electrolyte was measured with pH indicators as the specimen warmed up and the crack electrolyte melted. Resulting pH values for stress-corrosion cracks within several metals are given in Table 11.3.

Peterson and co-workers [10] also measured the pH within a stress-corrosion crack using the experimental arrangement shown in Fig. 11.8. By placing a pH electrode at the base of a crack, a lens tissue was wet by solution which traveled down the length of the stress-corrosion crack. Some of Peterson's results are also given in Table 11.3.

There are two modes by which stress-corrosion cracks can propagate within the metal. The crack path may be *intergranular*, that is, cracks grow along grain boundaries, or cracks may be *transgranular*, in which cracks travel across grain boundaries. Various alloys display various modes of crack growth and examples have been compiled by Scully [11]. Figure 11.9 shows an example of intergranular and transgranular crack growth on the same alloy but in different concentrations of the same electrolyte [12].

Table 11.3 pH within stress-corrosion cracks

Alloy	pH detected at crack front	Bulk solution	Sources
7075 Al	3.5	3.5% NaCl	Bown, Fujii, and Dahlberg [8]
0.45% C steel	3.7	3.5% NaCl	Sandoz, Fujii, and Brown [9]
AISI type 4340 steel	3.7	3.5% NaCl	Sandoz, Fujii, and Brown [9]
	3.5–3.9	3.5% NaCl (bulk pH adjusted between 2 and 10)	Smith, Peterson, and Brown [10]
17-4 PH steel (Fe-17 Cr-4 Ni precipitation-hardened)	3.7	3.5% NaCl	Sandoz, Fujii, and Brown [9]
Ti-8 Al-1 Mo-1 V	1.7	3.5% NaCl	Bown, Fujii, and Dahlberg [8]

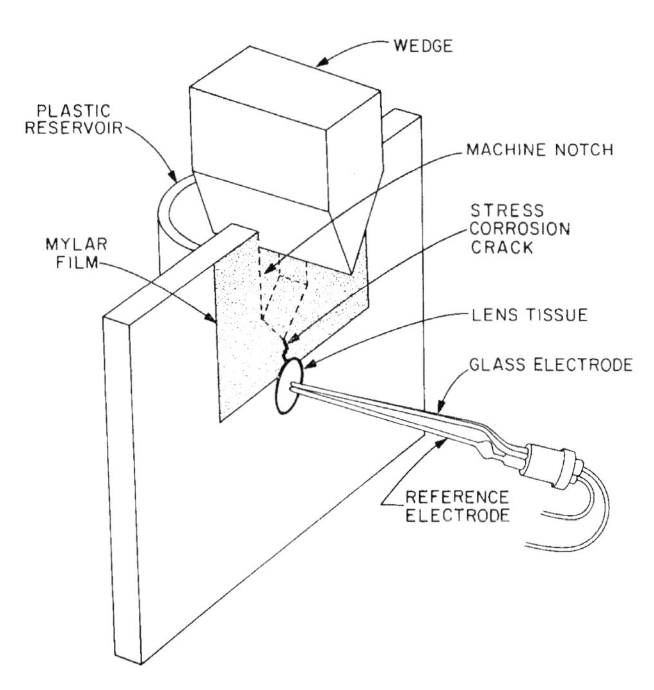

Fig. 11.8 An experimental arrangement for determining the pH at a crack tip [7, 10]. Reproduced by permission of © NACE International 1970

Fracture Mechanics and SCC

An important advance in the study of SCC was the application of fracture mechanics to the problem. Fracture mechanics is a relatively new discipline based largely upon the work of G.R. Irwin [13] at the Naval Research Laboratory in Washington, DC, and is used to treat fracture problems involving cracks in a quantitative manner. The application of fracture mechanics to SCC is due in large part to the work of B.F. Brown, M.H. Peterson, C.D. Beachem, and their co-workers [14–16], also at the Naval Research Laboratory.

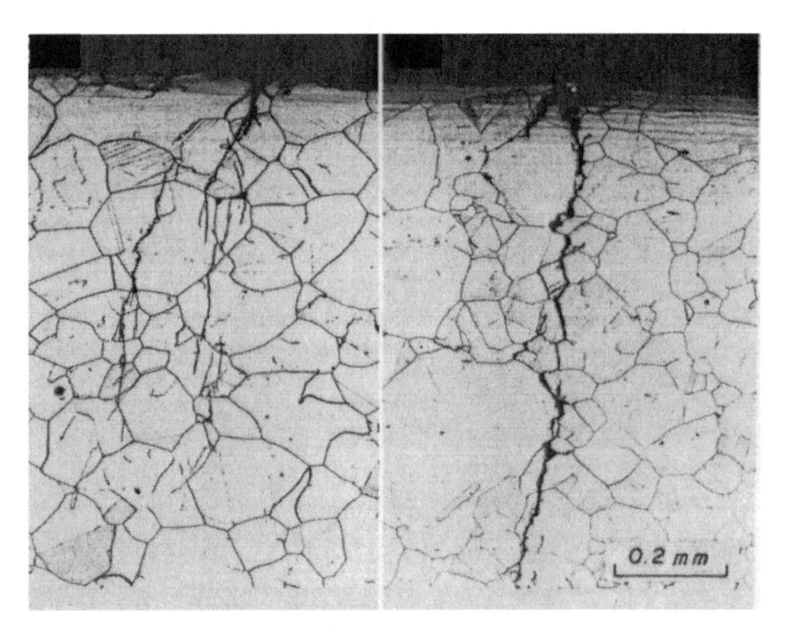

Fig. 11.9 Stress-corrosion cracking of sensitized type 304 stainless steel in (*left*) boiling 45% magnesium chloride solution and (*right*) in boiling 20% magnesium chloride solution [12]. Reproduced with permission of the Japan Institute of Metals

General Effect of a Flaw

Consider a homogeneous metal body as in Fig. 11.10(a) which is subjected to a tensile force P. The stress is given by $\sigma = P/A$ (where A is the cross–sectional area of the specimen) and is uniform throughout the specimen. The stress can be considered to be borne by lines of force (stress lines)

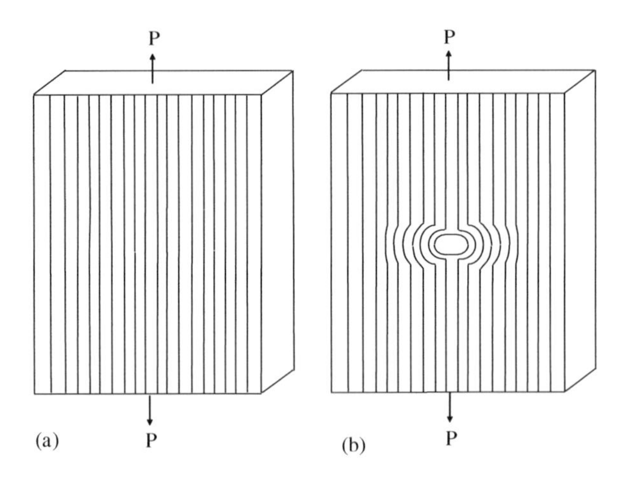

Fig. 11.10 (**a**) Schematic illustration showing lines of force in a body subject to a tensile load P. (**b**) Distortion of the lines of force when the body contains a center flaw. (The lines of force are shown only for the front face of the specimen but are considered to exist throughout the specimen thickness.) Redrawn from Tetelman and McEvily [4] by permission of John Wiley & Sons, Inc

which exist throughout the body, as shown in Fig. 11.10(a) [4]. These lines of force can be thought of as parallel columns of small area A' (with $A' \to 0$) which support the external load P. Because the body is uniform throughout, the lines of force exist with uniform spacing, as shown in Fig. 11.10(a).

But suppose there is a flaw in the structure, as shown in Fig. 11.10(b). Because there is no metal in the hole, the lines of force cannot continue through the hole, but must bend around the hole, as shown in Fig. 11.10(b). Thus, in the vicinity of the flaw (i.e., discontinuity), the lines of force (and thus the stresses) are highly concentrated. However, at distances removed from the discontinuity, the lines of force are more evenly spaced.

These considerations give rise to the concept of a stress intensity factor, K, which is defined as follows:

$$K = \frac{\sigma(\text{local})}{\sigma(\text{average})} \tag{5}$$

where σ (average) is the average stress in the bulk of the otherwise uniform body. Thus, the stress intensity factor K is a measure of how the stress is increased near a local discontinuity or flaw. The power of fracture mechanics is that stress-intensity factors have been derived for flaws of various geometries.

General Assumptions

The general assumptions which are made in the application of fracture mechanics to SCC are the following:

(1) Flaws (i.e., cracks) are inherently present in structures.
(2) These cracks exist in a linear elastic field (that is, the stress and strain are linearly related).
(3) The stress ahead of a crack can be calculated from the stress intensity factor K.
(4) The propagation of a crack leading to failure occurs above a critical stress intensity factor.

These ideas are discussed in more detail as we proceed into this chapter.

Fracture Mechanics and K_{I}

There are three modes by which a metal may fracture, called mode I, mode II, and mode III and these three modes are pictured in Fig. 11.11 [17]. The mode I fracture involves crack opening, mode II is edge sliding, and mode II is crack tearing, which has an additional sliding displacement superimposed on a mode I crack.

The mode I crack is the lowest energy mode and is the mode usually encountered in SCC. For a plane sharp defect as shown in Fig. 11.12, the stress components of the crack are given by [17]

$$\sigma_x = \frac{K_I}{\sqrt{2\pi r}} \cos\frac{\theta}{2}\left(1 - \frac{\theta}{2}\sin\frac{3\theta}{2}\right) \tag{6}$$

$$\sigma_y = \frac{K_I}{\sqrt{2\pi r}} \cos\frac{\theta}{2}\left(1 + \sin\frac{\theta}{2}\sin\frac{3\theta}{2}\right) \tag{7}$$

where K_{I} is the stress intensity factor for a mode I crack, r is the crack length measured from the crack tip, and θ is measured relative to the crack direction.

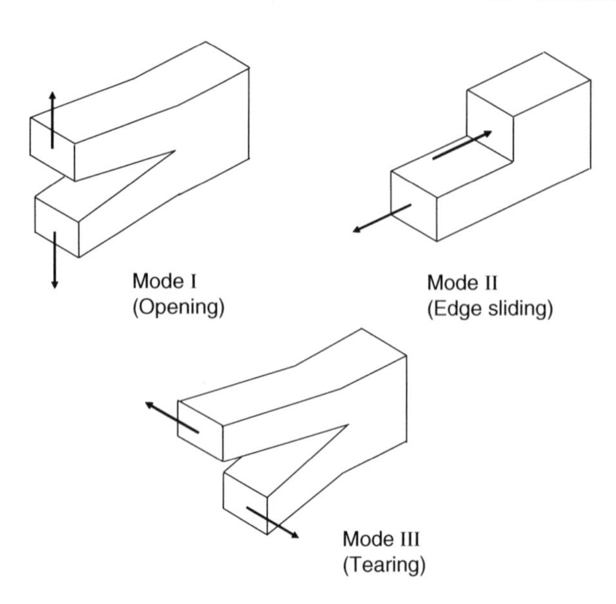

Fig. 11.11 Three basic modes of crack displacements [17]. Reproduced by permission, from [17], copyright ASTM International, 100 Barr Harbor Drive, West Conshohocken, PA 19428

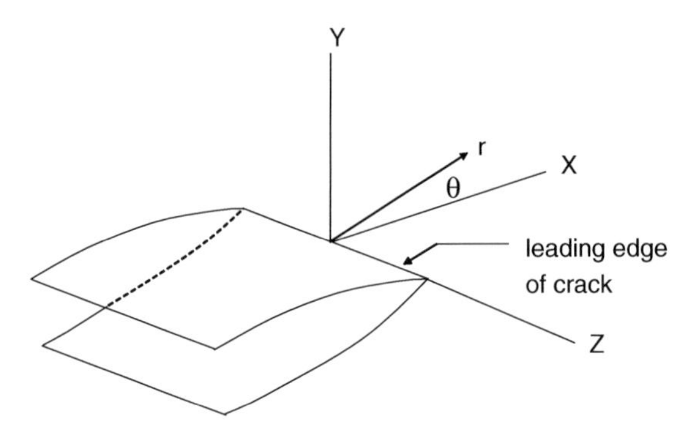

Fig. 11.12 Diagram of the leading edge of a mode I crack and the system of co-ordinates. The crack length is designated by r [17]. Reproduced by permission, from [17], copyright ASTM International, 100 Barr Harbor Drive, West Conshohocken, PA 19428

Stress Intensity Factors for Specific Types of Flaws

Expressions for the stress intensity factor K_{I} which appears in Eqs. (6) and (7) have been derived for various types of specific flaws [17, 18]. For example, for a center crack in an infinite sheet, as shown in Fig. 11.13,

$$K_{\mathrm{I}} = \sigma\sqrt{\pi a} \tag{8}$$

where a is half the crack length and σ is the average external stress. For a semi-infinite sheet of width W (but infinite in length),

$$K_{\mathrm{I}} = \sigma\sqrt{\pi a}\left(\frac{W}{\pi a}\tan\frac{\pi a}{W}\right)^{1/2} \tag{9}$$

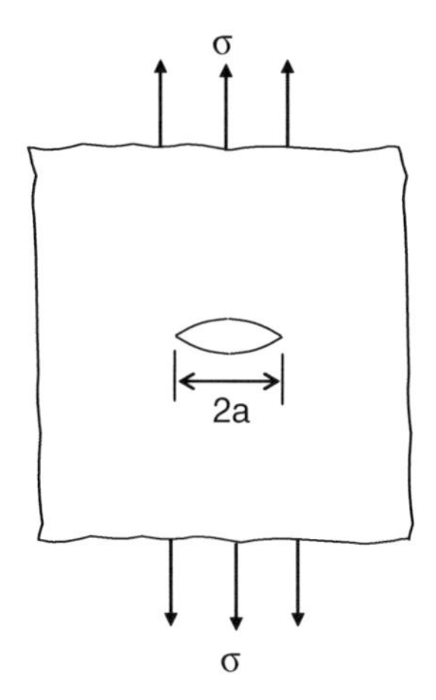

Fig. 11.13 A center crack in an infinite sheet

Two other configurations used in fracture mechanics are the single-edge notched specimen in Fig. 11.14 and the thumbnail-shaped surface crack in Fig. 11.15. For the single-edge notched specimen:

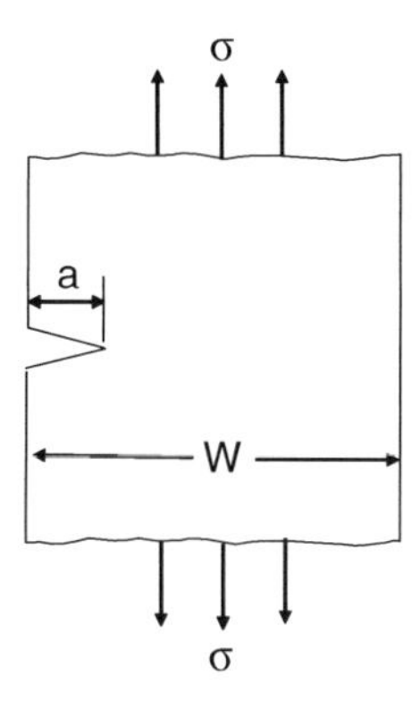

Fig. 11.14 A single-edge notched specimen

$$K_{\mathrm{I}} = \sigma\sqrt{\pi a}\,f(a/W) \tag{10}$$

where a is the depth of the notch, W is the specimen thickness, and $f(a/W)$ is a geometrical function given by [6, 19]

$$f(a/W) = 1.12 - 0.231\frac{a}{W} + 10.56\left(\frac{a}{W}\right)^2 - 21.72\left(\frac{a}{W}\right)^3 + 30.39\left(\frac{a}{W}\right)^4 \tag{11}$$

For the thumbnail-shaped surface crack:

$$K_{\mathrm{I}}^2 = \frac{1.2\pi\sigma^2 a}{\phi^2 - 0.212\left(\frac{\sigma}{\sigma_y}\right)^2} \tag{12}$$

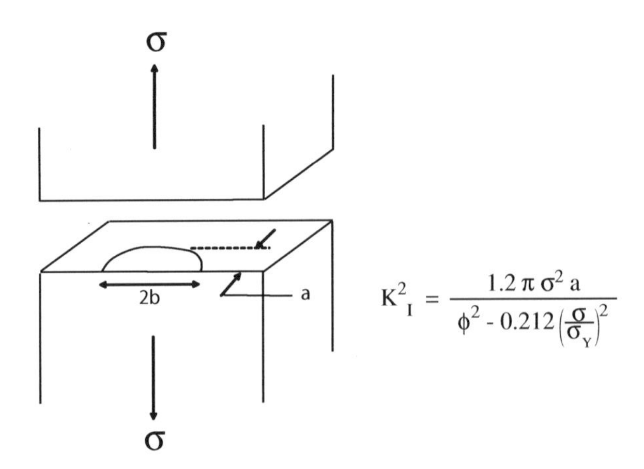

Fig. 11.15 Opened view of a tensile specimen with a thumbnail crack. The parameter ϕ is an integral function of *a*/*b*. For an infinitely long crack, $a/b \approx 0$ and $\phi^2 = 1.00$. For a typical thumbnail, $a/b = 0.5$ and $\phi^2 = 1.46$ [15]

where 2*b* is the crack length and *a* is now the crack depth, as shown in Fig. 11.15, and σ_Y is the yield stress of the alloy. The parameter ϕ is an integral function given by $\phi^2 = 1.00$ for an infinitely long surface crack ($a/b = 0$), and $\phi^2 = 1.46$ for a typical thumbnail crack ($a/b = 0.5$).

For each of these last two specimens, as well as for the center crack specimen, the stress intensity factor has the form

$$K_I = \sigma\sqrt{\pi\,a}f(G) \tag{13}$$

where *f*(*G*) is a factor specific to the particular geometry of the specimen and *a* is the critical dimension of the crack. Note that K_I increases with increasing crack length and with increasing stress σ.

Example 11.1: Calculate the stress intensity factor K_I in both English and metric units for an infinite sheet having a center crack of length 0.01 in., a load of 10,000 pounds, and a cross-sectional area of 0.50 in.^{2}.
Solution:

$$K_I = \sigma\sqrt{\pi\,a}$$

$$\sigma = \frac{10{,}000\,\text{lbs}}{0.05\,\text{in.}^2} = 20{,}000\frac{\text{lbs}}{\text{in.}^2}$$

$$a = \frac{0.01\,\text{in.}}{2} = 0.005\,\text{in.}$$

$$K_I = 20{,}000\,\frac{\text{lbs}}{\text{in.}^2}\sqrt{\pi(0.005\,\text{in.})}$$

$$K_I = 2{,}500\ \text{psi}\sqrt{\text{in.}} = 2.50\,\text{ksi}\sqrt{\text{in.}}$$

$$\sigma = \frac{(10{,}000\,\text{lbs})\left(\frac{454\,\text{g}}{\text{lb}}\right)\left(\frac{1\,\text{kg}}{1{,}000\,\text{g}}\right)}{(0.50\ \text{in.}^2)\left(\frac{2.54\,\text{cm}}{\text{in.}}\right)^2\left(\frac{1\,\text{m}}{100\,\text{cm}}\right)^2} = 1.41\times10^7\frac{\text{kg}}{\text{m}^2}$$

In metric units it is necessary to convert the mass per unit area to the force per unit area by using $F = mg$. Thus,

$$\sigma = \left(1.41 \times 10^7 \frac{\text{kg}}{\text{m}^2}\right)\left(\frac{9.8066\,\text{m}}{\text{s}^2}\right)$$

$$\sigma = 1.38 \times 10^8 \frac{\text{kg m}}{\text{m}^2\,\text{s}^2}$$

$$\sigma = 1.38 \times 10^8 \frac{\text{N}}{\text{m}^2}$$

$$\sigma = (1.38 \times 10^8\,\text{Pa})\left(\frac{1\text{MP}a}{1 \times 10^6\,\text{Pa}}\right) = 138\,\text{MPa}$$

Then,

$$K_\text{I} = 138\,\text{MPa}\sqrt{\pi(0.005\,\text{in.})\left(\frac{2.54\,\text{cm}}{\text{in.}}\right)\left(\frac{1\,\text{m}}{100\,\text{cm}}\right)} = 2.76\,\text{MPa}\,\sqrt{\text{m}}$$

The Cantilever Beam and K_{Iscc}

Stress-corrosion tests are often carried out in a pre-cracked cantilever beam experiment introduced by B. F. Brown and co-workers [14–16] at the Naval Research Laboratory (NRL). This test uses a bar of square or rectangular cross section containing a notch and a small crack and having the geometry shown in Fig. 11.16. In a typical experiment by Brown and Beachem [15], the metal specimen size was 13 cm long, 1 cm thick, and 2.5 cm high. One end of the specimen was clamped in a stand and the other end clamped to a loading arm, as shown in Fig. 11.16. The crack in the specimen was surrounded by a plastic cell containing a reservoir of electrolyte (usually 3.5% NaCl or natural seawater in the NRL experiments) and a load was applied at the end of a cantilever beam, as shown in Fig. 11.16.

The stress intensity factor for a given load P and crack length a (length of the notch plus the crack length in the specimen) is calculated from the specimen dimensions, the crack length, and the load P:

$$K_\text{I} = \frac{4.12\,M\left(\frac{1}{\alpha^3} - \alpha^3\right)}{BD^{3/2}} \tag{14}$$

where α is given by

$$\alpha = 1 - \frac{a}{D} \tag{15}$$

where a is the crack length (notch plus fatigue pre-crack), B is the specimen width, and the other dimensions are given in Fig. 11.16. The moment M is given by

$$M = PS \tag{16}$$

where P is the applied load and S is the distance between the crack and the point of application of the load (see Fig. 11.16). The formula for K_I for the cantilever beam specimen again has the same

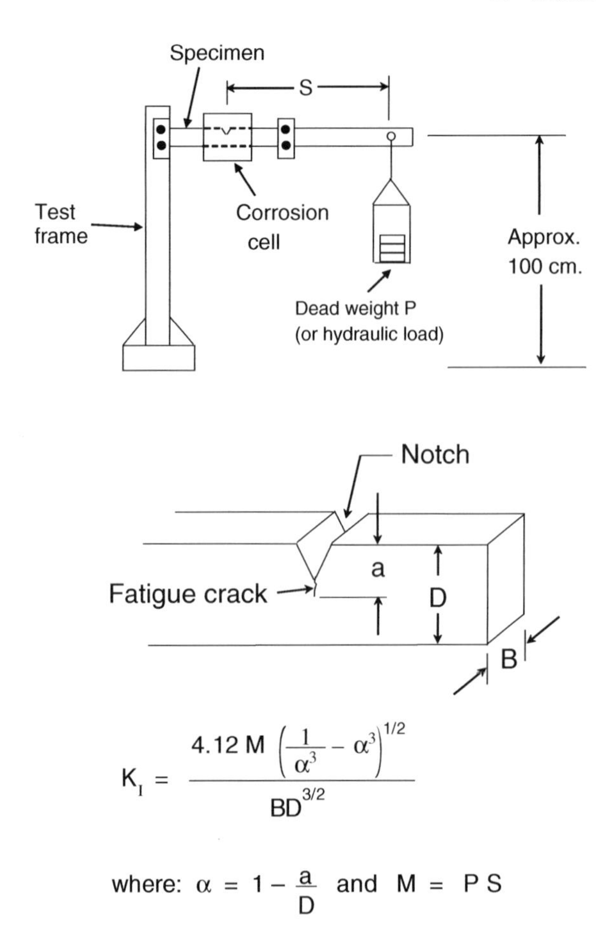

Fig. 11.16 (*Top*) Schematic diagram of a cantilever beam stress-corrosion test. (*Center*) Specimen details. (*Bottom*) Mathematical expression for the stress intensity factor K_I [14–16, 65]

general form as in Eq. (13). see Problem 11.3. Thus, the units for K_I are again ksi $\sqrt{\text{in.}}$ in English units or MPa $\sqrt{\text{m}}$ in metric units. The depth of the crack increases during the SCC test, and therefore the stress intensity factor also increases.

Figure 11.17 shows SCC cantilever beam data for AISI 4340 steel in 3.5% NaCl [15] in which K_I is plotted vs. the logarithm of the time to failure. Large values of K_I (large loads and/or large crack lengths) produce short times to failure, whereas smaller values of K_I (small loads and/or small crack lengths) lead to longer times to failure. Note that SCC does not occur in these alloys in 3.5% NaCl if K_I has a value smaller than 15 ksi $\sqrt{\text{in.}}$ Thus, this value of K_I is called the *critical stress intensity factor for stress-corrosion cracking* and is given the designation K_{Iscc}. The significance here is that values of K_I less than K_{Iscc} do not produce SCC, but values of greater than K_{Iscc} lead to SCC. The value of K_I which produces cracking in the dry specimen (in the absence of electrolyte) is called the *fracture toughness* and is given the symbol K_{Ic}. The value of K_{Ic} is also obtained or approached in the presence of electrolytes when there is a large applied load, so that the sample suffers fast mechanical fracture and there is little or no effect of the electrolyte in the presence of the large applied load.

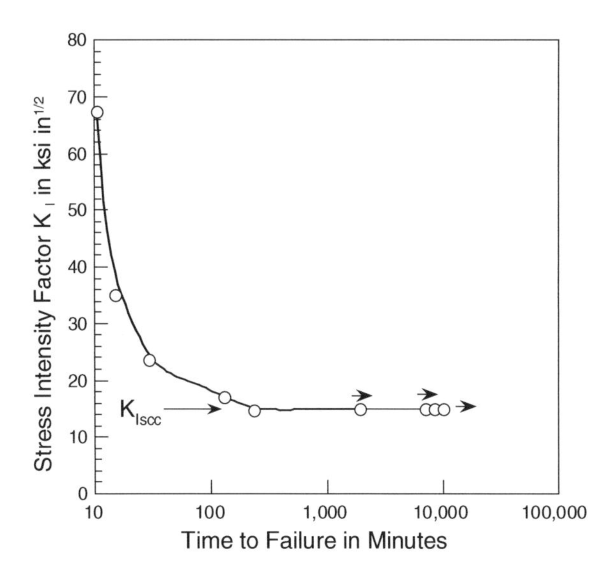

Fig. 11.17 K_{Iscc} tests for type 4340 steel in 3.5% NaCl for cantilever beam specimens [15]. The *arrows* associated with data points indicate no fracture within the time given. Reproduced by permission of Elsevier Ltd

K_{Iscc} as a Characteristic Property

Brown [14] showed that the value of K_{Iscc} in Fig. 11.17 as measured by the cantilever beam method was in excellent agreement with values measured for center-cracked (Fig. 11.13) or surface-cracked (Fig. 11.15) specimens. This agreement shows that the value of K_{Iscc} for a given alloy and environment is not a function of the test setup and K_{Iscc} is in fact considered to be a material property (for the given environment).

The stress intensity factor K_{Iscc} is a function of the alloy type, alloy composition, strength level of the alloy, and the nature of the electrolyte. Figure 11.18 shows that Mo additions increase K_{Iscc} values for Fe–Cr–Ni alloys in aqueous 22% NaCl solutions at 105°C [20]. (Recall from Chapters 9 and 10 that Mo also has a beneficial effect on the passivity and pitting resistance of Fe–Cr–Ni alloys.) Extensive tables of K_{Iscc} in 3.5% NaCl have been compiled for various titanium alloys [21, 22] and for various precipitation-hardened stainless steels [20]. The effect of the strength level of the alloy is considered in a later chapter.

For a given alloy K_{Iscc} is sometimes (but not always) lower for electrolytes than for non-electrolytes. Several examples are compiled in Table 11.4.

SCC Testing

There are two general approaches in conducting SCC tests. The first approach is to use pre-cracked specimens, as discussed above, and the method involves determination of the threshold stress intensity factor for SCC, as shown in Fig. 11.19(a). The crack velocity can also be measured as a function of the stress intensity factor, as shown in Fig. 11.19(b). (Sometimes the metal will fail before regions II and III appear).

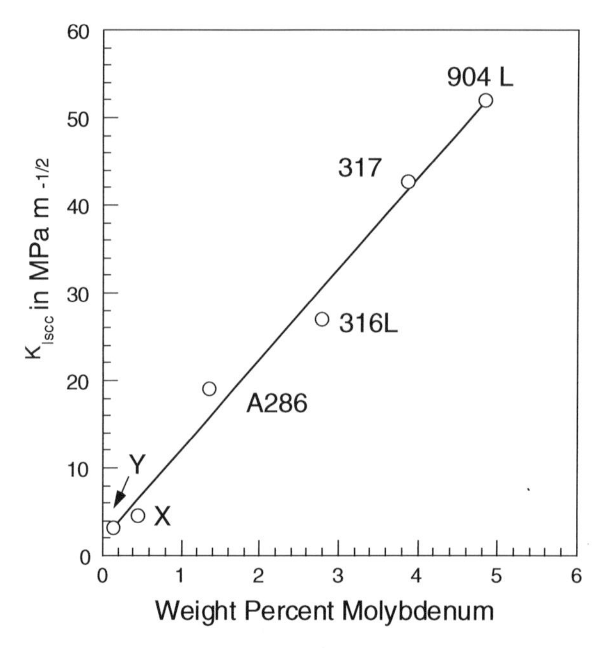

Fig. 11.18 Effect of molybdenum on the stress intensity factor of Fe–Cr–Ni–Mo alloys in 22% NaCl [20]. (The range of Cr is between 15 and 21%; the range of Ni is between 13 and 25%.) Reproduced by permission of John Wiley & Sons, Inc

Table 11.4 Comparison of K_{Iscc} for electrolyte solutions vs. non-electrolytes at approximately 25°C

Alloy	Electrolyte	ksi $\sqrt{\text{in.}}$	MPa $\sqrt{\text{m}}$	Non-electrolyte	ksi $\sqrt{\text{in.}}$	MPa $\sqrt{\text{m}}$
AISI 4340 steel [33] (220 ksi yield strength)	Seawater	10	11	Methanol	40	44
Uranium alloy [65]	3.5% NaCl	15	17	Water	21	23
Ti-8 Al-1 Mo-1 V [67, 68]	3.5% NaCl	18–23	20–25	CCl_4	18–23	20–25
	0.6 M NaCl	20	22	Methanol	14	15
	0.1 M LiCl	25	28	Glycerine	26	29

The second approach is to use smooth (but stressed) specimens which do not contain defects of a known size or shape, such as a U-bend specimens, as shown in Fig. 11.20. For smooth specimens, the SCC data are presented as the time to failure as a function of applied stress (*not* stress intensity). Such curves define a threshold stress level, σ_{th}, below which SCC does not occur, as shown in Fig. 11.19(c).

Several different types of metal specimens used in smooth or pre-cracked tests are shown in Fig. 11.20 [23]. More details on SCC test methods have been given by Sedriks [23]. The environments used in SCC tests should duplicate or simulate the actual environment in which SCC occurred; and as shown in Table 11.2, the SCC environments are often specific to the type of alloy.

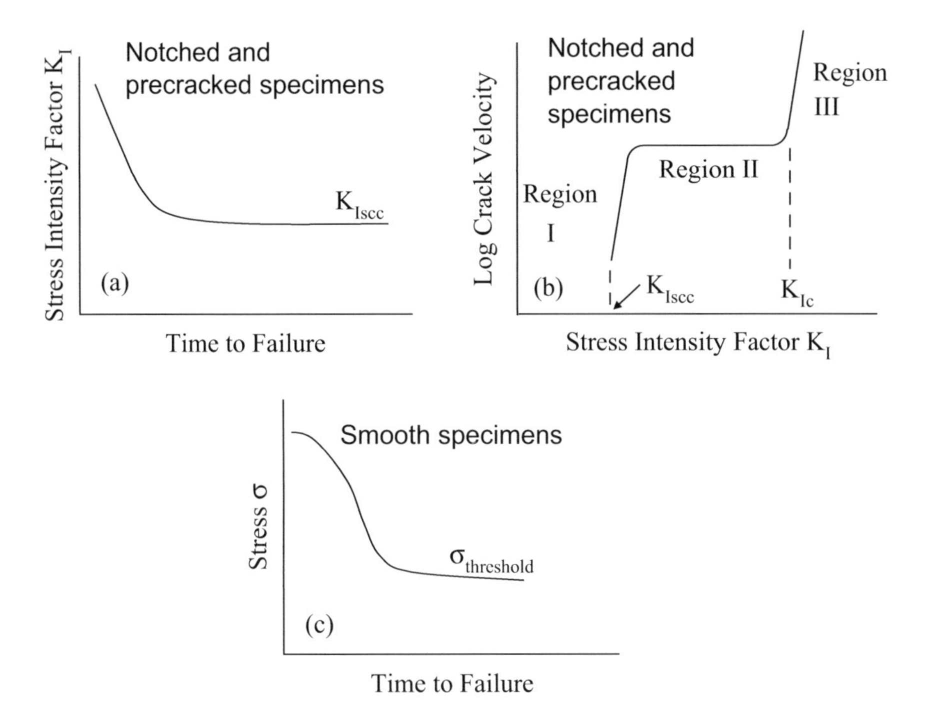

Fig. 11.19 Three different ways to present stress-corrosion cracking data [1, 6, 14]. K_{Iscc} refers to the critical stress intensity factor for stress-corrosion cracking and K_{Ic} to the value for purely mechanical rupture

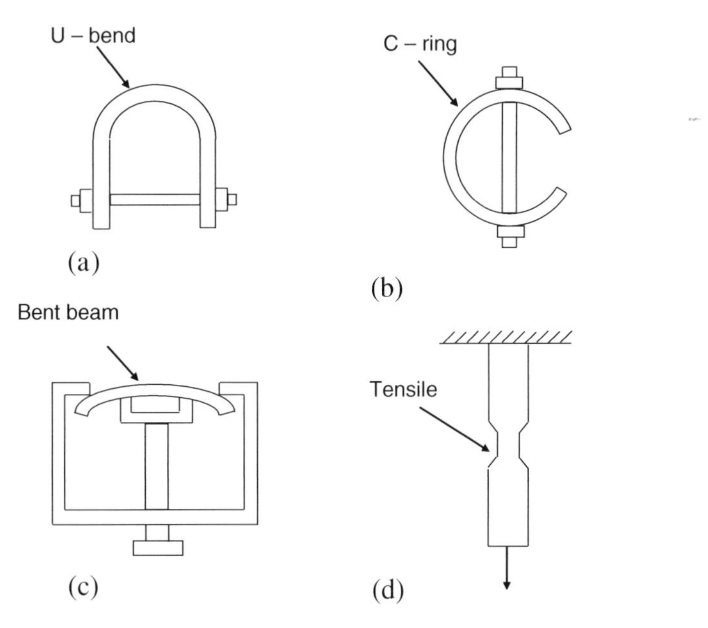

Fig. 11.20 Several different types of smooth specimens used in stress-corrosion cracking tests [1, 23]. None of these specimens contain pre-cracks. Reproduced by permission of NACE International

Interpretation of SCC Test Data

The fracture mechanics approach assumes that there is a critical flaw size which leads to SCC and to failure if the operating stress intensity factor exceeds a critical value, K_{Iscc}. In this approach, K_{Iscc} is used as a characteristic parameter of the metal/electrolyte system. But in smooth specimens which do not contain intentional flaws, a measure of the resistance to SCC is the *threshold stress*, σ_{th}, below which SCC does not occur. This section describes how to combine the SCC data from smooth and pre-cracked specimens in order to provide a fracture-safe set of operating conditions [1, 24].

For Al alloy 7079 (Al-4 Zn-3 Mg-0.5 Cu), for a thumbnail-shaped crack the value of K_{Iscc} was observed to be 4 ksi $\sqrt{\text{in}}$. (4.4 MPa $\sqrt{\text{m}}$) [1]. The threshold stress was found to be $\sigma_{\text{th}} = 8$ ksi (55 MPa) and the yield stress of the alloy was $\sigma_Y = 67$ ksi (460 MPa) [1]. Inserting these values of K_{Iscc} and σ_y into Eq. (12) and assuming a shallow surface crack ($a/b \approx 0$, $\phi^2 = 1.00$) gives

$$\sigma = \left(\frac{1.00}{0.2356\,a + 4.73 \times 10^{-5}} \right)^{1/2} \tag{17}$$

where σ is in ksi and b is in inches. Thus, for a flaw depth, a, the critical allowable stress, σ, can be calculated from Eq. (17). Results are given by the slanted line in Fig. 11.21. Any combination of b and σ which lies below this slanted line will produce a K_{I} less than K_{Iscc}, so that SCC will not occur. But because the yield stress of the alloy should not be exceeded, a safe region for the pre-cracked specimen is the region below the yield stress and the slanted line.

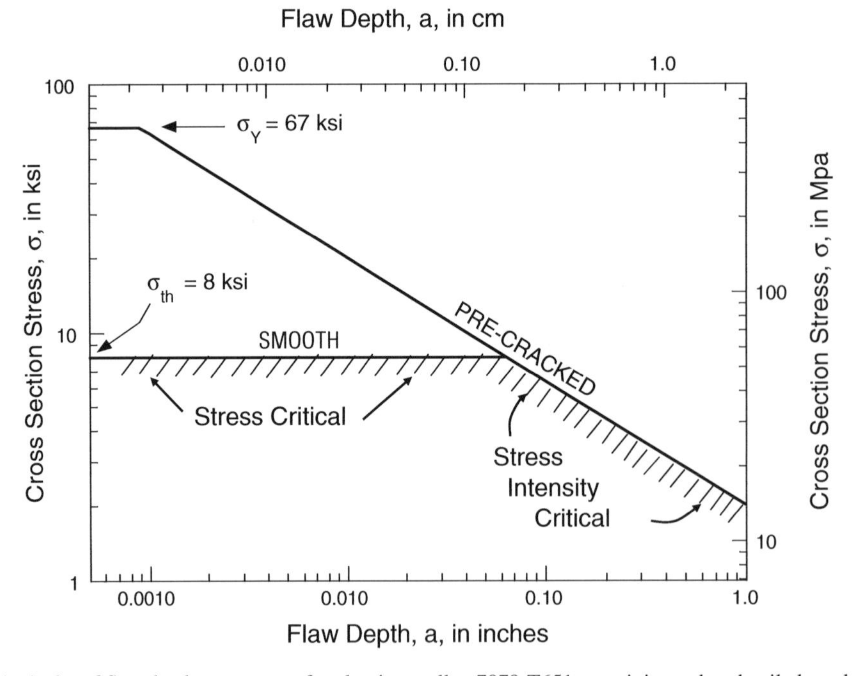

Fig. 11.21 A plot of flaw depth vs. stress σ for aluminum alloy 7079-T651 containing a thumbnail-shaped crack [1]. The parameter a refers to the depth of the crack, σ_Y is the yield stress of the alloy, and σ_{th} is the threshold stress for smooth specimens. This "safe-zone plot" shows where the alloy can be used safely for both smooth and pre-cracked specimens

However, for smooth specimens, the threshold strength σ_{th} should not be exceeded. Thus, a conservative estimate of the fracture-safe region is below the hatched line in Fig. 11.21. That is, we should take into account data for both the smooth and pre-cracked specimens. The intersection of the straight line for the threshold stress with the sloping line from fracture mechanics has sometimes been interpreted to give the inherent flaw size of a smooth specimen. Plots as in Fig. 11.21 are called safe-zone plots.

Note that the "stress critical" and "stress intensity critical" lines in Fig. 11.21 intersect at a flaw size of 0.06 in. (1.5 mm). Two interpretations regarding this flaw size are possible. First, this flaw size is the inherent flaw size resident in a "smooth sample." Second, other forms of corrosion such as pitting or intergranular corrosion first occur so as to produce a defect of this size, which in turn leads to stress-corrosion cracking.

Metallurgical Effects in SCC

For high-strength steels, K_{Iscc} depends on the strength level of the steel. Figure 11.22 shows that the value of K_{Iscc} for AISI 4340 steel in seawater decreases with increasing yield strength of the alloy. This means that any advantage in using a higher strength steel is negated in the presence of seawater, which increases the susceptibility to SCC [7, 16].

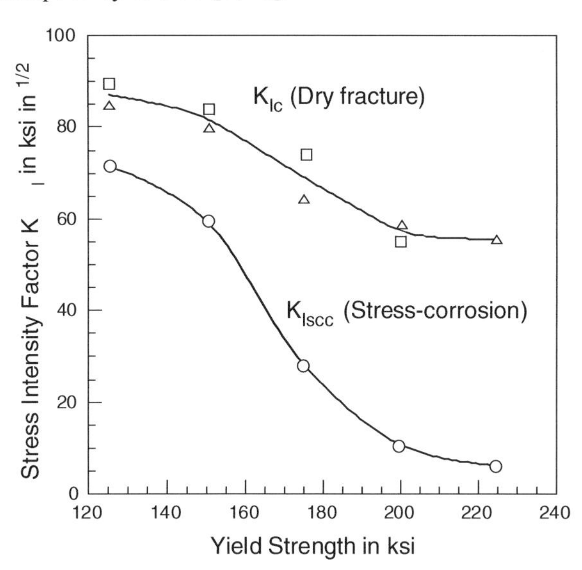

Fig. 11.22 Effect of the yield strength (yield stress) of type 4340 steel on the stress intensity factor for dry fracture K_{Ic} or for stress-corrosion cracking K_{Iscc} in flowing seawater [7, 16]. Reproduced by permission of © NACE International 1990

The resistance to SCC also depends on the shape of the metal grains and on the stressing direction. This is illustrated in Fig. 11.23, which shows the resistance to stress corrosion for Al alloy 7075 (Al-5.5 Zn-2.5 Mg-1.5 Cu) having three different grain shapes [24, 25]. For each case the specimens were stressed in directions parallel to (A) or perpendicular to (B), the principal grain axis. The stress-corrosion resistance was defined as the highest initial applied stress that did not cause SCC in 84 days exposure to 3.5% NaCl in an alternate immersion test. Figure 11.23 shows that the resistance to SCC was the highest where the most highly oriented grain was stressed parallel to the principal grain axis.

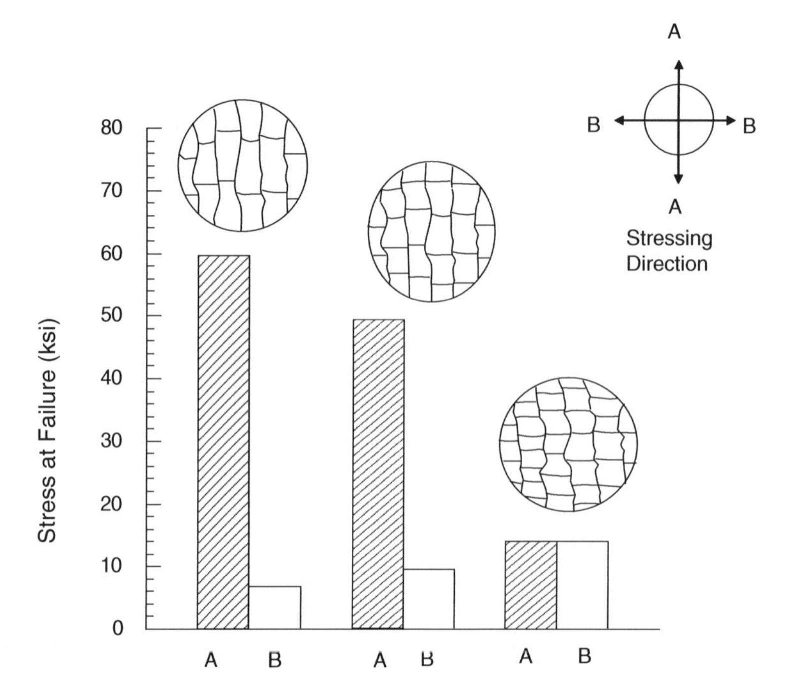

Fig. 11.23 Effect of grain shape and stress direction on the stress-corrosion cracking of Al alloy 7075 in a 3.5% NaCl alternate immersion test [25]. Reproduced by permission of NACE International

In addition to the grain shape, the grain size is also a factor in determining the resistance to SCC. A fine grain size has been shown to increase the resistance to SCC of carbon steels in solutions containing H_2S or nitrates [26], of brass in ammonia [27], and of stainless steels in boiling magnesium chloride [28]. The effect of a coarse grain size on increased susceptibility to SCC has been attributed to changes in the grain boundary composition and microstructure. That is, the total quantity of a precipitated phase that is responsible for SCC will become more enriched in the grain boundary as the grain becomes larger. (Mechanisms of SCC are discussed in a later section.) However, Parkins [29] suggested that the effect of grain size is explained by the well-known experimental observation that the yield strength of the alloy decreases with increasing grain size. (This is called the Hall and Petch effect [30], in which the barrier to slip across a grain increases with increasing grain size.)

Environmental Effects on SCC

So far we have considered stress-corrosion cracking largely from a mechanics point of view. The role of the electrolyte has been to lower the critical value of the stress intensity factor for SCC (K_{Iscc}) relative to the critical value in the absence of the electrolyte (K_{Ic}).

In general, environmental factors which influence SCC include the nature of the electrolyte, the electrode potential, the concentration of corrodent species, and the temperature.

Effect of Electrode Potential

The electrode potential is one of the major controlling parameters in SCC. Evidence is anecdotal in nature, having been accumulated though various case histories of SCC. For smooth specimens the electrode potential can affect the threshold stress or the time to failure (or equivalently, the cracking rate). For pre-cracked specimens the electrode potential can affect the value of K_{Iscc}.

Uhlig and Cook [31] conducted experiments on the effect of the electrode potential on the SCC of type 304 stainless steel in boiling magnesium chloride solutions. The time to failure of U-bend specimens was decreased by polarization in the anodic direction but the time to failure was increased dramatically by cathodic polarization. As shown in Fig. 11.24, there is a critical electrode potential below which (more negative than) SCC does not occur. Uhlig and Cook explained this behavior on the basis that the damaging Cl^- ion adsorbs on the plastically deformed metal surface above the critical potential, but desorbs below this potential. This effect is considered again later when mechanisms of SCC are discussed.

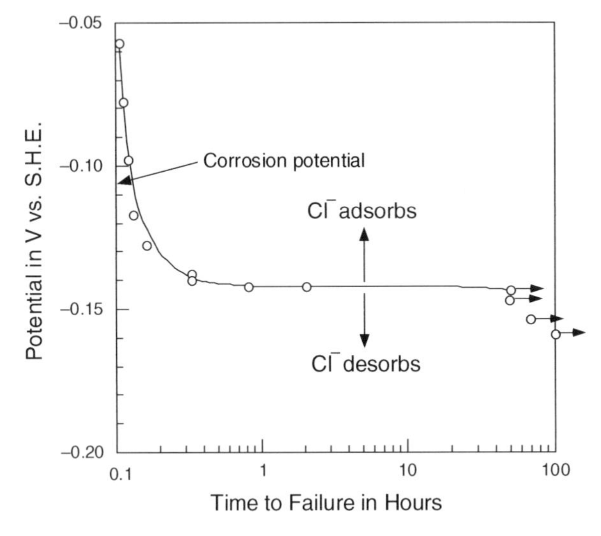

Fig. 11.24 Effect of electrode potential on the stress-corrosion cracking of type 304 stainless steel C-ring specimens in boiling magnesium chloride solutions [31]. Reproduced by permission of ECS – The Electrochemical Society

Other examples where the electrode potential affects the susceptibility to SCC or the cracking rate include the SCC of alloy 825 (a Ni–Cr–Fe–Mo alloy) in acidified solutions containing Cl^- ions and H_2S [32], the SCC of Fe-18 Cr-9 Ni stainless steel in boiling magnesium chloride solutions [20], and the SCC of various steels in chloride solutions [33, 20]. Beck [34] has shown that the crack growth rate of Ti-8Al-1Mo-1 V in 0.6 M halide solutions is a linear function of the electrode potential and that the effect is similar for Cl^-, Br^-, or I^- solutions.

The electrode potential also affects K_{Iscc} values. Fujii and Metzbower [35] found that the values for the high-strength steel HY 130 decreases with increasing negative potential, as shown in Table 11.5. Data compiled for various steels [6, 33] also indicate that cathodic protection decreases

Table 11.5 Effect of potential on K_{Iscc} values for HY-130 steel in 3.5% NaCl solutions [35]

Condition	E vs. Ag/AgCl (V)	K_{Iscc} (ksi $\sqrt{\text{in.}}$)	K_{Iscc} (MPa $\sqrt{\text{m}}$)
Freely corroding	–0.78	129	142
Coupled to Zn	–1.0	97	107
Coupled to Mg	–1.5	88	97

K_{Iscc} values. The effect of electrode potential on K_{Iscc} values for AISI 4340 steel in chloride solutions is shown in Fig. 11.25. It can be seen that there is a slight effect of electrode potential on K_{Iscc}. More dramatic changes in K_{Iscc} with electrode potential have been observed for other steels [36].

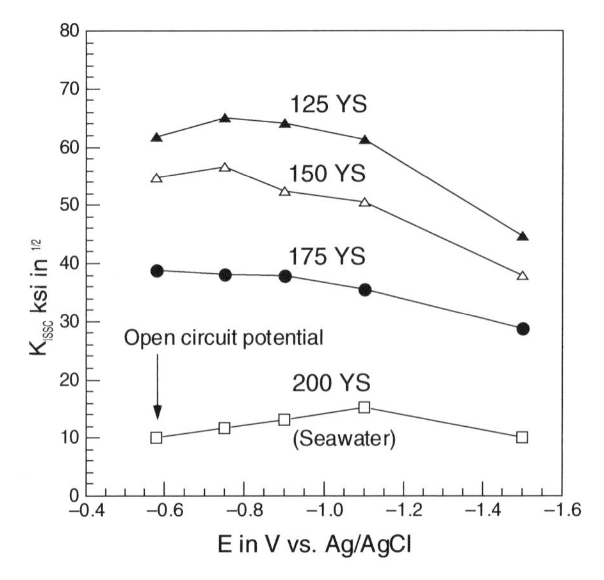

Fig. 11.25 Effect of electrode potential on the stress intensity factor of type 4340 steel of various yield strengths (YS) given in ksi. The solution is 3.5% NaCl (unless noted otherwise) [7, 33]

Effect of Cl^- Concentration and pH

In general, an increase in the concentration of a critical solution species causes an increase in the susceptibility to SCC, although relationships are not always simple. For example, an increase in Cl^- concentration causes an increase in the SCC susceptibility of type 304 stainless steel at 100°C, as shown in Fig. 11.26 [20]. But the effect of Cl^- is also related to the concentration of dissolved O_2, as shown in studies on the time to cracking of type 304 stainless steel [28]. Increasing the concentration of Cl^-, Br^-, or I^- ions increases the velocity of SCC cracks for Ti-8 Al-1 Mo-1 V [37].

The effect of pH depends on the particular system. For example, the lifetime of an Al-7 Mg alloy increases greatly for pH values greater than approximately 7 [25]. The lifetime of brass in ammoniacal copper sulfate also depended on pH, attaining a minimum lifetime at a pH of about 7.2 [38]. However, pH values between 1.05 and 11.5 had no effect on the SCC susceptibility of magnesium alloys in a salt solution [39].

It should be remembered, however, that in crack propagation it is the local pH within the crack tip that is important rather than the bulk pH. As noted in Table 11.3, various metals can generate locally acidic crack tip environments even though the bulk electrolyte is neutral.

Effect of Temperature

The effect of increase temperature is to increase the susceptibility to stress-corrosion cracking or to increase the rate of crack growth. Examples have been given by Sedriks [20] and by Marsh and Gerberich [36].

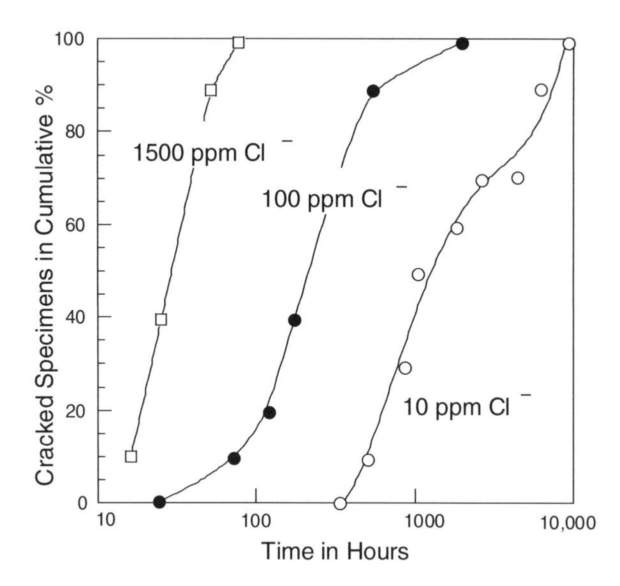

Fig. 11.26 Effect of chloride concentration on the SCC susceptibility of type 304 stainless steel [20]. Reproduced by permission of John Wiley & Sons, Inc

Mechanisms of SCC

Various mechanisms have been proposed to explain SCC. No single mechanism has been universally accepted and different mechanisms apply to different cases of SCC. The major mechanisms are

(1) Anodic dissolution
(2) Film rupture
(3) Stress-sorption cracking
(4) Hydrogen embrittlement

These mechanisms fall into two classifications: those which involve dissolution effects (the first two above) and those that involve mechanical effects (the last two above).

Anodic Dissolution

The anodic dissolution mechanism applies if there is an active anodic path established by microstructural heterogeneities in the metal. (This mechanism is also called the *electrochemical mechanism* of SCC or the *active path mechanism* of SCC.) As an example, $CuAl_2$ precipitates which form along the grain boundaries in an Al-4% Cu alloy render the grain boundaries anodic to the grain, as shown schematically in Fig. 11.27 [40]. Anodic dissolution of the grain boundary in the presence of an applied stress leads to intergranular SCC.

However, the anodic dissolution model cannot explain instances of transgranular SCC. For example, failures of stainless type 304 have occurred due to a phenomenon called *sensitization*. When austenitic stainless steels are heated through the temperature range 425–900°C (800–1650°F), chromium tends to combine with carbon to form chromium carbides. These carbides precipitate preferentially at grain boundaries, thus depleting chromium from the metal located in the grain boundaries. The grain boundaries thus become anodic to the grains, so that an active path exists

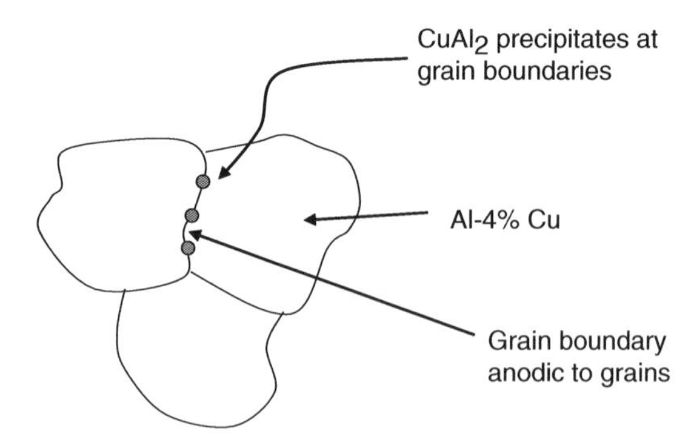

Fig. 11.27 Effect of $CuAl_2$ precipitates at grain boundaries on an Al-4% Cu alloy

along the grains for electrochemical dissolution. But the fracture mode has been observed to be transgranular rather than intergranular. (Proper heat treatment precludes sensitization of the stainless steel alloy [40].)

The electrochemical dissolution mechanism also cannot apply to cases where aluminum and titanium alloys undergo SCC in the presence of carbon tetrachloride, which is not an electrolyte.

Film Rupture Mechanism

The film rupture mechanism assumes that the oxide film on a stress-corrosion crack ruptures due to the stress in the film or near the crack tip. Bare metal near the crack tip is then exposed to the electrolyte within the crack, and the crack grows by anodic dissolution assisted by the applied stress. The process continues until the metal at the crack tip is repassivated, and the process then repeats itself, as illustrated in Fig. 11.28. This mechanism is supported by experimental observations of striations which are visible in the electron microscope and which are due to discontinuous cracking and arrest. The film rupture mechanism explains the SCC of brass in aqueous ammonia [41].

Stress-Sorption Cracking

According to the stress-sorption cracking mechanism, damaging ions weaken the cohesive bonds between metal surface atoms at the crack tip, as depicted in Fig. 11.29. This mechanism depends on the concept that the surface energy of a solid is reduced by chemisorption of a solution species and that the stress required for fracture is given by the Griffith criterion of crack formation in brittle solids [4]. For a crack in a thin plate,

$$\sigma_F = \left(\frac{2\gamma E}{\pi a}\right)^{1/2} \tag{18}$$

where σ_F is the stress required for brittle fracture, E is Young's modulus, γ is the surface free energy of the solid, and $2a$ is the crack length. Because adsorption of a solution species reduces γ, the stress σ_F required for fracture also decreases. (A common example of this phenomenon is the Rehbinder effect, in which a touch of saliva applied to a crack in a glass rod causes the glass to fracture at the crack more easily.)

Evidence for the stress-sorption mechanism was provided by experiments by Uhlig and Cook [31] on the effect of electrode potential on the SCC of type 304 stainless steel in boiling magnesium chloride solutions (referred to earlier). Uhlig and Cook [31] found that the addition of acetate or

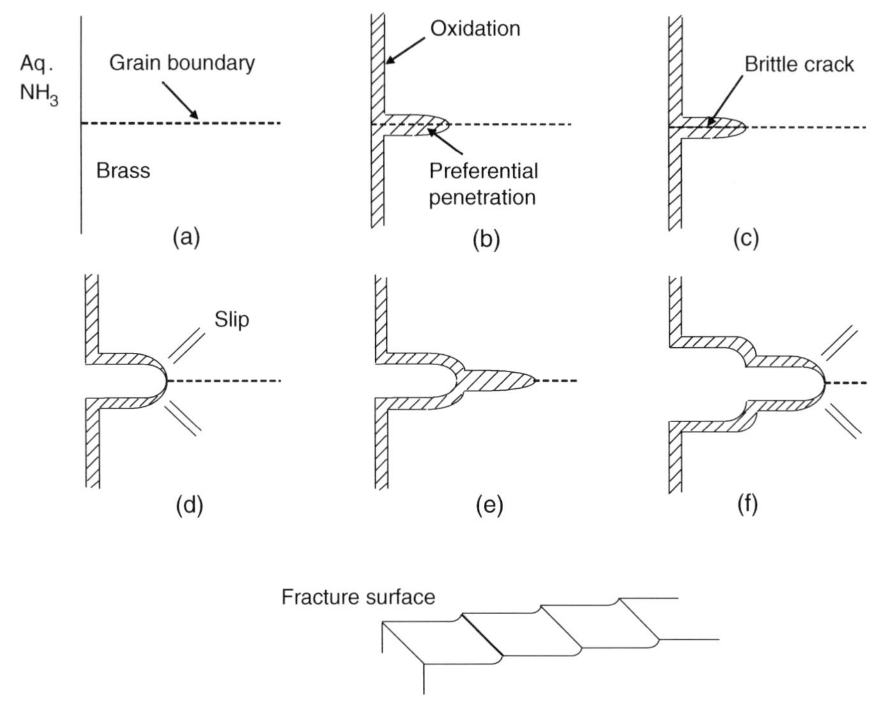

Fig. 11.28 Schematic illustration of the film rupture mechanism for stress-corrosion cracking. (**a**) The oxide film is considered to grow preferentially along a grain boundary (**b**) until it undergoes brittle fracture (**c**). The crack is arrested by slip in the substrate exposing metal to the environment (**d**). Further intergranular penetration then occurs (**e**) leading to further limited fracture (**f**). Crack propagation proceeds by repeated the cycles, and the resulting fracture surface shows evidence of discontinuous fracture [1, 41]

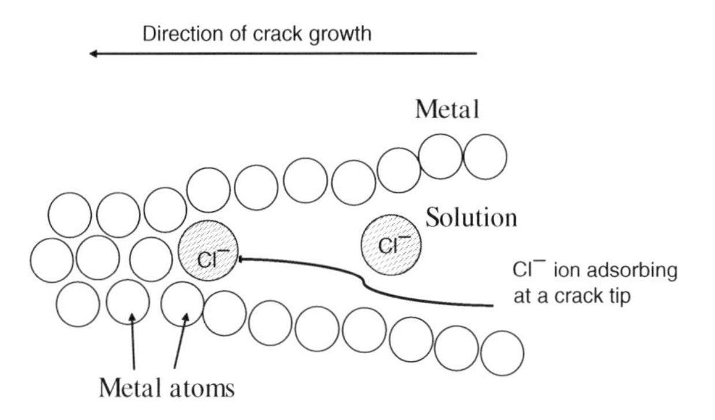

Fig. 11.29 Schematic diagram of the stress-sorption mechanism of SCC. An aggressive anion, such as Cl^-, adsorbs at or near the crack tip to weaken the metal–metal bond at the crack tip (this figure is drawn using 1.24 Å (0.124 nm) as the radius of a metal (Fe) atom and 1.81 Å (0.181 nm) as the radius of a chloride ion)

nitrate ions raised the critical potential for SCC, as shown in Table 11.6. These authors proposed that there is competitive adsorption between the inhibitive species (acetate or nitrate) and the aggressive species (chloride). Adsorption of the inhibitive species prevails when its concentration is sufficiently high. For 5% $NaNO_3$, for example, the corrosion potential lies below the critical potential for SCC, so that SCC does not occur at the open-circuit corrosion potential in this solution.

Table 11.6 Corrosion and critical potentials for cold-rolled type 304 stainless steel in $MgCl_2$ boiling at 130°C [31]

Anion addition	Corrosion potential E_{corr}(V vs. SHE)	Critical potential E_{crit}(V vs. SHE)	$E_{corr} < E_{crit}$?	SCC at E_{corr}?
None	–0.11	–0.145	No	Yes
0.1% sodium acetate	–0.12	–0.132	No	Yes
2% sodium acetate	–0.12	–0.116	Yes	No
None	–0.11	–0.145	No	Yes
2% sodium nitrate	–0.06	–0.090	No	Yes
5% sodium nitrate	–0.08	–0.070	Yes	No

The stress-sorption model also explains the SCC of alloys in non-electrolytes, such as aluminum alloys in carbon tetrachloride solutions.

Hydrogen Embrittlement

Hydrogen embrittlement refers to the loss of ductility or to the cracking in a metal caused by the entrance of hydrogen atoms into the metal lattice. (Hydrogen embrittlement is also called hydrogen-assisted cracking.) Hydrogen is a culprit due to several reasons. These include its small atomic size, its solubility in most metals, and its mobility (hydrogen atoms diffuse through most metals). In addition, hydrogen atoms can be present at the metal surface from several different sources. Hydrogen atoms arise from corrosion reactions or from various metal processing steps, such as melting or rolling, cleaning in acid solutions ("pickling"), or in electroplating from acid solutions.

Hydrogen atoms are produced by the reduction of hydrogen ions enroute to the evolution of hydrogen gas. The first step is

$$2\,H^+ + 2\,e^- \rightarrow 2\,H_{ads} \tag{19}$$

followed by either

$$2\,H_{ads} \rightarrow H_2 \tag{20}$$

or

$$H_{ads} + H^+ e^- \rightarrow H_2 \tag{21}$$

Equation (19) followed by Eq. (20) is called the *combination* mechanism of hydrogen evolution, while Eq. (19) together with Eq. (21) is called the *electrochemical* mechanism of hydrogen evolution [42]. Either mechanism involves the adsorption of hydrogen atoms, and some of these adsorbed H atoms can enter the metal rather than participating in Eqs. (20) or (21) (see also Chapter 7).

As shown by Brown and co-workers (Table 11.3), the electrolyte near the crack tip becomes acidified, so that hydrogen reduction and the production of hydrogen atoms is possible within the tip of the stress-corrosion crack. This possibility has been shown dramatically by Peterson et al. [10] for type 4340 steel using the experimental setup shown in Fig. 11.8. Potential–pH data for crack tip electrolytes formed in a 3.5% NaCl solution are superimposed on the partial Pourbaix diagram shown in Fig. 11.30. It can be seen that all data points lie below the "a" line for hydrogen evolution, so that it is possible for the reduction of hydrogen ions to occur within the crack tip.

The actual mechanism by which hydrogen embrittlement occurs is not entirely clear but may be due to several different effects. These are as follows:

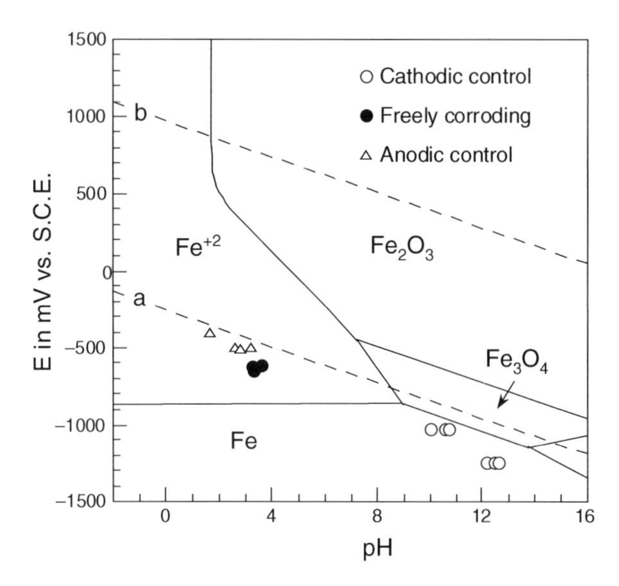

Fig. 11.30 Electrochemical conditions at a propagating stress-corrosion crack in type 4340 steel and superimposed on a Pourbaix diagram for Fe [7, 10]. Note that all experimental data lie below the "a" line for hydrogen evolution, so that the reduction of hydrogen ions is possible within the stress-corrosion crack. • Freely corroding, Δ external sample anodically polarized, o external sample cathodically polarized. (The experimental data were obtained using the device shown in Figure 11.8.) Reproduced by permission of © NACE International 1970

(1) An *internal pressure* is caused by the release of interstitial hydrogen atoms as molecular hydrogen gas within voids or flaws.
(2) *Brittle metal hydrides* are formed at the crack tip, and these brittle hydrides can be fractured to allow the crack to grow.
(3) Hydrogen diffuses to and adsorbs at the crack tip, *reducing the surface free energy*. (This is a form of stress-sorption cracking.)
(4) Hydrogen diffuses ahead of the crack tip and aids in *plastic deformation* of the metal in advance of the crack.

Experimental evidence for the plastic deformation mechanism has been provided by Beachem [43], who used the electron microscope to examine in detail the fracture surfaces on various carbon steels subjected to SCC in aqueous NaCl or charged with hydrogen gas. The appearance of regions of plastic-deformed metal ahead of the crack tip suggested that the presence of sufficiently concentrated hydrogen dissolved in the metal just ahead of the crack tip assisted the fracture process.

Anodic Dissolution vs. Hydrogen Embrittlement

It is possible to distinguish between anodic dissolution and hydrogen embrittlement by observing the effect of the electrode potential on the time to failure. Figure 11.31 compares schematically the mechanisms of anodic dissolution and hydrogen embrittlement [44] and also illustrates the expected effect of the electrode potential on the susceptibility to SCC. For the anodic dissolution mechanism, polarization in the anodic direction increases the rate of anodic dissolution and thus decreases the time to failure. For the hydrogen embrittlement mechanism, polarization in the cathodic direction increasingly generates more hydrogen atoms for penetration into the metal, and thus decreases the time to failure.

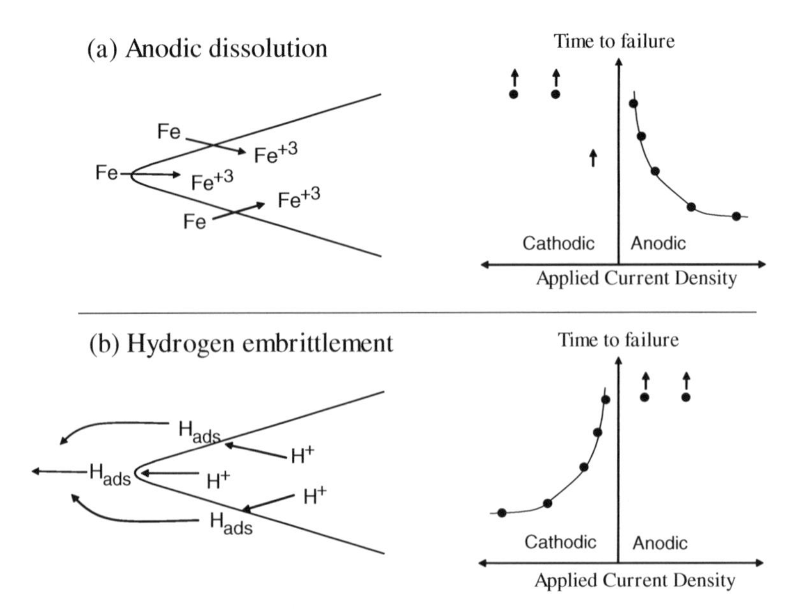

Fig. 11.31 Schematic comparison of the anodic dissolution mechanism (active path corrosion) of SCC and the hydrogen embrittlement mechanism of SCC. Adapted from Brown [44]

Figure 11.32 illustrates the effect of small impressed currents on the time to failure of a martensitic steel in an aqueous solution containing $(NH_4)_2S$ and acetic acid [44]. A small cathodic current decreases the time to fracture, while small anodic currents increase the fracture time. This behavior indicates that in the case of zero impressed current, the failure occurs by hydrogen embrittlement.

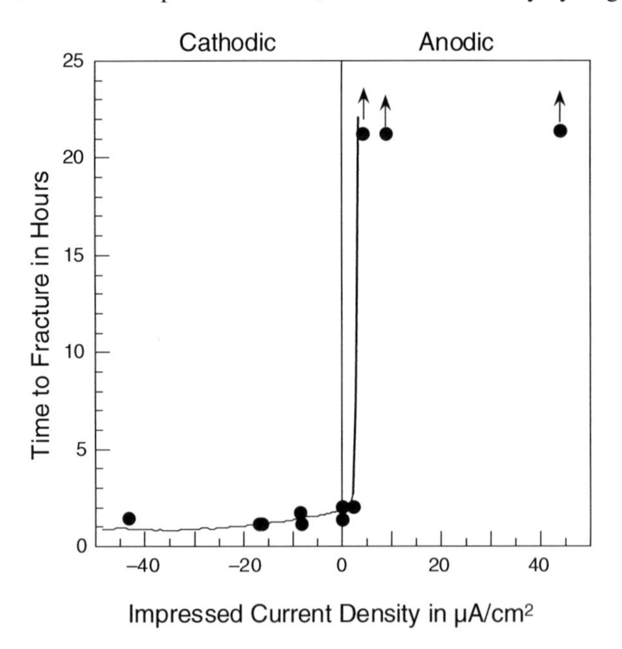

Fig. 11.32 Effect of impressed current on time to fracture of a martensitic steel in 5% acetic acid plus 0.1% ammonium sulfide. Redrawn from Brown [44]. This figure shows that the mechanism of stress-corrosion cracking was hydrogen embrittlement

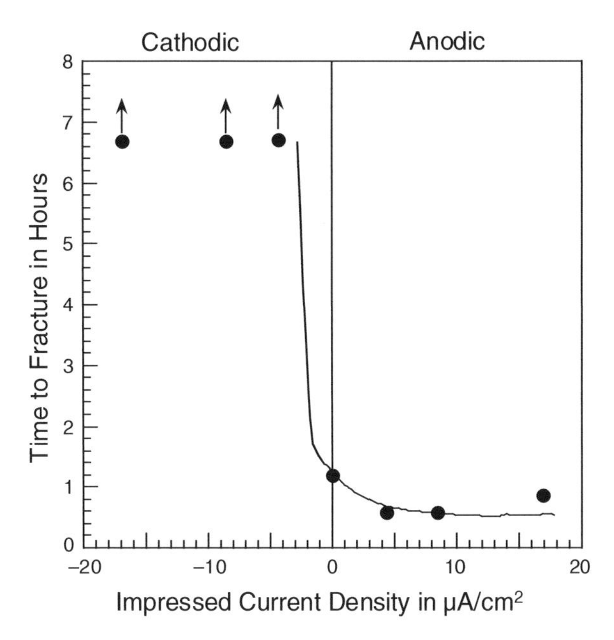

Fig. 11.33 Effect of impressed current on time to fracture of a martensitic steel in distilled water containing H_2S. Redrawn from Brown [44]. This figure shows that the mechanism of stress-corrosion cracking was anodic dissolution

Figure 11.33 shows the effect of small impressed currents on the same steel but in a different solution (H_2S in distilled water) [44]. Here a small cathodic current increases the time to failure, but a small anodic current decreases the time to failure, indicating that the SCC mechanism is anodic dissolution. (Not shown in the figure is that hydrogen embrittlement occurs at larger impressed cathodic currents.)

If it is suspected that hydrogen embrittlement is the cause of SCC, then hydrogen permeation measurements should be made. Hydrogen permeation can be determined by an electrochemical technique described elsewhere [45], and such experiments provide information as to hydrogen uptake as a function of electrode potential. Using such measurements, Wilde [46] showed that the SCC of a high-strength Fe-12 Cr-1 Mo steel in sodium chloride solutions was due to hydrogen embrittlement.

On a Universal Mechanism of SCC

Each of the different SCC mechanisms discussed in the previous sections has its own shortcomings. Some of these are listed in Table 11.7, which illustrate the point made earlier that no single mechanism of SCC can apply to all cases of SCC.

Protection Against Stress-Corrosion Cracking

Methods of preventing SCC are based upon our knowledge of the mechanisms of SCC and upon engineering experience. Preventative measures include the following [47]:

(1) Lowering the stress below the threshold stress for fracture or maintaining K_I below values of K_{Iscc}.

Table 11.7 Shortcomings in the various mechanisms of stress-corrosion cracking

Mechanism	Comment
Anodic dissolution	SCC of sensitized type 304 stainless steel often occurs transgranularly even though possible active paths may exist at grain boundaries
Film rupture	Requires extremely high local corrosion rates to explain crack growth rates
Stress sorption	Local plastic deformation would blunt a crack tip and eliminate the sharp tip required for this mechanism to apply. In addition, there is a competition for the bare metal surface between the adsorption of solution species and oxide film formation
Hydrogen embrittlement	The rate of hydrogen penetration cannot always be related to SCC susceptibility

(2) Eliminating the critical environmental species (if possible).
(3) Applying cathodic protection, but not if the mechanism of SCC is hydrogen embrittlement.
(4) Adding inhibitors to the solution.
(5) Materials selection. (For example, Ni additions to Fe-18 Cr stainless steels increase the time to failure in boiling magnesium chloride solutions [20] and Mo additions to Fe–Cr–Ni alloys raise the K_{Iscc} values in sodium chloride solutions, as shown in Fig. 11.18.)
(6) Short peening of the metal surface to put the surface in a state of compressive stress (rather than tensile stress) can improve the resistance to SCC [48]. (Shot peening is the controlled impingement on a metal surface of particles of steel, glass, or ceramics.)

Corrosion Fatigue

Environmental cracking under a cyclic load, as shown in Fig. 11.34 [49], is referred to as corrosion fatigue. Corrosion fatigue fracture surfaces often have characteristic striations due to the intermittent nature of crack growth. Environmental effects are not as specific as in SCC, and the crack growth is usual transgranular, whereas stress-corrosion cracks can be either transgranular or intergranular. The mechanism of corrosion fatigue is not well understood but involves the local slip of metal

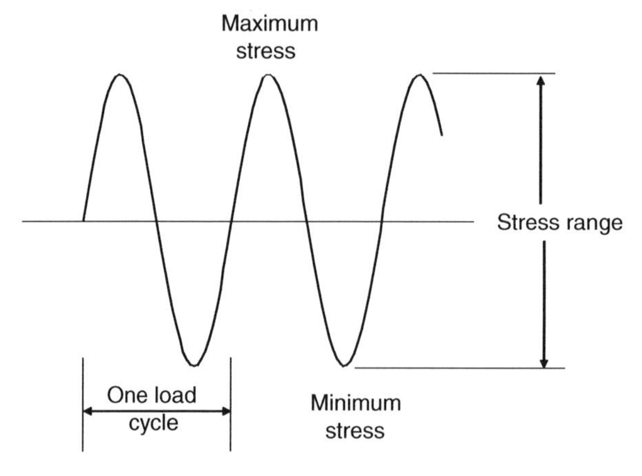

Fig. 11.34 Load cycling in a corrosion fatigue experiment

atoms within grains assisted by the corrosion process. The same mechanisms which apply to SCC also apply to corrosion fatigue [50] with the added complexity that the interaction between the environment and mechanical properties is more complicated in the presence of a cyclic load.

Corrosion Fatigue Data

Corrosion fatigue is characterized by plots of tensile stress (*S*) versus the number of cycles (*N*) to failure, as in Fig. 11.35, which are called *S–N* curves [51, 52]. The stress *S* is the maximum stress in the cycle. The definition of cycles to failure varies with individual researcher and can either be the initiation of a small visible macrocrack or final failure of the test specimen [52].

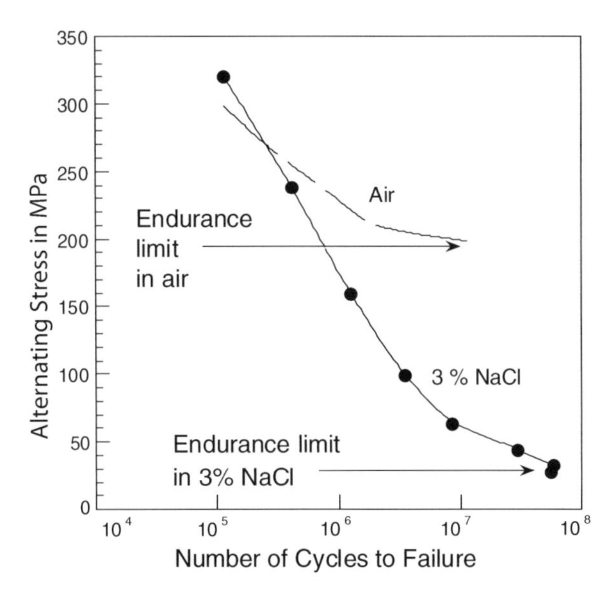

Fig. 11.35 *S–N* fatigue curves for a 0.21% C steel in air and in 3% NaCl [51]

Figure 11.35 shows *S–N* curves for a 0.21% C steel in air and in 3% NaCl [51]. In a given environment, failure does not occur if the stress is held below a critical value called the *endurance limit*. The effect of the environment is to lower the endurance limit and thus make failure easier. The *S–N* curves are not usually frequency dependent in air but are frequency dependent in a corrosive environment. Corrosion fatigue is more pronounced at lower frequencies because the crack spends more time in the local environment before the local electrolyte with its changes in solution chemistry is replaced due to the pumping action of the opening and closing of the crack.

A second way of presenting corrosion fatigue data is in a plot of the crack growth per cycle, $\mathrm{d}a/\mathrm{d}N$, versus the range of the stress intensity factor, ΔK, where $\Delta K = K_{\max} - K_{\min}$. This type of plot is shown schematically in Fig. 11.36 [52]. This figure shows that the effect of the corrosive environment is to increase the crack growth rate. In addition, the effect of the environment is most prominent at the low ΔK values and is least effective at high ΔK values where mechanics is dominant and where fast fracture occurs. Plots of $\mathrm{d}a/\mathrm{d}N$ vs. ΔK display a linear region:

$$\frac{\mathrm{d}a}{\mathrm{d}N} = C(\Delta K)^n \tag{22}$$

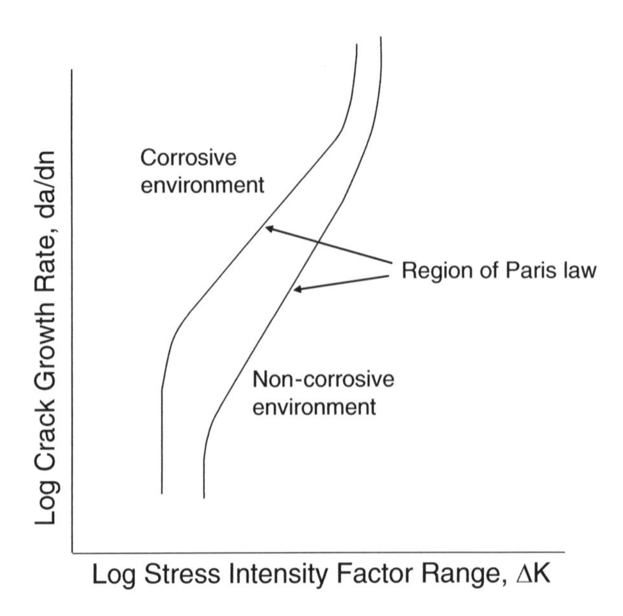

Fig. 11.36 Schematic plot of the fatigue crack growth per cycle, da/dN, vs. the range of the stress intensity factor, ΔK, where $\Delta K = K_{max} - K_{min}$ [52]

where C and n are constants to be determined experimentally. Equation (22) is called the Paris law [53]. The parameter n is a positive number and is usually between 2 and 4. For example, for the corrosion fatigue of several precipitation-hardened stainless steels in seawater, n has a value of approximately 2 [54].

Figure 11.37 shows experimental da/dN vs. ΔK curves in flowing seawater [55] for Al alloy 7475 (Al-5 Zn-2 Mg-1.5 Cu), 17-4 precipitation-hardened (PH) steel (Fe-16 Cr-4 Ni-3 Cu), and a high-strength steel HY-130 (Fe-4 Ni with minor amounts of Cr, Mo, and Mn). As shown in Fig. 11.37, at any given value of ΔK, the aluminum alloy has the greatest crack growth rate of the three alloys in flowing seawater.

Protection Against Corrosion Fatigue

Preventative measures are similar to those for stress-corrosion cracking. These include [47] reducing the stress on the component or annealing to relieve residual stress. Because corrosion fatigue is a surface phenomenon, surface finish is of major importance, and procedures such as polishing to smooth the surface eliminates stress concentration sites. As in the case of SCC, shot peening of the surface is also useful for protection against corrosion fatigue because the surface is left in a state of residual compression [52].

Corrosion fatigue resistance can also be improved by using corrosion inhibitors or coatings such as nitride coatings. Steels containing chromium and molybdenum (which form nitrides) can be treated by anhydrous ammonia to produce nitride surface layers having improved corrosion fatigue resistance. Ion implantation of nitrogen into titanium surgical implants has been found to increase the resistance to corrosion fatigue in simulated body fluids [56]. (Ion implantation as a surface modification technique is discussed in Chapter 16 on "Special Topics in Corrosion Science".)

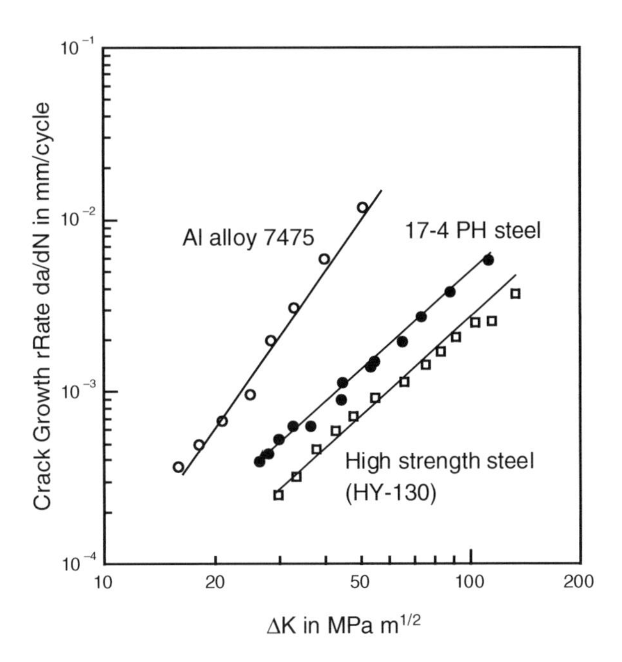

Fig. 11.37 Corrosion fatigue data for three different alloys in flowing seawater (see the text for alloy compositions). Redrawn from Ref. [55] by permission from ASTM International, 100 Barr Harbor Drive, West Coshohocken, PA 19428

Corrosion fatigue can also be minimized by design considerations such as eliminating crevices where corrosion can be initiated.

Cavitation Corrosion

Cavitation corrosion occurs when the flow of a corrosive liquid produces a localized low pressure which leads to the formation of bubbles in the liquid. These low pressures necessary to form bubbles can be caused by fluid flow across curved interfaces. The force *of* the collapse of these bubbles involves creation of shock waves and high-velocity microjects [57, 58]. The impact of these waves and jets against the metal surface causes a mechanical damage, which is intensified by the corrosive effect of the liquid.

The mechanical damage is caused by continuous and local bombardment of the vapor bubbles against the metal surface, as depicted in Fig. 11.38. The collapse of the bubbles against the metal surface destroys the passive film and removes underlying metal, as shown schematically in Fig. 11.38(a) – (c). The metal surface may repassivate, but the process is repeated as in Fig. 11.38(d)–(f). The appearance of cavitation is similar to pitting, but the pitted areas are closely spaced and the surface is roughened considerably.

The pressure ΔP required to maintain a spherical bubble of radius r in a liquid having a surface tension γ is given by the Young and Laplace equation [59]:

$$\Delta P = \frac{2\gamma}{r} \tag{23}$$

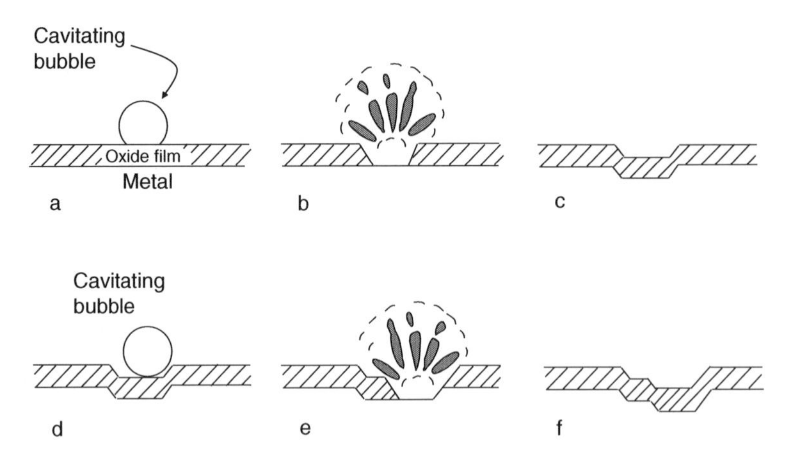

Fig. 11.38 Schematic drawing of cavitation damage. A cavitation bubble in (**a**) collapses and destroys the oxide film in (**b**) and the oxide film is reformed in (**c**). The process is repeated in steps (**d**)–(**f**). Redrawn from Ref. [47] with the permission of The McGraw-Hill Companies

The magnitude of this pressure is given in the following calculation.

Example 11.2: A laboratory vibrating device used to study cavitation corrosion produces spherical bubbles which are 20 μm in diameter [57]. Calculate the pressure difference across these bubbles.
Solution: The surface tension of water is 72 mJ/m^2. Thus,

$$\Delta P = \frac{2(72 \times 10^{-3}\mathrm{J/m^2})}{10 \times 10^{-6}\,\mathrm{m}}$$

$$\Delta P = 14{,}400\,\frac{\mathrm{J}}{\mathrm{m}^3}$$

But 1 J = 1 N m, so that

$$\Delta P = 14{,}400\,\frac{N}{\mathrm{m}^2} = 14{,}400\ \mathrm{Pa} = 0.0144\,\mathrm{MPa}$$

A small pressure of the order of 0.01 MPa is not sufficient to cause local deformation in a metal as, for example, in a typical aluminum alloy, where the yield stress is 460 MPa (Fig. 11.21). Thus, most of the mechanical damage to the alloy is produced by the hydrodynamic effects of the liquid due to shock and microjets rather than by the simple collapse of the bubbles. The total pressure inside a bubble is thus

$$P = P_{\mathrm{external}} + \frac{2\gamma}{r} \tag{24}$$

where P_{external} includes all the external effects on the bubble such as hydrodynamic effects plus the effect of atmospheric pressure.

Cavitation corrosion is studied using a device which vibrates at high frequencies to produce cavitating bubbles on the surface of a metal which is immersed in a liquid [60]. Engleberg and Yahalom [61] were among the earliest to demonstrate that there is an electrochemical component to cavitation damage. Figure 11.39 shows the effect of cavitation on the anodic polarization curve of a carbon steel in a potassium hydrogen phthalate buffer of pH 4.0. Cavitation had no effect on the active region

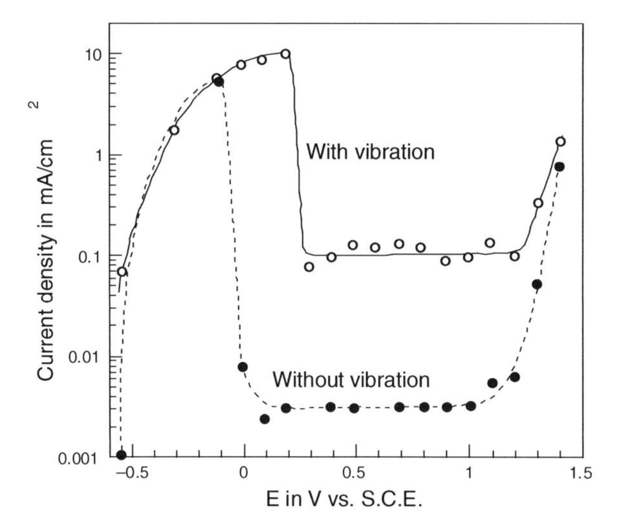

Fig. 11.39 Effect of cavitation on the anodic behavior of a carbon steel in a phthalate buffer of pH 4.0 [61]. Reproduced by permission of Elsevier Ltd

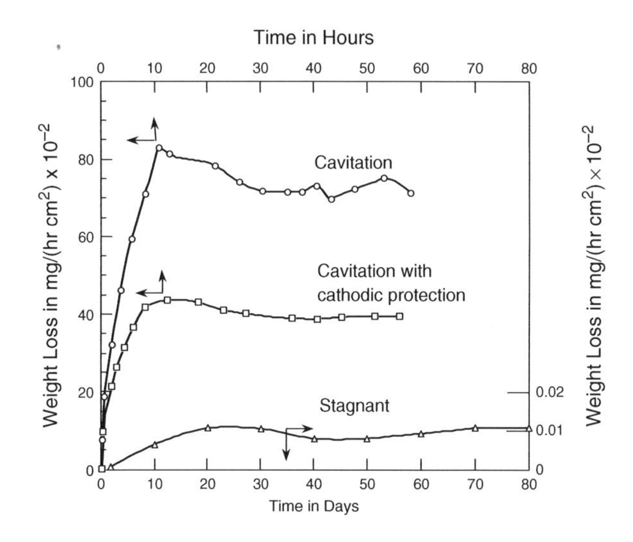

Fig. 11.40 Weight loss data for a nickel aluminum bronze alloy in seawater with and without cavitation [62]. Cathodic protection reduces the weight loss for the cavitating solution. Reproduced by permission of NACE International

of the polarization curve but increased the current density in the passive region by a factor of 30. Figure 11.40 shows rate loss data for a nickel aluminum bronze (80% Cu-9% Al-4.9% Fe-4.9% Ni-1.2% Mn) in seawater with and without cavitation [62]. The effect of cavitation is to increase the rate of metal loss, but this rate can be decreased by cathodic protection. Some investigators suggest that the role of cathodic protection is to provide hydrogen bubbles which cushion the shock wave produced by the cavitating bubbles [58].

Cavitation corrosion, like corrosion fatigue, is usually transgranular in nature [63]. In this regard the metallurgical microstructure of the alloy is important because cracks can be initiated by the pile-up of dislocations. To prevent or delay this process, it is necessary to produce a large number of very fine homogeneously dispersed obstacles, such as precipitates, grain boundaries, or needle-like phases. Therefore surfaces containing a fine dispersion of a hard particle in a more ductile matrix should improve the cavitation resistance. The high cavitation resistance of martensitic stainless steels has been attributed to the limitation of dislocation motion by the fine platelets of the martensitic phase [58, 63]. (Martensite is a hard and brittle carbon-containing solid phase found in certain steels.)

Cavitation corrosion can be minimized by several approaches [47, 58] including a change in design to minimize hydrodynamic flow conditions, by providing smooth surfaces to preclude sites for bubble initiation, by cathodic protection, and by alloy selection. Titanium alloys and Ni–Cr–Mo alloys which form stable passive films exhibit good resistance to cavitation corrosion [58]. Various compilations of cavitation corrosion data are available [47, 57, 64].

Erosion Corrosion and Fretting Corrosion

Erosion corrosion is the attack of a metal caused by the movement of a corrosive liquid against the metal surface. The velocity of the fluid plays an important role in erosion corrosion and increases in velocity generally lead to increases in erosion corrosion. Turbulent flow results in an increased contact of the liquid with the metal surface compared to laminar flow, as depicted in Fig. 11.41, and causes more damage than laminar flow. For instance, increased corrosion at the inlet ends of tubing is due to turbulent flow at the inlet side (as compared to laminar flow further down the pipe.)

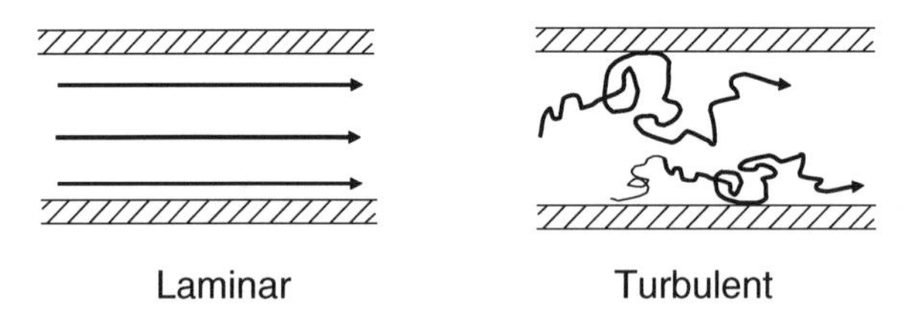

Fig. 11.41 Schematic diagram of laminar flow vs. turbulent flow in a pipe (The direction of flow is from *left* to *right*)

Protective measures [47] include design considerations, materials selection, or the addition of inhibitors. Design considerations include reduction of surface velocity and turbulence by measures such as larger radius elbows, larger pipe diameters, or gradual changes in flow channel dimensions. Materials selection involves the use of passive alloys, such as stainless steel or titanium, which have increased resistance to erosion corrosion relative to plain carbon steels [47].

Fretting corrosion is caused by the slight periodic motion of two surfaces rubbing against each other and is due to the combined effects of wear and corrosion. The mechanical damage is caused by the abrasive rupture of the oxide film and dislodgement of metal particles which themselves are eventually turned into oxide particles that accumulate between the two moving metal surfaces. These oxide debris particles act as abrasives which continue to add to the mechanical wear. With continuing mechanical motion, the passive film again breaks down, and the process repeats itself.

Fretting corrosion can be reduced by various measures [40, 47]. These include the use of lubricants which reduce friction and also act as corrosion inhibitors, reducing the load between fretting surfaces, the use of elastomer gaskets between the abrading surfaces, and the use of wear-resistant alloys (such as cobalt-based alloys).

The key scientific issues for both erosion corrosion and fretting corrosion involve passive film breakdown and repassivation (or lack thereof) in the presence of the mechanical disruption caused by either fluid flow or wear.

Problems

1. Explain why the parameter K_{Iscc}, which arises from fracture mechanics, can vary with the electrode potential.
2. Calculate the stress intensity factor K_I for a high-strength steel having a yield stress of 1,470 MPa and a thumbnail-shaped crack of depth a =1.0 mm when the depth of the crack is one-quarter the crack length and the alloy is loaded to one-half of its yield stress.
3. Show that K_I for a cantilever beam specimen as given in Fig. 11.16 has the units of MPa $\sqrt{m}$.
4. If a high-strength steel has a critical stress intensity factor of 20 ksi $\sqrt{in.}$ in seawater, what is the longest crack length that can be tolerated if the structure is loaded to a stress of 50 ksi. Assume a center crack in an infinite sheet.
5. The following data are for the stress-corrosion cracking of a titanium alloy in 3.5% NaCl [16]. The alloy had a yield stress of 104 ksi.

Time to fracture (min)	Stress intensity factor K_I(ksi $\sqrt{in.}$)
0.1	128
2.0	55
2.5	71
3.7	52
8.1	56
150	53
655	49
No break in 1,380 min	47

   (a) Plot these data to graphically determine the critical stress-intensity factor, K_{Iscc}.
   (b) For an edge crack which is 0.20 in. long (a = 0.20 in.) in a plate of the alloy which is 10.0 in. wide, what is the maximum external stress which can be applied without causing fracture in the 3.5% NaCl solution?
   (c) Suppose that the value of the external stress found in Part (b) above is applied to a specimen containing a thumbnail-shaped crack where the crack length is the same as in Part (b) above ($2b$ = 0.20 in.). The crack depth a is given by a/b = 0.5. Will fracture occur in this case?

6. Calculate the safe-zone plot for an aluminum alloy in a corrosive environment when the critical stress intensity factor is 13 MPa $\sqrt{m}$, the threshold value for fracture (for smooth specimens) is 48 MPa, and the yield stress of the alloy is 410 MPa. Assume a center crack in an infinite specimen. Indicate regions on the diagram where smooth specimens are immune to fracture, where cracked specimens are immune to fracture, and the overall fracture-safe zone.
7. Based on the effect of electrode potential on K_{Iscc} for HY-130 steel as given in Table 11.5, suggest a possible mechanism for stress-corrosion cracking in 3.5% NaCl.
8. The rate of crack growth (v) of a titanium alloy in 0.6 M NaCl at –300 mV SCE was observed to be 0.010 cm/s [34]. The crack had the geometry shown below [66]. Calculate the current density required for this crack growth rate if the crack growth is due to anodic dissolution. Assume that

the anodic current is due to the reaction:

$$Ti \rightarrow Ti^{3+} + 3e^-$$

The density of Ti is 4.5 g/cm^3 and the atomic weight is 47.87 g/mol.

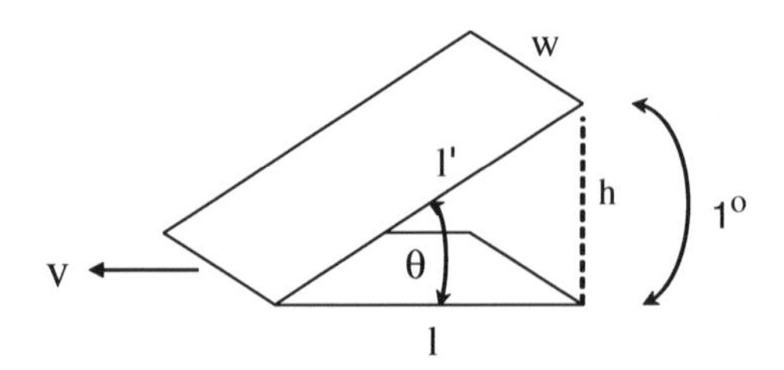

9. If the effect of an applied current on the time to failure in stress-corrosion cracking is as shown below, what can you say about the mechanism of SCC at the open-circuit potential (zero applied current)?

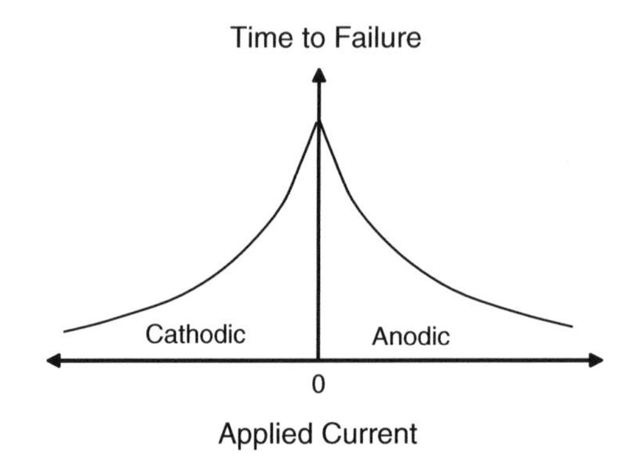

10. In what ways is corrosion fatigue similar to stress-corrosion cracking? In what ways do these two forms of localized corrosion differ?
11. For a precipitation-hardened stainless steel in seawater, the constants in the Paris Law, i.e., Eq. (22), are $C = 1.6 \times 10^{-9}$ (MPa)$^{-2}$ and $n = 2$ [54]. How many cycles will it take for a crack to grow by 1.0 mm if $\Delta K = 20$ MPa $\sqrt{m}$?
12. How are cavitation corrosion, erosion corrosion, and fretting corrosion similar to each other? How do they differ from each other?
13. Suppose that you have access to a high-velocity flume of a liquid which enables you to study erosion corrosion. Devise an experiment or test which will determine the effect of electrode potential on the total erosion-corrosion damage.

References

1. B. F. Brown, "Stress Corrosion Cracking Control Measures", U. S. Government Printing Office, Washington, DC (1977).
2. R. W. Staehle in "Fundamental Aspects of Stress Corrosion Cracking", R. W. Staehle, A. J. Forty, and D. van Rooyen, Eds., p. 3, NACE, Houston, TX (1969).

3. J. H. Brunton and M. C. Rochester in "Treatise on Materials Science and Technology", Vol. 16, C. M. Preece, Ed., p. 185, Academic Press, New York, NY (1979).
4. A. S. Tetelman and A. J. McEvily, Jr., "Fracture of Structural Materials", Chapters 1 and 2, John Wiley, New York, NY (1967).
5. B. F. Brown in "Stress Corrosion Cracking of Metals – A State of the Art", ASTM STP 518, p. 3, ASTM, Philadelphia, PA (1972).
6. B. F. Brown in "Stress-Corrosion Cracking in High Strength Steels and in Titanium and Titanium Alloys", B. F. Brown, Ed., p. 2, U. S. Government Printing Office, Washington, DC (1972).
7. B. F. Brown in "The Theory of Stress Corrosion Cracking in Alloys", J. C. Scully, Ed., p. 186, NATO, Brussels, Belgium (1971).
8. B. F. Brown, C. T. Fujii, and E. P. Dahlberg, *J. Electrochem. Soc.*, *116*, 218 (1969).
9. G. Sandoz, C. T. Fujii, and B. F. Brown, *Corros. Sci.*, *10*, 839 (1970).
10. J. A. Smith, M. H. Peterson, and B. F. Brown, *Corrosion*, *26*, 539 (1970).
11. J. C. Scully, "The Fundamentals of Corrosion", p. 177, Pergamon Press, Oxford (1990).
12. M. Kowaka and T. Kudo, *Trans. JIM*, *16*, 385 (1975).
13. G. R. Irwin and A. A. Wells, *Met. Revs.*, *10*, 223 (1965).
14. B. F. Brown, *Met. Revs.*, *13*, 171, (1968).
15. B. F. Brown and C. D. Beachem, *Corros. Sci.*, *5*, 745 (1965).
16. M. H. Peterson, B. F. Brown, R. L. Newbegin, and R. E. Groover, *Corrosion*, *23*, 142 (1967).
17. P. C. Paris and G. C. Sigh in "Fracture Toughness Testing and Its Applications", ASTM STP 381, p. 30, ASTM, Philadelphia, PA (1965).
18. H. Tada, P. C. Paris, and G. R. Irwin, "The Stress Analysis of Cracks Handbook", Del Research Corporation, St. Louis, MO (1973).
19. S. A. Meguid, "Engineering Fracture Mechanics", p. 134, Elsevier Applied Science, London (1989).
20. A. J. Sedriks, "Corrosion of Stainless Steels", Chapter 7, Wiley-Interscience, New York, NY (1996).
21. R. W. Judy, Jr. and R. J. Goode, "Stress-Corrosion Cracking Characteristics of Alloys of Titanium in Salt Water", NRL Report 6564, Naval Research Laboratory, Washington, DC, July 21 (1967).
22. R. W. Schutz in "Stress-Corrosion Cracking", R. H. Jones, Ed., p. 265, ASM International, Materials Park, OH (1992).
23. A. J. Sedriks, "Stress Corrosion Cracking Test Methods", NACE, Houston, TX (1990).
24. M. V. Hyatt and M. O. Speidel in "Stress-Corrosion Cracking in High Strength Steels and in Titanium and Titanium Alloys", B. F. Brown, Ed., p. 147, U. S. Government Printing Office, Washington, DC (1972).
25. D. O. Sprowls and R. H. Brown in "Fundamental Aspects of Stress Corrosion Cracking", R. W. Staehle, A. J. Forty, and D. van Rooyen, Eds., p. 466, NACE, Houston, TX (1969).
26. S. W. Ciaraldi in "Stress-Corrosion Cracking", R. H. Jones, Ed., p. 41, ASM International, Materials Park, OH (1992).
27. E. N. Pugh, J. V. Craig, and A. J. Sedriks in "Fundamental Aspects of Stress Corrosion Cracking", R. W. Staehle, A. J. Forty, and D. van Rooyen, Eds., p. 118, NACE, Houston, TX (1969).
28. R. M. Latanision and R. W. Staehle in "Fundamental Aspects of Stress Corrosion Cracking", R. W. Staehle, A. J. Forty, and D. van Rooyen, Eds., p. 214, NACE, Houston, TX (1969).
29. R. N. Parkins in "Fundamental Aspects of Stress Corrosion Cracking", R. W. Staehle, A. J. Forty, and D. van Rooyen, Eds., p. 361, NACE, Houston, TX (1969).
30. R. W. Hertzberg, "Deformation and Fracture Mechanics of Engineering Materials", p. 129, John Wiley & Sons, New York, NY (1996).
31. H. H. Uhlig and E. W. Cook, Jr, *J, Electrochem. Soc.*, *116*, 173 (1969).
32. N. Sridhar and G. Cragnolino in "Stress-Corrosion Cracking", R. H. Jones, Ed., p. 131, ASM International, Materials Park, OH (1992).
33. G. Sandoz in "Stress-Corrosion Cracking in High Strength Steels and in Titanium and Titanium Alloys", B. F. Brown, Ed., p. 79, U. S. Government Printing Office, Washington, DC (1972).
34. T. R. Beck in "The Theory of Stress Corrosion Cracking in Alloys", J. C. Scully, Ed., p. 64, NATO, Brussels, Belgium (1971).
35. C. T. Fujii and E. A. Metzbower "Stress-Corrosion Cracking in the HY-130 System", NRL Memorandum Report 2814, Naval Research Laboratory, Washington, DC, June (1974).
36. P. G. Marsh and W. W. Gerberich in "Stress-Corrosion Cracking", R. H. Jones, Ed., p. 64, ASM International, Materials Park, OH (1992).
37. M. J. Blackburn, W. H. Smyrl, and J. A. Feeny in "Stress-Corrosion Cracking in High Strength Steels and in Titanium and Titanium Alloys", B. F. Brown, Ed., p. 245, U. S. Government Printing Office, Washington, DC (1972).

38. J. A. Beavers in "Stress-Corrosion Cracking", R. H. Jones, Ed., p. 211, ASM International, Materials Park, OH (1992).
39. W. K. Miller in "Stress-Corrosion Cracking", R. H. Jones, Ed., p. 251, ASM International, Materials Park, OH (1992).
40. H. H. Uhlig and R. W. Revie, "Corrosion and Corrosion Control", Chapter 7, John Wiley, New York, NY (1985).
41. E. N. Pugh in "The Theory of Stress Corrosion Cracking in Alloys", J. C. Scully, Ed., p. 418, NATO, Brussels, Belgium (1971).
42. J. O'M. Bockris and A. K. N. Reddy, "Modern Electrochemistry", Vol. 2, p. 1233, Plenum Press, New York, NY (1977).
43. C. D. Beachem, *Metall. Trans.*, *3*, 437 (1972).
44. B. F. Brown, "Stress-Corrosion Cracking and Related Phenomena in High-Strength Steels", NRL Report 6041, Naval Research Laboratory, Washington, DC, November 6 (1963).
45. M. A. Devanathan and Z. Stachurski, *J. Electrochem.* Soc., *111*, 619 (1964).
46. B. E. Wilde, *Corrosion*, *27*, 326 (1971).
47. M. G. Fontana and N. D. Greene, "Corrosion Engineering", pp. 72, 88–115, McGraw-Hill, New York, NY (1978).
48. C. P. Dieport in "Impact Surface Treatment", S. A. Meguid, Ed., p. 86, Elsevier Applied Science, London, (1986).
49. D. W. Hoeppner in "Corrosion Fatigue: Chemistry, Mechanics, and Microstructure", A. J. McEvily and R. W. Staehle, Eds., p. 3, NACE, Houston, TX (1972).
50. D. J. Duquette in "Corrosion Mechanisms", F. Mansfeld, Ed., p. 367, Marcel Dekker, New York, NY (1987).
51. V. I. Pokhmurskii and A. M. Krohmalnyi in "Corrosion Fatigue", R. M. Parkins and Y. M. Kolotrykin, Eds., p. 54, The Metals Society, London (1983).
52. T. W. Crooker "Basic Concepts for Design Against Structural Failure by Fatigue Crack Propagation", NRL Report 7347, Naval Research Laboratory, Washington, DC, January 13 (1972).
53. P. Paris and F. Erdogan, *Trans. ASME, J. Basic Engineering, Series D*, *85*, 528 (1963).
54. A. J. Sedriks, "Corrosion of Stainless Steels", Chapter 8, Wiley-Interscience, New York, NY (1996).
55. F. D. Bogar and T. W. Crooker, *J. Testing Evaluation*, *7*, 155 (1979).
56. J. Yu, Z. J. Zhao, and L. X. Li, *Corrros. Sci.*, *35*, 587 (1993).
57. C. M. Preece in "Treatise on Materials Science and Technology", Vol. 16, C. M. Preece, Ed., p. 249, Academic Press, New York, NY (1979).
58. A. Karimi and J. L. Martin, *Inter. Metal Revs.*, *31*, 1 (1986).
59. A. W. Adamson, "Physical Chemistry of Surfaces", 3rd edition, p. 2, John Wiley, New York, NY (1976).
60. "Standard Test Method for Cavitation Erosion Using Vibratory Apparatus", ASTM Test Method G32-92, "1998 Annual Book of ASTM Standards", Volume 3.02, p. 103, ASTM, West Conshohocken, PA (1998).
61. G. Engelberg and J. Yahalom, *Corros. Sci.*, *12*, 649 (1972).
62. A. Al-Hashem, P. G. Caceres, W. T. Riad, and H. M. Shalaby, *Corrosion*, *51*, 331 (1995).
63. W. W. Hu, C. R. Clayton, and H. Herman, *Mater. Sci. Eng*, *45*, 263 (1980).
64. N. D. Tomashov, "Theory of Corrosion and Protection of Metals", p. 465, MacMillan Company, New York, NY (1966).
65. W. F. Czyrkalis and M. Levy, "Stress Corrosion Cracking of Uranium Alloys", AMMRC Report TR 73-54, Army Materials and Mechanics Research Center, Watertown, MA, December (1973).
66. T. R. Beck, *J. Electrochem. Soc.*, *115*, 890 (1968).
67. R. A. Bayles and D. A. Meyn in "Corrosion Cracking", V. S. Goel, Ed., p. 241, American Society of Metals, Metals Park, OH (1986).
68. W. H. Smyrl and M. J. Blackburn, *J. Mater. Sci.*, *9*, 777 (1974).

Chapter 12
Corrosion Inhibitors

Introduction

Several different methods of corrosion protection have been considered so far in this text. These methods can be classified as ones of changing the electrode potential of the metal surface or of changing the nature of the metal itself. Cathodic protection and anodic protection are two methods of corrosion control by changing the electrode potential. The nature of the metal can be changed through the use of metallic coatings (sacrificial or noble) or by material selection in which a corrosion-resistant alloy is chosen over a less corrosion-resistant one. Very often, however, material selection is not an option due to the requirements of retaining some other desirable property (such as mechanical strength), or is prohibited by cost. In such cases, altering the environment by the use of corrosion inhibitors then becomes a possible means of corrosion control.

A corrosion inhibitor is any chemical substance which when added to a solution (usually in small amounts) increases the corrosion resistance. Corrosion inhibitors modify electrochemical reactions by their action from the solution side of the metal/solution interface, and the increase in corrosion resistance can be measured by various parameters. In uniform corrosion, the rate of general corrosion decreases from some value i_o (without the inhibitor) to an inhibited value i, and the percent inhibition ($\% I$) is given by

$$\%I = \frac{i_o - i}{i_o} \times 100 \tag{1}$$

In pitting, the use of inhibitors raises the pitting potential but may also decrease the pit growth rate and the number of corrosion pits per unit area. In crevice corrosion, inhibitors usually decrease the propagation rate of crevice corrosion. In stress-corrosion cracking, inhibitors can increase the time to failure by increasing the time to crack initiation and/or by decreasing crack growth rates.

Corrosion inhibitors are used in various industrial applications, including potable (drinking) water, cooling water systems, automobile engine coolants, acid pickling (cleaning) solutions, to protect reinforcing steel bars in concrete, and in oil recovery and storage. Various applications and the inhibitors used are listed in Table 12.1. Inhibitors are also used in the surface treatment of metals to improve the corrosion resistance (chromates on aluminum alloys or galvanized steel) or to improve their paint adhesion (phosphates on autobody steel sheet). Corrosion inhibitors can also be incorporated into paints or organic coatings.

A growing interest in corrosion inhibitors is illustrated by the increasing number of published articles in the corrosion literature. Hackerman [5] noted the following:

E. McCafferty, *Introduction to Corrosion Science*, DOI 10.1007/978-1-4419-0455-3_12,

Table 12.1 Some industrial applications of corrosion inhibitors [1–4]

Application	Inhibitors
Potable (drinking) water	$CaCO_3$ deposition
	Silicates
	Polyphosphates
Recirculating cooling water	Chromates
	Nitrates
	Polyphosphates
	Silicates
	Zinc salts
	Benzotriazole (for copper)
Automotive engine coolant systems (ethylene glycol/water)	Borax–nitrite mixtures
	Sodium mercaptobenzothiazole
	Benzotriazole
High-chloride solutions (seawater, refrigerating brines)	Chromates
	Sodium nitrite
	Chromate–phosphate mixtures
Acid pickling (cleaning mixtures)	Various amines
	Pyridine
	Quinoline
	Mercaptans
	Phenylthiourea
Oil recovery	Primary, secondary, tertiary amines
	Diamines
	Amides
	Polyethoxylated amines
Steel-reinforced concrete	Calcium nitrite
	Sodium benzoate
Surface treatment of metals	
Aluminum aircraft components and galvanized steel	Chromates
Autobody steel sheet	Phosphates

- The first issue of *Chemical Abstracts* (published in 1907) contained no references to corrosion inhibition.
- The first such reference appeared in 1909 followed by a handful of references in each of the next few issues.
- The number of references on corrosion inhibition had grown to 280 by the 50th Edition (1956) and in 1990, the number of entries was voluminous.

This trend can be seen in Fig. 12.1, which records the number of listings for a computerized search using the ISI Web of Science ® and the keywords "corrosion and (inhibition or inhibitor)". As can be seen in Fig. 12.1, there is an explosion of entries for the time frame 1990–1999, and the number of entries for the 9 years from 2000 to 2008 is double that for the previous 10 years.

This chapter will draw on selective examples from the vast amount of work which has been done but will focus primarily on the fundamental aspects of corrosion inhibition.

The mechanisms of corrosion inhibition are different for acidic and neutral solutions. In acid solutions, natural oxide films initially present on the metal surface are dissolved away so that inhibitors then interact directly with the metal surface. In neutral solutions, the system is more complex because

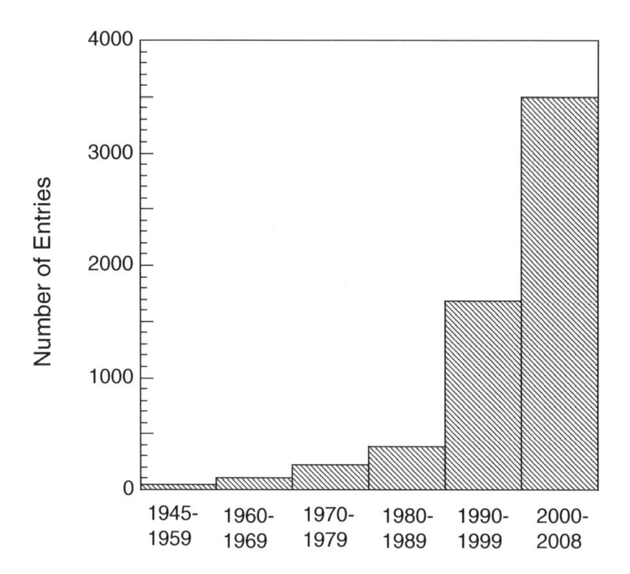

Fig. 12.1 Results for a computerized literature search for "corrosion inhibition" or "corrosion inhibitor"

the metal surface is oxide covered and also because dissolved oxygen may participate in the electrode reaction. Both types of solutions are addressed in this chapter. Most of this chapter deals with corrosion inhibition in aqueous solutions, although vapor-phase inhibitors are also considered.

Types of Inhibitors

Inhibitors can be classified in several different ways. The two main types are (i) adsorption inhibitors and (ii) film-forming inhibitors.

Adsorption inhibitors form a chemisorptive bond with the metal surface and impede ongoing electrochemical dissolution reactions. Most organic inhibitors are chemisorption-type inhibitors. For example, Fig. 12.2(a) shows schematically that an aliphatic organic amine has an electron pair on the nitrogen atom which is available for donation to the metal surface. In addition, the hydrocarbon tails of the molecule are oriented away from the interface toward the solution so that further protection is provided by the formation of an array of hydrophobic hydrocarbon tails located on adjacently adsorbed amines. This hydrophobic network serves to keep water molecules and aggressive anions, such as Cl^-, away from the metal surface, as shown in Fig. 12.2(b)

There are two types of film-forming inhibitors: (i) passivating inhibitors and (ii) precipitation inhibitors. As implied by their name, passivating inhibitors function by promoting the formation of a passive film on the surface. Passivating inhibitors may be oxidizing or non-oxidizing agents. (Oxidizing agents themselves are reduced in the process of oxidizing another molecule). Chromates are typical oxidizing inhibitors, and with iron or steels, the chromate ion is reduced to Cr_2O_3 or $Cr(OH)_3$ on the metal surface to produce a protective mixed oxide of chromium and iron oxides [6]. Adsorption is also important with oxidizing inhibitors because they are usually adsorbed on the metal surface prior to their reduction and formation of the passive film. Non-oxidizing passivators like benzoates, azelates, and phosphates also first adsorb on the surface before forming the passive film.

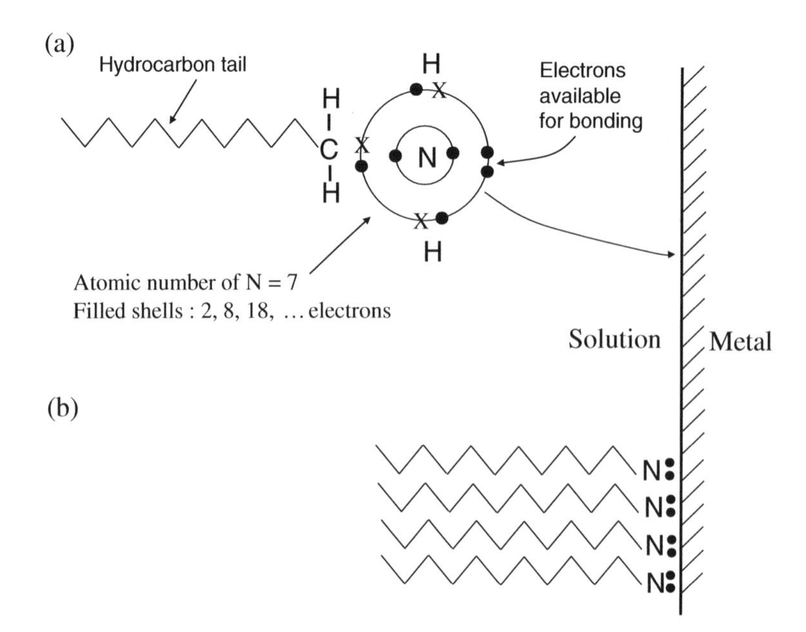

Fig. 12.2 (**a**) Chemisorption of amines at a metal surface. The *solid black dots* denote electrons belonging to the N atom, whereas the Xs refer to electrons from H or C atoms. (The size of the nitrogen atom is exaggerated relative to the rest of the molecule.) (**b**) Formation of a close-packed monolayer on the metal surface

With precipitation inhibitors, a precipitation reaction between cations of the corroding metal and the inhibitor deposits a three-dimensional barrier film on the metal surface. Such a film is formed when the solubility product is exceeded for the salt formed between the cations of the metal and the anions of the inhibitor. Phosphates and silicates are examples of precipitation-type inhibitors.

From another point of view, inhibitors can be classified as anodic, cathodic, or mixed inhibitors, depending on which partial electrochemical reaction is affected. As examples, chromates are anodic inhibitors, some phosphates are cathodic inhibitors, and most organic compounds are mixed inhibitors.

Acidic Solutions

Much of the information about the mechanisms of corrosion inhibitors has evolved from the study of acidic solutions. The reason is that the dissolution of air-formed surface films allows focus on the direct interaction between the inhibitor and the cleaned metal surface. Another simplification often made in mechanistic studies is that the solution is de-aerated so that oxygen does not participate in the electrode reaction and the cathodic reaction is restricted to hydrogen evolution. However, the effect of dissolved oxygen in acidic solutions does not appear to be important until the pH is as high as 2–4 [7].

Despite these simplifications, the general results for acidic solutions have considerable practical application. Corrosion inhibitors are used in a variety of acidic environments, including commercial pickling (cleaning) processes, in the storage and transport of acids, in oil and gas production, in the pulp and paper industry, and in the cleaning of corroded metal artifacts. In addition, corrosion inhibition in de-aerated acidic solutions serves as a scaled-up model for the inhibition of the propagation of localized corrosion processes, such as crevice corrosion, pitting, and stress-corrosion cracking, where the internal local electrolyte becomes both acidic and depleted in oxygen.

The most important act of the inhibitor in acid solutions is to chemisorb on the metal surface. The process of chemisorption and factors which affect it are discussed below.

Chemisorption of Inhibitors

Chemisorption involves an actual charge transfer or charge sharing between the inhibitor molecule and the metal surface. By interacting with metal surface atoms, the chemisorbed inhibitor interferes with metallic dissolution. The simplest picture is one of a blockage of active surface sites, but this view is not quite complete. Like all adsorbed species, chemisorbed molecules have a certain residence time at the surface and thus play a dynamic role by participating in a number of adsorption–desorption steps.

The main characteristics of chemisorption are the high heats of adsorption, persistence, and specificity. Table 12.2 compares the process of chemisorption with physisorption (i.e., physical adsorption), a more general type of adsorption but one involving weaker interactions.

Table 12.2 Comparison of chemical adsorption (chemisorption) and physical adsorption (physisorption) [8]

	Physisorption	Chemisorption
Type of electronic interaction	Van der Waals or electrostatic forces	Charge transfer or charge sharing
Reversibility	Adsorbed species readily removed by solvent washing	Adsorption is irreversible, more persistent
Energetics	Low heat of adsorption, <10 kcal/mol (40 kJ/mol)	Higher heat of adsorption, >10 kcal/mol (40 kJ/mol)
Kinetics	Rapid adsorption	Slow adsorption
Specificity	Adsorbed species relatively indifferent to identity of surface	Specific interaction, strong dependence on identity of surface

Chemisorption-type inhibitors usually contain N, S, or O atoms; and chemisorption occurs through the donation of electrons from these atoms to the metal surface, as shown in Fig. 12.2 for a primary amine. In a homologous series differing in the identity of the donor atom, the order of corrosion inhibition is

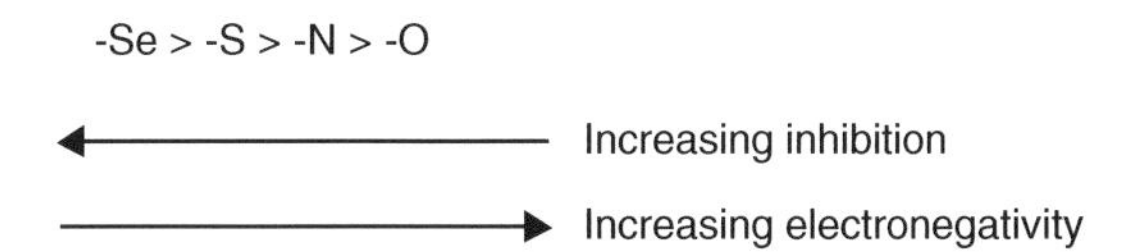

The order of corrosion inhibition is the reverse order of the electronegativity of these atoms. For instance, sulfur compounds are usually better corrosion inhibitors than are their nitrogen analogs because a S atom is less electronegative than a N atom (is less effective in drawing electrons to itself), so S is a better electron donor than is N. (It should be noted that sulfur and selenium compounds are less soluble than the corresponding nitrogen compounds).

Direct evidence for the chemisorption of corrosion inhibitors has been gained through many studies which have utilized various techniques such as radiotracer measurements, infrared spectroscopy, and modern surface analytical techniques such as X-ray photoelectron spectroscopy (XPS), among others.

Effect of Inhibitor Concentration

The inhibitor efficiency (% *I*) increases as the concentration of the dissolved inhibitor increases. Figure 12.3 shows polarization curves for iron in 6 M HCl containing various amounts of the diamine $NH_2(CH_2)_3NH_2$ [9, 10]. It can be seen in Fig. 12.3, as well as in Fig. 12.4, that the corrosion rate decreases (and % *I* increases) with increasing concentration of the diamine.

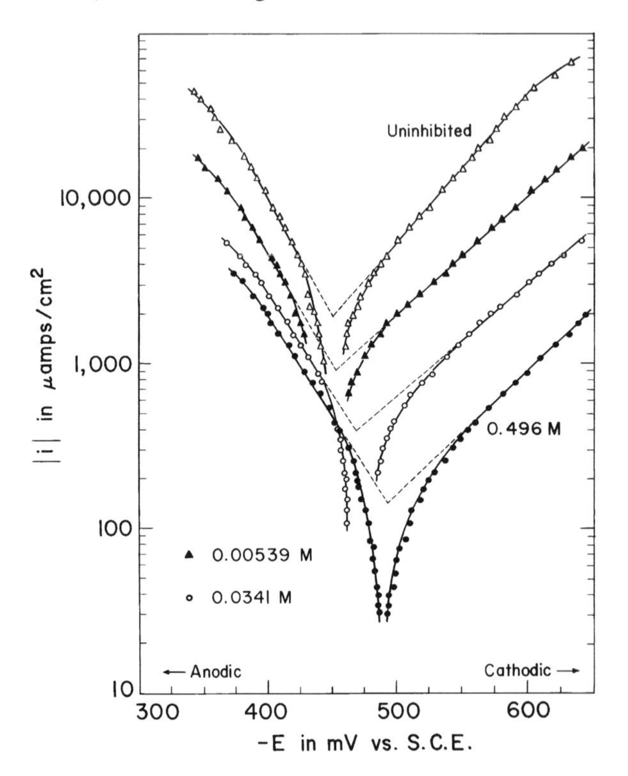

Fig. 12.3 Polarization curves for iron in 6 M HCl with NH_2-$(CH_2)_3$-NH_2 inhibitor. The *numbers* on the figure refer to the concentration of the diamine [9, 10]. Reproduced by permission of ECS – The Electrochemical Society

With increased concentration of inhibitor, the corrosion rate decreases because the adsorption of the inhibitor also increases, as shown in Fig. 12.5(a). Figure 12.5(a) illustrates schematically an *adsorption isotherm*, which relates the concentration of the inhibitor in solution to the amount of inhibitor taken up by the surface. For chemisorption-type inhibitors, the maximum fraction of the surface able to be covered is one monolayer (i.e., one molecular layer). When this coverage is reached, the corrosion rate attains its minimum value, and further increases in inhibitor concentration do not produce any further decrease in the corrosion rate, as shown in Fig. 12.5(b).

There are several different types of adsorption isotherms, and these are discussed later in this chapter.

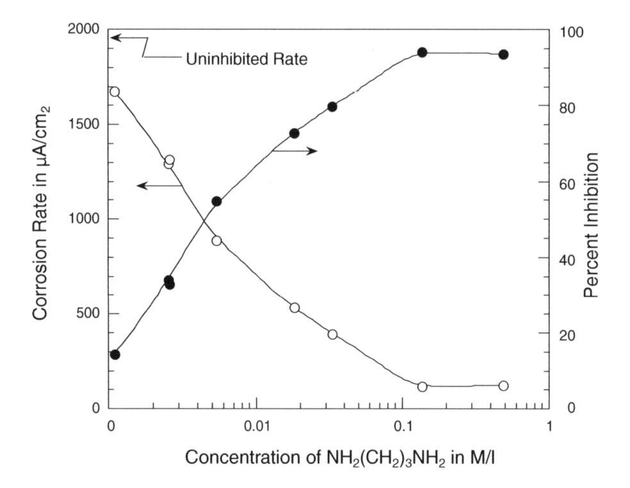

Fig. 12.4 Corrosion rate or percent inhibition for iron in 6 M HCl containing $NH_2(CH_2)_3NH_2$ [9, 10]

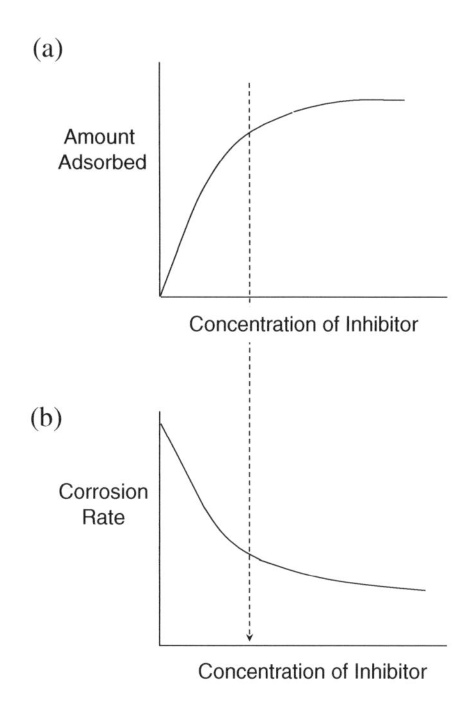

Fig. 12.5 Schematic diagram showing the correlation between (**a**) the adsorption isotherm of an inhibitor and (**b**) the corrosion rate

Chemical Factors in the Effectiveness of Chemisorbed Inhibitors

The chemisorption of organic molecules, and accordingly their effectiveness as corrosion inhibitors, is influenced by three principal chemical factors. These are (i) the electron donor ability of the molecule, (ii) the size of the molecule, and (iii) its solubility [9–13].

A first estimate of the ability of a molecule to act as an electron donor is given by its base strength. (Stronger bases are better electron donors than are weak bases). Organic bases are protonated in acidic solutions:

$$RNH_2 + H^+ \rightarrow RNH_3^+ \tag{2}$$

but as the amine approaches the metal surface, the molecule loses its proton so that the lone pair of electrons on the nitrogen atom can interact with the metal surface. The base strength is usually given in terms of the acid dissociation constant K_a for the reverse of Eq. (2):

$$K_a = \frac{[RNH_2][H^+]}{[RNH_3^+]} \tag{3}$$

or by

$$pK_a = \log \frac{1}{K_a} \tag{4}$$

The greater the value of pK_a, the greater the base strength, as shown in Appendix G. A greater base strength, in turn, means that there is a greater tendency for electron donation and chemisorption.

Figure 12.6 shows pK_a values for a series of aliphatic (straight chain) carboxylic acids and aliphatic amines. The pK_a values for the amines are much higher than those for the carboxylic acids. Thus, an aliphatic amine is a better Lewis base (electron donor) than is the corresponding carboxylic acid and accordingly is a better adsorption-type corrosion inhibitor.

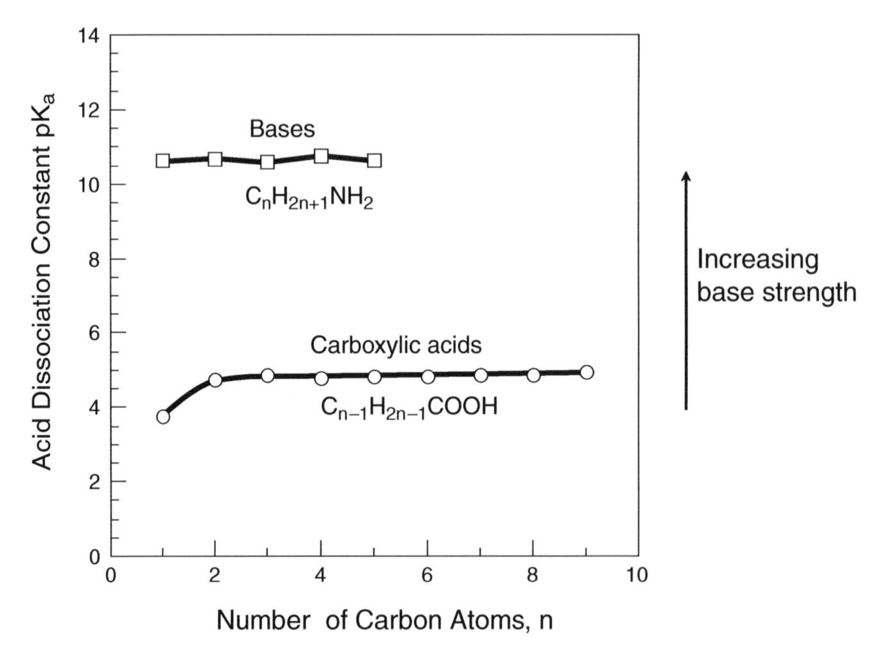

Fig. 12.6 The acid dissociation constant pK_a for a homologous series of amines and carboxylic acids. (Larger values of pK_a indicate a greater base strength)

A second factor in determining the effectiveness of a chemisorbed organic inhibitor is the molecular size. In general, the larger the molecular area, the better the inhibitor (other factors being equal). However, with very large molecules (such as polymers), there may be steric hindrance problems when a large molecule attempts to fit onto a surface already partially occupied with previously adsorbed molecules. Thus, complete coverage of the surface by large molecules may sometimes be difficult to achieve.

The third factor affecting chemisorption, and thus inhibition, is the solubility of the organic molecule. Less soluble molecules have a greater tendency to be adsorbed than soluble molecules. This is illustrated in Fig. 12.7, which depicts adsorption isotherms for compounds of two differing solubilities. For a fixed concentration (C^*), the less soluble compound is further along in its adsorption isotherm and has a correspondingly greater coverage of adsorbate on the metal surface.

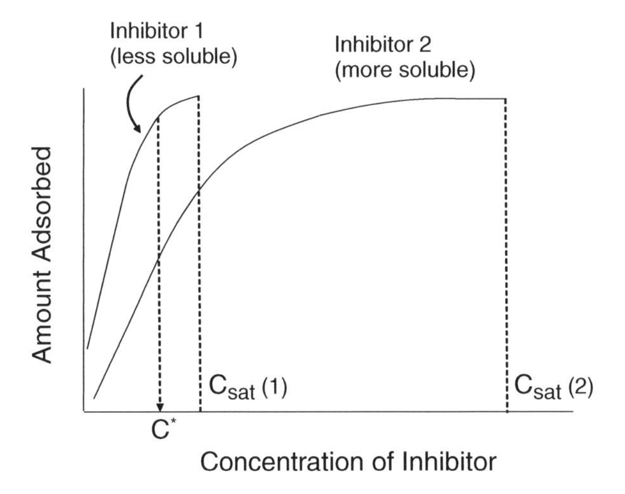

Fig. 12.7 Adsorption isotherms (schematic) for two compounds of differing solubility (C_{sat}). At a fixed concentration C^*, the compound of lesser solubility has the greater relative coverage on the metal surface

The interplay of these three factors is illustrated in a study on the corrosion inhibition of primary, secondary, and tertiary amines (RNH_2, R_2NH, and R_3N, respectively) in sulfuric acid [14]. Their results are shown in Fig. 12.8 for a fixed concentration of inhibitor (0.1 mol%). These results show the following:

(1) The inhibitor efficiency follows the order

$$R_3N > R_2NH > RNH_2$$

(2) Within a given series (primary or secondary or tertiary amines), the inhibition efficiency increases with increasing chain length.

Figure 12.8 also indicates pK_a values for various amines. There are only small differences in pK_a values in passing from a primary to secondary to tertiary amine. However, another measure of the base strength can be obtained though the use of Taft induction constants σ^* [15]. These constants indicate the electron-withdrawing or electron-donating ability of various constituent groups. Selected values of Taft induction constants are given in Table 12.3. Electron-withdrawing groups have positive values of σ^*, and electron-donating groups have negative values of σ^*. Taft constants apply to aliphatic compounds only (linear or cyclic).

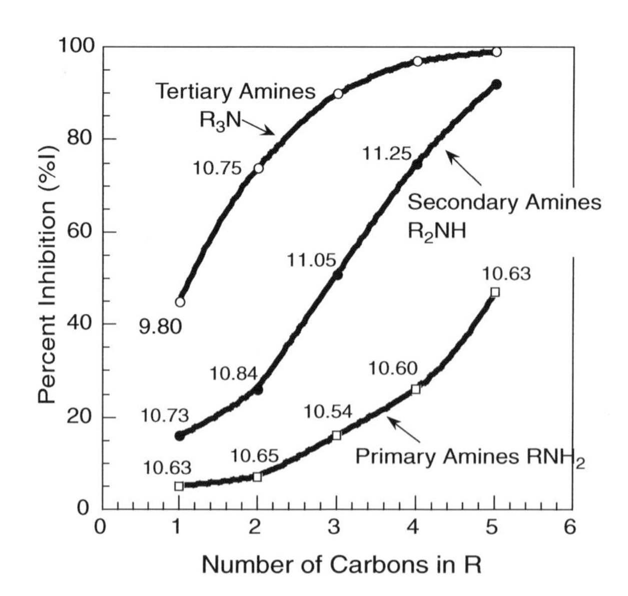

Fig. 12.8 The corrosion inhibition of primary (RNH_2), secondary (R_2NH), and tertiary amines (R_3N) in 1N sulfuric acid. (The *numbers* on the figure refer to pK_a values)

Table 12.3 Values of some Taft substituent constants σ^* [15]

Substituent	Taft σ^*
NO_2	+4.0
Cl	+2.96
COOH	+2.08
$CONH_2$	+1.68
SH	+1.68
CH_3CO	+1.65
OH	+1.34
NH_2	+0.62
C_6H_5	+0.60
H	+0.49
$ClCH_2CH_2$	+0.38
$C_6H_5CH_2$	+0.22
CH_3	0.00
C_2H_5	−0.10
n-C_3H_7	−0.12
n-C_4H_9	−0.13
Cyclohexyl	−0.15
n-C_5H_{11}	−0.25
$(CH_3)_3C$	−0.30

Example 12.1: Use the Taft induction constants in Table 12.3 to compare the electron-donating properties of $C_2H_5NH_2$, $(C_2H_5)_2NH$, and $(C_2H_5)_3N$

Solution: For $C_2H_5NH_2$:

$$\Sigma\sigma^* = \sigma^*(C_2H_5) + 2\sigma^*(H)$$
$$\Sigma\sigma^* = -0.10 + 2(0.49) = 0.88$$

For $(C_2H_5)_2NH$:

$$\Sigma\sigma^* = 2\sigma^*(C_2H_5) + \sigma^*(H)$$
$$\Sigma\sigma^* = 2(-0.10) + 0.49 = 0.29$$

For $(C_2H_5)_3N$:

$$\Sigma\sigma^* = 3\sigma^*(C_2H_5)$$
$$\Sigma\sigma^* = 3(-0.10) = -0.30$$

Thus, the electron-donating ability increases in the order

$$(C_2H_5)_3N > (C_2H_5)_2NH > C_2H_5NH_2$$

and the tertiary amine is the strongest base of the three.

Figure 12.9 shows the data in Fig. 12.8 re-plotted in terms of the summed Taft constants $\Sigma\sigma^*$ for each amine. It can be seen clearly that for a given homologous series (given n), the inhibition efficiency is greatest for the tertiary amine, which also has the greatest tendency to donate electrons.

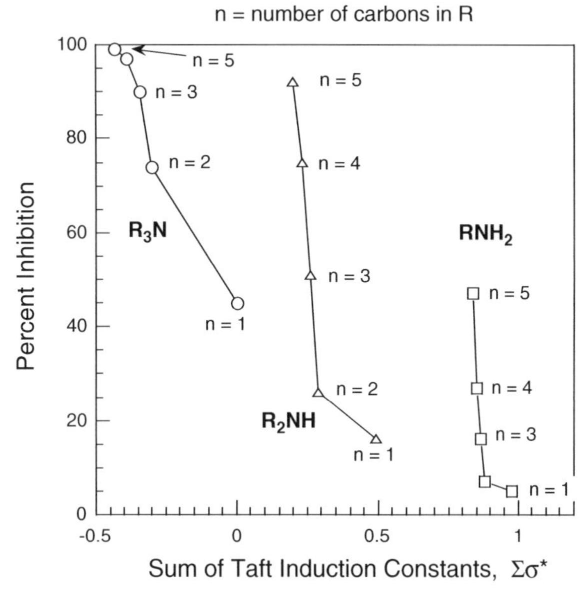

Fig. 12.9 Data in Fig. 12.8 re-plotted in terms of the summed Taft constants $\Sigma\sigma^*$. Drawn from data in [14] by permission of Plenum Press (Springer)

Within a given series (primary or secondary or tertiary amine), the inhibitor efficiency increases with increasing size of the molecule. The larger molecules cover a larger portion of the surface and also have decreased solubility as the molecular weight increases.

Involvement of Water

When a metal is immersed into an aqueous solution, the metal surface becomes covered with water molecules. In order to adsorb onto the metal surface, the organic molecule must first displace these adsorbed water molecules. The adsorption of organic molecules from aqueous solutions is thus a replacement reaction [16].

$$\mathrm{Org\,(soln)} + n\mathrm{H_2O\,(ads)} \rightarrow \mathrm{Org\,(ads)} + n\mathrm{H_2O\,(soln)}$$

where "soln" and "ads" refer to the aqueous and adsorbed phases, and n is the number of water molecules which must be desorbed from the metal surface to accommodate the organic molecule. If an aliphatic (straight-chain) amine adsorbs in the vertical configuration, as in Fig. 12.2, then two water molecules must first be desorbed because the cross-sectional areas are about 10 Å^2 for H_2O and 24 Å^2 for aliphatic amine.

The overall free energy of adsorption ΔG_{ads} involves both organic and water molecules:

$$\Delta G_{\mathrm{ads}} = \Delta G_{\mathrm{ads}}^{\mathrm{Org}} - n\Delta G_{\mathrm{ads}}^{\mathrm{H_2O}} \tag{5}$$

where $\Delta G^{\mathrm{org}}{}_{\mathrm{ads}}$ and $\Delta G^{\mathrm{H_2O}}$ are the free energies of adsorption per mole of the organic compound and water, respectively. See Problem 12.7. Free energies (and heats) of adsorption are less for electrode processes than for adsorption from the vapor phase because of this involvement of pre-adsorbed water. Adsorption from solution and from the vapor phase is related by a thermodynamic cycle [16], as shown in Fig. 12.10.

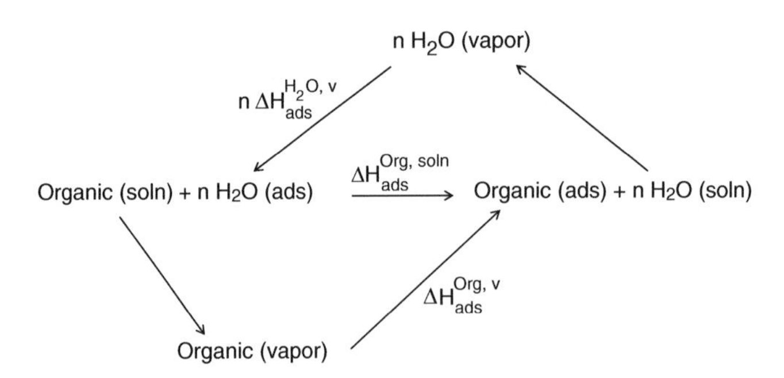

Fig. 12.10 Thermodynamic cycle relating the heat of adsorption of an organic molecule from an aqueous solution $\Delta H_{\mathrm{ads}}{}^{\mathrm{Org,\,soln}}$ and the heats of adsorption of the organic molecule $\Delta H_{\mathrm{ads}}{}^{\mathrm{Org,v}}$ and of water $\Delta H_{\mathrm{ads}}^{\mathrm{H_2O,v}}$ from the vapor phase [16]

Heats of adsorption in electrode processes are difficult to measure directly but can be calculated from such thermodynamic cycles or by fitting of adsorption data to special adsorption isotherms based on these cycles [17]. The free energies of adsorption of various organic molecules from solution onto metal surfaces are of the order of 4–7 kcal/mol ($\sim$30 kJ/mol) [16–19] rather than of the order of 25 kcal/mol ($\sim$100 kJ/mol), which is expected for adsorption directly from the vapor phase. See Table 12.4.

In addition to affecting the adsorption of corrosion inhibitors, water molecules also participate in charge transfer, i.e., corrosion reactions. The corrosion of iron in acidic solutions proceeds through a series of steps in which water molecules adsorbed on metal atoms aid in the dissolution of the metal atom, as will be seen in Chapter 16. Thus, replacement of adsorbed water by adsorbed organic molecules alters the kinetics of anodic dissolution reactions in acid solutions.

Table 12.4 Free energy of adsorption of various organic compounds from aqueous solutions onto metals

Organic compound	Metal	ΔG^{o}_{ads} (kJ/mol)	References
n-Decylamine	Nickel	−28.5	Bockris and Swinkels [16]
	Iron	−27.6	
	Copper	−30.5	
	Lead	−25.9	
n-Hexyl alcohol	Mild steel	−16.3	Smialowski and Wieczorek [17]
n-Hexyl amine		−20.5	
n-Octyl alcohol		−22.6	
n-Octyl amine		−23.8	
Naphthalene	Nickel	−25.1	Bockris et al. [18]
	Iron	−29.3	
	Copper	−29.3	
ω-Benzoyl alkanoic acid	Iron	−25.8	Kern and Landolt [19]
N-Ethyl morpholine		−28.7	

Competitive vs. Co-operative Adsorption

When the solution contains adsorbable anions, such as Cl^- (or I^-), organic inhibitors must compete with these ions for sites on the metal surface. This process of competitive adsorption is depicted in Fig. 12.11. In this type of adsorption, the protonated inhibitor loses its associated proton on entering the electrical double and chemisorbs by donating electrons to the metal, as discussed earlier.

The inhibitor can adsorb in another manner, as shown in Fig. 12.11. The protonated inhibitor can electrostatically adsorb onto the halide-covered surface through its hydrogen ion. Evidence for this latter type of adsorption is provided by cases in which the presence of certain anions increases the efficiency of an organic inhibitor. For instance, Fig. 12.12 illustrates a synergistic effect of *n*-decylamine and Br^- ions on the corrosion of mild steel in phosphoric acid [20]. The amine plus the bromide ion together are more effective than either alone in inhibiting corrosion. Other cases which imply co-operative adsorption between halide ions and organic inhibitors have been reported [21].

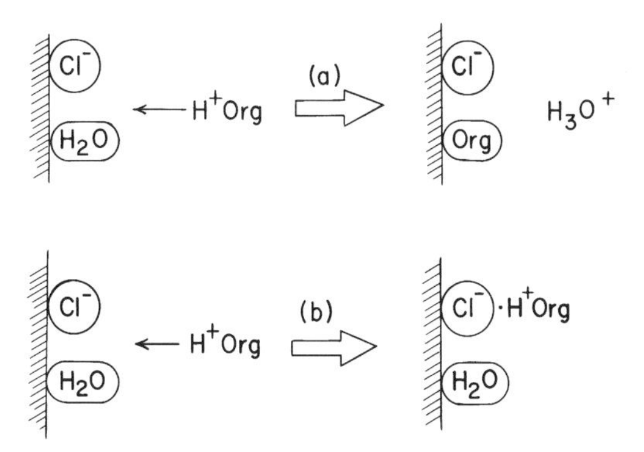

Fig. 12.11 (**a**) Competitive adsorption and (**b**) co-operative adsorption between a chloride ion and an organic inhibitor. Reproduced by permission of © NACE International 1974

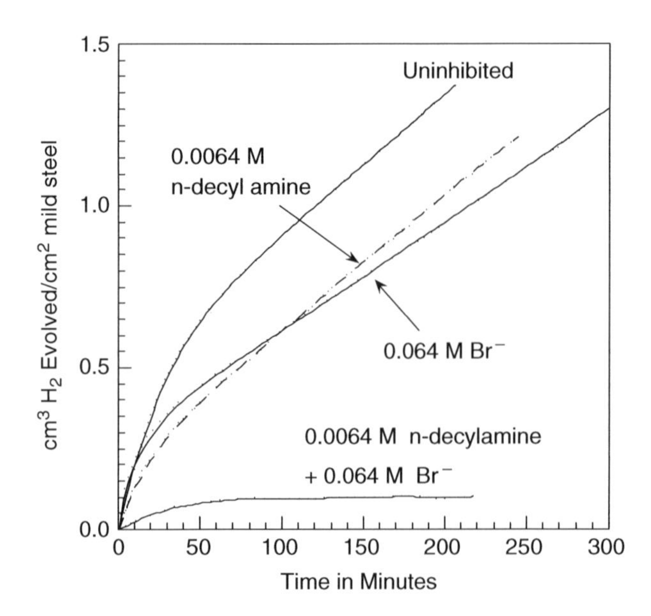

Fig. 12.12 Synergistic effect of Br^- ions and *n*-decylamine on the corrosion inhibition of mild steel in 3 M phosphoric acid [20]. Reproduced by permission of ECS – The Electrochemical Society

Effect of the Electrical Double Layer

The electrical double layer (edl), which was introduced in Chapter 3, is important in corrosion inhibition because it plays a role in the adsorption of organic molecules. Organic molecules must enter the electrical double layer to adsorb at the metal surface (for both competitive and co-operative adsorption). Intrusion of organic molecules into the edl changes both its composition and its structure. See Fig. 12.13. The adsorption of a long-chain organic molecule increases the thickness l of the edl and the presence of an organic molecule in the edl also decreases the effective dielectric constant ϵ of the edl. Thus the capacitance of the edl per unit area C_{dl} given by

$$C_{\mathrm{d}l} = \frac{\varepsilon}{l} \tag{6}$$

decreases with the adsorption of the organic molecule. In measuring C_{dl}, the faradaic processes, e.g., metal corrosion reactions, must be separated from the double layer properties. This can be done by using an experimental technique called the single current pulse method, in which the electrical double layer is charged quickly before the response due to slower faradaic processes is observed. More details as to this method are given in Fig. 12.14. The single pulse technique is effective in measuring the double layer capacitance, although double layer capacitances are frequently measured using AC impedance techniques. AC impedance techniques and their applications are treated separately in Chapter 14.

Figure 12.15 shows double layer capacitance–potential plots for iron in 6 M HCl with and without added $NH_2(CH_2)_{11}NH_2$ [9, 10]. The decrease in C_{dl} upon addition of the inhibitor shows that the diamine is adsorbed. Moreover, addition of the inhibitor leads to a general decrease in C_{dl} for potentials in both anodic and cathodic directions. Thus, the diamine is a mixed inhibitor because

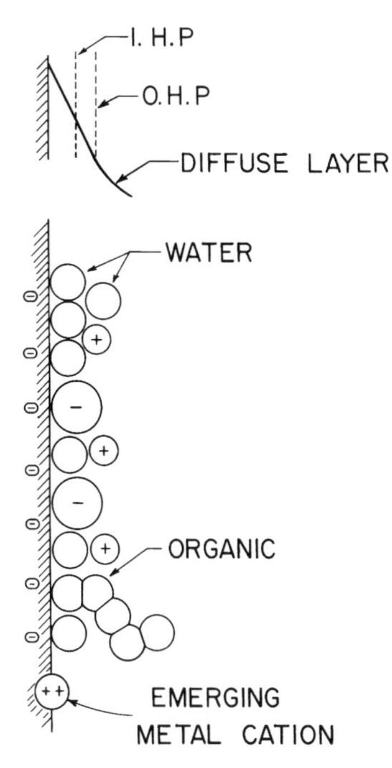

Fig. 12.13 The electrical double layer at the metal/solution interface showing the adsorption of an organic molecule. IHP refers to the inner Helmholtz plane and OHP refers to the outer Helmholtz plane

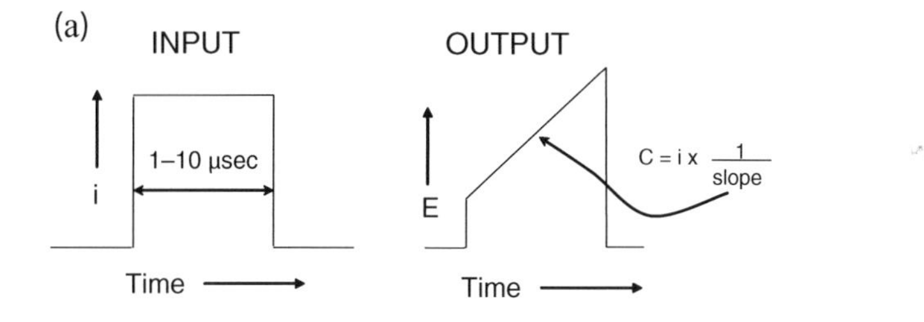

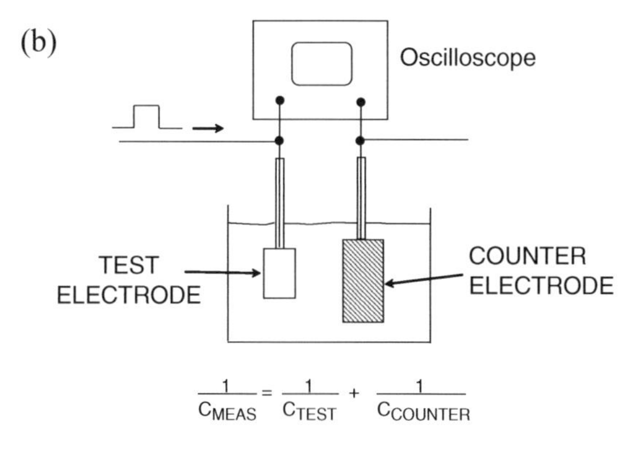

Fig. 12.14 (**a**) The single current pulse technique for measuring double layer capacitance. (**b**) The experimental setup

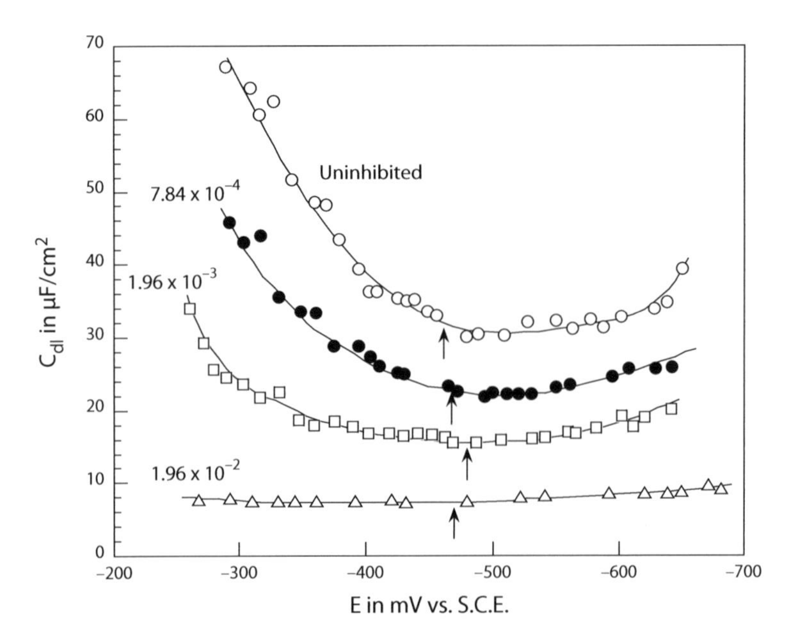

Fig. 12.15 Double layer capacitance of iron in 6 M HCl with $NH_2(CH_2)_{11}NH_2$ inhibitor. The *numbers* on the figure refer to the concentration of the diamine, and the arrows indicate the open-circuit corrosion potentials [9, 10]. Reproduced by permission of ECS – The Electrochemical Society

it affects both anodic and cathodic partial reactions. At cathodic overvoltages, C_{dl} rises due to the adsorption of H^+ ions, and at anodic overvoltages, it rises due to the adsorption of Cl^- ions.

The Potential of Zero Charge

As discussed in Chapter 3, the metal side of the metal/solution interface can have an excess or a deficiency of electrons. Thus, the metal surface can have a negative or a positive charge, respectively. In the former instance, positive charges are attracted to the surface and in the latter instance, negative charges will be favored. Between these two cases, there is an electrode potential at which the surface has zero charge, and this potential is called the *potential of zero charge* (pzc). The closer a metal is to its pzc, the easier it is for neutral organic molecules to adsorb onto the metal surface. This is especially true for those organic molecules which do not chemisorb strongly and would not be favored in competitive adsorption with other surface-active solution species, such as Cl^- ions.

Table 12.5 lists pzc values for various metals [22, 23]. These values are not always invariant quantities for any given metal and vary somewhat with the nature of the solution, especially if the solution contains adsorbable anions. But the idea is that above (more positive than) the pzc, the adsorption of negatively charged species is favored, and below (more negative than) the pzc, the adsorption of positively charged species is favored.

From Table 12.5, the pzc for iron is −0.40 V vs. SHE or −0.64 V vs. SCE. The open-circuit corrosion potential of iron in the HCl solutions containing the C_{11}-diamine is approximately −0.46 V vs. SCE (Fig. 12.15). Thus, the open-circuit corrosion potential for iron is more positive than the pzc

Table 12.5 Potentials of zero charge E_{pzc} for various metals [22, 23]

Metal	E_{pzc} (V vs. SHE)	Solution
Cadmium	−0.72	0.01–0.001 N NaF
	−0.70	0.005 N KCl
Zinc	−0.62 to −0.65	1 N H_2SO_4
	−0.60	0.02 N Na_2SO_4 + H_2SO_4
Lead	−0.56	0.01–0.001 N NaF
	−0.64	0.01 N Na_2SO_4
Aluminum	−0.52	0.01 M KCl
Chromium	−0.45	0.1 M NaOH
Tin	−0.46	0.001 N $KClO_4$
	−0.38	0.01 N KCl
Iron	−0.40	0.003 N HCl
	−0.37	0.001 N H_2SO_4
Cobalt	−0.32	0.02 N Na_2SO_4, pH 1
	−0.43	0.02 N Na_2SO_4, pH 3
Nickel	−0.21	0.02 N Na_2SO_4, pH 1
	−0.37	0.02 N Na_2SO_4, pH 5
		0.0015 N HCl
	−0.28	0.0015 N HBr
Copper	−0.05	1 N Na_2SO_4
	−0.03	0.02 N Na_2SO_4
	+0.007	0.01 N KCl
Gold	+0.15	1 N KCl
	+0.23	0.02 N H_2SO_4
		1 N $NaClO_4$ + 0.001 N
	+0.30	$HClO_4$

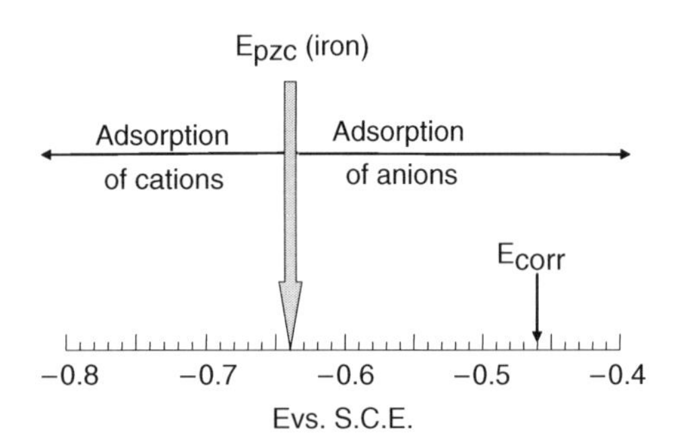

Fig. 12.16 Adsorption and the potential of zero charge. E_{pzc} refers to the potential of zero charge for iron and E_{corr} to the corrosion potential of iron in 6 M HCl containing the C_{12}-diamine

for iron. This means that there is a preponderance of negative ions (most likely Cl^-) on the solution side of the interface. See Fig. 12.16. Thus, at the open-circuit corrosion potential, co-operative adsorption of the protonated inhibitor is possible (in addition to competitive adsorption).

Effect of Molecular Structure

The molecular structure of the organic molecule is important in determining the effectiveness of adsorption-type corrosion inhibitors. This is because molecular structure influences the electron-donating ability of the molecule and accordingly its adsorption and effectiveness as an inhibitor. We have already begun to examine the role of molecular structure in earlier discussions of the effectiveness of primary vs. secondary vs. tertiary amines. There have been many studies on the effect of molecular structure on corrosion inhibition, and only a few examples are given here.

In general, aliphatic amines are better inhibitors than are aromatic amines. Heterocyclic amines, such as pyridine, are better inhibitors than are aromatic amines. The order of effectiveness for six-membered rings is [11]

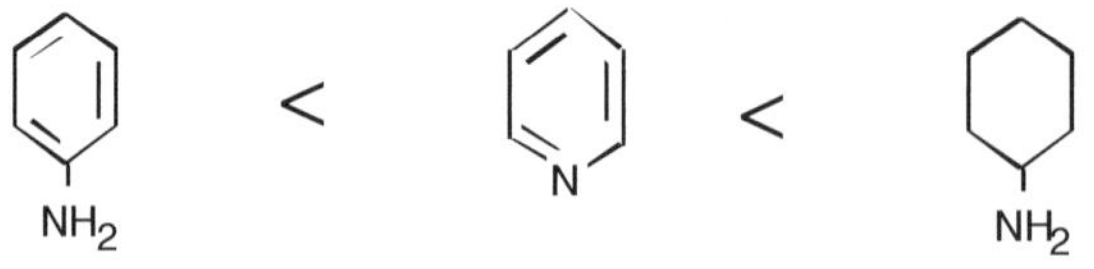

NH_2 < N < NH_2

aniline < pyridine < cyclohexylamine

The important factor here is that the base strength (i.e., the electron-donating ability) also increases in this order. (These three compounds have similar molecular areas and solubilities).

The nature of substituents on aromatic rings influences the availability of electrons for formation of chemisorptive bonds and thus are able to modify the effectiveness of organic inhibitors. This is because the substituents in ring compounds exert inductive effects on electron-donating centers, such as the $-NH_2$ group. Electron-withdrawing substituents, like –Cl, will decrease the charge density of electrons at the N atom resulting in increased adsorption and reduced protection. On the other hand, electron-donating substituents like $-CH_3$ will increase the availability of donor electrons at the adsorbable atom so as to result in increased protection. A $-CH_3$ group in the para position has a greater inductive effect than a $-CH_3$ group in the meta position. Thus, the order of inhibitor effectiveness for substituted pyridines is [13]

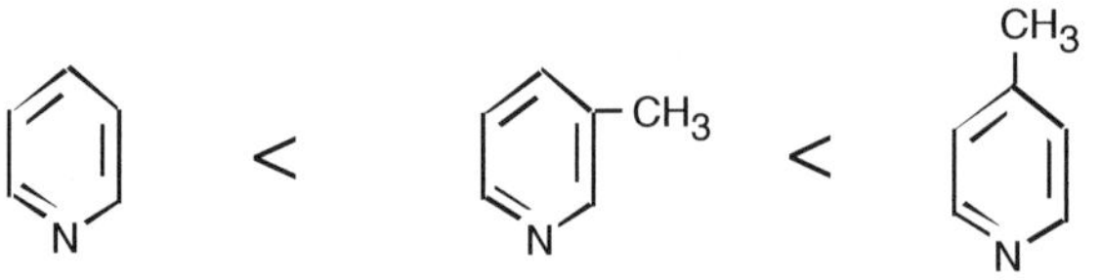

N < N, CH_3 < CH_3, N

Nobe and co-workers [24, 25] have shown that with ring-substituted organic molecules, the inhibitor efficiency is linearly related to the Hammett σ constant [15]. The Hammett σ constant is a measure of the ability of a substituted group on a ring to provide electrons (negative σ) or to withdraw electrons (positive σ) from a ring structure. (The Hammett σ is similar to the Taft σ^* except that the Hammett σ applies to ring structures only). Selected values of Hammett σ constants are listed in Table 12.6. Electron-withdrawing groups have positive values of σ, and electron-donating groups have negative values of σ.

Figure 12.17 shows that the corrosion rate of ring-substituted benzoic acids in acidic solutions decreases with the ability of the substituent group to donate electrons to the ring [24]. A similar effect was observed for substituted benzotriazole in which electron-donating groups also decreased the corrosion rate [25].

There is another way in which changes in molecular structure can change the inhibitor effectiveness, and that is by changes in spatial configuration, rather than in electron configuration. Studies with polymer amines have shown that very short chains of adsorbable repetitive groups improve

$$-\underset{}{\overset{\text{H}}{\overset{|}{\text{N}}}}-(CH_2)_2-\overset{\text{H}}{\overset{|}{\text{N}}}-$$

Table 12.6 Values of some Hammett substituent constants σ [15]

	Hammett σ	
Substituent	*Meta*	*Para*
CH_3	−0.07	−0.17
CH_2CH_3	−0.07	−0.15
$CH(CH_3)_2$	−0.07	−0.15
$C(CH_3)_3$	−0.10	−0.20
CN	+0.56	+0.66
COOH	+0.36	+0.43
CHO	+0.36	+0.22
$CONH_2$	+0.28	+0.36
CF_3	+0.43	+0.54
NH_2	−0.16	−0.66
NO_2	+0.71	+0.78
OH	+0.12	−0.37
OCH_3	+0.12	−0.27
SH	+0.25	+0.15
F	+0.34	+0.06
Cl	+0.37	+0.23
Br	+0.39	+0.23
I	+0.35	+0.28
H	0.00	0.00

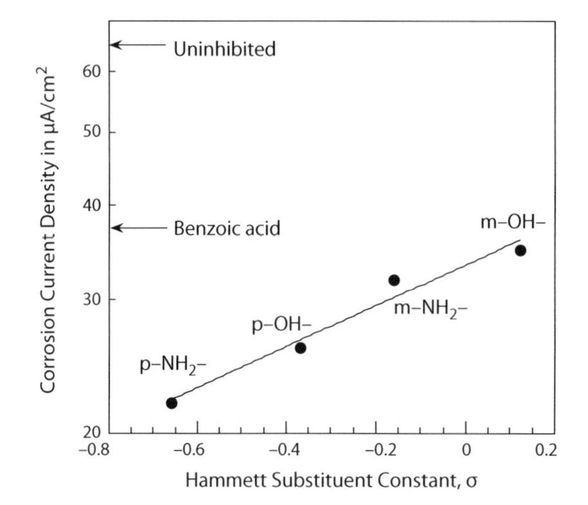

Fig. 12.17 Corrosion rates of inhibited iron in 0.5 M H_2SO_4 as a function of the Hammett sigma constant for ring-substituted benzoic acid [24]. The concentration of the inhibitor was 0.02 M in each case. Reproduced by permission of ECS – The Electrochemical Society

inhibitor effectiveness [26]. In a series of polymethylene amines, the order of inhibitor effectiveness follows the following order:

$$4\text{ units} > 3\text{ units} > 2\text{ units} > \text{monomer}$$

The effect of ring strain was shown in studies [27–29] which compared secondary amines and cyclic imines:

CH_3 CH_3 CH_2 — CH_2
$(CH_2)_m$ $(CH_2)_m$ ring closure → $(CH_2)_m$ $(CH_2)_m$
CH_2 CH_2 CH_2 CH_2
N N
H H

Both sets of compounds are secondary amines, have similar molecular weights (within 2 mass units), and similar molecular areas. Yet, the closed ring structures are more effective corrosion inhibitors than are their branched chain analogs. This structure effect is most pronounced for rings with 9–12 carbon atoms. This higher efficiency is due to the large strain in these rings, which is relieved with the formation of a strong chemisorptive bond with the metal surface.

Adsorption Isotherms

As discussed earlier, an adsorption isotherm relates the concentration of the inhibitor in solution to the amount of inhibitor taken up by the surface. The surface coverage of the inhibitor, in turn, is important for chemisorption-type inhibitors, as shown in Fig. 12.5.

There are several types of adsorption isotherms, but the two most important ones which pertain to corrosion inhibition are the Langmuir isotherm and the Temkin isotherm. The Langmuir isotherm is given by

$$\frac{\theta}{1-\theta} = KC \tag{7}$$

where θ is the fraction of the surface covered by the inhibitor of concentration C, and K is a constant (which includes the heat of adsorption of the inhibitor). The derivation of Eq. (7) is given in Appendix H. The Langmuir adsorption isotherm results if the adsorbent is energetically homogeneous (all adsorption sites are equivalent) so that the heat of adsorption of the inhibitor is independent of coverage. The test of the Langmuir isotherm is that plots of $\theta/(1-\theta)$ vs. C produce a straight line. (Equivalently, plots of the logarithm of $\theta/(1-\theta)$ vs. the logarithm of C also produce a straight line, as shown in Fig. 12.18 [30]).

Another test of the Langmuir adsorption isotherm follows from rearranging Eq. (7) as

$$\frac{C}{\theta} = \frac{1}{K} + C \tag{8}$$

A plot of C/θ vs. C will produce a straight line of unit slope if the Langmuir isotherm applies [31, 32].

As stated above, the Langmuir adsorption isotherm assumes that the heat of adsorption of the corrosion inhibitor ΔH_{ads} is independent of fractional coverage of the inhibitor. Often this relationship is not obeyed, but instead ΔH_{ads} decreases with increasing coverage, as shown schematically in Fig. 12.19. This decrease in ΔH_{ads} arises from two sources. First, the corroding metal is energetically heterogeneous due to its polycrystalline nature and to the presence of grain boundaries and defects (See Fig. 2.5). Thus, the most active sites are the first ones to interact with the inhibitor and are those which produce the greatest ΔH_{ads}. Less active sites interact at higher coverages and with

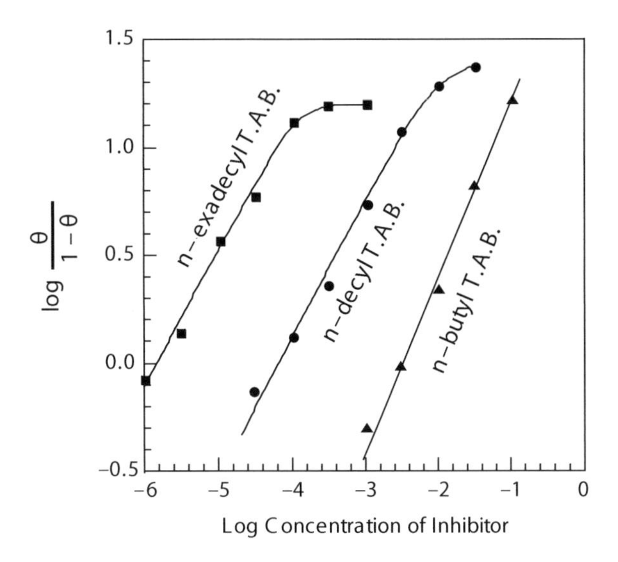

Fig. 12.18 Test of the Langmuir isotherm for various trimethylammonium bromides (TAB) compounds adsorbed on mild steel from 0.5 M H_2SO_4 [30]. Reproduced by permission of John Wiley & Sons Ltd

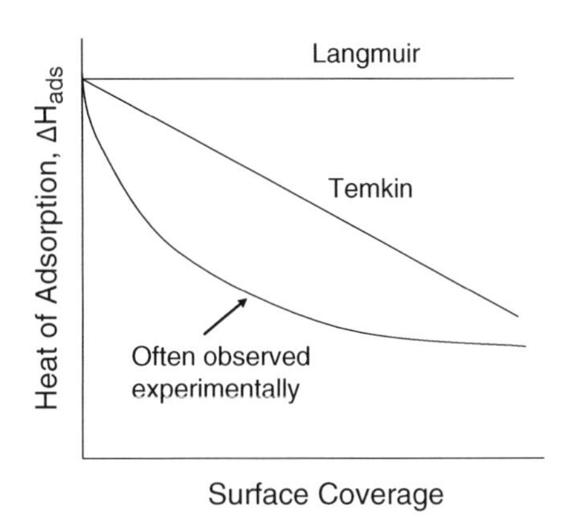

Fig. 12.19 Schematic diagram showing the heat of adsorption vs. coverage for Langmuir and Temkin adsorption isotherms

lesser values of ΔH_{ads}. Second, lateral repulsive interactions between adsorbed molecules (or ions) result in a decrease in ΔH_{ads} with increasing coverage.

The second major adsorption isotherm, the Temkin adsorption isotherm, allows for a linear decrease in the heat of adsorption with coverage.

$$\Delta H_{ads} = \Delta H^{o}_{ads} - r\theta \tag{9}$$

where $\Delta H^{o}{}_{ads}$ is the heat of adsorption at near-zero coverages and r is the Temkin parameter. When Eq. (9) is inserted into the Langmuir model in Eq. (7), the result is

$$\frac{\theta}{1-\theta} = K' C e^{-r\theta/RT} \tag{10}$$

Table 12.7 Comparison of Langmuir and Temkin adsorption isotherms

	Langmuir	Temkin
Assumption	Homogeneous surface	Heterogeneous surface
Heat of adsorption	Independent of coverage	Decreases linearly with coverage
Equation	$\frac{\theta}{1-\theta} = KC$ $\frac{C}{\theta} = \frac{1}{K} + C$	$\theta = \frac{2.303RT}{r} \log C$ $+ \frac{2.303RT}{r} \log K'$
Test of isotherm	Linear plot of $\theta/(1-\theta)$ vs. C Linear plot of θ/C vs. C	Plot of θ vs. log C gives a linear portion

where K' is a constant. At intermediate coverages ($0.3 < \theta < 0.7$), Eq. (10) gives

$$\theta = \frac{2.303\,RT}{r} \log C + \frac{2.303\,RT}{r} \log K' \tag{11}$$

as shown in Appendix I. Thus, a test of the Temkin adsorption isotherm is that a plot of coverage θ vs. the logarithm of inhibitor concentration C gives a straight line over a portion of the plot, as shown in Fig. 12.20. Table 12.7 compares Langmuir and Temkin adsorption isotherms.

Direct experimental determination of adsorption isotherms during the actual corrosion inhibition process is difficult to determine but can be done using radiotracer, infrared, or quartz crystal microbalance techniques. Alternately, the amount of adsorbed inhibitor can be inferred from double layer capacitance measurements or from corrosion current densities. These indirect methods yield relative, but not absolute, adsorption isotherms. The electrode surface can be modeled as being composed of two parallel capacitors for the inhibitor-bare and inhibitor-covered portions added on a coverage basis. The capacitance C is then

$$C = C_o(1 - \theta) + C_{sat}\theta \tag{12}$$

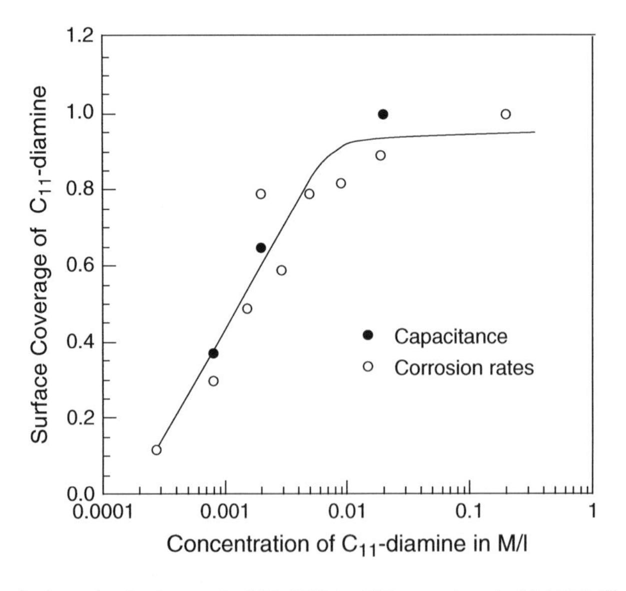

Fig. 12.20 Experimental adsorption isotherms for $NH_2(CH_2)_{11}NH_2$ onto iron in 6 M HCl [9, 10]

where C_o and C_{sat} are the capacitances (per unit area) of the inhibitor-free and inhibitor-covered surface, respectively, and θ is the coverage of the inhibitor. The value for C_{sat} is that for which further addition of the inhibitor causes no further decrease in capacitance. (While there is maximum adsorption at such a concentration, there may not be complete monolayer coverage, and this must be determined by other experimental methods). From Eq. (12)

$$\theta = \frac{C_o - C}{C_o - C_{sat}} \tag{13}$$

Similarly, the corrosion rate (i) can be given by the corrosion rates of two parallel reactions for the uninhibited (i_o) and inhibitor-covered (i_{sat}) surfaces. That is

$$i = i_o(1 - \theta) + i_{sat}\theta \tag{14}$$

or

$$\theta = \frac{i_o - i}{i_o - i_{sat}} \tag{15}$$

Adsorption isotherms given in Fig. 12.20 were determined using Eqs. (13) and (15). Many systems of organic inhibitors have been observed to obey Temkin isotherms [9, 10, 21, 26, 33–35], and not surprisingly so, given the heterogeneous nature of the surface of a corroding metal.

Nearly Neutral Solutions

The corrosion of metals in nearly neutral solutions (and open to the air) differs from that in acidic solutions for two reasons. First, in acid solutions, the metal surface is oxide free, but in neutral solutions, the surface is oxide covered. Second, in acid solutions, the main cathodic reaction is hydrogen evolution, but in air-saturated neutral solutions, the cathodic reaction is oxygen reduction.

Effect of Oxide Films

Corrosion inhibitors can interact strongly with a surface even if it is oxide covered. Table 12.8 compares the heats of adsorption ΔH_{ads} for several small organic molecules onto the surfaces of metals, oxides, or oxide films. The heats of adsorption are from the vapor phase so as to consider the energy of interaction between the adsorbate and the surface without the complicating effect of the solvent. The heats of adsorption in Table 12.8 are for near-zero coverages, so values of ΔH_{ads} are those which pertain before ΔH_{ads} begins to decrease with increasing coverage. It can be seen that in some cases, the heats of adsorption onto the oxide or oxide-covered surface are well within the range expected for a chemisorptive process. This strong interaction is due to the interaction of dipoles in the organic molecule with cations or anions present in the oxide surface. In addition, pi-bonding is also possible between benzene rings and the metal or its oxides.

Table 12.8 also shows that water molecules are strongly adsorbed onto oxide surfaces or oxide films. Thus, for oxide-covered surfaces, corrosion inhibitors in neutral solutions must also compete with water molecules for adsorption sites, just as in the case for bare metal surfaces in acid solutions.

Table 12.8 Heats of adsorption from the vapor phase at near-zero coverages for various small organic molecules or water on metals, oxides, or oxide-covered metal surfaces

Adsorbate	ΔH_{ads} on metal (kJ/mol)	ΔH_{ads} on metal oxide or oxide film (kJ/mol)	References
n-$C_3H_7NH_2$	−226 (on nickel)	−84 (on oxidized nickel)	Yu et al. [36]
	−105 (on copper)	−96 (on oxidized copper)	
n-C_3H_7OH	−347 (on nickel)	−138 (on oxidized nickel)	Yu et al. [36]
	−105 (on copper)	−67 (on oxidized copper)	
		−200 (on Al_2O_3)	Rossi et al. [37]
n-$C_3H_7NH_2$	−138 (on iron)	−188 (on oxidized iron)	Yu Yao [38]
n-$C_4H_9NH_2$	−144 (on iron)	−186 (on oxidized iron)	
n-$C_5H_{11}NH_2$	−165 (on iron)	−192 (on oxidized iron)	
n-$C_6H_{13}NH_2$	−172 (on iron)	−213 (on oxidized iron)	
$(n$-$C_3H_7)_2NH$	−128 (on iron)	−183 (on oxidized iron)	
$(C_2H_5)_3N$	−156 (on iron)	−207 (on oxidized iron)	
ϕ-CH_3[a]	–	−69 to −77 (on ZnO)	Nakazawa [39]
		−83 (on FeO)	
ϕ-OH[a]	–	−131 (on ZnO)	Nakazawa [39]
		−140 (on FeO)	
H_2O		−113 (on oxidized iron)	Yu Yao [38, 40]
		−134 (on α-Fe_2O_3)	
		−134 (on α-Al_2O_3)	
		−80 to −90 (on NiO)	Matsuda [41]
D_2O		−89 (on ZnO)	Nakazawa [39]
		−65(on FeO)	

[a]The symbol ϕ denotes the benzene ring.

Because the metal surface is oxide-covered, the role of the inhibitor in neutral solutions is to make oxide films protective and to keep them so. Oxidizing inhibitors, like chromates, react with the oxide-covered surface and are incorporated into the passive film. With non-oxidizing inhibitors, however, such chemical reactions do not occur. Phosphates, for example, are incorporated into the passive film by exchange with surface oxide or hydroxide ions [42]. Other inhibitors, such as sodium azelate, are not taken up as uniformly but instead function by repairing pores in oxide films by reacting there with dissolved cations to form insoluble local precipitates [43].

Chelating Compounds as Corrosion Inhibitors

Chelating compounds are organic molecules with at least two polar functional groups capable of ring closure with a metal cation. The functional groups may either be basic groups, such as $-NH_2$, which can form bonds by electron donation, or acidic groups, such as –COOH, which can co-ordinate after the loss of a proton. An example of a chelate is shown in Fig. 12.21. The chelating molecule may interact with metal ions which exist in the oxide film, or they may react with metal cations which are first produced by metallic dissolution. In the latter case, a complex of high molecular weight and low solubility is precipitated near the metal surface, and a barrier film is formed. Other chelating agents include mercaptoacetic acids, 8-hydroxyquinoline, and ethylenediaminetetraacetic acid. In each case, surface chelates are formed involving bonding between a surface cation and an oxygen, a sulfur, or a nitrogen atom in the chelating molecule, as in Fig. 12.21 [44].

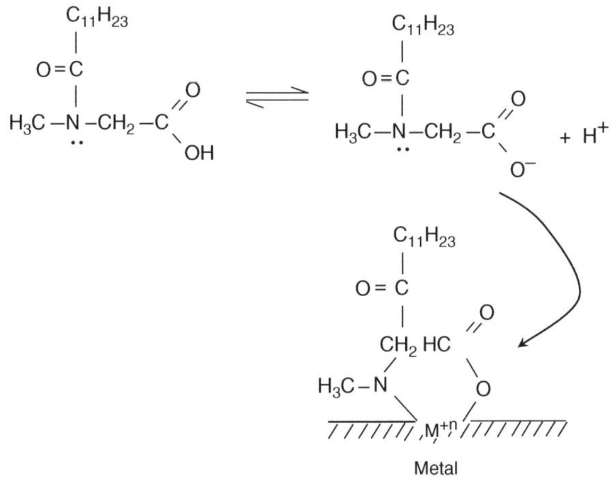

Fig. 12.21 A sarcosine-type surface chelate [44]. Reproduced by permission of NACE International

It is well known that in bulk solutions, five-membered chelate rings are the most stable [45], and it appears that this effect carries over to surface chelates as well. Steric requirements for surface chelation are more restrictive than those for the bulk solution. Not all compounds which are capable of forming chelates in the bulk solution are effective inhibitors. As an example, 2,2-bipyridine and 1,10-phenanthroline are classical chelating agents for ferrous ions in solution [45]. However, in cooling water studies, the former provided no protection, while the latter did [44]. The differences in corrosion inhibition are because the *trans*-2,2-bipyridine resists surface chelation, but the rigid structure of the 1,10-phenanthroline is favorable for surface chelation, as shown in Fig.12.22 [44].

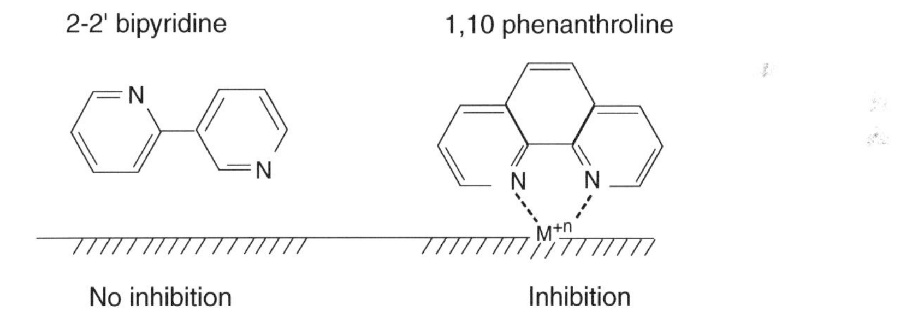

Fig. 12.22 Steric effects in surface chelation [44]. Reproduced by permission of NACE International

Chelating compounds have been studied as corrosion inhibitors for steel in industrial cooling waters [44, 46] and for zinc [47, 48] and aluminum [49, 50] in various environments. A current interest in chelate inhibitors has been prompted by the search to replace chromate inhibitors with compounds which are less toxic and less polluting.

Chromates and Chromate Replacements

Chromates (CrO_4^{2-}) are excellent corrosion inhibitors when added to aqueous solution or if used to treat metals to form corrosion-resistant "conversion" coatings. However, Cr^{6+} is a known carcinogen (the oxidation state of Cr is +6 in chromates). In addition, other toxic chemicals, such as cyanides and fluorides, are used in chromate conversion solutions. Thus, environmental concerns have been raised over the use of chromates and the disposal of their processing baths, and, accordingly, there is an increased interest in finding or developing replacements for chromate inhibitors. However, a lack of detailed fundamental understanding as to why chromates are such effective inhibitors has complicated the search for chromate replacements.

As mentioned in Chapter 6, chromates are passivating-type inhibitors and are reduced to form Cr_2O_3, which is incorporated into the passive film on the metal surface. The search for chromate replacements has taken several directions. These include studies on

- compounds which are analogous to chromates, such as molybdates (MoO_4^{2-}), tungstates (WO_4^{2-}), and vanadates (VO_4^{3-} or VO_3^{-}) [51–55]
- phosphates (PO_4^{3-}) [56, 57]
- cerium salts [58]
- "environment-friendly" inorganic coatings, such as hydrotalcite coatings (lithium aluminum hydroxylated carbonate) [59]
- diamond-like coatings [60]
- ion beam surface modification [61] (see Chapter 16)
- natural products as corrosion inhibitors [62, 63]

The use of natural products (biological molecules) as corrosion inhibitors is discussed later in this chapter.

Inhibition of Localized Corrosion

Pitting, crevice corrosion, and stress-corrosion cracking can all be mitigated by the use of corrosion inhibitors. In pitting corrosion, the use of an inhibitor usually raises the pitting potential, whereas in crevice corrosion, the inhibitor usually affects the propagation stage and reduces the current between the crevice metal and the component of the metal outside the crevice. In stress-corrosion cracking, inhibitors can increase the time to failure by increasing the time to initiation of cracks and/or by decreasing the crack growth rate. Inhibitors can also increase the value of K_{Iscc} so as to increase the resistance of specimens containing inherent flaws to SCC.

Examples of the inhibition of these various forms of localized corrosion are given below.

Pitting Corrosion

The effect of a corrosion inhibitor on pitting is usually to increase the pitting potential. This effect has been shown previously in Fig. 10.22, where it was seen that increasing additions of sulfate ion increase the pitting potential of 304 stainless steel in a chloride solution.

Two additional examples on the effect of capronate and chromate inhibitors on the pitting potential of iron are shown in Figs. 12.23 and 12.24 [64, 65]. For each inhibitor, the pitting potential increases linearly with the logarithm of the inhibitor concentration. That is:

$$\left(\frac{\partial E_{\text{pit}}}{\partial \log[\text{I}]}\right)_{[\text{A}]} = \text{constant} \tag{16}$$

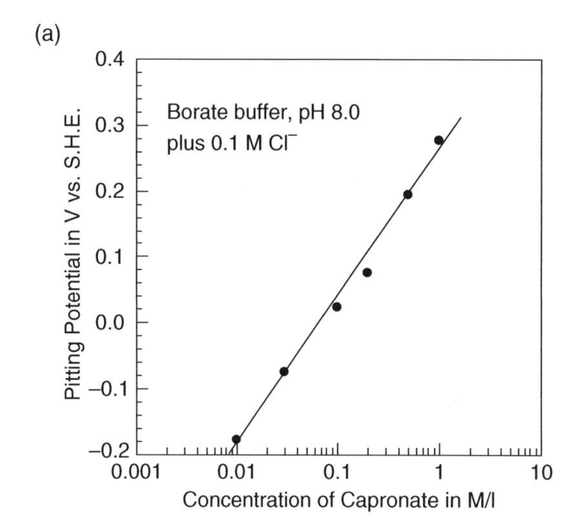

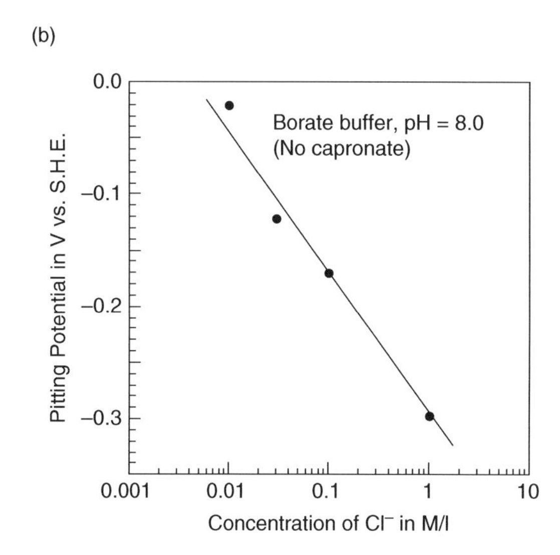

Fig. 12.23 (**a**) Pitting potential of iron as a function of capronate concentration (at a constant chloride concentration). (**b**) Pitting potential of iron as a function of chloride concentration (no inhibitor present). Redrawn from [64] by permission of Elsevier Ltd

where [I] refers to the concentration of inhibitive ion (capronate or chromate) and [A] refers to the concentration of the aggressive ion (chloride). Figures 12.23 and 12.24 also show that at a constant concentration of inhibitive ion (capronate or chromate), the pitting potential decreases with increasing concentration of aggressive ion (chloride), according to the following equation:

$$\left(\frac{\partial E_{\text{pit}}}{\partial \log [\text{A}]} \right)_{[\text{I}]} = \text{constant}' \qquad (17)$$

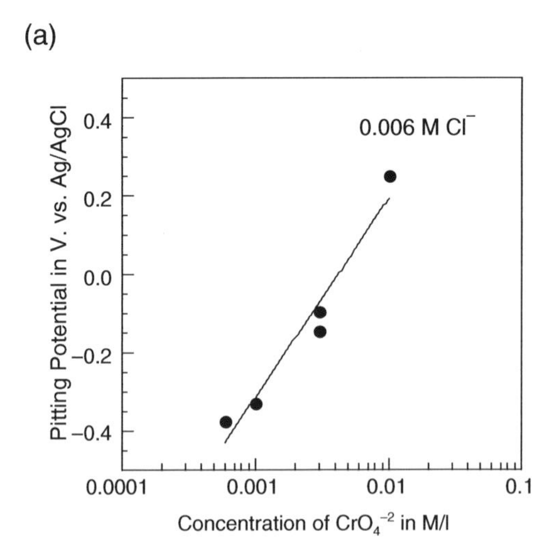

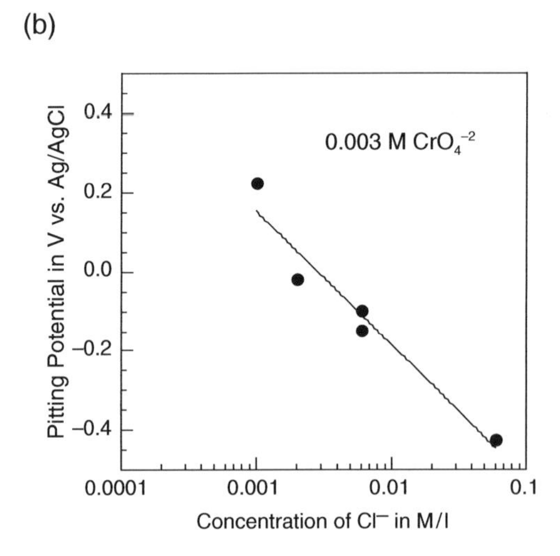

Fig. 12.24 (**a**) Effect of CrO_4^{2-} concentration on the pitting potential of iron (at constant Cl^- concentration). (**b**) Effect of Cl^- concentration on the pitting potential of iron (at constant CrO_4^{2-} concentration) [65]

The relationship in Eq. (17) is similar to that seen previously in Fig. 10.21 for aluminum or type 304 stainless steel in chloride solutions but in the absence of a corrosion inhibitor.

Equations (16) and (17) can be explained on the basis of competitive adsorption between the aggressive and the inhibitive ions [64–67]. That is, the aggressive and inhibitive ions compete for adsorption sites on the oxide-covered metal surface. If adsorption of the inhibitive ion prevails, then passive film formation ensues, the pit initiation process is stifled, and the surface must be polarized to a higher electrode potential to initiate passive film breakdown and pitting. (That is, the pitting potential is increased.)

The adsorption of either the inhibitive or the aggressive ions can each be described by Temkin adsorption isotherms, but Eq. (11) must first be modified to take into account the fact that the adsorbates are charged species. See Appendix J. For the aggressive ion:

$$\theta_A = \frac{2.303\,RT}{r_A}\log[A] + \frac{2.303\,RT}{r_A}\log K'_A + \frac{z_A\,FE}{r_A} \tag{18}$$

where E is the electrode potential and z_A is the charge on the aggressive ion. Similarly for the inhibitive ion

$$\theta_I = \frac{2.303\,RT}{r_I}\log[I] + \frac{2.303\,RT}{r_I}\log K'_I + \frac{z_I\,FE}{r_I} \tag{19}$$

where z_I is the charge on the inhibitive ion. Equations (18) and (19) combine to give

$$\begin{aligned}\left[\left(\frac{\theta_A}{\theta_I}\right)\frac{z_I}{r_I} - \frac{z_A}{r_A}\right]\frac{FE}{2.303RT} &= \frac{1}{r_A}\log[A] - \left(\frac{\theta_A}{\theta_I}\right)\frac{1}{r_I}\log[I] \\ &\quad + \frac{1}{r_A}\log K'_A - \left(\frac{\theta_A}{\theta_I}\right)\frac{1}{r_I}\log K'_I\end{aligned} \tag{20}$$

According to the competitive adsorption model of Uhlig and co-workers [66, 67], pitting initiates at sites on the oxide-covered metal surface at some critical ratio of $\theta_A/\theta_I = \theta_{crit}$. This does not mean that adsorption of the aggressive ion (usually chloride) is sufficient to cause pitting but that the ensuing processes of film breakdown and pit initiation occur when some critical amount of Cl^- has been taken up by the surface. Using a scanning Cl^--sensitive micro-electrode, Lin et al. [68] found that chloride ions accumulated in Cl^--rich islands on the surface of 18-8 stainless steel. Recall that in the pitting of aluminum, the amount of chloride taken up by the surface increases as the pitting potential is approached, as was discussed in Chapter 10. Thus, when $\theta_A/\theta_I = \theta_{crit}$, the electrode potential is the pitting potential E_{pit}:

$$\frac{\theta_A}{\theta_I} = \theta_{crit} \text{ at } E = E_{pit} \tag{21}$$

Taking differentials in Eq. (20) at a constant concentration of aggressive ion and subject to Eq. (21) gives:

$$\left(\frac{\partial E_{pit}}{\partial \log[I]}\right)_{[A]} = -\frac{2.303\,RT}{F}\,\frac{\theta_{crit}\left(\frac{r_A}{r_I}\right)}{\theta_{crit}\left(\frac{r_A}{r_I}\right)z_I - z_A} = \text{constant}' \tag{22}$$

The right-hand side of Eq. (22) is a constant so that Eq. (22) has the form of Eq. (16) and explains the straight line observed experimentally in Figs. 12.23(a) and 12.24(a).

Taking differentials in Eq. (20) at a constant concentration of inhibitor ion gives:

$$\left(\frac{\partial E_{pit}}{\partial \log[A]}\right)_{[I]} = \frac{2.303RT}{F}\,\frac{1}{\theta_{crit}\left(\frac{r_A}{r_I}\right)z_I - z_A} = \text{constant} \tag{23}$$

The right-hand side of Eq. (23) is also a constant so that Eq. (23) has the form of Eq. (17) and explains the straight lines observed experimentally in Figs. 12.23(b) and 12.24(b).

Moreover, the slopes in Fig. 12.23(a) and (b) have opposite signs, as predicted by Eqs. (22) and (23). The same holds for Fig. 12.24(a) and (b).

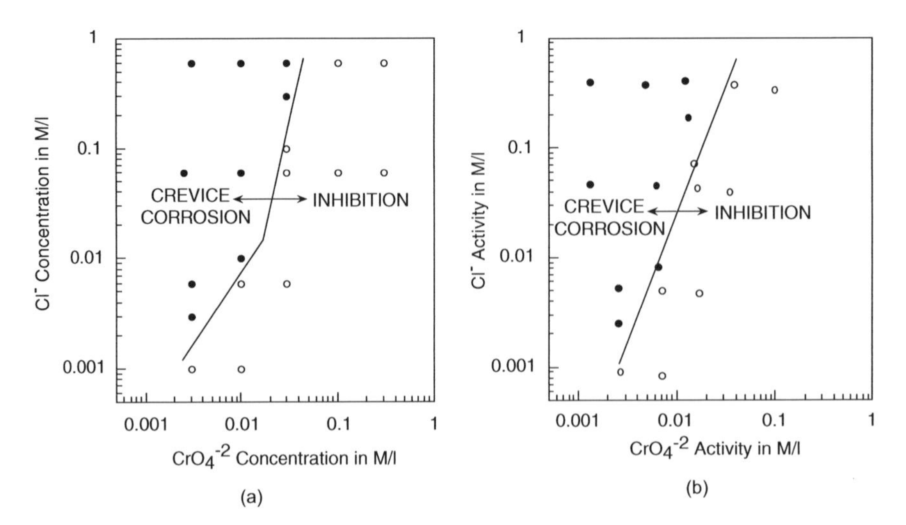

Fig. 12.25 (**a**) Crevice corrosion or inhibition of iron in 0.25-mm crevices as a function of the composition of the electrolyte. Note that for a fixed concentration of Cl^-, a certain minimum concentration of $CrO_4{}^{2-}$ is required for inhibition and that this minimum amount increases with increasing Cl^- concentration. (**b**) Same data re-plotted as a function of activities. Note that a single straight line separates the regions of corrosion and inhibition in part (**b**) [65]. Reproduced by permission of ECS – The Electrochemical Society

Crevice Corrosion

As discussed briefly in Chapter 10, chromates are also effective in reducing the crevice corrosion of iron in chloride solutions. Figure 10.11 has shown that for a given Cl^- concentration, the crevice corrosion of iron is reduced dramatically by using increasing amounts of inhibitor. For more dilute chloride solutions, a smaller concentration of chromate was required for protection, as was shown in Table 10.2. Figure 12.25 summarizes the crevice corrosion results for many solutions of different combinations of $CrO_4{}^{2-}$ and Cl^- concentrations. In all instances, a critical minimum concentration of $CrO_4{}^{2-}$ is required for inhibition, given a certain concentration of Cl^-. When the data in Fig. 12.25(a) are re-plotted on the basis of activities rather than concentrations, a single straight line results, as in Fig. 12.25(b). The borderline separating the regions of crevice corrosion and inhibition is given by:

$$\frac{\mathrm{d}\log a_{\mathrm{Cl}^-}}{\mathrm{d}\log a_{\mathrm{CrO}_4^{-2}}} = \text{constant}'' \tag{24}$$

For $CrO_4{}^{2-}/Cl^-$ solutions where crevice corrosion occurred, the electrode potential measured within the crevice was always between −0.620 and −0.660 V vs. Ag/AgCl, regardless of the concentration of the bulk electrolyte, as shown in Fig. 12.26. As shown in Table 10.6, the electrode potential of −0.620 to −0.660 V vs. Ag/AgCl is a value which is characteristic for iron in occluded corrosion cells.

Equation (20) can be applied to the inhibition of crevice corrosion if it is assumed that there is also a critical ratio θ_A/θ_I which leads to the breakdown of the passive film within the crevice. When $\theta_A/\theta_I = \theta_{crit}$, then $E = E_{active} = -0.640\ \mathrm{V} \pm -0.020\ \mathrm{V}$ vs. Ag/AgCl = constant. Taking differentials in Eq. (20) at constant E (after first replacing concentrations by activities) gives:

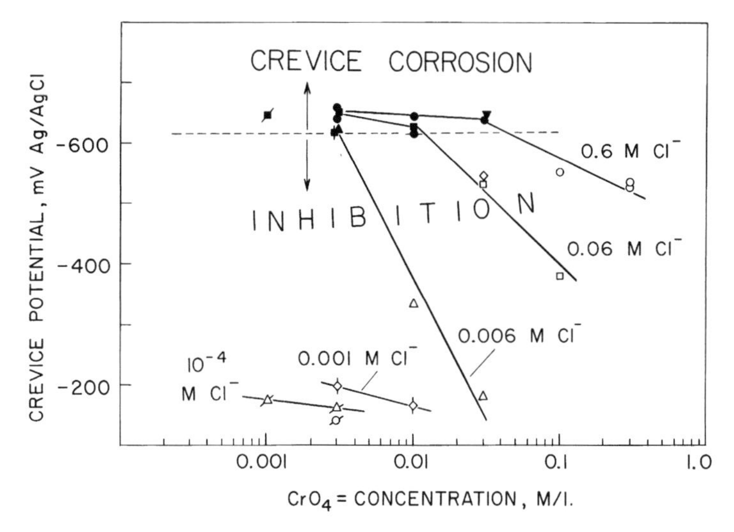

Fig. 12.26 Crevice corrosion of iron in chromate/chloride solutions always occurred when the electrode potential within the crevice was more negative than −0.62 V vs. Ag/AgCl. Crevice corrosion was inhibited at more positive internal electrode potentials [65]. Reproduced by permission of ECS – The Electrochemical Society

$$\left(\frac{\partial \log a_{\mathrm{A}}}{\partial \log a_{\mathrm{I}}}\right)_{E_{\mathrm{active}}} = \theta_{\mathrm{crit}} \frac{r_{\mathrm{A}}}{r_{\mathrm{I}}} = \text{constant}'' \tag{25}$$

which has the form of Eq. (24) and describes the borderline between regions of crevice corrosion and inhibition in Fig. 12.25(b). Details regarding derivation of Eq. (25) are left as an exercise in Problem 12.11.

Thus, inhibition of two forms of localized corrosion (pitting and crevice corrosion) is similar in that if a critical surface ratio of aggressive to inhibitive ion is exceeded, pitting ensues on open surfaces and active corrosion occurs within crevices.

Stress-Corrosion Cracking and Corrosion Fatigue

An interesting example on the inhibition of SCC is given by the effect of benzotriazole (BTA) on type 304 stainless steel in 1 M HCl [69].

Figure 12.27 shows that increasing amounts of BTA increase the K_{Iscc} values for single-edge notched specimens in 1 M HCl. Recall from Chapter 11 that

$$K_{\mathrm{I}} = \sigma\sqrt{\pi a} \cdot (\text{geometrical factor}) \tag{26}$$

so that increases in K_{I} mean that a larger flaw size (a) can be tolerated for a given stress σ.

In the same study [69], it was also shown that additions of BTA increased the time to failure for smooth tensile specimens, as shown in Fig. 12.28. Thus, additions of BTA were able to protect against SCC for either smooth specimens or pre-cracked specimens of type 304 stainless steel in 1 M HCl.

As stated earlier, corrosion inhibitors can reduce stress-corrosion cracking in various ways. Inhibitors can increase the time to failure by increasing the time to initiation of cracks and/or by

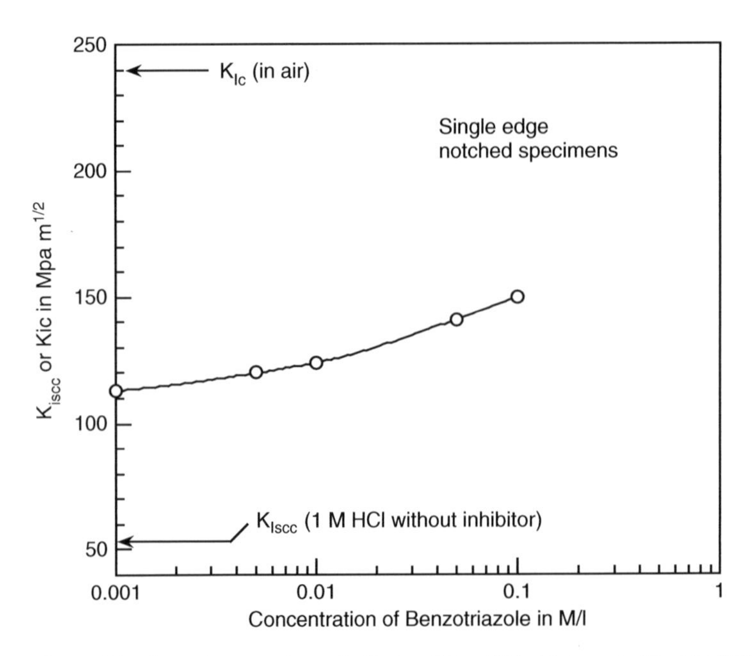

Fig. 12.27 Effect of benzotriazole on the stress intensity factor of type 304 stainless steel in 1 M HCl for single-edge notched specimens. Drawn from data in [69]

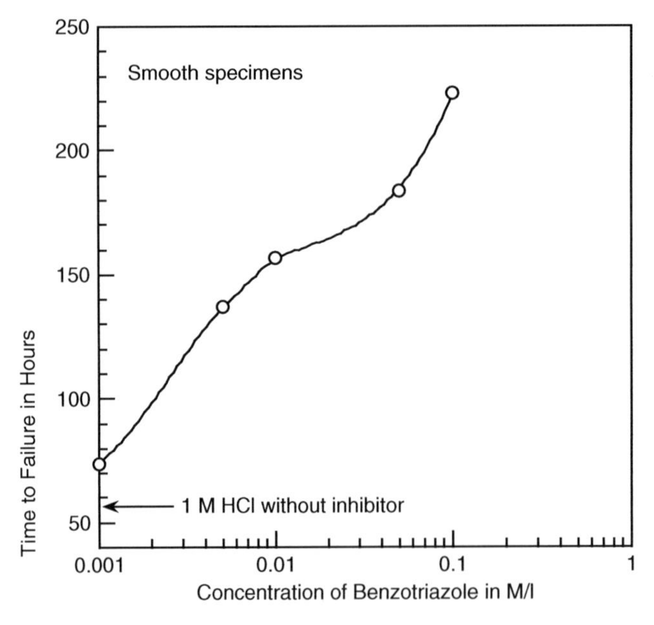

Fig. 12.28 Effect of benzotriazole on the time to fracture of type 304 stainless steel in 1 M HCl for smooth specimens not containing an intentional defect. Drawn from data in [69]

decreasing the crack growth rate. Inhibitors can also increase the value of K_{Iscc} so as to increase the resistance of specimens containing inherent flaws to SCC.

Corrosion inhibitors can also reduce corrosion fatigue damage. An example is given by the use of various inhibitors applied to the corrosion fatigue of a high-strength steel [70]. The environment was air of 90% relative humidity (maintained in an environmental chamber surrounding the crack). Thus, water vapor adsorbs and condenses within the crack tip to form an electrolyte. Various inhibitors were dissolved in appropriate solvents and were applied as droplets near the crack tip. Figure 12.29 shows corrosion fatigue plots of d*a*/d*N* vs. ΔK. It can be seen that several inhibitors were effective in reducing the crack growth rate.

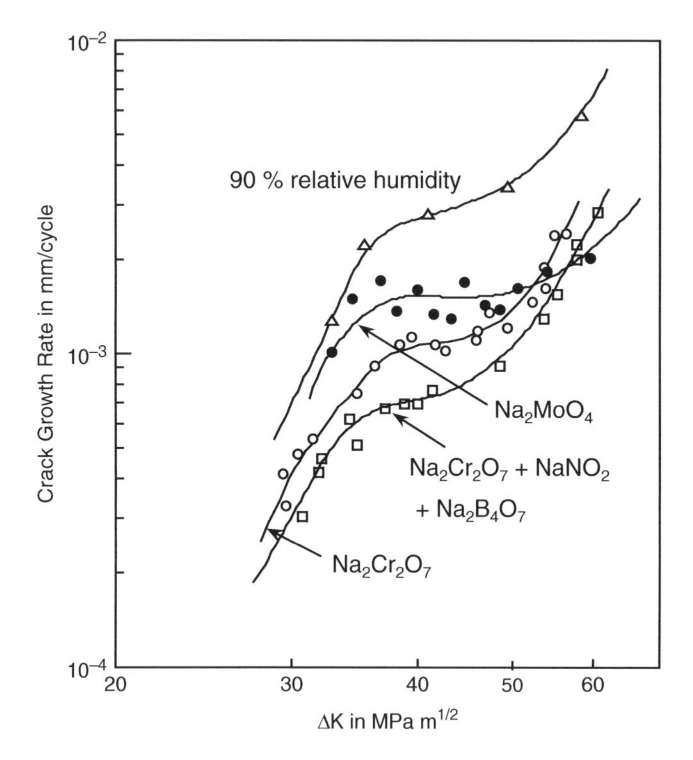

Fig. 12.29 Effect of various inhibitors on the corrosion fatigue crack growth rate of a high-strength steel in a humid environment [70]. Reproduced by permission of © NACE International 1980

In both stress-corrosion cracking and in corrosion fatigue (as well as in crevice corrosion), for the corrosion inhibitor to be effective it is essential that the inhibitor reaches the inside of the occluded corrosion cell (i.e., the crack tip or the crevice interior). This is accomplished most easily in the case of corrosion fatigue because the cyclic load causes the crack to open and close. This, in turn, produces a pumping action in the electrolyte which serves to assist the inhibitors to move from the bulk solution inward toward the crack tip.

New Approaches to Corrosion Inhibition

New approaches to corrosion inhibition have been spurred in part by the desire for environmentally acceptable corrosion inhibitors and in part by recent advances in surface modification by monomolecular films. The first of these two approaches, which is discussed below, is the use of natural products or biological molecules as non-toxic, biodegradable corrosion inhibitors.

Biological Molecules

The use of natural products as corrosion inhibitors is not new. Inhibitors of acid corrosion were known as far back as the Middle Ages by metal-working craftsmen, who used flour, bran, and yeast as pickling inhibitors [71]. In the early 1900s, molasses, vegetable oils, starch, tars, and oils were also used [71].

In recent years, attention has turned to various organic molecules, such as α-amino acids, which are components of natural products, and thus are environmentally acceptable. Amino acids are the molecular units that make up proteins, which in turn are the building blocks of biological systems. The formulas of some particular α-amino acids are shown in Fig. 12.30 [72]. All proteins are derived from 20 specific naturally occurring amino acids. In addition to their non-polluting properties, amino acids are attractive candidates as corrosion inhibitors because they contain adsorbable amino and carboxyl functional groups for attachment to metal surfaces. Amino acids and other natural products, such as peptides, catechols, vitamins, and siderophores (isolated from bacteria), have all been examined recently as corrosion inhibitors for different metals in various solutions [62, 63, 73–80].

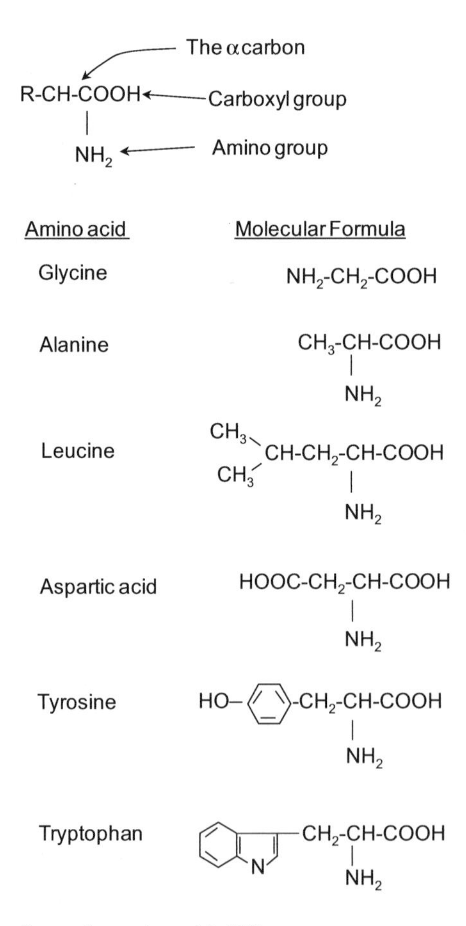

Fig. 12.30 Molecular formulas of several α-amino acids [72]

The application of natural products has been taken a step beyond the use of constituent molecules by using actual extracts from plant systems, such as tannins or tree barks [78–80], or from animal systems. An example of the latter is provided by the work of Hansen and co-workers [63, 76] on the properties of the adhesive protein of the common blue mussel *Mytilus edulis L.* The blue mussel is shown suspended by byssal threads from a glass plate in Fig. 12.31. These byssal threads terminate in adhesive plaques with which the mussel attaches itself to surfaces. The mussel adhesive protein (MAP), which has a molecular weight greater than 10,000, has been found to contain a repeating decapeptide that provides a variety of functional groups for attachment to bare or oxide-covered metal surfaces. These attachment possibilities include coulombic interactions, dipole–dipole interactions, hydrogen bonding through catechol groups, as well as co-ordination and chelation of metal ions or surface hydroxyls by the catecholic functional groups. These possibilities are shown in Fig. 12.32.

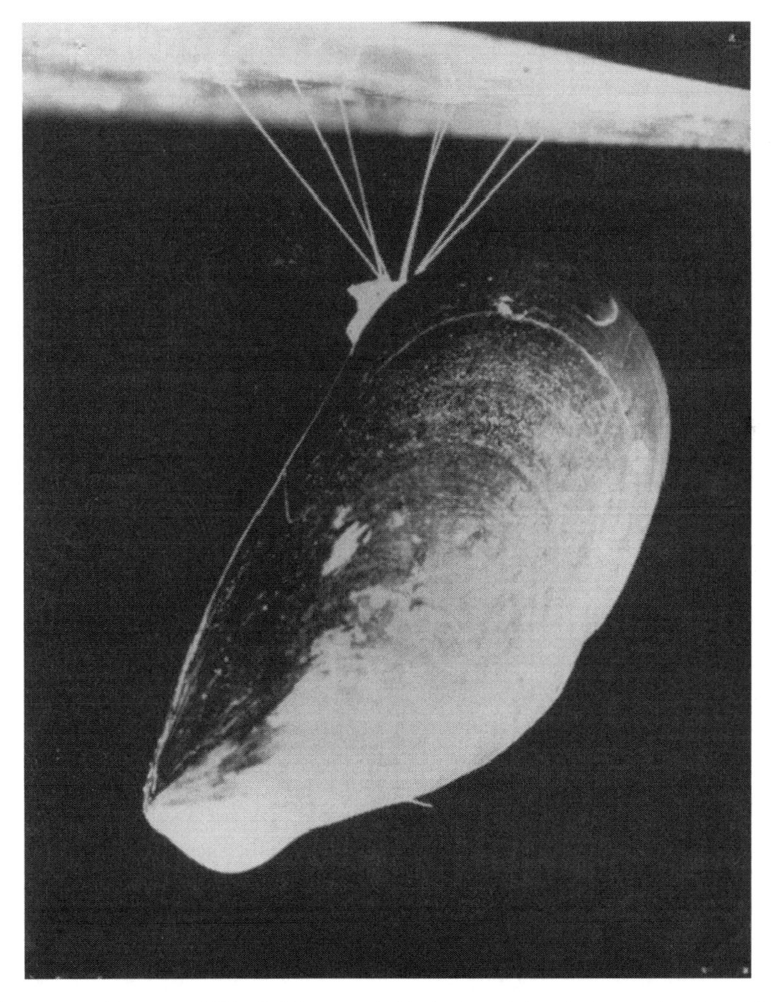

Fig. 12.31 The common blue mussel (*M. edulis*) suspended by its byssal threads from a glass plate. Courtesy of Prof. J. H. Waite, University of California at Santa Barbara

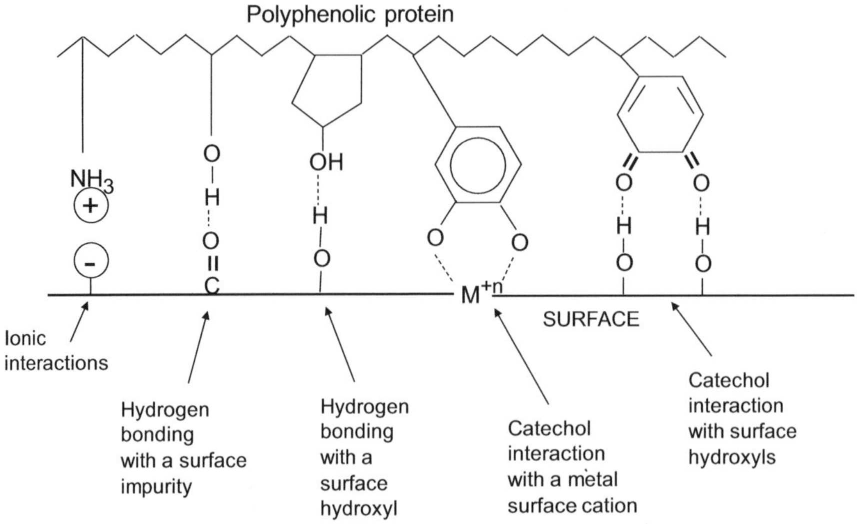

Fig. 12.32 Possible interactions between the adhesive protein isolated from the blue mussel and the oxide-covered aluminum surface [63, 76]. Reproduced by permission of ECS – The Electrochemical Society

Hansen and co-workers also utilized the complexing agent parabactin (Fig. 12.33) which was isolated from bacteria [62] and which also contains various surface-active groups, including catechols.

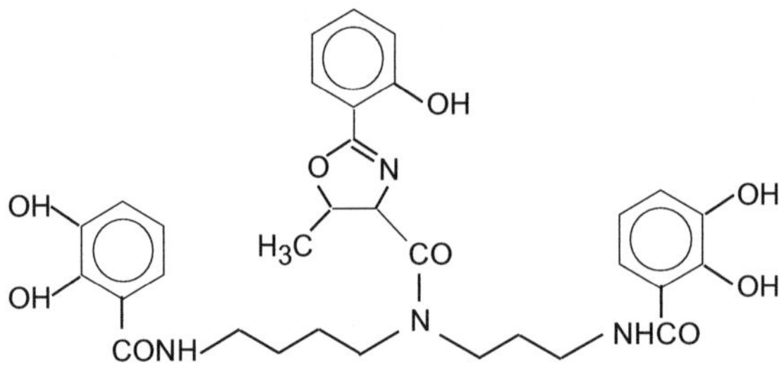

Fig. 12.33 Structural formula for the siderophore parabactin isolated from bacteria [62]. Reproduced by permission of ECS – The Electrochemical Society

Table 12.9 shows the results of pitting potential measurements for aluminum in 0.1 M NaCl. Both parabactin and MAP increase the pitting potential of aluminum. Figure 12.34 shows the effect of these molecules on the polarization resistance R_p (see Chapter 7) at the open-circuit corrosion

Table 12.9 Pitting potentials of aluminum in 0.1 M NaCl with and without natural products [76]

	E_{pit} (V vs. SCE)
Untreated aluminum	−0.700
Parabactin (1.3 × 10^{-5} M)	−0.640
Parabactin (1.0 × 10^{-4} M)	−0.600
Mussel adhesive protein (7.7 × 10^{-6} M, pH 2.5)	−0.600

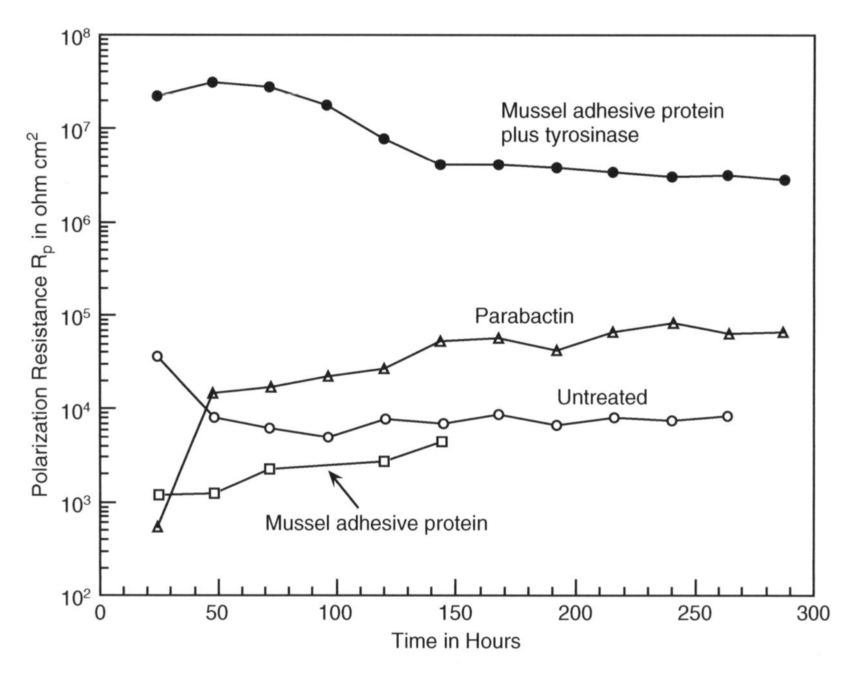

Fig. 12.34 Polarization resistance for aluminum samples treated with the mussel adhesive protein, the mussel adhesive protein plus tyrosinase, and the bacterial siderophore parabactin [63, 76]. Reproduced by permission of ECS – The Electrochemical Society

potentials. Parabactin and MAP with an enzyme added to cross-link the adhesive polymer were most effective in increasing R_p. These experiments are representative of the emerging possibilities on the use of natural products as new and novel corrosion inhibitors.

Langmuir–Blodgett Films and Self-assembled Monolayers

There are two novel approaches to corrosion inhibition which have attracted much recent interest. These are (i) Langmuir–Blodgett films and (ii) self-assembled monolayers. These approaches, which are related to each other, are borrowed from the field of surface chemistry. Both methods involve pre-treatment of the metal surface prior to its use in the corrosive electrolyte. The thickness of the inhibitor films is only one monolayer for self-assembled films and is one to several monolayers for Langmuir–Blodgett films.

Langmuir–Blodgett films are oriented insoluble monomolecular films of a substance like stearic acid, $CH_3(CH_2)_{16}COOH$, which contains a surface-active polar group, –COOH, and a non-polar hydrocarbon tail, $CH_3(CH_2)_{16}$–. The film is formed by first dissolving the compound in a suitable solvent and then casting it on a clean surface of water. After evaporation of the solvent, the polar end group of stearic acid resides in the water surface, and the hydrocarbon tail is oriented outward, as shown schematically in Fig. 12.35. The film is contained on a trough called a Langmuir film balance, which is fixed at one end and has a moveable barrier at the other end. By moving the adjustable barrier, the film can be compressed and force–area curves determined, as shown in Fig. 12.36. As the area available to the surface film is progressively decreased, the molecules in the film come into

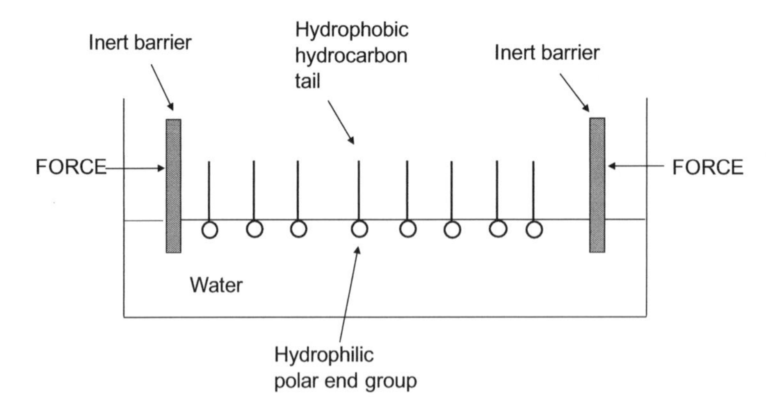

Fig. 12.35 Diagram of a Langmuir film balance. (The size of the molecule is greatly exaggerated relative to the rest of the drawing)

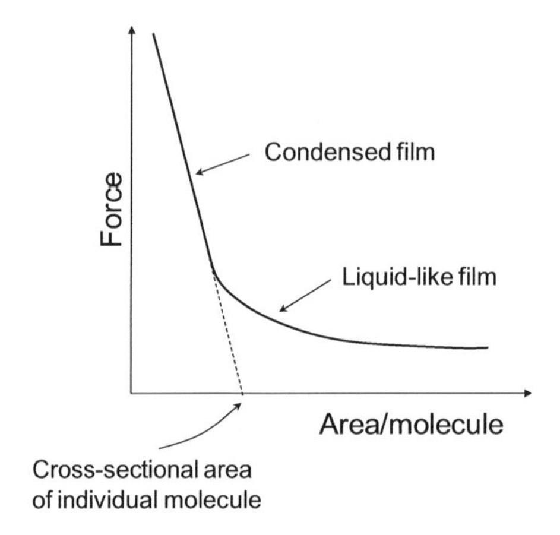

Fig. 12.36 Schematic force–area curves for a monomolecular film spread on a water surface in a Langmuir film balance

contact with each other, and the film is converted from a liquid-like film to a solid close-packed condensed film.

When a metal sample is withdrawn from a Langmuir film balance which contains a close-packed monolayer of a condensed film, the film is transferred to the metal surface, as shown in Fig. 12.37. For the initial emersion, the polar end groups are attached to the metal surface, and the hydrocarbon tails are oriented perpendicular to the surface, as in Fig. 12.37. Subsequent emersions can be carried out to build up the thickness of the film. The orientation is hydrocarbon tail-to-hydrocarbon tail for the second emersion and for all subsequent even-numbered emersions, and polar group-to-polar group for all odd-numbered emersions, as shown in Fig. 12.37.

Langmuir–Blodgett films of stearic acid have been recently deposited onto iron [81] and were observed to provide (short-term) protection in 0.1 M NaCl, as shown in Fig. 12.38. The effectiveness of the Langmuir–Blodgett film was observed to increase with the number of monolayers deposited. A similar effect was observed for Langmuir–Blodgett films of *N*-octadecylbenzidine on copper immersed in 3.4% NaCl [82].

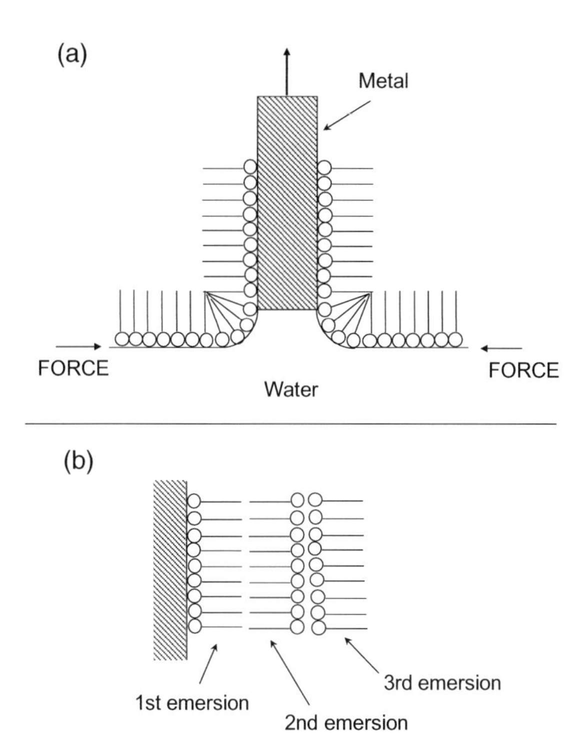

Fig. 12.37 (**a**) Transfer of a Langmuir–Blodgett film onto a metal surface by withdrawal of the metal specimen through a monomolecular film spread on a water surface. (**b**) Formation of multimolecular layers by successive withdrawals

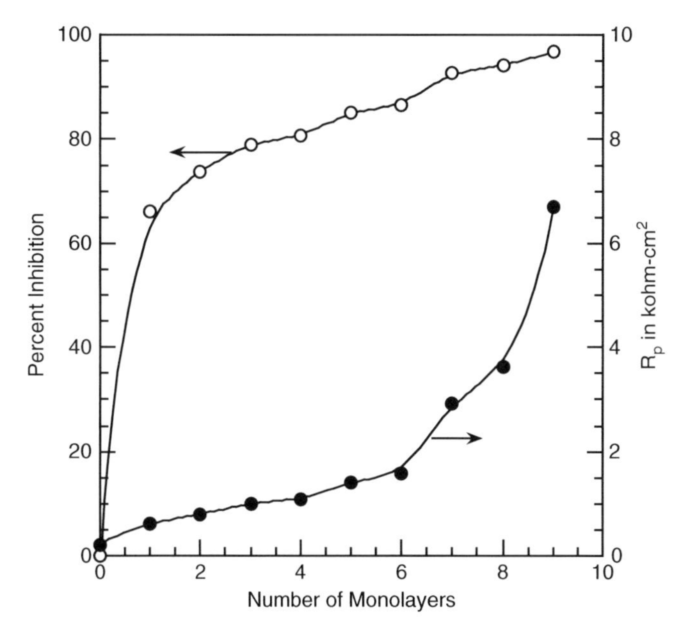

Fig. 12.38 Effect of the number of Langmuir–Blodgett layers on the percent inhibition and polarization resistance R_p of iron in 0.1 M NaCl after 30 min of exposure. Redrawn from [81] by permission of © NACE International 1995

A similar approach involves the use of self-assembled monolayers (SAM) in which a close-packed monomolecular oriented film is first formed by adsorption from an organic solvent onto a metal surface. The metal with its SAM intact is then immersed into the electrolyte of interest. Alkane thiols (of general formula R-SH) and related compounds have recently been shown to be effective in retarding the corrosion of iron or copper in various environments [83, 84]. With alkane thiols, the -SH group is adsorbed on the metal surface, and the hydrocarbon tail is oriented perpendicular thereto.

The idea of spread monolayers is not new. Benjamin Franklin delighted in spreading monolayers of vegetable oils on ponds, and Frau Pockels demonstrated in 1891 [85] that films could be confined by means of barriers. Langmuir and Blodgett developed the surface balance and the technique of transferring films to solid surfaces in the 1920s and the 1930s. It is only recently that these concepts have been applied to corrosion science.

Vapor-Phase Inhibitors

Inhibitors can be used to protect against atmospheric corrosion in gaseous atmospheres that contain moisture or other corrosive agents, such as Cl^-, H_2S, SO_2, and others. In such applications, the inhibitor is applied from the vapor phase and is called a vapor-phase inhibitor (VPI) or volatile corrosion inhibitor (VCI). These inhibitors volatilize and adsorb on all surfaces located in an enclosed space. Vapor-phase inhibitors are used to protect metal surfaces in storage or transport, as well as to protect electronic materials, such as circuit boards. Vapor-phase inhibitors can provide corrosion protection for periods ranging from months to years [86, 87].

Vapor-phase inhibitors can be either liquids or solids and are very often volatile amines. Some examples of VPIs are given in Table 12.10. (The chemical compositions of commercially available vapor-phase inhibitors are often proprietary in nature.) Liquids can be impregnated in papers or polymeric films, which can be used to package the components to be stored and protected. Solids can be contained in porous bags or sachets. The vapor pressure of the VPI must be high enough to provide a sufficient concentration of the inhibitor in the enclosed space but low enough to sustain an acceptable service life [88]. Solid-phase vapor-phase inhibitors have vapor pressures in the range 10^{-6} to 10^{-4} mmHg (approximately 10^{-4} to 10^{-2} Pa) [90].

The volatilization of the VPI is merely the means by which the inhibitor is transported to the metal to be protected. The inhibitor then functions by adsorption from the vapor phase onto the metal surface. The mechanism of inhibition by vapor-phase inhibitors in thin-layer condensed electrolytes is similar to that for bulk aqueous solutions in that the VPI forms an adsorptive bond with the metal surface or its oxide film. Figure 12.39 shows the corrosion current density for pure iron in electrolyte layers (containing SO_2) with and without a monolayer of adsorbed ammonium benzoate [91]. The VPI is most effective at the highest relative humidity where condensed electrolyte layers are formed.

Recent work on vapor-phase inhibitors has taken two directions. First, there is a renewed interest in the fundamental aspects of vapor-phase inhibition due to advances in instrumentation, such as the quartz crystal microbalance which is able to weigh microgram quantities and thus can determine adsorption isotherms for low-area metal samples. In addition, the Kelvin microprobe, which measures the work function, can map the spatial coverage of a VPI on a metal surface. The second recent direction regarding vapor-phase inhibitors is the search for environment-friendly vapor-phase inhibitors and for new applications.

Table 12.10 Some examples of vapor-phase inhibitors

Compound	Structure	Melting point (°C)	Metals protected
Cyclohexylamine	(cyclohexyl ring)–NH_2	−17.7	Mild steel [88]
Morpholine	$O(CH_2\text{-}CH_2)_2NH$ (ring: O, CH_2- CH_2, NH, CH_2- CH_2)	−4.9	Mild steel [88]
Benzotriazole	(benzene ring fused with N=N–NH ring)	98.5	Copper, aluminum, zinc [86]
Dicyclohexyl ammonium nitrate	(two cyclohexyl rings) $NH_2^+ NO_2^-$	154	Mild steel and aluminum [87]
n-Decylamine	$CH_3\text{-}(CH_2)_9\text{-}NH_2$	15	Mild steel and zinc [89]

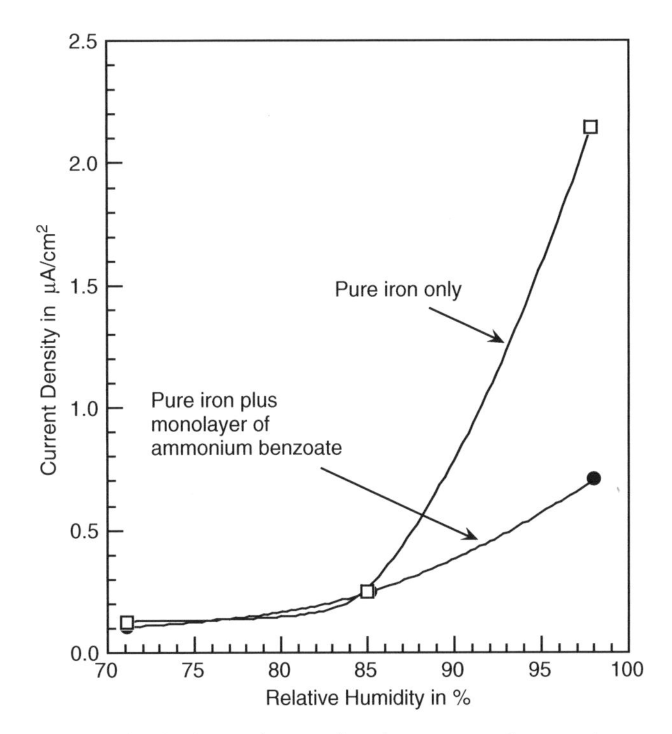

Fig. 12.39 Effect of a monomolecular layer of ammonium benzoate on the corrosion rate of iron in a humid atmosphere. Redrawn from [91] by permission of Elsevier Ltd

Problems

1. What type of tests or experiments would you set up to screen a set of several compounds for their effectiveness as corrosion inhibitors? What parameters should you attempt to measure?
2. The corrosion weight loss of mild steel immersed in a hydrochloric acid solution for 124 h was 1,229 mg/dm^2 for 1 M HCl and 7,066 mg/dm^2 for 4 M HCl [92]. In the presence of an aldehyde inhibitor, the weight losses over the same period were 153 mg/dm^2 for 1 M HCl and 2,860 mg/dm^2 for 4 M HCl. What was the percent inhibition in each case?
3. The following data were taken by measurement of double layer capacitances for benzotriazole on iron in sulfuric acid solution [31]. Show that these data obey a Langmuir adsorption isotherm.

Inhibitor concentration (mM)	Surface coverage, θ
0.21	0.69
0.42	0.84
2.08	0.97
4.16	1.00

4. The following data were taken [93] for the corrosion of mild steel in 0.5 M sulfuric acid containing various amounts of 5-hydroxyindole inhibitor. Calculate the fractional coverage of inhibitor for each concentration given below. Do the data better fit a Langmuir adsorption isotherm or a Temkin adsorption isotherm?

Concentration of inhibitor	Corrosion rate ($\mu A/cm^2$)
0	1,072
1.0×10^{-4}	879
5.0×10^{-4}	800
7.5×10^{-4}	772
1.0×10^{-3}	718
2.5×10^{-3}	568
5.0×10^{-3}	429
7.5×10^{-3}	354

5. Use the model of parallel reactions to show that if the minimum corrosion rate of an inhibited surface (i_{sat}) is much less than the corrosion rate of the uninhibited surface (i_o), then the fractional coverage of inhibitor is given by $\theta = (\% I)/100$.
6. In the following pairs of compounds, which compound in each group of two is expected to be the better inhibitor in acid solutions? Explain why in each case.

   (a) $CH_3(CH_2)_4NH_2$ or $CH_3(CH_2)_4SH$?
   (b) $CH_3(CH_2)_5COOH$ or $CH_3(CH_2)_{16}COOH$?
   (c) $CH_3(CH_2)_4NH_2$ or $H_2N(CH_2)_5NH_2$?
   The pK_a values are 10.63 for the monoamine and 10.25 for the first ionization of the diamine.

(d)

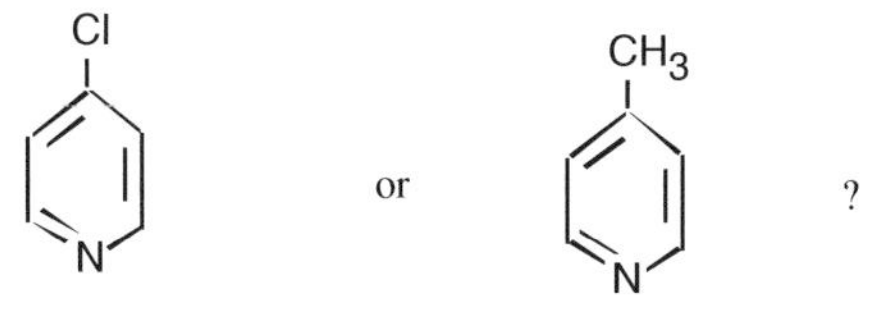

Cl

N

or

CH_3

N

?

(e)

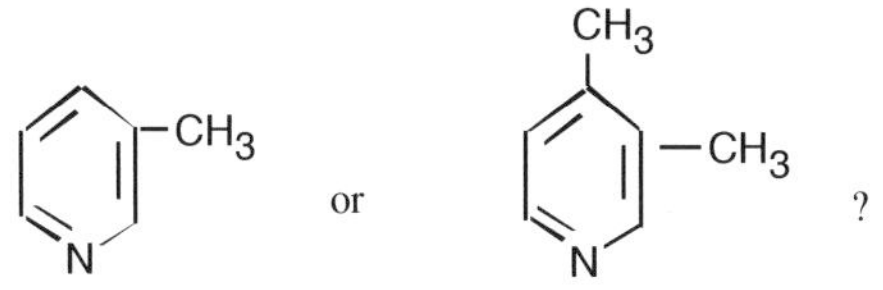

CH_3

N

or

CH_3

CH_3

N

?

7. For the adsorption of an organic molecule by replacement of previously adsorbed water molecules

$$\text{Org (soln)} + n\text{H}_2\text{O (ads)} \rightarrow \text{Org (ads)} + n\text{H}_2\text{O (soln)}$$

show that

$$\Delta G_{\text{ads}} = \Delta G^{\text{org}}_{\text{ads}} - n\Delta G^{\text{H}_2\text{O}}_{\text{ads}}$$

where $\Delta G^{\text{org}}{}_{\text{ads}}$ and $\Delta G^{\text{H}_2\text{O}}_{\text{ads}}$ refer to the free energy of adsorption per mole of the organic compound and water, respectively.

8. Chromate inhibitors are reduced to Cr_2O_3 on iron surfaces by the following reaction:

$$2\text{Fe} + 2\text{CrO}_4^{2-} + 4\text{H}^+ \rightarrow \text{Fe}_2\text{O}_3 + \text{Cr}_2\text{O}_3 + 2\text{H}_2\text{O}$$

for which the Nernst equation is

$$E = 1.437 + 0.0197 \log [\text{CrO}_4^{-2}] - 0.0394\,\text{pH}$$

The analogous reaction and the Nernst equation for tungstates are

$$2\text{Fe} + 6\text{WO}_4^{-2} + 12\text{H}^+ \rightarrow \text{Fe}_2\text{O}_3 + 3\text{W}_2\text{O}_5 + 6\text{H}_2\text{O}$$

$$E = 0.852 + 0.0591 \log [\text{WO}_4^{-2}] - 0.182\,\text{pH}$$

(a) Calculate the electrode potentials for each of these two reactions at pH 7.0 and $CrO_4{}^{2-}$ or $WO_4{}^{2-}$ concentrations of 1.0×10^{-3} M.

(b) What do the electrode potentials tell you about the spontaneity of each reaction under the given conditions and about the possibility of using tungstates as chromate replacements?

9. What are the similarities between the mechanisms of corrosion inhibition in acid solutions and in neutral solutions? What are the differences in mechanisms for these two types of solutions?
10. In some instances, very low concentrations of would-be corrosion inhibitors actually function as accelerators of corrosion until a larger surface coverage of the inhibitor is attained. Provide a possible explanation for this accelerating effect.

11. Show that the straight line in Fig. 12.25(b) which separates regions of crevice corrosion from regions of inhibition can be obtained from Eq. (20).

    HINT: Replace concentrations in Eq. (20) with activities and then take differentials at constant E ($E = E_{\text{active}}$) to get

$$\left(\frac{\partial \log a_{\text{A}}}{\partial \log a_{\text{I}}}\right)_{E_{\text{active}}} = \theta_{\text{crit}} \frac{r_{\text{A}}}{r_{\text{I}}} = \text{constant}'''$$

References

1. A. D. Mercer in "Corrosion", L. L. Shreier, Ed., Vol. 2, p. 18.1, Newnes-Butterworths, London (1976).
2. C. C. Nathan, Ed., "Corrosion Inhibitors", pp. 114, 126, 173, 196, NACE, Houston, TX (1974).
3. G. Butler and H. C. K. Ison, "Corrosion and Its Prevention in Waters", p. 157, Reinhold Publishing Company, New York (1966).
4. J. M. Gaidis and A. Rosenberg, *Mater. Perform.*, *43* (1), 48 (2004).
5. N. Hackerman, *Mater. Perform.*, *29* (2), 44 (1990).
6. E. McCafferty, M. K. Bernett, and J. S. Murday, *Corros. Sci.*, *28*, 559 (1988).
7. Z. Zembura, W. Ziolkowska, and H. Kolny, *Bull. Acad. Polon. Sci.*, *13* (7), 487 (1965).
8. N. Hackerman, *Corrosion*, *18*, 332t (1962).
9. E. McCafferty and N. Hackerman, *J. Electrochem. Soc.*, *119*, 146 (1972).
10. N. Hackerman and E. McCafferty in "Proceedings of the Fifth International Congress on Metallic Corosion", p. 542, NACE, Houston, TX (1974).
11. N. Hackerman and A. C. Makrides, *Ind. Eng. Chem.*, *46*, 523 (1954).
12. H. F. Finley and N. Hackerman, *J. Electrochem. Soc.*, *107*, 259 (1960).
13. R. C. Ayers, Jr. and N. Hackerman, *J. Electrochem. Soc.*, *110*, 507 (1963).
14. A. I. Altsybeeva, A. P. Dorokhov, and S. Z. Levin, *Protect. Metals*, *7*, 422 (1971)
15. J. A. Dean, Ed., "Lange's Handbook of Chemistry", p. 9.2, McGraw-Hill, New York (1999).
16. J. O'M. Bockris and D. A. J. Swinkels, *J. Electrochem. Soc.*, *111*, 736 (1964).
17. Z. Szklarska-Smialowska and G. Wieczorek, *Corros. Sci.*, *11*, 843 (1971).
18. J. O'M. Bockris, M. Green, and D. A. J. Swinkels, *J. Electrochem. Soc.*, *111*, 743 (1964).
19. P. Kern and D. Landolt, *Electrochim. Acta*, *47*, 589 (2001).
20. N. Hackerman, E. S. Snavely, Jr., and J. S. Payne, Jr., *J. Electrochem. Soc.*, *113*, 677 (1966).
21. Z. A. Iofa, Y. A. Nikorova, and V. V. Batrakov, "Proceedings of the Third International Congress on Metallic Corrosion", Vol. 2, p. 40, Moscow (1969).
22. I. L. Rozenfeld, "Corrosion Inhibitors", p. 115, McGraw-Hill, New York (1981).
23. R. S. Perkins and T. N. Andersen in "Modern Aspects of Electrochemistry", J. O'M. Bockris and B. E. Conway, Eds., Vol. 5, p. 203, Plenum Press, New York (1969).
24. A. Akiyama and K. Nobe, *J. Electrochem. Soc.*, *117*, 999 (1970).
25. N. Eldakar and K. Nobe, *Corrosion*, *33*, 128 (1977).
26. R. R. Annand, R. M. Hurd, and N. Hackerman, *J. Electrochem. Soc.*, *112*, 144 (1965).
27. N. Hackerman, R. M. Hurd, and R. R. Annand, *Corrosion*, *18t*, 37 (1962).
28. K. Aramaki and N. Hackerman, *J. Electrochem. Soc.*, *115*, 1007 (1968).
29. K. Aramaki, *J. Electrochem. Soc.*, *118*, 1553 (1971).
30. R. J. Meakins, *J. Appl. Chem.*, *13*, 339 (1963).
31. R. J. Chin and K. Nobe, *J. Electrochem. Soc.*, *118*, 545 (1971).
32. N. Eldakar and K. Nobe, *Corrosion*, *32*, 238 (1976).
33. M. A. Quraishi, J. Rawat, and M. Ajmal, *Corrosion*, *54*, 996 (1998).
34. A. E. Stoyanova, E. I. Sokolova, and S. N. Raicheva, *Corros. Sci.*, *39*, 1595 (1997).
35. M. S. A. Aal, S. Radwan, and A. El-Saied, *Br. Corros. J.*, *18*, 102 (1983).
36. Y.-F. Yu, J. J. Chessick, and A. C. Zettlemoyer, *J. Phys. Chem.*, *63*, 1626 (1959).
37. P. F. Rossi, M. Bassoli, G. Olivera, and F. Guzzo, *J. Therm. Anal.*, *41*, 1227 (1994).
38. Y.-F. Yu Yao, *J. Phys. Chem.*, *67*, 2055 (1963).
39. M. Nakazawa and G. A. Somorjai, *Appl. Surf. Sci.*, *84*, 309 (1995).

40. Y.-F. Yu Yao, *J. Phys. Chem.*, *69*, 3930 (1965).
41. T. Matsuda, T. Taguchi, and M. Nagao, *J. Therm. Anal.*, *38*, 1835 (1992).
42. J. G. N. Thomas, *Br. Corros. J.*, *5*, 41 (1970).
43. J. E. O. Mayne and C. L. Page, *Br. Corros. J.*, *7*, 115 (1972).
44. A. Weisstuch, D. A. Carter, and C. C. Nathan, *Mater. Prot. Perf.*, *10* (4), 11 (1971).
45. F. P. Dwyer and D. D. Mellor "Chelating Agents and Metal Chelates", Academic Press, New York (1964).
46. D. C. Zecher, *Mater. Perf.*, *15* (4), 33 (1976).
47. R. L. LeRoy, *Corrosion*, *34*, 98 (1978).
48. B. Müller, W. Kläger, and G. Kubitzki, *Corros. Sci.*, *39*, 1481 (1997).
49. F. Tirbonod and C. Fiaud, *Corros. Sci.*, *18*, 139 (1978).
50. Y. I. Kuznetsov and T. I. Bardasheva, *Russ J. Appl. Chem.*, *66*, 905 (1993).
51. B. D. Chambers, S. R. Taylor, and M. W. Kendig, *Corrosion*, *61*, 480 (2005).
52. C. B. Breslin, G. Treacy, and W. M. Carroll, *Corros. Sci.*, *36*, 1143 (1994).
53. A. A. O. Magalhaes, I. C. P. Margarit, and O. R. Mattos, *J. Electroanal. Chem.*, *572*, 433 (2004).
54. S. Z. El Abedin, *J. Appl. Electrochem.*, *31*, 711 (2001).
55. H. Guan and R. G. Buuchheit, *Corrosion*, *60*, 284 (2004).
56. V. F. Vetere and R. Romagnoli, *Br. Corros. J.*, *29*, 115 (1994).
57. S. M. Cohen, *Corrosion*, *51*, 71 (1995).
58. B. R. W. Hinton, D. R. Arnott, and N. E. Ryan, *Mater. Forum*, *9*, 162 (1986).
59. R. G. Buchheit, M. D. Bode, and G. E. Stoner, *Corrosion*, *50*, 205 (1994).
60. P. M. Natishan, E. McCafferty, J. W. Glesener, and A. A. Morrish in "Proceedings of the Fourth International Symposium on Diamond Materials", K. V. Ravi and J. P. Dismukes, Eds., p. 607, The Electrochemical Society, Pennington, NJ (1995).
61. E. McCafferty, P. M. Natishan, and G. K. Hubler, *Corros. Sci.*, *35*, 239 (1993).
62. E. McCafferty and J. V. McArdle, *J. Electrochem. Soc.*, *142*, 1447 (1995).
63. D. C. Hansen and E. McCafferty, *J. Electrochem. Soc.*, *143*, 115 (1996).
64. H.-H. Strehblow and B. Titze, *Corros. Sci.*, *17*, 461 (1977).
65. E. McCafferty, *J. Electrochem. Soc.*, *137*, 3731 (1990).
66. S. Matsuda and H. H. Uhlig, *J. Electrochem. Soc.*, *111*, 156 (1964).
67. H. H. Uhlig and J. R. Gilman, *Corrosion*, *20*, 289t (1964)
68. C.-J. Lin, R.-G. Du, and T. Nguyen, *Corrosion*, *56*, 41 (2000).
69. A. Devasenapathi and V. S. Raja, *J. Mater. Sci.*, *33*, 3345 (1998).
70. V. S. Argawala and J. J. De Luccia, *Corrosion*, *36*, 208 (1980).
71. I. M. Putilova, S. A. Balezin, and V. P. Barannik, "Metallic Corrosion Inhibitors", p. 1, Pergamon Press, New York (1960).
72. J. W. Hill and R. H. Petrucci, "General Chemistry", 2nd Edition, p. 980, Prentice Hall, Upper Saddle River, NJ (1999).
73. H. Ashassi-Sorkhabi, M. R. Majidi, and K. Seyyedi, *Appl. Surf. Sci.*, *225*, 176 (2004).
74. J. Telegdi, E. Kalman, and F. H. Karman, *Corros. Sci.*, *33*, 1099 (1992).
75. E. Mueller, C. S. Sikes, and B. J. Little, *Corrosion*, *49*, 829 (1993).
76. D. C. Hansen and E. McCafferty in "Environmentally Acceptable Inhibitors and Coatings", S. R. Taylor, H. S. Isaacs, and E. W. Brooman, Eds., p. 234, The Electrochemical Society, Pennington, NJ (1995).
77. I. Sekine, Y. Nakahata, and H. Tanabe, *Corros. Sci.*, *28*, 987 (1988).
78. G. D. Davis and J. A. Von Fraunhofer, *Mater. Perform.*, *42* (2), 56 (2003).
79. J. A. Von Fraunhofer and G. D. Davis, *Mater. Perform.*, *42* (12), 38 (2003).
80. G. Matamala, W. Smeltzer, and G. Droguett, *Corros. Sci.*, *42*, 1351 (2000).
81. W. Xing, Y. Shan, D. Guo, T. Lu, and S. Xi, *Corrosion*, *51*, 45 (1995).
82. A. Jaiswal, R. A. Singh, and R. S. Dubey, *Corrosion*, *57*, 307 (2001).
83. G. Grundmeier, C. Reinartz, M. Rohwerder, and M. Stratmann, *Electrochim. Acta*, *43*, 165 (1997).
84. Y. Yamamoto, H. Nishihari, and K. Aramaki, *J. Electrochem. Soc.*, *140*, 436 (1993).
85. W. A. Adamson, "Physical Chemistry of Surfaces", p. 99, John Wiley & Sons, New York (1976).
86. A. Subramananian, M. Natesan, V. S. Muralidharan, K. Balakrishnan, and T. Vasudevan, *Corrosion*, *56*, 144 (2000).
87. A. Wachter, T. Skei, and N. Stillman, *Corrosion*, *7*, 284 (1951).
88. E. Vuorinen, P. Ngobeni, G. H. van der Klashorst, W. Skinner, E. de Wet, and W. S. Ernst, *Br. Corros. J.*, *29*, 120 (1994).

89. E. M. Mora and J. M. Bastidas, *Corros. Prevent. Control*, *34* (6), 143 (1987).
90. I. L. Rozenfeld, "Corrosion Inhibitors", p. 302, McGraw-Hill, New York (1981).
91. A. Leng and M. Stratmann, *Corros. Sci.*, *34*, 1657 (1993).
92. M. N. Desai and M. B. Desai, *Corros. Sci.*, *24*, 649 (1984).
93. G. Moretti, G. Quartarone, A. Tassan, and A. Zingales, *Br. Corros. J.*, *31*, 49 (1996).

Chapter 13
Corrosion Under Organic Coatings

Introduction

Paints and organic coatings are used virtually everywhere in daily life to protect metal structures and components. Paints can be found on automobiles, trucks, planes, ships, trains, bridges, pipelines, industrial plants, and the exterior of storage tanks or vessels. In addition to protecting the underlying metal substrate from corrosion, paints also provide color and a pleasing decorative finish. In the area of electronics, organic coatings are used on circuit boards and components to provide protection against moisture, sulfur dioxide, carbon dioxide, and ionic contaminants in the atmosphere.

A paint film or organic coating protects a metal substrate from corrosion in two ways. First, the organic film or coating serves as a physical barrier to separate the metal substrate from the environment. Second, the organic film or coating may serve as a reservoir for corrosion inhibitors which resist corrosion. An extension of this second approach is provided by zinc-rich organic coatings which contain flakes of metallic zinc to provide added corrosion protection by galvanic protection when the organic coating has been breached.

Organic coatings are not perfect barriers and they can be penetrated by water, oxygen, and ions (such as chloride). When this happens, corrosion can occur beneath the organic coating at the organic coating/metal interface. The electrochemical processes that occur are similar to those which proceed in bulk solutions except that the reactions are confined to the narrow region beneath the organic coating. The small volumes of liquid which are involved in the early stages of the corrosion process produce extreme values of pH and ionic concentrations [1].

The local electrolyte which accumulates beneath the organic coating not only leads to metal dissolution but also weakens the adhesion of the organic coating to the metal. Thus, the processes of corrosion and de-adhesion both contribute to the failure of the protective system.

Organic coatings can also fail by non-electrochemical means. These include damage caused by mechanical impact or abrasion, cracking or crazing due to mechanical deformation, oxidation by ultraviolet radiation, and damage due to free-thaw cycles. In addition, all of the above processes allow slow access of the aqueous or ambient environment into the organic coating and thus also enhance the possibility of electrochemical corrosion at the interface between the organic coating and the metal.

Figure 13.1 shows an example of the failure of an organic coating.

E. McCafferty, *Introduction to Corrosion Science*, DOI 10.1007/978-1-4419-0455-3_13,

Fig. 13.1 An example of a paint failure leading to corrosion of the underlying substrate

Paints and Organic Coatings

A paint is an organic coating applied to a surface as a liquid, which subsequently dries or cures to form a protective film. Paint coatings are generally 25–1,000 μm in thickness [2].

A paint consists of at least three components [2–4]: (i) a *vehicle or binder*, (ii) a *pigment* suspended in the binder, and (iii) a *solvent*. In addition, there may be a fourth component, *special additives*, which have very specific purposes and are typically added to the paint in very small amounts. When the paint is applied and the solvent evaporates, the paint film forms the structure shown schematically in Fig. 13.2.

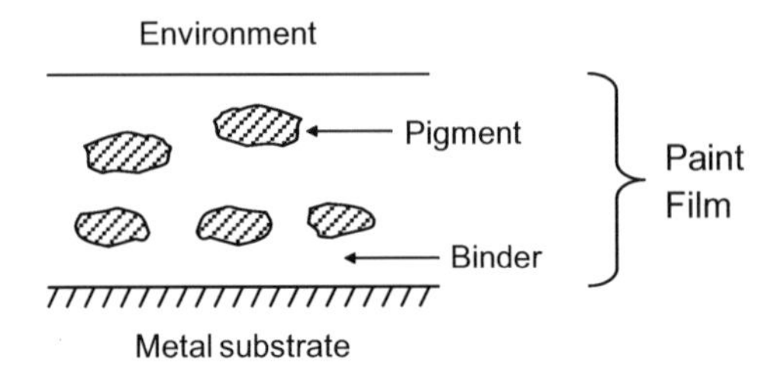

Fig. 13.2 Structure of a paint film upon drying of the solvent

Vehicles or binders are of various types, as listed in Table 13.1. The vehicle or binder is the film-forming agent in the paint and is the continuous polymeric phase in which all the other components are incorporated.

Pigments may be used to provide color to the paint film (e.g., Fe_2O_3 for red or TiO_2 for white) or to act as a corrosion inhibitor (e.g., $ZnCrO_4$). Pigments also include zinc powders or flakes (zinc-rich paints) which provide cathodic protection to the metal substrate when the paint film becomes permeated with water. Other pigments include fillers such as mica, which reinforce the paint film and present a more torturous path for the penetration of water and ions through the film.

Table 13.1 Some typical constituents of paints [2–4]

Vehicles	Pigments
Epoxy	Iron oxides
Urethane	Titanium dioxide
Acrylic	Clays
Vinyl	Silicates
Alkyd	Chromates
Silicones	Mica
	Zinc powder
Solvents	**Special additives**
Xylene	UV absorbers
Glycol ethers	Dispersing agents
Ketones	Mildew inhibitors
Mineral spirits	Fire-retarding agents
Water	Biocides for anti-fouling paints
	Abrasives for anti-slip paints

Solvents for paints are either volatile organic liquids or water. Oil-based paints require the use of organic solvents for thinning and cleanup of brushes or other applicators, so that the use of oil-based paints is being phased out due to environmental considerations.

Paint films often consist of various layers, as shown schematically in Fig. 13.3. The metal surface may first be treated with chromates or phosphates to form a conversion coating for improved corrosion resistance or paint adhesion. A primer may also be used prior to the application of the topcoat, with the topcoat itself having the structure in Fig. 13.2.

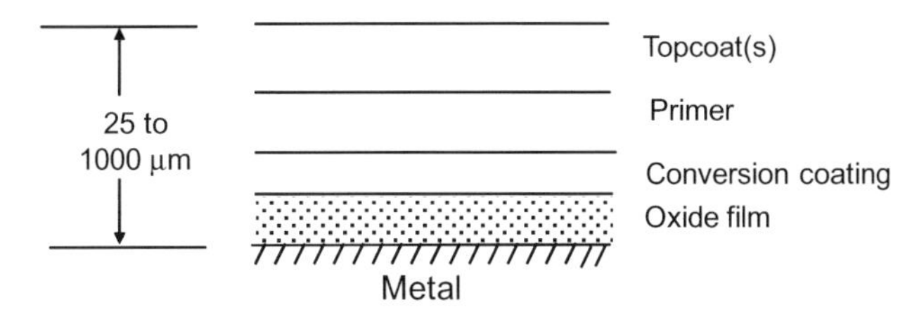

Fig. 13.3 Schematic diagram showing the various layers which may be present in a paint film

Paints are not generally used for protection below ground or in applications where mechanical damage may be a problem. Various types of organic coatings used in such applications include polyvinyl chloride, epoxies, urethanes, and polyesters; and the structural formulas of various organic compounds used in coatings are given in Fig. 13.4.

The primary purpose of a paint film or organic coating is to form a barrier between the metal substrate and its environment. However, an organic coating is not a perfect barrier but instead is like a membrane which can be penetrated by water, oxygen, and ions, as discussed below.

Underfilm Corrosion

Corrosion beneath an organic coating can occur only if an electrolyte exists at the organic coating/substrate interface. This electrolyte is provided by the penetration of water into the organic

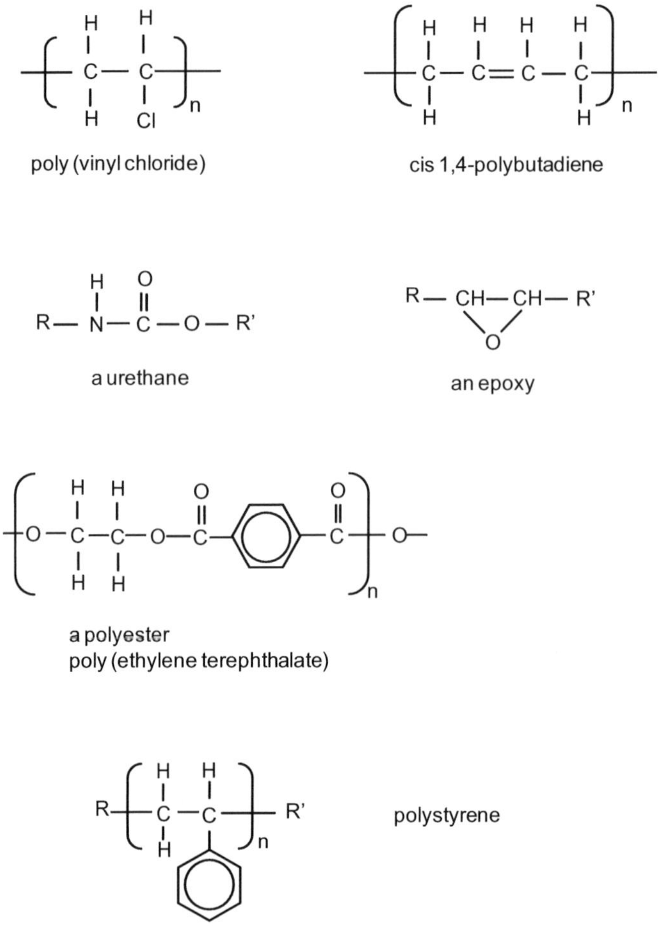

poly (vinyl chloride)

cis 1,4-polybutadiene

a urethane

an epoxy

a polyester
poly (ethylene terephthalate)

polystyrene

Fig. 13.4 Formulas of various organic compounds utilized in organic coatings for metals

coating and its subsequent accumulation beneath the coating. The presence of ions (such as chlorides) is also necessary in order to provide sufficient conductivity in the aqueous phase beneath the coating to allow local electrochemical reactions to occur.

Water Permeation into an Organic Coating

Water permeation into an organic coating occurs due to three effects [4]:

(1) Water diffuses into the coating due to a concentration gradient formed by immersion into an aqueous solution or exposure to a humid atmosphere.
(2) Water penetrates the film by osmosis due to an ionic concentration gradient which exists between the film (which may contain ionic impurities) and the environment.
(3) Water enters the film by capillary action through voids, pores, or cracks in the organic coating.

Penetration rates of water or gases into an organic coating are given in terms of a permeability coefficient P, which is the product of the usual diffusion coefficient D and the Henry's law constant K_H (also called the solubility coefficient). As shown in Chapter 8, the diffusion coefficient D pertains to Fick's first law:

$$J = -D\,\frac{dC}{dx} \tag{1}$$

where J is the flux past a plane and dC/dx the concentration gradient. For gases, it is easier to determine the partial pressure of the gas (P_{gas}) on each side of the polymer film (rather than its concentration) so that

$$J = -P\,\frac{dp_{gas}}{dx} \tag{2}$$

where P is the permeability coefficient. The concentration of a gas in the polymer film and its vapor pressure in the vapor phase are related by Henry's law:

$$C = K_H\,p_{gas} \tag{3}$$

where K_H is the Henry's law constant (see Chapter 8). For a linear concentration (or pressure) gradient, Eqs. (1)–(3) combine to give

$$P = DK_H \tag{4}$$

(See Problem 13.3). The permeability coefficient P should be independent of the coating thickness (for a homogeneous coating) because both D and K_H are independent of thickness. Permeability coefficients have the following units:

$$P = \frac{(\text{quantity of permeant}) \times (\text{film thickness})}{(\text{area}) \times (\text{time}) \times (\text{pressure drop across the film})} \tag{5}$$

(See Problem 13.4). However, permeability data are often expressed as transmission rates for a given thickness of coating. Experimentally determined transmission rates for water through various polymers or paint films are given in Table 13.2. It can be seen that all the polymers listed in Table 13.2 are permeable to water vapor and that the permeability coefficients vary greatly with the nature of the polymer. The flux of water through a free polymer film is typically of the order of 1–10 mg/cm^2 day for many classes of polymers [1, 5].

The transmission rate of a molecule through an organic coating depends on the thickness of the coating. If the transmission rate is J_1 through a coating of thickness L_1, then the transmission rate J_2 through a coating of the same material but having a different thickness L_2 is given by

$$\frac{J_2}{J_1} = \frac{L_1}{L_2} \tag{6}$$

as shown in Appendix K; that is, the transmission rate through a 100 μm coating is theoretically 50% of that through a 50 μm coating.

The transmission rate of water necessary to sustain the corrosion of steel at the base of the organic coating can be calculated as follows. The corrosion rate of unpainted steel in neutral electrolytes is 1–10 mils per year (mpy) [7, 8]. From the density of iron, a penetration rate of 10 mpy corresponds

Table 13.2 Transmission rates of water through organic coatings [5, 6]

Coating	Transmission rate in mg/cm^2 day	Film thickness in μm
Organic coating		
Epoxy/polyamide	15.5	10
Chlorinated rubber (plasticized)	9.5	10
Styrene acrylic latex	230	10
Vinyl chloride–vinylidene chloride copolymer	2.8	10
Polybutadiene	3	25
Polytetrafluoroethylene	0.48	25
Cellulose nitrate	63	25
Paint films		
Vinyl chloride–vinylidene chloride copolymer latex	6.8	10
Chlorinated rubber, unmodified	5.0	10
Coal tar epoxy	7.5	10
Acrylic water-borne primer	180	10
TiO_2-pigmented alkyd	65	10
Red lead-oil-based primer	54	10
Alkyd resin	2.3	100
Polyurethane	1.4	100
Phenolic resin	1.1	100

to a weight loss for iron of 9.8 × 10^{-6} moles/cm^2 day. The half-cell reactions in neutral electrolytes (see Chapter 2) are:

$$\text{Anodic:} \quad 2\,\text{Fe} \rightarrow 2\,\text{Fe}^{2+} + 4\,e^-$$

$$\text{Cathodic:} \quad \text{O}_2 + 2\text{H}_2\text{O} + 4\,e^- \rightarrow 4\,\text{OH}^-$$

$$\text{Overall:} \quad 2\,\text{Fe} + \text{O}_2 + 2\,\text{H}_2\text{O} \rightarrow 2\,\text{Fe}^{2+} + 4\,\text{OH}^-$$

From simple stoichiometry, the reaction of 9.8 × 10^{-6} moles/cm^2 day of Fe requires the same number of moles of H_2O, which in turn amounts to 0.18 mg H_2O/cm^2 day. (The transmission rate of H_2O needed to sustain the corrosion rate of 1 mpy is accordingly 0.018 mg H_2O/cm^2 day.)

As seen in Table 13.2, the transmission rate of water through various polymer films is several orders of magnitude higher than the transmission rate required for the corrosion of iron; see Table 13.3. This fact was first established by Mayne [9] and has been verified subsequently by other investigators [10, 11]. Thus, the permeation rate of H_2O through an organic coating or paint film is not the rate-determining step in the underfilm corrosion of steel substrates. However, the transmission of water is important in providing the electrolyte, in facilitating the diffusion of ions through the coating, and in contributing to the loss of adhesion of the organic coating, as is discussed later.

The data in Table 13.2 apply to free-standing films rather than to films attached to metal substrates. (That is, the polymer films were formed on a substrate such as glass and were then detached.) Thus, some reservation must be taken in applying permeability data for free films to films which are intact on metal substrates. However, Nguyen and co-workers [12] have more recently developed a technique based on Fourier transform infrared-multiple internal reflection (FTIR-MIR) for measuring in-situ water beneath organic coatings cast on metal substrates. This technique is able to detect molecular water which accumulates at the organic coating/metal interface. Results are shown in Fig. 13.5 for an epoxy coating applied to a silicon substrate. It can be seen that approximately 15 nm

Table 13.3 Comparison of required and operative permeation rates for corrosion to occur beneath coated iron specimens

Corrosion rate of open uncoated steel in neutral electrolytes [7, 8]	Required rate of supply of H_2O to sustain the corrosion rate[a]	Permeation rate of H_2O through organic coatings[b]	Required rate of supply of O_2 to sustain the corrosion rate[a]	Permeation rate of O_2 through organic coatings[c]
1 mil/year	$0.018 \frac{\text{mg } H_2O}{\text{cm}^2 \text{ day}}$		$0.016 \frac{\text{mg } O_2}{\text{cm}^2 \text{ day}}$	
		$0.5 \text{ to } 230 \frac{\text{mg } H_2O}{\text{cm}^2 \text{ day}}$		$0.002 \text{ to } 0.008 \frac{\text{mg } O_2}{\text{cm}^2 \text{ day}}$ (one value at $0.11 \frac{\text{mg } O_2}{\text{cm}^2 \text{ day}}$)
10 mils/year	$0.18 \frac{\text{mg } H_2O}{\text{cm}^2 \text{ day}}$		$0.16 \frac{\text{mg } O_2}{\text{cm}^2 \text{ day}}$	

[a]Based on the following reaction:
$2\,Fe + O_2 + 2\,H_2O \rightarrow 2\,Fe^{2+} + 4\,OH^-$
[b]From Table 13.2.
[c]From Table 13.4.

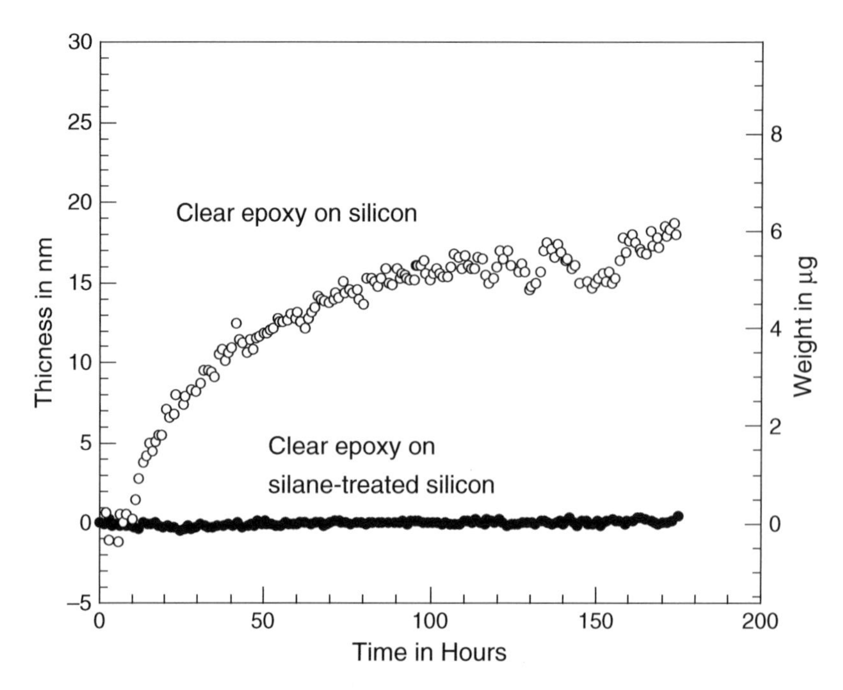

Fig. 13.5 Thickness and amount of water at the coating/substrate interface for clear epoxy coatings on silicon and on silane-treated silicon. The epoxy coating thickness was 130–140 μm. Redrawn from Nguyen et al. [12] with the permission of Elsevier Ltd

of water accumulates at the interface (corresponding to approximately 50 monolayers of water). Thus, this later evidence confirms the idea that electrolyte layers can be readily formed beneath organic coatings.

Nguyen and co-workers also found that the amount of water formed at the interface was much less for a silane-treated interface, as shown in Fig. 13.5. (Silanes are Si-containing molecules used to treat metals to produce hydrophobic surfaces.)

As mentioned earlier, the accumulation of water layers beneath a polymer film not only serves to provide an electrolyte for the operation of localized corrosion cells but also promotes the loss of adhesion of the organic coating. Silane-treated specimens displayed a greater adhesion strength than untreated samples after more than 600 h exposure to water [12].

Permeation of Oxygen and Ions into an Organic Coating

The rates of oxygen penetration through organic coatings are generally very much less than the rates for water, as shown in Table 13.4.

The transmission rate of O_2 necessary to support the corrosion of unpainted steel is calculated to be 0.016–0.16 mg O_2/cm^2 day. This value results from the earlier calculation that a corrosion rate of 10 mpy consumes 9.8×10^{-6} moles/cm^2 day of iron and thus requires half that amount of O_2 (from stoichiometry) or 4.9×10^{-6} moles/cm^2 day or 0.16 mg O_2/cm^2 day. (A corrosion rate of 1 mpy thus requires only 1/10 of that amount.) Most polymers in Table 13.4 have transmission rates of O_2 lower than that required for corrosion, as seen in Table 13.3. This observation has led some

Table 13.4 Transmission rates of oxygen through organic coatings [5, 10, 11]

Coating	Transmission rate in mg/cm^2 day	Film thickness in μm
Cellulose nitrate	0.106	100
Epoxy resin	7.3×10^{-3}	100
Vinyl chloride/vinyl acetate copolymer	7.5×10^{-3}	100
Chlorinated rubber	2.2×10^{-3}	100
Alkyd 15% poly (vinyl chloride)	6.8×10^{-3}	87
Epoxy/coal tar	4.1×10^{-3}	33
Epoxy/polyamide	6.4×10^{-3}	38
Polybutadiene	2.5×10^{-3}	25

investigators to suggest that the rate-determining step in underfilm corrosion is the slow permeation of O_2 through the organic coating [10, 11].

The rate of permeation of Cl^- ions is even less than the permeation rates for O_2, as seen in Table 13.5 [13]. The role of Cl^- ions in underfilm corrosion appears to be that of providing sufficient electrolyte conductivity for the operation of localized corrosion cells and to accelerate the anodic half-cell reaction when it has been initiated.

Table 13.5 Transmission rates of chloride through organic coatings [13]

Organic coating	Transmission rate in mg/cm^2 day	Film thickness in μm
Chlorinated rubber	6.00×10^{-3}	59
Vinyl	0.31×10^{-3}	63
Epoxy	0.90×10^{-3}	19
Alkyd/iron oxide	0.52×10^{-3}	62
Alkyd/red lead	0.20×10^{-3}	75
Epoxy/polyamide	0.21×10^{-3}	100

Breakdown of an Organic Coating

The sequence of events which occurs in the breakdown of an organic coating in the subcoating micro-environment is illustrated by the work of Ritter and Rodriguez [14]; see Figs. 13.6 and 13.7. These authors studied the corrosion of iron covered with a transparent cellulose nitrate coating, a coating selected for its ease of breakdown and observation. The sequence of events is as follows [14, 35]:

(1) The organic coating first becomes penetrated by water and oxygen molecules, as discussed above.
(2) Corrosion is initiated due to several possible causes, such as defects in the coating, mechanical rupture of the coating, or chemical rupture though generation of osmotic pressure at sites containing soluble salts.
(3) The substrate iron then goes into solution at a local anodic site:

$$Fe \rightarrow Fe^{2+} + 2e^-$$

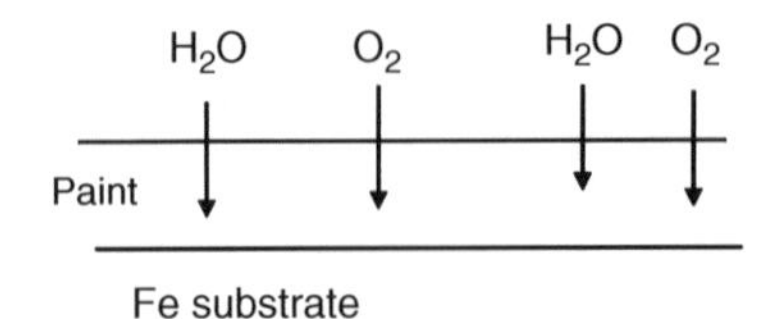

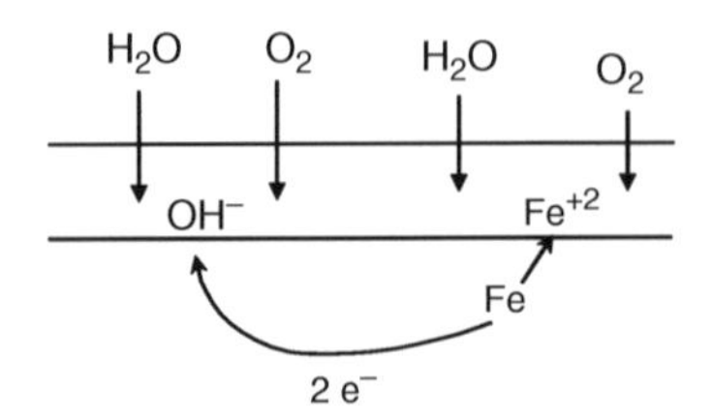

Fig. 13.6 The first two steps in the deterioration of an organic coating. After Ritter and Rodriguez [14] and reproduced by permission of NACE International

Acidic hydrolysis occurs so that the anodic site becomes locally acidic, as in the case of crevice corrosion and pitting (Chapter 10). Chloride ions then migrate into the local acidic environment, again as with crevice corrosion and pitting.

(4) A localized cathodic site is generated at a different location and the operation of the cathodic half-cell is fueled by O_2 and H_2O molecules which have previously permeated the organic coating.

$$O_2 + 2\,H_2O + 4\,e^- \rightarrow 4\,OH^-$$

The cathodic site becomes locally basic (alkaline) due to the production of these OH^- ions.

(5) The organic coating suffers disbonding from the metal substrate in the local alkaline environment. This disbonding and removal of the organic coating from the metal surface is called *cathodic delamination* and is discussed in more detail later.

(6) As the coating system continues to deteriorate, a breakthrough occurs between adjacent anodic and cathodic areas, and there is a catastrophic failure of the organic coating.

Cathodic delamination, as in step (5), is also a problem in coated systems which are protected by applied cathodic potential (cathodic protection). Areas on the metal adjacent to defects in the organic coating can suffer delamination in the presence of the applied cathodic potential.

Adhesion of Organic Coatings

The adhesion properties and corrosion behavior of an organic coating are related. If corrosion occurs beneath the organic coating, a loss of adhesion of the coating will ensue. On the other hand, if the coating does not possess good adhesion to the metal substrate, then undermining of the coating will lead to localized pockets of electrolyte which promote corrosion.

The relationship between adhesion and corrosion is seen schematically in Fig. 13.8.

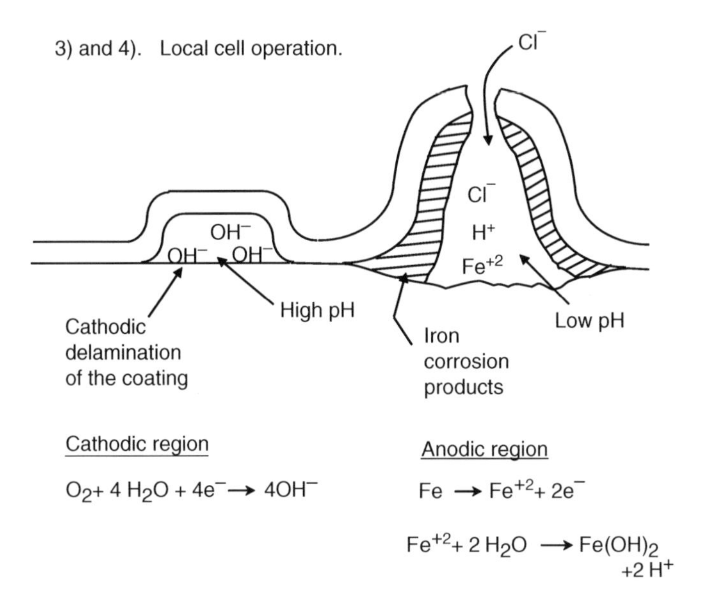

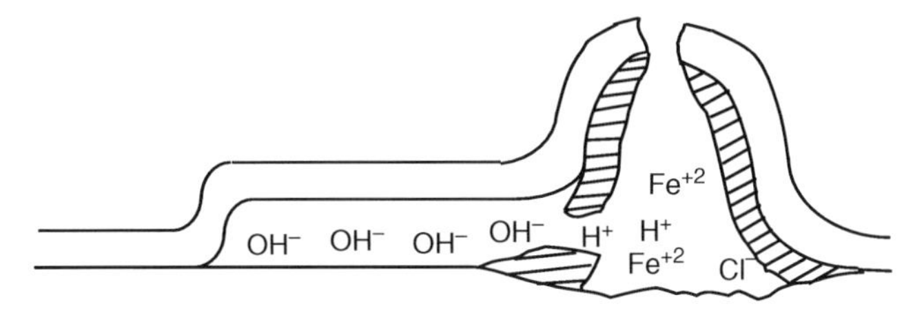

Fig. 13.7 Continued deterioration of an organic coating, according to Ritter and Rodriguez [14]. Reproduced by permission of NACE International

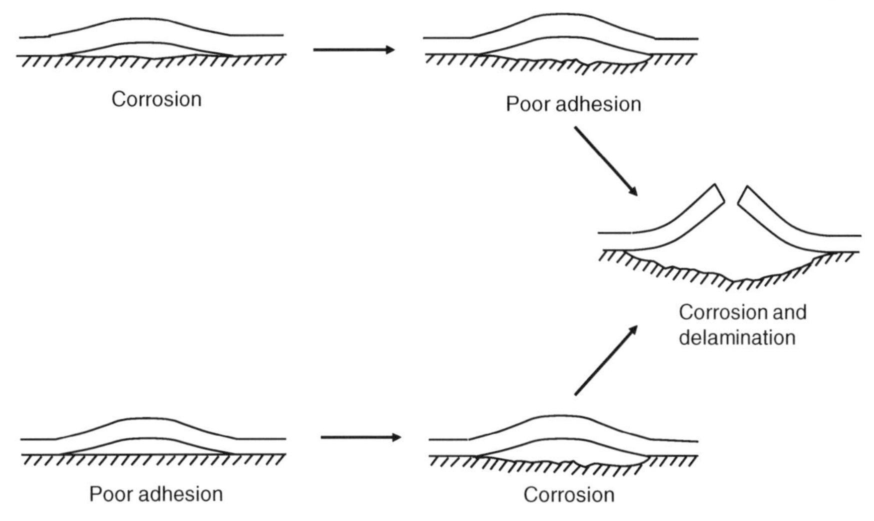

Fig. 13.8 Schematic illustration showing the relationship between poor adhesion and poor corrosion resistance of a coating

There are several fundamental theories of metal/polymer adhesion. These are as follows [15]:

(1) The *adsorption theory* (physical and chemical adsorption), in which the adhesive bond depends on adsorption forces between the two interfaces being joined
(2) The *chemical reaction theory*, in which a chemical reaction occurs between the polymer and the metal
(3) The *mechanical interlocking theory*, in which adhesion is due to the mechanical interlocking or keying of the polymer into cavities and pores of the metal surface
(4) The *electrostatic* theory, in which adhesion depends on the existence of an electrostatic charge between the two surfaces

Each of these last three mechanisms can operate under certain conditions, but the most pervasive mechanism is the first, in which the polymer adheres to the metal substrate (or, more properly, its oxide) because of interatomic and intermolecular bonds which are formed. These adsorption forces that exist across all boundaries include Lewis acid/Lewis base forces.

All metal surfaces to be coated with an organic polymer contain either the incipient air-formed oxide film or an oxide produced by a surface treatment (such as chromates). Thus, it is appropriate to consider the interaction of organic polymers with oxide films.

In Chapter 12 it was seen that many corrosion inhibitors are Lewis bases, which function by donation of electrons to the metal surface. Organic polymers can also be classified as Lewis bases (electron donors) or Lewis acids (electron acceptors). Poly(methyl methacrylate) (PMMA) is a typical basic polymer, as can be reasoned from its molecular structure, as shown in Fig. 13.9. The carbonyl oxygens of esters are basic (electron-donating) sites, which can form acid/base bonds with electron-accepting sites of Lewis acids.

Poly(vinyl chloride) (PVC) is an example of a polymer which is a Lewis acid. For PVC, it is the partially positive –CH portion of the vinyl group which is the electron-accepting site [16] and which can form acid/base bonds with electron-donating sites of Lewis bases; see Fig. 13.10.

Oxide films on metal surfaces have, in fact, acid/base properties, which will be discussed in more detail in Chapter 16, and the acidic or basic character of the oxide varies with the nature of the metal.

Figure 13.11 shows adhesion results for two organic polymers on a series of oxide-covered metals for which the oxide films have differing degrees of acidity (or basicity) [17]. Figure 13.11 shows that the adhesion strength of an acidic polymer (as measured by the peel strength) increases with increasing basicity of the oxide film. Conversely, the adhesion strength of the basic polymer PMMA (as measured by a pull-off test) increases with increasing acidity of the oxide film. These results illustrate the importance of Lewis acid/Lewis base interactions in the adhesion of organic polymers to metal surfaces.

However, the situation becomes more complicated when water molecules accumulate beneath the organic coating because water interferes with the adhesion between the organic coating and the metal substrate. The extent of adhesion in the presence of a liquid (usually water) is referred to as *wet adhesion* and is considered by Funke [18] to be the most important property of the organic coating. Two different reasons have been proposed for the loss of adhesion in the presence of water [4, 19]. These are as follows:

(1) Chemical disbonding due to the interaction of water molecules with covalent, hydrogen, or polar bonds between the organic polymer and the oxide film
(2) Mechanical disbonding due to forces caused by accumulation of water and osmotic pressure

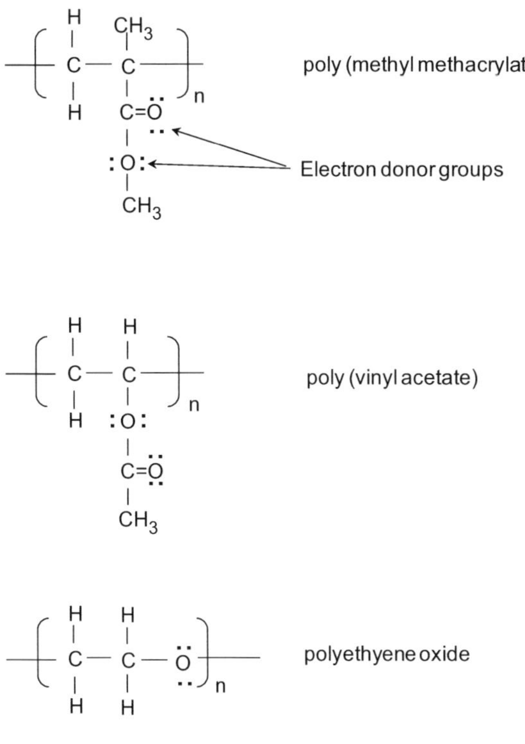

poly (methyl methacrylate)
Electron donor groups
poly (vinyl acetate)
polyethyene oxide

Fig. 13.9 Some examples of basic polymers. Electrons on the oxygen atom available for donation are indicated by *black solid dots*

δ+
Can interact with
electron-rich donor groups
poly (vinyl chloride)
δ-
poly (vinyl fluoride)
poly (vinylidene fluoride)

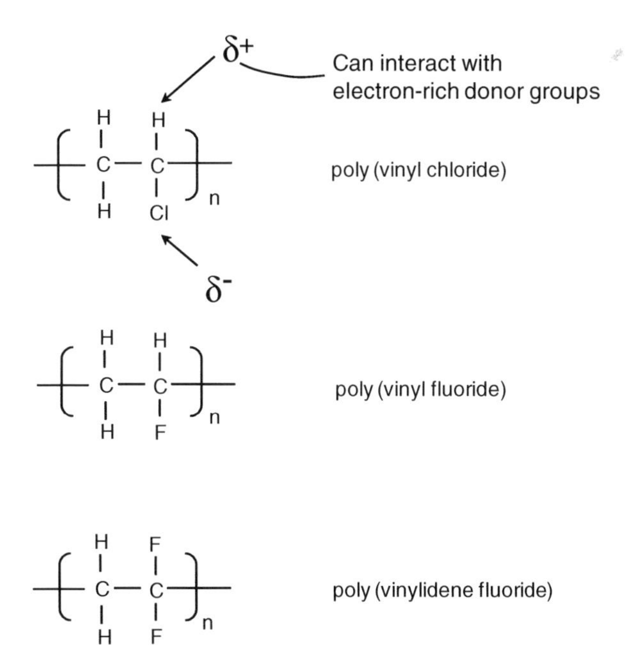

Fig. 13.10 Some examples of acidic polymers. Partial charges are denoted by δ^{+} and δ^{-}

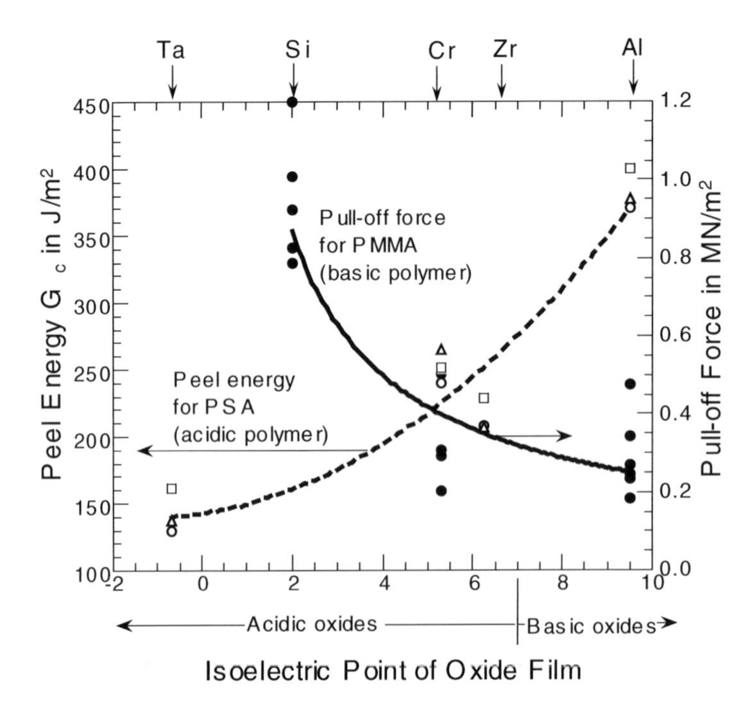

Fig. 13.11 Adhesion strength on various metals (indicated at *top*) of an acidic pressure-sensitive adhesive (PSA) and of a basic polymer, poly(methyl methacrylate) (PMMA) vs. the isoelectric point of the oxide film. For the PSA, the adhesion strength is measured by the peel energy, G_c, and for PMMA, the adhesion strength is measured by the pull-off force. The peel energy is shown for the various contact times prior to peel measurement: ○ 1 h, □ 10 h, △ 24 h [17]. Reproduced by permission of ECS – The Electrochemical Society

At present the mechanisms of wet adhesion (and its loss) are not understood very well and constitute a fertile area for further research and study.

Improved Corrosion Prevention by Coatings

Based on a mechanistic understanding of the breakdown of organic coatings, as discussed earlier, the following guidelines may be suggested to improve the durability of organic coatings. Most of the following guidelines are due to Leidheiser and Kendig [6]:

(1) *Reduce penetration by H_2O, O_2, and ions*. This can be achieved through the development of coatings with reduced permeability or by overcoating with another material.
(2) *Prevent electrons from reaching the reaction site*. This can be accomplished by providing a surface film between the metal and the organic coating which is a poor electron conductor.
(3) *Inhibit the anodic reaction* by pre-treatment of the surface or by including an anodic inhibitor in the coating.
(4) *Inhibit the cathodic reaction* by pre-treatment of the surface or by including a cathodic inhibitor in the coating.
(5) *Control the pH at anodic or cathodic sites* by including buffering agents in the coating.

(6) *Minimize anodic–cathodic breakthroughs.* A roughened metal surface will provide a torturous path for lateral diffusion of anodic and cathodic species. (A roughened surface will also increase the adhesion of the organic coating.)
(7) *Increase the adhesion of the polymer to the oxide film* by selecting Lewis acid polymers for basic oxide films and Lewis base polymers for acidic oxide films.

Filiform Corrosion

Filiform corrosion is the corrosion under organic coatings in the form of numerous narrow interconnected thread-like filaments. An example of filiform corrosion is shown in Fig. 13.12 [20]. This form of corrosion has been observed under organic coatings on aluminum, steel, magnesium, and galvanized steel [21, 22] and usually occurs under conditions of high relative humidity. Thus, filiform corrosion is a special type of atmospheric corrosion.

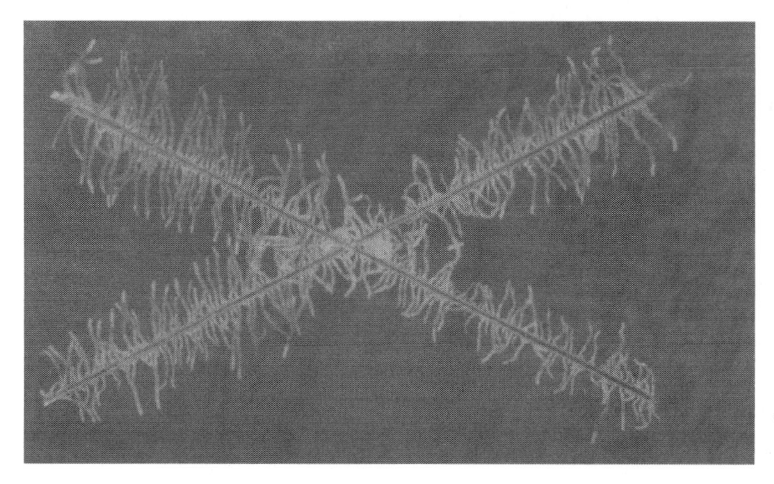

Fig. 13.12 Optical photograph of filiform corrosion under a paint film on Al 5001 which has been scribed with a cross-shaped scratch and exposed to vapors of 33% HCl [20]. Reproduced by permission of Elsevier Ltd

A schematic view of a filiform filament on iron is shown in Fig. 13.13. Filaments are usually of the order of 0.05–0.5 mm in thickness and can grow along the metal surface at the rate of 1 mm per day or less [1, 21–23]. The size of the filaments and their rate of growth is insensitive to the identity of the organic coating.

Filiform corrosion initiates at a break in the organic coating where a soluble ionic species is present; and once initiated, the filament travels in a relatively straight line for long distances. The filament propagates by a differential aeration cell in which oxygen enters the filament through its tail and diffuses toward the head of the filament, as shown in Fig. 13.13. The head of the filament is low in O_2 compared to the tail of the filament, so that the head of the filament is the primary anodic site. The head of the filament on iron is usually filled with a liquid solution and contains Fe^{2+} ions (this region is green in color for iron substrates). The pH is also lowest in this region due to acid hydrolysis, similar to the case of crevice corrosion. Ferrous ions produced in the anodic head diffuse toward the tail where they are oxidized to ferric ions and eventually become part of the dry solid corrosion product which fills the tail (reddish-brown in color).

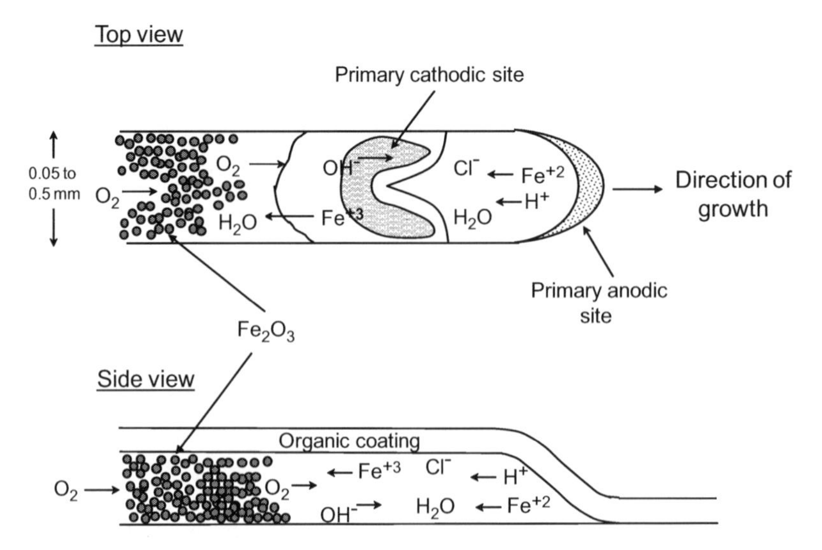

Fig. 13.13 Schematic illustration of filiform corrosion on steel. Redrawn from Ruggeri and Beck [23]. The filament propagates by a differential aeration cell in which oxygen enters the filament through its tail and diffuses toward the head of the filament. The head of the filament is low in O_2 compared to the tail of the filament, so that the head of the filament is the primary anodic site. Ferrous ions produced in the anodic head diffuse toward the tail where they are oxidized to ferric ions and precipitate as $Fe(OH)_3$, eventually becoming Fe_2O_3. Reproduced by permission of NACE International

Ruggeri and Beck [23] showed that oxygen diffuses through the filament tail in a set of experiments involving filament tails:

- The growth of a filament was stopped by cutting a filament tail across its width, peeling back the organic coating, and applying epoxy to the open portion of the tail so as to seal off its supply of oxygen.
- In control experiments, the filament continued to grow if epoxy was only applied to the tail (without cutting) or if the tail was only cut (but not sealed with epoxy).
- A filament which had been de-activated by sealing off the supply of oxygen could be re-activated by cutting open the tail to the atmosphere.

A characteristic of filament growth is that filaments rarely cross each other, although filaments may merge or divide [24]. Ruggeri and Beck [23] suggest that the reason filaments do not cross is that a growing head would encounter a new source of oxygen if it crossed another filament. This would cause changes in the positions of anodic and cathodic sites and would therefore change the direction of growth.

A more recent and finer scale investigation on filiform corrosion has been conducted by Leblanc and Frankel [25] using atomic force microscopy (for topographical measurements) and scanning Kelvin probe microscopy (for maps of the Volta potential distribution). (More about the scanning Kelvin probe is given later in this chapter.) This study generally supported the mechanism discussed above but with two new interesting observations. For a 300 nm thick epoxy coating on carbon steel, in addition to the predominantly cathodic region, local microscopic cathodes were also found in the nearby periphery of the filament head, suggesting that the organic coating in the filament head is disbonded by cathodic delamination.

In addition, Leblanc and Frankel [25] also observed voids at the edges of the tail beneath the organic coating. The authors suggest that these voids are caused by cathodic disbondment aided by mechanical prying of the epoxy film by the voluminous solid corrosion product. Leblanc and Frankel also suggest that such voids are involved in the filament growth process. That is, they provide fast channels for O_2 diffusion relative to transport through the solid tail and thus explain the fact that filaments deflect off tails without actually touching. When an active head approaches the tail of another filament, the advancing filament is exposed to a new source of oxygen contained in the void and the advancing filament changes the direction. The new direction responds to the sum of the two combined oxygen fluxes. This explanation supports the earlier ideas of Ruggeri and Beck [23] on filament crossing.

Corrosion Tests for Organic Coatings

There are a variety of test methods and measurements which can be used to assess or study the corrosion behavior of organic coatings [26, 27]. These methods include both accelerated standardized tests and experimental electrochemical measurements, but only a few methods are discussed here.

Accelerated Tests

The salt-fog test [28] is an accelerated exposure test (not limited to coatings) in which samples are exposed to a continuous spray of salt water in an enclosed high-humidity chamber. The sample size is typically 4 in. × 6 in., and organic coatings (with cut edges protected) are usually scribed with a large "X" and the performance of the coating is evaluated by its resistance to attack and de-adhesion along the scribe marks.

Other accelerated tests include alternate immersion cycles or exposure to various conditions of relative humidity, temperature, and ultraviolet light in a "weatherometer" [29]. Such tests, along with the salt-fog test, are useful in comparing or rating various organic coating systems but yield no mechanistic information.

Cathodic Delamination

The process of disbonding at cathodic sites is illustrated schematically in Fig. 13.14. Leidheiser and co-workers [30, 31] made extensive use of cathodic delamination tests under conditions of controlled electrode potential. In the cathodic delamination test, a small defect of the order of 1 mm diameter is punched into an organic coating, and the coated metal specimen is immersed into an electrolytic cell containing an ionic solution. The specimen is held at a constant electrode potential in the cathodic region, and replicate samples are removed at various times to determine the area of the coating which has been detached during the period of immersion. Typical experimental results are shown in Fig. 13.15 for the cathodic delamination of an alkyd coating at –0.8 V vs. SCE in aerated 0.5 M NaCl. After an initial delay time, the delamination rate is constant with time over the duration of the test.

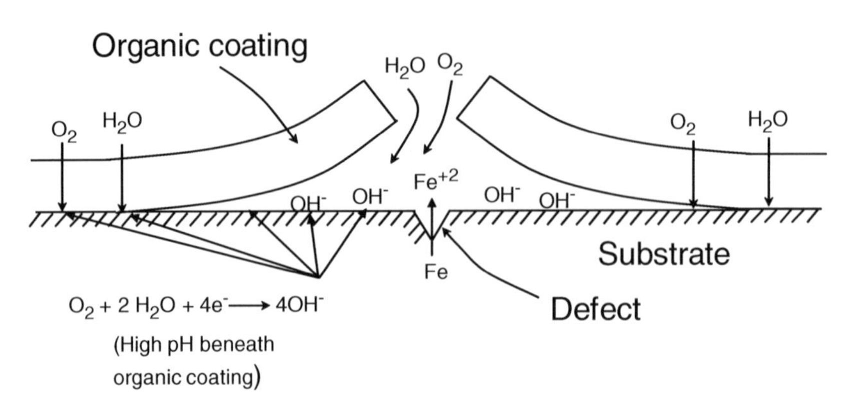

Fig. 13.14 Schematic illustration of cathodic delamination. When a defect triggers an anodic event, the cathodic reaction at other sites beneath the organic coating produces OH^- ions, which cause delamination of the coating

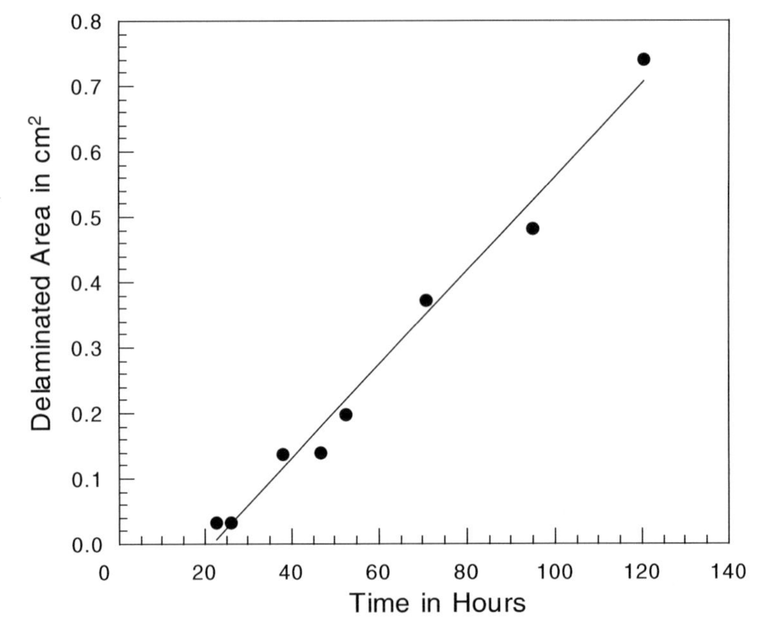

Fig. 13.15 The cathodic delamination at –0.8 V of a 142 μm thickness alkyd coating in aerated 0.5 M NaCl [31]. Reproduced by permission of Elsevier Ltd

Rates of cathodic delamination can be used to compare different coating systems because the delamination rate is sensitive to the nature of the organic polymer, the nature of the metal substrate, and the surface pre-treatment used prior to application of the organic coating. For a given system, the rate of cathodic delamination increases as follows [32]:

(1) the coating thickness decreases,
(2) the temperature increases,
(3) the electrode potential becomes more negative.

Delamination rates also depend on environmental factors, such as the nature of the cation in solution [30–32]. Delamination rates of polybutadiene coatings on steel in solutions of 0.5 M chlorides increase in the order

$$CaCl_2 < LiCl < NaCl < KCl < CsCl$$

Leidheiser and Wang [30] have correlated this latter effect to the diffusion coefficients of the various 0.5 M chloride solutions, as shown in Fig. 13.16. A similar effect has been observed by Stratmann et al. [33], who have suggested that delamination depends on the galvanic current between local anodes and cathodes and is possible only if cations are transferred from the cathode to the anode.

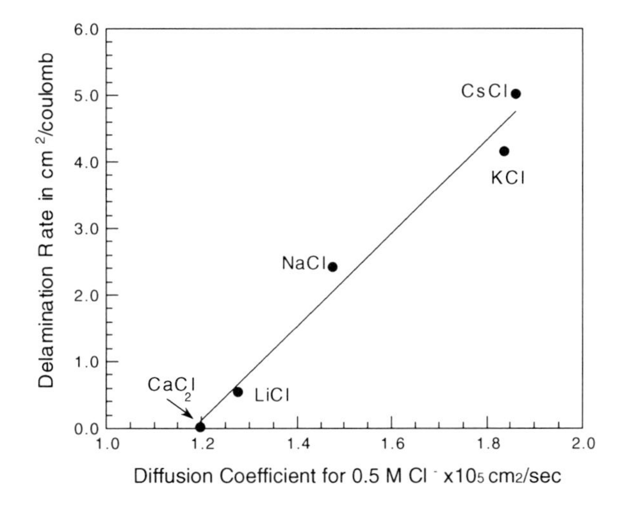

Fig. 13.16 The relationship between the cathodic delamination of polybutadiene coatings in various chloride solutions and the diffusion coefficient of the electrolyte. Reprinted from Ref. [30] by permission of the Federation of Societies for Coatings Technology

The mechanism of cathodic delamination is not completely understood at the microscopic level, but it is recognized that there are three possible modes of failure. These are (i) dissolution of the oxide film in the local alkaline environment which is formed, (ii) degradation of the polymer in the local alkaline environment, and (iii) loss of adhesion at the organic coating/oxide-coated metal interface [34, 35]. That is, the failure can occur within the oxide, within the polymer, or at the oxide/polymer interface, respectively.

Figure 13.17 shows five different planes along which delamination of the organic coating from the metal substrate can occur [1, 35]. Planes 3, 4, and 5 in Fig. 13.17 correspond to the three modes of cathodic delamination listed above.

Planes 1 and 2 in Fig. 13.17 are operative in cases of anodic undermining, which occurs when the major separation process is an anodic dissolution reaction occurring beneath the coating. Anodic undermining occurs beneath organic coatings on tin or aluminum [35].

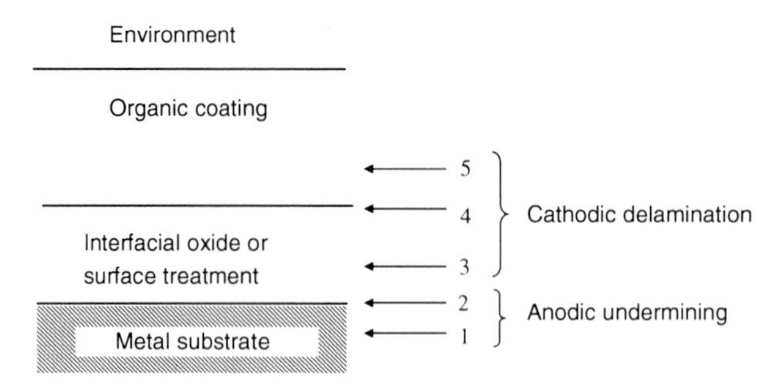

Fig. 13.17 Planes along which undermining of an organic coating may occur. After Leidheiser [35] and reproduced by permission of NACE International

AC Impedance Techniques – A Brief Comment

Alternating current (AC) impedance techniques have become more prevalent in studying the intrusion of water and ions into organic coatings and in comparing the performance of various organic coatings or surface treatments. In addition, AC impedance techniques can produce models of the solution/organic coating/metal system and thus yield mechanistic information about the protective system. The theory of AC impedance techniques and their application to various problems, including corrosion protection by organic coatings, is discussed in Chapter 14.

Recent Directions and New Challenges

Recent work in the area of paints and organic coatings has taken two directions. One is driven by industrial needs and the second direction is due to advanced instrumental techniques which are available to study the corrosion behavior of organic coatings.

Just as the search for new corrosion inhibitors has been triggered by environmental considerations, so too is there a similar effort regarding paint coatings [36]. Much recent work has been directed toward reducing the volatile organic content (VOC) level of paint systems. For example, studies at the Naval Research Laboratory have shown that zero- or low-VOC variants of epoxies, polyurea, and polyurethanes have promise as protective coatings [37].

A second current active research direction involves the development of organic coating systems which can be applied with minimum surface preparation over partially corroded metal surfaces. It is generally recognized that the most important factor affecting the lifetime of a paint is proper preparation of the metal surface prior to application of the paint film. Ideally, paints should be applied to bright metal surfaces and in the absence of condensed moisture. However, there is much interest in developing coating systems which are tolerant to partially corroded or contaminated surfaces. The presence of rust itself is not the major problem, but nests of aggressive ions (chloride or sulfate) contained in the rust near the interface promote corrosion if a paint film is applied over such an area [38]. Adhesion between a paint coating and rust is generally good, and failure is likely to occur within the brittle rust layer [39]. There is currently considerable research and development on chemical treatments for the removal of chloride ions from rusted surfaces and for the conversion of surface rust into a coherent protective layer over which additional paint coatings may be applied.

A recent experimental thrust in the study of corrosion behavior of organic coatings has made extensive use of the scanning Kelvin probe (SKP) [25, 33, 40] (mentioned earlier in this chapter).

The SKP measures the work function of a sample using a vibrating condenser [40] so that portions of the sample which have an intact organic coating display a different response than for areas where there is coating defect or breakthrough. By scanning across the surface, a two-dimensional map is obtained with a lateral resolution of approximately 1 μm or less. Figure 13.18 shows a scanning Kelvin probe analysis of a defect on an organic coating on Al alloy 2024-T3 [41]. The specimen was clad with a conversion coating, a primer, and a polyurethane coating. The specimen was scribed and exposed to 12 M HCl for 1 h and then to a salt fog before SKP analysis.

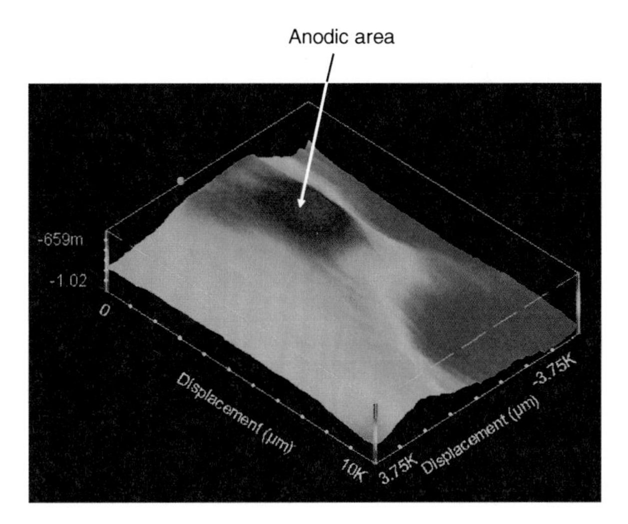

Fig. 13.18 Scanning Kelvin probe analysis of a defect on an organic coating on Al alloy 2024-T3 [41]. The specimen was scribed and exposed to 12 M HCl and to a salt fog before SKP analysis. Reproduced by permission of NACE International

There has also been much recent interest in the development of smart coatings, i.e., coatings which can detect the presence of a breakthrough and, in response, release a corrosion inhibitor at the required site. For example, Kendig and co-workers [42] have shown that conducting polyaniline films containing inhibitors for oxygen reduction are able to release the inhibitor at defects in the organic coating. The authors state that the inhibitor is released when there is a galvanic current between the conducting polymer and the bare metal surface.

In the future, improved and novel organic coatings will arise from increased environmental or performance demands, as well as from new mechanistic information gained from the continued use of advanced experimental techniques. It is also anticipated that the coatings field will draw on opportunities arising from the emerging areas of nanotechnology to develop a wide variety of smart organic coatings which can sense and repair corrosion damage.

Problems

1. How is the underfilm corrosion of an organic coating similar to crevice corrosion? How do these two forms of localized corrosion differ?
2. Filiform corrosion is a special type of underfilm corrosion. Would you expect the mechanism of detachment of the organic coating to be similar on a microscopic basis for both filiform corrosion and underfilm corrosion? Explain your answer.

3. When there is a linear concentration gradient across an organic coating, show that the permeability coefficient P is given by $P = DK_H$, where D is the diffusion coefficient and K_H is the Henry's law constant.
4. Show that the permeability coefficient P can have the units:

$$P = \frac{(\text{mol}) \times (\text{cm})}{(\text{cm}^2) \times (\text{s}) \times (\text{atm})}$$

5. As shown in Table 13.2, a 100 μm thick polyurethane film has a water transmission rate of 1.4 mg/cm^2 day:

   (a) Calculate the number of molecular layers of water which would be formed at the base of the organic coating after 1 day if all the water transmitted after 1 day accumulates there. The diameter of a water molecule is 0.30 nm.
   (b) Under the conditions of part (a), compare the thickness of the water layer to the thickness of the organic coating.

6. The transmission rate of through a 25 μm coating of polybutadiene was reported to be 3.0 mg/cm^2 day for water and 2.5 × 10^{-3} mg/cm^2 day for oxygen (5):

   (a) What is the limiting reactant for the cathodic reduction of oxygen on the underlying metal substrate?
   (b) What corrosion current density (in μA/cm^2) be sustained by these transmission rates if the underlying substrate is aluminum and the corrosion reaction is the following?

$$4\ \text{Al} + 3\ \text{O}_2 + 6\ \text{H}_2\text{O} \longrightarrow 4\ \text{Al}^{+3} + 12\ \text{OH}^-$$

7. Several investigators have measured the pH to be 14 at the front of a delaminating polymer. Leidheiser and Wang (30) conducted delamination experiments for polybutadiene on aluminum in 0.5 M NaCl solutions at an applied electrode potential of −1.35 V vs. SCE. (a) Is the oxide film on aluminum subject to dissolution under these conditions? Explain your answer. (b) What does your answer in part (a) mean in regard to the possible mechanism of delamination of polybutadiene on aluminum?
8. The transmission rate of water though a 10 μm thick coating of styrene acrylic latex is 230 mg/cm^2 day. What is the transmission rate of water through a 25 μm thick coating of the same material?
9. The oxide film on tantalum is acidic in nature. Which polymer would you expect to exhibit a greater (dry) adhesion to oxide-covered tantalum, poly (vinyl acetate) or poly (vinyl fluoride)? Explain your answer.
10. As shown in Table 13.2, polytetrafluoroethylene (PTFE) has a very low rate of water transmission. (a) Search the scientific literature or the Internet and list several other properties of fluoropolymers which make them attractive as organic coatings. (b) What is a drawback in the use of fluoropolymers as organic coatings?
11. (a) Would measuring the electrode potential of an underground pipeline protected by an organic coating be useful in monitoring the integrity of the coating? Explain your answer. (b) Would measuring the polarization resistance of a section the coated pipeline be useful for the same purpose? Explain your answer.

References

1. H. Leidheiser, Jr. in "Corrosion Mechanisms", F. Mansfeld, Ed., p. 165, Marcel Dekker, New York (1987).
2. K. B. Tator in "Corrosion and Corrosion Protection Handbook", P. A. Schweitzer, Ed., p. 355, Marcel Dekker, New York (1983).
3. C. H. Hare in "Uhlig's Corrosion Handbook", 2nd edition, R. W. Revie, Ed., p. 1023, John Wiley, New York (2000).
4. J. H. W. de Wit in "Corrosion Mechanisms in Theory and Practice", P. Marcus and J. Oudar, Eds., p. 581, Marcel Dekker, New York (1995).
5. N. L. Thomas, *Prog. Org. Coat.*, *19*, 101 (1991).
6. H. Leidheiser, Jr. and M. W. Kendig, *Corrosion, 32*, 69 (1976).
7. S. C. Dexter, "Handbook of Oceanographic Engineering Materials", p. 38, John Wiley, New York (1979).
8. H. H. Uhlig and W. R. Revie, "Corrosion and Corrosion Control", p. 96, John Wiley, New York (1985).
9. J. E. O. Mayne, *Corros. Technol.*, *1*, 286 (1954).
10. S. Guruviah, *J. Oil Col. Chem. Assoc.*, *53*, 669 (1970).
11. T. Haagen and W. Funke, *J. Oil Col. Chem. Assoc.*, *58*, 359 (1975).
12. N. Nguyen, E. Byrd, D. Bentz, and C. Lin, *Prog. Org. Coat.*, *27*, 181 (1996).
13. L. S. Hernandez, J. M. Miranda, and G. Garcia, *Corros. Prevention Control*, *45* (6), 181 (1998).
14. J. J. Ritter and M. J. Rodriguez, *Corrosion*, *38*, 223 (1982).
15. J. Schultz and M. Nardin in "Adhesion Promotion Techniques", K. L. Mittal and A. Pizzi, Eds., p. 1, Marcel Dekker, New York (1999).
16. F. M. Fowkes, D. O. Tischler, J. A. Wolfe, L. A. Lannigan, C. M. Ademu-John and M. J. Halliwell, *J. Polym. Sci., Polm. Chem. Ed.*, *22*, 547 (1984).
17. E. McCafferty, *J. Electrochem. Soc.*, *150*, B342 (2003).
18. W. Funke, *Prog. Org. Coat.*, *31*, 5 (1997).
19. H. Leidheiser, Jr. and W. Funke, *J. Oil Col. Chem. Assoc.*, *70 (5)*, 121 (1987).
20. J. L. Delplancke, S. Berger, X. Lefèbvre, D. Maetens, A. Pourbaix, and N. Heymans, *Prog. Org. Coat.*, *43*, 64 (2001).
21. G. M. Hoch in "Localized Corrosion", B. F. Brown, J. Kruger, and R. W. Staehle, Eds., p. 134, NACE, Houston, TX (1974).
22. W. H. Slabaugh, W. Dejager, S. E. Hoover, and L. L. Hutchison, *J. Paint Technol.*, *44*, 76 (1972).
23. R. T. Ruggeri and T. R. Beck, *Corrosion*, *39*, 452 (1983).
24. M. G. Fontana and N. D. Greene, "Corrosion Engineering", p. 46, McGraw-Hill Book Co., New York, NY (1978).
25. P. P. Leblanc and G. S. Frankel, *J. Electrochem. Soc.*, *151*, B105 (2004).
26. G. W. Walter, *Corros. Sci.*, *26*, 39 (1986).
27. J. N. Murray, *Prog. Org. Coat.*, *31*, 375 (1997).
28. ASTM Test B 117–97, "Standard Practice for Operating Salt Spray (Fog) Apparatus", "1998 Annual Book of ASTM Standards", Volume 3.02, p. 1, ASTM, West Conshohocken, PA (1998).
29. ASTM Active Standard G 154–06, "Standard Practice for Operating Fluorescent Light Apparatus for UV Exposure of Nonmetallic Materials", ASTM, West Conshohocken, PA (2006).
30. H. Leidheiser, Jr. and W. Wang, *J. Coatings Technol.*, *53* (672), 77 (1981).
31. H. Leidheiser, Jr., W. Wang, and L. Igetoft, *Prog. Org. Coat.*, *11*, 19 (1983).
32. U. Steinsmo and J. I. Skar, *Corrosion*, *50*, 934 (1994).
33. M. Stratmann, A. Leng, W. Fürbeth, H. Streckel, H. Gehmecker, and K.-H. Grosse-Brinkhaus, *Prog. Org. Coat.*, *27*, 261 (1996).
34. J. F. Watts. *J. Adhesion*, *31*, 73 (1989).
35. H. Leidheiser, Jr., *Corrosion*, *38*, 374 (1982).
36. "Environmentally Acceptable Inhibitors and Coatings", S. R. Taylor, H. S. Isaacs, and E. W. Broomen, Eds., The Electrochemical society, Pennington, NJ (1997).
37. M. V. Veazey, *Mater. Perfor.*, *42* (10), 22 (2003).
38. N. L. Thomas, *J. Oil Col. Chem. Assoc., 74*(3) 83 (1991).
39. N. L. Thomas, *J. Prot. Coat. Linings*, *6* (12) 63 (1989).
40. G. Grundmeier, W. Schmidt, and M. Stratmann, *Electrochim. Acta*, *45*, 2515 (2000).
41. D.C. Hansen, G.E. Grecsek and R.O. Roberts, "A Scanning Kelvin Probe Analysis of Aluminum and Aluminum Alloys", paper presented at NACE International 54th Annual Conference & Exposition, San Antonio, TX, April 25–30, 1999.
42. M. W. Kendig, M. Hon, and L. Warren, *Prog. Org. Coat.*, *47*, 183 (2003).

Chapter 14
AC Impedance

Introduction

The electrochemical techniques which have been considered so far in this text have been direct current techniques. This chapter describes a powerful technique [1, 2] which uses alternating current (AC). AC impedance techniques are not new, but prior to about 1980, AC measurements were generally made with a Schering bridge which requires manual balancing to null the bridge in order to measure the equivalent parallel resistance and capacitance. This method is tedious and time-consuming. With the advent of improved instrumentation under computer control and acquisition, AC impedance techniques have become an important method of making electrochemical measurements. AC impedance has been used to study the metal/solution interface, oxide films and surface treatments, and the corrosion behavior of organic coatings on metals. Each of these three applications is considered in this chapter.

In essence, the method consists of superimposing a small (10–20 mV) AC signal on the electrochemical system of interest and measuring the response of the system to this perturbation. When an electrochemical system is perturbed by an applied AC signal, the system relaxes to a new steady state; and the time τ required for this relation is known as the *time constant*. Sometimes τ is called the *relaxation time* by analogy to the process by which dipoles orient and relax in response to an alternating field [3, 4].

Relaxation Processes

Suppose that a liquid contains polar molecules which possess permanent dipoles, such as water molecules. In the absence of an external field, individual dipoles are oriented randomly in all directions, as shown in Fig. 14.1(a). However, in the presence of an alternating field, the dipoles will tend to orient themselves in the direction of the applied electric field, as illustrated in Fig. 14.1(b) and (c). With an alternating field, these water molecules must therefore turn back and forth in response to the changing direction of the electric field. The rotation of these dipoles, however, cannot respond perfectly to the electric field. This delayed response of a system to a change in external force is termed *relaxation*. Even if there are no permanent dipoles in the material, the presence of an electric field can induce a dipole, which then can respond to the alternating field and display a relaxation effect.

The AC impedance technique is also called *electrochemical impedance spectroscopy* (EIS). This is by analogy to conventional spectroscopy, in which the system displays a response at a characteristic frequency to an applied perturbation. In EIS, as will be seen later, plots of the imaginary part of

E. McCafferty, *Introduction to Corrosion Science*, DOI 10.1007/978-1-4419-0455-3_14,

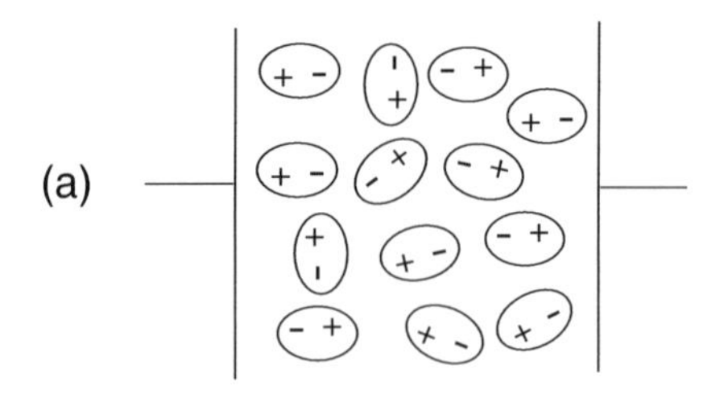

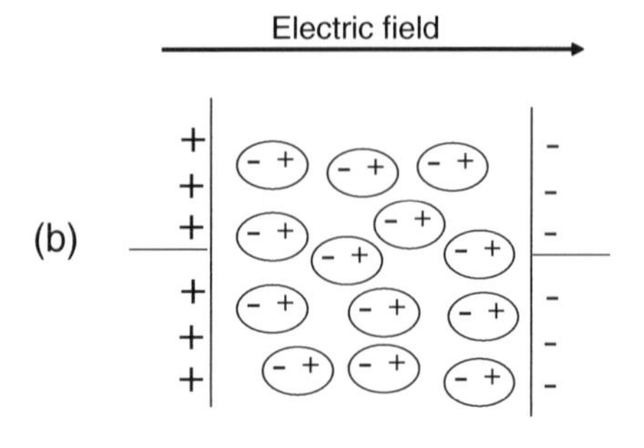

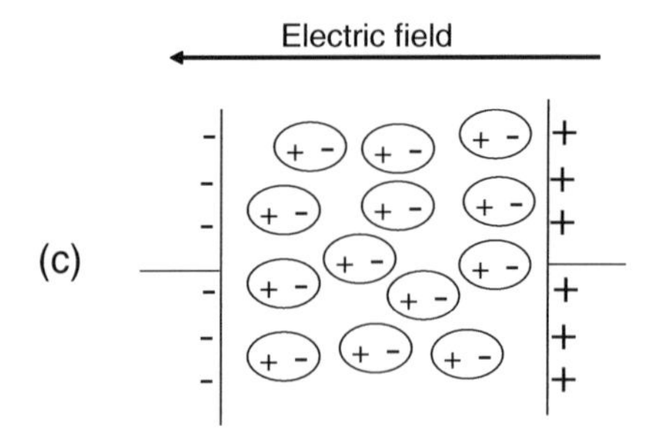

Fig. 14.1 (**a**) Random orientation of dipoles in the absence of an electric field. (**b**) and (**c**) Alignment of dipoles with the direction of an electric field

the impedance vs. the logarithm of the frequency produce a peak, analogous to usual observations in other types of spectroscopy .

Some types of spectroscopy and the characteristic frequencies involved are listed in Table 14.1.

Table 14.1 Several types of relaxation processes [5]

Type of relaxation	Frequency (Hz)	Portion of spectrum
Electronic transitions	10^{14}–10^{16}	Visible and ultraviolet
Vibrations of flexible bonds	10^{13}–10^{14}	Infrared
Rotation of small molecules	10^{11}–10^{13}	Far infrared
Rotation of polyatomic molecules	10^{9}–10^{11}	Microwave
Rotation of dipoles		
Liquid H_2O at 25°C [3, p. 128]	10^{10}	Microwave
Ice at 0°C [6]	10^{3}	Radiofrequencies
Physically adsorbed H_2O on α-Fe_2O_3 at 25°C [4]	10–10^{4}	Radiofrequencies
Electrochemical interfaces	10^{-2}–10^{5}	Radiofrequencies

In electrochemical systems, the relaxation time (in seconds) is given by

$$\tau = RC \tag{1}$$

where R is the resistance (in ohms) and C is the capacitance (in farads) of the system. The impedance is the resistance to the flow of an AC current:

$$Z = \frac{E}{I} \tag{2}$$

Both resistors and capacitors possess an impedance, as will be seen shortly.

Example 14.1: Show that the product RC has the units of seconds.

Solution:

$$RC = \text{ohms} \times \frac{\text{coulombs}}{\text{volts}}$$

or

$$RC = \text{ohms} \times \frac{(\text{amps})\,(\text{seconds})}{\text{volts}}$$

Then use $E = IR$, or

$$\text{volts} = (\text{amps})\,(\text{ohms})$$

so that

$$\frac{\text{amps}}{\text{volts}} = \frac{1}{\text{ohms}}$$

Use this in the expression above for RC to get

$$RC = \text{ohms} \times \frac{1}{\text{ohms}} \times \text{seconds}$$

or

$$RC = \text{seconds}$$

Experimental Setup

A block diagram of the experimental setup is shown in Fig. 14.2. The apparatus consists of an electrochemical cell, frequency generator, a frequency response analyzer (FRA), and a computer to control the experiment and to store the data. The electrode potential is controlled as usual with a potentiostat. The heart of the system is the FRA, which can measure the real and imaginary parts of the complex impedance. (The concept of a complex impedance is addressed below.) AC impedance measurements are usually made in the frequency range of 0.01–100,000 Hz (cycles/s), although this range can be extended. More detail on the experimental aspects of AC impedance techniques is given elsewhere [7, 8].

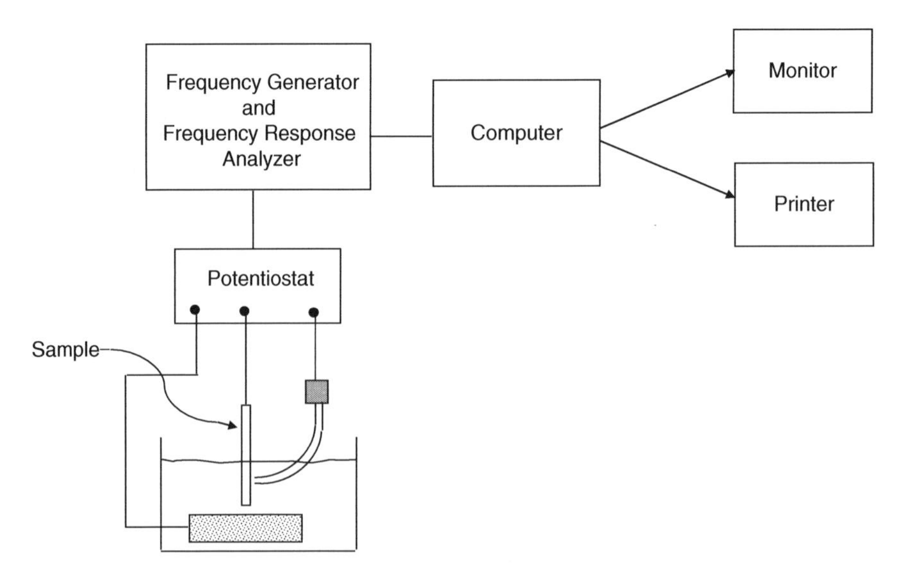

Fig. 14.2 A simple block diagram for the measurement of AC impedance by application of a sine wave and the use of a frequency response analyzer

Complex Numbers and AC Circuit Analysis

The analysis of AC electrical circuits utilizes complex numbers so it is useful to review some of their basic properties. A complex number z consists of an ordered pair of numbers, a and b, which are related by

$$z = a + \mathrm{j}\,b \tag{3}$$

where $\mathrm{j} = \sqrt{-1}$ (and is used here to avoid confusion with the current density i). The number a is called the *real part* of the complex number z and b is called the *imaginary part* (although both a and b themselves are real numbers). A complex number can be represented in the complex plane, as shown in Fig. 14.3. The absolute value of the complex number $|z|$ is given by its distance from the origin in the complex plane. Thus

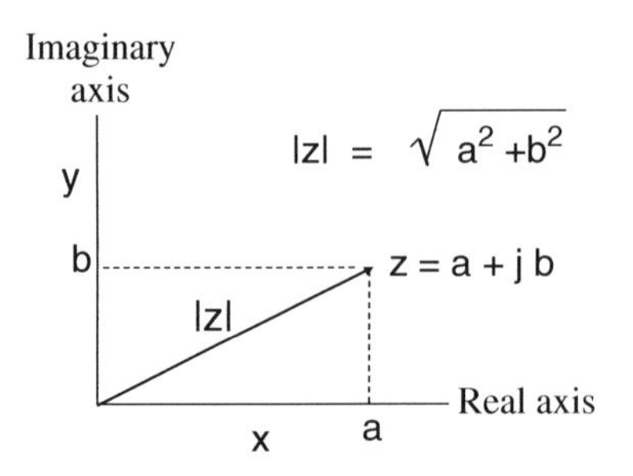

Fig. 14.3 Geometrical representation of a complex number

$$|z| = (a^2 + b^2)^{1/2} \quad (4)$$

The absolute value $|z|$ is sometimes called the modulus of z.

A useful identity in complex variables is Euler's equation:

$$e^{jx} = \cos x + j \sin x \quad (5)$$

and it is this equation which introduces complex variables into AC circuit analysis. The alternating voltage of an AC signal can be represented by a cosine (or sine) wave. We can write

$$E = E_0\, e^{j\omega t} \quad (6)$$

where E_0 is the amplitude of the voltage, t is the time, and $\omega = 2\pi f$ is the angular frequency, with f the frequency of the AC signal. Then from Eq. (5),

$$E = E_0\,(\cos \omega t + j \sin \omega t) \quad (7)$$

It is understood that to get the actual voltage, it is necessary to take the real part of Eq. (7):

$$E = \mathrm{Re}\,\{E_0(\cos \omega t + j \sin \omega t) = E_0(\cos \omega t) \quad (8)$$

Similarly, the current may also have a complex form:

$$I = I_0\, e^{j\omega(t-s)} \quad (9)$$

where δ is the phase angle. Because both the voltage and current can be written as complex numbers, it follows that the impedance Z given by

$$Z(\omega) = \frac{E(\omega,t)}{I(\omega,t)} \quad (10)$$

may also be a complex number. The impedance of a resistor R is simply the resistance itself:

$$Z_R = R \quad (11)$$

but the impedance of a capacitor is an imaginary number:

$$Z_C = \frac{1}{j\omega C} \quad (12)$$

where C is the value of the capacitance. Equation (12) is not obvious, so that its origin is given in Appendix L. Impedances in series or parallel follow the same rules as for resistances.

The Metal/Solution Interface

In Chapter 3 (Fig. 3.14), we discussed an equivalent circuit model for the metal/solution interface. This equivalent circuit, shown again in Fig. 14.4, contains a double-layer capacitance C_{dl} in parallel

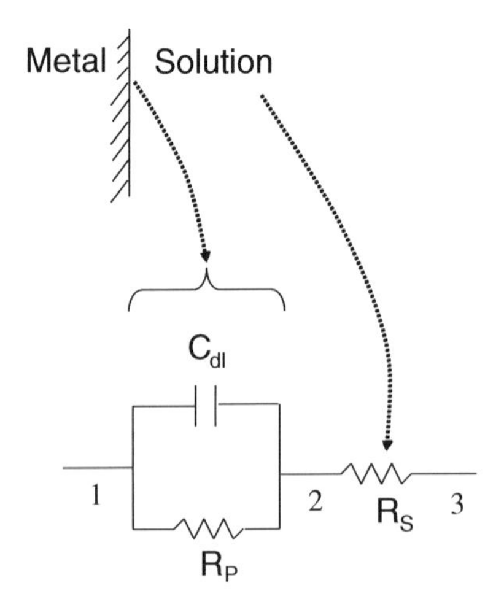

Fig. 14.4 Equivalent circuit for an electrical double layer (edl). C_{dl} is the capacitance of the edl, R_p is the polarization resistance, and R_s is the solution resistance

with the polarization resistance (i.e., charge transfer resistance) R_p. (R_p has the same meaning here as in Chapter 7). The parallel component of the circuit containing R_p and C_{dl} is in series with the solution resistance, R_s, which exists between the electrical double layer and the tip of the reference electrode.

Impedance Analysis

We can now perform an AC impedance analysis [9] of the assembly shown in Fig. 14.4. From Eqs. (11) and (12), the impedance Z_{12} between the points 1 and 2 is

$$\frac{1}{Z_{12}} = \frac{1}{R_P} + j\,\omega\, C_{dl} \tag{13}$$

or

$$Z_{12} = \frac{R_p}{1 + j\,\omega\, R_p\, C_{dl}} \tag{14}$$

The impedance Z between the points 1 and 3 in Fig. 14.4 is then

$$Z = R_S + \frac{R_P}{1 + j\,\omega R_P\, C_{dl}} \tag{15}$$

Multiplying the numerator and denominator in Eq. (15) by the term $(1 - j\omega R_p C_{dl})$ gives

$$Z = R_S + \frac{R_P}{1 + \omega^2 R_P^2\, C_{dl}^2} - j\,\frac{\omega\, R_P^2\, C_{dl}}{1 + \omega^2 R_P^2\, C_{dl}^2} \tag{16}$$

Writing the impedance Z as a complex number:

$$Z = Z' + \mathrm{j}\,Z'' \tag{17}$$

gives the real (Z') and imaginary (Z'') parts of the impedance to be

$$Z' = R_S + \frac{R_P}{1 + \omega^2 R_P^2 C_{dl}^2} \tag{18}$$

and

$$Z'' = -\frac{\omega R_P^2 C_{dl}}{1 + \omega^2 R_p^2 C_{dl}^2} \tag{19}$$

These last two equations can be combined by the elimination of ω. The result (after a little algebra) is

$$\left[Z' - \left(R_S + \frac{R_P}{2}\right)\right]^2 + (Z'')^2 = \left(\frac{R_P}{2}\right)^2 \tag{20}$$

Equation (20) is the equation of a semi-circle centered at (R_s + (R_p/2), 0), as shown in Fig. 14.5. The semi-circle intersects the Z'-axis at the values R_s and (R_s + R_p). The value of ω at the apex of the semi-circle is given by

$$\omega_{max} = \frac{1}{R_P C_{dl}} \tag{21}$$

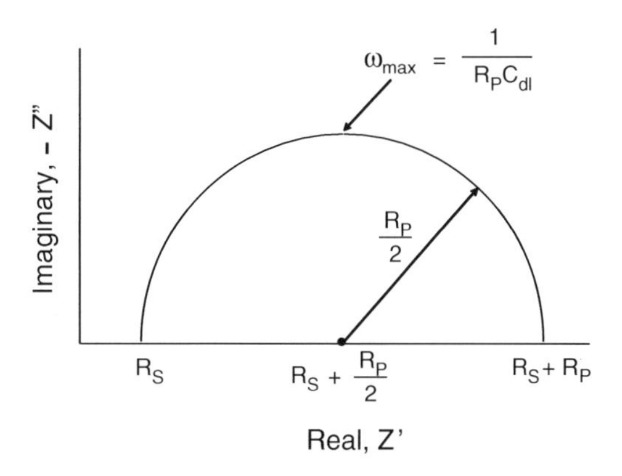

Fig. 14.5 Schematic diagram of complex-plane plot of the imaginary part of the impedance, Z'', vs. the real part of the impedance, Z'

The result in Eq. (21) follows from taking $dZ''/d\omega = 0$. See Problem 14.1. Then C_{dl} can be determined from Eq. (21) once R_p is known.

Thus, all three unknown quantities R_p, R_s, and C_{dl} can be determined from Fig. 14.5. Plots such as Fig. 14.5 are referred to by various workers as complex-plane plots, Argand diagrams, Nyquist plots, or Cole–Cole plots.

If the value of R_p is known, then the value of C_{dl} can also be determined from a linear plot of Z' vs. Z''/ω. The relationship is

$$Z' = R_s - \frac{Z''}{\omega R_P C_{dl}} \tag{22}$$

The slope of the plot is $-1/(R_p C_{dl})$ and the intercept is R_s. The origin of Eq. (22) is left as an exercise in Problem 14.2.

Additional Methods of Plotting Impedance Data

One of the disadvantages of the semi-circular plots in Fig. 14.5 is that the frequency dependence of the impedance does not appear on the diagram. In addition, the points are often distributed unevenly along the arc of the semi-circle. Finally, the maximum value of Z'' on the semi-circle and its corresponding frequency (used to obtain C_{dl}) may not actually correspond to a data point. These problems are illustrated in Fig. 14.6, which is the complex-plane plot for an interface for which $R_p = 1{,}000\ \Omega$, $R_s = 100\ \Omega$, and $C_{dl} = 10\ \mu F/cm^2$.

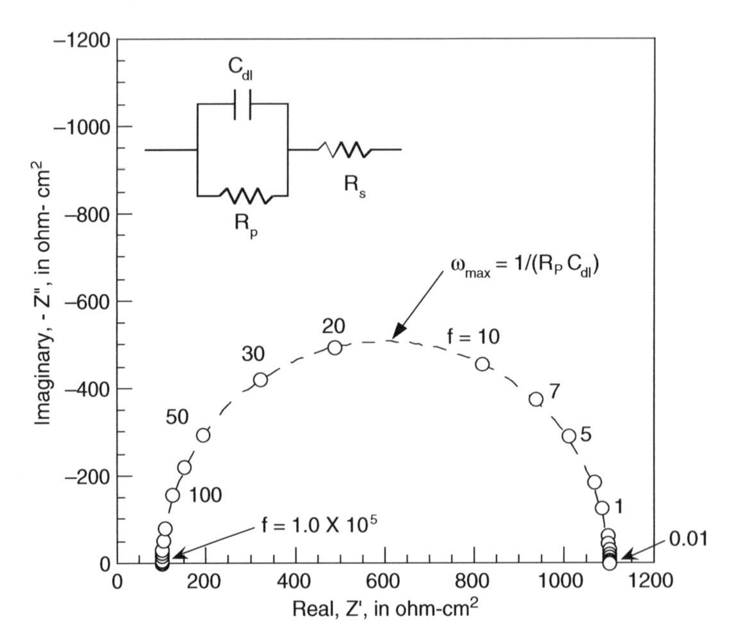

Fig. 14.6 Calculated complex-plane plot for the equivalent circuit shown in Fig. 14.4 with $R_p = 1{,}000\ \Omega$, $R_s = 100\ \Omega$, and $C_{dl} = 10\ \mu F/cm^2$ (frequencies in hertz are indicated on the figure)

(It should be noted that when C is expressed as a capacitance per unit area, i.e., as F/cm^2, then values of resistances and impedances are actually in Ω-cm^2 rather than in ohms. See Problems 14.3 and 14.4.)

An alternate method of displaying impedance data is a Bode plot, in which log $|Z|$ is plotted vs. log ω. The absolute value of the impedance Z follows from Eq. (4) as

$$|Z| = (Z'^2 + Z''^2)^{\frac{1}{2}} \tag{23}$$

Equations (18), (19), and (23) combine to give

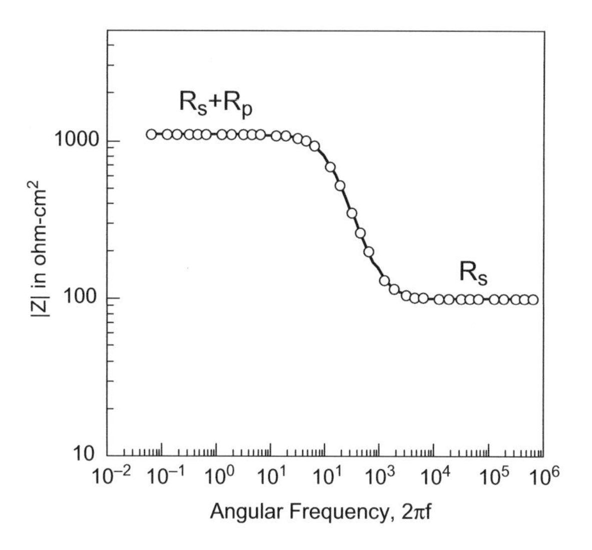

Fig. 14.7 Calculated Bode plot corresponding to the complex-plane plot in Fig. 14.6

$$|Z| = \left\{ \left(R_S + \frac{R_P}{1 + \omega^2 R_P^2 C_{dl}^2} \right)^2 + \left(\frac{\omega R_P^2 C_{dl}}{1 + \omega^2 R_P^2 C_{dl}^2} \right)^2 \right\}^{1/2} \tag{24}$$

The Bode plot corresponding to the complex-plane plot in Fig. 14.6 is given in Fig. 14.7. Resistances appear as horizontal lines in a Bode plot, and capacitances appear as slanted lines. It can be seen from Eq. (24) that for small values of ω

$$|Z| \to (R_S + R_P) \, \text{as} \, \omega \to 0 \tag{25}$$

For large values of ω

$$|Z| \to R_S \, \text{as} \, \omega \to \infty \tag{26}$$

To see this last result it is necessary to use L'Hospital's rule on Eq. (24). See Appendix M. The portion of the curve where the impedance is due largely to the capacitance C_{dl} has a slope equal to –1. See Problem 14.5.

Other methods of plotting impedance data include graphs of Z' or Z'' vs. ω, as shown in Fig. 14.8 for the equivalent circuit shown in Fig. 14.4. In addition, the phase angle δ can be plotted versus ω, where the phase angle δ is given by

$$\tan \delta = \frac{Z''}{Z'} \tag{27}$$

The most useful plot, however, has generally proven to be the Bode plot, and applications involving Bode plots are given in this chapter.

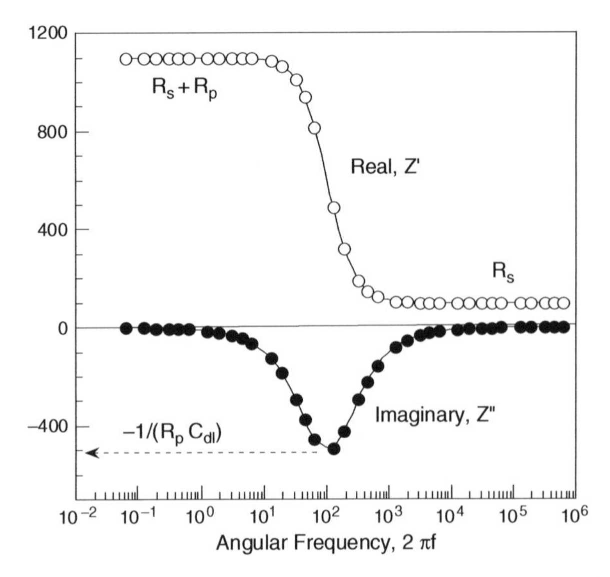

Fig. 14.8 Real and imaginary components of the complex impedance corresponding to the complex-plane plot in Fig. 14.6 and the Bode plot in Fig. 14.7

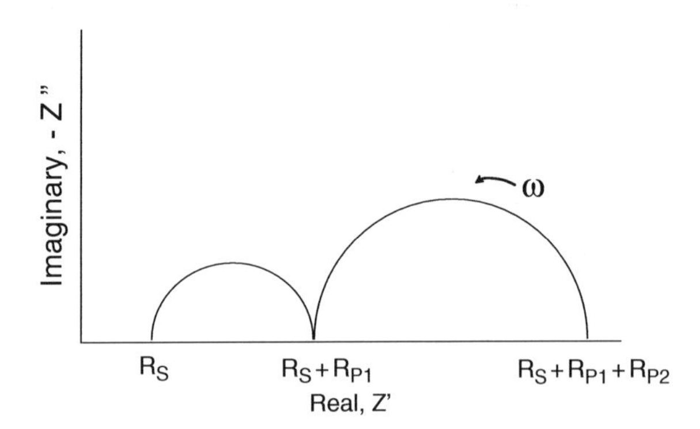

Fig. 14.9 Schematic diagram of a complex-plane plot for a system having two time constants

Multiple Time Constants and the Effect of Diffusion

An electrochemical system may contain more than one relaxation process and thus may have more than a single time constant. In such a case, the complex-plane plot will display more than one semi-circle, as shown schematically in Fig. 14.9 for the case of two time constants. (The Bode plot corresponding to Fig. 14.9 will contain three plateaus).

In addition, an electrochemical system may contain a component which is under diffusion control, for which [8]

$$Z = \frac{\sigma}{\omega^{1/2}} - \frac{j\sigma}{\omega^{1/2}} \tag{28}$$

where σ is the conductivity. The effect of diffusion appears at low frequencies in the complex-plane plot as a straight line 45° to the real axis, as shown schematically in Fig. 14.10. The effect of diffusion appears on the corresponding Bode plot as a slanted line with a slope of –1/2. See Problem 14.6. When diffusion effects are present in electrochemical systems, the equivalent circuit in Fig. 14.4 is modified by the inclusion of a diffusion component, known as the Warburg impedance, as shown in Fig. 14.11.

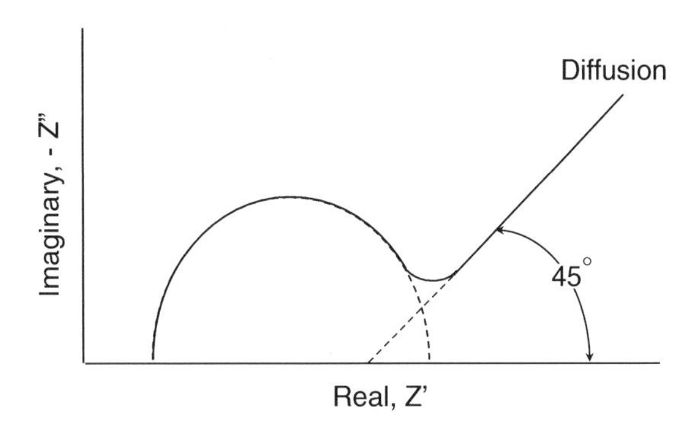

Fig. 14.10 Schematic diagram of a complex-plane plot for a system having a diffusion component

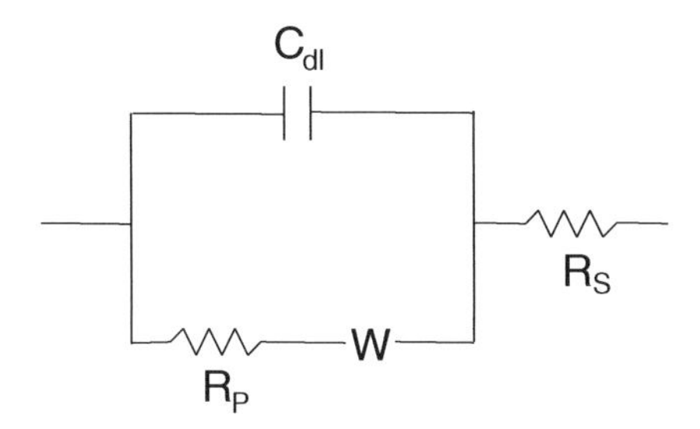

Fig. 14.11 Equivalent circuit for a electrochemical interface having a diffusion component. *W* represents the Warburg impedance due to diffusion

Kramers–Kronig Transforms

The Kramers–Kronig transforms are a set of integral equations which transform the real component of the impedance into the imaginary component [10–12], and vice versa. The Kramers–Kronig relationship have been expressed in various forms, but two of the most important are

$$Z'(\omega) - Z'(\infty) = \frac{2}{\pi} \int_0^{\infty} \frac{x Z''(x) - \omega Z''(\omega)}{x^2 - \omega^2} \, dx \tag{29}$$

and

$$Z''(\omega) = -\frac{2\omega}{\pi} \int_0^{\infty} \frac{Z'(x) - Z'(\omega)}{x^2 - \omega^2} \, dx \tag{30}$$

These transforms allow an independent check on the validity of the experimental impedance data. In addition, a useful relationship is [12]

$$R_P = \frac{2}{\pi} \int_0^\infty \frac{Z''(\omega)}{\omega} \, d\omega \tag{31}$$

If $Z''(\omega)$ is a symmetric function, as shown in Fig. 14.8, then Eq. (31) can be rewritten as

$$R_P = \frac{4}{\pi} \int_0^{\omega_{max}} \frac{Z''(\omega)}{\omega} \, d\omega \tag{32}$$

where ω_{max} is the angular frequency at which Z'' has its largest absolute value. If $Z''(\omega)$ can be expressed as a polynomial in ω of the form [11]

$$Z''(\omega) = a_0 + a_1\omega + a_2\omega^2 + a_3\omega^3 + \cdots + a_n\omega^n \tag{33}$$

then the integration in Eq. (32) is simple, and it is relatively easy to compute R_p from the imaginary part of the impedance (Z'') alone. Insertion of Eq. (33) into Eq. (32) and integrating gives

$$R_P = \frac{4}{\pi} \sum_i \left[a_0 \ln \omega + a_1\omega + \frac{a_2\omega^2}{2} + \frac{a_3\omega^3}{3} + \cdots + \frac{a_n\omega^n}{n} \right]_{\omega_i}^{\omega_{i+1}}$$

where the curve fitting of Z'' and the integration is done piecewise over i segments of the Z'' vs. ω plot [11]. This computation thus provides an independent check of the value of R_p because only Z'' is used in its determination. This value of R_p can be compared with the value of R_p as determined from either Bode plots or complex-plane diagrams (which use both Z' and Z'').

Application to Corrosion Inhibition

An example of the application of these principles is given by the corrosion inhibition of iron in acid solutions [13]. Figure 14.12 shows Bode plots for iron in de-aerated 1 M HCl with and without the addition of a biological molecule called aerobactin, the molecular formula of which is given in Fig. 14.13. The Bode plot in Fig. 14.12 shows that the addition of aerobactin increases the value of R_p. (Recall that the low-frequency limit of $|Z|$ gives (R_p + R_s).) As shown in Chapter 7, the corrosion rate decreases as R_p increases. The relationship given by Eq. (64) in Chapter 7 is

$$i_{corr} = \frac{1}{2.303\, R_P \left(\frac{1}{b_a} + \frac{1}{|b_C|} \right)} \tag{34}$$

(The term ($d\eta/di$) for small η in Chapter 7 is the same as R_p.) The actual corrosion rate can be calculated from R_p if the Tafel slopes b_a and b_c are known. See Problem 14.8.

The complex-plane plots corresponding to the Bode plots in Fig. 14.12 are shown in Fig. 14.14. The complex-plane plot also shows that the addition of aerobactin to 1 M HCl increases the value of R_p. (The semi-circle intersects the real axis at the values of R_s and (R_p + R_s).)

Figure 14.14 also shows that the complex-plane plot is not a perfect semi-circle but is instead a semi-circle with its center depressed below the real axis. Such complex-plane plots with depressed semi-circles, as shown in Fig. 14.14, are called Cole–Cole plots [10]. A depressed semi-circle results if there is a distribution of relaxation times τ around a most probable value $\tau_o = R_pC_{dl}$. This is tantamount to the distribution of capacitances around some central value C_{dl} [13, 14]. In such a case

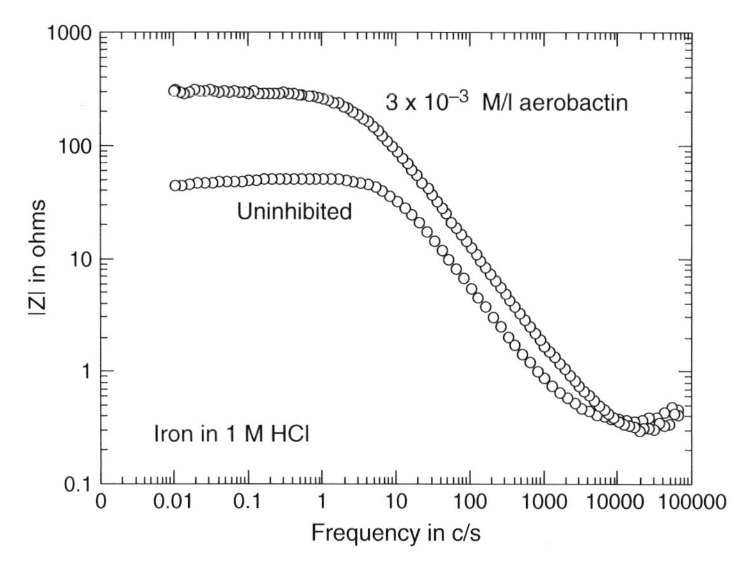

Fig. 14.12 Experimental Bode plots for iron in de-aerated 1 M HCl with and without the inhibitor aerobactin [13]. Sample area = 5.0 cm^2. Reproduced by permission of ECS – The Electrochemical Society

```
      CH3                                               CH3
      |                                                 |
      C=O                                             O=C
      |                                                 |
      N—OH                                           HO—N
      |                                                 |
     (CH2)4     O           COOH        O            (CH2)4
      |         ||          |           ||              |
  H—C—N — C—CH2— C —CH2—C — N—C—H
      |   |                 |                     |   |
  HOOC   H                 OH                    H   COOH
```

Fig. 14.13 Molecular formula for the inhibitor aerobactin [13]

Eq. (15) is replaced by [10]:

$$Z = R_S + \frac{R_P}{1 + (j\omega R_P C_{dl})^{1-\alpha}} \tag{35}$$

where α is a measure of the departure from ideality. The parameter α ranges from 0 to 1 and is zero for ideal behavior. The capacitance C_{dl} and the parameter α can be determined from the various arc chords U and V shown in Fig. 14.15. By calculating the ratio $|V/U|$ for each frequency, we obtain [4, 10, 13]:

$$\log\left|\frac{V}{U}\right| = (\alpha - 1)\log(2\pi f) + (\alpha - 1)\log(C_{dl} R_P) \tag{36}$$

The derivation of Eq. (36) is given in Appendix N. The value of C_{dl} follows from Eq. (36) and is given by

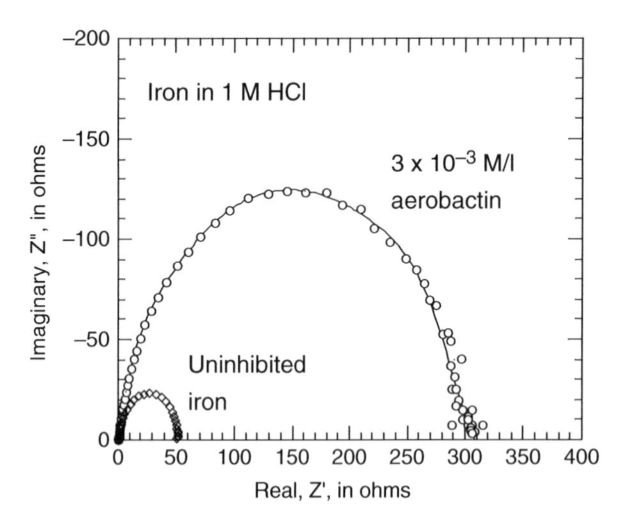

Fig. 14.14 Experimental complex-plane plot for iron in de-aerated 1 M HCl with and without the inhibitor aerobactin [13]. Sample area = 5.0 cm^2. Reproduced by permission of ECS – The Electrochemical Society

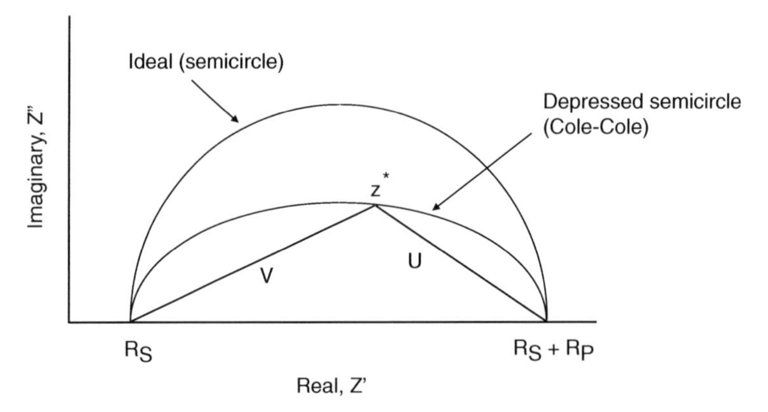

Fig. 14.15 Schematic complex-plane diagram for the impedance showing arc chords U and V for a given frequency [13]. The notation $z*$ indicates that z is a complex number. Reproduced by permission of ECS – The Electrochemical Society

$$C_{\mathrm{dl}} = \frac{10^{I/S}}{R_{\mathrm{P}}} \tag{37}$$

where I and S are the intercept and slope, respectively, of the plot of log $|V/U|$ vs. log ω.

Figure 14.16 shows the plot of Eq. (36) for iron in 1 M HCl containing 3×10^{-3} M aerobactin. The calculated value of C_{dl} was 56 μF/cm^2, and the value for uninhibited iron was 89 μF/cm^2, so that the double-layer capacitance decreased with adsorption of the inhibitor, as has been discussed in Chapter 12. The slope of the plot in Fig. 14.16 is $(\alpha-1) = -0.86$, or $\alpha = 0.14$.

In addition, polarization resistances were determined from Kramer–Kronig relations in Eqs. (31) and (34) for iron in 1 M HCl in the presence of various biological molecules. As shown in Table 14.2, there is good agreement between R_{p} values determined using the Kramers–Kronig relation and R_{p} values determined from Bode plots or complex-plane plots.

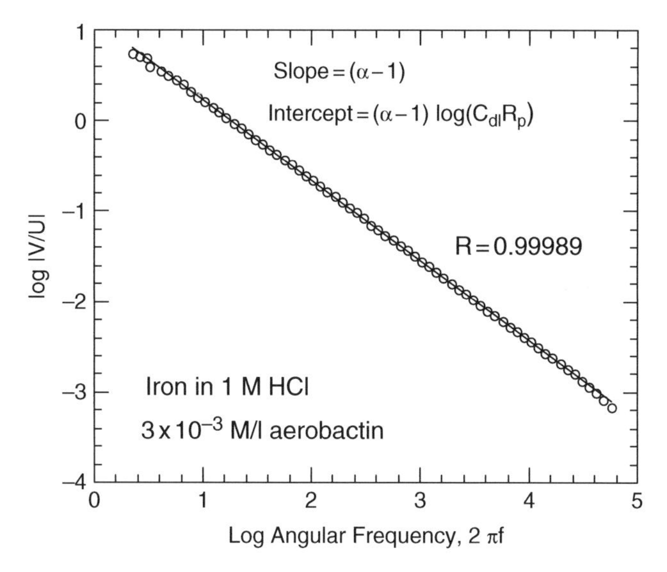

Fig. 14.16 A typical experimental arc chord plot used to determine the capacitance of the electrical double layer [13]. Reproduced by permission of ECS – The Electrochemical Society

Table 14.2 Polarization resistance (R_p) values determined by three different methods for iron in 1 M HCl with and without the addition of various biological molecules [13]

	R_p in Ωcm^2		
	Kramers–Kronig	Bode plot	Complex-plane plot
Uninhibited	20.4	20.3	20.4
Aerobactin 3.0×10^{-3} M	71.5	73.5	74.4
Rhodotorulic acid 1.2×10^{-3} M	75.5	71.6	71.4
Enterobactin 3.1×10^{-4} M	113	104	107
Parabactin 1.4×10^{-4} M	476	449	466

Organic Coatings

One of the most powerful applications of AC impedance techniques has been to assess the durability of organic coatings in aqueous media [1, 2, 15–17]. In this regard, AC impedance techniques have been used in two ways:

(1) to monitor the ingress of water (or ions) into the organic coating and
(2) to develop an equivalent circuit for the coating/substrate system and to relate the corrosion behavior to the properties of the elements of the equivalent circuit.

Kendig and Leidheiser [18] were among the earliest to use AC measurements to follow the change in capacitance and resistance of a coated metal in an aqueous solution (0.5 M NaCl). As time progressed, the capacitance of a polybutadiene coating increased and the resistance decreased due to the take-up of water and ions into the coating. (Recall from Chapter 13 that the transmission rate of water into an organic film is much greater than the transmission rate of chloride ions, so that the observed changes were most likely due to the ingress of water.) The basic idea is that the uptake of water modifies the dielectric constant ε of the coating and hence the capacitance C:

$$C = \frac{\varepsilon_0 \, \varepsilon \, A}{l} \tag{38}$$

where ε_o is the dielectric constant of free space (vacuum), A is the area, and l the thickness of the coating.

A later example is provided by the work of Bierwagen et al. [19] in which the absolute value of the impedance |Z| at 0.012 Hz was measured vs. time for various primer plus topcoat aircraft coatings in an alternate immersion test. The results are shown in Fig. 14.17. Coatings which were protective had values of |Z| at 0.012 Hz of 10^8–10^9 Ω over a period of 110 weeks. Non-protective coatings had an initial |Z| at 0.012 Hz of approximately 10^6 Ω and these values decreased to about 10^3 Ω in 20–40 weeks. Similar trends were observed for alkyd and epoxy marine coatings. Thus, the low-frequency values of |Z| can be used as an indicator of coating performance.

In the second approach, an equivalent circuit is used to represent the solution/coating/metal system. Figure 14.18 shows the equivalent circuit used frequently [1, 2, 16–19] to represent a polymer coated metal. The components of the circuit are:

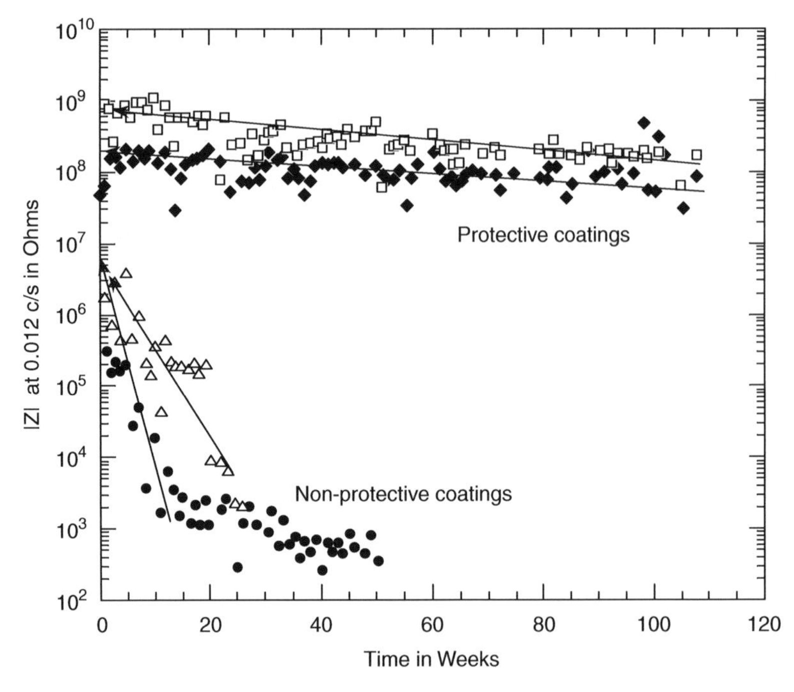

Fig. 14.17 |Z| at 0.012 Hz vs. exposure time for various aircraft coatings in a cyclic test involving exposure to UV light and alternate exposure and withdrawal from a salt fog consisting of 0.05% NaCl and 0.35% $(NH_4)_2SO_4$. □ Self-priming topcoat, ♦ Primer plus glossy topcoat, △ Primer plus flat topcoat • Yellow primer. Figure 14.17 redrawn from [19] by permission of Elsevier Ltd

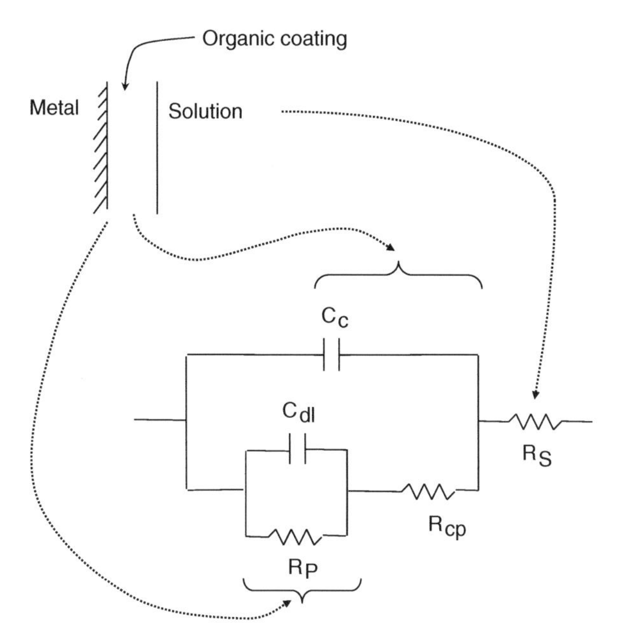

Fig. 14.18 Equivalent circuit for an organic coating on a metal substrate [1]. R_p is the polarization resistance at the metal/solution interface and C_{dl} is the capacitance of the electrical double layer at the metal solution interface. R_{cp} is the pore resistance of the coating and C_c the capacitance of the coating. R_s is the solution resistance.

C_c = coating capacitance
C_{dl} = double-layer capacitance at the metal/solution interface
R_p = polarization resistance at the metal/solution interface
R_{cp} = pore resistance of the coating
R_s = solution resistance

The pore resistance of the coating R_{cp} is of particular interest because it contains information regarding degradation of the protective properties of the coating.

By combining the impedances of the components of the equivalent circuit, the real and imaginary parts of the impedance are seen to be (after considerable algebra):

$$Z' = R_S + \frac{R_P + R_{cp} + \omega^2 (R_P C_{dl})^2 R_{cp}}{\left(1 - \omega^2 R_P C_{dl} R_{cp} C_c\right)^2 + \omega^2 \left(R_p C_{dl} + R_P C_c + R_{cp} C_c\right)^2} \tag{39}$$

and:

$$Z'' = -\frac{\omega\left[R_P^2 C_{dl}\left(1 + \omega^2 R_{cp}^2 C_{dl} C_c\right) + C_c \left(R_p + R_{cp}\right)^2\right]}{\left(1 - \omega^2 R_p C_{dl} R_{cp} C_c\right)^2 + \omega^2 \left(R_p C_{dl} + R_p C_c + R_{cp} C_c\right)^2} \tag{40}$$

The absolute value of the impedance |Z| then follows from inserting Eqs. (39) and (40) into Eq. (23). It can be seen that as $\omega \Rightarrow 0$, $|Z| \Rightarrow R_s + R_p + R_{cp}$. As $\omega \Rightarrow \infty$, $|Z| \Rightarrow R_s$. The latter result requires successive use of L'Hospital's rule.

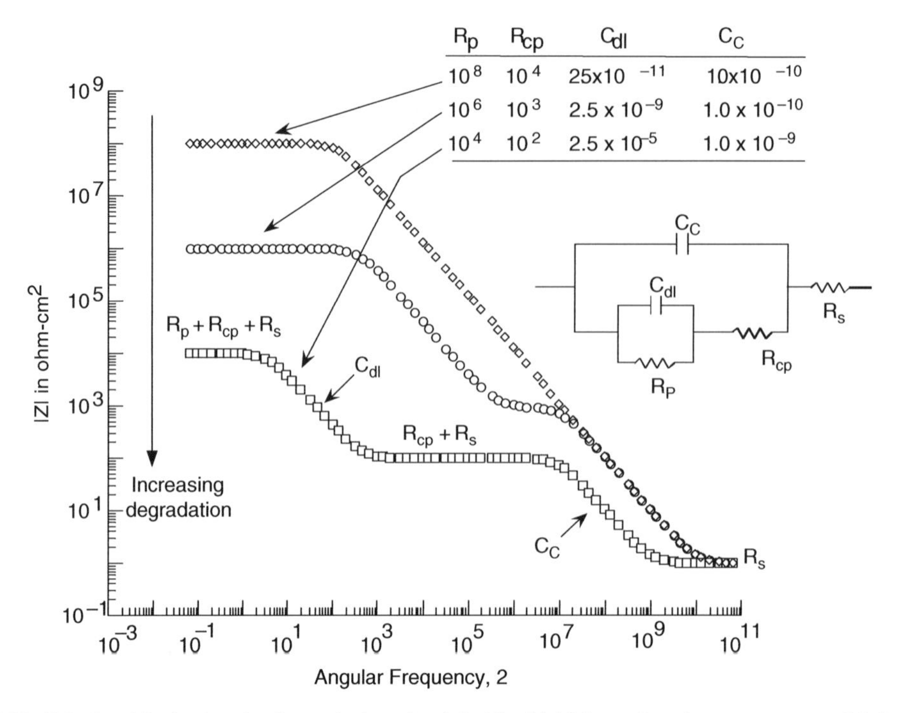

Fig. 14.19 Calculated Bode plots for the equivalent circuit in Fig. 14.18 for various input parameters. R is in Ω and C in F/cm^2

A typical Bode plot for an organic coating can be constructed by evaluating Z' and Z'' for representative values of the components of the equivalent circuit. If the organic coating has undergone little degradation, we can take $R_p = 10^8\ \Omega$, $R_{cp} = 10^4\ \Omega$, and $C_{dl} = 2.5 \times 10^{-11}$ F/cm^2 [1]. We can estimate C_c as in Example 14.2.

Example 14.2: What is the capacitance of a 35-μm thick organic coating having a dielectric constant of 4.0? (This example is taken from Mansfeld [1]).

Solution: The organic coating is considered to be a parallel plate capacitor, for which the capacitance is given by

$$C = \frac{\varepsilon_0\, \varepsilon\, A}{l}$$

where ε_0 is the dielectric constant of free space (8.85×10^{-14} F/cm), A the area, and l the thickness of the coating. Thus,

$$\frac{C}{A} = \frac{\left(8.85 \times 10^{-14} \mathrm{F\,cm^{-1}}\right)(4.0)}{\left(35 \times 10^{-6}\,\mathrm{m}\right)\left(100\,\frac{\mathrm{cm}}{\mathrm{m}}\right)} = 1.0 \times 10^{-10}\,\mathrm{F/cm^2}$$

Then, we can use $C_c = 1.0 \times 10^{-10}$ F/cm^2 and $R_s = 1\ \Omega$ along with the values of the other components of the equivalent circuit to calculate Z', Z'', and $|Z|$ from Eqs. (39), (40), and (23). These calculations produce the Bode plot in Fig. 14.19 (topmost curve).

If there is significant delamination of the coating and concomitant electrochemical activity beneath the organic coating, then R_p decreases and C_{dl} increases. The pore resistance of the coating R_{cp} also decreases. In addition C_c increases due to absorption of water which causes an increase in the effective dielectric constant of the organic coating. Figure 14.19 also shows the calculated Bode plot for the following set of parameters: $R_p = 10^4\ \Omega$, $R_{cp} = 10^2\ \Omega$, $C_c = 1.0 \times 10^{-9}$ F/cm^2, $C_{dl} = 2.5 \times 10^{-5}$ F/cm^2, and $R_s = 1\ \Omega$ (lowermost curve). The effect of coating degradation is to shift the Bode plot downward (and also to reveal more features in the plot). The Bode plot for an intermediate set of equivalent circuit parameters is also given in Fig. 14.19 (middle curve).

The low-frequency portion of Bode plots can be used to rate or compare the corrosion behavior of various organic coatings because the low-frequency limit contains the values ($R_{cp} + R_p$) which reflect the properties of the coating (R_{cp}) and of the polarization resistance (R_p). Actually, the low frequency limit also contains the value of the solution resistance, R_s, but this is constant for a given test solution and is also much less than ($R_{cp} + R_p$).

Figure 14.20 shows experimentally determined Bode plots [20] for three different epoxy coatings after 9 days immersion in an electrolyte used in atmospheric corrosion studies (8.5 mM $(NH_4)_2SO_4$ plus 26 mM NaCl). The Bode plots show that the electrocoat epoxy has the best corrosion resistance of the three coatings after the 9-day immersion period. The Bode plots in Fig. 14.20 also have the same general shapes shown in Fig. 14.19.

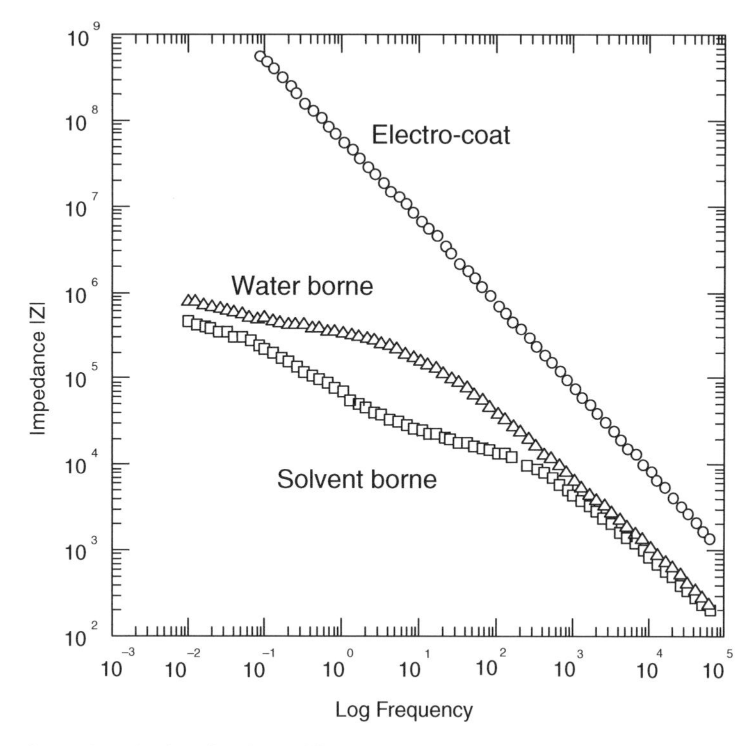

Fig. 14.20 Experimental Bode plots for three different epoxy coatings immersed 9 days in a solution containing 8.5 mM $(NH_4)_2SO_4$ and 26 mM NaCl. Redrawn from [20] with the permission of the American Chemical Society

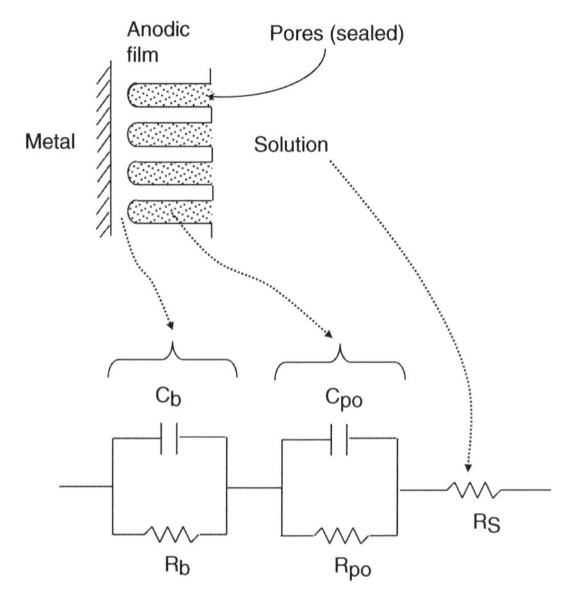

Fig. 14.21 Schematic diagram of an anodized coating and the corresponding equivalent circuit [2]. R_b and C_b refer to the barrier layer of the oxide, and R_{po} and C_{po} refer to the porous oxide layer. Reproduced by permission of Elsevier Ltd

Oxide Films and Surface Treatments

AC impedance techniques have also been useful in the study of oxide films, especially in the anodization of aluminum alloys [2, 21–24]. In Chapter 9, we discussed briefly the structure of oxide films on aluminum as formed by anodization in sulfuric acid. The anodic film consists of an inner barrier layer and an outer porous layer, as shown schematically in Fig. 14.21. (See also Fig. 9.22.) The porous outer layer can be sealed in hot water or in aqueous solutions such as dichromates.

An equivalent circuit frequently used to represent the oxide structure is also given in Fig. 14.21 [2]. This equivalent circuit contains contributions from the inner barrier layer (R_b and C_b) and from the outer porous layer (R_{po} and C_{po}). R_s is again the solution resistance. Some variations of this equivalent circuit have been used by other authors [21, 22].

By combining the impedances of the components of the equivalent circuit in Fig. 14.21, the real and imaginary parts of the impedance are given by:

$$Z' = R_S + \frac{R_b}{1 + \omega^2 R_b^2 C_b^2} + \frac{R_{po}}{1 + \omega^2 R_{po}^2 C_{po}^2} \tag{41}$$

and

$$Z'' = -\omega \left[\frac{R_b^2 C_b}{1 + \omega^2 R_b^2 C_b^2} + \frac{R_{po}^2 C_{po}}{1 + \omega^2 R_{po}^2 C_{po}^2} \right] \tag{42}$$

Then |Z| is determined by using Eqs. (41) and (42) in Eq. (23). For small values of ω,

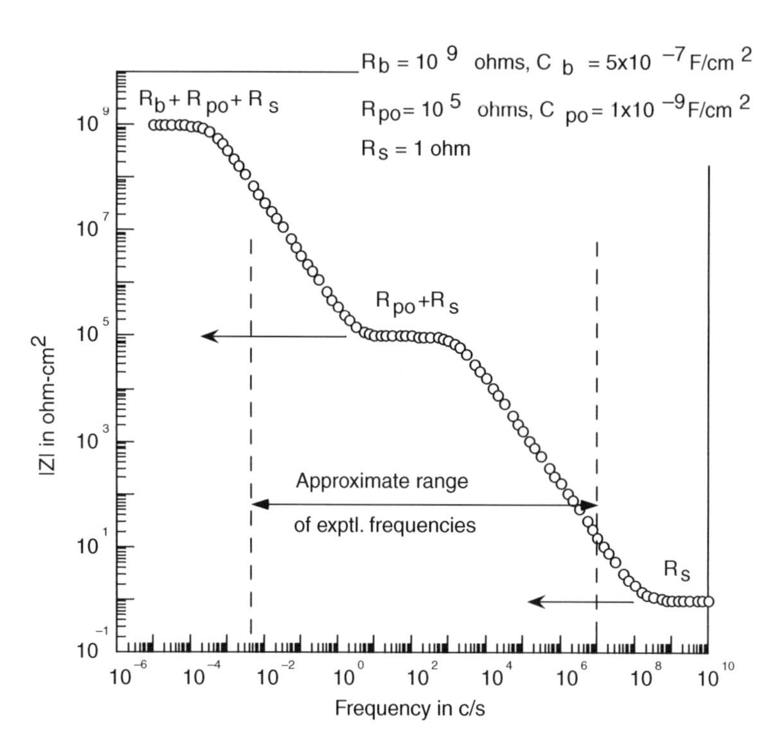

Fig. 14.22 Calculated Bode plot for the equivalent circuit in Fig. 14.21 with input parameters indicated on the figure

$$|Z| \rightarrow R_s + R_b + R_{po} \text{ as } \omega \rightarrow 0 \tag{43}$$

For large values of ω,

$$|Z| \rightarrow R_s \text{ as } \omega \rightarrow \infty \tag{44}$$

This last result again requires the use of L'Hospital's rule. Note that the low-frequency portion of the Bode plot is of particular interest because it contains information about the resistances of the two layers which comprise the anodic oxide film.

Figure 14.22 shows a Bode plot for representative values of the circuit components, as given by Mansfeld and Kendig [23]: $R_b = 10^9\ \Omega$, $C_b = 5 \times 10^{-7}$ F/cm^2, $R_{po} = 10^5\ \Omega$, $C_{po} = 10^{-9}$ F/cm^2, and $R_s = 1\ \Omega$. The exact shape of the curve, of course, depends on the values of the input parameters. Note that some features of the calculated Bode plot in Fig. 14.22 lie outside the range of applied experimental frequencies.

Figure 14.23 shows experimental Bode plots for aluminum alloy 2024 anodized in sulfuric acid with and without a dichromate seal [24]. It can be seen that the dichromate seal is effective in increasing the impedance $|Z|$ of the oxide film.

In order to compare the effectiveness of various anodization procedures and sealing treatments, Mansfeld and Kendig [24] defined a damage function D as

$$D = \log\left(\frac{|Z|_0}{|Z|_t}\right)_{0.1\text{ Hz}} \tag{45}$$

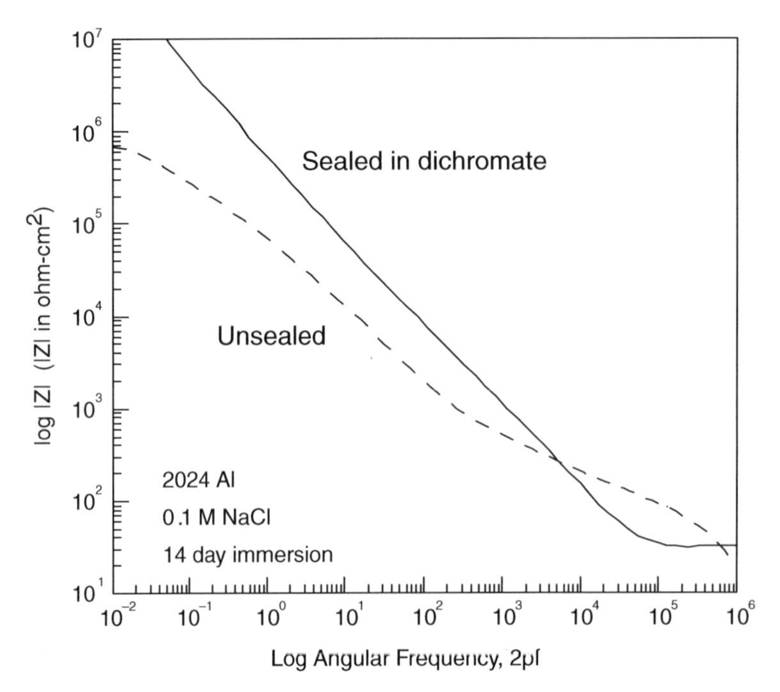

Fig. 14.23 Effect of a dichromate sealer on the Bode plots for anodized Al 2024 in 0.5 M NaCl. (Anodization was in sulfuric acid.) Redrawn from Mansfeld and Kendig [24] by permission of © NACE International (1985)

Table 14.3 Damage functions for anodized aluminum alloys after 7 days immersion in 0.5 M NaCl [24]

		Damage function for $t = 7$ days	
Aluminum alloy	Anodization	Unsealed	Sealed[a]
2024	Chromic acid	1.37	1.51
2024	Sulfuric acid	0.60	0.11
6061	Chromic acid	0.0	0.04
6061	Sulfuric acid	0.0	0.0
7075	Chromic acid	1.91	0.18
7075	Sulfuric acid	0.0	0.0

[a]Sealed in dichromate solutions.

This function D shows how the low-frequency impedance $|Z|$ changes as a function of immersion time (t) in the electrolyte. ($|Z|_0$ is the impedance after initial immersion). Oxide films which undergo breakdown will show a decrease in $|Z|$ with time, and thus an increase in D. Oxide films which are resistant to corrosion will not display decreases in $|Z|$, so that D will have zero or small values. Table 14.3 lists damage functions measured by Mansfeld and Kendig [24] for three different aluminum alloys. Sulfuric acid-anodized and chromic acid-anodized Al 6061 are so corrosion resistant that no corrosion occurred in 7 days, even in the absence of sealing. For Al 2024, sealing produced an improvement in the behavior for samples anodized in sulfuric acid, but not chromic acid. For Al 7075, sealing was beneficial.

Studies such as these show that AC impedance techniques can be used as a quality control tool in the surface treatment of aluminum alloys.

Concluding Remarks

As seen above, extensive use has been made of equivalent electrical circuits to characterize corrosion protection by inhibitors, by organic coatings, or by surface treatments such as anodization. Such equivalent circuits are based on best estimates of the physico-chemical properties of the system. However, a given equivalent circuit is not necessarily unique; and other possible equivalent circuits can be used. But if a given equivalent circuit is modified to take into account additional details, the new equivalent circuit can instead become unnecessarily complicated and unwieldy for analysis.

AC impedance instrumentation is a standard tool in the modern corrosion laboratory. Not only are these systems fully automated by computer control, but appropriate software is also available for data analysis. AC impedance has been successfully applied to problems involving the metal/solution interface, corrosion inhibition, anodization and surface treatments, and the corrosion behavior of organic coatings.

Problems

1. Show that Eq. (19) can be differentiated and $dZ''/d\omega$ set equal to zero to give

$$\omega_{\text{max}} = \frac{1}{R_{\text{p}}\, C_{\text{dl}}}$$

where ω_{max} is the frequency at which Z'' has its maximum absolute value.

2. In the text it was stated that for the equivalent circuit shown in Fig. 14.4, the term $1/(C_{\text{dl}}R_{\text{p}})$ can be determined from a linear plot of Z' vs. Z''/ω. Show that

$$Z' = R_{\text{s}} - \frac{Z''}{\omega\, C_{\text{dl}}\, R_{\text{p}}}$$

Hint: Combine Eqs. (18) and (19).

3. Show that the term ωRC is dimensionless when R is in ohms and C is in F.
4. (a) Use the fact that ωRC is dimensionless if R is in ohms and C is in F to show that the units of Z' and Z'' in Eqs. (18) and (19) are then each in ohms. (b) If C is in F/cm^2, show that R_{s}, R_{p}, Z', and Z'' are then each in Ω cm^2 for Eqs. (18) and (19).
5. Show that on a Bode plot of log $|Z|$ vs. log ω the slope for the impedance of a capacitor is –1.
6. (a) Show that a Warburg diffusion element appears on a complex-plane plot of $-Z''$ vs. Z' as a straight line 45° to the real axis. (b) Show that on a Bode plot of log $|Z|$ vs. log ω, the Warburg impedance gives a straight line with slope –1/2.
7. (a) Derive an expression for $|Z|$ for the following equivalent circuit. (b) Draw the Bode plot and complex-plane plot for $R = 1{,}000\ \Omega$ and $C = 1.0 \times 10^{-5}$ F/cm^2.

8. From the Bode plot in Fig. 14.12, calculate the corrosion rate for iron in 1 M HCl containing 3.0 $\times$ 10^{-3} M aerobactin. The anodic and cathodic Tafel slopes are 60 and –120 mV, respectively. The sample area is 5.0 cm^2.
9. Given the following impedance data for a metal immersed in an acid solution. Determine R_p, R_s, and C_{dl}.

f (Hz)	Z' (Ω)	Z'' (Ω)	f (Hz)	Z' (Ω)	Z'' (Ω)
0.01	2,010.0	−7.5397	50	15.614	−105.82
0.02	2,009.9	−15.079	70	12.868	−75.691
0.03	2,009.7	−22.617	100	11.406	−53.023
0.05	2,009.3	−37.686	200	10.352	−26.525
0.07	2,008.6	−52.742	300	10.156	−17.685
0.1	2,007.2	−75.291	500	10.056	−10.612
0.2	1,998.7	−149.94	700	10.029	−7.5799
0.3	1,984.7	−223.34	1,000	10.014	−5.3060
0.5	1,941.4	−364.06	2,000	10.004	−2.6530
0.7	1,879.8	−493.43	3,000	10.002	−1.7687
1.0	1,761.2	−660.17	5,000	10.001	−1.0612
2.0	1,285.2	−961.47	7,000	10.000	−0.75800
3.0	887.62	−992.56	10,000	10.000	−0.53060
5.0	449.32	−828.10	20,000	10.000	−0.26530
7.0	261.16	−662.81	30,000	10.000	−0.17687
10	141.49	−495.71	50,000	10.000	−0.10612
20	44.578	−260.71	70,000	10.000	−0.075800
30	25.517	−175.49	100,000	10.000	−0.075800

10. The figure below shows experimental Bode plots in a test solution for an epoxy coating containing no plasticizer or one of two candidate plasticizers [25]. (Figure reproduced by permission of Elsevier Ltd.) Does the presence of the plasticizer in the coating improve the corrosion protection of the coating? If so, which plasticizer is better? Explain your answers.

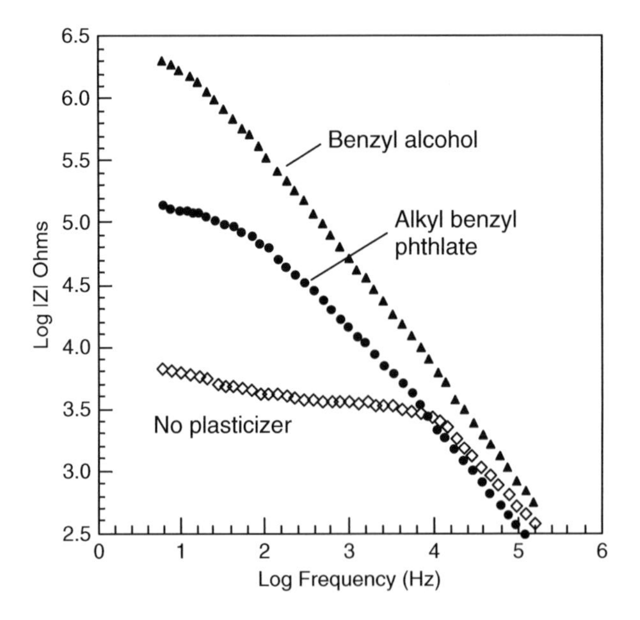

11. Calculate the capacitance per unit area of a barrier layer oxide film which is 200 Å thick and has a dielectric constant of 10.

References

1. F. Mansfeld, *J. Appl. Electrochemistry*, *25*, 187 (1995).
2. F. Mansfeld, *Electrochim. Acta*, *38*, 1891 (1993).
3. C. P. Smyth, "Dielectric Behavior and Structure", Chapters 1 and 2, McGraw-Hill, New York, NY (1955).
4. E. McCafferty, V. H. Pravdic, and A. C. Zettlemoyer, *Trans. Faraday Soc.*, *66*, 1720 (1970).
5. D. A. McQuarrie and J. D. Simon, "Physical Chemistry", p. 496, University Science Books, Sausalito, CA (1997).
6. R. P. Auty and R. H. Cole, *J. Chem. Phys.*, *20*, 1309 (1952).
7. J. R. Macdonald, Ed., "Impedance Spectroscopy", John Wiley & Sons, New York (1987).
8. A. J. Bard and L. R. Faulkner, "Electrochemical Methods", 2nd edition, p. 368, John Wiley & Sons, New York (2001).
9. F. Mansfeld, *Corrosion*, *36*, 301 (1981).
10. K. S. Cole and R. H. Cole, *J. Chem. Phys.*, *9*, 341 (1941).
11. D. D. Macdonald and M. Urquidi-Macdonald, *J. Electrochem. Soc.*, *132*, 2316 (1985).
12. M. Urquidi-Macdonald, S. Real, and D. D. Macdonald, *J. Electrochem. Soc.*, *133*, 2018 (1986).
13. E. McCafferty and J. V. McArdle, *J. Electrochem. Soc.*, *142*, 1447 (1995).
14. E. McCafferty, *Corros. Sci.*, *39*, 243 (1997).
15. A. Amirudin and D. Thierry, *Prog. Org. Coat.*, *26*, 1 (1995).
16. J. N. Murray, *Prog. Org. Coat.*, *31*, 375 (1997).
17. H. P. Hack and J. R. Scully, *J. Electrochem. Soc.* *138*, 33 (1991).
18. M. W. Kendig and H. Leidheiser, Jr., *J. Electrochem. Soc.*, *123*, 982 (1976).
19. G. Bierwagen, D. Tallman, J. Li, L. He, and C. Jeffcoat, *Prog. Org. Coat.*, *46*, 148 (2003).
20. R. L. Twite and G. P. Bierwagen in "Organic Coatings for Corrosion Control", ACS Symposium Series 689, G. P. Bierwagen, Ed., p. 308, American Chemical Society, Washington, DC (1998).
21. T. P. Hoar and G. C. Wood, *Electrochim. Acta.*, *7*, 333 (1962).
22. J. Hitzig, K. Jüttner, W. J. Lorentz, and W. Paatsch, *Corros. Sci.*, *24*, 945 (1984).
23. F. Mansfeld and M. W. Kendig, *J. Electrochem. Soc.*, *135*, 828 (1988).
24. F. Mansfeld and M. W. Kendig, *Corrosion*, *41*, 490 (1985).
25. J. M. McIntyre and H. Q. Pham, *Prog. Org. Coat.*, *27*, 201 (1996).

Chapter 15
High-Temperature Gaseous Oxidation

Introduction

So far this text has considered the corrosion of metals or alloys in aqueous environments. However, when a metal or alloy is exposed to an oxidizing gas, corrosion may occur in the absence of an electrolyte, especially at high temperatures. This phenomenon is sometimes called "dry corrosion" as opposed to "wet corrosion" which occurs in the presence of an aqueous electrolyte. Oxidizing gases include O_2, SO_2, H_2S, H_2O, and CO_2 but the most common oxidant is O_2.

High-temperature gaseous oxidation is important in many applications, such as in gas turbines, aircraft engines, petrochemical and power plants, and the extraction of metals from their ores. High-temperature oxidation usually concerns the temperature range of several hundred degrees centigrade to about 1,100°C or so. (The transition metals Cr, Fe, Ni, and Co melt at 1,890, 1,535, 1,453, and 1,495°C, respectively, but their oxides melt at higher temperatures. Aluminum melts at 660°C and aluminum oxide melts at about 2,000°C.) The oxide films formed in high-temperature oxidation reactions are much thicker than ordinary passive films and are often referred to as oxide layers or scales. The term "hot corrosion" is generally reserved for the attack of metals and alloys by molten salts or slags, but is not considered here.

Because this book is primarily concerned with aqueous corrosion, only a brief introduction to high-temperature oxidation is given here. For a more extensive treatment, specialized texts on high-temperature oxidation should be consulted [1–4].

Thermodynamics of High-Temperature Oxidation

The oxidation of a metal in air by oxygen proceeds by a chemical reaction such as

$$2\,\mathrm{Cu(s)} + \frac{1}{2}\,\mathrm{O_2\,(g)} \rightarrow \mathrm{Cu_2O\,(s)} \tag{1}$$

The standard free energy change for this reaction as written above is given by the standard free energy of formation, $\Delta G_f^{\,0}$, of Cu_2O because 1 mole of Cu_2O is formed from its elements. Tabulated values of $\Delta G_f^{\,0}$ exist as a function of temperature for various oxides [5, 6]. For instance, at 1,000 K, the free energy of formation of Cu_2O is −95.52 kJ/mole Cu_2O, so that at 1,000 K the oxidation reaction in Eq. (1) is spontaneous because the free energy change is negative. To compare the thermodynamic behavior of various oxides, it is useful to calculate their free energy change per mole of O_2. Thus, Eq. (1) can be rewritten as

E. McCafferty, *Introduction to Corrosion Science*, DOI 10.1007/978-1-4419-0455-3_15,

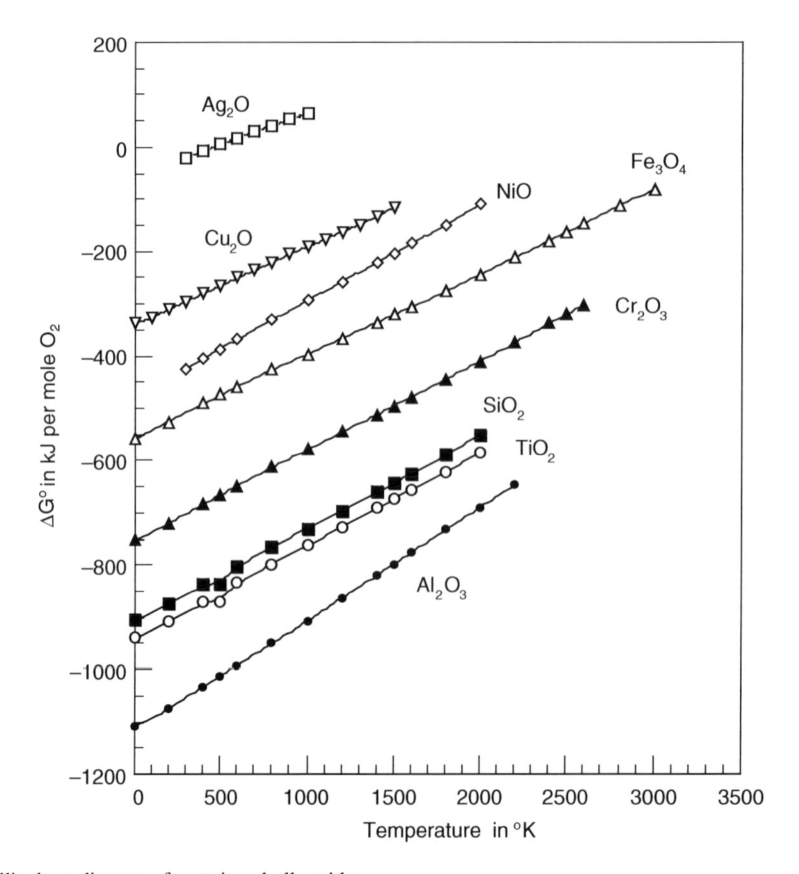

Fig. 15.1 Ellingham diagrams for various bulk oxides

$$4\,Cu(s) + O_2\,(g) \rightarrow 2\,Cu_2O\,(s) \tag{2}$$

and the free energy change for Eq. (2) at 1,000 K is 2(−95.52 kJ/mole Cu_2O) = −191.04 kJ/mole O_2.

Figure 15.1 shows the free energy change of Cu_2O in Eq. (2) as a function of temperature. The oxidation reaction is spontaneous at all temperatures shown, as indicated by the negative values for the change in free energy. The slope of the line of ΔG^0 vs. T gives the standard entropy change ΔS^0 from Eq. (8) in Chapter 4. That is

$$\Delta G^0 = \Delta H^0 - T\,\Delta S^0 \tag{3}$$

so that $d\Delta G^0/dT = -\Delta S^0$. The slope of ΔG^0 vs. T has a positive value, so that ΔS^0 is negative for the oxidation reaction. This means that the entropy decreases and the degree of order in the metal/oxygen system increases for the formation of a solid oxide from the metal and O_2 gas.

Ellingham Diagrams

Figure 15.1 also shows standard free energy changes for other oxides, with each case being calculated on the basis of 1 mole of O_2. For example, the oxidation of iron to Fe_3O_4 is written as

$$\frac{3}{2}\,\mathrm{Fe(s)} + \mathrm{O_2\,(g)} \rightarrow \frac{1}{2}\,\mathrm{Fe_3O_4\,(s)} \tag{4}$$

It can be seen that of the oxides given in Fig. 15.1, Al_2O_3 is the most stable because it has the largest decrease in free energy (at any given temperature) and Ag_2O is the least stable because it has the smallest decrease in free energy (at any given temperature). In fact, the oxidation of silver to Ag_2O is not spontaneous above 450 K where ΔG^0 for the oxidation reaction becomes positive.

Figure 15.1 also shows that the oxides of Fe and Ni (which are major components of engineering alloys) are less stable than the oxides of Cr, Al, and Si, which are often added as alloying elements for protection against high-temperature oxidation.

Diagrams as in Fig. 15.1 are called *Ellingham diagrams* [7, 8], and they have the same limitations as do Pourbaix diagrams in aqueous corrosion. That is, they contain information about the tendency of reactions to occur but do not take into account the kinetics of reactions. Ellingham diagrams can also be constructed for high-temperature reactions which form sulfides or carbides.

Although not shown in Fig. 15.1, there can be changes in the slope of a given curve if there is a phase change due to melting or a change in the structure of the oxide.

Equilibrium Pressure of Oxygen

The pressure of oxygen in equilibrium with solid can be calculated from an expression given in Chapter 4. That is

$$\Delta G^0 = -RT\,\mathrm{In}\,K \tag{5}$$

For Eq. (2), for example,

$$K = \frac{a^2\,(\mathrm{Cu_2O})}{a^4\,(\mathrm{Cu})\;P\,(\mathrm{O_2})} \tag{6}$$

where $a(Cu_2O)$ and $a(Cu)$ are the activities of Cu_2O and Cu, respectively, and $P(O_2)$ is the pressure of O_2 gas. But recall from Chapter 4 that the activities of solids are unity, so that Eq. (6) becomes

$$K = \frac{1}{P\,(\mathrm{O_2})} \tag{7}$$

and Eq. (7) then becomes

$$\Delta G^0 = RT\,\mathrm{In}\,P(\mathrm{O_2}) \tag{8}$$

Using $\Delta G^0 = -191.04$ kJ/mole in Eq. (8) gives $P(O_2) = 1.1 \times 10^{-10}$ atm at 1,000 K. At 1,000 K, pressures of O_2 below this value will not be sufficient to oxidize copper, whereas pressures greater than 1.1×10^{-10} atm will support oxidation.

Note that Eq. (5) can also be applied to Eq. (1), in which case a different K would be written, but the calculated value of $P(O_2)$ is the same as from Eq. (1); see Problem 15.2.

Table 15.1 gives equilibrium oxygen pressures for the oxidation of various metals at different temperatures. For each metal, the equilibrium pressure needed to support oxidation increases with temperature. All the equilibrium pressures are very low, except in the case of silver. For the oxidation of silver at 400 K, an oxygen pressure of 0.10 atm is required for oxidation; and this value is less than the partial pressure of oxygen in air (0.2) atm so that the oxidation of silver in air is thermodynamically favored in air at 400 K. However, at 500 K an oxygen pressure of 3.3 atm is required for oxidation so that the oxidation of silver in air is not thermodynamically favored at 500 K (or at higher temperatures). (Note that this argument pertains only to the tendency of oxidation to occur and not to the rate of oxidation.) There is a thermodynamic tendency for all the other metals in Table 15.1 to oxidize in air because the required oxygen pressures are very low.

Table 15.1 Equilibrium pressures of oxygen required for the oxidation of various metals at different temperatures

	O_2 pressure in atmospheres				
Oxidation reaction	400 K	500 K	1,000 K	1,500 K	2,000 K
$2Ag + 1/2O_2 \rightarrow Ag_2O$	0.10	3.3	2,320	–[a]	–[a]
$2Cu + 1/2O_2 \rightarrow Cu_2O$	2.6×10^{-37}	2.1×10^{-28}	1.1×10^{-10}	1.0×10^{-4}	–[a]
$Ni + \frac{1}{2} O_2 \rightarrow NiO$	–[b]	4.6×10^{-41}	4.6×10^{-16}	8.5×10^{-8}	1.4×10^{-3}
$3Fe + 2O_2 \rightarrow Fe_3O_4$	–[b]	–[b]	2.0×10^{-21}	6.8×10^{-12}	4.3×10^{-7}
$2Cr + 3/2O_2 \rightarrow Cr_2O_3$	–[b]	–[b]	5.4×10^{-31}	5.8×10^{-18}	1.9×10^{-11}
$Si + O_2 \rightarrow SiO_2$	–[b]	–[b]	7.2×10^{-39}	3.9×10^{-23}	4.5×10^{-15}
$Ti + O_2 \rightarrow TiO_2$	–[b]	–[b]	1.5×10^{-40}	3.4×10^{-24}	5.1×10^{-16}
$2Al + 3/2O_2 \rightarrow Al_2O_3$	–[b]	–[b]	–[b]	1.7×10^{-28}	9.9×10^{-19}

[a]The oxide is molten at this temperature.
[b]The value of the O_2 pressure is very low and is lower than in the entry to the right in the table.

Theory of High-Temperature Oxidation

The initial stages of oxidation involve the chemisorption of oxygen as O^{-2} ions on a clean metal surface to generate an electrical field at the surface. This electrical field then promotes the intrusion of oxidized metal atoms (cations) into the plane of adsorbed oxygen ions to produce a two-dimensional structure, which grows into an oxide film. These processes have been studied in detail at lower temperatures [4, 9] but at higher temperatures the initial stages of oxidation proceed very rapidly, so that it is the continued growth of the oxide film which is of paramount importance.

The growth of thicker oxide layers is described by the Wagner theory of oxidation [1–4, 10]. According to this model, oxide layers grow both from the metal substrate outward and from the outer oxide layer inward. Metal atoms are oxidized to metal cations at the metal substrate, and the cations which are produced diffuse outward through the oxide film, along with a concomitant outward migration of electrons produced by the oxidation reaction. At the outermost surface of the oxide film, O_2 gas is reduced to O^{-2} ions, which then diffuse inward through the oxide film. Thus situation is depicted in Fig. 15.2.

For the growth of an oxide of composition MO, the reactions are thus:

$$M \rightarrow M^{2+} + 2e^- \quad \text{at the metal/oxide Interface}$$

$$1/2\,O_2 + 2e^- \rightarrow O^{2-} \quad \text{at the metal/oxide Interface}$$

$$M + 1/2\,O_2 \rightarrow MO \quad \text{(overall reaction)}$$

Thus, the oxide layer must possess both an ionic and electronic conductivity for the oxide film to grow. The oxide layer serves as (i) an ionic conductor, (ii) an electronic conductor, (iii) an electrode for the reduction of oxygen (at the outer surface of the oxide), and (iv) a diffusion barrier through which metal cations and oxygen anions diffuse.

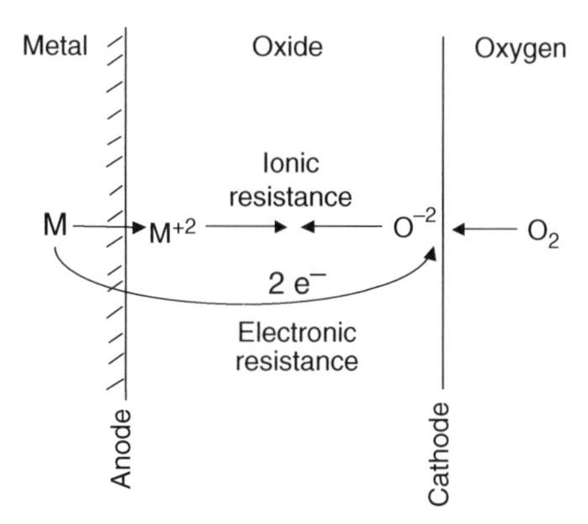

Fig. 15.2 The Wagner mechanism of high-temperature oxidation [10]. The oxide layer contains both an ionic and electronic resistance

The electronic resistivity of the oxide film is much higher than its ionic resistivity so that the ionic path is usually rate determining. Cations and anions do not diffuse with equal ease, and the diffusion of one or the other may be rate controlling. For the common base metals Fe, Ni, Cu, Cr, and Co, the rate-controlling step is the diffusion of cations outward but for the refractory metals Ta, Nb, Hf, Ti, and Zr, the rate-controlling step is the diffusion of O^{2-} anions inward [11]

Because the oxide layer grows by movement of ions (not atoms) through the film, the oxide film may be considered to be a solid electrolyte. This raises an interesting comparison between high-temperature oxidation and aqueous corrosion, as shown in Table 15.2.

The Wagner mechanism is not only applicable for high temperatures but also for lower temperature processes, such as the growth of rust layers in the natural atmosphere; see Problem 15.4.

Table 15.2 Comparison between electrochemical aspects of high temperature oxidation and aqueous corrosion [11]

	Method of corrosion control	
	Avoid contact between fixed anodes and fixed cathodes in a galvanic cell	Increase electrolyte resistance
Aqueous corrosion	Often possible, e.g., by inserting an inert barrier between anode and cathode	Often impossible because of fixed conditions
High-temperature oxidation	Impossible because metal and oxide are in contact	Possible, by alloying to reduce the ionic diffusivity

Oxidation Rate Laws

The kinetics of high temperature oxidation are usually studied in experiments which measure the gain in weight of a metal sample when an oxide is formed on the metal sample. The gain in weight is proportional to the thickness of the oxide which is formed. These measurements are made continuously using a microbalance assembly, with the sample placed in a furnace fitted with a thermocouple and temperature control.

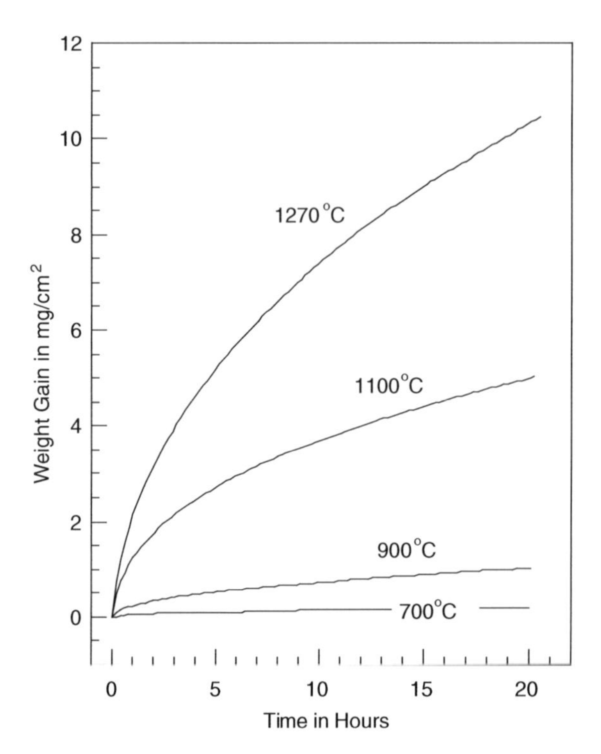

Fig. 15.3 Oxidation of high-purity nickel in oxygen at several different temperatures. Redrawn from Ref. [12] by permission of ECS – The Electrochemical Society

Figure 15.3 shows kinetic data [12] for the oxidation of annealed commercial purity nickel in oxygen at several different temperatures. At a given time, the total weight gain increases with temperature. Also, at a given time, the instantaneous rate of oxidation, as measured by the slope of the curve, increases with temperature.

There are three main laws which describe the kinetics of high-temperature oxidation. These are the (i) linear, (ii) parabolic, and (iii) logarithmic rate laws.

Linear Rate Law

In the linear rate law, the rate of oxidation is constant:

$$\frac{\mathrm{d}y}{\mathrm{d}t} = k \tag{9}$$

where y is the thickness of the oxide film. This equation holds when the reaction rate is constant at the metal/oxide interface, as, for example, when the gaseous reactant reaches the metal substrate through, cracks or pores in the oxide film. Equation (9) can be rewritten as $\mathrm{d}y = kt$, and integrated to give

$$y = kt + \mathrm{const} \tag{10}$$

where "const" is the constant of integration and can be determined from the initial conditions. If a clean metal surface is used in the rate study, then $y = 0$ at $t = 0$, so that const = 0. Otherwise, const is the initial thickness of the oxide film prior to high-temperature oxidation.

Metals which obey the linear rate law do not form protective oxides. An example is the oxidation of niobium in air at 1 atm between 600 and 1,200°C [13].

The oxide thickness in Eq. (10) can be replaced by the gain in weight per unit area, because the weight gain is proportional to the oxide thickness. (The constant in Eq. (10) will accordingly be different.) See Example 15.1. The same is true for the other two rate laws which follow.

Example 15.1: Show that for the linear rate law, the gain in weight per unit area, w, has the same form as in Eq. (10).

Solution: The weight gain per unit area, w, is related to the oxide thickness y by

$$w = \rho y \left(\frac{\text{g}}{\text{cm}^2} = \frac{\text{g}}{\text{cm}^3} \times \text{cm}\right)$$

where ρ is the oxide density. Thus, $y = w/\rho$, and substitution into Eq. (9) gives

$$\frac{1}{\rho}\frac{\text{d}w}{\text{d}t} = k$$

or

$$\frac{\text{d}w}{\text{d}t} = k\rho = k'$$

Thus, $\text{d}w = k'\text{d}t$, and integration gives

$$w = k't + \text{const}$$

which has the same form as Eq. (10).

Parabolic Rate Law

In the parabolic rate law, the diffusion of ions or electrons through the oxide is rate controlling, as illustrated in Fig. 15.2. The rate of oxidation is inversely proportional to the oxide thickness:

$$\frac{\text{d}y}{\text{d}t} = \frac{k'}{y} \tag{11}$$

Rewriting Eq. (11) as $y\text{d}y = k'\text{d}t$ and integrating gives

$$y^2 = 2k't + \text{const} = k_\text{p}t + \text{const} \tag{12}$$

where k_p is the parabolic rate constant. This rate law holds for protective oxides and is applicable to the high-temperature oxidation of many metals, including copper, nickel, iron, chromium, and cobalt [14]. The data in Fig. 15.3 follow the parabolic rate law.

Logarithmic Rate Law

At lower temperatures and for oxide films of 1,000 Å or less, the oxidation rate often follows a logarithmic law:

$$y = k'' \log(ct + 1) \tag{13}$$

where c is a constant to be determined from the experimental data. This rate law is semi-empirical, but is generally thought to result if the rate of oxidation is controlled by the transfer of electrons across the oxide film. The transfer of electrons can be rate controlling because of the thinner oxide films involved. The logarithmic rate law is obeyed for the initial oxidation of many metals, including copper, iron, zinc, nickel, and others [14].

Comparison of Rate Laws

The three main rate laws for oxidation are compared in Table 15.3. These three rate laws are also compared in Fig. 15.4, which is plotted with arbitrary values of $k = k' = k'' = 1$, $c = 100$, and the constants of integration taken as zero. Figure 15.4 shows that the linear rate law leads to a non-protective oxide whereas the parabolic and logarithmic rate laws pertain to protective oxides in which the rate of growth (and hence the rate of weight gain) decreases with time. The oxide film approaches a limiting thickness in the case of the logarithmic rate law.

Table 15.3 The three main rate laws for high-temperature oxidation

Rate law	Equation	Test for fit
Linear	$y = kt + \text{const}$	y vs. t is linear
Parabolic	$y^2 = 2k't + \text{const}$	y^2 vs. t is linear
Logarithmic	$y = k'' \log(ct + 1)$	y vs. $\log t$ is linear (for $ct > 1$)

The Wagner Mechanism and the Parabolic Rate Law

The Wagner mechanism of oxidation shown in Fig. 15.2 gives rise to a parabolic rate law [15–17]. This rate law results from assuming that the growth of the oxide film is proportional to a current I which flows through the oxide film:

$$\frac{dy}{dt} = BI \tag{14}$$

where B is a proportionality constant given by

$$B = \frac{M}{A\, d_{oxide} F} \tag{15}$$

where M is the gram-equivalent weight of the oxide, A is the area of the oxide, d_{oxide} the oxide density, and F is the Faraday. (B has the units of cm/C, and accordingly, dy/dt has the units of cm/s.) The corrosion current is given by

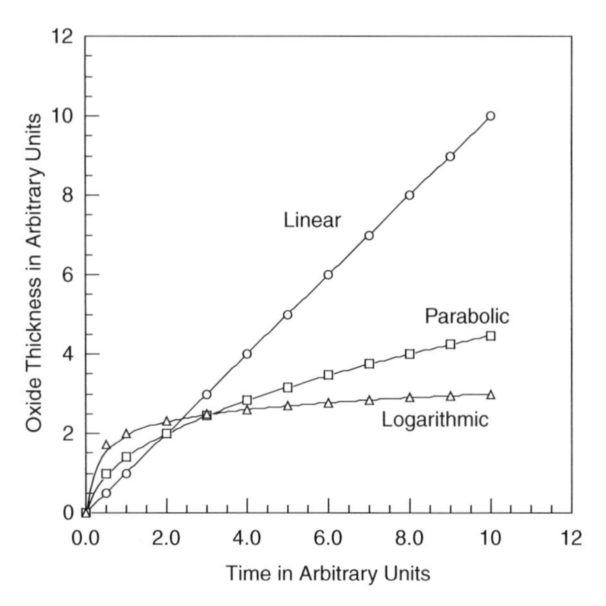

Fig. 15.4 Illustration of the three main rate laws. The linear rate law is plotted with $k = 1$ in Eq. (10) and the parabolic rate law is plotted with $k' = 1$ in Eq. (12), with the constants of integration taken to be zero in each case. The logarithmic rate law is plotted with $k'' = 1$ and $c = 100$ in Eq. (13)

$$I = \frac{E}{R} \tag{16}$$

where E is the emf across the oxide electrolyte and R is the oxide resistance. The emf across the oxide film can be determined from the free energy change ΔE for the oxidation reaction and the relationship $\Delta G = -nFE$. The ionic resistance and electronic resistance of the oxide layer exist in series because both must operate for oxidation to occur. Thus, the total resistance R of the oxide is given by

$$R = \left(\frac{1}{\kappa_{\text{ionic}}} + \frac{1}{\kappa_{\text{elecronic}}}\right)\frac{y}{A} \tag{17}$$

where κ_{ionic} and $\kappa_{\text{electronic}}$ are the ionic and electronic oxide conductivities, respectively. Equation (17) can be rewritten as

$$R = \left(\frac{\tau_a + \tau_c + \tau_e}{\tau_a + \tau_c}\frac{1}{\kappa} + \frac{\tau_a + \tau_c + \tau_e}{\tau_e}\frac{1}{\kappa}\right)\frac{y}{A} \tag{18}$$

where κ is the conductivity of the oxide and τ_a, τ_c, and τ_e are the transference numbers for the anion, cation, and electron, respectively. (The transference number is the fraction of current carried by a given species.) Then, since $\tau_a + \tau_c + \tau_e = 1$, Eq. (18) can be rewritten as

$$R = \left(\frac{1}{(\tau_a + \tau_c)\ \tau_e}\right)\frac{y}{\kappa A} \tag{19}$$

Equations (14), (15), (16), and (19) combine to give

$$\frac{\mathrm{d}y}{\mathrm{d}t} = \frac{M}{d_{\text{oxide}} F}\left[(\tau_a + \tau_c)\ \tau_e\,\kappa E\right]\frac{1}{y} \tag{20}$$

or

$$y\,dy = \frac{M}{d_{\text{oxide}}F}\,[(\tau_a + \tau_c)\ \tau_e\,\kappa E]\ dt \tag{21}$$

After integration,

$$y^2 = \frac{2M}{d_{\text{oxide}}F}\,[(\tau_a + \tau_c)\ \tau_e\,\kappa E]\ t + \text{const} \tag{22}$$

which has the form of a parabolic expression:

$$y^2 = k_p t + \text{const} \tag{23}$$

where

$$k_p = \frac{2M}{d_{\text{oxide}}F}\,[(\tau_a + \tau_c)\ \tau_e\,\kappa E] \tag{24}$$

The parabolic rate constant can also be expressed as a normalized rate constant by dividing y^2 by $2M/d_{\text{oxide}}$. Then Eq. (22) becomes

$$\frac{y^2}{\frac{2M}{d_{\text{oxide}}}} = \frac{1}{F}\,[(\tau_a + \tau_c)\ \tau_e\,\kappa E]\ t + \text{const}' \tag{25}$$

or

$$\frac{y^2}{\frac{2M}{d_{\text{oxide}}}} = k_{\text{rational}}\ t + \text{const}' \tag{26}$$

where k_{rational} is called the rational rate constant and is given by

$$k_{\text{rational}} = \frac{1}{F}\,[(\tau_a + \tau_c)\ \tau_e\,\kappa E] \tag{27}$$

The rational rate constant has the units equiv $\text{cm}^{-1}\ \text{s}^{-1}$.

Equations (24) or (27) provide a means for calculating parabolic rate constants from oxide parameters. Some calculated values compiled by Kubachewski and Hopkins [1] are listed in Table 15.4.

Table 15.4 Calculated and measured rational rate constants for high-temperature oxidation [1]

Metal	Oxidant	Temperature in °C	Rational rate constant (equiv/cm/s)	
			Calculated	Experimental
Cu	I_2 (gas)	195	3.8×10^{-10}	3.4×10^{-10}
Ag	Br_2 (gas)	200	2.7×10^{-11}	3.8×10^{-11}
Cu	O_2, $P = 8.3 \times 10^{-2}$ atm	1,000	6.6×10^{-9}	6.2×10^{-9}
Cu	O_2, $P = 3.0 \times 10^{-4}$ atm	1,000	2.1×10^{-9}	2.2×10^{-9}
Co	O_2, $P = 1$ atm	1,000	1.25×10^{-9}	1.05×10^{-9}
Co	O_2, $P = 1$ atm	1,350	3.15×10^{-8}	3.65×10^{-8}

The close agreement between calculated and experimental values for the rational rate constant lends support to the Wagner mechanism of oxidation.

Effect of Temperature on the Oxidation Rate

The oxidation rate increases with temperature according to the Arrhenius equation:

$$k = A' \exp^{-\frac{\Delta E_{\text{act}}}{RT}} \tag{28}$$

where k is the rate constant, R is the gas constant, T is the temperature in degrees Kelvin, and A' is a constant. The quantity ΔE_{act} is the activation energy and is the energy barrier which must be overcome for the reaction to occur. Taking logarithms and differentiating gives

$$\frac{d \log k}{d\left(\frac{1}{T}\right)} = -\frac{\Delta E_{\text{act}}}{2.303\,R} \tag{29}$$

The activation energy ΔE_{act} can be determined from the slope of a plot of log k vs. $1/T$. The integrated form of Eq. (29) is

$$\log \frac{k_2}{k_2} = \frac{\Delta E_{\text{act}}}{2.303\,R}\left(\frac{1}{T_1} - \frac{1}{T_2}\right) \tag{30}$$

Equation (29) also holds for lower temperature processes. (Recall the free energy barrier $\Delta G^{\neq}$ for aqueous corrosion processes in Chapter 7.)

Defect Nature of Oxides

Oxides are not perfect crystals but often possess various types of defects. Two of the most important defects in regard to high-temperature oxidation are point defects due to vacancies and interstitials. In an oxide, vacancies are missing ions and interstitials are extra ions which exist in the crystal lattice. There can be cationic and anionic vacancies and cationic and anionic interstitials. Figure 15.5 illustrates an interstitial cation and an anion vacancy. (The combination of an interstitial plus a vacancy of the same component is called a *Frenkel defect*. The combination of a cation vacancy and an anion vacancy is called a *Schottky defect.*)

Vacancies are important in high-temperature oxidation because they provide a means for diffusion of cations or anions across an oxide film, as in the Wagner mechanism of oxidation. Diffusion occurs by a vacancy mechanism if an ion in a normal lattice position jumps to an adjacent unoccupied position (i.e., a vacancy). This is illustrated schematically in Fig. 15.6a. This movement or jump of an ion in turn creates a vacancy in its wake, so that the process can be continued. If instead the point defect is an interstitial ion, then diffusion occurs by a slightly different mechanism in which an interstitial ion moves from one interstitial position to an adjacent interstitial position, as shown in Fig. 15.6b.

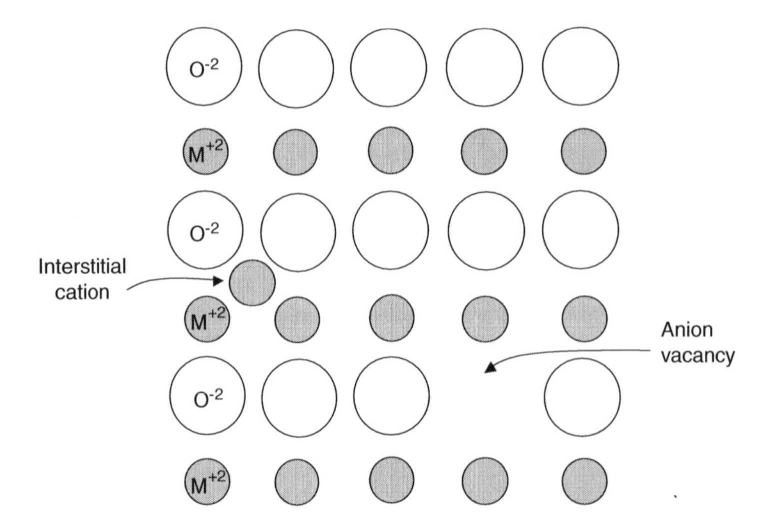

Fig. 15.5 Illustration of an interstitial cation and an anion vacancy

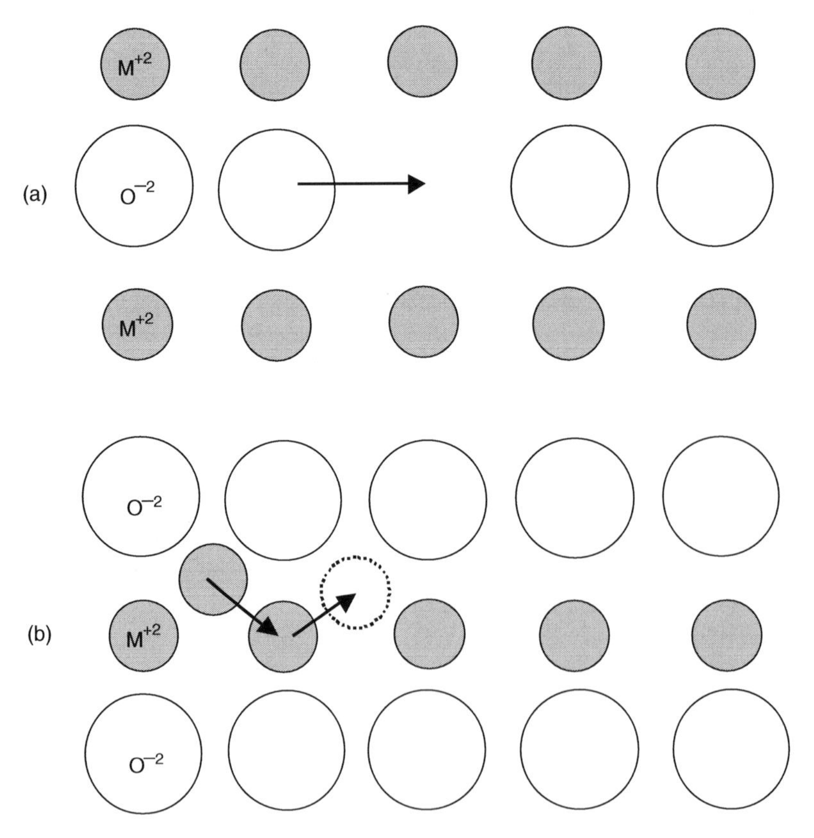

Fig. 15.6 Schematic illustration of ionic diffusion by (**a**) anion diffusion by means of vacancies and (**b**) cation diffusion by movement of interstitial cations

Semiconductor Nature of Oxides

The existence of point defects gives rise to the semiconductor nature of oxides. for example, suppose that ZnO contains interstitial Zn^{2+} ions, as illustrated in Fig. 15.7. Then for charge neutrality, two extra electrons must be added to the oxide structure (from the metal) for each interstitial Zn^{2+} cation. Thus, the current in the oxide film is carried by electrons, so that ZnO is an n-type semiconductor. Ionic transport occurs by interstitial diffusion.

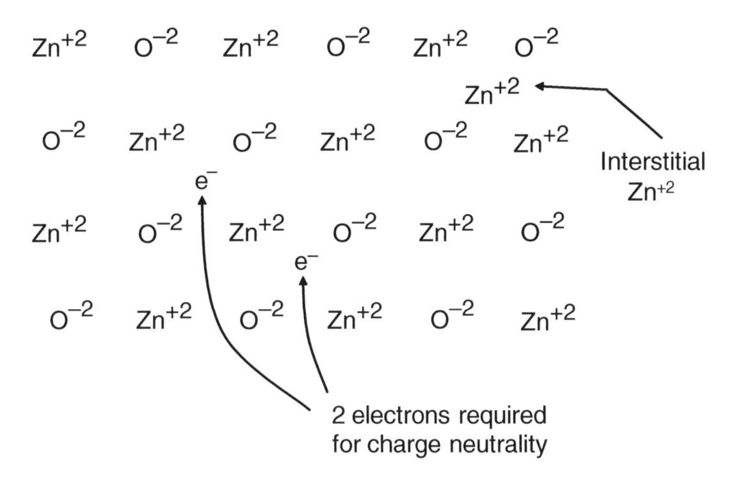

Fig. 15.7 Schematic illustration of an n-type oxide semiconductor containing interstitial cations

If instead the point defect is a cation vacancy, as illustrated in Fig. 15.8 for NiO, then the oxide contains a deficiency of positive charges (due to the missing Ni^{2+} ions). Then for charge neutrality, two electron holes must be added to the oxide structure for each Ni^{2+} vacancy. Thus, the current in the oxide film is carried by electron holes, so that NiO is a p-type semiconductor and ionic transport occurs by diffusion of vacancies.

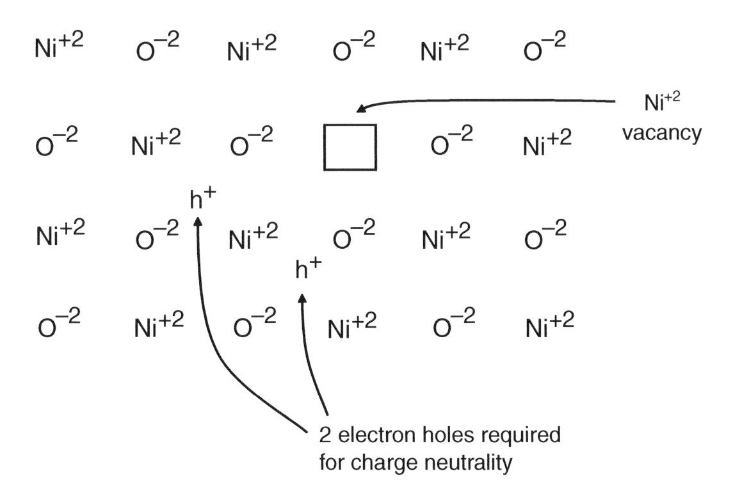

Fig. 15.8 Schematic illustration of a p-type oxide semiconductor containing cation vacancies

By similar arguments it can be seen that O^{2-} vacancies produce a deficiency of negative charges, so that electrons must be added for electroneutrality, thus producing an n-type semiconductor. Figure 15.9 shows schematically that oxygen vacancies can be formed at high temperatures by the loss of oxide ions to the gaseous phase.

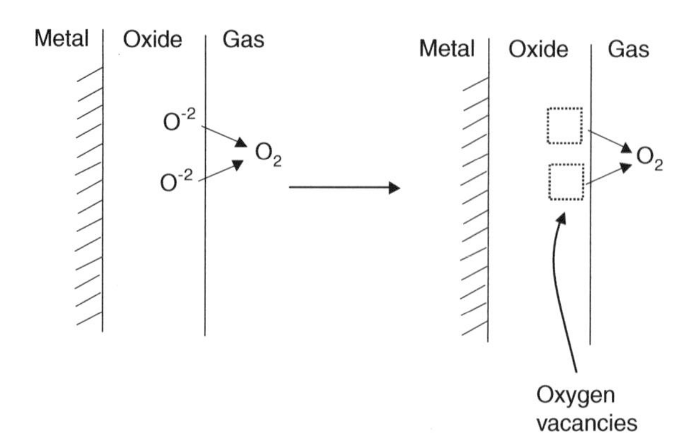

Fig. 15.9 Schematic illustration showing the loss of two oxide ions (O^{2-}) to the gas phase to form two oxygen vacancies in the oxide layer

The last possibility, i.e., interstitial anions, produces an excess of negative charges which are balanced by electron holes, thus giving a p-type semiconductor. However, examples of interstitial anions, e.g. O^{2-}, are not common due to the large size of the anion. These concepts are summarized in Table 15.5.

Table 15.5 Types of oxide semiconductors

Type of oxide semi-conductor	Defect	Requirement for charge neutrality	Charge carriers	Ionic transport	Examples [4, 18]
n-Type	Interstitial cations [extra (+) charge]	Add electrons	Electrons	Interstitial	ZnO, CdO
n-Type	Anion vacancies [missing (−) charge]	Add electrons	Electrons	Vacancies	Al_2O_3, TiO_2, Fe_2O_3, ZrO_2
p-Type	Cation vacancies [missing (+) charge]	Add holes	Holes	Vacancies	NiO, FeO, Cr_2O_3, Fe_3O_4
p-Type	Interstitial anions [extra (−) charge]	Add holes	Holes	Interstitial	UO_2

Hauffe Rules for Oxidation

The effect of various foreign ions on the rates of high-temperature oxidation can be explained on the basis of the semiconductor nature of the oxide layer. The effects of various solute ions on high-temperature oxidation rates are called the Hauffe rules for oxidation.

Consider an *n-type oxide film with interstitial cations*, such as ZnO with Zn^{2+} interstitial ions, as described earlier. The equilibrium which exists between ZnO and its interstitial cations is

$$Zn_i^{2+} + 2e^- + \frac{1}{2}\ O_2 \rightarrow ZnO \tag{31}$$

where Zn_i^{2+} represents an interstitial Zn^{2+} ion. By the law of mass action,

$$\frac{[ZnO]}{\left[Zn_i^{2+}\right]\left[e^-\right]^2 P(O_2)^{1/2}} = K \tag{32}$$

where $[Zn_i^{2+}]$ is the concentration of interstitial Zn^{2+} ions, $[e^-]$ is the concentration of electrons, [ZnO] is the concentration (activity) of ZnO, $P(O_2)$ is the partial pressure of oxygen, and K is the equilibrium constant.

The activity of solid ZnO is unity so that Eq. (32) becomes

$$\frac{1}{\left[Zn_i^{2+}\right]\left[e^-\right]^2 P(O_2)^{1/2}} = K \tag{33}$$

At a constant partial pressure of O_2, Eq. (33) gives

$$\left[Zn_i^{2+}\right]\left[e^-\right]^2 = \frac{1}{K\,P(O_2)^{1/2}} = K' \tag{34}$$

Suppose that one Zn^{2+} ion in the normal lattice is replaced with one Li^+ ion, as shown in Fig. 15.10. Then, for charge neutrality, one electron must be lost. Then as the concentration of electrons is decreased, the concentration of interstitial Zn^{2+} ions must increase, according to Eq. (34). But an increased concentration of interstitial cations means that the rate of interstitial diffusion will increase. This is because the rate of diffusion is proportional to the concentration gradient according to Fick's first law of diffusion, as given in Eq. (4) in Chapter 8. An increased rate of diffusion in turn yields an increased rate of oxidation, according to the Wagner theory of oxidation.

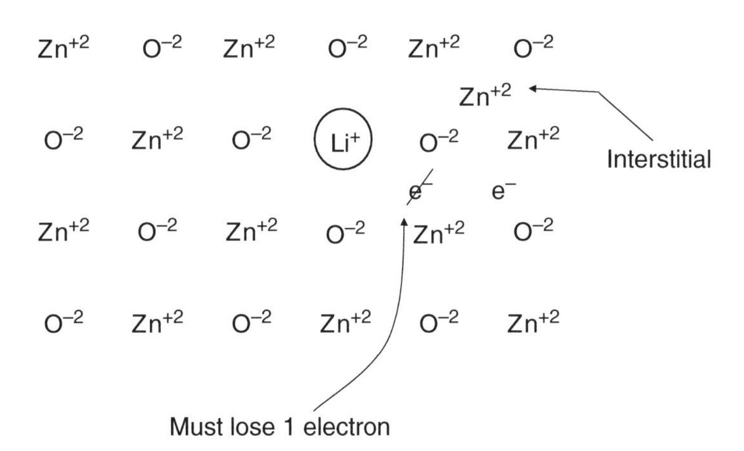

Fig. 15.10 Schematic illustration of a ZnO n-type semiconductor containing interstitial Zn^{2+} ions and in which a normal Zn^{2+} lattice ion is replaced by a lattice ion (Li^+) of a lower oxidation number

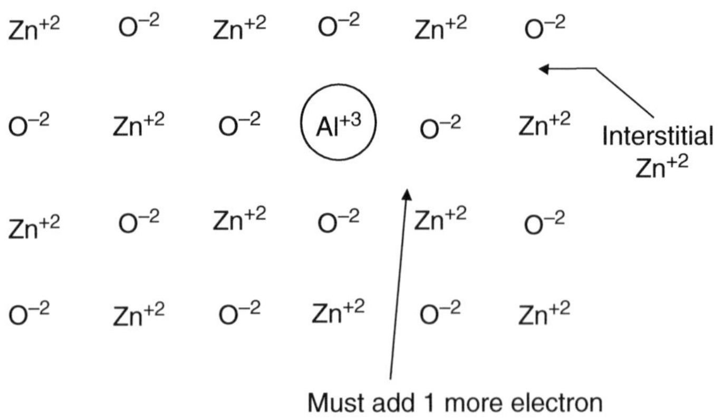

Fig. 15.11 Schematic illustration of a ZnO n-type semiconductor containing interstitial Zn^{2+} ions and in which a normal Zn^{2+} lattice ion is replaced by a lattice ion (Al^{3+}) of a higher oxidation number

If the ZnO film contains a solute ion of oxidation number greater than +2, e.g., Al^{3+} as shown in Fig. 15.11, then for charge neutrality one electron must be added for each Al^{3+} solute ion. According to Eq. (34), as the concentration of electrons increases, the concentration of interstitial Zn^{2+} ions decreases. This means that there is a decreased rate of ionic diffusion, and accordingly a decreased oxidation rate. See Table 15.6 for a compilation of these trends.

Table 15.6 Hauffe rules for oxidation

Type of oxide semiconductor	Defect	Example	Oxidation number of solute ion relative to oxide cation	Change in defect concentration and ionic diffusion	Change in oxidation rate
n-Type	Interstitial cations	ZnO	Lower	Increase	Increase
			Higher	Decrease	Decrease
	Anion vacancies	ZrO_2	Lower	Increase	Increase
			Higher	Decrease	Decrease
p-Type	Cation vacancies	NiO	Lower	Decrease	Decrease
			Higher	Increase	Increase
	Interstitial anions	UO_2	Lower	Decrease	Decrease
			Higher	Increase	Increase

If instead of interstitial cations, as in n-type semiconductors as described above, suppose that we consider a *p-type oxide semiconductor* (e.g., NiO) in which the point defects are *cationic vacancies.* The equilibrium which exists between the nickel ion vacancies and the NiO lattice is given by

$$\mathrm{NiO} + V_{\mathrm{Ni}}'' + 2\,h^{+} \rightarrow \frac{1}{2}\mathrm{O}_2 \tag{35}$$

where V_{Ni}'' represents a nickel ion vacancy of effective charge –2 (the prime refers to an effective negative charge) and h^{+} refers to an electron hole. By the law of mass action

$$\frac{P(\mathrm{O}_2)^{1/2}}{[\mathrm{NiO}]\ [V_{\mathrm{Ni}}'']\ [h^{+}]^2} = K \tag{36}$$

or

$$\left[V_{\mathrm{Ni}}''\right] \left[h^+\right]^2 = \frac{P(\mathrm{O}_2)^{1/2}}{K} = K' \tag{37}$$

at a constant partial pressure of oxygen.

Suppose next that one Ni^{2+} ion in the normal lattice position is replaced by one Li^+ ion, as shown schematically in Fig. 15.12. Then for charge neutrality, one electron must be lost, or equivalently, one electron hole must be added. As the concentration of electron holes increases, then the concentration of metal ion vacancies decreases, as per Eq. (37). This decrease produces a decrease in the rate of ionic diffusion and hence a decrease in the oxidation rate.

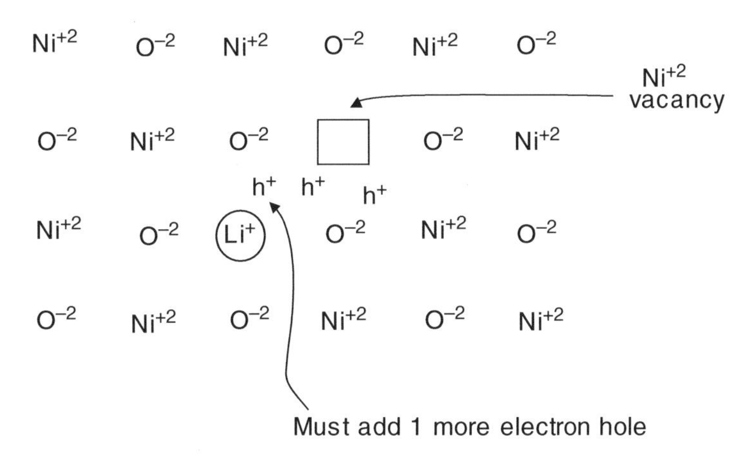

Fig. 15.12 Schematic illustration of a NiO p-type semiconductor containing oxygen vacancies and in which a normal Ni^{2+} lattice ion is replaced by a lattice ion (Li^+) of a lower oxidation number

If the NiO lattice contains a substitutional solute ion having an oxidation number higher than that of Ni^{2+}, e.g., Cr^{3+}, then by a similar argument as just given, for each Cr^{3+} ion substituted for a Ni^{2+} ion, one electron must be added (or one h^+ is lost). According to Eq. (37), a decrease in the concentration of electron holes yields an increase in the concentration of metal ion vacancies and hence an increase in the rate of ionic diffusion and an increase in the oxidation rate.

This trend has been observed for the oxidation of Cr–Ni alloys in which chromium is added progressively to nickel. Table 15.7 lists parabolic rate constants as compiled by Uhlig [14] for the high-temperature oxidation of Ni-Cr alloys. For dilute alloys, the increased addition of Cr to Ni results in an increase in the oxidation rate. This can be explained on the basis that increased alloying with Cr results in increased concentrations of Cr^{3+} in the oxide film which, according to the arguments above, leads to an increase in the oxidation rate. For higher alloying amounts of Cr, however, the oxidation rate decreases as the oxide becomes predominantly Cr_2O_3 instead of NiO. (Recall from the Ellingham diagram in Fig. 15.1 that Cr_2O_3 is more stable than NiO.)

So far we have considered only cation defects in regard to the Hauffe rules for oxidation. Consider next an *n-type oxide having anion vacancies*, such as ZrO_2. The equilibrium between an oxygen atom in the lattice and an oxygen vacancy is:

$$V_{\mathrm{O}}'' + 2e^- + \frac{1}{2}\mathrm{O}_2 \rightarrow \mathrm{O}_{\mathrm{lattice}} \tag{38}$$

Table 15.7 Oxidation of nickel alloyed with chromium at 1,000°C and 1 atm O_2 [14]

Wt% Cr	Parabolic rate constant ($g^2/cm^4/s^3$)
0	3.1×10^{-10}
0.3	14×10^{-10}
1.0	26×10^{-10}
3.0	31×10^{-10}
10.0	1.5×10^{-10}

where V_O'' represents an oxygen vacancy of effective charge +2. (The dots in V_O'' represent effective positive charges). Thus,

$$\frac{[O_{lattice}]}{\left[V_O''\right]\left[e^-\right]^2 P(O_2)^{1/2}} = K \tag{39}$$

or,

$$\left[V_O''\right]\left[e^-\right]^2 = \frac{1}{K\,P(O_2)^{1/2}} = K' \tag{40}$$

at a constant partial pressure of O_2.

If one Zr^{4+} ion in the normal lattice is replaced by an ion of lower oxidation state, say Al^{+3}, then one electron must be lost for charge neutrality; see Fig. 15.13. According to Eq. (40), a decrease in the concentration of electrons produces an increase in the concentration of oxygen vacancies, which in turn leads to an increase in the rate of diffusion and an increase in the oxidation rate. If Zr^{4+} ions are replaced by ions of a higher oxidation state, the opposite trend results.

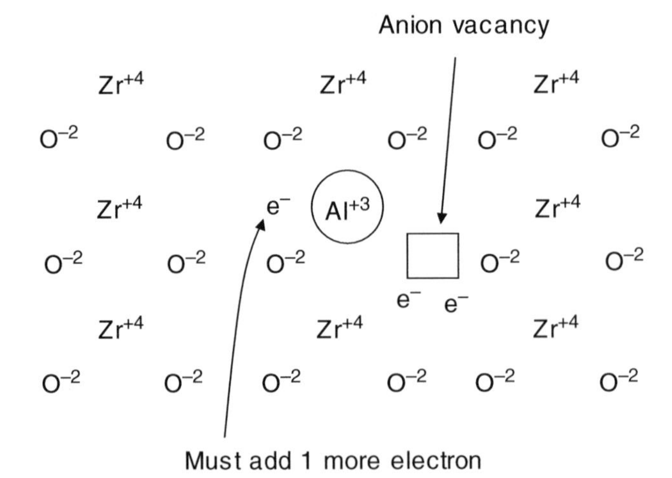

Fig. 15.13 Schematic illustration of a ZrO_2 n-type semiconductor containing oxygen vacancies and in which a normal Zr^{4+} lattice ion is replaced by a lattice ion (Al^{3+}) of a lower oxidation number

The last case to be considered is a *p-type oxide semiconductor with interstitial anions*. The equilibrium between an interstitial oxygen ion O_i^{2-} and the lattice is

$$O_i^{-2} + 2h^+ = \frac{1}{2}O_2 \tag{41}$$

The effect of solute ions is left as an exercise in Problem 15.11.

Table 15.6 compiles the effects of various solute ions on the high-temperature oxidation rates of the different types of oxide semiconductors.

Effect of Oxygen Pressure on Parabolic Rate Constants

The rate of parabolic oxidation is a function of the O_2 partial pressure. The general relationship, as will be seen below, is

$$k_p = A\,P\,(O_2)^{1/n} \tag{42}$$

where k_p is the parabolic rate constant and A is a constant, and n is negative for an n-type oxide and positive for a p-type oxide. The case of an n-type oxide containing interstitial cations is shown below.

For an n-type oxide MO with interstial cations, the equilibrium between the oxide and its interstitial cations is given by

$$M_i^{+2} + 2\,e^- + \frac{1}{2}\ O_2 \rightarrow MO \tag{43}$$

which is a generalization of Eq. (31). Thus,

$$\frac{1}{\left[M_i^{+2}\right]\left[e^-\right]^2\ P(O_2)^{1/2}} = K \tag{44}$$

But $[M_i^{+2}] = 2\,[e^-]$, so that substituting for $[e^-]$ into Eq. (44) gives the result

$$\left[M_i^{+2}\right] = K'\,P(O_2)^{-1/6} \tag{45}$$

For n-type oxides with interstitial cations, the rate of ionic diffusion is proportional to the concentration of interstitial cations (see Table 15.5). The rate of oxidation is in turn proportional to the rate of diffusion, so that the result is

$$k_p\,\alpha\,P(O_2)^{-1/6} \tag{46}$$

If an electron associates itself with an interstitial cation, then Eq. (43) is modified to

$$M_i^{+} + e^- + \frac{1}{2}\ O_2 \rightarrow MO \tag{47}$$

and by a similar argument it can be shown that

$$k_p\,\alpha\,P(O_2)^{-1/4} \tag{48}$$

Similar arguments can be made for n-type oxides with anion vacancies by using Eq. (38), for p-type oxides with cation vacancies by using Eq. (35), and for p-type oxides with interstitial anions by using Eq. (41). The results are given in Table 15.8.

Table 15.8 Effect of oxygen pressure on the parabolic rate constant

Type of oxide semiconductor	Equilibrium	Mechanism of diffusion	Dependence of parabolic rate constant on $P(O_2)$
n-Type with interstitial cations	Eq. (31) or (43)	Interstitial cations	$P(O_2)^{-1/6}$ or $P(O_2)^{-1/4}$
n-Type with anion vacancies	Eq. (38)	Anion vacancies	$P(O_2)^{-1/6}$
p-Type with cation vacancies	Eq. (35)	Cation vacancies	$P(O_2)^{1/6}$ or $P(O_2)^{1/4}$
p-Type with interstitial anions	Eq. (41)	Interstitial anions	$P(O_2)^{1/6}$

Non-uniformity of Oxide Films

Oxide films formed at high temperature are often not uniformly homogeneous in composition or structure. (That is, the solid electrolyte is not always homogeneous as are the usual electrolytes in aqueous corrosion.) An example is provided by the oxidation of iron, which possesses multiple oxidation states. As shown in Fig. 15.14, the oxide film consists of three distinct layers, FeO, Fe_3O_4, and Fe_2O_3 [11]. The phase richest in iron (FeO) exists nearest the metal surface and the phase richest in oxygen (Fe_2O_3) exists nearest the source of oxygen. The various voids in the oxide phase in Fig. 15.13 are attributed to the collapse of cation vacancies.

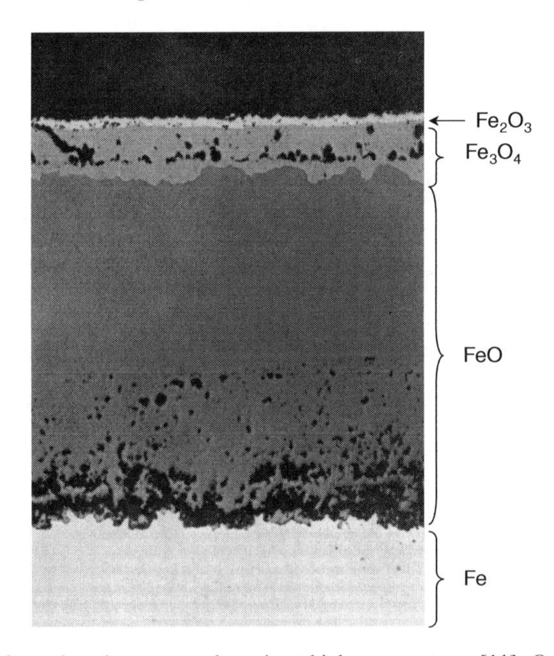

Fig. 15.14 Oxide layers formed on iron exposed to air at high temperatures [11]. Original photomicrograph by C.T. Fujii of the Naval Research Laboratory, Washington, DC and reproduced by permission of The McGraw-Hill Companies

Voids have also been observed in other instances, such as in the high-temperature oxidation of niobium [11]. (Nb_2O_5 has a Pilling–Bedworth ratio of 2.66.) These voids result in a non-protective oxide in which oxygen can diffuse through the voids to locations closer to the metal surface where the oxygen reduction reaction can occur. Thus, the effective thickness of the oxide film is reduced by the presence of the voids.

Protective vs. Non-protective Oxides

Based on the various factors which affect high temperature oxidation and on the properties of oxides, several generalizations may be drawn regarding the protective nature of oxide films.

Pilling–Bedworth Ratio

The Pilling–Bedworth ratio R given by

$$R = \frac{\text{Molecular volume of oxide}}{\text{Molecular volume of metal}} \quad (49)$$

as was discussed in Chapter 9 on passivity. It was seen that the Pilling–Bedworth ratio is one factor in determining whether or not an oxide forms a protective passive film. These considerations especially apply to high-temperature oxidation because the thickness of the oxide layer is much greater than that for ambient temperatures. This means that Pilling–Bedworth ratios taken from values for bulk oxides are more reliable for high-temperature oxide layers than for the thinner passive oxide films formed at lower temperatures.

A value of $R < 1$ once again means that the metal produces insufficient oxide to cover the metal surface, and the oxide is not protective. A value of R equal to or slightly greater than 1 is the optimal Pilling–Bedworth ratio for producing a protective oxide film. Values much greater than 1 introduce compressive strengths which lead to fracture and spalling of the oxide layer. However, the Pilling–Bedworth ratio is only one factor in determining the oxidation resistance of metals at high temperatures, as seen below.

Properties of Protective High-Temperature Oxides

To be protective against high-temperature oxidation, an oxide should have the following properties [11]:

(1) a coefficient of expansion approximately equal to that of the metal substrate,
(2) a good adherence to the metal substrate,
(3) a high melting point,
(4) a low vapor pressure,
(5) good high-temperature plasticity so as to resist fracture,
(6) low electrical conductivity or low diffusion coefficients for metal cations or oxygen anions,
(7) a Pilling–Bedworth ratio of approximately 1.

The actual behavior of metals and alloys must be measured experimentally. As in the case of passive films at ambient temperatures, it is not possible to predict the exact behavior of alloys from bulk oxide properties. Alloying additions of Cr, Al, and Si, however, have been generally found to be effective in forming high-temperature protective oxides.

Most high-temperature engineering alloys form chromia (Cr_2O_3) or alumina (Al_2O_3) scales. Fe-Cr, Ni-Cr, and Co-Cr alloys exhibit low oxidation rates when the Cr concentration is 15–30 wt.% [4]. Alloy 625 is another example of an alloy which provides a chromia-forming oxide. Alloy 625 has a nominal composition of 60 Ni-20 Cr-10 Mo-5 Fe (Table 9.11) and is used frequently in aqueous corrosion applications but was originally developed as an alloy for high-temperature applications. Alumina-forming oxides arise from alloys containing 5–7 wt.% Al and greater than 10 wt.% Cr [4]. A compilation of various high-temperature alloys is given by Khanna [4].

Problems

1. Based on thermodynamic considerations alone, what would be the effect on the stability of the oxide film formed by high-temperature oxidation if iron was alloyed with aluminum?
2. For the following reaction:

$$4\text{Cu (s)} + O_2 \text{ (g)} \rightarrow 2\,Cu_2O \text{ (s)}$$

   (a) Write an expression for the equilibrium constant.
   (b) Calculate the equilibrium pressure of O_2 from the equilibrium constant given that ΔG^0 is –191.04 kJ/mole at 1,000 K for the reaction as written above.
   (c) How does this pressure of O_2 compare with that calculated in the text for the reaction:

$$2\text{Cu (s)} + \frac{1}{2}\,O_2 \text{ (g)} \rightarrow Cu_2O \text{ (s)}$$

3. The free energy of formation of MgO is –469.888 kJ/mole at 1,200 K. Calculate the equilibrium pressure of O_2 required for the oxidation of Mg to MgO at this temperature. Is oxidation possible in air at this temperature?
4. In a study on the long-term growth of rust layers formed on a high tensile steel in the natural atmosphere, Horton et al. [19] found that atmospheric dusts become deposited on the air surface of the rust and then become enveloped within the rust layer. What does this observation suggest about the mechanism of growth of rust layers at ambient temperatures in the natural atmosphere?
5. (a) Show that for logarithmic oxidation, the rate of oxide growth dy/dt is inversely proportional to the time for large reaction times. (b) Then show that the oxidation rate approaches 0 for a large reaction time.
6. Smeltzer et al. [20] found that the oxidation rate for Ni–Al alloys at 1,298 K increases with an increase in the amount of Al (between 0.1 and 3.0 wt.%).

   (a) Is this increase in the oxidation rate for Ni-Al alloys with increasing Al content consistent with the thermodynamic trends in the Ellingham diagram?
   (b) Is this increase in oxidation rate consistent with the Hauffe rules for oxidation?

7. The parabolic rate constant in the preceding problem was determined to be 3.5×10^{-10} $g^2\ cm^{-4}\ s^{-1}$ for a Ni-0.1 wt% Al alloy at 1,298 K.

   (a) Calculate the weight gain at 1,298 K after 20 h of oxidation.
   (b) Calculate the oxide thickness (in μm) after 20 h oxidation assuming that the oxide is predominantly NiO having a density of 6.67 g/cm^3.

8. Lithium-aluminum alloys are used in aerospace applications where a reduction in weight is desired. On the basis of the Hauffe rules for oxidation, what would be the effect of solute Li^+ ions on the parabolic oxidation rate of aluminum?

9. The following data are taken from a study [21] on a Co-10% Cr alloy oxidized in 100 torr O_2 at 1,100°C. Which one of the three major rate laws best fits these data?

Exposure time (h)	Weight gain (mg/cm^2)
1.0	14.0
2.0	20.1
3.0	24.9
4.0	28.7

10. In a study [21] on the high-temperature oxidation of cobalt, the activation energy was determined to be 146 kJ/mole for the oxidation of pure cobalt at 100 torr O_2 pressure. The parabolic rate constant at 950°C was determined to be 25 mg^2 cm^{-4} h^{-1}. What is the weight gain of the cobalt sample in mg/cm^2 after 5 h of oxidation at 1,200°C?
11. Uranium dioxide, UO_2, is a p-type oxide with interstitial anions. Explain the effect of Al^{3+} solute ions on the rate of high-temperature oxidation. Explain the effect of Ta^{5+} solute ions on the oxidation rate.
12. For the oxidation of Cu to Cu_2O at a temperature of 1,000°C and a pressure of 100 mmHg, the oxide conductivity is 4.8 Ω/cm, the transference numbers are $\tau_a + \tau_c = 4 \times 10^{-4}$, and the cell emf is $E = 0.337$ V [1]. Calculate the rational rate constant for parabolic oxidation. How does your answer compare with the experimentally determined value of 7×10^{-9} equiv cm^{-1}s^{-1}

References

1. O. Kubaschewski and B. E. Hopkins, "Oxidation of Metals and Alloys", Butterworths, London (1962).
2. K. Hauffe, "Oxidation of Metals", Plenum Press, New York (1965).
3. P. Kofstad, "High Temperature Corrosion", Elsevier Applied Science, London (1988).
4. A. S. Khanna, "Introduction to High Temperature Oxidation and Corrosion", ASM International, Materials Park, OH (2002).
5. M. W. Chase, Jr., Ed., "NIST-JANAF Thermochemical Tables", National Institute of Standards and Technology, Gaithersburg, MD (1998).
6. J. P. Coughlin, "Contributions to the Data on Theoretical Metallurgy", Vol. XII, US Government Printing Office, Washington, DC (1954).
7. D. R. Gaskell, "Introduction to Metallurgical Thermodynamics", p. 272, Hemisphere Publishing Corp., Washington, DC (1981).
8. F. D. Richardson and J. H. Jeffes, *J. Iron Steel. Inst.*, *160*, 261 (1948).
9. D. F. Mitchell and M. J. Graham in "High Temperature Corrosion", R. A. Rapp, Ed., p. 18, National Association of Corrosion Engineers, Houston, TX (1983).
10. C. Wagner, *Z. Physik. Chem.*, *B21*, 25 (1933).
11. M. G. Fontana and N. D. Greene, "Corrosion Engineering", Chapter 11, McGraw-Hill, New York (1978).
12. D. Caplan, M. J. Graham, and M. Cohen, *J. Electrochem. Soc.*, *119*, 1205 (1972).
13. K. Hauffe, "Oxidation of Metals", p. 237, Plenum Press, New York (1965).
14. H. H. Uhlig and W. R. Revie, "Corrosion and Corrosion Control", Chapter 10, John Wiley, New York (1985).
15. O. Kubaschewski and B. E. Hopkins, "Oxidation of Metals and Alloys", p. 82, Butterworths, London (1962).
16. T. P. Hoar and L. E. Price, *Trans. Faraday Soc.*, *34*, 867 (1938).
17. N. D. Tomashov, "Theory of Corrosion and Protection of Metals", p. 74, MacMillan Co., New York (1966).
18. G. Wranglen, "An Introduction to Corrosion and Protection of Metals", Chapter 11, Chapman and Hall, London (1985).

19. J. B. Horton, W. C. Hahn, and J. F. Libsch, "Proceedings of the Third International Congress on Metallic Corrosion", Vol. IV, p. 401, Moscow (1969).
20. W. W. Smeltzer, H. M. Hindam, and F. A. Elrefaie in "High Temperature Corrosion", R. A. Rapp, Ed., p. 251, NACE, Houston, TX (1983).
21. P. K. Kofstad and A. Z. Hed, "Proceedings of the Fourth International Congress on Metallic Corrosion", p. 196, NACE, Houston, TX (1972).

Chapter 16
Selected Topics in Corrosion Science

Introduction

This chapter addresses selected topics in corrosion science. These topics can each form the basis for more extended treatments, but in large part they complement the body of material included in previous chapters. These special topics include (i) the application of electrode kinetics to corrosion processes, (ii) potential and current distribution, (iii) large structures and scaling rules, (iv) acid–base properties of oxide films, and (v) surface modification by directed energy beams (i.e., by ion implantation or laser-surface techniques).

Electrode Kinetics of Iron Dissolution in Acids

Throughout this text, for the dissolution of iron, we have written

$$Fe \rightarrow Fe^{2+} + 2e^- \tag{1}$$

This equation pertains to the overall dissolution reaction but does not describe the mechanism by which the reaction proceeds. In acid solutions, there are two experimental variable parameters which control the anodic current density, i_a. These are the pH (i.e., the concentration of hydrogen ions) and the electrode potential, E. Thus

$$i_a = f\left(\left[H^+\right], E\right) \tag{2}$$

where $[H^+]$ is the concentration of hydrogen ions in solution. Accordingly, we can experimentally observe two parameters which describe how i_a varies with changes in either $[H^+]$ or E. These parameters are

$$z_{H^+} = \left(\frac{\partial \log i_a}{\partial \log\left[H^+\right]}\right)_E \tag{3}$$

and

$$b_a^{-1} = \left(\frac{\partial \log i_a}{\partial \log E}\right)_{[H^+]} \tag{4}$$

where z_{H^+} is called the reaction order with respect to the H^+ ions and b_a is the usual Tafel slope (see Chapter 7).

E. McCafferty, *Introduction to Corrosion Science*, DOI 10.1007/978-1-4419-0455-3_16,

Various investigators have reported two different sets of kinetic data for the steady-state dissolution of iron in sulfuric or perchloric acids, as shown in Table 16.1. One group of investigators [1, 2] reports anodic Tafel slopes of 40 mV while a second group [3, 4] reports values of 30 mV. In addition, two different values of z_{H^+} have been observed, i.e., -1 vs. -2. Both groups of investigators have observed that dissolved ferrous ions do not affect the anodic dissolution of iron, i.e., $z_{Fe^{2+}} = 0$.

Table 16.1 Kinetic data for the anodic dissolution of iron in acid solutions

Solution	Investigator	b_a (mV)	z_{H^+}	z_{Cl^-}
Sulfuric acid	Bockris [1]	40	−1	–
	Kelly [2]	40	−1	–
	Heusler [3, 4]	30	−2	–
Hydrochloric acid	Lorenz [7, 8]	60	−1	−1
	McCafferty [9]	60	−1	−1
Concentrated acidic chloride solutions	McCafferty [9]	60	+1.8	+1
	Lorenz [8]	100	+1	+0.6
	Nobe [10]	100	+1	+1

The observation that z_{H^+} is a negative number (-1 or -2) is at first glance an astonishing result. This means that at *constant electrode potential*, the H^+ ion does not catalyze the anodic reaction, but instead, it is the OH^- ion which promotes anodic dissolution. The concentration of OH^- ions is very small in acid solutions ($[OH^-] = 10^{-13}$ M at pH 1.0), but OH^- ions are generated by the dissociation of adsorbed water molecules, as shown in the following mechanisms.

Bockris–Kelly Mechanism

Bockris et al. [1] and Kelly [2] proposed the following mechanism to explain their experimental results:

$$Fe + H_2O \underset{-1}{\overset{1}{\rightleftharpoons}} Fe(H_2O)_{ads} \tag{5}$$

$$Fe(H_2O)_{ads} \underset{-2}{\overset{2}{\rightleftharpoons}} Fe(OH^-)_{ads} + H^+ \tag{6}$$

$$Fe(OH^-)_{ads} \underset{-3}{\overset{3}{\rightleftharpoons}} Fe(OH)_{ads} + e^- \tag{7}$$

$$Fe\,(OH)_{ads} \xrightarrow{4} (FeOH)^+ + e^- \text{ (RDS)} \tag{8}$$

$$(FeOH)^+ + H^+ \xrightarrow{5} Fe^{2+}\,(aq) + H_2O \tag{9}$$

The overall reaction in steps 1–5 is $Fe \longrightarrow Fe^{2+} + 2e^-$. The mechanism above takes into account that the electrode surface contains adsorbed water molecules, adsorbed hydroxyl ions, and the surface intermediate $FeOH_{ads}$. The anodic reaction thus proceeds by two consecutive one-electron transfer reactions. The rate-determining step (RDS) occurs in step 4.

In the steady state, the rate of appearance of each surface intermediate is equal to its rate of consumption. That is,

$$\beta \frac{\mathrm{d}\theta_{\mathrm{i}}}{\mathrm{d}t} = 0 \tag{10}$$

where β is a proportionality constant which relates the surface coverage θ to the surface concentration in mol/cm^2. Thus

$$\beta \frac{\mathrm{d}\theta_1}{\mathrm{d}t} = \bar{k}_1 (1 - \theta_T) - \bar{k}_{-1}\beta\theta_1 - \bar{k}_2\beta\theta_1 + \bar{k}_{-2}\left[\mathrm{H}^+\right]\beta\theta_2 = 0 \tag{11}$$

$$\beta \frac{\mathrm{d}\theta_2}{\mathrm{d}t} = \bar{k}_2\beta\theta_1 - \bar{k}_{-2}\left[\mathrm{H}^+\right]\beta\theta_2 - \bar{k}_3\beta\theta_2 + \bar{k}_{-3}\beta\theta_3 = 0 \tag{12}$$

$$\beta \frac{\mathrm{d}\theta_3}{\mathrm{d}t} = \bar{k}_3\beta\theta_2 - \bar{k}_{-3}\beta\theta_3 - \bar{k}_4\beta\theta_3 = 0 \tag{13}$$

where $\bar{k}_i$ are the electrochemical rate constants and θ_1, θ_2, and θ_3 are the surface coverages of $Fe(H_2O)_{ads}$, $FeOH^-{}_{ads}$, and $FeOH_{ads}$, respectively; and $\theta_T = \theta_1 + \theta_2 + \theta_3$. The current density is carried in steps 3 and 4. Thus

$$\frac{i_a}{F} = \bar{k}_3\beta\theta_2 - \bar{k}_{-3}\beta\theta_3 + \bar{k}_4\beta\theta_3 \tag{14}$$

where F is the Faraday. Each rate constant k_i (bar) is related to the electrode potential E by

$$\bar{k}_i = k_i \mathrm{e}^{\pm n_i FE/2RT} \tag{15}$$

where n_i is the number of electrons transferred in the ith step. (The plus and minus signs refer to the anodic and cathodic directions, respectively.) In addition, in Eq. (15) it is assumed that the transfer coefficient $\alpha = 1/2$ (see Chapter 7).

The solution to Eqs. (11) – (15) subject to $(1 - \theta_T) \approx (1 - \theta_1)$ and $k_{-3} >> k_4$ is (*after considerable algebra*) [2]

$$i_a = 2F \frac{k_2 k_3 k_4 \beta}{k_{-2} k_{-3}\left[\mathrm{H}^+\right]} \left(\frac{k_1}{k_1 + k_{-1}\beta}\right) \mathrm{e}^{3FE/2RT} \tag{16}$$

Thus,

$$\log i_a = \log\,(\text{constants}) - \log\left[\mathrm{H}^+\right] + \frac{3FE}{2.303\,(2RT)} \tag{17}$$

Taking appropriate partial derivatives gives

$$z_{\mathrm{H}^+} = \left(\frac{\partial \log i_a}{\partial \log\left[\mathrm{H}^+\right]}\right)_E = -1 \tag{18}$$

and

$$b_a = \left(\frac{\partial E}{\partial \log i_a}\right)_{[\mathrm{H}^+]} = \frac{2}{3}\left(\frac{2.303RT}{F}\right) = 0.040\ \mathrm{V} \tag{19}$$

(Recall from Chapter 7 that 2.303 $(RT/F) = 0.0591$ V.)

The results in Eqs. (18) and (19) agree with the experimental observations of Bockris [1] and Kelly [2], as shown in Figs. 16.1 and 16.2.

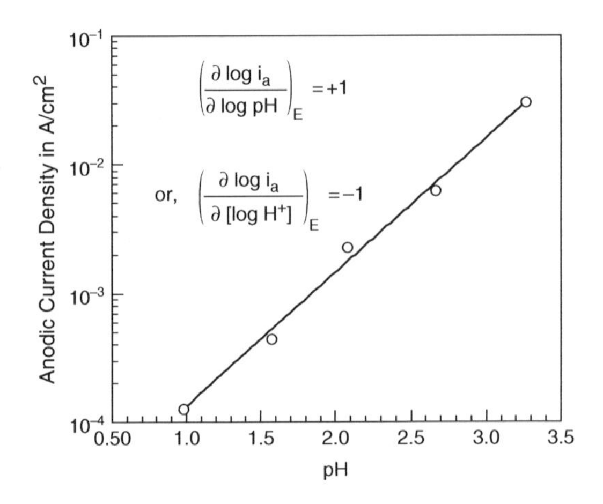

Fig. 16.1 Anodic current density of iron as a function of pH at constant electrode potential ($E = -0.300$ V vs. SHE) in $H_2SO_4 + Na_2SO_4$ solutions [2]. Reproduced by permission of ECS – The Electrochemical Society

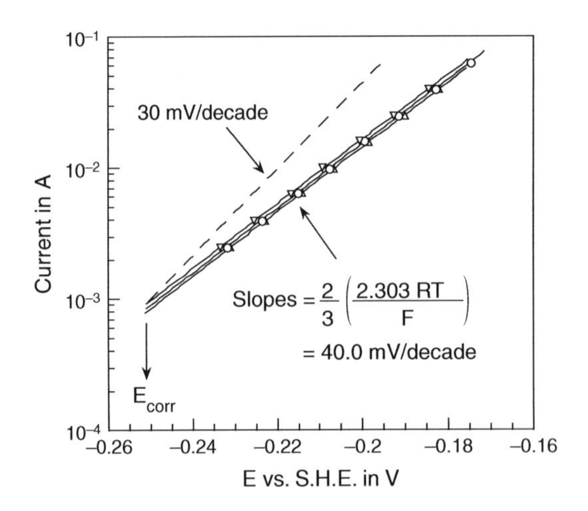

Fig. 16.2 Anodic polarization of iron in H_2-saturated 0.05 M H_2SO_4 at 30°C [2]. (The slope for a 30 mV Tafel slope is given by the *dotted line*.) Reproduced by permission of ECS – The Electrochemical Society

Heusler Mechanism

The Heusler mechanism (also called the catalyzed mechanism) is somewhat similar to the Bockris–Kelly mechanism. The first three steps are the same in each mechanism, but step 4 in the Bockris–Kelly mechanism is replaced by the following two steps:

$$\mathrm{Fe} + \mathrm{Fe(OH)_{ads}} \underset{-3a}{\overset{3a}{\rightleftharpoons}} [\mathrm{Fe(FeOH)_{ads}}] = \text{catalyst} \tag{20}$$

and

$$[\mathrm{Fe(FeOH)_{ads}}] + \mathrm{OH^-} \underset{-4}{\overset{4}{\rightleftharpoons}} (\mathrm{FeOH})^+ + (\mathrm{FeOH})_{\mathrm{ads}} + 2\mathrm{e}^- \tag{21}$$

The reaction then continues as in step 5 of the Bockris–Kelly mechanism. The surface species $\mathrm{Fe(FeOH)_{ads}}$ serves simply as a catalyst for step 4. In the Heusler mechanism it is assumed that most of the anodic reaction is carried out by step 4. With this assumption, then the overall reaction is again- $\mathrm{Fe} \rightarrow \mathrm{Fe}^{2+} + 2e^-$, and

$$i_a = 2F\bar{k}_4\beta\theta_{\text{catalyst}}\,[\mathrm{OH}]^- \tag{22}$$

By setting up a system of steady-state equations as was done for the Bockris–Kelly mechanism, it can be shown that

$$i_a = 2F\left(\frac{k_1}{k_1 + k_{-1}\beta}\right)\frac{k_2k_3k_4}{k_{-2}k_{-3}}\left(\frac{KK_{\mathrm{w}}}{[\mathrm{H}^+]^2}\right)\mathrm{e}^{2FE/RT} \tag{23}$$

where K is the equilibrium constant for the formation of the catalyst in step 3a, i.e., Eq. (20), and K_{w} is the dissociation constant for water. Equation (23) leads to

$$z_{\mathrm{H}^+} = \left(\frac{\partial \log i_a}{\partial \log[\mathrm{H}^+]}\right)_E = -2 \tag{24}$$

and

$$b_a = \left(\frac{\partial E}{\partial \log i_a}\right)_{[\mathrm{H}^+]} = \frac{1}{2}\left(\frac{2.303RT}{F}\right) = 0.030\ \mathrm{V} \tag{25}$$

One of the criticisms of the Heusler mechanism is that a simultaneous two-electron transfer in step 4 is less likely than two consecutive one-electron transfers, as in the Bockris–Kelly mechanism. However, Eqs. (24) and (25) explain the experimental results observed by Heusler.

Reconciliation of the Two Mechanisms

Lorenz and co-workers [5, 6] have shown that either mechanism is possible depending on the microstructure of the iron surface. On iron surfaces having a high density of active sites such as crystal planes, grain boundaries, and dislocations, the Heusler mechanism occurs. But on an electropolished and smooth iron surfaces having a low-surface activity, the Bockris–Kelly mechanism is favored. The Heusler mechanism (30 mV) can be induced on smooth surfaces (40 mV) by cold working to increase the density of surface sites. In addition, the Bockris–Kelly mechanism (40 mV) can occur on iron initially having a high density of active sites (30 mV) if the iron is first annealed to relieve high energy sites.

In addition, the adsorption of an organic molecule on an iron surface having a low density of active sites (a 40 mV surface) can transform the surface into one with a high density of active sites (a 30 mV surface). But the opposite effect does not occur.

These various relationships are illustrated schematically in Fig. 16.3.

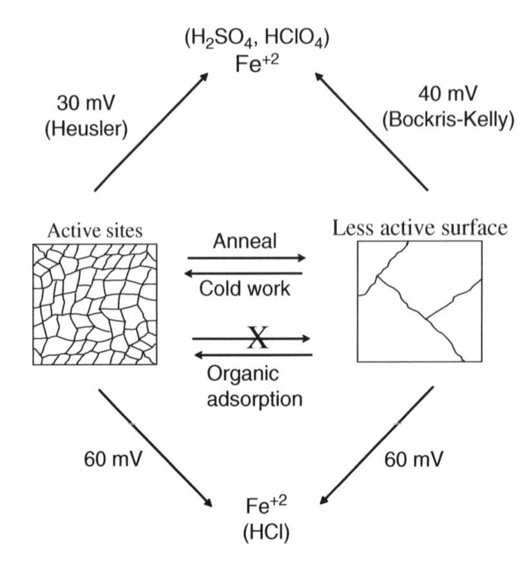

Fig. 16.3 Summary of anodic Tafel slopes (given on the figure) for the dissolution of iron in sulfuric acid, perchloric acid, or hydrochloric acid solutions

Additional Work on Electrode Kinetics

The mechanism of iron dissolution changes slightly in acidic chloride solutions, in which Cl^- ions compete with water molecules for surface sites. Details are given elsewhere [7–9], but it is generally found that $z_{H^+} = -1$ (as in chloride-free solutions), b_a is usually 60 mV, and $z_{Cl^-} = -1$. This last result means that at constant electrode potential, Cl^- ions inhibit the anodic dissolution of iron (rather than promoting anodic dissolution).

In highly concentrated acidic chloride solutions, however, both z_{H^+} and z_{Cl^-} are positive [8–10] indicative of a dramatic change in the mechanism of dissolution. These positive values mean that the H^+ and Cl^- ions act co-operatively to promote iron dissolution. (Recall co-operative adsorption from Chapter 12). See Table 16.1 for a compilation of these kinetic parameters for iron in acid solutions.

Detailed electrode kinetic studies have also been carried out for the dissolution of various other metals in acid solutions, including nickel, cobalt [11], indium [12], and titanium [13]. In addition, electrode kinetics have been applied to the dissolution of iron and other metals in neutral or basic solutions. But when oxide films are formed, the electrode kinetic approach finds limited utility, and electrode kinetic studies should be complemented by other electrochemical measurements (such as AC impedance) and by surface analysis of the oxide films.

Distribution of Current and Potential

In many electrochemical systems, the corrosion current and electrode potential are not distributed uniformly over the metal surface. Instances where there is instead a non-uniform distribution of current and potential include galvanic couples, cathodic protection systems, atmospheric corrosion under thin layers of electrolyte, and localized corrosion in pits, crevices, and stress-corrosion cracks. Two examples involving current and potential distribution [14] are given here.

Laplace's Equation

The concept of the electrostatic potential ϕ was introduced in Chapter 3. Recall that ϕ is defined at any point in the electrolyte, including "just outside" the electrode surface. The electrostatic potential ϕ obeys Laplace's equation [15]:

$$\nabla^2 \phi = 0 \tag{26}$$

where ∇ is the differential operator $(\partial/\partial x + \partial/\partial y + \partial/\partial z)$. Equation (26) applies if the solution is electroneutral (the usual situation), if there are no chemical reactions in the bulk electrolyte which produce or consume ions, and if there are no concentration gradients in the bulk of the solution. The derivation of Laplace's equation is given in Appendix O.

Circular Corrosion Cells

Many corrosion systems can be modeled by a co-planar circular geometry, as shown in Fig. 16.4. Examples include galvanic corrosion between dissimilar metals, pitting, crevice corrosion under O-rings or washers, corrosion under barnacles, or corrosion under dust particles in atmospheric corrosion.

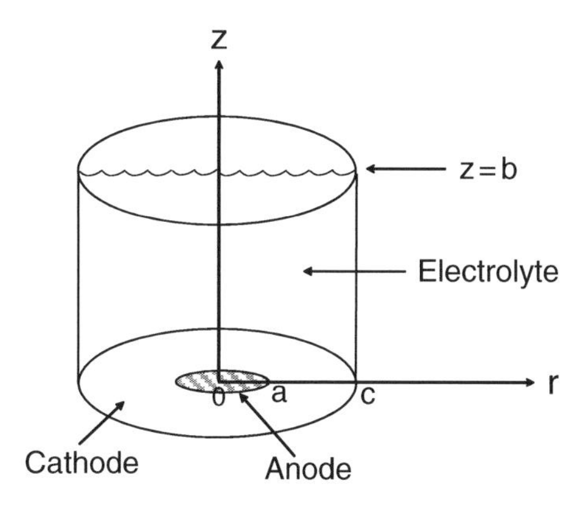

Fig. 16.4 Coplanar concentric geometry in a corrosion cell. Reproduced by permission of ECS – The Electrochemical Society

In cylindrical co-ordinates, Laplace's equation is

$$\frac{\partial^2 \phi(r,z)}{\partial r^2} + \frac{1}{r}\frac{\partial \phi(r,z)}{\partial r} + \frac{\partial^2 \phi(r,z)}{\partial z^2} = 0 \tag{27}$$

The approach is to solve $\phi(r, z)$ in Eq. (27) subject to appropriate boundary conditions and then to evaluate the local current density at the metal surface, $i(r, 0)$ from Ohm's law for electrolytes:

$$i(r, 0) = -\sigma \left[\frac{\partial \phi(r, z)}{\partial z} \right]_{z=0} \tag{28}$$

where σ is the electrolyte conductivity. Boundary conditions are that no current flows across the planes $r = 0$ (due to symmetry) or across the outer boundaries of the electrolyte at $r = c$ and $z = b$. Thus,

$$\left[\frac{\partial \phi(r, z)}{\partial z} \right]_{r=0} = \left[\frac{\partial \phi(r, z)}{\partial z} \right]_{r=c} = \left[\frac{\partial \phi(r, z)}{\partial z} \right]_{z=b} = 0 \tag{29}$$

The last boundary condition involves electrochemical conditions at the metal surface. Following Wagner [16] and Waber [17, 18], it is assumed that there is an extended linear dependence between the electrode potential, E, and the current density, i, as shown in Fig. 16.5. This is a good approximation in many cases, especially in thin-layer electrolytes where polarization curves display an extended linear region for both anodic and cathodic processes [19]. As pointed out by Stern and Geary [20], in bulk electrolytes the combined effects of concentration polarization and ohmic polarization may interfere with activation polarization so that a very short Tafel region is observed. Such cases often give straight line segments in i vs. E. Thus, this model invokes linearity over an extended potential range and just not in a pre-Tafel region close to the corrosion potential.

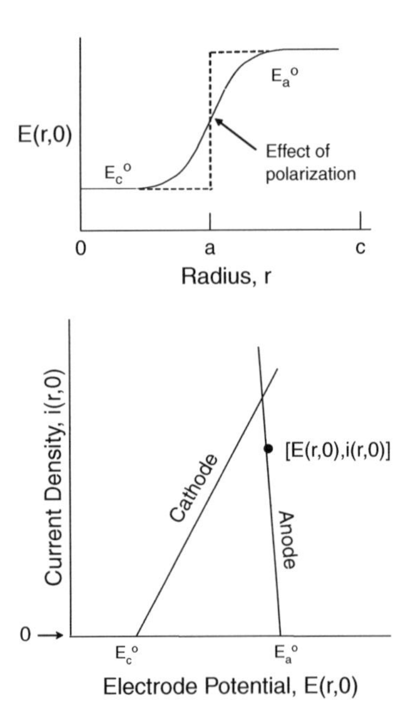

Fig. 16.5 Electrode potentials and extended linear polarization curves [14]. E_a^0 and E_c^0 refer to the open-circuit potential of anode and cathode, respectively. Reproduced by permission of ECS – The Electrochemical Society

Thus, the Wagner polarization parameters are defined as

$$L_a = \sigma \left| \frac{dE}{di} \right|_a \quad \text{for the anode} \tag{30a}$$

and

$$L_c = \sigma \left| \frac{dE}{di} \right|_c \quad \text{for the cathode} \tag{30b}$$

where σ is the conductivity of the electrolyte. Just outside the electrode surface,

$$E(r,0) = V' - \phi(r,0) \tag{31}$$

where V' is a constant which includes all the various differences in electrostatic potential across the extra interfaces which are introduced in the measurement of an electrode potential, E, as discussed in Chapter 3. From Fig. 16.5,

$$\frac{di}{dE} = \frac{i(r,0)}{E(r,0) - E_a^0} \tag{32}$$

Equations (28), (30a), (31), and (32) combine to give

$$\phi(r,0) - L_a \left[\frac{\partial \phi(r,z)}{\partial z} \right]_{z=0} = V' - E_a^0 \tag{33}$$

Similarly for the cathode,

$$\phi(r,0) - L_c \left[\frac{\partial \phi(r,z)}{\partial z} \right]_{z=0} = V' - E_c^0 \tag{34}$$

Equations (33) and (34) thus complete the boundary conditions. The solution of Laplace's equation, i.e., Eq. (27) subject to the boundary conditions in Eqs. (29), (33), and (34) is [14]:

$$\begin{aligned} E(r,0) = {} & \left(\frac{a}{c}\right)^2 E_a^0 + \left(\frac{c^2 - a^2}{c^2}\right) E_c^0 \\ & + \frac{2a}{c^2}(L_a - L_c) \sum_{n=1}^{\infty} C_n \sinh\left(x_n \frac{b}{c}\right) J_1\left(x_n \frac{a}{c}\right) \\ & - \sum_{n=1}^{\infty} C_n \cosh\left(x_n \frac{b}{c}\right) J_0\left(x_n \frac{r}{c}\right) \end{aligned} \tag{35}$$

where sinh and cosh are the hyperbolic functions given by $\sinh x = (e^x - e^{-x})/2$ and $\cosh x = (e^x + e^{-x})/2$, J_o and J_1 are Bessel functions of order 0 and 1, respectively, and x_n are the values where $J_1(x) = 0$. (Note that the parameter V′ in Eqs. (33) and (34) has vanished enroute to the solution given by Eq. (35).) The coefficients C_n are generated by a series of n equations, the form of which is given elsewhere [14]; but in brief, these equations contain the cell dimensions, the Wagner polarization parameters L_a and L_c, the quantities E_a^0 and E_c^0, and Bessel functions.

The derivation of Eq. (35) is too detailed to include in this text but is given elsewhere [14]. The major point to be taken here is that the potential distribution in Eq. (35) depends both on geometrical factors (a, b, and c) and on electrochemical factors (L_a, L_c, E_a^0 and E_c^0). The current distribution follows from applying Eq. (28) to Eq. (35). The result is [14]

$$\frac{i\,(r,0)}{\sigma} = \frac{1}{c}\sum_{n=1}^{\infty} C_n x_n \sinh\left(x_n \frac{h}{c}\right) J_0\left(x_n \frac{r}{c}\right) \tag{36}$$

The total anodic current is obtained from the local anodic current density $i(r, 0)$ by

$$I_a = \int_{r=0}^{a}\int_{\theta=0}^{2\pi} i\,(r,0)\, r\,\mathrm{d}r\,\mathrm{d}\theta \tag{37}$$

The result is [14]

$$\frac{I_a}{\sigma} = 2\pi\, a \sum_{n=1}^{\infty} C_n \sinh\left(x_n \frac{b}{c}\right) J_1\left(x_n \frac{a}{c}\right) \tag{38}$$

Parametric Study

Figure 16.6 shows the electrode potential $E(r, 0)$ for $L_a = 1$ cm and $L_c = 10$ cm for a corrosion cell with different electrolyte thicknesses. The coefficients C_n were computed up to $C_n = 100$. It can be

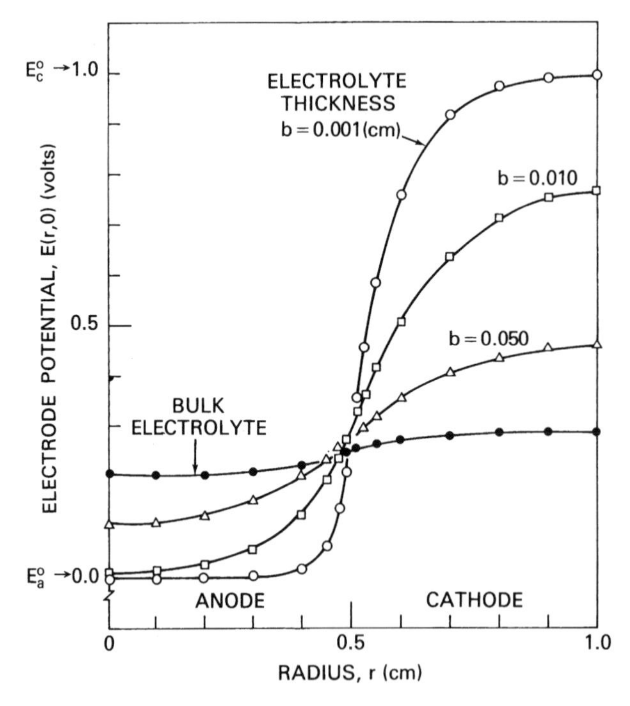

Fig. 16.6 Distribution of electrode potential in a concentric circular cell for $L_a = 1$ cm and $L_c = 10$ cm for different electrolyte thicknesses b [14]. (Anode radius $a = 0.5$ cm, cathode radius $c = 1.0$ cm, $E_a{}^0 = 0.00$ V, and $E_c{}^0 = 1.00$ V.) Reproduced by permission of ECS – The Electrochemical Society

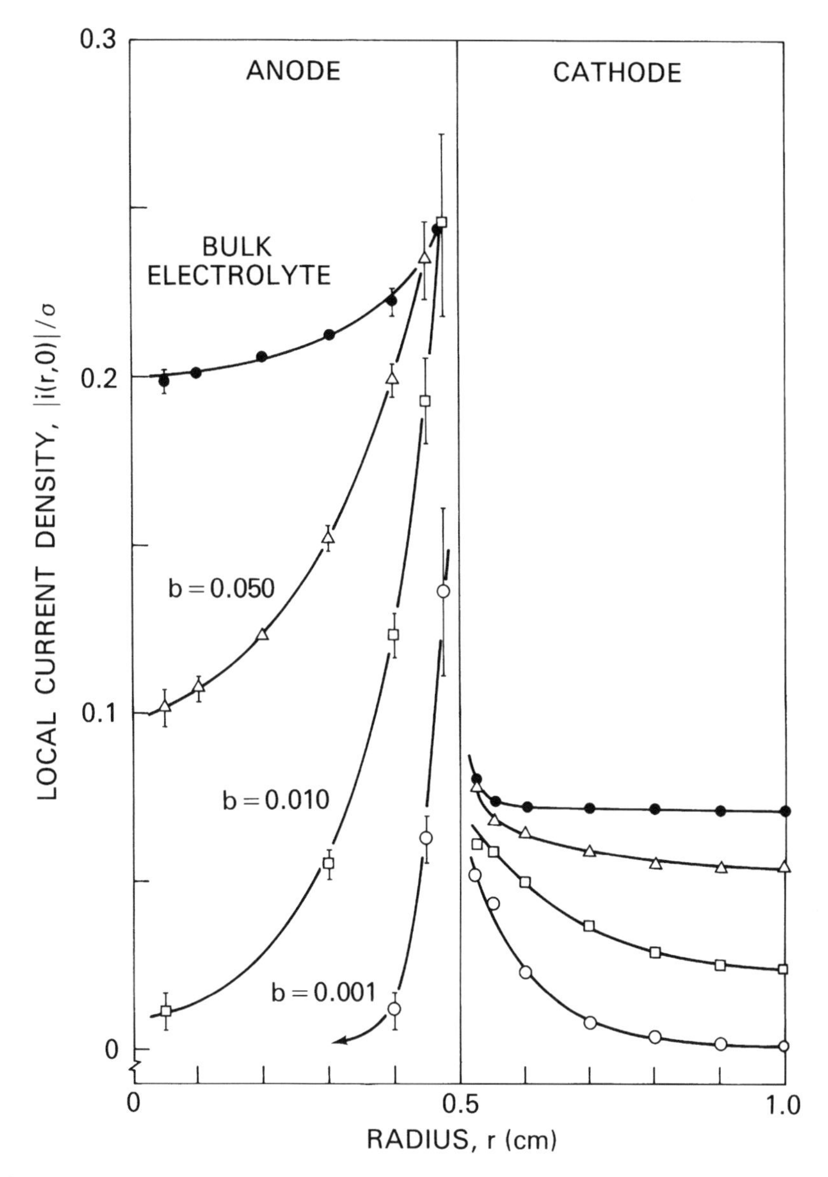

Fig. 16.7 Current distribution corresponding to the potential distribution given in Fig. 16.6 [14]. (Anode radius $a =$ 0.5 cm, cathode radius $c =$1.0 cm, $E_a^0 = 0.00$ V, and $E_c^0 = 1.00$ V.) The electrolyte thickness b in centimeters is given on the figure. Reproduced by permission of ECS – The Electrochemical Society

seen that the electrode potential is distributed almost uniformly for the bulk electrolyte (large values of b), but the potential distribution becomes increasingly non-uniform with decreasing thickness of the electrolyte layer.

The corresponding current distributions are shown in Fig. 16.7. The local current densities were calculated from Eq. (36) with $n = 100$–125. For the thin layers of electrolyte, there is a pronounced geometry effect in which the corrosion attack is concentrated near the anode/cathode juncture.

Figure 16.8 shows the total anodic current calculated from Eq. (38) for two different combinations of L_a and L_c. In both cases, the total current approaches values for the bulk electrolyte for an

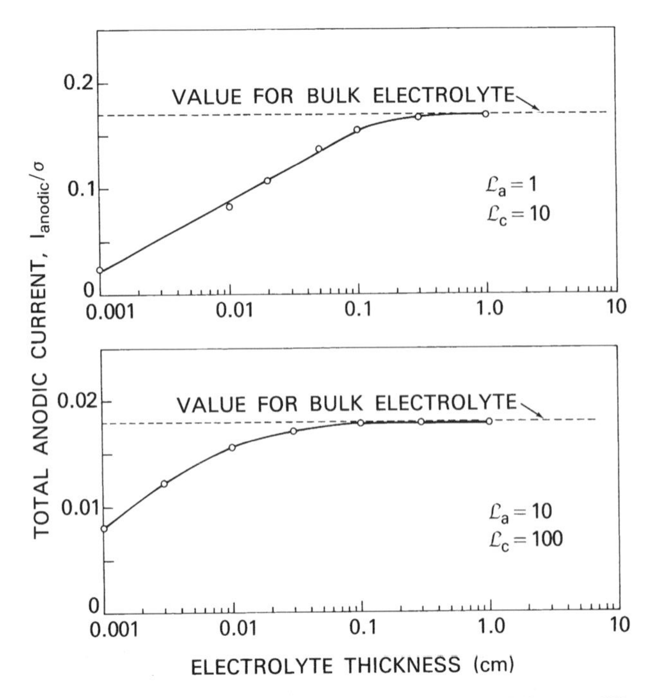

Fig. 16.8 Total anode current computed as a function of electrolyte thickness b for two different combinations of L_a and L_c [14]. (Anode radius $a = 0.5$ cm, cathode radius $c = 1.0$ cm, $E_a^0 = 0.00$ V, and $E_c^0 = 1.00$ V.) Reproduced by permission of ECS – The Electrochemical Society

electrolyte thickness of about 0.1 cm (or about 1,000 μm). This value is consistent with experimental results for iron in thin-layer electrolytes open to the air where the corrosion reaction is under cathodic control. For atmospheric corrosion under thin layers of 0.1 M NaCl, the cathodic polarization curve for iron was similar to that for the bulk electrolyte for an electrolyte thickness greater than 330 μm [19]; see Fig. 16.9. For iron in 0.6 M NaCl, the thickness of the diffusion layer for oxygen reduction was calculated in Chapter 8 to be 0.065 cm (650 μm). Thus, the cathodic polarization curve for iron in 0.6 M NaCl will be the same as in the bulk electrolyte for an electrolyte thickness of 650 μm or greater.

Application to the Experiments of Rozenfeld and Pavlutskaya

Rozenfeld and Pavlutskaya [19] have measured the potential distribution across an iron circular anode in contact with a copper disc in 0.1 M NaCl for both bulk and 165 μm thickness electrolytes. Their experimental results, which are shown in Figs. 16.10 and 16.11, provide an experimental comparison for calculations made using the model described above. Figure 16.12 shows cathodic polarization curves as measured by Rozenfeld [19] for copper in 0.1 M NaCl. For copper in bulk 0.1 M NaCl, the cathodic curve is nearly linear over an overvoltage of about 1 V, after which the curve shifts considerably, as the reaction moves from the region of oxygen reduction into that of hydrogen evolution. A tangent approximation was drawn to the linear portion of the curve, and the value of E_c^0 was replaced by the value E_c', which is given by the intersection of the tangent approximation and the potential axis. A similar treatment was done for the polarization curve for copper in 165 μm 0.1 M NaCl, which also displays an extended linear range. Values of L_c were calculated from Eq. (30b) using the slopes of the cathodic polarization curves and handbook values for the

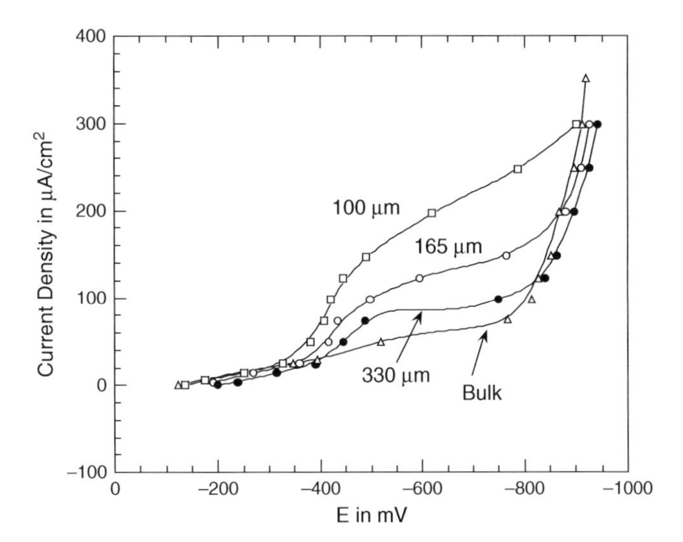

Fig. 16.9 Cathodic polarization curves of iron in 0.1 M NaCl for different electrolyte thicknesses (electrolyte thicknesses are shown on the figure). Redrawn from Rozenfeld [19] by permission of © NACE International (1972)

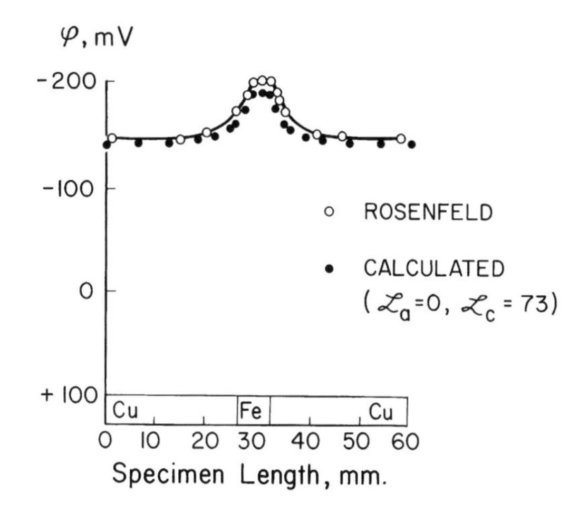

Fig. 16.10 Comparison of the calculated potential distribution [14] and experimental data of Rozenfeld [19] for circular electrodes in bulk 0.1 M NaCl. Reproduced by permission of ECS – The Electrochemical Society

electrolyte conductivity. The anodic polarization curves for iron in bulk and 165 μm 0.1 M NaCl were analyzed similarly, and Wagner polarization parameters are listed in Table 16.2.

The potential distributions calculated from Eq. (35) for the bulk and 165 mm electrolytes are given in Figs. 16.10 and 16.11. It can be seen that there is good agreement between the calculations and experimental data.

Large Structures and Scaling Rules

The previous examples of systems where there is a non-uniform distribution of electrode potential and current have concerned electrochemical cells of small size, but there are many practical cases

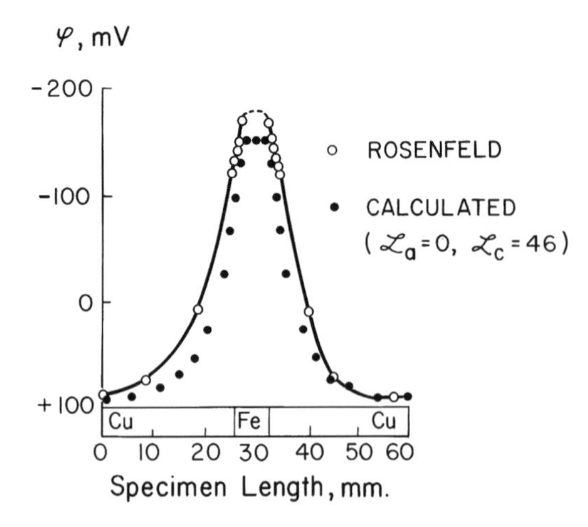

Fig. 16.11 Comparison of the calculated potential distribution [14] and experimental data of Rozenfeld [19] for circular electrodes in 165 μm 0.1 M NaCl. Reproduced by permission of ECS – The Electrochemical Society

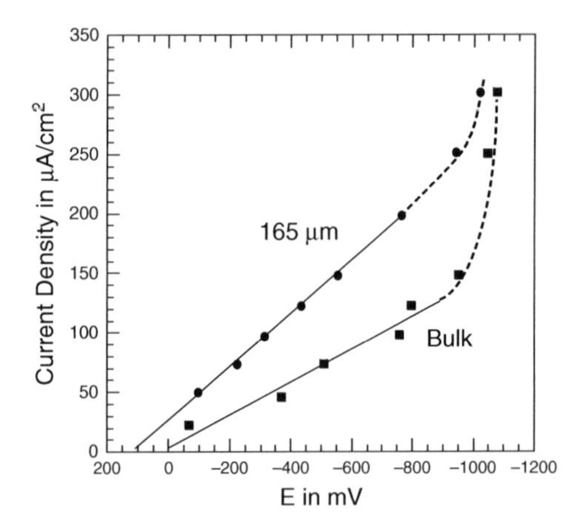

Fig. 16.12 Cathodic polarization of copper in 0.1 M NaCl at two different electrolyte film thicknesses [19]. Reproduced by permission of © NACE International (1972)

Table 16.2 Experimental parameters from Rozenfeld data [19] used to calculate the potential distribution for an iron disc in a copper annulus

Electrolyte thickness	L_a (iron) cm	L_c (copper) (cm)	E_a' (V)	E_c' (V)
Bulk	0	73	−0.19	0.00
165 μm	0	46	−0.15	+0.10

where much larger sizes are involved. These include the cathodic protection of ships, bridges, offshore platforms, or buried pipelines, as well as galvanic corrosion problems in large-scale structures, such as between a ship's propeller and hull.

With such complicated geometries, analytical solutions of the type discussed above are difficult if not impossible. Instead, numerical solutions of Laplace's equation are employed using techniques such as the boundary element method, the finite element method, or the finite difference method.

Standard texts on numerical analysis techniques should be consulted for a description of these methods. The application of the boundary element method to corrosion studies has been discussed by various authors [21, 22].

Modeling of the Cathodic Protection System of a Ship

An example of computer modeling of the potential distribution on a large structure is provided by the work of Lucas et al. [23], who analyzed the performance of a cathodic protection system on the hull of an aircraft carrier hull. The overall length of this type of vessel is 317 m (approximately 1,000 ft), and the geometry modeled has the shape shown in Fig. 16.13. Additional corrosion protection for the painted hull is provided by an impressed cathodic current protection (ICCP) system, in which an electrical current is provided to anodes by external power supplies (rather than by the sacrificial action of zinc anodes).

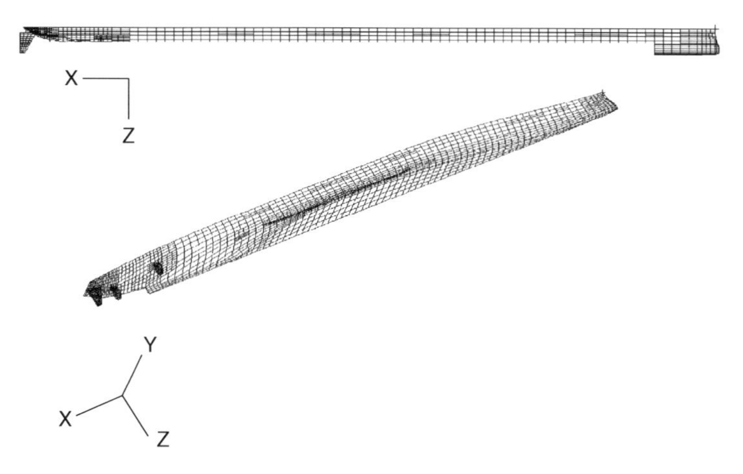

Fig. 16.13 The shape of the hull of a vessel for modeling of potential distribution [23]. Figure courtesy of Virginia G. DeGiorgi, Naval Research Laboratory, Washington D.C., and reproduced by permission of Elsevier Ltd

The ICCP system for the hull consisted of 17 anodes (platinum), placed as follows. Two pairs (four anodes) were located in the fore region of the hull, four anode pairs (eight anodes) were located in the mid-region, and two anode pairs and a single mid-line anode (five anodes) were located in the stern region. A certain degree of damaged paint was assumed (either 2.8% or 15%) of the hull surface.

Laplace's equation, i.e., Eq. (26) was solved using the boundary element method, using experimental electrochemical polarization parameters determined in previous studies. The computational results are given in Fig. 16.14, which shows the potential profile along the hull centerline. Potential values of –1.0 V vs. Ag/AgCl (or more negative values) indicate that the ICCP system is operating properly.

Figure 16.14 also shows experimentally measured electrode potentials at various locations along a physical scale model of the hull. These results were obtained after first constructing a scale model of the carrier hull. Then, zinc anodes were placed on the scale model at positions analogous to those in the actual full-sized hull, and electrode potentials of the immersed hull were measured vs. a Ag/AgCl reference electrode at precise fixed locations on the model hull.

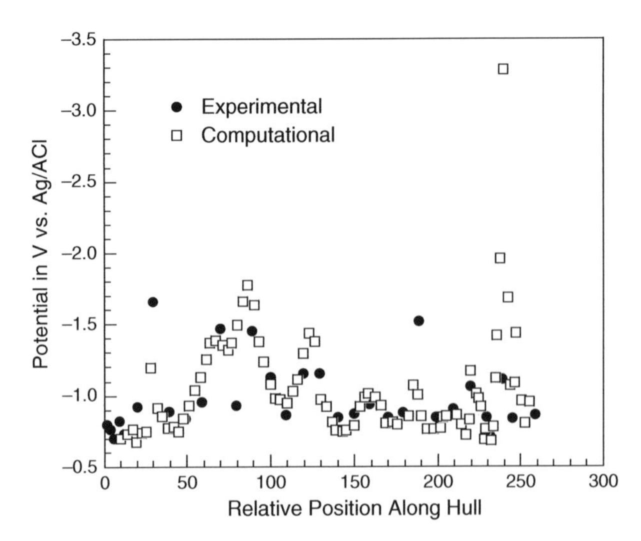

Fig. 16.14 Potential profiles along the centerline of the hull of a US Navy CVN aircraft carrier for a paint damage of 15% of the hull surface. Redrawn from Ref. [23] by permission of Elsevier Ltd

In order to construct a successful scale model, it is necessary not only to scale down the physical dimensions of the original object but also to scale down the conductivity of the electrolyte, as is discussed below.

Scaling Rules

It is often desirable to scale large structures down to a smaller size in order to conduct experiments in the laboratory (either at bench-top or at a pilot plant scale). However, when scaling down the geometrical size of an electrolytic cell, the current distribution in the scale model is not accurately reproduced unless the electrolyte conductivity is also scaled, as described below.

A suitable scaling parameter is the Wagner number, W_a, given by [24]

$$W_a = \frac{R_{\mathrm{act}}}{R_{\mathrm{ohm}}} \tag{39}$$

where R_{act} is the activation resistance and R_{ohm} is the ohmic resistance. The W_a number indicates whether the current density is under kinetic control and therefore relatively uniform, or whether it is under ohmic control and therefore more non-uniform, being dependent on the cell configuration.

The ohmic resistance is given by

$$R_{\mathrm{ohm}} = \frac{\ell}{\sigma A} \tag{40}$$

where σ is the conductivity of the electrolyte, A is the area, and ℓ is a characteristic length which depends on the cell configuration. (For parallel plate electrodes, for example, ℓ is the distance between electrodes.)

For activation polarization, from the Tafel equation,

$$\eta_{\text{act}} = a + b \log i \tag{41}$$

or

$$\eta_{\text{act}} = a + \frac{b}{2.303} \ln i = a + b' \ln \frac{I}{A} \tag{42}$$

where b′ is the usual Tafel slope divided by 2.303 and the current density $i = I/A$. Thus,

$$R_{\text{act}} = \frac{\partial \eta_{\text{act}}}{\partial I} = \frac{b'}{I} \tag{43}$$

Use of Eqs. (40) and (43) in Eq. (39) gives

$$W_a = \frac{\sigma\, b'}{\ell \dfrac{I}{A}} = \frac{\sigma\, b'}{\ell i} \tag{44}$$

In comparing two electrochemical cells of differing sizes, say an original structure and its scale model, we need

$$\frac{W_a\,(\text{model})}{W_a\,(\text{structure})} = \frac{\dfrac{\sigma\,(\text{model})\; b'\,(\text{model})}{\ell\,(\text{model})\; i\,(\text{model})}}{\dfrac{\sigma\,(\text{structure})\; b'\,(\text{structure})}{\ell\,(\text{structure})\; i\,(\text{structure})}} = 1 \tag{45}$$

That is, to have the current distribution in the original structure reproduced in the scale model, we need the same Wagner number in each case. The Tafel parameters b' are not dependent on the size of the system and thus are the same in each case. That is,

$$b'\,(\text{model}) = b'\,(\text{structure}) \tag{46}$$

To preserve the current density in the scale model, i.e., in order to have i(model) = i(structure) we need

$$\frac{\dfrac{\sigma\,(\text{model})}{\ell\,(\text{model})}}{\dfrac{\sigma\,(\text{structure})}{\ell\,(\text{structure})}} = 1 \tag{47}$$

Because the characteristic length ℓ is different in each case, we accordingly need to adjust the electrolyte conductivity in the scale model cell by

$$\sigma\,(\text{model}) = \sigma\,(\text{structure})\,\frac{\ell\,(\text{model})}{\ell\,(\text{structure})} \tag{48}$$

For a scaled-down model, ℓ (model) is less than ℓ (structure), so we need to reduce the electrolyte conductivity as per Eq. (48). See Problem 16.3 for an illustration of this concept.

Acid–Base Properties of Oxide Films

The properties of oxide films are important in a number of corrosion phenomena, including passivity and its breakdown, corrosion inhibition by organic molecules, and the adhesion of organic coatings. In previous chapters, we have briefly mentioned that the acid–base properties of oxide film are important in the pitting of passive films (Chapter 10) and in the adhesion of an organic coating onto an oxide-covered metal surface (Chapter 13).

Surface Hydroxyl Groups

It is well known that oxide surfaces or oxide-covered metals exposed to either the ambient environment or immersed in aqueous solutions terminate in an outermost layer of hydroxyl groups due to their interaction with water molecules [25, 26]. The reaction of a surface oxide with a water molecule is shown in Fig. 16.15. Using quantitative X-ray photoelectron spectroscopy, the concentration of surface hydroxyls in the air-formed oxide films on various metals has been found to be 6–20 OH groups/nm^2 [27]. The depth of the hydroxylated region extends 5–8 Å into the oxide. Although the surface hydroxyl layer is quite thin, it nonetheless has a large influence on the properties of the oxide film, especially on its surface charge.

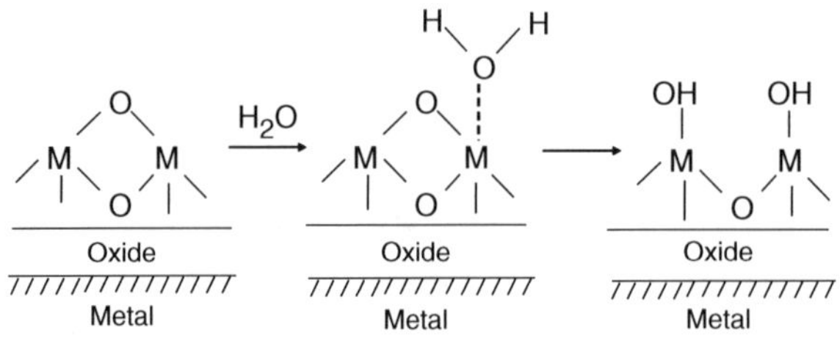

Fig. 16.15 Interaction of a water molecule with the oxide film on a metal

In aqueous solutions, the surface hydroxyl groups may remain undissociated. In such an instance, the oxide surface is said to be at its isoelectric point (IEP) and will have a net surface charge of zero. This will occur only if the pH of the aqueous solution has the same value as the isoelectric point of the oxide. It is more likely, however, that the oxide film will be charged. If the pH is less than the isoelectric point, the surface will acquire a positive charge:

$$-\mathrm{MOH_{surf}} + \mathrm{H^+_{(aq)}} \rightleftharpoons -\mathrm{MOH^+_{2\,surf}} \tag{49}$$

where M denotes a surface site occupied by a metal cation. If the pH is greater than the isoelectric point, the surface will acquire a negative charge:

$$-\mathrm{MOH_{surf}} + \mathrm{OH^-} \rightleftharpoons -\mathrm{MO^-_{surf}} + \mathrm{H_2O}$$

or

$$-\mathrm{MOH_{surf}} \rightleftharpoons -\mathrm{MO^-_{surf}} + \mathrm{H^+}\ (\mathrm{aq}) \tag{50}$$

In the above reactions the surface species $-MOH_2^+$ is a Bronsted acid because it is a proton donor (in the reverse direction of Eq. (49)), and $-MO^-$ is a Bronsted base in Eq. (50) (reverse direction) because it is a proton acceptor. The surface species -MOH is amphoteric, being a Bronsted base in Eq. (49) (forward direction) and a Bronsted acid in Eq. (50) (forward direction).

Nature of Acidic and Basic Surface Sites

A broadened view of acids and bases has been given by G.N. Lewis, in which a Lewis acid is an electron acceptor and a Lewis base is an electron donor [28, 29]. This definition allows a wide variety of molecules, ions, and compounds to be classified as Lewis acids or Lewis bases. With this in mind, the following lists the types of acid sites on metal oxide surfaces:

(1) unhydroxylated metal ions, $M^{+\delta_s}$
(2) protonated surface hydroxyls, $-MOH_2{}^+{}_{surf}$

$$-MOH_{2\ surf}^{+} \rightarrow -MOH_{surf} + H^{+}\ (aq)$$

(3) surface hydroxyls, $-MOH_{surf}$

$$-MOH_{surf} \rightarrow -MO_{surf}^{-} + H_{(aq)}^{+}$$

where δ_s is the partial charge on a metal surface ion. The first of these acid sites above is a Lewis acid because the metal ion $M^{+\delta_s}$ can accept electrons [30,31].

The basic sites on metal oxides are [30, 31]

(4) unhydroxylated oxygen anions, $O^{-\gamma_s}$
(5) dissociated surface hydroxyls, $-MO^-{}_{surf}$

$$-MO^{-}{}_{surf} + H_{(aq)}^{+} \rightarrow -MOH_{surf}$$

(6) surface hydroxyls, $-MOH_{surf}$

$$-MOH_{surf} + H_{(aq)}^{+} \rightarrow -MOH_{2\ surf}^{+}$$

where γ_s is the partial charge a surface oxygen anion. The first of these basic sites above is a Lewis base because the oxygen anion $O^{-\gamma_s}$ can be an electron donor.

While all these different types of acid or basic sites are possible, the acid–base properties of the oxide are usually dominated by the behavior of the hydroxyl groups.

Isoelectric Points of Oxides

The isoelectric point of an oxide or oxide film can be measured in several different ways. One method is to measure the zeta potential, i.e., the potential drop across the diffuse part of the electrical double layer, as a function of the pH of the electrolyte. This can be done by streaming potential measurements [32, 33], in which the aqueous phase is forced to flow across a fixed solid. The zeta potential goes through zero at the isoelectric point of the solid surface.

Another method is to measure contact angles of aqueous droplets on the oxide-covered metal surface [34–37]. In this approach, contact angles are measured at the hexadecane/aqueous solution interface, as shown in Fig. 16.16. The pH of the aqueous phase is varied, and the contact angle is observed to go through a maximum at the isoelectric point of the oxide film, as shown in Fig. 16.17. Details are given elsewhere [35–37].

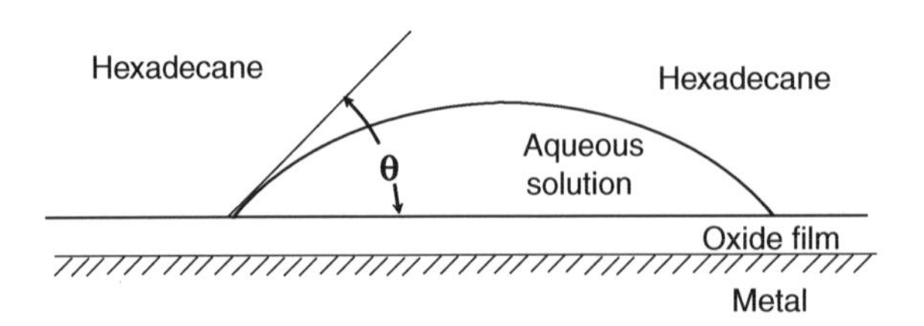

Fig. 16.16 Schematic diagram showing the contact angle of an aqueous droplet in a two-liquid–solid system. The isoelectric point of the oxide film is determined by measuring the contact angle as a function of the pH of the aqueous droplet

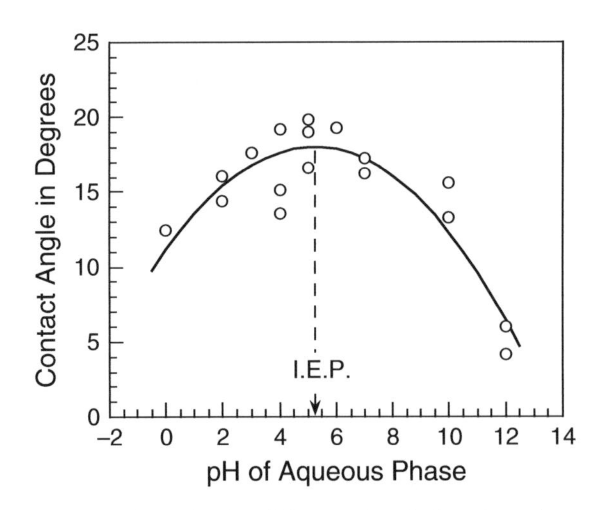

Fig. 16.17 Contact angles for a chromium-plated steel (ferrotype) at the hexadecane/aqueous solution interface, as a function of the pH of the aqueous phase [35, 36]. The maximum in the contact angle gives the isoelectric point (IEP) of the solid surface, i.e., of the oxide film on chromium. Reproduced by permission of Elsevier Ltd

Table 16.3 lists the isoelectric points (IEP) for various bulk oxides and oxide films. The isoelectric point is a function of the identity of the individual cation in the M-OH bond. Oxides with an IEP less than 7 are acidic oxides and oxides with an IEP greater than 7 are basic oxides. It can be seen that there is a wide range in the IEP values for metals. For instance, tantalum which is a passive metal has an oxide which is acidic and aluminum which is used in many applications in neutral solutions has an oxide which is basic. Much more data are available on bulk oxides than on intact oxide films, but in general, the IEP of an oxide film is similar to that for the stand-alone bulk oxide [37, 38].

Table 16.3 Isoelectric points of oxides and oxide films

Oxide	Isoelectric point	References
Oxide film on Ta	−0.7	McCafferty and Wightman [36, 37]
IrO_2	0.5	Ardizzone and Trasatti [56]
MoO_3	1.8−2.1	Compiled by Natishan [49]
SiO_2	1.8–2.2	Compiled by Natishan [49]
Ta_2O_5	2.8	Compiled by Natishan [49]
Oxide film on 316 stainless steel	4.2–5.2	Kallay et al. [57]
MnO_2	4.7–4.8	Tamura et al. [58]
Oxide film on Ti	2	Kallay et al. [59]
	4.2 (revised value)	McCafferty et al. [60]
TiO_2	5.0–6.5	Ardizzone and Trasatti [56]
SnO_2	5–6	Arai et al. [61]
Oxide film on Cr	5.2–5.3	McCafferty and Wightman [36, 37]
CeO_2	5.2–6.1	Hsu et al. [62]
ZrO_2	5.5–6.3	Compiled by Natishan [49]
Cr_2O_3	6.2–6.3	Compiled by Natishan [49]
NiO	7.3	Moriwaki et al. [63]
	8.2–8.6	Kokarev et al. [64]
Fe_2O_3	7.5	Yates and Healy [65]
	8.7	Moriwaki et al. [63]
		Smith and Salman [66]
CuO	8.5–9.5	Compiled by Kosmulski [67]
Al_2O_3	9.0–9.4	Compiled by Natishan [49]
Oxide film on Al	9.5	McCafferty and Wightman [36, 37]
ZnO	9.2–10.3	Compiled by Natishan [49]
Oxide film on iron	9.8	Kurbatov [68]
	10.0	Simmins and Beard [69]
CdO	10.3	Janusz [70]
MgO	12.4	Compiled by Natishan [49]

Surface Charge and Pitting

Figure 16.18 illustrates the surface charge character of several oxides. As seen in the diagram, in a solution of pH 7, the surface of aluminum oxide (and oxide-covered aluminum) consists of acidic groups (positively charged $=AlOH_2^+$) onto which Cl^- ions will adsorb. As discussed in Chapter 10, aluminum is susceptible to pitting corrosion; and the adsorption of Cl^- ions is the first step in the pitting process. However, for other oxides, such as tantalum oxide, the surface consists of basic groups (negatively charged TaO^-), so that Cl^- adsorption and its subsequent penetration through the oxide film is less favored. Thus, it is necessary to polarize tantalum further in the positive direction to initiate the pitting process. Hence, tantalum has a higher (more positive) pitting potential than aluminum.

Figure 16.19 shows pitting potentials as a function of the isoelectric point of the oxide for various metals in 1 M NaCl [38]. It can be seen that the pitting potential increases as the oxide IEP decreases, in accordance with the surface charge character of the various oxides.

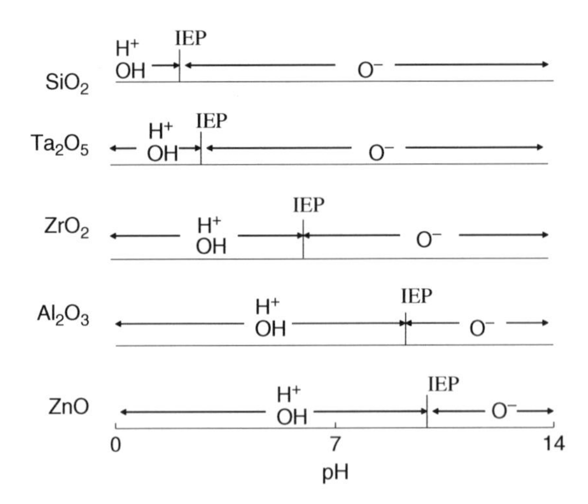

Fig. 16.18 Surface charge character of various oxides as a function of the pH of the aqueous solution. IEP refers to the isoelectric point of the oxide. Reproduced by permission of ECS – The Electrochemical Society

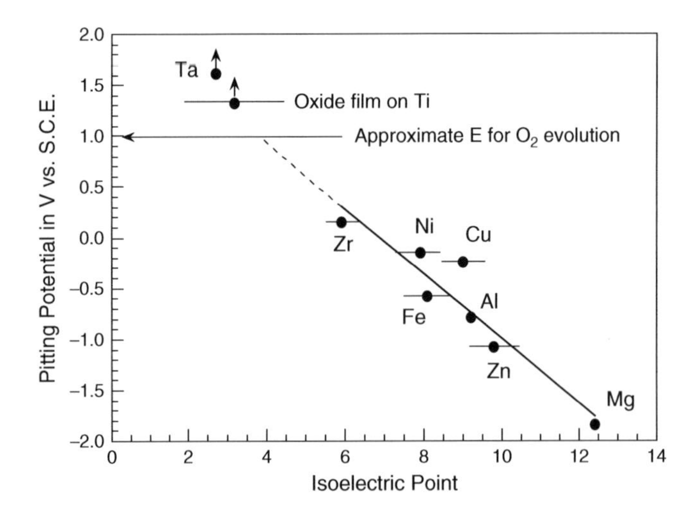

Fig. 16.19 Pitting potentials of oxide-covered metals in 1 M NaCl vs. the isoelectric point (IEP) of the oxide [38, 49]. Note that the least-squares *slanted line* intersects the potential for O_2 evolution at an IEP of about 4. This means that metals having isoelectric points below 4 have pitting potentials (in neutral solutions) above the potential for O_2 evolution. Reproduced by permission of ECS – The Electrochemical Society

Pitting Potential of Aluminum as a Function of pH

The pitting potential of aluminum is independent of pH for pH values between 4 and 8 [38–40], as shown in Fig. 16.20. Over this pH range, the oxide-covered aluminum surface consists of $-AlOH_2^+$ groups, as discussed above. Adsorption of Cl^- is favored, and pitting proceeds at the same critical potential for each pH.

As shown in Fig. 16.20, the pitting potential increases with pH for pH values at and above the isoelectric point (9.5). This is because Cl^- adsorption is less favored on either $-AlOH$ or $-AlO^-$

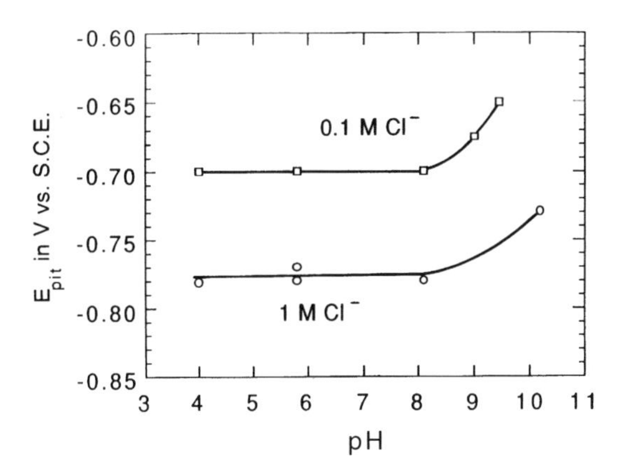

Fig. 16.20 Pitting potentials for aluminum as a function of pH in 0.1 and 1 M NaCl solutions [38, 39]. Reproduced by permission of Elsevier Ltd

surface groups, so that it is necessary to move the electrode potential to more positive values to first adsorb chloride ions and then to drive them into interior of the passive film [41–43].

More details on the pitting of aluminum have been given previously in Chapter 10.

Surface Modification by Directed Energy Beams

In recent years there has been much interest in improving the corrosion behavior of metals and alloys by surface modification techniques, such as ion implantation and laser-surface processing. Surface alloying (as opposed to bulk alloying) has the appeal of conserving expensive, scarce, or critical materials by concentrating them in the near-surface region where they are required for applications such as corrosion protection, wear resistance, or catalytic performance.

Ion implantation and laser processing are two methods of surface modification which not only conserve the amounts of alloying elements required but also offer the possibility of tailoring the surface without sacrificing bulk physical or mechanical properties. In addition, both ion implantation and laser processing are able to produce novel surface alloys or microstructures unattainable by conventional procedures.

Both ion implantation and laser processing use directed energy beams. Ion implantation is a technique for modifying the surface composition of a metal by bombarding it in vacuum with a high-energy beam of ions. Laser processing uses a high-power laser beam to heat a thin surface region with the underlying bulk metal providing rapid resolidification. An introduction is given here to both ion implantation and laser-surface processing. Only a few examples of corrosion research using these methods are given, although it should be realized that both methods have now been used quite extensively in research studies.

Ion Implantation and Related Processes

Ion implantation is a process by which virtually any element can be injected into the surface of a solid material to selected depths and concentrations by means of a beam of high-energy ions (usually tens to hundreds of kiloelectron volts) striking a target mounted in a vacuum chamber. A typical

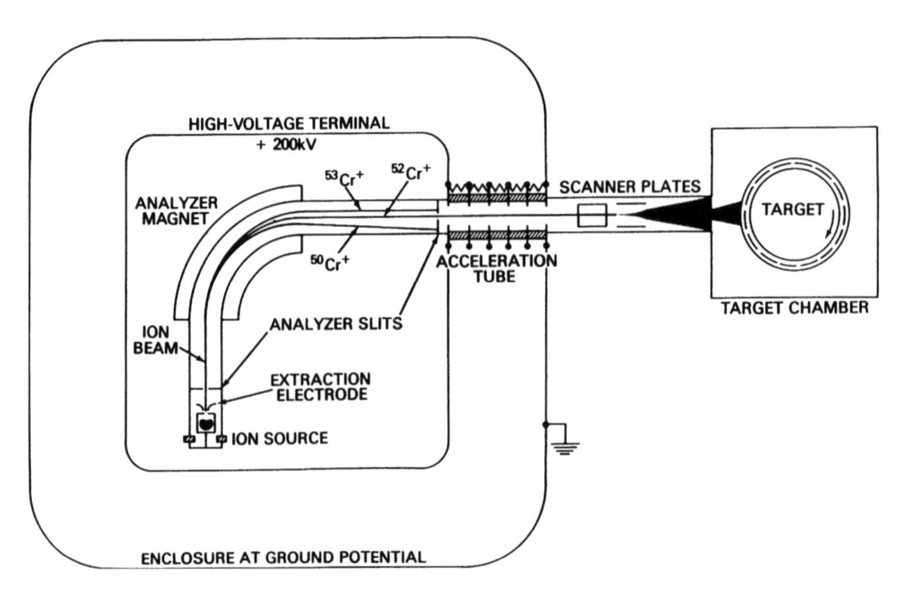

Fig. 16.21 A typical ion implantation apparatus. Ions of a chemical species are obtained by bombarding an appropriate material with an electron beam to create a plasma. An electric field at the extraction electrode extracts the ions from the plasma. The ions are then focused and accelerated with magnetic and electric fields before bombarding the material to be surface modified [44]

ion implantation system is shown schematically in Fig. 16.21 [44]. Atoms of a selected chemical element are ionized by collisions with electrons in an electrical discharge in a gas at a low pressure in the ion source. The ions are electrostatically extracted though an orifice into a high-vacuum region to a moderate energy, are further separated by isotope mass, are accelerated to the desired energy, and are focused. Farther down the line the beam is rastered to ensure that a uniform distribution of ions is implanted laterally into the target metal surface. Areas to be implanted are typically several square inches, although larger areas can be implanted by rotation of the sample through the ion beam.

The bombarding ions lose energy in collisions with substrate electrons and atoms. Penetration depths of tens to thousands of angströms are achieved before incident ions lose all their energy and come to a rest in the solid. The depth of penetration depends on the accelerating voltage, the mass of the ion, and the atomic mass of the substrate. The maximum concentration of implanted ions is generally located beneath the substrate surface, and the depth–concentration profile follows a Gaussian distribution. The surface concentration of implanted ions that can be achieved ranges from extremely dilute alloys to 50 at.% alloys.

Table 16.4 summarizes the various characteristics of the ion implantation process and Table 16.5 lists advantages of the ion implantation approach. One of the major disadvantages of ion implantation is in the shallow depth of the surface-modified region.

Two related extensions of ion implantation are ion beam mixing and ion beam-assisted deposition, each of which can extend the depth of the modified region.

In *ion beam mixing*, a sputter-deposited or sputter-evaporated thin film of thickness several hundreds to thousands of angströms is induced to intermix with the substrate using collisional cascades generated by an implanted ion, which may be either an inert ion (e.g., a noble gas) or may be an

Table 16.4 Ion implantation parameters

Implanted elements	Virtually any element can be implanted
Ion energy	Normally 2–200 keV
Implantation depth	Varies with ion energy, ion species, and host material. Ranges from 100 to 1,000 Å
Range distribution	Approximately Gaussian. Choice of energies allow tailored depth distribution
Concentration	From trace amounts up to 50 at.%
Host material	Any solid material can be implanted

Table 16.5 Advantages of ion implantation

1.	No sacrifice of bulk properties
2.	Rare, expensive, or critical materials can be conserved by concentrating them in the surface
3.	Solid solubility limit can be exceeded
4.	Metastable phases and amorphous phases can be produced which are often unattainable by conventional alloying techniques
5.	No change in sample dimensions or grain sizes
6.	Depth–concentration distribution is controllable (within limits)

ion of one of the desired constituents of the surface alloy. Ion beam mixing retains all the advantages of ion implantation (provided that complete mixing can be achieved) and in addition allows the possibility of producing a thicker surface alloy than can be provided by ion implantation alone.

Ion beam-assisted deposition (IBAD) is a process which combines physical vapor deposition (PVD) and ion implantation. The result is to produce coatings which are much thicker (of the order of microns) than the surface alloys produced by ion implantation. In addition, IBAD coatings have a greater adhesion than PVD coatings due to the partial mixing of the coating with the substrate by the energetic ion source.

Applications of Ion Implantation

Modification of the corrosion behavior of metal surfaces by ion implantation using accelerated ion beams was first achieved in the 1970s in England in collaborative efforts between scientists at the University of Manchester Institute of Science and Technology and Salford University. Concerted efforts were also established about the same time at various research institutions, notably the Harwell Laboratory in England, the University of Heidelberg in Germany, and the Naval Research Laboratory in the United States. The use of ion implantation to improve the corrosion behavior of iron and steel, stainless steels, and aluminum for the period 1985–2000 has been treated in a review article [45]. Between 1985 and 2000, 35 different elements were implanted into iron, stainless steels, or aluminum in regard to corrosion studies, as shown in Fig. 16.22.

Ion implantation can be used to alter either the anodic or cathodic half-cell reaction. A good example of the first approach is given by the early implantation studies of Ashworth et al. [46] who were the first to show that the ion implantation of chromium into iron produced passive Fe–Cr surface alloys similar to the corresponding conventional bulk alloys. Cr-implantation decreased the critical anodic current density for passivation in an acetate buffer but had little or no effect on the cathodic polarization curves.

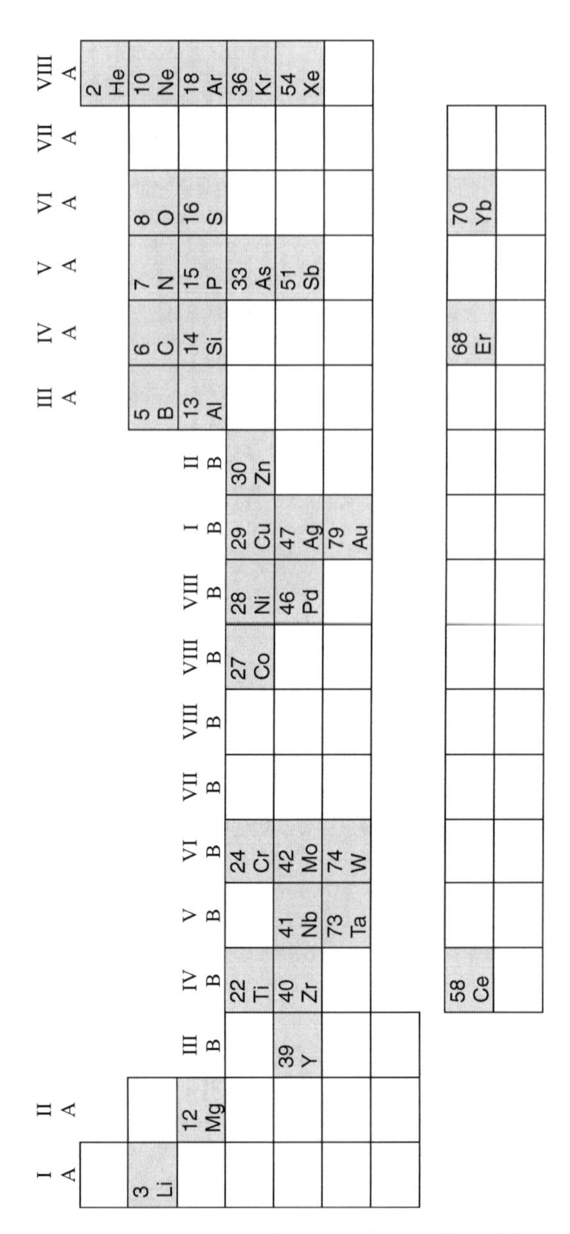

Fig. 16.22 Elements which have been implanted into iron, low alloy steels, stainless steels, or aluminum (1985–2000) and their locations on the periodic table (group numbers and atomic numbers are also shown). Reproduced by permission of © NACE International

One of the earliest examples which used ion implantation to modify corrosion behavior by alteration of the cathodic process is provided by work [47, 48] on the corrosion of titanium in hot acids. Ion implantation of Pd into titanium produced *surface alloys* which decreased the corrosion rate in a manner similar to that for Pd–Ti *bulk alloys*.

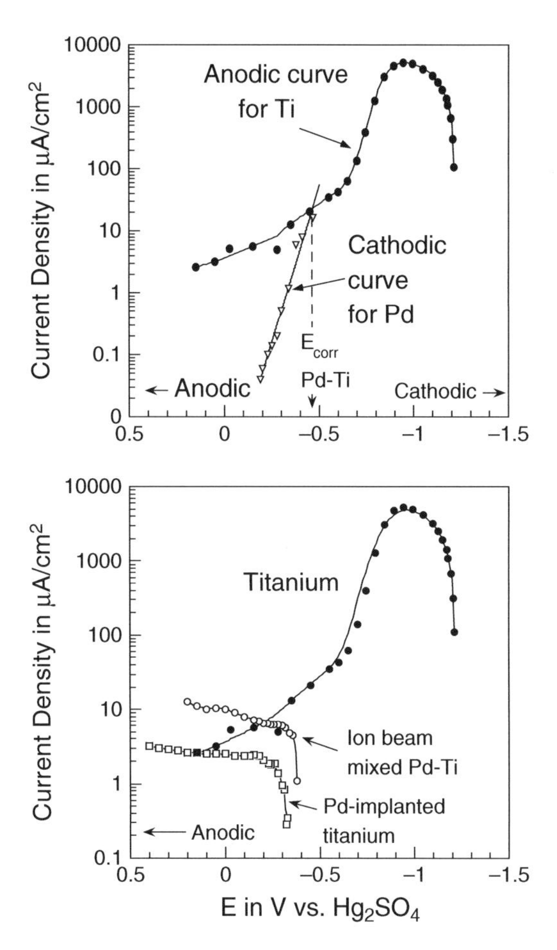

Fig. 16.23 (*Top*): Experimental anodic polarization curve for titanium in boiling 1 M H_2SO_4 and the cathodic polarization curve for palladium. The corrosion potential for a Pd–Ti alloy is indicated on the figure for alloy having equal areas of Pd and Ti. (*Bottom*): Anodic polarization curves for titanium, ion-beam mixed Pd–Ti, and for Pd-implanted titanium [47, 48]. Reproduced by permission of Elsevier Ltd

Figure 16.23 shows anodic polarization curves for titanium, Pd–implanted titanium, and ion beam mixed Pd-Ti surface alloys in boiling 1 M H_2SO_4. It can be seen that titanium can be passivated but only after extensive anodic polarization before undergoing an active/passive transition. (Enroute to the passive state, the titanium surface suffers considerable corrosion damage). By contrast, each of the two Pd–Ti surface alloys self-passivates, and each displays low anodic current densities.

This behavior is similar to that for Pd–Ti bulk alloys, as has been discussed in Chapter 9 in the section "Passivity by Alloying with Noble Metals"; see Figs. 9.45 and 9.46. Hydrogen reduction predominates on palladium sites rather than on titanium so that the resulting mixed potential for the surface alloy (or bulk alloy) occurs in the passive region for titanium rather than in its active region.

Ion implantation has also been used to increase the pitting resistance of aluminum. As shown in Fig. 16.24, implantation of Mo or Si into aluminum resulted in increases in the pitting potential in 0.1 M NaCl [49]. (Implantation of Al into aluminum had no effect on the pitting potential,

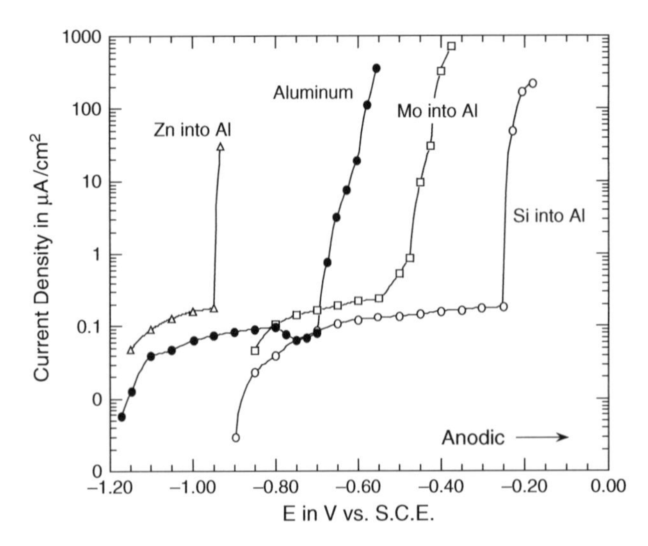

Fig. 16.24 Anodic polarization curves in de-aerated 0.1 M NaCl for unimplanted aluminum and for aluminum implanted with various ions [49]. Reproduced by permission of ECS – The Electrochemical Society

thus showing that the changes in pitting potential are chemical effects rather than physical effects introduced by the damage of the implantation process.)

XPS surface analysis has shown that the implanted ions are contained in the oxide film as well as in the underlying metal. This means that some of the aluminum–oxygen ionic bonds in the oxide film have been replaced with bonds formed between oxygen ions and the implanted cation. That is, the properties of the oxide film are modified by the ion implantation process.

Increases (or decreases) in pitting potential by ion implantation can be explained by two effects [37–39, 49, 50]. Ion implantation can change (i) the surface charge on the oxide film and (ii) the concentration of oxygen vacancies in the film. The first effect is related to the adsorption of Cl^- ions on the oxide surface and the second effect to the transport of Cl^- ions through the oxide film. Each of these two effects is discussed below.

Figure 16.25 shows the pitting potential of ion-implanted aluminum as a function of the isoelectric point (IEP) of the oxide of the implanted ions. (There is some scatter in the data because the various samples did not all receive the same dose of implanted ions.) Implantation into aluminum of metals such as Mo, Si, Nb, Ta, Cr, and Zr (all with oxides having a isoelectric point lower than that of Al_2O_3) resulted in an increase in the pitting potential, i.e., an increase in the resistance to pitting attack. Conversely, implantation with Zn or Li (whose oxides have a higher isoelectric point than that of Al_2O_3) decreased the pitting potential.

Ion implantation changes the surface charge of the oxide, as has been shown elsewhere [36, 37]. It has been shown that the IEP of aluminum is 9.5, of tantalum is –0.7, and of Ta-implanted aluminum is 5.0 [36, 37]. In each case the metals are actually oxide-covered so that the IEP values refer to the oxide films. Thus, the effect of implantation of Ta into aluminum is to change the surface charge at pH 7 from positive $AlOH_2^+$ groups to negative AlO^- or TaO^- surface groups, as per Fig. 16.18, so that the adsorption of Cl^- ions is less favored on the Ta-implanted surface.

Once initiated, corrosion pits were observed to propagate through the thickness of the ion-implanted region. Thus, the role of the implanted ion appears to be that of modifying the surface chemistry of the oxide film rather than providing an underlying substrate which is resistant to propagation in the occluded cell environment.

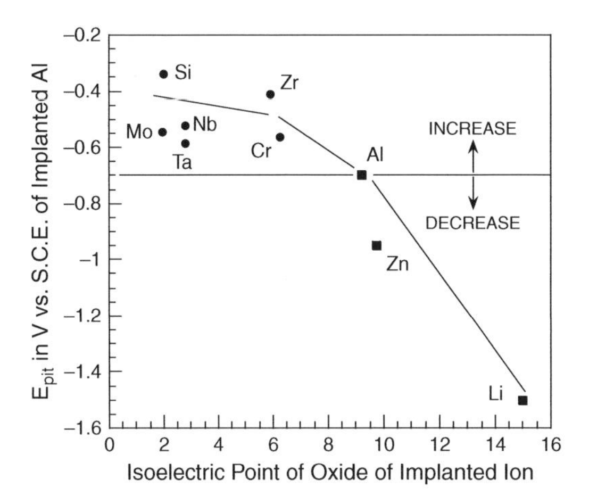

Fig. 16.25 Pitting potentials of ion-implanted aluminum binary surface alloys in 0.1 M NaCl [37, 38]. Reproduced by permission of ECS – The Electrochemical Society

The fact that the IEP of oxide-covered Ta-implanted aluminum is intermediate between the values for oxide-covered Ta and oxide-covered Al is in agreement with studies on bulk oxides which show that the IEP of a mixed oxide falls between the IEP values of its component oxides [51]. Thus, surface charge considerations, such as in Fig. 16.18, can be used as a guide to select implanted elements to be ion implanted into a given metal substrate.

The second effect of ion implantation is to change the concentration of oxygen vacancies in the oxide film. As discussed in Chapter 10, one of the possible mechanisms by which Cl^- ions penetrate passive films is by transport through oxygen vacancies in the film. Thus, an increase in the concentration of oxygen vacancies will assist the breakdown of the passive film and will result in a decrease in the pitting potential. The effect of solute ions in the oxide film can be reasoned by arguments similar to Hauffe's rules for oxidation, which were discussed in Chapter 15. The oxide film on aluminum is an n-type semiconductor with anion (oxygen) vacancies (Table 15.5). Ion implantation of Zr^{4+} ions, for example, will decrease the concentration of oxygen vacancies according to Table 15.6. (The oxidation number of Zr^{4+} is higher than that of Al^{3+}). A decrease in the concentration of oxygen vacancies in turn means an decreased transport of Cl^- ions through the oxide film, a decreased tendency for passive film breakdown, and an increase in the pitting potential, as observed in Fig. 16.25. By contrast, ion implantation of Zn^{2+} into the aluminum oxide film will have the opposite effect on the concentration of oxygen vacancies, according to Table 15.6, and the pitting potential will decrease, as shown in Fig. 16.25.

Laser-Surface Processing Techniques

There are several different laser techniques which can be used to process metal surfaces for improved corrosion resistance.

In *laser surface melting*, a laser beam melts a thin surface layer and the underlying bulk metal provides self-quenching, with cooling rates up to 10^7 K/s. The melting and rapid solidification can improve the corrosion resistance of some alloys by eliminating or minimizing phase separations so as to produce a more chemically homogeneous surface.

Laser surface alloying consists of melting the surface of a metal, adding known amounts of other metals, mixing these components, and alloying them to resolidify. This process produces a surface layer with chemical composition and properties which are different from the substrate material. The surface alloy is also metallurgically bonded to the substrate and has a high degree of adhesion.

Laser melt–particle injection consists of melting a shallow pool on the surface of a metal which is translated under the laser beam and of blowing particles into the melt pool from a nearby fine nozzle. The injected particles may melt completely to form a surface alloy upon resolidification or they may melt only partially so as to be built up as a coating.

In *laser consolidation of coatings*, a coating previously applied by a process such as flame spraying or plasma spraying is laser remelted to remove residual porosity. (The corrosion behavior of laser-melted titanium coatings has been discussed in Chapter 5.)

A typical experimental apparatus used for surface processing using kilowatt power continuous lasers is shown in Fig. 16.26 [52]. The focusing element is a single spherical mirror which is used to obtain a sharply focused laser beam. Test specimens are mounted on a rotating turntable, swept

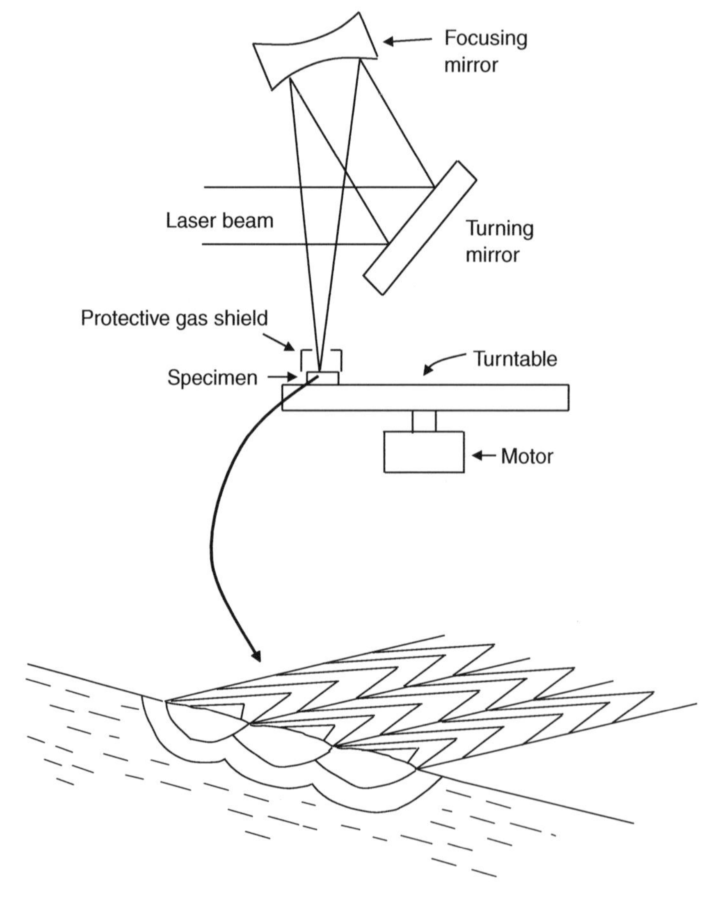

Fig. 16.26 (*Top*): Schematic diagram of a continuous-wave high-power laser-processing apparatus [52]. (*Bottom*): Schematic diagram showing melt stripes, surface ripples, and chevron marks in the laser-surface-melted region Reproduced by permission of ECS – The Electrochemical Society

though the focused beam, and protected during the processing by a flowing helium gas shield. The laser beam rapidly melts a small volume of metal, which subsequently rapidly solidifies by conduction of heat to the bulk specimen. Each pass results in the processing of a ribbon of material, typically 0.25 mm wide and 0.1 mm deep. The width and depth of such a ribbon is varied by changing the processing conditions, such as sweep speed, laser power, or spot diameter. Complete coverage of a large surface is obtained by using successive passes spaced a fraction of a pass width from one another. Table 16.6 lists typical processing conditions for surface modification for corrosion applications [53].

Table 16.6 Typical laser-processing conditions using a high-power continuous-wave laser for surface modification for corrosion applications [53]

Output power	7.5 kW
Spot diameter	0.15 mm
Average power density	4×10^7 W/cm^2
Sweep speed	50 cm/s
Interaction time	0.6 ms
Time metal is molten	1 ms
Melt depth	50 μm
Melt width	260 μm
Cooling rate	10^7 K/s

Applications of Laser-Surface Processing

Only two examples of corrosion research are given here, one involving laser surface melting and a second involving laser surface alloying. (Chapter 5 has already considered an additional application on laser consolidation of coatings.)

Figure 16.27 shows a cross section of laser-melted 304 stainless steel sample. The overlapping melt stripes can be clearly seen in the figure, which also shows that the growth in the resolidified melt is epitaxial, extending from grains in the base metal. Table 16.7 shows the effect of laser-processing

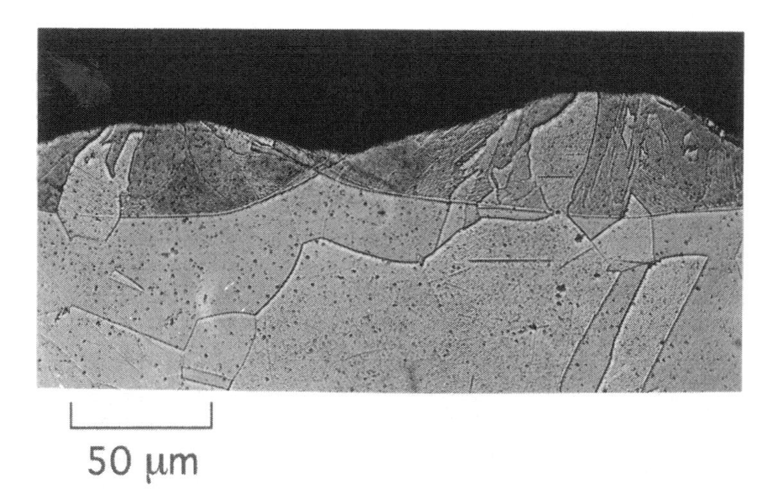

Fig. 16.27 Cross section of a laser-surface-melted 304 stainless steel showing the overlapping melt stripes and characteristic surface ripples due to laser processing. The epitaxial resolidification of the laser-melted region can also be seen [52]. Reproduced by permission of ECS – The Electrochemical Society

Table 16.7 Pitting potentials for 304 stainless steel in 0.1 M NaCl at various anodic sweep rates [52]

	Pitting potential in V vs. SCE[a]		
	1 mV/min	10 mV/min	100 mV/min
Type 304 stainless steel	+0.28	+0.28	+0.28
Laser-processed type 304 stainless steel	+0.50	+0.48	+0.80
	+0.58	+0.50	
	+0.80	+0.80	
		+0.80	

[a]Each entry in the table represents a separate experiment.

on pitting potentials obtained in 0.1 M NaCl [52]. For the laser processed stainless steels, samples were polished past the surface ripples but only partially through the melt zone to provide a smooth finish. There is some scatter in the pitting potentials of the laser-processed stainless steel, but the effect of laser processing is clearly to increase the pitting potential by 200–500 mV.

The reason for the increase in pitting potential is due to the redistribution or possible elimination of sulfide inclusions from the stainless steel. The role of sulfide inclusions in contributing to the breakdown of passivity on stainless steels has been discussed in Chapter 10, and an SEM micrograph of a typical sulfide inclusion on type 304 stainless steel has been shown previously in Fig. 10.36. These MnS particles were not detected after laser processing, and sulfur levels were reduced, as determined by electron microprobe analysis. Thus, the sulfide inclusions were melted during laser processing and are either eliminated or are redistributed into a larger number of smaller particles [52].

The preparation and characterization of laser-alloyed stainless steels was first carried out by groups at the Naval Research Laboratory [52, 53] and at Rockwell International [54, 55]. McCafferty and Moore [52, 53] laser alloyed Mo into 304 stainless steel to produce Fe–Cr–Ni–Mo surface alloys. Two types of alloys were prepared: Fe-18 Cr-10 Ni-3 Mo, which is within the specifications for bulk 316 stainless steel (which is often used instead of 304 stainless steel when improved resistance to pitting is needed), and Fe-19 Cr-12 Ni-9 Mo. These surface alloys were produced by sputter depositing up to 30 alternating layers of pure Mo and Ni–Cr alloys on a type 304 stainless substrate. (The Ni–Cr layers were provided to maintain the Ni and Cr levels in the surface alloy upon melting and mixing.) Details are given elsewhere [52, 53]. These specimens were laser surface melted to an initial depth of 150 μm followed by a second laser processing at a shallower depth of 100 μm two or more times to further homogenize the composition of the resulting surface alloy.

Figure 16.28 shows a cross section through the 3% Mo surface alloy. (The sample had been etched to bring out its structure.) The outlines of the various melt passes can be shown in the figure. Figure 16.29 shows electron microprobe traces taken across the 3% Mo surface alloy. It can be seen that Mo is distributed evenly throughout the surface alloy, except at the bottom of the laser-melted region. Molybdenum was also distributed uniformly throughout the 9% Mo alloy.

The anodic behavior of the two surface alloys in 0.1 M NaCl is shown in Fig. 16.30. The pitting potential of the 3% Mo surface alloy is similar to that for bulk 316 stainless steel (which contains 2–3% Mo). The 9% Mo surface alloy did not undergo pitting up to potentials of oxygen evolution. The compositions and the pitting potentials are summarized in Table 16.8.

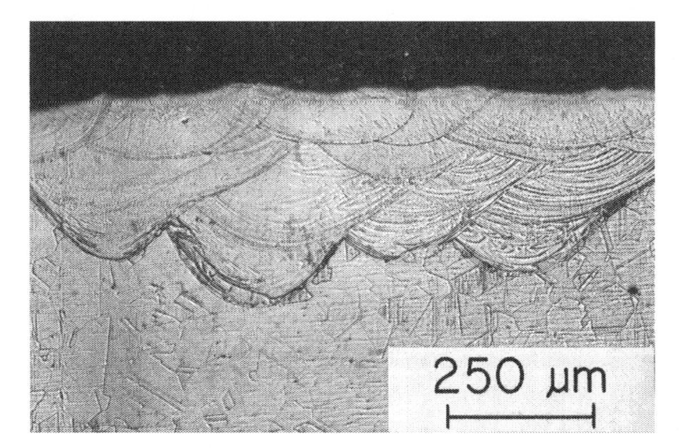

Fig. 16.28 Cross section of 3% Mo laser-surface alloy showing overlapping sets of melt passes due to deep alloying and shallower homogenization [52]. Reproduced by permission of ECS – The Electrochemical Society

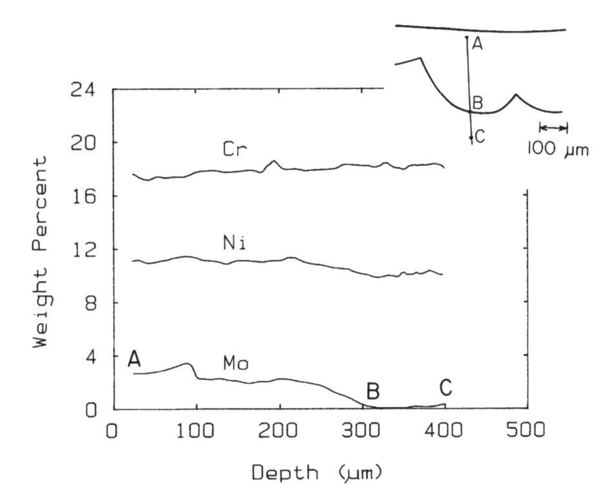

Fig. 16.29 Electron microprobe traces across the 3% Mo surface alloy. The *inset* to the *right* shows the outline of the laser-surface alloy [52]. Reproduced by permission of ECS – The Electrochemical Society

Thus, the 3% Mo surface alloy prepared by laser surface alloying exhibits both a surface composition and a pitting potential equivalent to type 316 stainless steel, but conserves the amount of Mo required by restricting its presence to the near-surface region. The rapid solidification feature of laser alloying enables production of a surface stainless steel containing 9% Mo, an amount in excess of that attainable by conventional alloying techniques. The improved pitting resistance of the 9% Mo surface alloy is comparable to that normally obtained with higher alloys containing larger amounts of Cr and Ni. Thus, the use of 9% Mo surface alloy also minimizes the amount of Cr and Ni required, thus conserving these alloying elements in addition to conserving Mo.

Comparison of Ion Implantation and Laser–Surface Processing

These two surface modification techniques are complementary in nature. The depth of the surface-modified region is greater with laser processing, but the surface is roughened with characteristic

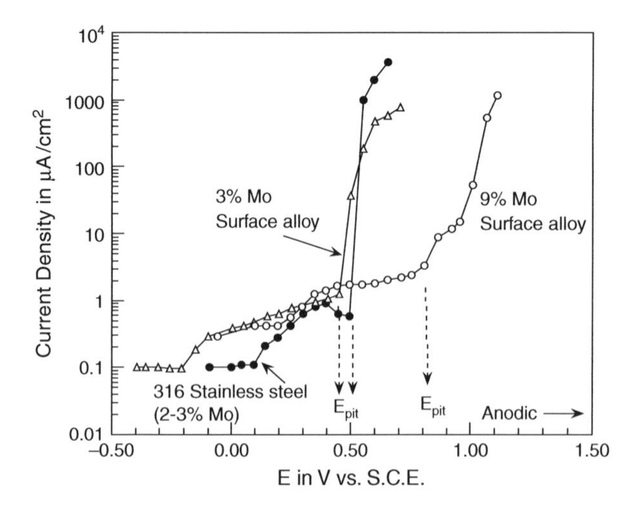

Fig. 16.30 Anodic polarization curves of Fe–Cr–Ni–Mo alloys in de-aerated 0.1 M NaCl [52]. Reproduced by permission of ECS – The Electrochemical Society

Table 16.8 Summary of composition of Fe–Ni–Cr–Mo alloys and pitting potentials in 0.1 M NaCl [52]

Alloy	Cr	Ni	Mo	E_{pit} in V vs. SCE
304 stainless steel	18–20	8–10	0	+0.300
316 stainless steel	16–18	10–14	2–3	+0.550
3% Mo surface alloy	18.9	9.1	3.7	+0.500
9% Mo surface alloy	19.2	11.7	9.6	Did not pit

surface ripples. With ion beam processing, there is no effect on surface topography, although the modified region is much shallower. Ion implantation, its related techniques, and laser surface processing are still primarily research techniques, but each offers the possibility of a new approach to corrosion protection.

Problems

Electrode kinetics

1. Given the following corrosion mechanism for a univalent metal ion

$$\mathrm{M} + \mathrm{H_2O} \underset{-1}{\overset{1}{\rightleftharpoons}} \mathrm{M\,(H_2O)_{ads}}$$
$$\mathrm{M\,(H_2O)_{ads}} \underset{-2}{\overset{2}{\rightleftharpoons}} \mathrm{M}\left(\mathrm{OH^-}\right)_{\mathrm{ads}} + \mathrm{H^+}$$
$$\mathrm{M}\left(\mathrm{OH^-}\right)_{\mathrm{ads}} \overset{3}{\rightarrow} \mathrm{M\,(OH)_{ads}} + e^- \quad \text{(slow step)}$$
$$\mathrm{M\,(OH)_{ads}} + \mathrm{H^+} \overset{4}{\rightarrow} \mathrm{M^+\,(aq)} + \mathrm{H_2O}$$

   (a) write an expression for the anodic current density i_a in terms of $\theta(\mathrm{OH^-})$, where $\theta(\mathrm{OH^-})$ is the surface coverage of the species $\mathrm{M(OH^-)_{ads}}$

(b) Set up a system of steady-state equations for each intermediate species and solve for $\theta(OH^-)$ in terms of the individual rate constants Assume that most of the surface is covered by adsorbed water molecules
(c) What is the Tafel slope b_a for this mechanism?
(d) What is the reaction order z_{H^+} for this mechanism?

Current and potential distribution

2. Suppose that a zinc-coated metal fastener is screwed flush into a nickel sheet and that a 300 μm thick layer of electrolyte condenses from a marine atmosphere in a circular droplet, as shown below:

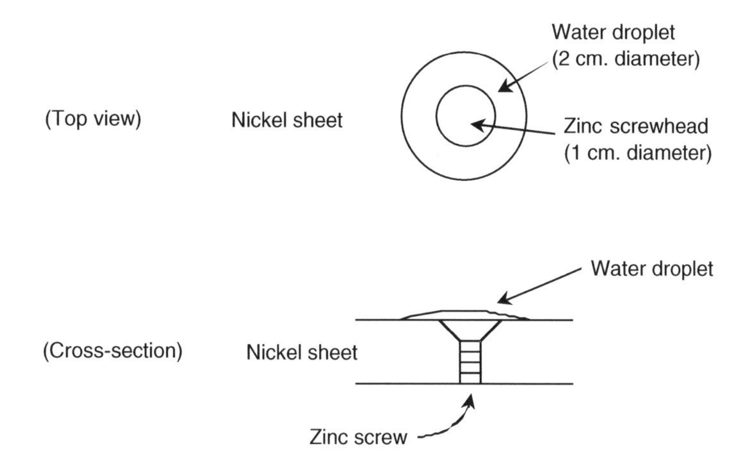

The anodic polarization curve for zinc in the thin-layer electrolyte gives an extended linear region with $|dE/di| = 20\ \Omega\ cm^2$ and the intercept of the linear region with the zero current axis is $E_a' = 0.0$ V. Similarly, the cathodic polarization curve for nickel in the thin-layer electrolyte gives an extended linear region with $|dE/di| = 200\ \Omega\ cm^2$ and intercept $E_c' = 1.0$ V. The conductivity of the electrolyte is $0.05\ \Omega^{-1}\ cm^{-1}$. What is the approximate local current density at the center of the screw head?

Hint: Use Fig. 16.7.

Scaling effects

3. Lucas and co-workers [23] constructed a physical scale model of an aircraft carrier using a scale of 1/100 in a study of the distribution of the electrode potential at various points along the hull of the model. The conductivity of seawater in which the aircraft carrier operates is $0.05\ \Omega^{-1}\ cm^{-1}$. What conductivity of chloride solution should be used in the scaled-down model system?

Acid–base properties of oxide films

4. Given the isoelectric points in Table 16.3, what is the surface charge (positive or negative) of each of the following oxides when immersed in a solution of pH 7.0?

   Molybdenum (Mo)
   Zirconium (Zr)
   Cadmium (Cd)

5. Repeat the above problem for pH 11.0.
6. Would a negatively charge corrosion inhibitor be electrostatically adsorbed onto an oxide-covered surface of zinc at a solution pH of 8.0?

Ion implantation

7 In ion implantation, a typical dose of the implanted ions is 1.0×10^{16} ions/cm^2. The implanted ions are distributed into the host metal substrate throughout a depth of approximately 100 Å in a Gaussian distribution. Suppose that Cr is implanted as Cr^{3+} ions into an iron substrate and that 10% of the dose of 1×10^{16} Cr^{3+} ions/cm^2 is retained in the outermost iron surface. Assume that the metal surface contains only implanted Cr^{3+} ions and Fe host atoms. The radius of an Fe atom is 1.24 Å (124 pm).

   (a) What is the surface concentration of Cr^{3+} in atomic percent if all of the implanted ions reside in the outermost iron surface?
   (b) How does this surface concentration of Cr^{3+} compare with the critical composition needed in Fe-Cr alloys for passivity?

8. Based on the data in Table 16.3, name one ion which could be implanted into pure copper to improve its pitting potential in a chloride solution of pH 6.0? Explain why? What assumption(s) arc you making?
9. (a) Based on a surface charge argument, what is the expected effect on the pitting potential in a neutral chloride solution if Si is implanted into iron?
   (b) Based on a defect concentration argument, what is the expected effect on the pitting potential in a neutral chloride solution if Si is implanted into iron?
   (c) What conclusion can be drawn?

10. Name and discuss one drawback of the ion implantation approach as a method of corrosion protection.

Laser-surface modification

11. Laser-surface melting and the resulting local rapid resolidification impart a compressive stresses to the near-surface region. What is the effect of this local stress on corrosion fatigue?
12. Name and discuss one drawback of the laser-processing approach in regard to corrosion protection.

References

1. J. O'M. Bockris, D. Drazic, and A. R. Despic, *Electrochim. Acta*, *4*, 325 (1961).
2. E. J. Kelly, *J. Electrochem. Soc.*, *112*, 124 (1965).
3. K. E. Heusler, *Z. Elektrochem.*, *62*, 582 (1958).
4. K. F. Bonhoeffer and K. E. Heusler, *Z. Elektrochem.*, *61*, 122 (1957).
5. G. Eichkorn, W. J. Lorenz, L. Albert, and H. Fischer, *Electrochim. Acta*, *13*, 183 (1968).
6. F. Hilbert, Y. Miyoshi, G. Eichkorn, and W. J. Lorenz, *J. Electrochem. Soc.*, 118, 1919 (1971).
7. W. J. Lorenz, *Corros. Sci.*, *5*, 121 (1965).
8. N. A. Darwish, F. Hilbert, W. J. Lorenz, and H. Rosswag, *Electrochim. Acta*, *18*, 421 (1973).
9. E. McCafferty and N. Hackerman, *J. Electrochem. Soc.*, *119*, 999 (1972).
10. H. C. Kuo and K. Nobe, *J. Electrochem. Soc.*, *125*, 853 (1978).
11. W. J. Lorenz in " Corrosion Mechanisms", F. Mansfeld, Ed., p. 1, Marcel Dekker, New York, NY (1987).

12. B. Miller and R. E. Visco, *J. Electrochem. Soc.*, *115*, 251 (1968).
13. E. J. Kelly in "Proceedings of the Fifth International Congress on Metallic Corrosion", p. 137, NACE, Houston, TX (1974).
14. E. McCafferty, *J. Electrochem. Soc.*, *124*, 1869 (1977).
15. J. Newman in "Advances in Electrochemistry and Electrochemical Engineering", Vol. 5, C. W. Tobias, Ed., p. 87, Interscience, New York, NY (1967).
16. C. Wagner, *J. Electrochem. Soc.*, *98*, 116 (1951).
17. J. T. Waber, *J. Electrochem. Soc.*, *101*, 271 (1954).
18. J. T. Waber and B. Fagan, *J. Electrochem. Soc.*, *103*, 64 (1956).
19. I. L. Rozenfeld, "Atmospheric Corrosion of Metals", pp. 60–90, NACE, Houston, TX (1972).
20. M. Stern and A. L. Geary, *J. Electrochem. Soc.*, *104*, 56 (1957).
21. R. S. Munn in "Computer Modeling in Corrosion", ASTM STP 1154, R. S. Munn, Ed., p. 215, ASTM, Philadelphia, PA (1992).
22. V. G. DeGeorgi, E. D. Thomas II, and A. I. Kaznoff in Reference 21, p. 265.
23. V. G. DeGiorgi, E. D. Thomas II, and K. E. Lucas, *Engr. Anal. with Boundary Elements*, *22*, 41 (1998).
24. U. Landau, Lecture Notes, "Workshop on Electrochemical Engineering", Chapter 12, Case Western University, Cleveland, OH (2004).
25. W. Stumm, "Chemistry of the Solid-Water Interface", Chapter 2, John Wiley, New York (1992).
26. J. C. Bolger in "Adhesion Aspects of Polymeric Coatings", p. 3, K. L. Mittal, Ed., Plenum Press, New York (1983).
27. E. McCafferty and J. P. Wightman, *Surface Interface Anal.*, *26*, 549 (1998).
28. W. F. Luder, *Chem. Revs.*, *27*, 547 (1940).
29. G. N. Lewis, *J. Franklin Inst.*, *226*, 293 (1938).
30. K. Aramaki, *Corrosion Engineering*, *45*, 733 (1996).
31. T. Yamanaka and K. Tanabe, *J. Phys. Chem.*, *80*, 1723 (1976).
32. D. Fairhurst and V. Ribitsch in "Particle Size Distribution II", ACS Symposium Series No, 472, T. Provder, Ed., p. 337, American Chemical Society, Washington, DC (1991).
33. C. Bellmann, A. Opfermann, H.-J. Jacobasch, and H.-J. Adler, *Fresnius J. Anal. Chem.*, *358*, 255 (1997).
34. L.-K. Chau and M. D. Porter, *J. Colloid Interface Sci.*, 145, 283 (1995).
35. E. McCafferty and J. P. Wightman, *J. Colloid Interface Sci.*, *194*, 344 (1997).
36. E. McCafferty and J. P. Wightman, *J. Adhesion Sci. Technol.*, *13*, 1415 (1999).
37. E. McCafferty, *J. Electrochem. Soc.*, *146*, 2863 (1999).
38. E. McCafferty, *J. Electrochem. Soc.*, *150*, B342 (2003).
39. E. McCafferty, *Corros. Sci.*, *37*, 481 (1995).
40. W. M. Carroll and C. B. Breslin, *Br. Corros. J.*, *26*, 255 (1991).
41. A. Kolics, A. S. Besing, P. Baradlai, R. Haasch, and A. Wieckowski, *J. Electrochem. Soc.*, *148*, B251 (2001).
42. P. M. Natishan, W. E. O'Grady, E. McCafferty, D. E. Ramaker, K. Pandya, and A. Russell, *J. Electrochem. Soc.*, *146*, 1737 (1999).
43. S. Y. Yu, W. E. O'Grady, D. E. Ramaker, and P. M. Natishan, *J. Electrochem. Soc.*, *147*, 2952 (2000).
44. G. K. Hubler, "Ion Beam Processing", NRL Memorandum Report 5928, Naval Research Laboratory, Washington, DC (1987).
45. E. McCafferty, *Corrosion*, *57*, 1011 (2001).
46. V. Ashworth, D. Baxter, W. A. Grant, and R. P. M. Proctor, *Corrosion Sci.*, *16*, 775 (1976).
47. G. K. Hubler and E. McCafferty, *Corros. Sci.*, *20*, 103 (1980).
48. E. McCafferty, P. M. Natishan, and G. K. Hubler, *Corros. Sci.*, *30*, 209 (1990).
49. P. M. Natishan, E. McCafferty, and G. K. Hubler, *J. Electrochem. Soc.*, *135*, 321 (1988).
50. E. McCafferty, *Corros. Sci.*, *45*, 1421 (2003).
51. P. Roy and D. W. Fuerstenau, *Surf. Sci.*, *30*, 487 (1972).
52. E. McCafferty and P. G. Moore, *J. Electrochem. Soc.*, *133*, 1090 (1986).
53. E. McCafferty and P. G. Moore in "Laser Surface Treatment of Metals", C. W. Draper and P. Mazzoldi, Eds., p. 263, Martinus Nijhoff Publishers, The Netherlands (1986).
54. J. B. Lumsden, D. S. Gnanamuthu, and R. J. Moores in "Corrosion of Metals Processed by Directed Energy Beams", C. R. Clayton and C. M. Preece, Eds., p. 129, AIME, Warrendale, PA (1982).
55. J. B. Lumsden, D. S. Gnanamuthu, and R. J. Moores in "Fundamental Aspects of Corrosion Protection by Surface Modification", E. McCafferty, C. R. Clayton, and J. Oudar, Eds., p. 122, The Electrochemical Society, Pennington, NJ (1984).
56. S. Ardizzone and S. Trasatti, *Adv. in Colloid Interface Sci.*, *64*, 173 (1996).

57. N. Kallay, D. Kovacevic, I. Dedic, and V. Tomasic, *Corrosion*, *50*, 598 (1994).
58. H. Tamura, T. Oda, M. Nagayama, and R. Furuichi, *J. Electrochem. Soc.*, *136*, 2782 (1989).
59. N. Kallay, Z. Torbic, M. Golic, and E. Matijevic, *J. Phys. Chem.*, *95*, 7028 (1991).
60. E. McCafferty, J. P. Wightman, and T. F. Cromer, *J. Electrochem. Soc.*, *146*, 2849 (1999).
61. T. Arai, D. Aoki, Y. Okaba, and M. Fujihara, *Thin Solid Films*, *273*, 322 (1996).
62. W. P. Hsu, L. Rönnquist, and E. Matijevic, *Langmuir*, *4*, 31 (1988).
63. H. Moriwaki, Y. Yoshikawa, and T. Morimoto, *Langmuir*, *6*, 847 (1990).
64. G. A. Kokarev, V. A. Kolesnikov, A. F. Gubin, and A. A. Korabonav, *Soviet Electrochem.*, *18*, 407 (1982).
65. D. E. Yates and T. W. Healy, *J. Colloid Interface Sci.*, *52*, 222 (1975).
66. G. W. Smith and T. Salman, *Canadian Met. Quart.*, *5* (2), 93 (1966).
67. M. Kosmulski, *Langmuir*, *13*, 1615 (1997).
68. G. Kurbatov, E. Darque-Ceretti, and M. Aucouturier, *Surf. Interface Anal.*, *18*, 811 (1992).
69. G. W. Simmons and B. C. Beard, *J. Phys. Chem.*, *91*, 1143 (1987).
70. W. Janusz, *J. Colloid Interface Sci.*, *145*, 119 (1991).

Chapter 17
Beneficial Aspects of Corrosion

Introduction

Throughout this text we have considered corrosion to be a troublesome or dangerous process. And indeed it is! Chapter 1 has discussed the impact of corrosion on health and safety, on wastage of materials, and on economic loss.

Figure 17.1 shows a humorous look at the forces of corrosion. Table 17.1 lists some quotations about corrosion [1–3]. As seen in the table, corrosion (rusting) does not fare well and has been compared to disease, stagnancy, obsession with material wealth, and even to envy. Personal degradation by tribological forces is deemed to be preferred over corrosion.

But there can also be some advantageous aspects to corrosion if corrosion is managed in a controlled manner [4, 5]. Some benefits of corrosion include the following:

Rust Is Beautiful

Rust layers possess an attractive reddish-brown hue. Recall Fig. 1.6, which shows a picture of the attractive rust-colored giant watering can, which graces an outdoor garden center near Alexandria, Virginia.

"Weathering" steels have this same general appearance. Weathering steels are low alloy steels which form attractive-looking rust layers upon exposure in the natural atmosphere. More importantly, these rust layers are also tightly adherent and provide corrosion protection. An example has already been given in Fig. 7.2.

The chemical composition of a typical weathering steel is given in Table 17.2.

Weathering steels have been used in the construction of buildings, bridges, and utility towers. Because of their low corrosion rate, which is established after about 1 year, and with their pleasing appearance, weathering steels have been touted as requiring no painting.

Copper Patinas Are Also Beautiful

The beautiful blue-green patinas which form on copper roofs and statues are due to the initial corrosion of copper to form an oxide film, which then provides further corrosion protection to the underlying copper. Example can be found in the blue-green roofs of hotels, public buildings, churches, and private homes (usually porch roofs), as well as on copper or bronze statues.

E. McCafferty, *Introduction to Corrosion Science*, DOI 10.1007/978-1-4419-0455-3_17,

Fig. 17.1 A humorous look at the forces of corrosion. Reprinted with permission from the comic strip "Hagar the Horrible" by Chis Browne, *Washington Post*, August 15, 2006. By permission of Hagar © King Features Syndicate

Table 17.1 Some quotations regarding corrosion [1–3]

"Time will rust the sharpest sword".
Sir Walter Scott (1771–1832)

"It is better to wear out than to rust out".
Richard Cumberland (1631–1792)

"When I rest, I rust".
German proverb

"Iron rusts from disuse, stagnant water loses its purity, even so does inaction sap the vigor of the mind".
Leonardo da Vinci (1452–1519)

"Keep up your bright swords, for the dew will rust them".
William Shakespeare (1564–1616)

"Where moth and rust doth corrupt and where thieves break through to steal".
Matthew, New Testament, 6:19–20

"As iron is eaten with rust, so are the envious consumed by envy".
Antistenes (c. 455 BC to c. 365 BC)

"Usura* rusteth the chisel
It rusteth the craft and craftsman".
Ezra Pound (1885–1972)

"Rust Never Sleeps"
Album recorded in 1979 by Neil Young, North American folk-rock artist

*Pound's use of the word usura (usury) refers to the obsession with wealth or money.

Table 17.2 Composition of weathering steels compared to two other structural steels [6]

	Composition (%)							
Type of steel	C	Mn	P	S	Si	Cu	Ni	Cr
Structural carbon steel	0.17	0.57	0.019	0.050	0.043	0.05	0.02	0.02
Structural copper steel	0.18	0.49	0.024	0.034	0.025	0.32	0.02	0.02
Early Cor-Ten steel	0.09	0.30	0.16	0.035	0.93	0.42	0.03	1.1
Later Cor-Ten steel	0.06	0.48	0.11	0.030	0.54	0.41	0.51	1.0

Cathodic Protection

Cathodic protection by a sacrificial anode is another example of beneficial corrosion. When a zinc anode is electrically connected to an iron structure, as in Fig. 5.5, the zinc, which has the more negative open-circuit potential, becomes the anode in the iron/zinc couple. Thus, the sacrificial zinc anode is allowed to intentionally corrode, but in so doing, it provides protection to the more valuable structure iron (or steel), which may be a bridge, a pipeline, or a ship.

Electrochemical Machining

Electrochemical machining utilizes controlled anodic dissolution (i.e., corrosion) between a work-piece (anode) and a "tool" (cathode) in an electrolytic cell. The shape of the machined work piece is controlled by moving the tool electrode through the electrolyte in a prescribed path.

Metal Cleaning

Scale-covered or rusted metals can be cleaned prior to surface treatment by immersing them in acid solutions ("pickling").

Etching

In studies on the metallography of metals or alloys, specimens are etched in appropriate solutions (often acids) to allow the specimen to corrode in a controlled manner in order to bring out the grain boundary structure.

Batteries

Batteries physically separate the anode and cathode in order to provide an electromotive force between the two electrodes. In lead acid batteries (automobile batteries), the reactions are as follows:

$$\text{Anode:} \quad Pb + H_2SO_4 \rightarrow PbSO_4 + 2H^+ + 2e^-$$

$$\text{Cathode:} \quad PbO_2 + H_2SO_4 + 2H^+ + 2e^- \rightarrow PbSO_4 + 2H_2O$$

$$\text{Overall:} \quad Pb + PbO_2 + 2H_2SO_4 \rightarrow 2PbSO_4 + 2H_2O$$

Thus, in order for this battery to operate, the lead anode must undergo oxidation (corrosion). Six cells are used in series to produce a total cell voltage of approximately 12 V. (This battery can be recharged, a process which reverses the half-cell reactions above.)

Passivity

Chapter 9 has discussed in much detail the protection of metals or alloys by passive films. These passive films arise from the controlled corrosion of the substrate metal to form a protective oxide film.

Anodizing

Anodizing is a special form of passivation in which an oxide film of tens to hundreds of microns in thickness is formed anodically in an electrochemical cell. Such anodized oxide coatings provide improved corrosion protection, increased abrasion, or enhanced adhesion of paints or other organic coatings.

Titanium Jewelry and Art

Titanium is a gray-colored metal, but it is often used in jewelry and art pieces. This is because titanium can be colored by the process of anodizing to produce oxide films of various beautiful colors, including yellow, blue, green, and violet. The color of the oxide film depends on the applied voltage and on the thickness of the film. Specimens can be anodized by immersion into an electrochemical cell, or selected areas can be anodized with a brush which brings the electrolyte and the cell to the metal. In this case, a brush tip or sponge is saturated with the electrolyte, and the wetted brush or sponge is in contact with a metal cathode (which is insulated). The desired pattern and color is then "painted" onto the titanium surface; see Fig. 17.2.

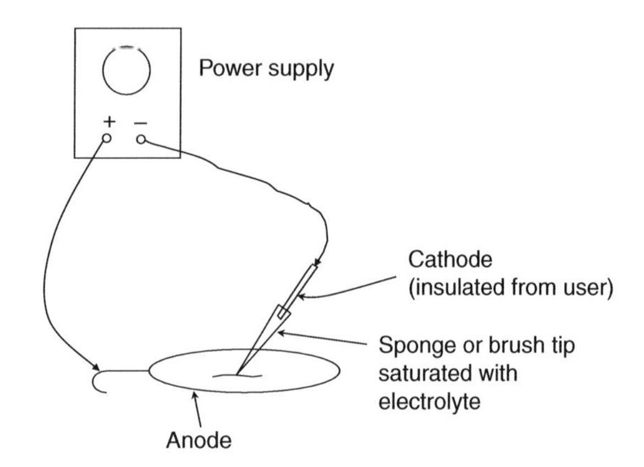

Fig. 17.2 Schematic diagram of the anodizing of a titanium earring using a brush tip

Caution to Inexperienced Artisans:

There is a real danger of getting an electric shock. Be careful, and don't let corrosion throw you for a loop!

References

1. "Familiar Quotations: John Bartlett", 16th edition, J. Kaplan, Ed., Little, Brown, and Company, Boston, MA (1992).
2. "Dictionary of Quotations", B. Evans, Delacorte Press, New York (1968)
3. Canto XLV in "Selected Poems of Ezra Pound", A New Directions Paperback, New York (1957).

4. H. Leidheiser, Jr., "Corrosion: sometimes good is mostly bad", *C. & E. News*, April 5, p. 78 (1965).
5. P. F. King in "Corrosion and Corrosion Protection", R. P. Frankenthal and F. Mansfeld, Eds., p. 285, The Electrochemical Society, Pennington, NJ (1981).
6. C. P. Larrabee and S. K. Coburn, "First International Congress on Metallic Corrosion", p. 276, Butterworths, London (1962).

Answers to Selected Problems

Chapter 2

1. (a) Physical; (b) – (d) physical plus environmentally assisted.
2. Anodic: $Zn\ (s) \rightarrow Zn^{2+}\ (aq) + 2e^-$
   Cathodic: $2H^+\ (aq) + 2e^- \rightarrow H_2\ (g)$
   and: $O_2\ (g) + 2H_2O\ (l) + 4e^- \rightarrow 4OH^-$
3. Anodic: $Fe\ (s) \rightarrow Fe^{2+}\ (aq) + 2e^-$
   and: $Cr(s) \rightarrow Cr^{3+}\ (aq) + 3e^-$
   Cathodic: $O_2\ (g) + 2H_2O\ (l) + 4e^- \rightarrow 4OH^-$
4. (a) Anodic: $Cu\ (s) \rightarrow Cu^{2+}\ (aq) + 2e^-$
   Cathodic: $O_2\ (g) + 2H_2O\ (l) + 4e^- \rightarrow 4OH^-$
   (b) Precipitation occurs.
   (c) Decreasing the pH increases the concentration of H^+ ions and decreases the concentration of OH^- ions so that the equilibrium is shifted to the left.
6. 93.1 mA/cm^2.
7. (a) 3.97×10^{-4} $g/(cm^2\ h)$; (b) 0.363 mA/cm^2.
8. (a) 7.40×10^{-3} A/m^2; (b) 8.23×10^{-3} mm/year; (c) 61 years.
9. 5.9 mils thickness required.
10. 0.664 mA/cm^2.
11. Hf has the larger atomic weight and thus the larger weight loss.
12. The final pH is 14.4. Figure 2.11 shows that this will cause an increase in the corrosion rate of Al.
13. (a) Anodic areas usually appear near the point of the nail. (b) The bent part of the nail is an anodic area because straining the metal perturbs metal atoms from their equilibrium lattice positions.

Chapter 3

1. The number of water molecules involved in primary hydration is 0.86% of the total water molecules.
2. 2.5×10^7 V/cm.
3. (a) 75 $\mu F/cm^2$; (b) 56 $\mu F/cm^2$; (c) 30 $\mu F/cm^2$.
4. $V = PD_{M1/S} - PD_{ref/S}$.
6. (a) $E = -0.380$ V vs. SCE; $E = -0.360$ V vs. Ag/AgCl; (c) $E = -0.454$ V vs. $Cu/CuSO_4$.

E. McCafferty, *Introduction to Corrosion Science*, DOI 10.1007/978-1-4419-0455-3,

8. (a) False; (b) True; (c) True.
9. (a) False; (b) True; (c) False.
10. The double-layer capacitance C_{dl} decreases as the corrosion rate decreases, and thus can be used to monitor the corrosion rate.
11. The double-layer capacitance per unit area C_{dl} is given by $C_{dl} = \varepsilon/d$. As an organic molecule adsorbs, ε decreases and the thickness d of the edl increases, so that C_{dl} decreases.

Chapter 4

2. (a) 1364 cal/mol; 5708 J/mol.
3. (a) $E = +0.235$ V vs. SHE. (b) The electrode potential for the SCE electrode is + 0.242 V vs. SHE, so there is only 7 mV difference between the two.
4. $E = -0.337$ V vs. SHE.
5. $E = +0.209$ V vs. SHE.
6. The reaction is spontaneous.
7. (a) The overall reaction is spontaneous in the standard state; (b) $E^0 = +0.656$ V vs. SHE; (c) $E = 0.656 + 0.0295 \log [MoO_4^{2-}] - 0.0.0591$ pH; (d) the overall reaction to form MoO_2 is spontaneous, under the given conditions; and (e) Spontaneity of forming MoO_2 depends on the experimental conditions, i.e., pH and MoO_4^{2-} concentration.
8. μ^0 (Sn^{2+} (aq)) = – 26,634 J/mol = –6365 cal/mol.
9. $E = -2.394$ V vs. SHE.
10. $[Pb^{+2}] = 0.036$ M.
11. We cannot add the two given values of E^0. We must write $\Delta G^0 = -nFE^0 = \Sigma(\nu_i\mu_i{}^0$ (products)) $- \Sigma(\nu_i\mu_i{}^0$ (reactants)) for each of the two given half-cell reactions to calculate $\mu^0(Fe^{2+})$ and $\mu^0(Fe^{3+})$ for use in calculating E^0 (Fe^{3+}/Fe). The result is E^0 (Fe^{+3}/Fe) = –0.041 V vs. SHE.
12. In the solid metal, a metal atom is confined to a certain lattice position. In the aqueous solution, the metal cation can move through the solution, so the disorder increases. Thus, the entropy increases.

Chapter 5

1. (a) The nickel electrode is the anode and will corrode; (b) cell emf is +0.570 V; (c) Ni (s) + Cu^{2+} (aq) → Cu (s) + Ni^{2+} (aq).
2. (a) The aluminum electrode is the anode and will corrode; (b) cell emf is +1.391 V; (c) 2Al (s) + $3Ni^{2+}$ (aq) → $2Al^{3+}$ (aq) + 3Ni (s).
3. $[Cd^{2+}]/[Fe^{2+}] = 0.032$.
4. (a) Pb in contact with 0.01 M Pb^{2+} is the anode. (b) Cell emf is +0.05 V.
5. (a) Zn is more negative in the emf series and in the galvanic series for seawater. (b) Sn is more negative than Cu in the emf series, but both have approximately the same potential in seawater. (c) Ni is more negative than Ag in the emf series, but both have approximately the same potential in seawater. (d) Ti is more negative than Fe in the emf series, but is more positive in the galvanic series for seawater.
6. (a) Mg can be used to cathodically protect steel in seawater. (b) Al can be used to cathodically protect steel in seawater. (c) Either Be or Cd can be used to cathodically protect steel in seawater.
7. Tin will be the anode and will corrode to protect the underlying steel substrate.

8. (a) Aluminum will be the anode. (b) No galvanic effect. (c) Potentials are near each other so that a galvanic effect is not likely. (d) Aluminum will be the anode.
9. The electrode potential of the second-phase particle Al_2CuMg is more negative than the aluminum matrix, so the second-phase particle will preferentially corrode. However, the aluminum matrix has an electrode potential more negative than the second-phase particle Al_3Fe, so that the aluminum matrix will corrode in a galvanic couple between Al and Al_3Fe.
10. Cadmium is a sacrificial coating, and Cd will corrode if a break develops in the coating down to the steel substrate.
11. If a break develops in the stainless steel coating and extends down to the plain steel substrate, the steel substrate will galvanically corrode.
12. (a) If a break develops down to the Cr layer, Cr will corrode. (b) If the break continues down to the steel substrate, the steel substrate will corrode.
13. Case B is worse. This situation involves a large area cathode connected to a smaller area anode, so that the anodic current density will be increased at the surface of the steel nut and bolt. Case B is the less troublesome situation of a small cathode connected to a larger area anode.
14. Waterline corrosion below the surface is possible due to the operation of an differential oxygen cell.

Chapter 6

1. (a) Mild steel lies just within a region of corrosion; zinc lies in a region of corrosion. (b) Coupled mild steel lies in a region of immunity; coupled zinc lies in a region of corrosion. (c) These results are consistent with predictions based on the galvanic series for seawater.
2. Corrosion of silver at pH 5 occurs between approximately 0.45 and 1.1 V vs. SHE.
3. The region of corrosion as Pd^{2+} ions lies above the "a" line for H_2 evolution. Thus, H_2 is not stable in this region of Pd corrosion, so the reaction of Pd with H^+ ions to form H_2 is not thermodynamically favored.
4. (b) Three methods of protection are (i) increase the electrode potential into a region of passivity (anodic protection), (ii) make the electrode potential more negative to move into a region of immunity (cathodic protection), (iii) increase the pH to move into a region of passivity.
5. 96.5 min.
6. $E^0 = -1.550$ V vs. SHE.
10. $E = 1.311 + 0.0197 \log [CrO_4^{2-}] - 0.0985$ pH
11. (a) Ti (s) + H_2O (l) $\rightarrow$ TiO (s) + $2H^+$ (aq) + $2e^-$; 2TiO (s) + H_2O (l) $\rightarrow$ Ti_2O_3 + $2H^+$ (aq) + $2e^-$; Ti_2O_3+ H_2O (l) $\rightarrow$ $2TiO_2$ (s) + $2H^+$ (aq) + $2e^-$ (b) $dE/dpH = -0.0591$ for each of the three reactions.
12. The Pourbaix diagram for Mg is similar in general shape to that for Al or Zn (Figs. 6.1 and 6.7).
14. The electrode potential must be more positive than -0.177 V vs. SHE.
15. The data point lies below the "a" line for H_2, so that H_2 gas is stable under the experimental conditions and is likely the gas which was observed.
16 Nb and Ta have similar Pourbaix diagrams. Nb and Ta appear in the same column in the Periodic Table and are expected to have similar chemical behavior. (See Fig. 9.24.)

Chapter 7

1. (a) 0.411 g/cm^2 h; (b) 123 mA/cm^2.
2. (a) 71.3 mg/cm^2 year. (b) The corrosion rate decreases with time. (c) The rust layer becomes more protective with time.
3. (a) 7.5 × 10^{-3} M/L; (b) 0.10 g/cm^2; (c) 41.9 mL (STP)/cm^2.
4. 3.6 mg/cm^2 h.
5. The weight fractions of metals in the alloy are Fe 0.058, Cr 0.162, Mo 0.153, Ni 0.546. Their cation fractions in solution are Fe 0.057, Cr 0.100, Mo 0.130, Ni 0.712. Thus, Ni is preferentially corroded, while the others are not.
6. $b_a = 0.0591$ V.
8. (a) $E_{\text{corr}} = -0.459$ V vs. SCE. (b) $i_{\text{corr}} = 24$ μA/cm^2.
9. (a) The overvoltage is given by $\eta = 0.0591/n$ (in V), where n is the number of electrons transferred. If $n = 1$, $\eta = 0.0591$ V. If $n = 2$, $\eta = 0.0296$ V. (b) $\eta = 0.118/n$ (in V), where n is the number of electrons transferred. If $n = 1$, $\eta = 0.118$ V. If $n = 2$, $\eta = 0.0591$ V.
10. $\log i_{\text{corr}} = (1/2) \log (A_c/A_a) +$ constant.
11. (a) $i_{\text{corr}} = 1.2 \times 10^{-5}$ A/cm^2 (b) $E_{\text{corr}} = -0.03$ V
12. 1.6 μA/cm^2.
13. $b_a = 0.073$ V.

Chapter 8

1. (a) 0.0173 cm; (b) 2005 jumps/s.
2. 3.7 h.
3. 5.6 h.
4. 35 s (using 0.065 cm as the thickness δ of the diffusion layer from Example 8.2).
5. For $\delta = 0.05$ cm, $i_L = 3.9 \times 10^{-5}$ A/cm^2. For $\delta = 0.1$ cm, $i_L = 1.9 \times 10^{-5}$ A/cm^2. (Using $D = 2.0 \times 10^{-6}$ cm^2/s and the concentration of dissolved $O_2 = 8.0$ mg/L from Fig. 8.5.)
6. (b) Increasing the flow velocity increases the limiting cathodic current density i_L for the reduction of oxygen and hence increases the corrosion rate. (c) The concentration of dissolved O_2 decreases, so i_L decreases and so does the corrosion rate.
7. The following assumptions must hold: (i) the only cathodic reaction is the reduction of O_2, (ii) the O_2 reduction reaction is under diffusion control, and (iii) the anodic and cathodic polarization curves intersect in the region of the limiting cathodic current density.
8. (b) $E_{\text{corr}} = -0.070$ V vs. SHE $= -0.31$ V vs. SCE. (c) From the galvanic series for seawater, $E_{\text{corr}} = -0.1$ to -0.2 V vs. SCE.
9. (a) 11 μA/cm^2; (b) 3.1 g/m^2 day.
10. The electrolyte in contact with the hotter end of the metal will contain less dissolved O_2 (Fig. 8.4). Thus, an oxygen concentration cell will be set up, with the hotter end of the metal becoming the anode.
11. 238 μA/cm^2.

Chapter 9

1. 0.94–1.91 monolayers of O^{-2}.
2. The calculated potential for the reduction of γ-Fe_2O_3 to Fe^{2+} is −0.214 V vs. SHE., in good agreement with the experimental value for the first plateau. The calculated potential for the

reduction of Fe_3O_4 to Fe is −0.534 V vs. SHE., in good agreement with the experimental value for the second plateau.

3. Tantalum. Based on the Pourbaix diagrams in Fig. 9.24, tantalum is either passive or immune over a large range of pH and electrode potential.
4. $E = -0.672$ V vs. SHE; not in agreement with the experimentally observed Flade potential.
6. 15.9 at.% Ni. The electron configuration theory is based on the concept of an adsorbed film and does not take into account the properties of the oxide film.
7. The effective molecular weight of the oxide is 136.17 g/mol. The effective atomic weight of the metal is 55.27 g/mol. The effective value of $n = 1.74$. The density of the oxide is 6.67 g/cm^3. The density of the metal is 7.0 g/cm^3 (given). $R = 1.49$, so that the oxide film is protective.
8. The addition of Mo to Fe–Ni alloys is predicted to be detrimental. This trend is in agreement with the text, which states that Mo is beneficial only when Cr is present.
9. (a) $i_{corr} \approx 400$ μA/cm^2; (b) 0 to +0.3 V vs. SCE; (c) 30 μA/cm^2; (d) 9.5×10^{-4} cm penetration; (e) at −0.7 V vs. SCE, $i_{anodic} \approx 30$ μA/cm^2.
10. (a) The anodic and cathodic polarization curves intersect in a region of active corrosion. (b) The anodic and cathodic curves intersect in a region of passivity. (c) The situation is unstable and the system oscillates between a state of active corrosion and one of passivity.

Chapter 10

1. Pitting and crevice corrosion are similar in their propagation stages. Each of these two forms of localized corrosion results in the formation of an internal solution of low pH and concentrated in Cl^- ions and in metal cations. Pitting and crevice corrosion differ in their initiation stages. Crevice corrosion initiates by means of a differential O_2 cell, whereas pitting involves the localized breakdown of an oxide film (by one of several different mechanisms).
3. (c) Both the 1.0 cm^2 crevice and the 2.0 cm^2 crevice have the same weight loss, but the smaller sample has the greater weight loss per unit area.
6. The pitting potential is $E_{pit} = -0.70$ V vs. S.C.E.
7. The current density within the pit is approximately 1,900 A/cm^2.
8. (a) The pitting potential for pure Al in 0.6 M Cl^- is −0.74 V vs. S.C.E. = −0.72 V vs. Ag/AgCl. (b) The pitting potential of Al alloys in seawater from Table 10.5 is −0.88 V vs. Ag/AgCl. (c) Agreement is fair. Differences may be caused by the following reasons: (i) Use of 0.6 M Cl^- does not satisfactorily simulate seawater, which contains other ions as well as organic matter. (ii) Short-term laboratory tests do not represent data obtained over a much longer period of immersion. (iii) We have considered laboratory data for pure Al rather than for the Al alloy of interest.
9. (b) Safe ranges are more negative than −0.7 V vs. SCE (immunity) or between −0.35 V and −0.58 V vs. SCE. (perfect passivity).
10. Alloy B (but not alloy A) can be used instead of 304 stainless steel. Alloy B has a higher pitting potential in 1 M HCl than does the other two alloys and has a critical current density for passivation similar to that for 304 stainless steel and lower than that of alloy A.
11. With increased temperature the rate increases for all individual steps which lead to pitting. This includes the rate of Cl^- transfer within the oxide film, the rate of oxide thinning, and the rate of electrochemical reactions at the metal/oxide interface. Thus, the metal can be driven to a lower electrode potential to cause pitting to occur.

Chapter 11

2. $K_I = 38$ MPa $\sqrt{m}$.
4. Crack length $2a = 0.10$ in. = 2.5 mm.
5. (a) $K_{Iscc} = 45$ ksi $\sqrt{in.}$; (b) $\sigma = 51$ ksi; (c) $K_I = 19$ ksi $\sqrt{in.}$ is less than K_{Iscc} so that fracture does not occur.
6. Smooth specimens do not fracture below the threshold yield stress $\sigma_{th} = 48$ MPa. For pre-cracked specimens, the stress is given by $\sigma = 3.103/\sqrt{a}$. Pre-cracked specimens do not fracture below the intersection of this line and the yield stress, $\sigma_Y = 410$ MPa. The safe-zone region lies below the intersection of the lines $\sigma = 3.103/\sqrt{a}$ and $\sigma_{th} = 48$ MPa.
7. Making the electrode potential more negative decreases K_{Iscc} (or increases the susceptibility to stress-corrosion cracking). (If K_{Iscc} decreases, the critical flaw size also decreases for the same applied stress σ.) Hydrogen embrittlement is a possible mechanism of stress-corrosion cracking because the alloy is more susceptibile with increasing negative potentials, where more cathodic charging with hydrogen gas can occur.
8. 1.2 A/cm^2.
9. The time to failure decreases for both cathodic and anodic currents. Thus, the mechanism for stress-corrosion cracking must be a mixed-mode mechanism, involving both hydrogen embrittlement and anodic dissolution within the crack.
11. 1,563 cycles.

Chapter 12

2. For 1 M HCl, $\%I = 88\%$; for 4 M HCl, $\%I = 60\%$.
3. A plot of C/θ vs. C gives a straight line of unit slope.
4. Calculate θ from $\theta = (i_0 - i)/(i_0 - i_{sat})$. Then, a plot of θ vs. log C gives a straight line over a portion of the plot, indicative of a Temkin isotherm. Various tests for the Langmuir isotherm fail.
6. (a) The S-containing compound is better because the S atom is a better electron donor than the N atom. (b) $CH_3(CH_2)_{16}COOH$ is better because it has the longer chain length and thus presents a greater barrier to the aqueous solution when a close-packed monolayer is formed on the metal surface. (c) Both compounds have similar base strengths, but the diamine (second compound) is better because it offers more attachment possibilities to the metal surface. Either of the two $-NH_2$ groups can adsorb with the molecule in the vertical configuration or both $-NH_2$ groups can adsorb with the molecule being in the flat configuration. (d) The $-CH_3$ compound (second compound) is better. This is because the p-CH_3 group is an electron donor and provides electrons to the ring. The Hammet sigma value $\sigma(p\text{-}CH_3) = -0.17$. The Cl-containing compound is an electron acceptor and withdraws electrons from the ring. The Hammet sigma value for the Cl group is $\sigma(p\text{-}Cl) = +0.23$. (e) The second compound is better. It contains two electron-donating groups, so that the electron density on the N atom is higher for the second compound.
8. (a) For the conditions given, $E = +1.103$ V vs. SHE for chromate reduction, and $E = -0.153$ V vs. SHE For tungstate reduction, (b) under the given conditions, chromate reduction is spontaneous, but tungstate reduction is not, by $\Delta G = -nFE$.
9. The corrosion of metals in nearly neutral solutions differs from that in acidic solutions for two reasons. (1) In acid solutions, the metal surface is oxide-free, but in neutral solutions, the surface is oxide-covered. (2) In acid solutions the main cathodic reaction is hydrogen evolution,

but in air-saturated neutral solutions, the cathodic reaction is oxygen reduction. In each case the inhibitor must interact with the metal or oxide-covered metal surface. For acid solutions, adsorption alone may be sufficient to cause inhibition. But in nearly neutral solutions, the inhibitor must be able to enhance the protective properties of the oxide film. Possibilities include (i) formation of surface chelates, (ii) incorporation into the film of an oxidizing inhibitor (such as CrO_4^{2-}), which itself is reduced to Cr_2O_3, (iii) incorporation of the inhibitor into the oxide film or into its pores, (iv) formation of a precipitate on the oxide film.

10. If the inhibitor does not completely cover the metal surface, there can be a difference in potential between the covered sites and the uncovered sites. This difference can cause a galvanic effect, so that the uncovered sites act as local anodes, and thus localized corrosion may occur.

Chapter 13

1. When an anodic site is activated beneath an organic coating, the local solution becomes acidified and chloride ions migrate into the vicinity of the site, as in crevice corrosion. The cathodic reaction (O_2 reduction) also occurs beneath the organic coating. This is different than the usual case of crevice corrosion, in which the principal cathodic reaction takes place outside the crevice. The major damage to the organic coating is caused in alkaline regions (cathodic) rather than in the acidic region (anodic). This is different than the usual case of crevice corrosion where the main corrosion damage occurs within the crevice (anodic region).
2. Filiform corrosion is different in its geometric appearance than the usual type of underfilm corrosion, but the mechanism of detachment is similar in both cases. Cathodic regions beneath the coating exist in each case and can give rise to cathodic delamination.
5. (a) 33,000 layers of H_2O molecules, (b) 9.9% of the thickness of the organic coating contains H_2O molecules.
6. (a) In 1 day the coating will transmit 1.67×10^{-4} mol H_2O, as compared to 7.81×10^{-8} mol O_2. To use up all the H_2O, we would need 8.4×10^{-5} moles O_2, but we have much less. Thus, O_2 is the limiting reactant. (b) 0.35 $\mu A/cm^2$.
7. (a) The data point lies in the region of the Pourbaix diagram for aluminum where Al goes into solution as AlO_2^-. (b) Thus, the oxide film is not stable so that dissolution of the oxide is a possible mechanism of delamination. However cathodic delamination due to failure of the polymer in the alkaline environment cannot be ruled out.
8. 92 mg/cm^2 day.
9. From Figs. 13.9 and 13.10, poly(vinyl acetate) is a basic polymer and poly(vinyl fluoride) is an acidic polymer. Thus, the basic polymer, poly(vinyl acetate), would be expected to have the greater adhesion to the acidic oxide.
10. (a) Desirable properties are that polytetrafluoroethylene (PTFE) has good corrosion resistance in most solutions and is also a hydrophobic polymer which is not wetted by aqueous solutions. In addition, PTFE has a low coefficient of friction and is a good electrical insulator. (b) An undesirable property is that when drastically overheated, PTFE can release toxic gases. In addition, perfluorooctanoic acid, which is used to make PFTE, is a carcinogen. It is also difficult to get PTFE to adhere to the metal substrate without using certain adhesion promoters. (Adhesion promoters are often bifunctional chemical agents which act as molecular bridges between two chemically different materials. Adhesion promoters improve adhesion by chemically bonding to both types of materials simultaneously.)

11. (a) Measuring the electrode potential of an underground coated system would be useful in finding holidays (breakthroughs) in the coating. (b) Similarly, measuring the polarization resistance of specimens of the coated metal could give information on the integrity of the coating. Values of R_p would decrease if there is a breakthrough in the coating.

Chapter 14

7. $|Z| = [R^2 + 1/(\omega^2 C^2)]^{1/2}$.
8. 12 μA/cm^2.
9. $R_p = 1{,}910\ \Omega$; $R_s = 10\ \Omega$, $C_{dl} = 3.1 \times 10^{-5}$ F.
10. The low-frequency value of $|Z|$ is the parameter of interest because it contains the values of the polarization resistance R_p and the pore resistance R_{cp} of the coating. $|Z| \rightarrow R_p + R_{cp} + R_s$ as $\omega \rightarrow 0$. At low frequencies, $|Z|$ is larger when either plasticizer is contained in the organic coating. Benzyl alcohol is the better of the two plasticizers.
11. 4.4×10^{-7} F/cm^2.

Chapter 15

1. Based on thermodynamics, there will be an increase in the stability of the oxide. When iron is alloyed with aluminum, the oxide will be a mixed oxide consisting of Fe_2O_3 and Al_2O_3, with Al_2O_3 being more stable than Fe_2O_3, per the Ellingham diagram.
2. (a) $K = 1/P_{O_2}$; (b) $P_{O_2} = 1.0 \times 10^{-10}$;(c) the two results are the same.
3. $P_{O_2} = 1.2 \times 10^{-41}$ atm, so that oxidation is possible in air at 1,200 K.
4. The observation that dust particles become enveloped within the rust layer means that the rust layer grows both from the outside oxide surface inward and from the metal surface outward. This is similar to the Wagner mechanism of high-temperature oxidation.
6. (a) According to the Ellingham diagram, the addition of Al to Ni should result in a more stable alloy. But the oxidation rate increases with Al content, in disagreement with the Ellingham diagram. (b) From Table 15.5, NiO is a p-type oxide with cation vacancies. From the Hauffe rules in Table 15.6, a solute of higher oxidation number (Al^{3+}) results in an increase in oxidation rate, in agreement with the Hauffe rules.
7. (a) 5.0×10^{-3} g/cm^2 after 20 h; (b) 7.5 μm.
8. According to Table 15.5, Al_2O_3 is an n-type oxide with anion vacancies. The addition of a solute like Li^+ with an oxidation number lower than that of Al^{3+} causes an increase in the oxidation rate, according to the Hauffe rules in Table 15.6.
9. The parabolic rate law gives the best fit.
10. 37.7 mg/cm^2.
11. (a) Replacing U^{4+} with Al^{3+} reduces the oxidation rate. (b) Replacing U^{+4} with Ta^{+5} increases the oxidation rate.
12. $k_{rational} = 6.7 \times 10^{-9}$ equiv cm^{-1}s^{-1}.

Chapter 16

1. (a) $i_a/F = k_3\theta_{OH^-}$;(b) $k_1k_2/\{(k_1 + k_{-1}\beta)\ (k_{-2}[H^+])\}$;(c) $b_a = 0.118$ V; (d) $z_{H^+} = -1$.
2. 2.5 mA/cm^2.

3. $5.0 \times 10^{-4} \Omega^{-1} \text{cm}^{-1}$.
4. Surface charges at pH 7.0 are MoO_3 negative, ZrO_2 negative, CdO positive.
5. Surface charges at pH 11.0 are MoO_3 negative, ZrO_2 negative, CdO negative.
6. At pH 8.0, the oxide-covered surface has a positive charge. Thus, a negatively charged corrosion inhibitor can be adsorbed at pH 8.0.
7. (a) The concentration of implanted Cr^{3+} ions at the surface is 33 at.%. (b) This concentration is higher than the critical concentration of 13 at.% Cr required for passivity in Fe–Cr binary alloys.
8. (a) At pH 6.0, the surface of oxide-covered copper has a positive charge. We want to implant copper with a metal whose oxide will have a negative charge at pH 6.0 in order to replace a portion of the positively charge surface with negative charges (to minimize Cl−uptake). Thus, any oxide having an isoelectric point less than 6.0 will do. Possibilities are Ce, Ti, Ta, Si, and Mo. (b) We are assuming that the implanted ion is contained in the oxide film and that the isoelectric point for an oxide film is similar to that for bulk oxides.
9. (a) In neutral solutions, the surface of oxide-covered Fe is positively charged, and surface of oxide-covered Si is negatively charged. Thus implantation of Si into Fe will convert a portion of the positively charged surface into negative charges. The uptake of Cl^- will be reduced, so protection against pitting will be increased. (b) Fe_2O_3 is an n-type semiconductor with anion vacancies (Table 15.6). Insertion of Si^{4+} ions will decrease the concentration of anion vacancies (Table 15.6). This reduces ionic diffusion through the oxide film so that protection against pitting will be increased. (c) Each argument is consistent with the other, so we can conclude that implantation of Si into Fe provides increased resistance to pitting.
10. Ion implantation has the following limitations: (i) the depth of the modified region is shallow (hundreds to thousands of angstroms); (ii) ion implantation is a line-of-sight process so that recesses in the sample geometry cannot be easily ion implanted; (iii) large-sized samples cannot be processed readily.
11. The resulting stress is a compressive stress which is beneficial to protection against corrosion fatigue.
12 Laser processing produces a rough surface containing surface ripples which must be machined if a smoother surface finish is desired. The depth of modified region is thicker than in the case of ion implantation, but is still limited to the near-surface of the order of hundreds of microns deep.

Appendix A
Some Properties of Various Elemental Metals

Element	Symbol	Atomic number	Atomic weight (g/mol)[a]	Density (g/cm^3)[a]	Oxidation number of most common aqueous ions[b, c]	
					Mono-atomic	Oxyions
Lithium	Li	3	6.941	0.534	+1	
Beryllium	Be	4	9.012	1.85	+2	
Boron	B	5	10.811	2.34		+3 (BO_3^{3-}) +3 ($B_4O_7^{2-}$)
Sodium	Na	11	22.990	0.97	+1	
Magnesium	Mg	12	24.305	1.74	+2	
Aluminum	Al	13	26.982	2.70	+3	+3 (AlO_2^-)
Silicon	Si	14	28.086	2.3296		+4 (SiO_3^{2-})
Potassium	K	19	39.098	0.89	+1	
Calcium	Ca	20	40.078	1.54	+2	
Scandium	Sc	21	44.956	2.99	+3	
Titanium	Ti	22	47.867	4.506	+2, +3, +4	+4 (TiO_3^{2-})
Vanadium	V	23	50.942	6.0	+2, +3	+5 (VO_4^{3-})
Chromium	Cr	24	51.996	7.15	+2, +3	+6 (CrO_4^{2-})
Manganese	Mn	25	54.938	7.3	+2, +3	+7 (MnO_4^-)
Iron	Fe	26	55.845	7.87	+2, +3	+2 (FeO_2^{2-})
Cobalt	Co	27	58.933	8.86	+2, +3	+2 (CoO_2^{2-})
Nickel	Ni	28	58.693	8.90	+2	+2 (NiO_2^{2-})
Copper	Cu	29	63.546	8.96	+1, +2	+2 (CuO_2^{2-})
Zinc	Zn	30	65.409	7.134	+2	+2 (ZnO_2^{2-})
Gallium	Ga	31	69.723	5.91	+2, +3	+3 (GaO_3^{3-})
Germanium	Ge	32	72.64	5.3234	+2, +4	+4 (GeO_3^{2-})
Rubidium	Rb	37	85.468	1.53	+1	
Strontium	Sr	38	87.62	2.64	+2	
Yttrium	Y	39	88.906	4.47	+3	
Zirconium	Zr	40	91.224	6.52	+4	+4 (ZrO_3^{2-})
Niobium	Nb	41	92.906	8.57	+3	
Molybdenum	Mo	42	95.94	10.2	+3	+6 (MoO_4^{2-})
Technetium	Tc	43	98	11	+2	+7 (TcO_4^-)
Ruthenium	Ru	44	101.07	12.1	+2, +3	+6 (RuO_4^{2-})
Rhodium	Rh	45	102.906	12.4	+1, +2, +3	+6 (RuO_4^{2-})

Element	Symbol	Atomic number	Atomic weight (g/mol)[a]	Density (g/cm^3)[a]	Oxidation number of most common aqueous ions[b, c]	
					Mono-atomic	Oxyions
Palladium	Pd	46	106.42	12.0	+2	+2 (PdO_2^{2-})
Silver	Ag	47	107.868	10.5	+1, +2	
Cadmium	Cd	48	112.411	8.69	+2	+2 (CdO_2^{2-})
Indium	In	49	114.818	7.31	+1, +2, +3	+3 (InO_2^{-})
Tin	Sn	50	118.710	7.287	+2, +4	+4 (SnO_3^{2-})
Antimony	Sb	51	121.760	6.68		+3 (SbO_2^{-}) +5 (SbO_3^{-})
Cesium	Cs	55	132.905	1.93	+1	
Barium	Ba	56	137.327	3.62	+2	
Lanthanum	La	57	138.905	6.15	+3	
Cerium	Ce	58	140.116	6.770	+3, +4	
Hafnium	Hf	72	178.49	13.3	+4	
Tantalum	Ta	73	180.948	16.4	+3	+5 (TaO_3^{-})
Tungsten	W	74	183.84	19.3		+6 (WO_4^{2-})
Rhenium	Re	75	186.207	20.8	+3	+6 (ReO_4^{2-}) +7 (ReO_4^{-})
Osmium	Os	76	190.23	22.587		+6 (OsO_4^{2-})
Iridium	Ir	77	192.217	22.562	+3	+6 (IrO_4^{2-})
Platinum	Pt	78	195.084	21.5	+2	+6 (PtO_4^{2-})
Gold	Au	79	196.967	19.3	+1, +3	+3 (AuO_3^{3-})
Mercury	Hg	80	200.59	13.534	+1, +2	
Thallium	Tl	81	204.383	11.8	+1, +3	
Lead	Pb	82	207.2	11.3	+2, +4	+2 (PbO_2^{2-}) +4 (PbO_3^{2-})
Bismuth	Bi	83	208.980	9.79	+3	
Polonium	Po	84	209	9.20	+2	+4 (PoO_3^{2-})
Radium	Ra	88	226	5	+2	

[a]"*Handbook of Chemistry and Physics*", CRC Press (online Edition) (2002).
[b]M. Pourbaix, "*Atlas of Electrochemical Equilibrium*", National Association of Corrosion Engineers, Houston, TX (1974).
[c]P. Vanysek, "*Electrochemical Series*" in "*Handbook of Chemistry and Physics*", CRC Press (online Edition) (2002).

Appendix B
Thermodynamic Relationships for Use in Constructing Pourbaix Diagrams at High Temperatures

Consider the thermodynamic cycle shown in Fig. 6.19. At constant pressure:

$$C_p^{\text{react}}(T - 298) + \Delta H_{\text{T}} + C_p^{\text{prod}}(298 - T) - \Delta H_{298} = 0 \quad \text{(B1)}$$

where $C_{\text{p}}^{\text{prod}}$ and $C_{\text{p}}^{\text{react}}$ refer to the heat capacities at constant pressure of products and reactants, respectively. Thus,

$$\Delta H_T - \Delta H_{298} = (C_p^{\text{prod}} - C_p^{\text{react}})(T - 298) \quad \text{(B2)}$$

or:

$$\Delta \text{H}_{\text{T}} - \Delta \text{H}_{298} = \Delta \text{C}_{\text{p}}(\text{T} - 298) \quad \text{(B3)}$$

where $\Delta C_p = C_p^{\text{prod}} - C_p^{\text{react}}$. When the products and reactants are in their standard states, then Eq. (B3) becomes:

$$\Delta H_T^0 - \Delta H_{298}^0 = \Delta C_p^0 (T - 298) \quad \text{(B4)}$$

or:

$$\Delta H_T^0 - \Delta H_{298}^0 = \int_{298}^{T} \text{d}(\Delta H) = \int_{298}^{T} \Delta C_p^0 \text{d}T) \quad \text{(B5)}$$

Because $dS = \text{d}H/T$ at constant pressure,

$$\Delta S_T^0 - \Delta S_{298}^0 = \int_{298}^{T} \frac{d(\Delta H)}{T} = \int_{298}^{T} \Delta C_p^0 \, \text{d} \ln T) \quad \text{(B6)}$$

Then

$$\Delta G_T^0 = \Delta H_T^0 - T \, \Delta S_T^0 \quad \text{(B7)}$$

and

$$\Delta G_{298}^0 = \Delta H_{298}^0 - 298 \, \Delta S_{298}^0 \quad \text{(B8)}$$

combine to give:

$$\Delta G_T^0 - \Delta G_{298}^0 = (\Delta H_T^0 - \Delta H_{298}^0) - T(\Delta S_T^0 - \Delta S_{298}^0) - (\Delta T)\Delta S_{298}^0) \tag{B9}$$

where $\Delta T = T - 298$. Use of Eqs. (B5) and (B6) in Eq. (B9) gives the result:

$$\Delta G_T^0 = \Delta G_{298}^0 + \int_{298}^{T} \Delta C_P^0 \, \mathrm{d}T - T \int_{298}^{T} \Delta C_P^0 \, \mathrm{d}\ln T - (\Delta T)\Delta S_{298}^0 \tag{B10}$$

Equation (B10) is the expression for relating the standard free energy change at some elevated temperature T to the standard free energy change at 25°C. Equation (B10) is the same as Eq. (30) in Chapter 6. The heat capacities required in Eq. (B10) are either measured or estimated. Ionic entropies are usually estimated by the empirical correlation method of Criss and Cobble [B1, B2] which relates entropies at elevated temperatures to entropies at 298 K.

References

B1. C. M. Criss and J. W. Cobble., *J. Am. Chem. Soc.*, *86*, 5385 (1964).
B2. J. W. Cobble, *J. Am. Chem. Soc.*, *86*, 5394 (1964).

Appendix C
Relationship Between the Rate Constant and the Activation Energy for a Chemical Reaction

For the general reaction

$$A + B \rightarrow [AB]^{\neq} \rightarrow \text{products} \tag{C1}$$

where $[AB]^{\neq}$is the activated complex; see Fig. 7.14. The rate of the reaction depends on the concentration of the activated complex and its rate of passage over the energy barrier; That is,

$$\text{rate of reaction} = \begin{pmatrix}\text{concentration}\\ \text{of complex}\end{pmatrix} \times \begin{pmatrix}\text{rate of passage}\\ \text{over energy barrier}\end{pmatrix} \tag{C2}$$

When the activated complex is poised at the top of the energy barrier, its vibrational energy ($h\nu$) is just equal to the thermal energy (kT). That is, $h\nu = kT$, where h is Planck's contant, ν is the frequency of vibration of the complex, and k is Boltzmann's constant. Thus, $\nu = (kT/h)$ is the rate of passage of activated complexes over the energy barrier. Then, Eq. (C2) becomes

$$\text{rate} = [AB^{\neq}]\frac{kT}{h} \tag{C3}$$

The equilibrium constant for the formation of the activated complex is

$$K^{\neq} = \frac{[AB^{\neq}]}{[A][B]} \tag{C4}$$

or

$$[AB^{\neq}] = K^{\neq}[A][B] \tag{C5}$$

Also, $\Delta G^{\neq} = -RT \ln K^{\neq}$ gives

$$K^{\neq} = e^{-\Delta G^{\neq}/RT} \tag{C6}$$

Use of Eqs. (C5) and (C6) in Eq. (C3) gives

$$\text{rate} = [A][B]\frac{kT}{h}e^{-\Delta G^{\neq}/RT} \tag{C7}$$

But from classical kinetics,

$$\text{rate} = (\text{rate constant})\,[\text{A}]\,[\text{B}] \tag{C8}$$

Comparison of Eqs. (C7) and (C8) gives the result

$$\text{rate constant} = \frac{kT}{h} e^{-\Delta G^{\neq}/RT} \tag{C9}$$

which is the same as Eq. (6) in Chapter 7.

Appendix D
Random Walks in Two Dimensions

We want to calculate the root mean square distance $\sqrt{< d^2 >}$ which a particle can move after N two-dimensional steps. Suppose that after $N-1$ jumps, an atom or ion has the co-ordinates x_{N-1} and y_{N-1}, as shown in Fig. D.1. As described in Chapter 8, let the jump distance be l units and suppose that the particle can jump in any of the eight discrete directions shown in Fig. D.2.

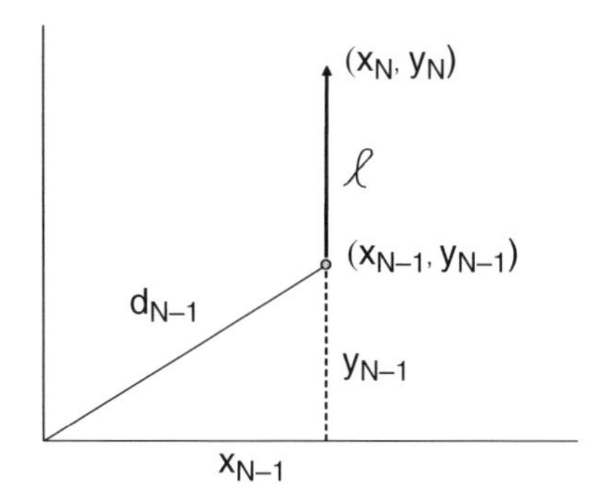

Fig. D.1 A particle jumps from a given site to a new site one jump distance in the North direction

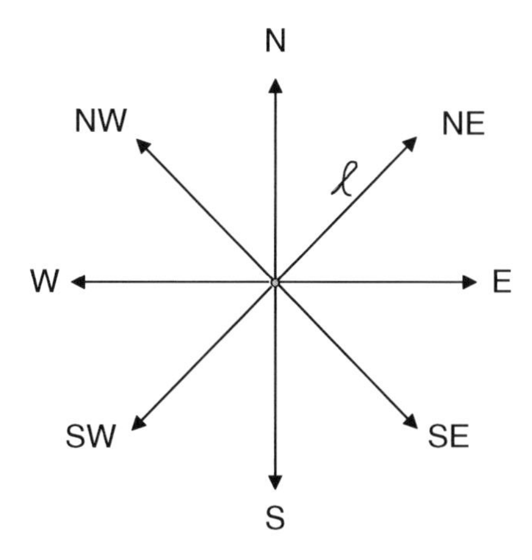

Fig. D.2 Diagram showing the possible jump directions considered in this example of a two-dimensional walk

Suppose that the particle jumps from co-ordinates (x_{N-1}, y_{N-1}) in the north direction (N), as shown in Fig. D.1. Then,

$$x_N = x_{N-1} \tag{D1}$$

and

$$y_N = y_{N-1} + l \tag{D2}$$

Squaring both sides of these two equations gives

$$X_N^2 = X_{N-1}^2 \tag{D3}$$

and

$$Y_N^2 = Y_{N-1}^2 + 2\,l\,y_{N-1} + l^2 \tag{D4}$$

Adding Eqs. (D3) and (D4) gives

$$x_N^2 + y_N^2 = x_{N-1}^2 + y_{N-1}^2 + 2l y_{N-1} + l^2 \tag{D5}$$

Equation (D5) can be written as

$$d_N^2 = d_{N-1}^2 + w^2 \tag{D6}$$

where

$$d_{N-1}^2 = x_{N-1}^2 + y_{N-1}^2 \tag{D7}$$

and

$$w^2 = 2\,l\,y_{N-1} + l^2 \tag{D8}$$

Instead of jumping from co-ordinates (x_{N-1}, y_{N-1}) in the north direction (N), suppose that the particle jumps in the northeast direction (NE). Then, from simple geometry, as shown in Fig. D.3:

$$x_N = x_{N-1} + \frac{l}{\sqrt{2}} \tag{D9}$$

$$y_N = y_{N-1} + \frac{l}{\sqrt{2}} \tag{D10}$$

Squaring both sides in Eqs. (D9) and (D10) and adding gives

$$x_N^2 + y_N^2 = x_{N-1}^2 + y_{N-1}^2 + \frac{2\,l}{\sqrt{2}}(x_{N-1} + y_{N-1}) + l^2 \tag{D11}$$

which again has the form of Eq. (D6), except now that

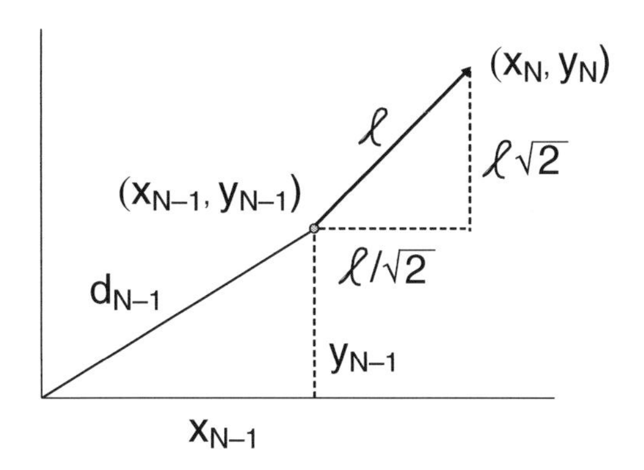

Fig. D.3 The particle jumps from the second location in Fig. D.1 to a new site one jump distance in the Northeast direction

$$w^2 = \frac{2\,l}{\sqrt{2}}\,(x_{N-1} + y_{N-1}) \,+\, l^2 \tag{D12}$$

Thus, we can continue around the compass in Fig. D.2 and generate a set of equations (as per Eq. (D6)), where w^2 has the form for each case as given in Table D.1.

Table D.1 Forms for w^2 in $d^2{}_N = d^2{}_{N-1} + w^2$

Direction of jump	w^2
N	$w^2 = 2\,ly_{N-1} + l^2$
NE	$w^2 = \frac{2\,l}{\sqrt{2}}(x_{N-1} + y_{N-1}) + l^2$
E	$w^2 = 2\,lx_{N-1} + l^2$
SE	$w^2 = \frac{2\,l}{\sqrt{2}}(x_{N-1} - y_{N-1}) + l^2$
S	$w^2 = -2\,ly_{N-1} + l^2$
SW	$w^2 = \frac{2\,l}{\sqrt{2}}(-x_{N-1} - y_{N-1}) + l^2$
W	$w^2 = -2\,lx_{N-1} + l^2$
NW	$w^2 = \frac{2\,l}{\sqrt{2}}(-x_{N-1} + y_{N-1}) + l^2$

The average value of w^2 should be l^2. From Table D.1, it can be seen that adding up the various values of w^2 and dividing by 8 gives

$$\overline{w^2} = \frac{8\,l^2}{8} = l^2 \tag{D13}$$

Thus, starting at the origin and making one jump gives

$$d_1^2 = d_0^2 + l^2 = l^2 \tag{D14}$$

After the second jump

$$d_2^2 = d_1^2 + l^2 = 2\,l^2 \tag{D15}$$

After the third jump

$$d_3^2 = d_2^2 + l^2 = 3\,l^2 \tag{D16}$$

Thus, in general,

$$d_N^2 = Nl^2 \tag{D17}$$

or, the expected root-mean square value of d is

$$< d^2 >= Nl^2 \tag{D18}$$

which is Eq. (5) in Chapter 8. Thus,

$$\sqrt{< d^2 >} = \sqrt{Nl^2} \tag{D19}$$

It can be seen that if the jump direction is considered for a larger number of jump directions that in turn approach a continuum, the result will be the same as in Eq. (D19).

Appendix E
Uhlig's Explanation for the Flade Potential on Iron

Uhlig assumed that the following surface reaction was responsible for the passivation of iron:

$$Fe(s) + 3H_2O\,(1) \rightarrow Fe(O_2{\cdot}O)_{ads} + 6H^+(aq) + 6e^- \tag{E1}$$

where $Fe(O_2{\cdot}O)_{ads}$ refers to the chemisorbed monolayer with a second layer of adsorbed O_2 molecules. The approach is to calculate the standard free energy change ΔG^0 for Eq. (E1), and then to get E^0 from $\Delta G^0 = -nFE^0$. The standard free energy change for Eq. (E1) is

$$\Delta G^0 = \mu^0\,(Fe(O_2{\cdot}O)_{ads}) + 6\mu^0(H^+(aq)) - \mu^0((Fe(s)) - 3\mu^0((H_2O(1)) \tag{E2}$$

Recall that $\mu^0(H^+(aq)) = 0$ and $\mu^0((Fe(s)) = 0$. The value for $\mu^0((H_2O\ (l))$ has been tabulated, but the value for $\mu^0(Fe(O_2.O)_{ads})$ is not available and must be calculated. Uhlig calculated $\mu^0(Fe(O_2.O)_{ads})$ by considering the following surface reaction:

$$Fe\,(s) + \frac{3}{2}O_2 \rightarrow Fe\,(O_2{\cdot}O)_{ads} \tag{E3}$$

Uhlig then used

$$\Delta G^0_S = \Delta H^0_S - T\Delta S^0_S \tag{E4}$$

where the subscripts refer to the surface reaction in Eq. (E3). With experimental data reported in the surface chemistry literature for the heat of adsorption $\Delta H_s{}^0$ and entropy of adsorption $\Delta S_s{}^0$ of oxygen on iron, Eq. (E4) gives

$$\Delta G^0_S = \left(\frac{-75{,}000\,\text{cal}}{\text{mol}\,O_2}\right)\left(\frac{3}{2}\text{mol}\,O_2\right) - 298^\circ\left(\frac{-46.2\,\text{cal}}{\text{deg mol}\,O_2}\right)\left(\frac{3}{2}\text{mol}\,O_2\right) \tag{E5}$$

or, $\Delta G_s{}^0 = -\ 91{,}848$ cal per mole of $Fe(O_2{\cdot}O)_{ads}$. This is also the value for $\mu^0(Fe(O_2.O)_{ads})$ because Eq. (E3) involves the formation of $Fe(O_2{\cdot}O)_{ads}$ from its elements.

Use of $\Delta G_s{}^0 = -\ 91{,}848$ cal for $Fe(O_2.O)_{ads}$ in Eq. (E2) along with the value of $-56{,}690$ cal/mole for liquid water gives

$$\Delta G^0 = (1\,\text{mol}\,Fe(O_2{\cdot}O)_{ads})\left(\frac{-91{,}848\,\text{cal}}{\text{mol}}\right) - (3\,\text{mol}\,H_2O)\left(\frac{-56{,}690\,\text{cal}}{\text{mol}}\right)$$

or, $\Delta G^0 = 78{,}222$ cal mol in Eq. (E1). Then,

$$\Delta G^0 = -nFE^0 \tag{E6}$$

gives

$$(78{,}222\,\text{cal})\left(\frac{4.184\,\text{J}}{\text{cal}}\right) = -(6\,\text{equiv})\left(\frac{96{,}500\,\text{coul}}{\text{equiv}}\right)E^0 \tag{E7}$$

or, $E^0 = -\ 0.57$ V vs. SHE. But this value is for the oxidation reaction and the standard electrode potential is for the reduction reaction, so that $E^0 = +\,0.57$ V vs. SHE. This value is in good agreement with the Flade potential of +0.58 V for Franck's data for iron in sulfuric acid.

Appendix F
Calculation of the Randic Index $X(G)$ for the Passive Film on Fe–Cr Alloys

From Chapter 9, we begin with a hexagonal graph of Cr_2O_3 which contains Fe^{3+} ions substituted for some Cr^{3+} ions to form a mixed oxide, $x\mathrm{Fe_2O_3}\cdot(1-x)\ \mathrm{Cr_2O_3}$. The original hexagonal graph G_0 for Cr_2O_3 contains N vertices (Cr^{3+} ions) and (3/2) N edges (O^{2-} ions). In the new graph G for the mixed oxide D edges have been deleted one Fe^{3+} substituted per edge deletion. From Eq. (18) in Chapter 9:

$$\frac{D}{N} = \frac{x}{1-x} \tag{F1}$$

where x is the mole fraction of Fe_2O_3 in the mixed oxide.

To calculate $X(G)$, we would need to know which edges have been deleted from G_0. However, the process of edge deletion is random, so we follow the procedure used by Meghirditchian [F1] in analyzing the network of a silica glass. For $X(G)$, we use its expected value $E[X(G)]$ given by

$$E[X(G)] = \sum_{i,j} (ij)^{-0.5} E[A_{ij}^{(G)}] \tag{F2}$$

where $A_{ij}^{(G)}$ is the number of edges in G connecting vertices of degrees i and j. We calculate the expected value of $E[A_{ij}^{(G)}]$ following Meghirditchian [F1].

Edge deletion in the hexagonal network G_0 can be treated by considering two adjacent vertices k and l in G_0 and the connecting edge (kl), as shown in Fig. F.1. Suppose that after edge deletion, the edge (kl) remains undeleted, but the degrees of vertices k and l become i and j, as also shown in Fig. F.1. To do this, we

(1) delete $(3 - i)$ edges incident with the vertex k,
(2) delete $(3 - j)$ edges incident with vertex l, and
(3) delete $(D - 6 + i + j)$ edges from the remaining $(3/2)N - 5$ edges (See Table F.1).

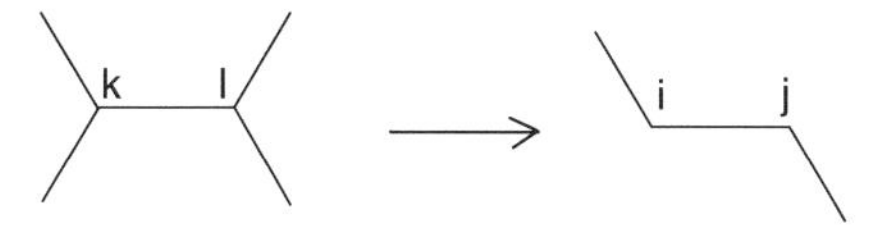

Fig. F.1 Edge deletion in a hexagonal network. After edge deletion, the vertices of degree k and l become i and j (shown here for the case where i and j are both two)

Table F.1 Edge deletion in a hexagonal array

Type edge	Number of edges	Degree of vertex after edge deletion	Possibilities	Number edges deleted	Ways to delete edges
kl	1	----	----	----	----
Incident with *k* 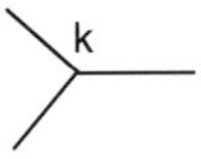	2	*i*	$i = 2$ 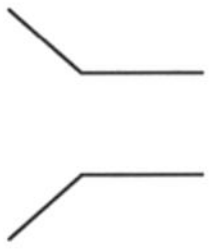	$3 - i$	$\binom{2}{i-1}$
			$i = 1$ 	$3 - i$	$\binom{2}{i-1}$
Incident with *l* 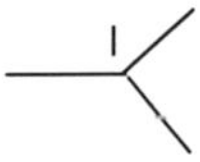	2	*j*	$j = 2$ 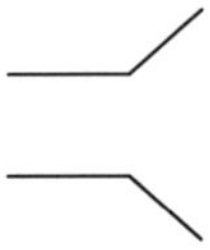	$3 - j$	$\binom{2}{j-1}$
			$j = 1$ 	$3 - j$	$\binom{2}{j-1}$
Remaining edges	$\frac{3}{2}N - 5$			$D - 6 + i + j$	$\binom{\frac{3}{2}N - 5}{D - 6 + i + j}$

The probability P_{ij} of this occurrence is [F1, F2]

$$P_{ij} = \frac{\binom{2}{i-1}\binom{2}{j-1}\binom{\frac{3}{2}N - 5}{D - 6 + i + j}}{\binom{\frac{3}{2}N}{D}} \tag{F3}$$

Recall that

$$\binom{a}{b} = \frac{a!}{b!\,(a-b)!}$$

From the properties of factorial numbers and for large N and D, Eq. (F3) becomes (after some algebra)

$$P_{ij} = \frac{\binom{2}{i-1}\binom{2}{j-1}\left(\frac{\frac{3}{2}N-D}{D}\right)^{i+j-1}}{\left(\frac{\frac{3}{2}N}{D}\right)^{5}} \tag{F4}$$

which is the probability of any single edge (kl) in G_0 becoming an edge of G joining vertices of degrees i and j. Then the expected value for the total number of edges in G joining vertices of degrees i and j is

$$E[A_{ij}^{(G)}] = A(G_0)P_{ij} \tag{F5}$$

where $A(G_0)$ is the number of edges in G. Thus

$$E[X(G)] = A(G_0)\sum_{j=1}^{3}\sum_{i=1}^{3}(ij)^{-0.5}P_{ij} \tag{F6}$$

or

$$E[X(G)] = A(G_0)\sum_{j=i}^{3}\sum_{i=1}^{3}(ij)^{-0.5}\alpha_{ij}P_{ij} \tag{F7}$$

where

$$\alpha_{ij} = \begin{cases} 2 \text{ if } i \neq j \\ 1 \text{ if } i = j \end{cases}$$

Use of Eqs. (F1) and (F4) in Eq. (F7) with $A(G_0) = (3/2)N$ gives

$$E[X(G)] = \frac{\left(\frac{1}{3}\right)^5 N}{(1-x)^5}(3-5x)\{24\,x^4 + 24\sqrt{2}\,x^3(3-5x) + 4(\sqrt{3}+3)\,x^2\,(3-5x)^2 + 2\sqrt{6}\times(3-5x)^{\,3} + \tfrac{1}{2}\,(3-5x)^4\} \tag{F8}$$

References

F1. J. J. Meghirditchian, J. Am. Chem. Soc., *113*, 395 (1991).

F2. E. McCafferty, *Electrochem. and Solid State Lett. 3*, 28 (2000).

Appendix G
Acid Dissociation Constants pK_a of Bases and the Base Strength

Organic bases are protonated in acid solutions. For a secondary amine, for instance,

$$R_2NH^+ \rightleftharpoons R_2N + H^+ \tag{G1}$$

where R_2N denotes the depronated amine and R_2NH^+ its conjugate acid, with R referring to an organic substituent. (If one of the R′s is an H atom, then the amine is a primary amine). The acid dissociation constant for Eq. (G1) is

$$K_a = \frac{[R_2N]\,[H^+]}{[R_2NH^+]} \tag{G2}$$

and

$$pK_a = \log \frac{1}{K_a} \tag{G3}$$

The stronger the conjugate acid, the greater its tendency to produce protons. Thus, the stronger the acid, the larger the ratio $[R_2N]/[R_2NH^+]$, and the larger the value of K_a. Large values of K_a correspond to small values of pK_a.

Thus, the smaller the pK_a, the greater the acid strength. Conversely, the greater the pK_a, the stronger base and greater the tendency to donate electrons.

Appendix H
The Langmuir Adsorption Isotherm

Suppose that a species A adsorbs from the aqueous phase onto a metal surface

$$\text{A (aqueous)} \underset{k_{-1}}{\overset{k_1}{\rightleftharpoons}} \text{A (adsorbed)} \tag{H1}$$

where k_1 and k_{-1} refer to the rate constants in the forward (adsorption) and reverse (desorption) directions.

The process is represented by the free energy diagram in Fig. H.1(also see Chapter 7). In Fig. H.1, $\Delta G_1^{\neq}$ is the height of the free energy barrier in the forward direction, $\Delta G_{-1}^{\neq}$ is the corresponding height in the reverse direction, and ΔG_{ads} is the change in free energy due to adsorption of the species A.

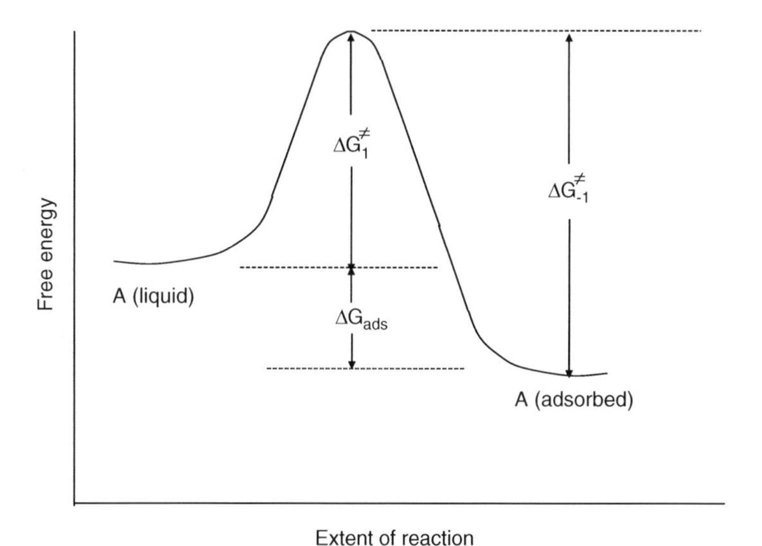

Fig. H.1 Free energy diagram for the adsorption of a species A from the aqueous phase onto a metal surface. $\Delta G_1^{\neq}$ is the height of the free energy barrier in the forward direction, $\Delta G_{-1}^{\neq}$ is the corresponding height in the reverse direction, and ΔG_{ads} is the change in free energy due to adsorption of the species A

The process of adsorption requires a vacant site on the metal surface (i.e., a site not already occupied by an adsorbed molecule or ion). From absolute reaction rate theory, as discussed in Chapter 7, the rate of adsorption (the forward direction) is

$$\text{rate forward} = k_1\,(1-\theta)\,Ce^{-\,\Delta G^{\neq}_{-1}/RT} \tag{H2}$$

where θ is the surface coverage of the adsorbed species and C is the concentration of species A in the liquid phase. The rate of desorption (the reverse direction) is given by

$$\text{rate reverse} = k_{-1}\theta e^{-\,\Delta G^{\neq}_{-1}/RT} \tag{H3}$$

At equilibrium the two rates are equal so that

$$k_{-1}\,\theta\,e^{-\,\Delta G^{\neq}_{-1}/RT} = k_1\,(1-\theta)\,Ce^{-\,\Delta G^{\neq}_{-1}/RT} \tag{H4}$$

Thus,

$$\frac{\theta}{1-\theta} = \frac{k_1}{k_1}Ce^{(\Delta G^{\neq}_{-1}-\Delta G^{\neq}_{1})/RT} \tag{H5}$$

But from Fig. H.1,

$$\Delta G^{\neq}_{-1} - \Delta G^{\neq}_{1} = \Delta G_{\text{ads}} \tag{H6}$$

Use of Eq. (H6) in Eq. (H5) gives

$$\frac{\theta}{1-\theta} = \frac{k_1}{k_{-1}}e^{\Delta G_{\text{ads}}/RT}C \tag{H7}$$

Then we can write $\Delta G_{\text{ads}} = \Delta H_{\text{ads}} - T\,\Delta S_{\text{ads}}$, so that Eq. (H7) becomes

$$\frac{\theta}{1-\theta} = \frac{k_1}{k_{-1}}e^{-\Delta S_{\text{ads}}/R}e^{\Delta H_{\text{ads}}/RT}C \tag{H8}$$

or

$$\frac{\theta}{1-\theta} = K_1\,e^{\Delta H_{\text{ads}}/RT}C \tag{H9}$$

where $e^{-\Delta S_{\text{ads}}/R}$ has been incorporated into the constant K_1.

In the Langmuir model, the metal surface is assumed to be homogeneous everywhere. In addition, it is assumed that there are no lateral interactions between adsorbed molecules. (If there were lateral interactions, then the heat of adsorption would vary with the surface coverage θ.) Thus, ΔH_{ads} is independent of the surface coverage θ. Then Eq. (H9) becomes

$$\frac{\theta}{1-\theta} = KC \tag{H10}$$

where

$$K = K_1\,e^{\Delta H_{\text{ads}}/RT} \tag{H11}$$

Equation (H10) is the same as Eq. (7) in Chapter 12.

Appendix I
The Temkin Adsorption Isotherm

The Temkin adsorption isotherm removes the restriction in the Langmuir model that the heat of adsorption is independent of surface coverage θ. The Langmuir adsorption isotherm can be written as

$$\frac{\theta}{1-\theta} = K_1\,\mathrm{e}^{\Delta H_{\mathrm{ads}}/RT}\,C \tag{I1}$$

where the terms have the same meaning as in Eq. (H9).

In the Temkin model, the heat of adsorption is assumed to decrease linearly with surface coverage θ; that is

$$\Delta H_{\mathrm{ads}} = \Delta H_{\mathrm{ads}}^{0} - r\theta \tag{I2}$$

where $\Delta H_{\mathrm{ads}}{}^{0}$ is the initial heat of adsorption (at near-zero coverages) and r is the Temkin parameter. Insertion of Eq. (I2) into Eq. (I1) gives

$$\frac{\theta}{1-\theta} = K_1\,e^{\Delta H^0/RT}\,e^{-r\theta/RT}\,C \tag{I3}$$

or

$$\frac{\theta}{1-\theta} = K'\,\mathrm{e}^{-r\theta/RT}\,C \tag{I4}$$

where

$$K' = K_1\,e^{\Delta H^0/RT} \tag{I5}$$

Taking natural logarithms in Eq. (I4) gives

$$\ln\left(\frac{\theta}{1-\theta}\right) + \frac{r\theta}{RT} = \ln K' + \ln C \tag{I6}$$

Next we compare the terms $\ln[\theta/(1-\theta)]$ and $r\theta/RT$. This is done in Table I.1 for typical values of r of 10–20 kcal/mol. As seen in Table I.1,

$$\ln\left(\frac{\theta}{1-\theta}\right) << \frac{r\theta}{RT}$$

for $0.3 < \theta < 0.7$. Then Eq. (I6) simplifies to

$$\frac{r\theta}{RT} = \ln K' + \ln C \tag{I7}$$

or

$$\theta = \frac{2.303\,RT}{r}\log C + \frac{2.303\,RT}{r}\log K' \tag{I8}$$

which is Eq. (11) in Chapter 12.

Table I.1 Comparison of the terms ln $[\theta/(1-\theta)]$ and $r\theta/RT$ in the derivation of the Temkin adsorption isotherm

θ	$\ln\frac{\theta}{1-\theta}$	$\frac{r\theta}{RT}$	
		$r = 10$ kcal/mole	$r = 20$ kcal/mole
0.1	−2.198	1.69	3.38
0.2	−1.386	3.38	6.76
0.3	−0.847	5.08	10.1
0.4	−0.405	6.78	13.6
0.5	0.000	8.47	16.9
0.6	0.405	10.0	20.2
0.7	0.847	11.9	23.8
0.8	1.386	13.6	27.2
0.9	2.198	15.3	30.6

Appendix J
The Temkin Adsorption Isotherm for a Charged Interface

If a molecule or ion adsorbs at a charged interface in the presence of an electrode potential,

$$\text{A (aqueous)} \underset{k_{-1}}{\overset{k_1}{\rightleftharpoons}} \text{A (adsorbed)}$$

then the process is represented by the free energy diagram in Fig. J.1, where the solid lines in Fig. J.1 apply to the open-circuit potential E_0.

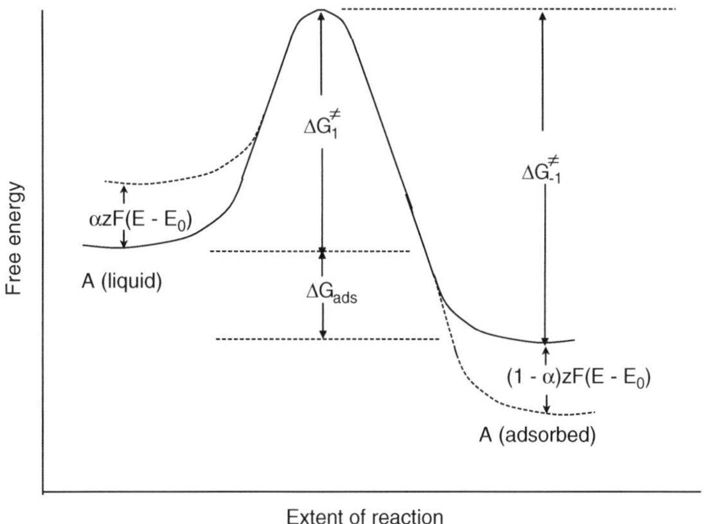

Fig. J.1 Free energy diagram for the adsorption of a species A from the aqueous phase onto a metal surface in the presence of an electrode potential. The *solid lines* refer to the open-circuit potential E_0, and the *dotted lines* to a polarized potential E. $\Delta G_1^{\neq}$ is the height of the free energy barrier in the forward direction at potential E_0, $\Delta G_{-1}^{\neq}$ is the corresponding height in the reverse direction at potential E_0, and ΔG_{ads} is the change in free energy due to adsorption of the species A at potential E_0

Suppose that the adsorbed species has a negative charge, and that the electrode potential is changed from the open-circuit potential E_0 to a more positive potential E. Then the height of the free energy barrier will be decreased for adsorption (the forward direction) and increased for desorption (the reverse direction), as shown by the dotted lines in Fig. J.1.

The rate of adsorption in the forward direction at the open-circuit potential E_0 is

$$\text{rate forward} = k_1\,(1-\theta)[A]\,e^{-\Delta G_1{}^{\neq}/RT} \tag{J1}$$

(see Eq. (H2) in Appendix H). Under the new applied potential E, the rate of adsorption in the forward direction is

$$\text{rate forward} = k_1\,(1-\theta)\,[A]\;e^{-[\Delta G^{\neq}-\alpha zF(E-E_0)]/RT} \tag{J2}$$

where z is the charge on the adsorbing species, α is the symmetry factor, and $\alpha ZF(E-E_0)$ represents the coulombic interaction of the charge with the electric field. (If the decrease in the free energy barrier in the forward direction is equal to the increase in the free energy barrier in the reverse direction, then α = 0.5.)

Similarly, the rate of desorption in the reverse direction under potential E is

$$\text{rate reverse} = k_{-1}\,\theta\,\mathrm{e}^{-\left[\Delta G^{\neq}_{-1}+(1-\alpha)zF(E-E_0)\right]/RT} \tag{J3}$$

At equilibrium under the potential E, the rate in the forward direction is equal to the rate in the reverse direction, or

$$k_{-1}\,\theta\,\mathrm{e}^{-\left[\Delta G^{\neq}_{-1}+(1-\alpha)zF(E-E_0)\right]/RT} = k_1\,(1-\theta)\,[A]\,\mathrm{e}^{-\left[\Delta G^{\neq}_{1}-\alpha ZF(E-E_0)\right]/RT} \tag{J4}$$

Thus,

$$\frac{\theta}{1-\theta} = \frac{k_1}{k_{-1}}[A]\,\mathrm{e}^{(\Delta G^{\neq}_{-1}-\Delta G^{\neq}_{1})/RT}\;\mathrm{e}^{-zFE_0/RT}\;\mathrm{e}^{zFE/RT} \tag{J5}$$

As in Appendix H,

$$\Delta G^{\neq}_{-1} - \Delta G^{\neq}_{-1} = \Delta G_{\text{ads}} \tag{J6}$$

and writing $\Delta G_{\text{ads}} = \Delta H_{\text{ads}} - T\,\Delta S_{\text{ads}}$ gives

$$\frac{\theta}{1-\theta} = \frac{k_1}{k_{-1}}[A]\,\mathrm{e}^{\Delta H_{\text{ads}}/RT}\,\mathrm{e}^{-\Delta S_{\text{ads}}/R}\;\mathrm{e}^{-zFE_{\text{o}}/RT}\;\mathrm{e}^{zFE/RT} \tag{J7}$$

As in Appendix I, we write

$$\Delta H_{\text{ads}} = \Delta H^0_{\text{ads}} - r\theta \tag{J8}$$

where $\Delta H_{\text{ads}}{}^0$ is the initial heat of adsorption (at near-zero coverages) and r is the Temkin parameter. Then

$$\frac{\theta}{1-\theta} = \frac{k_1}{k_{-1}}[A]\,\mathrm{e}^{\Delta H^0_{\text{ads}}/RT}\,\mathrm{e}^{-r\theta/RT}\;\mathrm{e}^{-\Delta S_{\text{ads}}/R}\;\mathrm{e}^{-ZFE_{\text{o}}/RT}\;\mathrm{e}^{zFE/RT} \tag{J9}$$

$$\frac{\theta}{1-\theta} = K_2\,[A]\,\mathrm{e}^{-r\theta/RT}\;\mathrm{e}^{zFE/RT} \tag{J10}$$

where

$$K_2 = \frac{k_1}{k_{-1}} \mathrm{e}^{\Delta H^0_{\mathrm{ads}}/RT} e^{-\Delta S_{\mathrm{ads}}/R} \mathrm{e}^{-zF E_0/RT} \tag{J11}$$

Equation (J10) has a form similar to that of Eq. (I4) in Appendix I, and the remainder of the treatment is similar to that in Appendix I. Taking logarithms in Eq. (J10) gives

$$\ln\left(\frac{\theta}{1-\theta}\right) + \frac{r\theta}{RT} = \ln K_2 + \ln\,[A] + \frac{zFE}{RT} \tag{J12}$$

As shown in Appendix I,

$$\ln\left(\frac{\theta}{1-\theta}\right) << \frac{r\theta}{RT}$$

for $0.3 < \theta < 0.7$. Then Eq. (J12) becomes

$$\theta = \frac{2.303\,RT}{r} \log\,[A] + \frac{2.303\,RT}{r} \log\,K_2 + \frac{zFE}{r} \tag{J13}$$

which is Eq. (18) in Chapter 12.

Appendix K
Effect of Coating Thickness on the Transmission Rate of a Molecule Permeating Through a Free-Standing Organic Coating

Consider the case where the exterior of the coating is in contact with concentration C_0 of permeating molecule. The concentration profiles with time within an organic coating of thickness L_1 are given in Fig. K.1. The concentration of permeating molecule at the underside boundary of the organic film is zero because the permeating molecules exit the film at that distance.

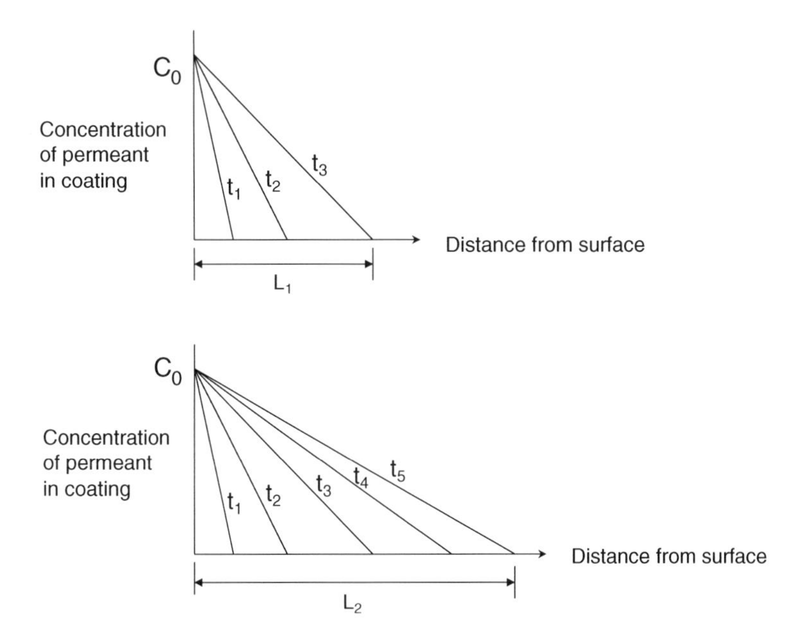

Fig. K.1 Concentration profiles (at different times) for a permeating molecule in free-standing organic films of thickness L_1 (*top*) and L_2 (*bottom*)

From Ficks first law of diffusion, the transmission rate J_1 of the permeating molecule in the coating of thickness L_1 is

$$J_1 = -D\frac{\mathrm{d}C}{\mathrm{d}x} \tag{K1}$$

or

$$J_1 = -D\left(\frac{0 - C_0}{L_1}\right) \tag{K2}$$

For a second coating of identical material but of a different thickness L_2,

$$J_2 = -D\left(\frac{0 - C_0}{L_2}\right) \tag{K3}$$

Equations (K2) and (K3) combine to give:

$$\frac{J_2}{J_1} = \frac{L_1}{L_2} \tag{K4}$$

Appendix L
The Impedance for a Capacitor

The voltage across a capacitor C is

$$E(t) = \frac{1}{C} Q(t) \tag{L1}$$

where $Q(t)$ is the charge given by

$$Q(t) = I(t)t \tag{L2}$$

Use of Eq. (L2) in Eq. (L1) gives

$$E(t) = \frac{1}{C} I(t)t \tag{L3}$$

so that

$$\frac{\mathrm{d}\,\mathrm{E(t)}}{\mathrm{d}t} = \frac{1}{C} I(t) \tag{L4}$$

and

$$I(t) = C \frac{\mathrm{d}\,\mathrm{E(t)}}{\mathrm{d}t} \tag{L5}$$

The impedance is given by

$$Z_C = \frac{E(t)}{I(\mathrm{t})} = \frac{E(t)}{C \dfrac{d\,E(t)}{\mathrm{d}t}} \tag{L6}$$

The voltage $E(t)$ is given by Eq. (6) in Chapter 14, that is

$$E = E_0\,\mathrm{e}^{\mathrm{j}\omega \mathrm{t}} \tag{L7}$$

Use of Eq. (L7) in Eq. (L6) gives

$$Z_C = \frac{E(t)}{I(\mathrm{t})} = \frac{E_0 \mathrm{e}^{\mathrm{j}\omega t}}{\mathrm{C}\mathrm{j}\omega E_0^{\mathrm{j}\omega t}} \tag{L8}$$

or

$$Z_C = \frac{1}{\mathrm{j}\omega C} \tag{L9}$$

Reference

L1. G. H. Hostetter, "Fundamentals of Network Analysis", p. 162, Harper & Row, New York, NY (1980).

Appendix M
Use of L'Hospital's Rule to Evaluate |Z| for the Metal/Solution Interface for Large Values of Angular Frequency ω

For the model of the electrical double layer shown in Fig. 14.4, we have

$$|Z| = \left\{ \left(R_S + \frac{R_p}{1 + \omega^2 R_p^2 C_{dl}^2} \right)^2 + \left(\frac{\omega R_p^2 C_{dl}}{1 + \omega^2 R_p^2 C_{dl}^2} \right)^2 \right\}^{1/2} \tag{M1}$$

which is Eq. (24) in Chapter 14. For $\omega \to \infty$, the first term in the parentheses on the right-hand side of Eq. (M1) approaches $(R_s)^2$, but we need to apply L'Hospital's rule to the second term on the right-hand side of Eq. (M1). Thus,

$$\lim_{\omega \to \infty} \left(\frac{\omega R_p^2 C_{dl}}{1 + \omega^2 R_p^2 C_{dl}^2} \right) = \lim_{\omega \to \infty} \left(\frac{\frac{d}{d\omega} \left(\omega R_p^2 C_{dl} \right)}{\frac{d}{d\omega} \left(1 + \omega^2 R_p^2 C_{dl}^2 \right)} \right) \tag{M2}$$

$$\lim_{\omega \to \infty} \left(\frac{\omega R_p^2 C_{dl}}{1 + \omega^2 R_p^2 C_{dl}^2} \right) = \lim_{\omega \to \infty} \left(\frac{R_p^2 C_{dl}}{2\omega R_p^2 C_{dl}} \right) \tag{M3}$$

or

$$\lim_{\omega \to \infty} \left(\frac{\omega R_p^2 C_{dl}}{1 + \omega^2 R_p^2 C_{dl}^2} \right) = 0 \tag{M4}$$

Thus,

$$\lim_{\omega \to \infty} |Z| = |R_S^2|^{1/2} = R_S \tag{M5}$$

Appendix N
Derivation of the Arc Chord Equation for Cole–Cole plots

Complex plane plots of Z'' vs. Z' are sometimes not perfect semi-circles but instead are depressed semi-circles, as in Fig. N.1. (This figure is the same as Fig. 14.15).

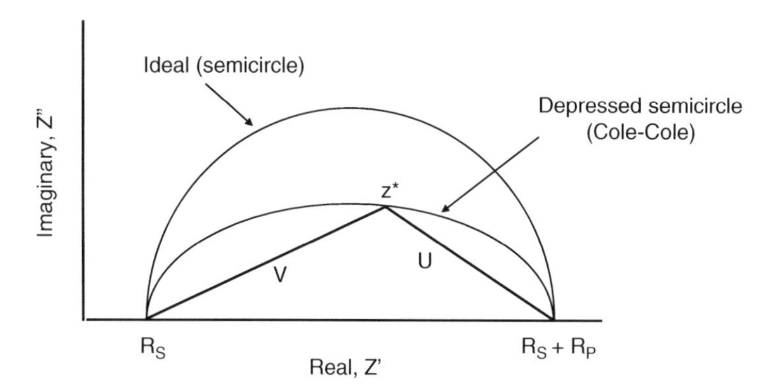

Fig. N.1 Schematic complex-plane diagram for the impedance showing arc chords U and V for a given frequency (N2). The notation z^* indicates that z is a complex number

Such depressed semi-circles are called Cole–Cole plots. For the electrical double layer at a metal/solution interface, these plots result if there is a distribution of relaxation times τ around a most probable value $\tau_0 = R_{dl}C_p$. This is equivalent to the distribution of capacitances around some central value C_{dl}:

$$Z = R_S + \frac{R_P}{1 + \left(j\omega R_p C_{dl}\right)^{1-\alpha}} \tag{N1}$$

where α is a measure of the departure from ideality. The parameter α ranges from 0 to 1 and is zero for ideal behavior.

The vectors U and V in Fig. N.1 are given by

$$V = Z^* - R_s \tag{N2}$$

$$U = Z^* - (R_S + R_p) \tag{N3}$$

where Z^* is the complex impedance. From Fig. N.1,

$$V + U = (R_S + R_p) - R_S = R_p \tag{N4}$$

Then Eq. (N1) becomes

$$V + U = \frac{V + U}{1 + (\mathrm{j}\omega R_{\mathrm{p}} C_{\mathrm{dl}})^{1-\alpha}} \tag{N5}$$

or

$$V + V(\mathrm{j}\omega R_{\mathrm{p}} C_{\mathrm{dl}})^{1-\alpha} = V + U \tag{N6}$$

so that

$$\frac{V}{U} = \frac{1}{(\mathrm{j}\omega R_{\mathrm{p}} C_{\mathrm{dl}})^{1-\alpha}} \tag{N7}$$

or

$$\frac{V}{U} = (\mathrm{j}\omega R_{\mathrm{p}} C_{\mathrm{dl}})^{\alpha-1} \tag{N8}$$

Taking absolute values on each side

$$\left|\frac{V}{U}\right| = \left|\mathrm{j}^{(\alpha-1)}\right| (\omega R_{\mathrm{p}} C_{\mathrm{dl}})^{\alpha-1} \tag{N9}$$

Then use is made of the property that any complex number can be written in polar form:

$$z = a + \mathrm{j}b = r\,\mathrm{e}^{\mathrm{j}\theta} \tag{N10}$$

For the special case where $z = \mathrm{j}$,

$$\mathrm{j} = 1 \cdot \mathrm{e}^{\mathrm{j}\frac{\pi}{2}} \tag{N11}$$

$$\mathrm{j}^{(\alpha-1)} = \mathrm{e}^{(\mathrm{j}\frac{\pi}{2})(\alpha-1)} \tag{N12}$$

or

$$\mathrm{j}^{(\alpha-1)} = \cos\frac{\pi}{2}(\alpha - 1) + \mathrm{j}\,\sin\frac{\pi}{2}(\alpha - 1) \tag{N13}$$

With

$$|z| = \sqrt{a^2 + b^2} \tag{N14}$$

$$\left|\mathrm{j}^{(\alpha-1)}\right| = \sqrt{\cos^2\left[\frac{\pi}{2}(\alpha - 1)\right] + \sin^2\left[\frac{\pi}{2}(\alpha - 1)\right]} = 1 \tag{N15}$$

Use of this result in Eq. (N9) gives

$$\left|\frac{V}{U}\right| = (\omega R_{\mathrm{p}} C_{\mathrm{dl}})^{\alpha-1} \tag{N16}$$

and

$$\log \left| \frac{V}{U} \right| = (\alpha - 1) \log \omega + (\alpha - 1) \log (C_{dl} R_p) \tag{N17}$$

which is the desired result.

References

N1. E. McCafferty, V. H. Pravdic, and A. C. Zettlemoyer, *Trans. Faraday Soc.*, *66*, 1720 (1970).
N2. K. S. Cole and R. H. Cole, *J. Chem. Phys.*, *9*, 341 (1941).
N3. E. McCafferty and J. V. McArdle, *J. Electrochem. Soc.*, *142*, 1447 (1995).

Appendix O
Laplace's Equation

The flux N_i of each dissolved species in an electrolyte arises from three effects. These are: (a) the motion of charged species in an electric field (*migration*), (b) *diffusion* due to a concentration gradient, and (c) *convection* due to bulk motion of the fluid. These three contributions give [O1]

$$N_i = \underbrace{-z_i \mu_i F C_i \nabla\phi}_{\text{migration}} \underbrace{- D_i \nabla C_i}_{\text{diffusion}} + \underbrace{C_i v}_{\text{convection}} \tag{O1}$$

where

N_i = ionic flux of species i
z_i = charge of species i
μ_i = mobility of species i
F = Faraday's constant
C_i = concentration of species i
ϕ = electrostatic potential
D_i = diffusion coefficient of species i
v = fluid velocity
∇ = differential operator $(\partial/\partial x + \partial/\partial y + \partial/\partial z)$

The current density i is given by

$$i = F\sum_i z_i N_i \tag{O2}$$

The material balance of each species is

$$\frac{\partial C_i}{\partial t} = -\nabla N_i + R_i \tag{O3}$$

where R_i is the rate of production of species i due to chemical reactions in the bulk of the electrolyte. If ions are not generated in the bulk, then $R_i = 0$.

The condition of electroneutrality in the solution is

$$\sum_i z_i C_i = 0 \tag{O4}$$

Multiplying Eq. (O3) by z_i gives

$$\frac{\partial(z_i C_i)}{\partial t} = -\nabla z_i N_i \tag{O5}$$

(with $R_i = 0$). Then, summing over all species gives

$$\frac{\partial}{\partial t}\sum_i z_i C_i = -\nabla \sum_i z_i N_i \tag{O6}$$

But from Eq. (04), $\sum_i z_i C_i = 0$, so that Eq.(06) gives

$$\nabla \sum_i z_i N_i = 0 \tag{O7}$$

Then, Eq. (O2) can be written as

$$\nabla i = \nabla\left(F\sum_i z_i N_i\right) = F\sum_i z_i N_i \tag{O8}$$

Using Eq. (O7) in Eq. (O8) gives

$$\nabla i = 0 \tag{O9}$$

This result will be used below. Use of Eq. (O1) in Eq. (O2) gives

$$i = F\sum_i z_i\left(-z_i\,\mu_i F C_i \nabla\phi - D_i \nabla C_i + C_i v\right)$$

or

$$i = -F^2\nabla\phi\sum_i z_i^2 \mu_i C_i - F\sum_i z_i D_i \nabla C_i + Fv\sum_i z_i C_i \tag{O10}$$

By virtue of electroneutrality, the last term is zero. If there are no concentration gradients in the bulk of the solution, $\nabla C_i = 0$. Thus, Eq. (O10) becomes

$$i = -\sigma\,\nabla\phi \tag{O11}$$

where σ is the electrolyte conductivity given by

$$\sigma = -F^2\sum_i z_i^2 \mu_i C_i \tag{O12}$$

We can operate on Eq. (O11) as follows:

$$\nabla i = -\sigma\,\nabla^2\phi \tag{O13}$$

But $\nabla i = 0$ by Eq. (O9), so that Eq. (O13) gives

$$\nabla^2 \phi = 0 \tag{O14}$$

which is Laplace's equation.

Reference

O1. J. Newman, in "Advances in Electrochemistry and Electrochemical Engineering", C. W. Tobias, Ed., Vol. 5, p. 87, Interscience Publishers, New York, NY (1967).

Index